OMICS TECHNOLOGIES AND BIO-ENGINEERING

OMICS TECHNOLOGIES AND BIO-ENGINEERING
Towards Improving Quality of Life

VOLUME 1
EMERGING FIELDS, ANIMAL AND MEDICAL BIOTECHNOLOGIES

Edited by

DEBMALYA BARH
Institute of Integrative Omics and Applied Biotechnology (IIOAB), Nonakuri, Purba Medinipur, India
Laboratório de Genética Celular e Molecular, Departamento de Biologia Geral, Instituto de Ciências Biológicas (ICB), Federal University of Minas Gerais, Belo Horizonte, Minas Gerais, Brazil

VASCO AZEVEDO
Laboratório de Genética Celular e Molecular, Departamento de Biologia Geral, Instituto de Ciências Biológicas (ICB), Federal University of Minas Gerais, Belo Horizonte, Minas Gerais, Brazil

ACADEMIC PRESS
An imprint of Elsevier

Academic Press is an imprint of Elsevier
125 London Wall, London EC2Y 5AS, United Kingdom
525 B Street, Suite 1800, San Diego, CA 92101-4495, United States
50 Hampshire Street, 5th Floor, Cambridge, MA 02139, United States
The Boulevard, Langford Lane, Kidlington, Oxford OX5 1GB, United Kingdom

Notices
Knowledge and best practice in this field are constantly changing. As new research and experience broaden our
understanding, changes in research methods, professional practices, or medical treatment may become necessary.

Practitioners and researchers must always rely on their own experience and knowledge in evaluating and using
any information, methods, compounds, or experiments described herein. In using such information or methods
they should be mindful of their own safety and the safety of others, including parties for whom they have a
professional responsibility.

To the fullest extent of the law, neither the Publisher nor the authors, contributors, or editors, assume any liability
for any injury and/or damage to persons or property as a matter of products liability, negligence or otherwise, or
from any use or operation of any methods, products, instructions, or ideas contained in the material herein.

British Library Cataloguing-in-Publication Data
A catalogue record for this book is available from the British Library

Library of Congress Cataloging-in-Publication Data
A catalog record for this book is available from the Library of Congress

ISBN: 978-0-12-804659-3

For Information on all Academic Press publications
visit our website at https://www.elsevier.com/books-and-journals

Working together
to grow libraries in
developing countries

www.elsevier.com • www.bookaid.org

Publisher: John Fedor
Acquisition Editor: Rafael Teixeira
Editorial Project Manager: Mariana Kuhl
Production Project Manager: Kiruthika Govindaraju
Cover Designer: Victoria Pearson

Typeset by MPS Limited, Chennai, India

Dedication

We dedicate this book to our next generation.

Estella   Vasco-Jean   Laura Emília   Antônio Luiz   Shaurya Shree

Contents

6. Omics Approaches in Forensic Biotechnology: Looking for Ancestry to Offence

SYED B. NIZAMI, SAYYADA Z. HASSAN KAZMI,
FATIMA ABID, MUSTAFEEZ M. BABAR, ANEEQA NOOR,
NAJAM-US-SAHAR S. ZAIDI, SAMI U. KHAN,
HUMNA HASAN, MOHSIN ALI AND ALVINA GUL

7. Biotechnology and Bioengineering in Astrobiology: Towards a New Habitat for Us

SAMEEN R. IMADI, MUSTAFEEZ M. BABAR, SAMI U.
KHAN, HUMNA HASAN, MOHSIN ALI AND ALVINA GUL

8. Lab-on-a-Chip Technology and Its Applications

BURAK YILMAZ AND FAZILET YILMAZ

9. Robotics and High-Throughput Techniques

HUMNA HASAN, MUHAMMAD HASSAN SAFDAR,
SANA ZAHID, MARIA BIBI AND ALVINA GUL

10. 3D Printing Technologies and Their Applications in Biomedical Science

SYEDA M. BAKHTIAR, HINA A. BUTT, SHUJA ZEB,
DARRAK M. QUDDUSI, SAIMA GUL AND ERUM
DILSHAD

11. Next-Generation Sequencing and Data Analysis: Strategies, Tools, Pipelines and Protocols

PABLO H.C.G. DE SÁ, LUIS C. GUIMARÃES,
DIEGO A. DAS GRAÇAS, ADONNEY A. DE
OLIVEIRA VERAS, DEBMALYA BARH, VASCO
AZEVEDO, ARTUR L. DA COSTA DA SILVA AND
ROMMEL T.J. RAMOS

19. Medical Biotechnology: Techniques and Applications

PHUC V. PHAM

20. Tissue Engineering: Towards Development of Regenerative and Transplant Medicine

MUSTAFA S. ELITOK, ESRA GUNDUZ, HACER E. GURSES
AND MEHMET GUNDUZ

21. Therapeutic Aspects of Stem Cells in Regenerative Medicine

ALOK MISHRA AND MUKESH VERMA

22. Genetic Engineering: Towards Gene Therapy and Molecular Medicine

SHAILENDRA DWIVEDI, PURVI PUROHIT, YOGESH
MITTAL, GARIMA GUPTA, APUL GOEL, RAKESH C.
VERMA, SANJAY KHATTRI, PRAVEEN SHARMA, SANJEEV
MISRA AND KAMLESH K. PANT

23. Biotechnology for Biomarkers: Towards Prediction, Screening, Diagnosis, Prognosis, and Therapy

DIPALI DHAWAN

24. Omics Approaches in In Vitro Fertilization

AIMAN TANVEER, NEHA MALVIYA AND DINESH YADAV

25. Safety and Ethics in Biotechnology and Bioengineering: What to Follow and What Not to

ANJANA MUNSHI AND VANDANA SHARMA

List of Contributors

Duygu Abbasoğlu Anadolu University, Eskişehir, Türkiye

Zain U. Abedin University of Strathclyde, Glasgow, United Kingdom

Fatima Abid The University of Lahore, Islamabad, Pakistan

Muhammad J. Abid The University of Lahore, Lahore, Pakistan

Mohsin Ali National University of Sciences and Technology, Islamabad, Pakistan; National University of Sciences and Technology, Shifa Tameer-e-Millat University, Shifa, Pakistan

Vasco Azevedo Federal University of Minas Gerais, Belo Horizonte, Minas Gerais, Brazil

Mustafeez M. Babar Shifa College of Pharmaceutical Sciences, Shifa Tameer-e-Millat University, Islamabad, Pakistan

Syeda M. Bakhtiar Capital University of Science and Technology, Islamabad, Pakistan

Debmalya Barh Federal University of Minas Gerais, Belo Horizonte, Minas Gerais, Brazil; Institute of Integrative Omics and Applied Biotechnology (IIOAB), Purba Medinipur, India

Yasemin Baskın Dokuz Eylul University, İzmir, Türkiye

Attya Bhatti National University of Sciences and Technology (NUST), Islamabad, Pakistan

Maria Bibi Quaid-I-Azam University, Islamabad, Pakistan

Hina A. Butt Capital University of Science and Technology, Islamabad, Pakistan

Gülay Büyükköroğlu Anadolu University, Eskisehir, Türkiye

Gizem Çalıbaşı Dokuz Eylul University, İzmir, Türkiye

Artur L. da Costa da Silva Federal University of Pará, Belém, Pará, Brazil

Kenny da Costa Pinheiro Federal University of Pará, Belém, Pará, Brazil

Diego A. das Graças Federal University of Pará, Belém, Pará, Brazil

Ayşe Banu Demir Dokuz Eylul University, İzmir, Türkiye

Adonney A. de Oliveira Veras Federal University of Pará, Belém, Pará, Brazil

Pablo H.C.G. de Sá Federal University of Pará, Belém, Pará, Brazil

Dipali Dhawan PanGenomics International Pvt. Ltd., Ahmedabad, Gujarat, India

Erum Dilshad Capital University of Science and Technology, Islamabad, Pakistan

Devrim Demir Dora Akdeniz University, Antalya, Türkiye

Shailendra Dwivedi All India Institute of Medical Sciences (AIIMS), Jodhpur, India; King George Medical University (KGMU), Lucknow, India

Mustafa S. Elitok Turgut Ozal University, Ankara, Turkey; Yunus Emre Mahallesi, Ankara, Turkey

Serpil Eraslan Bogazici University, İstanbul, Türkiye

Yağmur Esemen Eastern Mediterranean University (EMU), Mersin, Türkiye

Preetam Ghosh Virginia Commonwealth University, Richmond, VA, United States

Apul Goel   King George Medical University (KGMU), Lucknow, India
Mohan L. Gope   Ajmera Infinity, Bangalore, India
Rajalakshmi Gope   NIMHANS, Bangalore, India
Luis C. Guimarães   Federal University of Pará, Belém, Pará, Brazil
Alvina Gul National University of Sciences and Technology, Islamabad, Pakistan; National University of Sciences and Technology, Shifa Tameer-e-Millat University, Shifa, Pakistan
Saima Gul   Capital University of Science and Technology, Islamabad, Pakistan
Esra Gunduz   Turgut Ozal University, Ankara, Turkey;   Yunus Emre Mahallesi, Ankara, Turkey
Mehmet Gunduz   Turgut Ozal University, Ankara, Turkey;   Yunus Emre Mahallesi, Ankara, Turkey
Sibel Gunes   Eskisehir Osmangazi University, Eskisehir, Turkey
Garima Gupta   All India Institute of Medical Sciences (AIIMS), Jodhpur, India
Hacer E. Gurses   Turgut Ozal University, Ankara, Turkey;   Yunus Emre Mahallesi, Ankara, Turkey
Humna Hasan   Quaid-I-Azam University, Islamabad, Pakistan
Candan Hızel   OPTI-THERA Inc., Montreal, QC, Canada
Sameen R. Imadi   National University of Sciences and Technology, Islamabad, Pakistan
Rija Irfan   National University of Sciences and Technology, Islamabad, Pakistan
Peter John   National University of Sciences and Technology (NUST), Islamabad, Pakistan
Rommel Thiago Jucá Ramos   Federal University of Pará, Belém, Pará, Brazil
Sayyada Z. Hassan Kazmi   University of Karachi, Karachi, Pakistan
Rohini Keshava   Bangalore University, Bangalore, India
Azka Khan   National University of Sciences and Technology, Islamabad, Pakistan
Sami U. Khan   University of Haripur, Khyber Pakhtunkhwa, Pakistan
Sanjay Khattri   All India Institute of Medical Sciences (AIIMS), Jodhpur, India
Neha Malviya   D.D.U Gorakhpur University, Gorakhpur, India
Alok Mishra   National Institutes of Health (NIH), Bethesda, MD, United States
Sanjeev Misra   All India Institute of Medical Sciences (AIIMS), Jodhpur, India;   King George Medical University (KGMU), Lucknow, India
Rohan Mitra   NIMHANS, Bangalore, India
Yogesh Mittal   All India Institute of Medical Sciences (AIIMS), Jodhpur, India
Anjana Munshi   Central University of Punjab, Bathinda, India
Joseph J. Nalluri   Virginia Commonwealth University, Richmond, VA, United States
Syed B. Nizami   University of Karachi, Karachi, Pakistan
Aneeqa Noor   National University of Sciences and Technology, Shifa Tameer-e-Millat University, Shifa, Pakistan
Ayşe Feyda Nursal   Hitit University, Çorum, Türkiye
Filiz Özdemir   Anadolu University, Eskisehir, Türkiye
Kamlesh K. Pant   All India Institute of Medical Sciences (AIIMS), Jodhpur, India
Phuc V. Pham   University of Science, VNUHCM, Ho Chi Minh city, Vietnam
Purvi Purohit   All India Institute of Medical Sciences (AIIMS), Jodhpur, India
Darrak M. Quddusi   Capital University of Science and Technology, Islamabad, Pakistan
Rommel T.J. Ramos   Federal University of Pará, Belém, Pará, Brazil

Muhammad Hassan Safdar Quaid-I-Azam University, Islamabad, Pakistan

Ayla Eker Sariboyaci Eskisehir Osmangazi University, Eskisehir, Turkey

Behiye Şenel Anadolu University, Eskisehir, Türkiye

Murat Sevimli Eskisehir Osmangazi University, Eskisehir, Turkey

Tugba Sevimli Eskisehir Osmangazi University, Eskisehir, Turkey

Adeena Shafique National University of Sciences and Technology, Islamabad, Pakistan

Praveen Sharma All India Institute of Medical Sciences (AIIMS), Jodhpur, India

Vandana Sharma Indraprastha Apollo Hospitals, New Delhi, India

Nida A. Syed National University of Sciences and Technology (NUST), Islamabad, Pakistan

Aiman Tanveer D.D.U Gorakhpur University, Gorakhpur, India

Şükrü Tüzmen Eastern Mediterranean University (EMU), Mersin, Türkiye; Arizona State University (ASU), Phoenix, AZ, United States

Onur Uysal Eskisehir Osmangazi University, Eskisehir, Turkey

Mukesh Verma National Institutes of Health (NIH), Rockville, MD, United States

Rakesh C. Verma UP Rural Institute of Medical Sciences & Research Saifai, Etawah, India

Kinza Waqar National University of Sciences and Technology, Islamabad, Pakistan

Dinesh Yadav D.D.U Gorakhpur University, Gorakhpur, India

Sangeeta Yadav D.D.U Gorakhpur University, Gorakhpur, India

Burak Yılmaz Turgut Ozal University, Ankara, Turkey

Fazilet Yılmaz Turgut Ozal University, Ankara, Turkey

Muhammad A. Zahid Shifa Tameer-e-Millat University, Islamabad, Pakistan

Sana Zahid Quaid-I-Azam University, Islamabad, Pakistan

Najam-us-Sahar S. Zaidi National University of Sciences and Technology, Shifa Tameer-e-Millat University, Shifa, Pakistan

Shuja Zeb Capital University of Science and Technology, Islamabad, Pakistan

About the Editors

Vasco Azevedo is graduated from veterinary school of the Federal University of Bahia in 1986. He obtained his Master (1989) and Ph.D. (1993) degrees in microbial genetics from the Institut National Agronomique Paris-Grignon (INAPG) and Institut National de la Recherche Agronomique (INRA), France, respectively. He did his Postdoctoral research (1994) at the Department of Microbiology, School of Medicine, University of Pennsylvania, United States. Since 1995, he is a professor at the Federal University of Minas Gerais (UFMG), Brazil. In 2004, Prof. Azevedo won the Livre-docência contest at the University of São Paulo, which is considered as the best university in Brazil. Livre-docência is a degree awarded by the
Higher Education Department of Brazil through a public examination open only to the doctoral degree holders and is a recognition to a superior quality of teaching and research. In 2017, Prof. Azevedo defended his third thesis to become a Doctorate in Bioinformatics from the UFMG. He is a also a Fellow of Brazilian Academy of Sciences. He has published 380 research articles, 3 books, and 29 book chapters. Prof. Azevedo is expert in bacterial genetics, genomics, transcriptome, proteomics, and development of new vaccines and diagnostics against infectious diseases. He is pioneer in genetics of Lactic Acid Bacteria and *Corynebacterium pseudotuberculosis* in Brazil.

Debmalya Barh, M.Sc. (applied genetics), M.Tech. (biotechnology), M.Phil. (biotechnology), Ph.D. (biotechnology), Ph.D. (bioinformatics), Post-Doc (bioinformatics), PGDM, is a honorary Principal Scientist at the Institute of Integrative Omics and Applied Biotechnology (IIOAB), India—a virtual global platform of multidicipliary research and advocacy. He is blended with both academic and industrial research and has more than 12 years bioinformatics and personalized diagnostic/medicine industry experience where his main focus is to translate academic research into high value commercial products for common mans' reach. He has published more than 150 articles in reputed international journals and has edited 15
cutting-edge omics—related reference books published by Taylor & Francis, Springer, Elsevier, etc. He has also coauthored 30 + book chapters. He also frequently reviewes articles for international journals like Nature Publications, Elsevier, BMC Series, PLoS One, etc.. Due to his significant contribution in the field, he has been recognized by Who's Who in the World and Limca Book of Records.

PART I

EMERGING FIELDS

Overview and Principles of Bioengineering: The Drivers of Omics Technologies

Dinesh Yadav, Aiman Tanveer, Neha Malviya and Sangeeta Yadav

D.D.U Gorakhpur University, Gorakhpur, India

1.1 INTRODUCTION

1.1.1 Science of "Omics"

Innovations in technologies have led to the better understanding of the principles of bioengineering by providing potential tools for comprehensive studies. "Omics" in general reflects totality and, when added as suffix to any specific scientific field, provides an insight with a broader spectrum (Malviya et al., 2016). Genomics is one of the earlier established sciences of omics for comprehensive analysis of the genome, once the whole-genome sequence of an organism is deciphered. Though several omics exists (as mentioned in Table 1.1), genomics, proteomics, metabolomics, transcriptomics, and glycomics are the major branches of omics (Fig. 1.1).

The growth and by-products productivity, as accomplished by bioengineering, can be modulated by intervention of omics technologies. The tools of omics technologies have paved the way to bioengineering in in vivo as well as in cell cultures (Alyass et al., 2015).

Omics Technologies and Bio-engineering: Towards Improving Quality of Life
DOI: https://doi.org/10.1016/B978-0-12-804659-3.00001-4

3

TABLE 1.1 Brief Description of Different Branches of Science of Omics

Branches of science of "omics"	Description
GENE (DNA) LEVEL	
Genomics	Comprehensive study of genome of a particular organism
Structural genomics	Aims to determine the 3D structures of every proteins encoded by the genome using high-throughput technologies comprising of both experimental and modeling approaches
Functional genomics	Aims to determine the putative functions of the genes in totality at different levels of transcription, translation along with providing an insight into the complexity of gene regulations
Comparative cenomics	Aims to analyze different genomes to focus on the similarity and variability of diverse genomics features from evolutionary point of view. The availability of whole-genome sequences is a prerequisite
ORFeomics	Focuses on the open reading frame (ORF) of a particular genome in totality and involves tools of bioinformatics
Epigenomics	Comprehensive study of epigenetic modifications associated with gene expression in regulation for a particular genome. It is mainly manifested in the form of DNA methylation and histone modifications at genome level
Matagenomics	High-throughput genomics technologies associated with analyzing the genetic material obtained directly from the diverse environmental samples. It is considered to be one of the best approach to utilize the untapped microbial diversity due to limitation of cultivation-based methods as source for novel enzymes and metabolites
Pharmacogenomics	Intervention of genomics in analyzing the drug response for a patient to optimize drug therapy with the sole purpose to eliminate the side effects and enhance the efficacy of a particular drug for a patient
Exomics	Comprehensive analysis of exons, coding for the protein region in whole genome of an organism. Whole-exome sequencing is gaining importance as compared to whole-genome sequencing to get an insight into the molecular basis of diseases and phenotypic traits
Enviromics	Aims to analyze the total environmental factors associated with the diseased state of an organism as a part of genetic epidemiological studies
Nutrigenomics	High-throughput genomic technologies for understanding the molecular level interactions between nutrients with the genome with an ultimate goal to develop personalized nutrition based on genotype
RNA LEVEL	
Transcriptomics	Comprehensive study of the whole RNA transcript encoded by the genome under specific conditions or in a specific cell utilizing high-throughput technologies such as microarray. It can also be referred to as part of the functional genomics which focus on the gene expression especially on mRNA (transcript). Expressed sequence tag (EST) sequencing, microarray, serial analysis of gene expression (SAGE), and RNA sequencing (RNA-seq) are important tools of transcriptomics
PROTEIN LEVEL	
Proteomics	Aims to study the structural and functional aspects of total proteins of an organism or system using high-throughput technologies. Because of the spatial and temporal variability of the proteome of an organism, the proteomics study is quite complicated as compared to genomics, which remains more or less constant for an organism. Proteomics is further subdivided into expression proteomics, proteogenomics, and structural proteomics for carrying out specific proteomic studies

(Continued)

TABLE 1.1 (Continued)

Branches of science of "omics"	Description
Secretomics	Part of proteomics studies associated exclusively with analyzing the total secreted proteins of a cell, tissue, or organism. These secreted proteins are known to be associated with diverse physiological processes and is gaining importance in cancer biology research for the discovery of appropriate biomarkers. Two-dimensional gel electrophoresis and mass spectrometry (MS) are important tools for secretomics
Allergenomics	Aims to investigate the putative proteinous allergens using high-throughput proteomics technologies. The quantitative and qualitative estimation of antigens based on its interaction with the IgE antibodies of a patient is an important step in allergenomics studies

METABOLITE LEVEL

Metabolomics	Comprehensive study of metabolites of a cell, tissue, organ, or organism, manifested as intermediates and product of metabolism. This provides an insight into the repertoire of small molecule metabolites such as hormones, signaling molecules, and secondary metabolites possessed by an organism at a particular time, as it is highly dynamic system. The integration of genomics, proteomics, and metabolomics is a prerequisite for understanding the cellular biology in totality. Nuclear magnetic resonance (NMR) spectroscopy and MS are widely used tools for analysis of the metabolites. The metabolic responses of living systems to pathophysiological stimuli are often referred as "metabonomics"

CELL AND CELLULAR SYSTEM LEVEL

Cytomics	Aims to investigate the systems, subsystems, and other functional system of a cell in totality, revealing the structural and functional diversity of an organism
Physiomics	Comprehensive analysis of cell physiological features derived from the genes and proteins of an organism. The bioinformatics-based construction of gene networks, metabolic pathway profiling, and RNAi are some of tools for physiomics
Phenomics	Associated with understanding the phenotype of an organism by analyzing the physical and biochemical traits under the influence of genetic and environmental changes. Developing appropriate tools for quantitative and qualitative estimation of the phenomes is still a major challenge. Due to temporal and spatial variations of the phenotypes, complete characterization of the phenome is not possible as compared to complete genome characterization using tools of genomics
Embryomics	Aims to investigate the diverse cell types associated with embryogenesis, giving an insight into the location and development history of the cells in the embryo. This is closely associated with the development of regenerative medicine
Toponomics	Associated with understanding the network of proteins and other biomolecules of a cell or tissue (toponome) utilizing the tools of system biology, molecular biology, genomics, proteomics, and metabolomics. Imaging cycle microscopy is often used for elucidation of the spatial organization

MISCELLANEOUS

Foodomics	Comprehensive analysis of diverse components of food and nutrition using high-throughput technologies, in light of improving the human health. The term "nutrigenomics" is a part of the foodomics. The major omics, i.e., genomics, transcriptomics, proteomics, and metabolomics, are extensively applied in foodomics for better understanding of food and nutrition

(Continued)

TABLE 1.1   (Continued)

Branches of science of "omics"	Description
Exposomics	Aims to studies the total environmental influences during the life course of a human, which might be contributing to human diseases. The estimation of environmental effects is comparatively difficult to assess than the genetic factors related to diseases
Connectomics	Omics approach for analyzing the organism's complex nervous system using tools such as diffusion MRI and 3D electron microscopy revealing the complex connections especially in brain and eye
Bibliomics	Bioinformatics intervention revealing the totality of the biological text information. The term was introduced by European Bioinformatics Institute (EBI). There have been substantial developments of databases, potential tools, and softwares for analysis of the sequences in the recent years in bioinformatics
Interferomics	Basically a system biology approach for investigating the life processes at the stages between the posttranscription and pretranslation. The determination of events during this period is crucial for gene expression and regulation and is generally carried out by tools of RNA interference
Glycomics	Omics-based approach for analyzing the total glycan structures of a given cell type or organisms. The complexity of the sugars structures, biosynthetic pathways are investigated using tools such as MS, high-performance liquid chromatography (HPLC), matrix-assisted laser desorption/ionization-time-of-flight–mass spectroscopy (MALDI-TOF-MS), X-ray crystallography, and NMR. It is an integral part of glycobiology
Kinomics	Comprehensive study of kinases associated with phosphorylation-based signaling, known to be closely associated with functioning of cells in physiological and pathological states. Kinome represents the totality of kinases and kinase signaling and is generally investigated by tools such as RNA interference, MS, phage-based assays, fluorescence resonance energy transfer, and antibody array
Lipidomics	Omics approach for assessing the total cellular lipids of biological system and is gaining importance owing to its association with the metabolic diseases such as obesity, stroke, hypertension, and diabetes. It is often a part of metabolomics and generally analyzed by tools such as MS, NMR, and bioinformatics tools
Mechanomics	An interface between biology and biomechanics associated with studies related to mechanical stimuli influencing the biological processes at various levels ranging from cell, tissue, organ or molecular level. The various in vivo mechanical stimuli such as shear flow, tensile stretch mechanical compression influencing biological processes is accessed using transcriptomics and proteomics approaches (Wang et al., 2014)
Regulomics	Aims to investigate the regulatory elements associated with gene expression and gene regulation of an organism. The availability of genome sequence is a prerequisite and more emphasis is on understanding the distribution and type of transcription factors of an organism. The tools of genomics and proteomics are applied in regulomics
Metallomics	Omics approaches exclusively for comprehensive study of metal containing biomolecules of an organism using tools of proteomics. It has relevance in drug discovery based on the assessment of toxicity of the metal containing biomolecules
Phytochemomics	Comprehensive analysis of the phytochemicals, their impact on human health, potential for preventing aging, and protecting from several diseases. This is a growing area based on the growing popularity of herbal drugs in the recent years. The omics-based approaches, namely proteomics and metabolomics, are extensively applied for phytochemomics

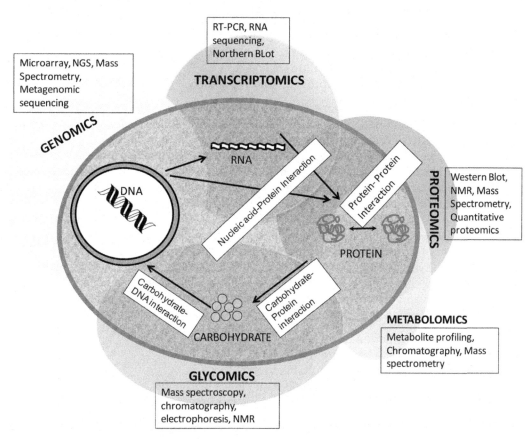

FIGURE 1.1   Bacterial cell showing interplay between the different branches of omics along with the relevant tools and their involvement in the bacterial gene flow.

1.2 GENOMICS

Genomics is a branch of science associated with a comprehensive study of genome of a particular organism. The innovations in sequencing technologies in the recent years have led to the sequencing of several genomes providing substantial importance to the genomics for better understanding of life processes. The genomics aims for

- Decoding the biological information from the whole-genome sequences
- Assessing the total number of genes present in an organism
- Elucidating organization of genes on the chromosomes
- Revealing complexity of genome with reference to distribution of different types of sequences
- Providing an insight into the regulatory sequences present in the genome
- Predicting putative functions of the genes

- Analyzing the genome-wide comparative studies of sequenced genome from evolutionary point of view
- Providing accessibility for genetic manipulation for desired features
- Better understanding of the central dogma of life
- Better understanding of gene expression and regulation

Bioengineering provides a means through which the entire genotype of the biological entity can be altered. Since genes carry all the genetic information stored in them, the genes can be easily moved from one organism to the other, to modulate the target organism depending on our need. Engineering a gene encompasses the basic theme of fragmenting the desired gene and reassembling into the organism of interest. This has proven to be of immense interest ranging from cloning of the gene in unicellular organism to gene therapy in humans. If we are interested in bioengineering of a gene, it can be manipulated inside a living bacterial host or other feasible cell lines. Bacterial cells are the most common host for genetic manipulation as they are easy to culture, relatively cheap to maintain, have faster rate of growth, and have a large number of mutant strains available for multidimensional study of the desired gene (Neus and Villaverde, 2013; Weeks Donald et al., 2016).

The most conventional method for bioengineering is genetic engineering. The basic strategies for production of recombinant proteins are as follows:

- Identification and isolation of the gene of interest
- Selection and designing of appropriate vector
- Ligation of gene of interest into suitable vector
- Transformation/transfection of recombinant clones in suitable host cells
- Selection of recombinants and optimization of transgene integration in host cell

There have been substantial developments at each step of cloning processes and polymerase chain reaction (PCR) is one major breakthrough technologies applied extensively for in vitro amplification of desired genes, provided the sequence information is available (Demain and Vaishnav, 2009). The important microorganisms explored as cell factories for the production of recombinant proteins using the genetic engineering approach along with their unique features and some products are summarized in Table 1.2 (Fisher et al., 2014).

Innovations in sequencing technologies leading to deciphering of several genome sequences have expedited the science of genomics. The traditional method of DNA sequencing popularly known as first generation sequencing technique, proposed by Sanger (enzymatic, dideoxy, or chain termination method) and Maxam and Gilbert (chemical method), has undergone several modifications over the years increasing the speed, precision, and read length of the sequences. The automation of Sanger's method was further upgraded as automated DNA sequencing method with a major change from gel-based to capillary-based analysis. The next-generation sequencing (NGS), which reflects the second, third, and fourth generation sequencing technologies, is currently very popular and highly efficient for genome sequencing. Some of the NGS techniques are pyrosequencing (Roche 454), ABI SOLiD, HeliScope, Ion Torrent PGM, Pacific Biosciences' real-time single molecule sequencing (PacBioRS), combinatorial probe-anchor ligation (cPAL), Oxford Nanopore, etc. The benefits of advanced sequencing technologies include

TABLE 1.2 List of Microorganisms Used for Recombinant Protein Production

Group	Main Species	Main Features	Examples
Proteobacteria	*Caulobacter crescentus Rhodobacter sphaeroides Pseudoalteromonas haloplanktis Pseudomonas fluorescens Halomonas elongata Chromohalobacter salexigens Escherichia coli*	Fast expression, high yield, amenable to culture and genome modifications, cheaper, production of mass culture is fast, easy purification of fusions, high production of membrane proteins, improved protein folding	Hematopoietic necrosis virus, capsid proteins, human nerve growth factor, human granulocyte colony-stimulating factor, β-lactamase, nucleoside diphosphate kinase, alkaline phosphatase, levanfructotransferase, human granulocyte colony-stimulatory factor, insulin-like growth factor, cholera toxin b
Actinobacteria	*Streptomyces lividans Streptomyces griseus Nocardia lactamdurans Mycobacterium smegmatis Corynebacterium glutamicum Corynebacterium ammoniagenes*	Efficient secretion, posttranslational modificationsHigh-level production and secretion	Trypsin, lysine-6-aminotransferaseHsp65-hIL-2 fusion protein, protein-glutaminase, pro-transglutaminase
Firmicutes	*Bacillus subtilis Bacillus brevis Bacillus megaterium Bacillus licheniformis Lactococcus lactis Lactobacillus plantarum Lactobacillus casei*	High-level production and secretionGenetically well characterized, so easy to manipulateHighly developed transformation and gene replacement technologies; rapid growth, safe according to US FDA, cost-effective recovery	β-Galactosidase, disulfideisomeraseAntibodies, subtilisin, neutral protease, minor serine protease, bacillopeptidase F, metalloesterase, fibronectin-binding protein A, internalin A, β-galactosidase
Yeasts	*Saccharomyces cerevisiae, Hansenula, Pichia pastoris, Candida, Torulopsis*	High yield, durable products, cheap, high-density growth, high productivity, fast growth in chemically defined media, product processing similar to mammalian cells such as protein folding and protein glycosylation	γ-Interferon, erythropoietin, human chorionic gonadotropin, tissue plasminogen activator, interferons, insulin, interleukin, human serum albumin, tumor necrosis factor, tetanus toxin

enhanced speed, accuracy, reduction in cost, increase read length, reduced amount of initial DNA required, less chance of contamination, automation, etc.

Microarray is a high-throughput technique for determination of expression level of large number of genes at a time or for whole-genome typing. Numerous DNA sequences known as probe or spots are attached to the solid surface. The DNA samples to be analyzed are labeled with fluorescent probes and washed against the DNA microarray chip. The complementary DNA sequences of the target bind to the DNA of the chip and are subsequently detected with the help of fluorescent detectors and other imaging techniques. In addition to this core method, there are several modifications of the microarray technique, depending upon its uses in different applications. Microarray can also be used for comparative analysis of different genomes by comparative hybridization. Specific microbes, such as pathogens or microbes in food, can be detected by GeneID microarray. Microarray followed by chromatin immunoprecipitation assay can be used in identifying the DNA sequence bound to any given protein. Single nucleotide polymorphism can also be typed with microarray. Most recent modifications are tiling array, double-stranded B-DNA microarrays, double-stranded Z-DNA microarrays, and multistranded DNA microarrays.

The tools of genomics can be applied in medical field for diseases diagnostics, gene therapy, and somatic cell therapy. In agriculture identification of potential genes associated with desirable agronomic traits can be done from the sequenced genomes and utilized in transgenic technology for developing biotic and abiotic stress-tolerant crops.

1.3 TRANSCRIPTOMICS

Transcriptomics is a comprehensive analysis of whole sets of transcripts for a particular cell, tissue, organ, or whole organism corresponding to a particular time or developmental stages or may be under some specific physiological conditions. Transcriptomics aims for the following:

- Functional annotation of the genome
- Providing an insight into transcriptional structure of the genes
- Deciphering transcriptional start site
- Elucidating splicing
- Understanding posttranslational modifications
- Cataloguing all types of transcripts (mRNA, tRNA, rRNA, siRNA, noncoding RNAs, etc.)
- Providing an insight into the complexity of development processes or to understand diseases
- Elucidating differentially expressed genes

It provides an overview regarding the genetic content and the regulatory system in the cell. The wild-type cells, whether bacterial cell or cell line, possess the same genome content. Based on different applications and conditions, different genes are expressed, resulting in different patterns of gene expression in different organisms. There are cellular regulatory machineries through which the gene expression can be switched on and off. Great deal of work has been performed on the eukaryotes as it contains both introns and exons which through splicing provides better spectrum of the total transcriptome. Trancriptome study of prokaryotes is lagging due to the absence of the 3'-end poly(A) tail, which is considered to be a signature of mature mRNA in eukaryotes. Data related to the expression level of the genes in the given genome, genome profiling, comparative expression levels between different experimental data sets, and effect of different parameters on gene expression can be assessed by the help of transcriptomics tools.

Commonly used techniques for transcriptome study are expressed sequence tag (EST)-based methods, SAGE, hybridization-based microarray, real-time PCR, NGS-based RNA-sequencing (RNA-seq) methods, RNA interference, and bioinformatics tools for transcriptomes analysis. The methodology basically involves RNA isolation, purification, quantification, cDNA library construction, and high-throughput sequencing. The important considerations for selection of tools of transcriptomics are cost effectiveness, sensitivity, high throughput, and minimal concentration of starting RNA.

Sequencing of the ESTs gives an overview regarding the expression level of the gene. Large-scale EST sequencing projects have been applied on the model organisms such as *Arabidopsis*, *Drosophila*, and human. EST databases are also available which provide reference for the expression profile. Although it is helpful in studying gene abundance and novel gene discovery, it is highly expensive and time-consuming technique. The cDNA or EST microarrays are available whereby cDNA sequences or ESTs are attached on the microchip. The technique is similar to DNA microarray, but the sample here are the fluorescently labeled cDNA molecules. Being a cost-effective and rapid technique, it helps in the study of relative abundance between the genes. SAGE is another method, which is similar to EST sequencing. It is advantageous over EST because only short "tags" of about only 15 bases are sequenced. The short fragments generated are then joined together and sequenced. A pool of cDNA can be subjected to high-throughput NGS known as RNA-seq for quantification, discovery of novel ESTs, and profiling of RNAs.

DNA chip, gene chip, biochip, or microarray is a collection of DNA, cDNA, oligonucleotides spots attached to a solid support such as glass, silicon chip forming an array for the purpose of expression profiling; monitoring expression levels of thousands of genes simultaneously and is a hybridization-based method. Several innovations regarding the fabrication of chips, controlling hybridization conditions, and computational analysis of data generated have been achieved over the years. The high-density chip allows relatively low-cost gene-expression profiling, once the fabrication of chip has been achieved. The major limitation of this technology is that genome sequence information is a prerequisite and also higher background inherent of hybridization technique.

The quantitative real-time PCR (qRT-PCR) is a type of PCR preferred for reliable quantification of low-abundance mRNA or low-copy transcripts for transcriptomics studies. The major advantages of this technique are its high sensitivity, better reproducibility, and wide dynamic quantification range. It facilitates gene expressions and regulation studies even in a single cell based on its exponential amplification ability. The availability of diverse types of fluorescence monitoring system attached with the PCR resulted in its popularity for gene-expression studies. The problems of nonspecific amplification, formation of primer-dimers are some of the limitations of qRT-PCR.

RNA-seq is one of the advanced high-throughput technology for transcriptomics. RNA-seq, also known as whole-transcriptome shotgun sequencing, utilizes NGS tools. The advantages of RNA-seq are that it does not rely on the availability of the genome sequence, has no upper quantification limits, shows high reproducibility, and possesses a large dynamic detection range.

The stem cells and cancer cells transcriptomes are emerging field of interest which can provide an insight into the processes of cellular differentiation and carcinogenesis.

Transcriptomics can be used in in vitro fertilization for proper embryo selection. Similarly, in biomarker discovery, it can be used in assessing the safety of drugs or chemical risk.

1.4 PROTEOMICS

The study and characterization of entire set of proteins present in cell, organ, or organism at a particular time is referred to as proteomics. Due to alternative splicing and post-translational modifications, there are vast arrays of protein content in all living cells. Therefore, this diverse and complex proteome needs to be carefully evaluated to get an insight into the cell physiology and other metabolic processes. To overcome this complexity, many proteomic techniques have come up facilitating large-scale, high-throughput analyses for the detection, identification, and functional investigation of proteome. The basic procedures involved in proteomic study are the isolation of total protein from the cell, its separation, and quantification (Kellner, 2000). The proteomics aims for the following:

- Comprehensive and global study of expression of genetic information at protein level
- Providing an insight into three dimensional structures of proteins
- Assessing protein—protein or DNA—protein interactions
- Elucidating protein structure and function relationship
- Assessing distribution of proteins in different cellular and subcellular components of the cell
- Elucidating the molecular basis of patho(physiological) processes
- Quantitative protein expression profile of a cell, an organism, or a tissue under defined conditions
- Facilitating genome annotation
- Providing an insight into alternative splicing
- Elucidating posttranslational modifications

The tools of proteomics can be applied for diverse applications as listed in Table 1.3.

Liquid chromatography (LC) followed by MS is a reliable and powerful tool for proteomics study. It helps in characterization of both native and recombinant proteins. However, the requirement of very concentrated samples and difficulty in handling problematic class of proteins, such as membrane proteins, large-sized and complex folded proteins, and the disulfide rich entities, are the major disadvantages of this technique. To overcome these shortcomings, sample has to be enriched, instrumentation must be improved, and the resulting data should be carefully processed. In addition, vast arrays of other proteomic techniques such as gel-based and gel-free techniques are also available. Gel-based techniques are the commonly used one-dimensional and two-dimensional polyacrylamide gel electrophoresis. Fluorescence two-dimensional difference gel electrophoresis is the modification of two-dimensional polyacrylamide gel electrophoresis wherein the protein samples are fluorescently labeled with Cy2, Cy3, or Cy5 prior to second dimensional electrophoresis. This makes it highly sensitive and quantitative data can also be inferred.

TABLE 1.3   Applications of Proteomics

1.	Discovery of protein biomarkers
2.	Study of tumor metastasis
3.	Field of neurotrauma
4.	Renal disease diagnosis
5.	Neurology by investigating brain protein groups in neurodegeneration
6.	Fetal and maternal medicine
7.	Urological cancer research
8.	Autoantibody profiling for the study and treatment of autoimmune disease
9.	Cardiovascular research
10.	Diabetes research
11.	Nutrition research

Among the gel-free techniques are the isotope-coded affinity tag (ICAT). As the name indicates, the protein is labeled with the ICAT reagent. The labeled protein after digestion with trypsin and separation by chromatographic procedures is detected with tandem MS and quantitative estimation is done by LC peak areas. But this technique is difficult to work with acidic proteins and works selectively for proteins with high cysteine content. To overcome the limitation of ICAT a newer technique called stable isotope labeling with amino acids in cell culture has emerged. Here the cells are directly labeled with the radiolabeled isotopes which makes the study of differential expression highly convenient. Isobaric tag for relative and absolute quantitation is a high-throughput technique which allows relative quantification of the protein samples. Samples are multiplexed and need to be separated before MS analysis. Multidimensional protein identification technology is an integration of three different techniques. Site-specific tryptic digestion of the protein is performed and the peptides are separated by HPLC. From the HPLC the peptides enter MS column. The MS data is analyzed for the constituent proteins. Similar to DNA and cDNA microarrays there are proteins and antibody microarrays available for the proteomic-based studies.

Mass spectrometry is the key technique in all the proteomic analysis. In a spectrophotometer, first the protein samples are ionized and gas phase ions are produced, these ions are then separated according to their mass to charge ratio and finally the ions are detected for generation of mass spectrophotometry data. The three main parts of a mass spectrophotometer are ion source, mass analyzer, and ion detector. Major ionization sources are MALDI and electrospray ionization, and four major mass analyzers are TOF, ion trap, quadrupole, and Fourier transform ion cyclotron which are mainly applied for protein identification and characterization.

1.5 METABOLOMICS

It is a comprehensive systematics quantitative assessment of all metabolites in a metabolome under specified conditions. In general, metabolites are substances produced in or by biological processes. It can be either primary metabolite or secondary metabolites. It includes peptides, oligonucleotides, sugars, nucleosides, organic acid, ketones, aldehydes, amines, amino acids, lipids, steroids, alkaloids, drugs, xenobiotics, etc. (Fiehn, 2002). Metabolomics aims for:

- Deciphering the molecular network regulation
- Assessing pathway discovery
- Understanding functional genomics
- Analyzing material samples to gain an understanding of their biochemical composition and structure
- Providing mechanistic insight into diverse type of biological processes
- Unraveling connecting link between genotypes and phenotypes
- Assisting increased metabolic fluxes into valuable biochemical pathways using metabolic engineering
- Investigating the cause of biological effects such as plant−pathogen interactions, plant−environment interactions

The metabolomics tools find significant applications in the study of human diseases and oncobiology. These metabolites are used as biomarkers in several diseases such as male infertility, lung cancer, Alzheimer's disease, respiratory disease, Huntington's disease, kidney cancer, and renal cell carcinoma.

The basic methodology of metabolomics involves extraction from the targeted biological tissues/samples, separation of metabolite, and detection followed by identification and quantification (Liao et al., 1996). The basic tools of metabolomics associated with separation are various types of chromatographies such as gas chromatography (GC), HPLC, and capillary electrophoresis (CE), while metabolites can be detected based on different physical properties such as mass, rotation/vibration, absorbance and emittance of light (hv), spin, volatility, size, charge, and hydrophobicity. The methods applicable for the detection of the metabolites are MS, infrared spectrometry, UV-visible spectrometry, fluorescence spectrometry, nuclear magnetic resonance (NMR), GC, LC, CE, and size exclusion chromatography.

1.6 GLYCOMICS

Glycomics deals with a comprehensive study of structural and functional aspects of carbohydrates, which play a significant role in the storage and transmission of biological information associated with diverse types of physiological and pathophysiological processes (Furukawa et al., 2013). It is also considered as a branch of metabolomics and often referred to as glycobiology. Glycans are oligosaccharides or polysaccharides present either in free form or attached to other biomolecules. Contrary to DNA, RNA, and protein, glycans are not encoded through template biosynthetic pathway and also lack proofreading mechanism. Glycosyltransferase is involved in glycan biosynthesis while epimerase, sulfotransferase, etc., are glycan-modifying enzymes. These enzymes are required for site-specific attachment to proteins and lipids to form glycoproteins, proteoglycans, and glycolipids. Glycans attached to proteins are responsible for folding, stability, regulating protein structure and function. Glycoproteins, proteoglycans and glycolipids are diverse and each cell has its own type. According to the data, 70% of therapeutic proteins approved by European and US regulatory agencies are glycoproteins. A large variety of glycans located on the cell surface is associated with vital functions and influences the cell—cell adhesion, immune response, pathogenesis events, cellular process such as development, differentiation, and cellular alteration leading to cancer. The glycomics aims for the following:

- Glycan expression analysis
- Elucidating the diversity of glycan structures
- Elucidating the physiological and pathophysiological roles of glycans located on the surface
- Determining the complexity of glycan with reference to composition, branching, linkages, and anomericity
- Deciphering the potential of glycans as biomarkers for diseases associated with cellular alterations
- Elucidating the N- or O-linked glycans derived from glycoproteins, proteoglycans, and glycosphinolipids, which are important cell surface glycoconjugates

One of the basic techniques to study glycans is high-resolution MS in which even small quantities of glycans can be characterized. The N-glycans of the glycoproteins can be separated and sequenced separately, while the glycolipids can be directly sequenced without separation of the lipid component. Another glycan, glycosaminoglycans, is difficult to analyze through MS due to its large size. Through MS, multiple glycans of any subtype can be profiled. However, mass spectrophotometer cannot perform analysis of complex samples containing multiple glycans of different subtypes. With the help of MS, the mode of binding between the protein and glycans in the glycoproteins can also be studied.

Although MS provides detailed information of glycan, it cannot be used to screen large number of samples at a time. For high-throughput studies, lectin and antibody arrays are available. This is similar to DNA microarray, where lectins (or glycan-specific antibodies) are attached on the array chips instead of DNA spots. The distribution of glycans in different cells and tissues can be routinely analyzed with the help of lectins and glycan-specific antibodies. These act as probe to profile the expression pattern of different glycans. The glycans can also be metabolically labeled with the help of imaging reagents and visualized under the fluorescent microscope. It helps in the real-time imaging of the total glycome content in any cell or tissue.

Many publications are focused on N- and O-glycans as disease markers (de Leoz et al., 2008). Alterations in the degree of branching and levels of sialylation and fucosylation in N-glycans have been reported as a consequence of many diseases such as cancer of prostate, breast, liver, ovary, pancreas, and gastric. Cancer is signified through changes in the branching of N-glycans, truncation of O-glycans, linkage, and acetylation (Varki et al., 2009). The tools of glycomics can be utilized for several clinical applications for disease diagnosis, detection of cancer, and personalized medicine. As therapeutic agents, glycan-integrated nanomaterials can be used as vaccine against viral and bacterial infections and cancer cells.

1.7 OMICS-DRIVEN BIOENGINEERING

1.7.1 Cellular Engineering

The cellular engineering is mainly associated with the modifications and alterations of host cells for enhancing the yield and quality of the expressed protein. There are several approaches for altering the cellular processes such as silencing or overexpressing of targeted genes and altering expression of genes using microRNA (miRNA). Gene silencing is the major approach for cellular engineering. RNA silencing is a novel gene regulatory mechanism that limits the transcript level by either suppressing transcription or by activating a sequence-specific RNA degradation process (RNAi). miRNAs are noncoding RNAs which regulate gene expression and further cell physiology. RNA interference is being used to silence multiple genes of different cellular pathways for optimum productivity of bioengineered proteins. RNA interference can be used to target several features such as apoptosis, glycosylation, and enzymes (e.g., dihydrofolatereductase). Once miRNA sequences are identified through complete genome sequence, it can be either overexpressed or silenced to achieve desired regulation of gene expression. New advanced tools that are currently used for gene editing are zinc finger nucleases, modified meganucleases, hybrid DNA/RNA oligonucleotides, Transcription Activator like (TAL) effector nucleases, and modified Clustered Regularly Interspaced Short Palindrome Repeats (CRISPR)/Cas9. These tools have the ability to target one specific DNA sequence in genome, which leads to double-stranded DNA break. DNA repair to these breaks leads to gene knockouts or gene replacement by homologous recombination.

1.7.2 Enzyme Engineering

Enzymes are considered to be one of the major products of biotechnology with immense industrial applications and are known to expedite the rates of a wide range of biochemical reactions along with exquisite specificity. There exist several enzymes especially produced from the microbial sources having diverse application in all sphere of life. The growing need of industrially important enzymes has led to the development of innovations mainly protein engineering—based technologies for developing novel biocatalysts with desired features. The advent of recombinant DNA technology and omics-based technologies, such as genomics and proteomics, has expedited the development of novel biocatalysts in the recent years. The microbial enzymes have been commercially produced by fermentation technology utilizing either submerged fermentation or solid-state fermentation. There is a need for intervention of omics-based technologies at different steps of enzyme technology from production and manipulation of enzymes for desired features to search for novel sources by utilizing the untapped microbial diversity. Protein engineering techniques aim at modifying the sequence of a protein, and hence its structure, to create enzymes with improved functional properties such as stability, specific activity, inhibition by reaction products, and selectivity toward nonnatural substrates (Bloom et al., 2005). For past few decades, biocatalysts have been successfully exploited for the synthesis of various complex drug intermediates, speciality chemicals, and even commodity chemicals in the pharmaceutical, chemical, and food industries due to their inherent ability to catalyze reactions with high velocity and unmet specificity under a variety of conditions, as well as their potential as a greener alternative to chemical catalysts. The role of protein engineering is to overcome the limitations of natural enzymes as biocatalysts and engineer process−specific biocatalysts. Some of the tools of protein engineering applied for enzyme manipulation are discussed as follows.

Site-directed mutagenesis is an important tool in enzyme engineering targeting for specific and desired changes in the gene sequences coding for the concerned enzyme. The basic mechanism involves synthesis of an oligonucleotide primer. The desired mutation sites are incorporated into the primer sequence. This may be single-point mutation, multiple-point mutation, or insertion or deletion of the targeted DNA. Next, the DNA polymerase is used to amplify the rest of the DNA sequence. The final DNA sequence after amplification contains the incorporated mutant site which is then transformed in the host cell and cloned. The mutants are confirmed and selected by DNA sequencing.

Directed evolution of enzymes is a well-established high-throughput technology with potential for desired manipulation of enzymes. It is a process which mimics natural evolution carried out in laboratory and involves iterative strategy. The procedure begins by determining a target biomolecule, metabolic pathway, or organism, and a desired phenotypic goal. A diverse library of mutants is generated in vivo or in vitro through methods that mirror the strategies of traditional evolution: introduction of random mutations in the genetic material and/or "sexual" gene recombination. A high-throughput screening or selection method is used to identify improved progeny among the library, which are subsequently used as parents in a second round of the cycle. The process is repeated until the phenotypic goal is achieved, or when no further improvement of the phenotype is observed despite repeated iterations. Using tools of directed evolution, attempts have been made to enhance catalytic activity of enzyme, improve the stability over a wide range of temperature and pH, alter the specificity, expand or limit substrate specificity, etc.

The search for potential microbial sources for enzymes has been always a limitation in enzyme research as the prerequisite is to develop suitable media for culturing of the microbes in lab conditions. Since only less than 1% of the microbial diversity has been explored based on ability to culture in lab conditions, the remaining 99% is still to be explored. The relatively newer concept of "metagenomics approach" seems to have immense potential for searching novel sources for potential enzymes. In metagenomics approach, random environmental samples are picked and the DNA is directly isolated. The genomic library of isolated metagenomic DNA is prepared and screened by either structure-based or function-based screening procedures. Since then large number of enzymes have been discovered through this approach.

The deciphering of whole-genome sequences of microbes, known to be potential sources of industrially important enzymes, resulted in genome-wide identification and bioinformatics-based characterization of putative genes coding for the targeted enzymes. Cloning and expression of genes of targeted enzymes could be in-silico analyzed for three-dimensional structural predictions and validation along with deciphering potential catalytic sites prior to subjecting it to wet-lab experimentation (Damborsky and Brezovsky, 2014). The tools of genomics and proteomics can be applied for characterization of enzymes and appropriate enzyme engineering can be targeted based on it.

1.7.3 Metabolic Engineering

Metabolic engineering is basically meant for the production of chemicals, fuels, pharmaceuticals, and medicine by altering the metabolic pathways. This method involves useful alteration of metabolic pathways to better understand and utilize the cellular pathways. Metabolic engineering is motivated by commercial applications by which we can improve the developing strains for production of useful metabolites. This method requires overexpression or downregulation of certain proteins in a metabolic pathway, such that the cell produces a new product. First step for successful engineering requires the complete understanding of host cell for genetic modifications. The effect of genetic manipulation on growth should also be examined. The genetic manipulation may negatively effect on metabolic burden. For example, genetic manipulation for overexpression of phosphoenolpyruvate inhibits heat-shock response and nitrogen regulation. To achieve the product several methods, such as elimination of competitive pathway and toxic byproduct and expression of a heterologous enzyme to improve the synthesis of products, are used for the modification of the host cell. In classical metabolic engineering approach, genetic manipulation is based on the prior knowledge of enzyme network pathway and its kinetics; on the other hand, in inverse metabolic engineering the environmental or genetic conditions are considered for the desired phenotype for genetic manipulation. A number of approaches are used for secondary metabolites improvement using bacteria in metabolic engineering, such as (1) heterologous expression of gene clusters, (2) regulatory networks pathway engineering, (3) gene insertion or deletion, and (4) stimulation using certain substrates. Therefore, metabolic engineering is currently used as cell factories for production of amino acids, biofuels, pharmaceuticals, bioplastics, platform chemicals, silk proteins, etc.

1.8 BIOINFORMATICS INTERVENTION IN OMICS

The outcome of sequencing projects resulted in substantial increase in the sequence databases globally, demanding the intervention of web-based bioinformatics for proper storage, retrieval, and analysis of sequences. The huge data generated in the field of genomics, proteomics, transcriptomics, metagenomics, and metabolomics needs bioinformatics tools, browsers, or softwares for quick, reliable, and efficient analysis. Several bioinformatics tools have been developed for omics as listed in Table 1.4.

1.9 APPLICATIONS OF OMICS

1.9.1 Food and Agriculture Sector

There have been several innovations for improvement of crop over the years mainly targeting for enhancing the yield or improving the quality. The classical breeding, molecular

TABLE 1.4 List of the Important Bioinformatics Tools/Softwares/Browsers for Science of Omics

Softwares/Browsers name	Website Address	Description
GENOMICS		
UCSC Genome Browser	genome.ucsc.edu/cgi-bin/	Provides a rapid display of portion of genomes along with predicted genes, ESTs, mRNAs, CpG islands, assembly gaps and coverage, chromosomal bands, mouse homologies, etc.
BPhyOG	http://cmb.bnu.edu.cn/BPhyOG	Interactive server for genome-wide inference of bacterial phylogenies based on overlapping genes
"Genomic Repeat Element Analyzer for Mammals" (GREAM)	http://resource.ibab.ac.in/GREAM	For analysis, screening, and selection of potentially important mammalian genomic repeats such as transposons, retro-transposons, and other genome-wide repetitive elements
OrthoVenn	http://probes.pw.usda.gov/OrthoVenn or http://aegilops.wheat.ucdavis.edu/OrthoVenn	Useful for genome-wide comparisons and visualization of orthologous clusters. OrthoVenn provides coverage of vertebrates, metazoa, protists, fungi, plants, and bacteria
Prokaryotic Sequence homology Analysis Tool (PSAT)	http://nwrce.org/cgi-bin/psat/select.cgi	A web tool to compare genomic neighborhoods of multiple prokaryotic genomes
Cancer Genomic Data Server (CGDS)	http://www.cbioportal.org/index.do	The portal currently contains data from 126 cancer genomics studies
BPhyOG	http://cmb.bnu.edu.cn/BPhyOG	Interactive server for genome-wide inference of bacterial phylogenies based on overlapping genes
CVTree3	http://tlife.fudan.edu.cn/cvtree3/	Web server for whole-genome-based and alignment-free prokaryotic phylogeny and taxonomy
g:Profiler	http://biit.cs.ut.ee/gprofiler/	Web server for functional interpretation of gene
GEneSeT	http://bioinfo.vanderbilt.edu/webgestalt/	Designed for functional genomic, proteomic, and large-scale genetic studies
NuST (nucleoid survey tools)	lgm.upmc.fr/nust/	For the analysis of the aggregation of specific gene sets along the chromosome, at different observation scales
PROTEOMICS		
UFO	http://ufo.gobics.de	Ultra-fast functional profiling of whole-genome protein sequences
PIQMIe	http://piqmie.semiqprot-emc.cloudlet.sara.nl	Web server for semi-quantitative proteomics data management and analysis
ProteomeAnalyst	webdocs.cs.ualberta.ca/	Web server built to predict protein properties, such as general function, in a high-throughput fashion

(Continued)

TABLE 1.4 (Continued)

Softwares/Browsers name	Website Address	Description
iHOP	http://www.ihop-net.org/UniPub/iHOP/	For protein interactions
PIC	http://pic.mbu.iisc.ernet.in/	Protein interaction calculator
String	http://string-db.org/	Web resource of known and predicted protein–protein interactions
Maspectras	http://genome.tugraz.at/maspectras/	For management and analysis of proteomics LC–MS/MS data
SRMAtlas	http://www.srmatlas.org/	Targeted proteomics assays to detect and quantify yeast proteins in complex proteome digests by MS
MRMaid	www.mrmaid.info	Quantification of specific proteins by MS/MS
pProRep	http://www.php.net	An integrating electrophoretic and mass spectral data from proteome analyses into a relational database
ELASPIC	http://elaspic.kimlab.org	Proteome-wide structure-based prediction of mutation effects on protein stability and binding affinity
myProMS	http://bioinfo.curie.fr/myproms	Server for management and validation of MS-based proteomic data
TRANSCRIPTOMICS		
T-Rex	http://genome2d.molgenrug.nl	Transcriptome analysis web server for RNA-seq expression data
NFFinder	http://nffinder.cnb.csic.es	For searching similar transcriptomics experiments in the context of drug repositioning
Trinotate	http://trinotate.github.io	Automatic functional annotation of transcriptomes
TRAPID	http://bioinformatics.psb.ugent.be/webtools/trapid	An online tool for the fast and efficient processing of assembled RNA-seq transcriptome data
ExPASy	https://www.expasy.org/	Provides access to scientific databases and software tools in different areas including proteomics, genomics, phylogeny, systems biology, population genetics, transcriptomics, etc.
METAGENOMICS		
WebMGA	http://weizhongli-lab.org/metagenomic-analysis	Web server for fast metagenomic sequence, provides users with rapid metagenomic data analysis using fast and effective tools
METAGENassist	http://www.metagenassist.ca	Comprehensive web server for comparative metagenomics

(Continued)

TABLE 1.4 (Continued)

Softwares/Browsers name	Website Address	Description
CoMet	http://comet.gobics.de	For comparative functional profiling of metagenomes
MG-RAST	metagenomics.anl.gov	Platform for metagenomes providing quantitative insights into microbial populations based on sequence data
EBI Metagenomics	https://www.ebi.ac.uk/ metagenomics/	Is an automated pipeline for the analysis and archiving of metagenomic data
myPhyloDB	http://www.myphylodb.org	Web server for the storage and analysis of metagenomic data
METABOLOMICS		
MetaboAnalyst	http://www.metaboanalyst.ca/	For processing and statistical analysis of metabolomics data
BIOSPIDER	www.biospider.ca	Web server for automating metabolome annotations
MBROLE	http://csbg.cnb.csic.es/mbrole/	Provide the user with information on metabolite interactions with proteins
GLYCOMICS		
GlycoViewer	http://www.systemsbiology.org.au/ glycoviewer	For visual summary and comparative analysis of the glycome
GLYCOSCIENCES.de	http://www.glycosciences.de/	Contains bioinformatics resources for glycobiology and glycomics. Applications are related to carbohydrate 3D structures, MS, or NMR

breeding, and genomics-assisted breeding reflect the developments of technologies. The omics-driven agriculture has great potential for identifying the genes showing potential agronomics traits, developing appropriate molecular markers, applying tools of forward and reverse genetics and also for developing genetically modified or biotech crops with desired features. The genomics provides information regarding the genetic blueprint for making changes at the genetic level. Molecular markers can be identified with the help of genomics, and their association with the different genes can be exploited for genetic improvement. The genome-wide association studies can be made to evaluate the correlation between a specific genetic trait and a marker allele, the selected markers are then used for prediction of the breeding values. Once the breeding values of the genome-wide markers are estimated, they can be substantially applied in crop-improvement programs. Contrary to this strategy, instead of targeting markers, whole-genome sequence can be explored for genetic improvement using traditional breeding or using transgenic approaches.

Genomics-driven biomarkers are also useful in the study of wild habitats and their conservation. These can also be used for assessment of genetic diversity. Similarly, using "metagenomics" approach one can utilize the unexplored microbial diversity available in different environmental samples such as soil, air, and water for fishing out novel sources for enzymes of metabolites. Intervention of omics is also feasible for identification of contamination in food and relevant molecular markers can be typed for easy identification of pathogens.

1.9.2 Health Sector

The genomics-based information of pathogens and human can facilitate the development of suitable drugs. The effect of drugs on individual may vary, based on the variation of the sequences, and needs to be addressed based on the individual genome sequences. This has led to the concept of personalized medicine. The omics study encompasses all the aspects of information driven from a body related to the genome, transcriptome, proteome, and metabolome. All these parameters can be studied as personalized omics for an individual. In addition, a relatively newer microbiomics providing detailed parameter of all the microorganisms, both good and bad in our body, can be utilized for developing appropriate healthcare technology. Using omics-based technology one can have better understanding about the causative agent for the targeted diseases, can facilitate typing of disease symptoms, can assist in developing relevant biomarkers, etc. There has been substantial improvement in the disease diagnosis, pathogen detection, understanding the physiological and pathological conditions of patients, getting an insight into the cause of genetic or hereditary diseases, and developing appropriate strategies for enhancing life expectancy of human. There have been efforts for omics-based studies for cancer and other diseases over the years.

1.9.3 Environmental Sector

The effect of chemicals on the environment can be monitored by omics technologies. Through the integrated approach of different omics-based methods, the ill-effect of these agents can be monitored. The biomarker for toxicity of specific chemical reagent can be typed to study its mechanism of action and also the risk associated with each hazard-causing chemical. Environmental omics is still in preliminary stage and improvement must be made for characterization of mixture of toxic samples and development of technologies and assays to monitor its impact (Yue Ge et al., 2013). This will help in better understanding of environmental toxicity and sustainable recovery of the environment from health hazards. Omics-based approaches can be used for understanding the interactions of environmental and genetic factors, toxicity mechanism, and short- or long-term effects of environmental chemicals on human health. Using meta-omics, there is a possibility of assessment of microbial communities for identifying novel enzymes associated with degradation of environmental pollutants, developing novel biomarkers for environmental risk monitoring, etc.

1.10 CONCLUSION

Integrating the omics technologies will assist in better understanding of life processes. Over the years, there have been substantial improvements in technological innovations relevant to different branches of omics. The availability of whole-genome sequences has expedited the growth of genomics, proteomics, and metabolomics along with bioinformatics intervention, and has led to the development of appropriate tools for analysis of genome, proteome, and metabolome efficiently. The extrapolation of the omics information is for the development of strategies for improvement in different sector of life such as agriculture, health, industry, and environment. Efforts have been made to upgrade the tools of omics by enhancing the speed and accuracy, making it cost-effective and easy to operate and analyze the results. In the recent years, based on the success of genomics, proteomics, and metabolomics there have been reports of repertoire of omics dealing exclusively with specific areas. There is an immediate need for developing appropriate links between diverse omics based on the emergence of huge data generated by omics technologies.

References

Alyass, A., Turcotte, M., Meyre, D., 2015. From big data analysis to personalized medicine for all: challenges and opportunities. BMC Med. Genomics 8, 33.

Bloom, J.D., Meyer, M.M., Meinhold, P., Otey, C.R., MacMillan, D., Arnold, F.H., 2005. Evolving strategies for enzyme engineering. Curr. Opin. Struct. Biol. 15, 447–452.

Damborsky, J., Brezovsky, J., 2014. Computational tools for designing and engineering enzyme. Curr. Opin. Chem. Biol. 19, 8–16.

de Leoz, M.L.A., An, H.J., Kronewitter, S., Kim, J., Beecroft, S., Vinall, R., et al., 2008. Glycomic approach for potential biomarkers on prostate cancer: profiling of N-linked glycans in human sera and pRNS cell lines. Dis. Markers 25, 243–258.

Demain, A.L., Vaishnav, P., 2009. Production of recombinant proteins by microbes and higher organisms. Biotechnol. Adv. 27, 297–306.

Fiehn, O., 2002. Metabolomics-the link between genotypes and phenotypes. Plant Mol. Biol. 48, 155–171.

Fisher, A.K., Freedman, B.G., Bevan, D.R., Senger, R.S., 2014. A review of metabolic and enzymatic engineering strategies for designing and optimizing performance of microbial cell factories. Comput. Struct. Biotechnol. J. 11, 91–99.

Furukawa, J., Fujitani, N., Shinohara, Y., 2013. Recent advances in cellular glycomic analyses. Biomolecules 3, 198–225.

Ge, Y., Wang, D.-Z., Chiu, J.-F., Cristobal, S., Sheehan, D., Silvestre, F., et al., 2013. Environmental omics: current status and future directions. J. Integr. Omics 3 (2), 75–87.

Kellner, R., 2000. Proteomics-concepts and perspective. Fresenius J. Anal. Chem. 366, 517–524.

Liao, J.C., Hou, S.Y., Chao, Y.P., 1996. Pathway analysis, engineering, and physiological considerations for redirecting central metabolism. Biotech. Bioeng. 52, 129–140.

Malviya, N., Tanveer, A., Yadav, S., Yadav, D., 2016. Emerging tools and approaches to biotechnology in the omics era. In: Khan, M.S., Khan, I.A., Barh, D. (Eds.), Applied Molecular Biotechnology: The Next Generation of Genetic Engineering. CRC Press, Taylor and Francis Group, Boca Raton, FL, pp. 1–29.

Neus, F.M., Villaverde, A., 2013. Bacterial cell factories for recombinant protein production; expanding the catalogue. Microb. Cell Factories 12, 113.

Varki, A.C.R., Esko, J.D., Freeze, H.H., Stanley, P., Bertozzi, C.R., Hart, G.W., et al., 2009. Essentials of Glycobiology, second ed. Cold Spring Harbor Laboratory Press, Cold Spring Harbour, NY.

Wang, J., Lu, D., Mao, D., Long, M., 2014. Mechanomics: an emerging field between biology and biomechanics. Protein Cell 5 (7), 518–531.

Weeks Donald, P., Spalding Martin, H., Yang, B., 2016. Use of designer nucleases for targeted gene and genome editing in plants. Plant Biotechnol. J. 14 (2), 483–495.

CHAPTER

2

Omics Approaches Towards Transforming Personalized Medicine

Dipali Dhawan

PanGenomics International Pvt. Ltd., Ahmedabad, Gujarat, India

2.1 INTRODUCTION

The increasing knowledge on genetic variations and their effect on the therapeutic regime have enabled various researchers and clinicians to utilize this information for better patient management. A number of monogenic polymorphisms in gene coding for drug metabolizing enzymes and drug targets have been reported in the literature (Evans and Relling, 1999; Vesell, 1984). A large portion of interindividual variability in drug response has been attributed to genetic variations. However, the applicability of pharmacogenetics has been more prominent in research and academics as compared to clinics (Crews, et al., 2011; Nelson et al., 2011; Pulley et al., 2012; O'Donnell et al., 2012; Cavallari and Nutescu, 2012). There are a number of applications of pharmacogenetics that are summarized in Fig. 2.1. The idea of tailored therapeutics will take a few more years to be completely applied to the clinics and enable efficacious treatment to the patients.

2.2 PERSONALIZED MEDICINE

Personalized medicine is a very old concept that was coined almost hundred years ago. With newer technologies and availability of genome data, it has been possible to screen individuals to enable the identification of genetic variants affecting diagnosis, prognosis, and theranosis of any disorder. There are a number of variants that affect the drug metabolism and response which serves as a basis for pharmacogenetics to enable personalized medicine. The sequencing of the human genome has enabled the translation of precision medicine from the research phase to the clinical phase. There are a number of tools that

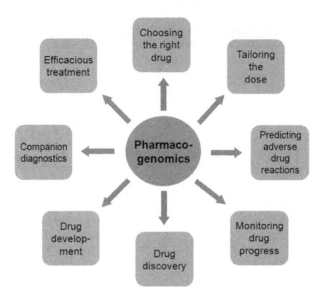

FIGURE 2.1   Applications of pharmacogenomics.

have been developed to help in the smooth application of this science to personalize treatment. The Genome-wide Association Studies have enabled the identification of genetic variants for different traits of various populations. The knowledge from these studies has eased the path towards tailored dosing for efficacious treatment. There are several terms being used interchangeably with personalized medicine, including "pharmacogenetics," "precision medicine," "theranosis," and "tailored medicine."

2.3 PHARMACOGENOMICS OF VARIOUS DISORDERS

2.3.1 Cancer

The applicability of pharmacogenomics in cancer is the highest because the therapeutic window for the cancer drug is narrow, the time of treatment for the patient is critical, and the cost of treatment is very high.

6-Mercaptopurine (6-MP) and Thiopurine methyltransferase (TPMT): Acute lymphoblastic leukemia (ALL) is the most commonly observed childhood malignancy (Pui and Evans, 2006). 6-MP is the most well prescribed therapy for ALL; however, it leads to myelosuppression as a side effect (Lennard et al., 1987). The genetic testing of *TPMT* has a dual advantage as it is useful in determining the risk of toxicity and the response in terms of minimal residual disease posttreatment (Stanulla, et al., 2005). It has been recommended by the FDA to identify individuals who are homozygous for nonfunctional TPMT alleles or have low or intermediate TPMT activity, and dose determination is done on the basis of this test. Dosing recommendations of mercaptopurine based on the genotype and phenotype have been suggested by the Clinical Pharmacokinetics Implementation Consortium

(Relling et al., 2011). About 3%−14% individuals who are heterozygous for nonfunctional *TPMT* alleles are at a significantly higher risk of toxicity due to mercaptopurine as compared to individuals with two functional alleles (Relling et al., 2011). *TPMT*2, *TPMT*3A, *TPMT*3B*, and *TPMT*3C* are the nonfunctional alleles associated with reduced levels of TPMT activity (McLeod and Siva, 2002).

Irinotecan and UDP glucuronosyltransferase 1 (UGT1A1): Irinotecan is prescribed to patients suffering from rhabdomyosarcoma and other solid tumors. The enzyme UGT1A1 metabolizing irinotecan is highly polymorphic and leads to irinotecan-induced neutropenia (Innocenti et al., 2009; Hoskins et al., 2007). There is a common insertion/deletion in the *UGT1A1* promoter TATA box, which reduces the transcriptional activity of *UGT1A1*. The efficiency of the translation is inversely correlated to the number of tandem TA repeats present in the individual (5, 6, 7, 8 alleles) (Beutler et al., 1998). The most commonly observed are the 6 and 7 repeats. 6 repeats has been classified as *UGT1A1*1*, whereas 7 repeats has been classified as *UGT1A1*28* (Hall et al., 1999). There is a strong evidence for the role of *UGT1A1*28* for severe toxicity (Innocenti et al., 2004; Iyer et al., 2002; Rouits et al., 2004; Marcuello et al., 2004; Toffoli et al., 2006; Carlini et al., 2005; Font et al., 2003; Han et al., 2006; Massacesi et al., 2006). The importance of this genetic testing has resulted in the label change of this drug by the FDA, making genetic testing mandatory for irinotecan (Innocenti and Ratain, 2006).

5-Fluorouracil (5-FU) and dihydropyrimidine dehydrogenase (DPYD): 5-FU, used in the treatment of most of the solid tumors, is metabolized by dihydropyrimidine dehydrogenase (DPD). (van Kuilenburg 2004; Heggie et al., 1987). DPD activity is variable among different individuals, and it is reported that 3%−5% of individuals are deficient or show low DPD activity (Etienne et al., 1994; Lu et al., 1993). The gene *DPYD* is highly polymorphic, and one of the variant allele *DPYD*2A* (c.1905 + 1G > A, also called as IVS14 + 1G > A) has a significant effect on toxicities, including leucopenia and severe mucositis, observed due to 5-FU (Schwab et al., 2008). This variant results in skipping of exon 14 and results in a nonfunctional enzyme leading to toxicities (Wei et al., 1996; Vreken et al., 1996). Various other studies have also reported lethal 5-FU toxicities in patients with DPD deficiency (van Kuilenburg et al. 2003; van Kuilenburg et al. 2001; Johnson and Diasio, 2001; Milano et al., 1999). Furthermore, other concomitant drugs may also interact with 5-FU metabolism and hence affect the risk profile for *DPYD* variants. The label for 5-FU mentions a warning for DPD-deficient patients not to use this drug; however, genetic testing is not mandatory for it (Yen and McLeod, 2007; Ciccolini et al., 2010).

2.3.2 Cardiovascular Diseases

The ability to diagnose and treat various cardiovascular disorders has improved owing to the better understanding of individual characteristics like genetic variations. Examples of some of these cardiovascular applications have been mentioned below:

Warfarin: Warfarin is widely prescribed for the prevention and treatment of thromboembolic diseases. However, it has varied responses in different individuals due to genetic polymorphisms. The two most well-studied genes for understanding this variable response are *CYP2C9* (Cytochrome P450 2C9) and *VKORC1* (vitamin K epoxide reductase complex subunit 1), The single nucleotide polymorphisms (SNPs) in these two genes are responsible for more than one-third of the variation observed in warfarin therapeutic outcome (Wadelius et al., 2007; Gage et al., 2008; Klein et al., 2009; Caldwell et al., 2008; Wadelius et al., 2009). To understand whether genotype information could enable the reduction in the incidence rates of hospitalizations occurring due to adverse effects of warfarin, the Medco-Mayo Warfarin Effectiveness Study was conducted with almost 4000 individuals (Epstein et al., 2010). About 900 patients who were starting warfarin therapy were genotyped for *CYP2C9* and *VKORC1*. Their genotype data were sent to their doctors and outcomes were measured in comparison with 2700 controls who were not genotyped. Two external control groups from a different set were also followed during the study, one synchronous with the genotyped group and the other synchronous with the control group. A 31% reduction of hospitalization was observed after a 6-month follow-up in the genotyped patients in comparison to the controls (Epstein et al., 2010). The results of such studies suggest the importance of pharmacogenomic applications in cardiovascular medicine.

Clopidogrel: The standard therapeutic regime for acute coronary syndrome patients includes aspirin and clopidogrel (Anderson et al., 2007; Kushner et al., 2009). However, there is interindividual variability observed in patients prescribed clopidogrel therapy, due to genetic variations present in *CYP2C19* (Cytochrome P450 2C19). The most commonly observed and major contributor towards this variation is the *CYP2C19*2* variant (Gurbel et al., 2003; Angiolillo et al., 2007; Hulot et al., 2006). TRITON-TIMI 38 (Trial to Assess Improvement in Therapeutic Outcomes by Optimizing Platelet Inhibition with Prasugrel Thrombolysis in Myocardial Infarction 38), AFIJI (Appraisal of Risk Factors in Young Ischemic Patients Justifying Aggressive Intervention), and FAST-MI (French Registry of Acute ST-Elevation and Non-ST-Elevation Myocardial Infarction) are the three largest studies with patients on clopidogrel therapy. All the three studies showed a significant correlation between reduced function alleles and higher rates of cardiovascular death and myocardial infarction (Mega et al., 2009; Simon et al., 2009; Collet et al., 2009). Based on these findings, the US FDA approved a label change for clopidogrel to include a "boxed warning" which mentioned that individuals carrying reduced function CYP2C19 alleles which are termed as poor metabolizers will experience diminished effectiveness of the drug at standard dosing and hence alternative therapies must be considered in such patients.

2.3.3 Infectious Diseases

There are a lot of infectious diseases prevalent all over the world, and treatment regimes are different in various regions. The most well-studied examples of pharmacogenomics in the perspective of infectious diseases include malaria, tuberculosis, and HIV.

Amodiaquine for Malaria: Amodiaquine is a well-prescribed drug for malaria; however, it induces adverse reactions in individuals harboring variations in Cytochrome P450 2C8 (*CYP2C8*), Cytochrome P450 1A1 (*CYP1A1*), and Cytochrome P450 1B1 (*CYP1B1*) (Li et al., 2002; Kerb et al., 2009). Reduced *CYP2C8* activity leads to impaired metabolism, thereby leading to hepatotoxicity and agranulocytosis (Dai et al., 2001). Among the various different alleles of CYP2C8, *2, *3, and *4 are the ones with low activity (Gil and Berglund, 2007). CYP2C8*2 has shown a sixfold reduced intrinsic clearance of Amodiaquine in vitro as compared to the wild-type allele (*1) (Gil and Berglund, 2007; Parikh et al., 2007). Also CYP2C8*3 has been shown to have no detectable Amodiaquine metabolism activity, and these individuals have an increased risk of side effects due to decreased clearance of Amodiaquine (Parikh et al., 2007).

Isoniazid for Tuberculosis: Isoniazid has been widely used for the treatment of tuberculosis. Isoniazid gets activated by N-acetyltransferase (NAT2) which is highly polymorphic. The *NAT2* genotype carried by the individual determines the early bactericidal activity of isoniazid. Hence, the mean bactericidal activity is higher in slow acetylators as compared to rapid acetylators (Donald et al., 2004). Pharmacogenomics has paved the way for isoniazid treatment after prior genotyping of the NAT2 allele. The *NAT2*4* allele is completely active, and the individuals with these alleles are called rapid acetylator (wild-type). These individuals can be administered drugs according to the recommended therapeutic doses. However, the individuals with one active and one inactive allele (heterozygous) are termed intermediate acetylators and require a lower therapeutic dose. The slow acetylators are the individuals who have mutations leading to *NAT2*5A*, *NAT2*5B*, and *NAT2*6A* (Cascorbi and Roots, 1999). The individuals who are slow acetylators are at a greater risk of drug-induced side effects as there is reduced drug clearance in these individuals. The treatment regime can be decided based on the number of alleles—none, one, or two carried by the patient. This would help in avoiding treatment failure or disease relapse and thereby enable better patient management in a life-threatening disease like tuberculosis (Weiner et al., 2003).

Efavirenz and Indinavir for HIV therapy: Efavirenz, a non-nucleoside reverse transcriptase inhibitor, undergoes metabolism by Cytochrome P450 2B6 (CYP2B6) (Ward et al., 2003). CYP2B6 has been reported to be highly polymorphic with polymorphisms including $415A > G$, $516G > T$, $136A > G$, $296G > A$, $785A > G$, $419G > A$, and $1172T > A$ being the most important ones affecting gene expression (Lamba et al., 2003; Lang et al., 2004; Zukunft et al., 2005). Various studies have reported CNS toxicity with the accumulation of efavirenz in the plasma (Haas et al., 2004; Rodriguez-Novoa et al. 2005, Marzolini et al., 2001). Another important drug, Indinavir, a protease inhibitor, is metabolized by Cytochrome P450 3A5 (CYP3A5), which is highly polymorphic (Paulussen et al., 2000). The most important variant that has been well studied, CYP3A5 (6986G), leads to aberrant splicing of the transcripts and hence very low CYP3A5 protein levels (Kuehl et al., 2001). This leads to higher drug levels in the system and an increased risk towards kidney toxicities (Anderson et al., 2004).

2.3.4 Immune Disorders

Autoimmune disorders require multiple lifelong drugs, and it is difficult to understand which drug will be best suited to the patient. Pharmacogenomics enables in the identification of the correct drug in the patient, hence being more effective and avoiding adverse side effects in these patients. Examples of some applications in immune disorders have been mentioned below:

Azathioprine: Azathioprine has been the first line of therapy for autoimmune diseases such as rheumatoid arthritis, multiple sclerosis, and lupus (Remy, 1963). About 1%−2%

people on standard doses suffer from severe or fatal myelosuppression, whereas 10%−20% people have moderately severe myelosuppression (McLeod et al., 2000). This myelosuppression is caused by genetic variants in the *TPMT* gene. Homozygosity for low-activity *TPMT* variants (e.g., *TPMT*3A*) occurs in 0.5%−1% of individuals, and a standard dose of azathioprine in homozygous low activity *TPMT* variants like *TPMT*3A* may lead to severe or fatal myelosuppression, whereas 5%−10% of people carrying low-activity *TPMT* variants show lowered inactivation of azathioprine and are at a risk of moderately severe myelosuppression (McLeod et al., 2000; Weinshilboum and Sladek, 1980; Collie-Duguid et al., 1999). Due to these severe effects of azathioprine, the FDA, in 2003, revised the drug label of azathioprine to spread awareness about the risk of myelosuppression in individuals with low TPMT activity and to provide genetic testing prior to treatment (FDA, 2003).

Codeine: Codeine is the most well-prescribed opioid analgesic for pain relief in patients suffering from autoimmune diseases (Ytterberg et al., 1998; Kimura et al., 2006). There are reports of side effects, such as respiratory depression, drowsiness, fatigue, and constipation, in patients who are on codeine (Kimura et al., 2006). Codeine is also reported to show adverse reactions in neonates who received the metabolized drug in the form of morphine from the breast milk fed by the mothers on codeine. Codeine is metabolized by Cytochrome P450 2D6 (CYP2D6), and variants in this gene lead to poor, normal, or ultrarapid metabolism in individuals (Sindrup and Brosen, 1995). Hence, genetic testing for *CYP2D6* variants for individuals before prescribing therapy would enable in curbing adverse drug reactions by prescribing alternate medications such as acetaminophen or non-steroidal antiinflammatories, or a reduced dose of codeine.

2.4 APPLICATIONS OF PHARMACOGENOMICS

2.4.1 Personalized Medicine

There is a considerable interindividual variation in response to a particular therapy due to various reasons, namely genetics, epigenetics, environmental factors, and other factors such as gender, age, and drug−drug interaction (Meyer et al., 2013). This variability leads to a number of differences among people, which are functionally important to understand and analyze namely disease susceptibility, drug response, response to diet and exercise, and many others. Pharmacogenomics has revolutionized the field of medicine by utilizing new generation of "omics" technologies for the better understanding of disease risk and response to therapeutics. The Hercep Test/trastuzumab (Herceptin) from Dako and Genentech/Roche for the detection of Her2-positive subset of breast cancer patients is one of the best cited examples of pharmacogenomics in personalized medicine. Another well-cited example is of imatinib (Glivec, Novartis), which is a synthetic tyrosine kinase inhibitor used for the treatment of chronic myeloid leukemia and gastrointestinal stromal tumors (Nair, 2010).

2.4.2 Drug Discovery

The entire process of drug discovery and development is very elaborate and extremely time-consuming. It involves the discovery of the drug molecule, and its development followed by animal and clinical studies. After this lengthy process, out of the thousands of drug molecules screened, very few enter clinical trials, and probably one or two are approved for commercialization. This entire process involves a lot of money as well (Dickson and Gagnon, 2004). Some of the drugs approved in this manner may even fail after reaching the market due to adverse effects of the drug.

A number of drug targets have been identified after the sequence of the human genome has become available. This includes enzymes and receptors that account for major proportion of drugs that have been developed for various different disorders. A list of some of the enzymes and receptors for which targeted drugs have been developed are listed in Tables 2.1 and 2.2.

To save time, money, and efforts of people involved in the drug discovery process, it is wise to invest in a genetic test that could determine along with the drug discovery and clinical trials whether the drug would be efficacious in a particular population. This information can help in selecting appropriate target patients for the drug and also helps in preventing adverse side effects (Abraham and Adithan, 2001). Recently, drug development programs targeted towards rare form of cystic fibrosis (Ramsey et al., 2011) and the use of an exon-skipping antisense compound for Duchenne muscular dystrophy (Eteplirsen) are examples of genetically guided drug discovery. The drugs designed with a pharmacogenomic approach have a predetermined efficacy, and the chances of such a drug failing in preclinical and clinical studies due to problems of efficacy will be minimal.

2.5 COMPANION DIAGNOSTICS

The US FDA, in 2014, issued a document to outline the oversight and approval of companion diagnostics. This document is used to identify patients who are likely to benefit or suffer from adverse events from a therapeutic and tailoring the dose according to the safety and effectiveness of the drug (US Food and Drug Administration, 2014). Most of the FDA approved companion diagnostics include molecular assays to detect DNA. The companion diagnostic is commonly termed as a "gatekeeper" to enable treatment decisions (Jørgensen, 2013) and is related to the indication, safety, and efficacy of the therapeutic linked to it. Companion diagnostics appears to offer a set of tools to prevent the failure of drugs during trials or after the launch in the market (Fig. 2.2). This will enable safer and more efficacious therapeutics with genotype- and/or phenotype-focused therapeutics. The advent of newer technologies and improved knowledge on the complexity and variability of human genome has forced the pharmaceutical companies to opt for companion diagnostics and rethink about their current business model (Liebman, 2008). It is very evident that tailored therapy is preferable to the blockbuster model, thereby offering more efficacious and safer drugs to the patient pool (Culbertson et al., 2007). This process also speeds up the approval process as all the necessary efficacy studies have already been done on various populations to justify the approval of the product.

Most of the pharmaceutical companies have started investing in companion diagnostics including Roche, Merck, Astra Zeneca, Pfizer, Amgen, Biogen, Bristol Myers Squibb,

TABLE 2.1 Target Enzymes, Approved Drugs, and Activity of Drugs

Type of Enzyme	Approved Drug	Activity of Drug
OXIDOREDUCTASES		
Aldehyde dehydrogenase	Disulfiram	Inhibitor
Monoamine oxidases	Tranylcypromine, moclobemide	MAO-A inhibitor
	Tranylcypromine	MAO-B inhibitor
Cyclooxygenases	Acetylsalicylic acid, profens, acetaminophen and dipyrone (as arachidonylamides)	COX1 inhibitor
	Acetylsalicylic acid, profens, acetaminophen and dipyrone (as arachidonylamides)	COX2 inhibitor
Vitamin K epoxide reductase	Warfarin, phenprocoumon	Inhibitor
Aromatase	Exemestane	Inhibitor
Lanosterol demethylase (fungal)	Azole antifungals	Inhibitor
Lipoxygenases	Mesalazine	Inhibitor
	Zileuton	5-Lipoxygenase inhibitor
Thyroidal peroxidase	Thiouracils	Inhibitor
Iodothyronine-5′ deiodinase	Propylthiouracil	Inhibitor
Inosine monophosphate dehydrogenase	Mycophenolate mofetil	Inhibitor
HMG-CoA reductase	Statins	Inhibitor
5α-Testosterone reductase	Finasteride, dutasteride	Inhibitor
Dihydrofolate reductase (bacterial)	Trimethoprim	Inhibitor
Dihydrofolate reductase (human)	Methotrexate, pemetrexed	Inhibitor
Dihydrofolate reductase (parasitic)	Proguanil	Inhibitor
Dihydroorotate reductase	Leflunomide	Inhibitor
Enoyl reductase (mycobacterial)	Isoniazid	Inhibitor
Squalene epoxidase (fungal)	Terbinafin	Inhibitor
$\Delta 14$ reductase (fungal)	Amorolfin	Inhibitor
Xanthine oxidase	Allopurinol	Inhibitor
4-Hydroxyphenylpyruvate dioxygenase	Nitisinone	Inhibitor
Ribonucleoside diphosphate reductase	Hydroxycarbamide	Inhibitor

(Continued)

TABLE 2.1 (Continued)

Type of Enzyme	Approved Drug	Activity of Drug
TRANSFERASES		
Protein kinase C	Miltefosine	Inhibitor
Bacterial peptidyl transferase	Chloramphenicol	Inhibitor
Catecholamine-O-methyltransferase	Entacapone	Inhibitor
RNA polymerase (bacterial)	Ansamycins	Inhibitor
Reverse transcriptases (viral)	Zidovudine	Competitive inhibitors
	Efavirenz	Allosteric inhibitors
DNA polymerases	Acyclovir, suramin	Inhibitor
GABA transaminase	Valproic acid, vigabatrin	Inhibitor
Tyrosine kinases	Imatinib	PDGFR/ABL/KIT inhibitor
	Erlotinib	epidermal growth factor receptor (EGFR) inhibitor
	Sunitinib	VEGFR2/PDGFRβ/KIT/FLT3
	Sorafenib	VEGFR2/ PDGFRβ/RAF
Glycinamide ribonucleotide formyl transferase	Pemetrezed	Inhibitor
Phosphoenolpyruvate transferase (MurA, bacterial)	Fosfomycin	Inhibitor
Human cytosolic branched-chain aminotransferase	Gabapentin	Inhibitor
HYDROLASES (PROTEASES)		
Aspartyl proteases (viral)	Saquinavir, indinavir	HIV protease inhibitor
HYDROLASES (SERINE PROTEASES)		
Unspecific	Aprotinine	Unspecific inhibitors
Bacterial serine protease	B-lactams	Direct inhibitor
Bacterial serine protease	Glycopeptides	Indirect inhibitor
Bacterial lactamases	Sulbactam	Direct inhibitor
Human antithrombin	Heparins	Activator
Human plasminogen	Streptokinase	Activator
Human coagulation factor	Factor IX complex, Factor VIII	Activator
Human factor Xa	Fondaparinux	Inhibitor

(Continued)

I. EMERGING FIELDS

TABLE 2.1   (Continued)

Type of Enzyme	Approved Drug	Activity of Drug
HYDROLASES (METALLOPROTEASES)		
Human ACE	Captopril	Inhibitor
Human HRD	Cilastatin	Inhibitor
Human carboxypeptidase A (Zn)	Penicillamine	Inhibitor
Human enkephalinase	Racecadotril	Inhibitor
HYDROLASES (OTHER)		
26S proteasome	Bortezomib	Inhibitor
Esterases	Physostigmine	AChE inhibitor
	Obidoxime	AChE reactivators
	Caffeine	PDE inhibitor
	Amrinon, milrinone	PDE3 inhibitor
	Papaverine	PDE4 inhibitor
	Sildenafil	PDE5 inhibitor
	Valproic acid	HDAC inhibitor
	Carbamezapine	HDAC3/HDAC7 inhibitor
Glycosidases (viral)	Zanamivir, oseltamivir	α-glycosidase inhibitor
Glycosidases (human)	Acarbose	α-glycosidase inhibitor
Lipases	Orlistat	Gastrointestinal lipases inhibitor
Phosphatases	Cyclosporin	Calcineurin inhibitor
	Lithium ions	Inositol polyphosphate phosphatase inhibitor
GTPases	6-Thio-GTP (azathioprine metabolite)	Rac1 inhibitor
Phosphorylases	Bacitracin	Bacterial C55-lipid phosphate dephosphorylase inhibitor
LYASES		
DOPA decarboxylase	Carbidopa	Inhibitor
Carbonic anhydrase	Acetazolamide	Inhibitor
Histidine decarboxylase	Tritoqualine	Inhibitor
Ornithine decarboxylase	Eflornithine	Inhibitor
Soluble guanylyl cyclase	Nitric acid esters, molsidomine	Activator

(Continued)

TABLE 2.1 (Continued)

Type of Enzyme	Approved Drug	Activity of Drug
ISOMERASES		
Alanine racemase	D-Cycloserine	Inhibitor
DNA gyrases (bacterial)	Fluoroquinolones	Inhibitor
Topoisomerases	Irinotecan	Topoisomerase I inhibitor
	Etoposide	Topoisomerase II inhibitor
$\Delta 8, 7$ isomerase (fungal)	Amorolfin	Inhibitor
LIGASES (ALSO KNOWN AS SYNTHASES)		
Dihydropteroate synthase	Sulfonamides	Inhibitor
Thymidylate synthase (fungal and human)	Fluorouracil	Inhibitor
Thymidylate synthase (human)	Methotrexate, pemetrexed	Inhibitor
Phosphofructokinase	Antimony compounds	Inhibitor
mTOR	Rapamycin	Inhibitor
Haem polymerase (Plasmodium)	Quinoline antimalarials	Inhibitor
1,3-β-D-glucansynthase (fungi)	Caspofungin	Inhibitor
Glucosylceramide synthase	Miglustat	Inhibitor

Adapted from Imming, P., Sinning, C., Meyer, A., 2006. Drugs, their targets and the nature and number of drug targets. Nat Rev Drug Disc 5, 821–834.

and Eli Lilly. The most active area in companion diagnostics is Oncology, as the therapeutic window is narrow and the time of treatment is critical in such patients. Some popular examples of therapeutics for which companion diagnostics were developed include Herceptin (Genetech), Tarceva (OSI Pharmaceuticals/Genetech), Iressa (Astra Zeneca), and Erbitux (Imclone/Bristol Myers Squibb). In the recent past FDA approved the companion diagnostic called the Qiagen therascreen KRAS RGQ PCR kit for metastatic colorectal cancer patients to be treated with the drug Vectibix. The mutations in the KRAS gene render the drug Vectibix as ineffective in the treatment of colorectal cancer. Hence, this was an important companion test to identify patients who would benefit from treatment with this drug. A list of companion diagnostic markers used along with cancer therapeutics has been listed in Table 2.3. Hence, with the growing trend of companion diagnostics, it will be easy to identify patients who will benefit from a particular drug, identify patients who will be at an elevated risk of side effects after treatment with a particular drug, and/or monitor the course of treatment to tailor the dose in the patients for achieving maximum safety and effectiveness.

TABLE 2.2 Receptors, Approved Drugs, and Activity of Drugs

Type of Receptor	Approved Drug	Drug Target
GABA receptors	Barbiturate	Barbiturate binding site
	Benzodiazepines	Benzodiazepine binding site agonists
	Flumazenil	Benzodiazepine binding site antagonists
Acetylcholine receptors	Pyrantel (of Angiostrongylus), levamisole	Nicotinic receptor agonists
	Alcuronium	Nicotinic receptor stabilizing antagonists
	Suxamethonium	Nicotinic receptor depolarizing antagonists
	Galantamine	Nicotinic receptor allosteric modulators
Glutamate receptors (ionotropic)	Memantine	NMDA subtype antagonists
	Acamprosate	NMDA subtype expression modulators
	Ketamine	NMDA subtype phenylcyclidine binding site antagonists
G-protein-coupled receptors	Pilocarpine	Muscarinic receptor agonists
	Tropane derivatives	Muscarinic receptor antagonists
	Darifenacine	Muscarinic receptor M_3 antagonists
Adenosine receptors	Adenosine	Agonists
	Lignans from valerian	Adenosine A_1 receptor agonists
	Caffeine, theophylline	Adenosine A_1 receptor antagonists
	Caffeine, theophylline	Adenosine A_{2A} receptor agonists
Adrenoceptors	Adrenaline, noradrenaline, ephedrine	Agonists
	Xylometazoline	α_1- and α_2-receptors agonists
	Ergotamine	α_1-receptor agonists
	Methyldopa (as methylnoradrenaline)	α_2-receptor, central agonists
	Isoprenaline	β-adrenoceptor antagonists
	Propanolol, atenolol	β_1-receptor antagonists
	Salbutamol	β_2-receptor agonists
	Propanolol	β_2-receptor antagonists
Angiotensin receptors	Sartans	AT_1-receptors antagonists
Calcium-sensing receptor	Strontium ions	Agonists
	Cinacalcet	Allosteric activators

(Continued)

TABLE 2.2 (Continued)

Type of Receptor	Approved Drug	Drug Target
Cannabinoid receptors	Dronabinol	CB1- and CB2-receptors agonists
Cysteinyl-leukotriene receptors	Montelukast	Antagonists
Dopamine receptors	Dopamine, levodopa	Dopamine receptor subtype direct agonists
	Apomorphine	D_2, D_3 and D_4 agonists
	Chlorpromazine, fluphenazine, haloperidol, metoclopramide, ziprasidone	D_2, D_3 and D_4 antagonists
Endothelin receptors (ET_A, ET_B)	Bosentan	Antagonists
$GABA_B$ receptors	Baclofen	Agonists
Glucagon receptors	Glucagon	Agonists
Glucagon-like peptide-1 receptor	Exenatide	Agonists
Histamine receptors	Diphenhydramine	H_1-antagonists
	Cimetidine	H_2-antagonists
Opioid receptors	Morphine, buprenorphine	μ-opioid agonists
	Naltrexone	μ-, κ- and δ-opioid antagonists
	Buprenorphine	κ-opioid antagonists
Neurokinin receptors	Aprepitant	NK_1 receptor antagonists
Prostanoid receptors	Misoprostol, sulprostone, iloprost	Agonists
Prostamide receptors	Bimatoprost	Agonists
Purinergic receptors	Clopidogrel	P_2Y_{12} antagonists
Serotonin receptors	Ergometrine, ergotamine	Subtype-specific (partial) agonists
	Buspirone	$5-HT_{1A}$ partial agonists
	Triptans	$5-HT_{1B/1D}$ agonists
	Quetiapine, ziprasidone	$5-HT_{2A}$ antagonists
	Granisetron	$5-HT_3$ antagonists
	Tegaserode	$5-HT_4$ partial agonists
Vasopressin receptors	Vasopressin	Agonists
	Terlipressin	V_1 agonists
	Desmopressin	V_2 agonists
	Oxytocin	OT agonists
	Atosiban	OT antagonists

(Continued)

I. EMERGING FIELDS

TABLE 2.2   (Continued)

Type of Receptor	Approved Drug	Drug Target
CYTOKINE RECEPTORS		
Class I cytokine receptors	Pegvisomant	Growth hormone receptor antagonists
	Erythropoietin	Erythropoietin receptor agonists
	Filgrastim	Granulocyte colony stimulating factor agonists
	Molgramostim	Granulocyte-macrophage colony stimulating factor agonists
	Anakinra	Interleukin-1 receptor antagonists
	Aldesleukin	Interleukin-2 receptor agonists
TNFα receptors	Etanercept	Mimetics (soluble)
INTEGRIN RECEPTORS		
Glycoprotein IIb/IIIa receptor	Tirofiban	Antagonists
RECEPTORS ASSOCIATED WITH TYROSINE KINASE		
Insulin receptor	Insulin	Direct agonists
Insulin receptor	Biguanides	Sensitizers
NUCLEAR RECEPTORS (STEROID HORMONE RECEPTORS)		
Mineralocorticoid receptor	Aldosterone	Agonists
	Spironolactone	Antagonists
Glucocorticoid receptor	Glucocorticoids	Agonists
Progesterone receptor	Gestagens	Agonists
Estrogen receptor	Estrogens	Agonists
	Clomifene	(Partial) antagonists
	Fulvestrant	Antagonists
	Tamoxifen, raloxifene	Modulators
Androgen receptor	Testosterone	Agonists
	Cyproterone acetate	Antagonists
Vitamin D receptor	Retinoids	Agonists
ACTH receptor agonists	Tetracosactide (also known as cosyntropin)	Agonists
NUCLEAR RECEPTORS (OTHER)		
Retinoic acid receptors	Isotretinoin	RAR-α agonists
	Adapalene, isotretinoin	RAR-β agonists
	Adapalene, isotretinoin	RAR-γ agonists

(Continued)

TABLE 2.2   (Continued)

Type of Receptor	Approved Drug	Drug Target
Peroxisome proliferator-activated receptor (PPAR)	Fibrates	PPAR-α agonists
	Glitazones	PPAR-γ agonists
Thyroid hormone receptors	L-Thyroxine	Agonists

Adapted from Imming, P., Sinning, C., Meyer, A., 2006. Drugs, their targets and the nature and number of drug targets. Nat Rev Drug Disc 5, 821–834.

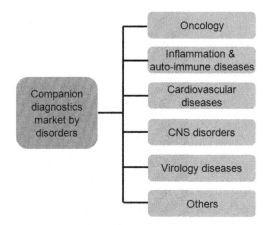

FIGURE 2.2   Companion diagnostics market by disorders.

2.6 LIMITATIONS OF PHARMACOGENOMICS

The field of pharmacogenomics has progressed a lot in the recent years; however, there are a lot of challenges that are yet to be resolved for personalized medicine to be smoothly put into clinical practice (Roden et al., 2006). Some important challenges faced by personalized medicine are discussed below.

Diseases and the response to therapeutics can be multigenic and can further be affected by environmental factors such as lifestyle, nutrition, and age. Hence, the complete understanding of a complex biological response would be difficult if all or most of these parameters are not monitored. High throughput and quick technologies need to be developed to enable screening of whole genomes to understand the effect of genes on disease risk and drug response. Ethical and privacy issues with genetic information need to be handled with caution as it may have implications for insurance companies. Cost-effective methods for SNP genotyping and expression studies need to be designed. Updated information about new genetic diagnostic tests must be provided to health care

TABLE 2.3   List of Approved Companion Diagnostics (Oncology)

Drug Name	Test Name	Developed By	Intended Use
Imatinib mesylate	KIT mutation detection by PCR	ARUP Laboratories	Gleevec eligibility in Aggressive Systemic Mastocytosis (ASM)
Imatinib mesylate	PGFRB FISH	ARUP Laboratories	Gleevec Eligibility in Myelodysplastic Syndrome/Myeloproliferative Disease (MDS/MPD)
Tagrisso (osimertinib)	Cobas epidermal growth factor receptor (EGFR) mutation test v2	Roche Molecular Systems, Inc.	Qualitative detection of mutations of the EGFR for non-small-cell lung cancer (NSCLC) for eligibility to osimertinib
Iressa (gefitinib)	Therascreen EGFR RGQ PCR kit	Qiagen Manchester Ltd	Qualitative detection of exon 19 and exon 21 mutations in EGFR for NSCLC for eligibility to gefitinib
Xalkori (crizotinib)	VENTANA ALK (D5F3) CDx Assay	Ventana Medical Systems, Inc.	Qualitative detection of the anaplastic lymphoma kinase protein in NSCLC for eligibility to crizotinib
Erbitux (cetuximab); Vectibix (panitumumab)	The Cobas KRAS mutation test	Roche Molecular Systems, Inc.	Erbitux or Vectibix eligibility in colorectal cancer (CRC) tumor by detection of KRAS mutations
Lynparza (olaparib)	BRAC Analysis CDx	Myriad Genetic Laboratories, Inc.	Lynparza eligibility for breast cancer by detection of mutations in BRCA1 and BRCA2
Erbitux (cetuximab); Vectibix (panitumumab)	DAKO EGFR PharmDx Kit	Dako North America, Inc.	Identifying CRC patients eligible for cetuximab or panitumumab by qualitative immunohistochemistry (IHC) kit for EGFR expression
Gilotrif (afatinib)	Therascreen EGFR RGQ PCR Kit	Qiagen Manchester, Ltd.	Qualitative detection of mutations in EGFR for NSCLC for eligibility to gefitinib
Herceptin (trastuzumab)	INFORM HER-2/NEU	Ventana Medical Systems, Inc.	Qualitative determination of Her-2/Neu gene amplification by fluorescence in situ hybridization (FISH) for eligibility to trastuzumab
Herceptin (trastuzumab)	INSITE HER-2/NEU KIT	Biogenex Laboratories, Inc.	Semiqualitative determination of Her-2/Neu overexpression by IHC for eligibility to trastuzumab
Mekinist (tramatenib); Tafinlar (dabrafenib)	THxID BRAF Kit	bioMerieux Inc.	Qualitative detection of the BRAF V600E and V600K in human melanoma for treatment with tramatenib or dabrafenib

(Continued)

TABLE 2.3 (Continued)

Drug Name	Test Name	Developed By	Intended Use
Tarceva (erlotinib)	Cobas EGFR Mutation Test	Roche Molecular Systems, Inc.	Qualitative real-time PCR test for detection of EGFR mutations in NSCLC for treatment with erlotinib
Zelboraf (vemurafenib)	COBAS 4800 BRAF V600 Mutation Test	Roche Molecular Systems, Inc.	Qualitative detection of BRAF V600E using real-time PCR for treatment with vemurafenib

From http://www.fda.gov/MedicalDevices/ProductsandMedicalProcedures/InVitroDiagnostics/ucm301431.htm.

providers for better patient management. Large population studies must be conducted to understand the ethnic variations and their effects on the disease risk and/or therapeutic outcome.

2.7 FUTURE PROSPECTS

With the progress in the field of omics, it is expected that the next few years will see a better translation of pharmacogenetics from research to clinics. Currently there are limited applications in a few critical areas of medicine; however, it is growing progressively. Pharmacogenetics has enabled well-timed and efficient drug dosage selection.

2.8 CONCLUSION

There are a number of applications of pharmacogenetics, and although there are a few challenges, these can be overcome by improved study designs and validation of studies in different populations. Hence, pharmacogenetics shows immense promise in improving the safety and effectiveness of therapeutics for patients of different disorders.

References

Abraham, B.K., Adithan, C., 2001. Genetic polymorphism of CYP2D6. Indian J. Pharmacol. 33, 147–169.

Anderson, J.L., Adams, C.D., Antman, E.M., Bridges, C.R., Califf, R.M., Casey Jr., D.E., et al., American College of Cardiology; American Heart Association Task Force on Practice Guidelines (Writing Committee to Revise the 2002 Guidelines for the Management of Patients With Unstable Angina/Non ST-Elevation Myocardial Infarction); American College of Emergency Physicians; Society for Cardiovascular Angiography and Interventions; Society of Thoracic Surgeons; American Association of Cardiovascular and Pulmonary Rehabilitation; Society for Academic Emergency Medicine 2007. ACC/AHA 2007 guidelines for the management of patients with unstable angina/non−ST-elevation myocardial infarction: a report of the American College of Cardiology/American Heart Association Task Force on Practice Guidelines (writing committee to revise the 2002 guidelines for the management of patients with unstable angina/non ST-elevation myocardial infarction): developed in collaboration with the American College of Emergency Physicians, the Society for

Cardiovascular Angiography and Interventions, and the Society of Thoracic Surgeons: endorsed by the American Association of Cardiovascular and Pulmonary Rehabilitation and the Society for Academic Emergency Medicine. Circulation 11, e148–e304.

Anderson, P., Schuetz, E., Fletcher, C., 2004. CYP3A5 and MDR1 (P-gp) polymorphisms in HIV-infected adults: associations with indinavir concentrations and antiviral effects. In: 11th Conference on Retroviruses and Opportunistic Infections, Feb 8–11, San Francisco.

Angiolillo, D.J., Fernandez-Ortiz, A., Bernardo, E., Alfonso, F., Macaya, C., Bass, T.A., et al., 2007. Variability in individual responsiveness to clopidogrel: clinical implications, management, and future perspectives. J. Am. Coll. Cardiol. 49, 1505–1516.

Beutler, E., Gelbart, T., Demina, A., 1998. Racial variability in the UDP-glucuronosyltransferase 1 (UGT1A1) promoter: a balanced polymorphism for regulation of bilirubin metabolism?. Proc. Natl. Acad. Sci. U.S.A. 95, 8170–8174.

Caldwell, M.D., Awad, T., Johnson, J.A., Gage, B.F., Falkowski, M., Gardina, P., et al., 2008. CYP4F2 genetic variant alters required warfarin dose. Blood 111, 4106–4112.

Carlini, L.E., Meropol, N.J., Bever, J., Andria, M.L., Hill, T., Gold, P., et al., 2005. UGT1A7 and UGT1A9 polymorphisms predict response and toxicity in colorectal cancer patients treated with capecitabine/irinotecan. Clin. Cancer Res. 11 (3), 1226–1236.

Cascorbi, L., Roots, I., 1999. Pitfalls in N-acetyl transferase 2 genotyping. Pharmacogenetics 9 (1), 123–127.

Cavallari, L., Nutescu, E.A., 2012. A Team Approach to Warfarin Pharmacogenetics. Pharmacy Practice News, New York, McMahon Publishing.

Ciccolini, J., Gross, E., Dahan, L., Lacarelle, B., Mercier, C., 2010. Routine dihydropyrimidine dehydrogenase testing for anticipating 5-fluorouracil-related severe toxicities: hype or hope?. Clin. Colorectal. Cancer 9, 224–228.

Collet, J.P., Hulot, J.S., Pena, A., Villard, E., Esteve, J.B., Silvain, J., et al., 2009. Cytochrome P450 2C19 polymorphism in young patients treated with clopidogrel after myocardial infarction: a cohort study. Lancet 373, 309–317.

Collie-Duguid, E.S., Pritchard, S.C, Powrie, R.H., Sludden, J., Collier, D.A., Li, T., et al., 1999. The frequency and distribution of thiopurine methyltransferase alleles in Caucasian and Asian populations. Pharmacogenetics 9, 37e42.

Crews, K.R., Cross, S.J., McCormick, J.N., Baker, D.K., Molinelli, A.R., Mullins, R., et al., 2011. Development and implementation of a pharmacist-managed clinical pharmacogenetics service. Am. J. Health Syst. Pharm. 68 (2), 143–150.

Culbertson, A.W., Valentine, S.J., Naylor, S., 2007. Personalized medicine: technological innovation and patient empowerment or exuberant hyperbole? Drug Discov. World (Summer) 16–31.

Dai, D., Zeldin, D.C., Blaisdell, J.A., Chanas, B., Coulter, S.J., Ghanayem, B.I., et al., 2001. Polymorphisms in human CYP2C8 decrease metabolism of the anticancer drug paclitaxel and arachidonic acid. Pharmacogenetics 11 (7), 597–607.

Dickson, M., Gagnon, J.P., 2004. The cost of new drug discovery and development. Discov. Med. 4, 172–179.

Donald, P.R., Sirgel, F.A., Venter, A., Parkin, D.P., Seifart, H.I., van de Wal, B.W., et al., 2004. The influence of human N-acetyltransferase genotype on the early bactericidal activity of isoniazid. Clin. Infect. Dis. 39 (10), 1425–1430.

Epstein, R.S., Moyer, T.P., Aubert, R.E., O Kane, D.J., Xia, F., Verbrugge, R.R., et al., 2010. Warfarin genotyping reduces hospitalization rates: results from the MM-WES (Medco-Mayo Warfarin Effectiveness Study). J. Am. Coll. Cardiol. 55, 2804–2812.

Etienne, M.C., Lagrange, J.L., Dassonville, O., Fleming, R., Thyss, A., Renée, N., et al., 1994. Population study of dihydropyrimidine dehydrogenase in cancer-patients. J. Clin. Oncol. 12 (11), 2248–2253.

Evans, W.E., Relling, M.V., 1999. Pharmacogenomics: translating functional genomics into rational therapeutics. Science 286 (5439), 487–491.

Font, A., Sanchez, J.M., Taron, M., Martinez-Balibrea, E., Sánchez, J.J., Manzano, J.L., et al., 2003. Weekly regimen of irinotecan/docetaxel in previously treated non-small cell lung cancer patients and correlation with uridine diphosphate glucuronosyltransferase 1A1 (UGT1A1) polymorphism. Invest. New Drugs 21 (4), 435–443.

Food and Drug Administration, 2003. Pediatric Oncology Subcommittee of the Oncologic Drugs Advisory Committee. http://www.fda.gov/ohrms/dockets/ac/03/minutes/3971M1.htm.

Gage, B.F., Eby, C., Johnson, J.A., Deych, E., Rieder, M.J., Ridker, P.M., et al., 2008. Use of pharmacogenetic and clinical factors to predict the therapeutic dose of warfarin. Clin. Pharmacol. Ther. 84, 326–331.

Gil, J., Gil Berglund, E., 2007. CYP2C8 and antimalaria drug efficacy. Pharmacogenomics 8 (2), 187–198.

Gurbel, P.A., Bliden, K.P., Hiatt, B.L., O'Connor, C.M., 2003. Clopidogrel for coronary stenting: response variability, drug resistance, and the effect of pretreatment platelet reactivity. Circulation 107, 2908–2913.

Haas, D., Ribaudo, H., Kim, R., Tierney, C., Wilkinson, G.R., Gulick, R.M., et al., 2004. Pharmacogenetics of efavirenz and central nervous system side effects: an Adult AIDS Clinical Trials Group study. AIDS 18, 2391–2400.

Hall, D., Ybazeta, G., Destro-Bisol, G., Petzl-Erler, M.L., Di Rienzo, A., et al., 1999. Variability at the uridine diphosphate glucuronosyltransferase 1A1 promoter in human populations and primates. Pharmacogenetics 9 (5), 591–599.

Han, J.Y., Lim, H.S., Shin, E.S., Yoo, Y.K., Park, Y.H., Lee, J.E., et al., 2006. Comprehensive analysis of UGT1A polymorphisms predictive for pharmacokinetics and treatment outcome in patients with non-small-cell lung cancer treated with irinotecan and cisplatin. J. Clin. Oncol. 24 (15), 2237–2244.

Heggie, G.D., Sommadossi, J.P., Cross, D.S., Huster, W.J., Diasio, R.B., 1987. Clinical pharmacokinetics of 5-fluorouracil and its metabolites in plasma, urine, and bile. Cancer Res. 47, 2203–2206.

Hoskins, J.M., Goldberg, R.M., Qu, P., Ibrahim, J.G., McLeod, H.L., 2007. UGT1A1*28 genotype and irinotecan-induced neutropenia: dose matters. J. Natl. Cancer Inst. 99, 1290–1295.

Hulot, J.S., Bura, A., Villard, E., Azizi, M., Remones, V., Goyenvalle, C., et al., 2006. Cytochrome P450 2C19 loss-of-function polymorphism is a major determinant of clopidogrel responsiveness in healthy subjects. Blood 108, 2244–2247.

Imming, P., Sinning, C., Meyer, A., 2006. Drugs, their targets and the nature and number of drug targets. Nat. Rev. Drug Disc 5, 821–834.

Innocenti, F., Kroetz, D.L., Schuetz, E., Dolan, M.E., Ramírez, J., Relling, M., et al., 2009. Comprehensive pharmacogenetic analysis of irinotecan neutropenia and pharmacokinetics. J. Clin. Oncol. 27, 2604–2614.

Innocenti, F., Ratain, M.J., 2006. Pharmacogenetics of irinotecan: clinical perspectives on the utility of genotyping. Pharmacogenomics 7, 1211–1221.

Innocenti, F., Undevia, S.D., Iyer, L., Chen, P.X., Das, S., Kocherginsky, M., et al., 2004. Genetic variants in the UDP-glucuronosyltransferase 1A1 gene predict he risk of severe neutropenia of irinotecan. J. Clin. Oncol. 22, 1382–1388.

Iyer, L., Das, S., Janisch, L., Wen, M., Ramírez, J., Karrison, T., et al., 2002. UGT1A1*28 polymorphism as a determinant of irinotecan disposition and toxicity. Pharmacogenomics J. 2 (1), 43–47.

Johnson, M.R., Diasio, R.B., 2001. Importance of dihydropyrimidine dehydrogenase (DPD) deficiency in patients exhibiting toxicity following treatment with 5-fluorouracil. Adv. Enzyme Regul. 41, 151–157.

Jørgensen, J.T., 2013. Companion diagnostics in oncology—current status and future aspects. Oncology 85, 59.

Kerb, R., Fux, R., Morike, K., Kremsner, P.G., Gil, J.P., Gleiter, C.H., et al., 2009. Pharmacogenetics of antimalarial drugs: effect on metabolism and transport. Lancet Infect. Dis. 9 (12), 760–774.

Kimura, Y., Walco, G.A., Sugarman, E., Conte, P.M., Schanberg, L.E., 2006. Treatment of pain in juvenile idiopathic arthritis: a survey of pediatric rheumatologists. Arthritis Rheum. 55, 81e5.

Klein, T.E., Altman, R.B., Eriksson, N., Gage, B.F., Kimmel, S.E., Lee, M.T., et al., 2009. Estimation of the warfarin dose with clinical and pharmacogenetic data. N. Engl. J. Med. 360, 753–764.

Kuehl, P., Zhang, J., Lin, Y., Lamba, J., Assem, M., Schuetz, J., et al., 2001. Sequence diversity in CYP3A promoters and characterization of the genetic basis of polymorphic CYP3A5 expression. Nat. Genet. 27, 383–391.

Kushner, F.G., Hand, M., Smith Jr., S.C., King III., S.B., Anderson, J.L., Antman, E.M., et al., 2009. 2009 focused updates: ACC/AHA guidelines for the management of patients with ST-elevation myocardial infarction (updating the 2004 guideline and 2007 focused update) and ACC/AHA/SCAI guidelines on percutaneous coronary intervention (updating the 2005 guideline and 2007 focused update)—a report of the American College of Cardiology Foundation/American Heart Association Task Force on Practice Guidelines. Circulation. 120, 2271–2306.

Lamba, V., Lamba, J., Yasuda, K., Strom, S., Davila, J., Hancock, M.L., et al., 2003. Hepatic CYP2B6 expression: gender and ethnic differences and relationship to CYP2B6 genotype and CAR (constitutive androstane receptor) expression. J. Pharmacol. Exp. Ther. 307, 906–922.

Lang, T., Klein, K., Richter, T., Zibat, A., Kerb, R., Eichelbaum, M., et al., 2004. Multiple novel non-synonymous CYP2B6 gene polymorphisms in Caucasians: demonstration of phenotypic null alleles. J. Pharmacol. Exp. Ther. 311, 34−43.

Lennard, L., Van Loon, J.A., Lilleyman, J.S., Weinshilboum, R.M., 1987. Thiopurine pharmacogenetics in leukemia: correlation of erythrocyte thiopurine methyltransferase activity and 6-thioguanine nucleotide concentrations. Clin. Pharmacol. Ther. 41, 18−25.

Li, X.Q., Bjorkman, A., Andersson, T.B., Ridderstrom, M., Masimirembwa, C.M., 2002. Amodiaquine clearance and its metabolism to N-desethylamodiaquine is mediated by CYP2C8: a new high affinity and turnover enzyme-specific probe substrate. J. Pharmacol. Exp. Ther. 300 (2), 399−407.

Liebman, M.N., 2008. Personalized medicine—end of the blockbuster? Pharma. Focus Asia 9, 4−8.

Lu, Z.H., Zhang, R.W., Diasio, R.B., 1993. Dihydropyrimidine dehydrogenase-activity in human peripheral-blood mononuclear-cells and liver—population characteristics, newly identified deficient patients, and clinical implication in 5-fluorouracil chemotherapy. Cancer Res. 53 (22), 5433−5438.

Marcuello, E., Altés, A., Menoyo, A., Del Rio, E., Gómez-Pardo, M., Baiget, M., 2004. UGT1A1 gene variations and irinotecan treatment in patients with metastatic colorectal cancer. Br. J. Cancer 91, 678−682.

Marzolini, C., Telenti, A., Decosterd, L., Greub, G., Biollaz, J., Buclin, T., 2001. Efavirenz plasma levels can predict treatment failure and central nervous system side effects in HIV-1-infected patients. AIDS 15, 71−75.

Massacesi, C., Terrazzino, S., Marcucci, F., Rocchi, M.B., Lippe, P., Bisonni, R., et al., 2006. Uridine diphosphate glucuronosyltransferase 1A1 promoter polymorphism predicts the risk of gastrointestinal toxicity and fatigue induced by irinotecan-based chemotherapy. Cancer 106 (5), 1007−1016.

McLeod, H.L., Krynetski, E.Y., Relling, M.V., Evans, W.E., 2000. Genetic polymorphism of thiopurine methyltransferase and its clinical relevance for childhood acute lymphoblastic leukemia. Leukemia 14, 567e72.

McLeod, H.L., Siva, C., 2002. The thiopurine S-methyltransferase gene locus—implications for clinical pharmacogenomics. Pharmacogenomics 3 (1), 89−98.

Mega, J.L., Close, S.L., Wiviott, S.D, Shen, L., Hockett, R.D., Brandt, J.T., et al., 2009. Cytochrome P-450 polymorphisms and response to clopidogrel. N. Engl. J. Med. 360, 354−362.

Meyer, U.A., Zanger, U.M., Schwab, M., 2013. Omics and drug response. Annu. Rev. Pharmacol. Toxicol. 53, 475−502.

Milano, G., Etienne, M.C., Pierrefite, V., Barberi-Heyob, M., Deporte-Fety, R., Renée, N., 1999. Dihydropyrimidine dehydrogenase deficiency and fluorouracil-related toxicity. Br. J. Cancer 79, 627−630.

Nair, S.R., 2010. Personalized medicine: striding from genes to medicines. Perspect. Clin. Res. 1 (4), 146−150.

Nelson, D.R., Conlon, M., Baralt, C., Johnson, J.A., Clare-Salzler, M.J., Rawley-Payne, M., 2011. University of Florida clinical and translational science institute: transformation and translation in personalized medicine. Clin. Transl. Sci. 4 (6), 400−402.

O'Donnell, P.H., Bush, A., Spitz, J., Danahey, K., Saner, D., Das, S., et al., 2012. The 1200 patients project: creating a new medical model system for clinical implementation of pharmacogenomics. Clin. Pharmacol. Ther. 92, 446−449.

Parikh, S., Ouedraogo, J.B., Goldstein, J.A., Rosenthal, P.J., Kroetz, D.L., 2007. Amodiaquine metabolism is impaired by common polymorphisms in CYP2C8: implications for malaria treatment in Africa. Clin. Pharmacol. Ther. 82 (2), 197−203.

Paulussen, A., Lavrijsen, K., Bohets, H., Hendrickx, J., Verhasselt, P., Luyten, W., et al., 2000. Two linked mutations in transcriptional regulatory elements of the CYP3A5 gene constitute the major genetic determinant of polymorphic activity in humans. Pharmacogenetics 10, 415−424.

Pui, C.H., Evans, W.E., 2006. Treatment of acute lymphoblastic leukemia. N. Engl. J. Med. 354, 166−178.

Pulley, J.M., Denny, J.C., Peterson, J.F., Bernard, G.R., Vnencak-Jones, C.L., Ramirez, A.H., et al., 2012. Operational implementation of prospective genotyping for personalized medicine: the design of the Vanderbilt PREDICT project. Clin. Pharmacol. Ther. 92, 87−95.

Ramsey, B.W., Davies, J., McElvaney, N.G., Tullis, E., Bell, S.C., Dřevínek, P., et al., 2011. A CFTR potentiator in patients with cystic fibrosis and the G551D mutation. N. Engl. J. Med. 365, 1663−1672.

Relling, M.V., Gardner, E.E., Sandborn, W.J., Schmiegelow, K., Pui, C.H., Yee, S.W., et al., 2011. Clinical Pharmacogenetics Implementation Consortium guidelines for thiopurine methyltransferase genotype and thiopurine dosing. Clin. Pharmacol. Ther. 89 (3), 387−391.

Remy, C.N., 1963. Metabolism of thiopyrimidines and thiopurines. S-methylation with S-adenosylmethionine transmethylase and catabolism in mammalian tissues. J. Biol. Chem. 238, 1078e84.

Roden, D.M., Altman, R.B., Benowitz, N.L., Flockhart, D.A, Giacomini, K.M., Johnson, J.A., et al., 2006. Pharmacogenetics Research Network. Pharmacogenomics: challenges and opportunities. Ann. Intern. Med. 145 (10), 749–757.

Rodriguez-Novoa, S., Barreiro, P., Rendon, A., Jimenez-Nacher, I., Gonzalez-Lahoz, J., Soriano, V., 2005. Influence of 516G > T polymorphisms at the gene encoding the CYP450-2B6 isoenzyme on efavirenz plasma concentrations in HIV-infected subjects. Clin. Infect. Dis. 40, 1358–1361.

Rouits, E., Boisdron-Celle, M., Dumont, A., Guérin, O., Morel, A., Gamelin, E., 2004. Relevance of different UGT1A1 polymorphisms in irinotecan-induced toxicity: a molecular and clinical study of 75 patients. Clin. Cancer. Res. 10, 5151–5159.

Sarepta Therapeutics Announces Eteplirsen Meets Primary Endpoint of Increased Novel Dystrophin and Achieves Significant Clinical Benefit on 6-Minute Walk Test After 48 Weeks of Treatment in Phase IIb Study in Duchenne Muscular Dystrophy, October 03, 2012. http://finance.yahoo.com/news/sarepta-therapeutics-announces-eteplirsen-meets-110000135.html.

Schwab, M., Zanger, U.M., Marx, C., Schaeffeler, E., Klein, K., Dippon, J., et al., German 5-FU Toxicity Study Group 2008. Role of genetic and nongenetic factors for fluorouracil treatment-related severe toxicity: a prospective clinical trial by the German 5-FU Toxicity Study Group. J. Clin. Oncol. 26, 2131–2138.

Simon, T., Verstuyft, C., Mary-Krause, M., Quteineh, L., Drouet, E., Meneveau, N., et al., 2009. Genetic determinants of response to clopidogrel and cardiovascular events. N. Engl. J. Med. 360, 363–375.

Sindrup, S.H., Brosen, K., 1995. The pharmacogenetics of codeine hypoalgesia. Pharmacogenetics 5, 335e46.

Stanulla, M., Schaeffeler, E., Flohr, T., Cario, G., Schrauder, A., Zimmermann, M., et al., 2005. Thiopurine methyltransferase (TPMT) genotype and early treatment response to mercaptopurine in childhood acute lymphoblastic leukemia. JAMA 293, 1485–1489.

Toffoli, G., Cecchin, E., Corona, G., Russo, A., Buonadonna, A., D'Andrea, M., et al., 2006. The role of UGT1A1*28 polymorphism in the pharmacodynamics and pharmacokinetics of irinotecan in patients with metastatic colorectal cancer. J. Clin. Oncol. 24 (19), 3061–3068.

U.S. Food and Drug Administration, 2014. In Vitro Companion Diagnostic Devices. Guidance for Industry and Food and Drug Administration Staff. Docket No. FDA-2011-D-0215.

van Kuilenburg, A.B., 2004. Dihydropyrimidine dehydrogenase and the efficacy and toxicity of 5-fluorouracil. Eur. J. Cancer 40, 939–950.

van Kuilenburg, A.B., Baars, J.W., Meinsma, R., Gennip, A.H., 2003. Lethal 5-fluorouracil toxicity associated with a novel mutation in the dihydropyrimidine dehydrogenase gene. Ann. Oncol. 14 (2), 341–342.

van Kuilenburg, A.B.P., Muller, E.W., Haasjes, J., Meinsma, R., Zoetekouw, L., Waterham, H.R., et al., 2001. Lethal outcome of a patient with a complete dihydropyrimidine dehydrogenase (DPD) deficiency after administration of 5-fluorouracil: frequency of the common IVS14 + 1G > A mutation causing DPD deficiency. Clin. Cancer Res. 7 (5), 1149–1153.

Vesell, E.S., 1984. New directions in pharmacogenetics: introduction. FASEB J. 43 (8), 2319–2325.

Vreken, P., Vankuilenburg, A.B.P., Meinsma, R., Smit, G.P., Bakker, H.D., De Abreu, R.A., et al., 1996. A point mutation in an invariant splice donor site leads to exon skipping in two unrelated Dutch patients with dihydropyrimidine dehydrogenase deficiency. J. Inherited Metabolic Dis. 19 (5), 645–654.

Wadelius, M., Chen, L.Y., Eriksson, N., Bumpstead, S., Ghori, J., Wadelius, C., et al., 2007. Association of warfarin dose with genes involved in its action and metabolism. Hum. Genet. 121, 23–34.

Wadelius, M., Chen, L.Y., Lindh, J.D., Eriksson, N., Ghori, M.J., Bumpstead, S., et al., 2009. The largest prospective warfarin-treated cohort supports genetic forecasting. Blood 113, 784–792.

Ward, B., Gorski, J., Jones, D., Hall, S., Flockhart, D., Desta, Z., 2003. The cyto-chrome P450 2B6 (CYP2B6) is the main catalyst of efavirenz primary and secondary metabolism: implication for HIV/AIDS therapy and utility of efavirenz as a substrate marker of CYP2B6 catalytic activity. J. Pharmacol. Exp. Ther. 306, 287–300.

Wei, X.X., McLeod, H.L., McMurrough, J., Gonzalez, F.J., Fernandezsalguero, P., 1996. Molecular basis of the human dihydropyrimidine dehydrogenase deficiency and 5-fluorouracil toxicity. J. Clin. Invest. 98 (3), 610–615.

Weiner, M., Burman, W., Vernon, A., Benator, D., Peloquin, C.A., Khan, A., et al., 2003. Low isoniazid concentrations and outcome of tuberculosis treatment with once-weekly isoniazid and rifapentine. Am. J. Respir. Crit. Care Med. 167 (10), 1341–1347.

Weinshilboum, R.M., Sladek, S.L., 1980. Mercaptopurine pharmacogenetics: monogenic inheritance of erythrocyte thiopurine methyltransferase activity. Am. J. Hum. Genet. 32, 651e62.

Yen, J.L., McLeod, H.L., 2007. Should DPD analysis be required prior to prescribing fluoropyrimidines? Eur. J. Cancer 43, 1011–1016.

Ytterberg, S.R., Mahowald, M.L., Woods, S.R., 1998. Codeine and oxycodone use in patients with chronic rheumatic disease pain. Arthritis Rheum. 41, 1603e12.

Zukunft, J., Lang, T., Richter, T., Hirsch-Ernst, K.I., Nussler, A.K., Klein, K., et al., 2005. A natural CYP2B6 TATA Box polymorphism (-82T-> C) leading to enhanced transcription and relocation of the transcriptional start site. Mol. Pharmacol. 67, 1772–1782.

Further Reading

Haga, S.B., Thummel, K.E., Burke, W., 2006. Adding pharmacogenetics information to drug labels: lesson learned. Pharmacogenet. Genomics 16 (12), 847–854.

Omics Approaches in Marine Biotechnology: The Treasure of Ocean for Human Betterments

Fatima Abid[1], Muhammad A. Zahid[2], Zain U. Abedin[3], Syed B. Nizami[4], Muhammad J. Abid[5], Sayyada Z. Hassan Kazmi[4], Sami U. Khan[6], Humna Hasan[7], Mohsin Ali[8] and Alvina Gul[8]

[1]The University of Lahore, Islamabad, Pakistan [2]Shifa Tameer-e-Millat University, Islamabad, Pakistan [3]University of Strathclyde, Glasgow, United Kingdom [4]University of Karachi, Karachi, Pakistan [5]The University of Lahore, Lahore, Pakistan [6]University of Haripur, Khyber Pakhtunkhwa, Pakistan [7]Quaid-I-Azam University, Islamabad, Pakistan [8]National University of Sciences and Technology, Islamabad, Pakistan

3.1 INTRODUCTION

Marine biotechnology emerged in 1980s, currently gaining momentum in the areas of genetics, marine food, marine energy, and human health with the introduction of fascinating analytical techniques of genomics, proteomics, transcriptomics, nutrigenomics, and metabolomics. Marine biotechnology is in its infancy opening doors for the exploration of a biodiversified marine world by the use of innovative omics techniques (Qin et al., 2012). With an increasing awareness on effectiveness of natural products, the pharmaceuticals emergence has increased dedicated to marine biotechnology (Qin et al., 2011).

The nature at epigenetic level is greatly complex. Therefore, for a better understanding of biological processes such as the metabolic pathways and physiological processes, the study of proteomics must be complemented with genomic studies. The diversified marine pool holds a treasure of massive biological molecules encapsulated in marine organisms. The

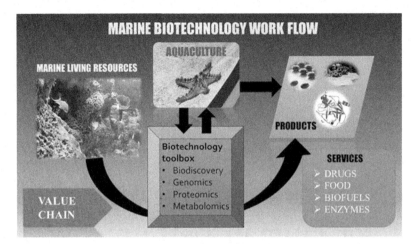

FIGURE 3.1 The global marine biotechnology specificity lies in the marine products and services obtained from unique marine living resources with the aid of biotechnological tools.

detection of an undiscovered pool of proteins having different physiochemical properties requires a validated method of analysis. Due to the lack of recent advances, there is no single method available for analysis of all these properties. However, nuclear magnetic resonance (NMR) and mass spectrometry (MS) have gained superiority among all available analytical techniques. It is, therefore, significant to unlock all the biological markers to make a lucid understanding of all biological pathways and their functions (Hochstrasser et al., 2002).

With biotechnological and biological advancements, efficient aquacultures have now come into existence. With optimization of omics tools, such as transcriptomics and proteomics, thousands of marine invertebrate transcripts and proteins have been identified. The ocean is a sustainable bioenergy source for the production of biofuels, thereby presenting a challenge. Moreover, the ocean world also presents a platform for discoveries of novel drugs, new opportunities in treatment, and diagnostic tools for the betterment of human health. Hence, marine-derived nutraceuticals have become a major focus in domains of human well-being. Fig. 3.1 provides a summary of the integration of omics tools in the marine biotechnology process.

3.2 IMPACT OF OMICS ON MARINE BIOTECHNOLOGY

Marine biotechnology is in its early developmental phase, based on unmapped and diversity of aquatic system. Omics is one of the rapidly growing technologies transforming the domains of biotechnological and biomedical research. It has transitioned the research approach from a single gene to the whole organism (Hollywood et al., 2006; Joyce and Palsson, 2006). With the introduction of genomics, the pharmacogenomics, agro-biotechnology, medical diagnostics, and aqua-biotechnology have started to flourish. The identification of marine species has been pronounced with the aid of nuclear and mito-chondrial genome studies. It is a very dynamic and powerful research tool for discovering the evolution of marine life which is helping in the sequencing of complete genome and understanding of their function (Titilade and Olalekan, 2015). The metagenomics study is still in infancy not yet applied on marine ecology on a broad scale (Venter et al., 2004; DeLong et al., 2006; Sogin et al., 2006). The recent advancements in genomic laboratory tools and development in oceanography have resulted in the establishment of a better understanding of relationship existing between biodiversified marine world and its impact on the ecosystem (Worm et al., 2006). Under genomics, the structure and sequence of organism's genome (DNA) is identified, which undoes the issues related to functions of the ecosystem, climatic change, and loss of biodiversity.

Metatranscriptomics, however, provides a new insight into the ocean world. To find a relationship between the changing environmental conditions and microbial functions, the high-throughput screening and sequencing techniques have evolved with metagenomics for the study of metatranscriptomics. With the emergence of metatranscriptomics, it has become possible for the scientists to discover the impact of changing conditions on the metabolic activity of marine organisms by revealing both known and unknown transcripts in natural habitat (Frias-Lopez et al., 2008; Gilbert et al., 2008).

Proteomics is another analytical technique for the analysis of marine organisms' functional genome (Schweder et al., 2008). With proteomics, it has become possible to investigate the bacterial response to stress conditions as well as to starvation. It allows knowing marine bacterial physiology and genome analysis.

3.3 OMICS-DRIVEN MARINE BIOTECHNOLOGY

In this century, advances in the field of biotechnology have opened up a new world of biological sciences not yet mapped. It has widened up new horizons of biotechnological research, thereby expanding the prospects in areas of industrial, medical, and agricultural applications. The evergrowing developments in the scientific research tools have permitted an explosive growth in molecular biology allowing to manipulate, analyze, and apply biological processes in environmental, health, and industrial domains.

The ocean world, extending from coasts to abysses, holds an enormous microbial biosphere: an unlocked resource of biotechnological approaches and interests (Luna, 2015). The mid-1990s osmic revolution, a field of molecular biology, has addressed the objects of "ome" such as genomics, proteomics, transcriptomics, nutrigenomics, and metabolomics. These osmics-driven technologies aim to quantify and characterize the pools of all biological molecules that structurally, functionally, and dynamically translate into organism(s). Omics ("ome" derived from a Greek word meaning "whole," "all," or "complete") on a broad scale thus focus on understanding the complex life in encapsulated "omes." This holistic approach has diverse and important applications in the field of biotechnology. Table 3.1 provides a summary of the applications of omics techniques in the field of marine biotechnology.

3.3.1 Genomics

Assessment of genetic annotation and sequencing is essential to the biotechnological potential of marine organisms. DNA sequencing through Sanger sequencing method has

TABLE 3.1 Applications of Omics Tools in Blue Biotechnology

Omics Approach	Applications in Marine Biotechnology
Genomics	Determination of biological diversity
	Development of Global Ocean Sampling campaign to investigate the genomics of marine organisms
	Understanding of marine microbial consortia
Proteomics	Preclinical drug safety evaluation
	Development of marine-based therapeutics
Transcriptomics	Study of relative gene expressions and exploitation of these studies for improving biological efficiency
Nutrigenomics	Understanding of food webs
	Development of patient-specific nutritional support
Metabolomics	Development of therapeutic enzymes
	Understanding of underwater metabolic processes and the way to enhance their efficiency
	Identification of biological markers

taken breathtaking progress due to the technical advances in sequencing technologies (Metzker, 2010). About half of the 1000 prokaryotic genomes being sequenced and annotated are of medical and industrial relevance. Until now no systematic phylogenetic genome sequence has been done. A large reservoir of nonsequenced genomes of undiscovered proteins still exists. Once the sequencing and assembly of the genome of organisms are done, the identification of gene-coding regions and the establishment of their annotations become the critical step (Stein, 2001). Subsequent sequence recognition through computational analysis determines the functional gene content (Siezen and Van Hijum, 2010).

Genomes of more than 150 marine bacteria have been sequenced, and among all the biological entities in the marine environment, viruses are the most commonly found microbial species in the aquatic ecosystem. A unique gene pool along with the molecular architecture of marine viruses has been determined through metagenomics surveys. The marine viral diversity estimated by the metagenome analysis indicates differences in the genetic makeup of hundreds of thousands of different biological species, with genes completely varying from other forms of life.

The genomics era of marine eukaryotes, comprising microalgae, macroalgae, and protozoa, is still embryonic because of the high cellular complexity and a massive genome size. However, the sequencing of these genomes provides a novel information and an insight into the concept of species and ecological niche (Pedrós-Alió, 2010). The genome of not more than 30 microalgae has been done with size ranging from 15 to 165 Mb. They are highly complexed single-cell marine organisms having human-like chromosomal and mitochondrial DNA. The algae also hold chloroplast DNA. With possession of a complex nucleus, there is a transfer of genetic material between the organelles via vertical/endo-symbiotic gene transfer and horizontal gene transfer.

Marine cyanobacteria are grouped into *Prochlorococcus* and *Synechococcus*. The grouping of these two closely linked genera is done on the basis of 16S rRNA, responsible for two-third oceans' photosynthesis. However, *Prochlorococcus* is also taken as the most abundantly existing photosynthetic organisms (Partensky et al., 1999). Genome sequencing of three strains of *Prochlorococcus* (Rocap et al., 2003; Dufresne et al., 2003) showed that all the three strains belong to the same species with a variation in their genome sizes. The strains SS120 and MED4 had a genome of 1.7 Mb, whereas the strain MIT9313 had a genome size of 2.4 Mb. This difference in sizes reflects the stability of a strain in its specific environment such as the MED4 strain is highly adapted to low nutrient content with the availability of high light intensity and hence resides at the ocean surface, and the MIT9313 strain is adapted to intermediate ocean depths and SS1220 lives at lower depths of the ocean. The genomic study of these marine cyanobacteria demonstrates how the evolution, taxonomy, and ecology along with the physiology of microorganisms alter.

Metagenomics libraries of various communities, if given, can be used for relative gene abundance comparison and traced back to particular environmental conditions. Various metagenomics approaches are conducted for sequencing. One approach allows the novel genes discovery, irrespective of their origin. It involves the environmental DNA fragmentation (into 3 kb size) with subsequent shotgun cloning into conventional vectors. Without screening, the DNA fragments are sequenced. Using the procedure of shotgun cloning, a sample of more than 100,000 genes was revealed, many of them without known functions (Venter et al., 2004). Another approach involves the fragmentation and cloning of large environmental DNAs of genes found in the genome. For example, if one clone has 16S rRNA gene, the bacterium from where it came can be identified. This approach was used to discover the function of an existing novel gene. Another strategy for sequencing of deep ocean samples involves the use of PCR. This technique is used in couple with 454, a Life Sciences Company technology, for 16S rRNA gene samples (Sogin et al., 2006). The magnitude of known marine samples and its diversity have doubled with the existence of these new sequencing and cloning techniques. A large collection of rare taxa that contributes in making a complete biodiversity and a dozen of abundant taxa, forming bacterial communities with diversity of sequences in a sample, makes diversity of that ecosystem (Pedrós-Alió and Simó, 2002)

3.3.2 Proteomics

Metagenome studies provide a set of DNA sequences, reflecting genetic makeup of living organisms. For biological processes to be understood at the molecular level, the genetic studies must be complemented with the proteomic studies. With the existence of vast variety of proteins having dissimilar physiochemical properties, it has become difficult to find a single method for protein analysis. However, MS is considered one of the superior proteomic tools and has become the vital element with the ability to partially characterize and quickly identify low abundant proteins. A proteome analysis comprehensively indicates the role of abundant organisms taking part in processes of nutrient cycling (Hochstrasser et al., 2002).

For the identification of protein of interest, right sample preparation followed by the selection of appropriate sensitive ionization technique are basic requirements. To study phototrophic marine organisms (bacteria), a mass spectroscopic protocol was developed for relevant biological protein identification (Stapels et al., 2004). Sensitive proteomic techniques can assess the entirety of abundant proteins; however, metaproteomics potential is still in its rudimentary phase (Wilmes and Bond, 2006; Maron et al., 2007). In 2004, from a sewage mud, Wilmes and Bond both investigated prokaryotic group by using 2-D protein gel electrophoresis and MS (Wilmes and Bond, 2006). A metaproteome study of Chesapeake Bay marine microbial community was conducted without its prior DNA being sequenced. However, the information resulting from this proteome analysis was limited (Kan et al., 2005). New strategies are needed to be designed to deal with the difficulty of metaproteomes. From the previous metaproteomic studies, it is indicated that a collateral comprehensive metagenome analysis is needed for an effective metaproteomic investigation. With the help of 454/Roche Pyrosequencing method, large-scale DNA sequencing is done (Margulies et al., 2005). This method is useful for the sequencing of DNA with 200 nucleotides; however, with current optimization, it is capable of providing extensive sequencing information of up to 400 nucleotides. For future complex marine microbial communities, the proteomic study with cell fractionation methods appears to be a powerful technique in dealing with diverse microbial consortia.

For the future study of unknown protein function, a comprehensive proteome analysis directed at cellular physiology is needed. However, a wide-ranging information on marine ecosystem can emerge from functional information available on proteins' physiological role.

3.3.3 Transcriptomics

Every sequenced genome of unicellular marine eukaryote has provided a new insight into the aquatic world (Armbrust et al., 2004; Read et al., 2013). An obvious alternate way to generate reference database other than genetic modeling is transcriptomics (Worden and Allen, 2010). To assemble reference database, transcriptomes are the building blocks from pure cultures, thereby generating a large number of coding sequences from known organisms. Transcriptomic data, however, should not be taken as an alternate of gene sequencing.

Marine Microbial Eukaryotic Transcriptome Sequencing Project (MMETSP) is intended to provide a significant base for integrating eukaryotic microbes into marine ecology by generating more than 650 functionally annotated, assembled, and publicly available transcriptomes, whereas the transcriptomes come from the most ecologically abundant and significant marine eukaryotes. Many eukaryotic groups are not covered by MMETSP transcriptomes because of the lack of lineages in the oceans with the difficulty in cultivating the lineages, and some of them are characterized poorly in molecular data (Keeling et al., 2014).

Transcriptomics sequences mRNAs from the microbial community without previous knowledge about gene expression. The protocols for prokaryotic mRNAs are generally difficult than eukaryote mRNAs because it lacks the poly (A) tails have comparatively short half-lives (Liang and Pardee, 1992; Brawerman, 1993). The study of environmental transcriptomics provides information on the natural ecosystem: bacterial mRNA dynamics and composition (Poretsky et al., 2005). The study can provide a deeper insight to microbial biogeochemistry and provide a way to better understand quantitative gene expression study (Poretsky et al., 2009).

3.3.4 Nutrigenomics

In the detection of various physiological stresses caused by nutrition or any disease, expression of genes is used as one of the biomarkers. But in the case of marine living organisms, the gene expression approach is less commonly used (Spitz et al., 2015). To meet the need for fish products, the growth of related industries has greatly increased. Gene expressions related to marine nutrition can be modified with the desire to introduce different components of feeds. Nutrigenomics is the study of nutrients as feeding signals manipulating the metabolite production. The sensory cellular systems detect these signals thereby affecting the protein and gene expression. The feeding components have, therefore, a massive effect on the safety of nurtured aquatic life as well as on related gene expression. Aquaculture expansion requires sustainability and protection from fish stock depletion along with the improvement in feed for marine culture for better health and well-being. Immune response to new diets can be interpreted by molecular biological approaches. Fig. 3.2 represents the integration of nutrigenomics approach for the exploitation of rich source of nutrition from the marine organisms.

Various genes are used as markers to understand dietary responses. In ubiquitin proteasome, a protein degradation pathway, the enzyme ubiquitin ligase serves as a marker to

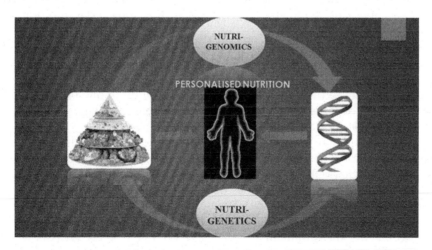

FIGURE 3.2 A representation of how nutrigenomics define the genetic variation of organisms/individual towards the determination of the optimal diet.

interpret fish performance and health. Under food-deprived and proinflammatory conditions, both genes Atrogin-1 and MuRF are increased. The expression of NF-kB mitochondrial activators' genes, MULAN and Mul1b, is also increased under such conditions.

Intestinal tissues are the most sensitive sites, which come into direct contact with fed food having the highest regulated transcriptional response. In numerous fish feed ratio schemes, many responses of biological processes such as stress-immunogenic responses could be different related to intestinal functions interacting with degradation of proteins, transport and metabolism of lipids, and mitochondrial energy generation pathways. Genes related to the immunogenic responsive proteins have shown different expressions in response to modifications in diet. The role of physiology in response to nutritional modifications is evident from changes in expression of genes and different transcriptomic response consequently altering the metabolic pathways and the physiological response (Exadactylos, 2014; Tacchi, 2011).

3.3.5 Metabolomics

Metabolomics is the study of metabolomes where metabolomes are low-molecular weight endogenous metabolites found in fluids, cells, or tissues. The metabolomics application to natural anthropogenic stresses in characterizing organisms' metabolic responses comes under environmental metabolomics. For the assessment of functions and determining organisms' health at a molecular level, this approach is beneficial. This novel omics approach is rapidly capable of measuring thousands of metabolites and hence useful in identifying useful biological markers and metabolic processes. Metabolomics have been applied to an aquatic system in the study of toxicity, disease, fish embryogenesis (Viant 2003), and environmental stress (Viant et al. 2003).

Embryos of Chinook salmon (*Oncorhynchus tshawytscha*) (Viant et al., 2006b) and medaka (*Oryzias latipes*) (Viant et al., 2006a) have been used in the study of toxic metabolomics. In fish and invertebrate the effects of disease such as a bacterial infection in Atlantic salmon (*Salmo salar*) on metabolome have been studied (Solanky et al. 2005).

Metabolomics is different from other omics approaches that are genomics, transcriptomics, and proteomics, in the way it responds to stressful and natural conditions. Unlike genes, transcripts, and proteins, the metabolites are compounds with diverse physiochemical properties, and due to this reason, there is no availability of one particular bioanalytical technique capable of measuring all types of metabolites. The two most commonly used techniques in metabolomics are NMR and mass spectrometry. However, both techniques lack the ability to completely measure all types of metabolites. About 1000 metabolites can be detected using MS, whereas up to 100 metabolites can be measured using NMR. Therefore, the researchers should abreast themselves with technological advancements able to cover a vast variety of metabolomes (Johnson et al., 2007).

The metabolomics study is very crucial for understanding the relationship of metabolites with organisms' physiology. For example, the energy status of a living organism can be determined by the measurement of ATP and glycogen levels (Wasser et al. 1996). Similarly, glutathione and ascorbate molecules serve as biological markers for cellular redox reactions (Kristal et al. 1998). An organism's reproductive status can be known by detection of the two steroidal molecules ketotestosterone and estradiol (Noaksson et al. 2004). Metabolomics weighs an advantage on other omics techniques over assessment of multispecies, which ultimately helps in the overall assessment of environmental stresses on the ecosystem.

3.4 CHALLENGES AND FUTURE OPPORTUNITIES

Advancements in the "omics" technology have provided a strong toolset thereby allowing to explore all phenotypic and functional gene networks and proteins of all cells and organisms. Omics technology has now transformed from genetic and protein listing into in-depth disease-specific meta-genetic analyses, pathway mapping and modifications, along with analyses of protein—protein interactions (Ning and Lo, 2010). This technology is nowadays massively employed by the marine scientists and the oceanographers providing an opportunity in quantifying and characterizing the ocean nature and in determining the function and distribution of biological organisms in ocean ecosystem (DeLong, 2009; Gilbert and Dupont, 2011; Giovannoni and Vergin, 2012). The tools encompassing omics science are multifarious, rapidly growing, and complex. It requires the huge capable data analytical tools and construction of sundry federated catalogs. In this arena sequencing new-generation technologies, its economics and applications to a large extent have become a big challenge for the marine science community (Gilbert et al., 2014). The primary limiting factor is insufficient analytical tools availability and intercomparison as well as visualization of environmental, oceanographic, and omics data sets. This has created a barrier between the integration of physical—biogeochemical models development and emerging omics databases.

3.4.1 Marine Food

The oceans hold a huge unused and underutilized source of food; a biodiverse bank of environmental niches. As a result of massive fisheries, the number of fish stocks has become endangered. To alleviate the effects of overfishing and escalate the replenishment of fish/wild stock, "Marine biotechnology" by captive stock breeding and selection can secure the healthy marine food. Aquaculture, therefore, needs to become more effective and efficient to meet the challenges of growing seafood and its supply to the market. As a result of mutual understanding of industry and research community, the focus on improving product quality, development of sustainable practices, and introduction of novel species for rigorous cultivation have markedly increased. Rapid biotechnological and biological advancements have yielded an efficient aquaculture.

In carnivorous, aquaculture fish meal and fish oil are artificial diet derived from wild sector. Shifts in diet profile induce an impact on fish health, immunity, and metabolism, which is closely monitored by molecular tools of liver-enzyme biomarkers, metabolic and immune parameters. The addition of carotenoid substances imparts natural color and dietary benefits in the fish. The biological functions such as tissue regeneration, growth, and reproduction are also improved. The healthier diet provided to the farmed species benefit not only the fish but also the consumers, and it serves as a nutritional carrier beneficial for the humans. Currently, however, commercial aquaculture is facing problems with the reproduction, larval development, health–disease management, growth, nutrition, and environmental interaction. For optimal production, provision of necessary physical conditions such as light, pressure, and temperature at deeper water surfaces and weather conditions is an important and a foremost challenge for the farming oceans. Supply of feed to increase the target organism is another challenge posed to the local biosphere balance. Therefore, for the sustainable development of aquaculture, the application of biotechnological and molecular tools is important. Nowadays the concern of consumers regarding GMOs has increased. In some farmed species, by manipulation of growth-controlling hormone genes, the rate of growth has increased to 30-fold.

Development in the food science has led to an improvement in health sector such as the consumption of omega-3 fatty acids EPA and DHA from marine food is very beneficial. Omega-3 polyunsaturated fatty acids produce a synergistic effect with Taurine component found in seafood. The proteinaceous fish part has largely been ignored from its health benefits. Dietary modifications provide an opportunity to produce seafood with added functional substances. Innovative genomics and molecular biology–based methods can be established in the near future to allow production of marine food capable of synthesizing fatty acids and that are genetically modified (Querellou et al., 2010).

3.4.2 Marine Energy

Marine biotechnology can contribute significantly to the emerging need of energy supply on a global scale. There are many different approaches to achieve this goal. For example, Microbial Enhanced Oil Recovery (MEOR) can help to improve the life of mature oil reservoirs effectively. MEOR is a bio-based approach to improve the efficiency of retrieval of fossil oil reserves by manipulating the structure or function of microbes in oil reservoirs. The ecophysiology of the oil reservoir can be changed by complementing different strategies. In situ stimulation of growth and metabolic activity of indigenous microbes can be promoted by injecting nutrients. Alternatively, exogenous microbes can be injected, which are adapted to oil reservoir conditions and are capable to produce desired effects. The results of MEOR trials are still controversial and their benefits are debatable. However, technical and economic feasibility and environmental safety of MEOR is one of the present challenges.

Ocean itself is an untapped and sustainable source of bioenergy. The production of biofuel with microalgae poses one option to harvest this huge potential, and there are many more examples related to the production of bioenergy from marine organisms/bacteria (Felix, 2010).

A great deal of research has already been done, which confirms microalgae and seaweeds as possible resources for the production of renewable fuels. In addition, indigenous seaweeds consist of a valuable reserve for food and cosmetics businesses in some

European countries like France. However, thousands of tons of undesirable seaweeds are contaminating the coasts and regularly need to be carried. Despite various trials, it appears that bioconversion of this large amount of seaweeds is technically feasible but not profitable. Creating seaweeds especially for biofuel production has also been recommended but its practicability remains dubious (Matsunaga et al., 2008).

Microalgae are capable of accumulating large quantities of hydrophobic compounds that can be converted into biodiesel. The production of biodiesel from microalgal acyl glycerides is mainly the focus of much interest. The concentration of lipids in microalgae varies between 10% and 60%. The buildup of high concentrations of lipids in globules mostly takes place as a result of stress such as that induced by the limitation of nitrogen. However, any benefit gained through an increase in oil content is usually counterbalanced by lower production rate induced by the same stress, and therefore using the entire algal biomass is important for achieving a good balance of energy (Steele et al., 2008).

Based on certain small-scale projects, it is estimated that a theoretical production of 20,000–80,000 L of oil per hectare can be achieved annually from microalga culture. Although present technology limits the efficiency of up to 20,000 L/ha per annum, it is considerably higher than production from earthly crops. The annual production of oil from palm and rapeseed is 6000 and 1500 L/ha, respectively. In comparison to terrestrial crops, there is no dispute about algae as an energy crop contending with food crops, and this is because the productivity of algal biomass is higher than for earthly crops and more importantly because microalgae can be grown in closed systems with close to complete reprocessing of nutrients and water. For cultivation of microalgae, areas that are not suitable for agriculture can be used. The cultivation of microalgae for the generation of bioenergy is a great challenge and one of the key aims for Marine Biotechnology in the current century. Detailed challenges are diversified by the need of an enormous upscaling for commercial production, and hence a highly multidisciplinary approach is needed. The success will be dependent on the overall efficiency at all levels of the production chain (Coscia and Oreste, 2003). Following are the main challenges that are foretold:

- Examination and understanding of the biodiversity of microalgae at the molecular level and on a global scale.
- Attainment of full sustainability of the whole production chain in terms of regional and global impact.
- Processing and harvesting of large amounts of microalgae for the production of the optimal mix of bioenergy and bioproducts.
- Taking advantage of the physiological potential of microalgae to produce biofuels with the bioengineering tools of the 21st century.
- Achievement of a net energy gain along with the whole production chain necessary to convert microalgal biomass into biofuels.

3.4.3 Human Health

Oceans and seas affect our health and well-being. Over thousands of years since the ancient times, oceans have served humans in different ways. It has interacted with humans in serving food, connecting people around the globe, providing living and in generation of economy. Intervention into the marine ecosystem or coastal damage due to natural disaster has led to a negative impact on human health. However, this interaction with marine environment may also have beneficial effects on human life (Bowen et al., 2006; Fleming et al., 2006; Walsh et al., 2011; and Bowen et al., 2014) The food web of aquatic ecosystem constitutes the photosynthetic organisms that produce oxygen by fixing carbon, for example, cyanobacteria, planktonic microalgae, and macrophytes. However, under some situations due to the abundance of various taxa, harm is caused to other organisms and humans. Harmful algal blooms (HABs) is a term used to describe adverse and harmful consequences of various taxa perceived by humans on the marine ecosystem. The HABs pose a threat to have beneficial effects onbenefits provided by seas and oceans on tourism, food supplies, recreation, and ultimately the human well-being. More than 300 microalgal species are involved in harmful damages to human health, among 100 of which cause lethal effects (Berdalet et al., 2015).

Toxicity results from direct contact with water contaminated with toxin, from inhalation, or by ingestion of toxin-contaminated seafood. However, adverse HABs' outcomes have reduced due to awareness. Designing of artificial beach and modifications in harbors' water circulation have pronounced HABs' incidence (Hallegraeff and Bolch, 1992; Anderson et al., 2002; Davidson et al., 2014). However, a number of marine-derived products are marketed in collaboration with industries. This is because of the discovery of pharmaceutically active biological compounds by the use of novel technologies. Cosmeceuticals, nutraceutical, and functional food, for example, have found tremendous industrial applications. Fig. 3.3 provides an image of various species of marine organisms that have proved to be of potential benefit to humans.

FIGURE 3.3   Representation of a biodiversified marine world.

3.5 CONCLUSION

Based on the benefits and the challenges presented, it is evident that marine or blue biotechnology is an emerging field of biotechnology with an immense potential in the field of medical, nutritional, electronic, and pharmaceutical sciences. Increased efforts need to be made to understand the structure, physiology, and chemistry of these organisms to exploit the agents that serve as an effective bioresource. World's marine ecosystems are underexplored and underutilized. A challenge, therefore, remains for the biotechnologists, physicists, and policy makers to develop means to study this resource for the development of industrially viable processes and products. By harnessing the marine biomaterials, new applications in the development of pharmaceuticals, industrial enzymes, nutritional products, and biosensors can be developed. Blue biotechnology can, hence, potentially be exploited by the enterprise sector for finding new means to fulfill the therapeutic and commercial gap. This would provide an insight into the development of sustainable exploitation of marine resources for the human betterment.

References

Anderson, D.M., Glibert, P.M., Burkholder, J.M., 2002. Harmful algal blooms and eutrophication: nutrient sources, composition, and consequences. Estuaries 25 (4), 704–726.

Armbrust, E.V., Berges, J.A., Bowler, C., Green, B.R., Martinez, D., Putnam, N.H., et al., 2004. The genome of the diatom Thalassiosira pseudonana: ecology, evolution, and metabolism. Science 306 (5693), 79–86.

Berdalet, E., Fleming, L.E., Gowen, R., Davidson, K., Hess, P., Backer, L.C., et al., 2015. Marine harmful algal blooms, human health and wellbeing: challenges and opportunities in the 21st century. J Mar Biol Assoc UK 2015, 1–31.

Bowen, R., Depledge, M., Carlarne, C., Fleming, L.E. (Eds.), 2014. Seas, Society and Human Wellbeing. Wiley, Publishers, UK.

Bowen, R.E., Halvarson, H., Depledge, M.H., 2006. The oceans and human health. Mar Pollut Bull 53, 631–639.

Brawerman, G., 1993. mRNA degradation in eukaryotic cells: an overview. In: Belasco, J.G., Brawerman, G. (Eds.), Control of Messenger RNA Stability. pp. 149–159.

Coscia, M.R., Oreste, U., 2003. New developments in marine biotechnology. Mar. Biotechnol. 5 (3), 213.

Davidson, K., Gowen, R.J., Harrison, P.J., Fleming, L.E., Hoagland, P., Moschonas, G., 2014. Anthropogenic nutrients and harmful algae in coastal waters. J Environ Manag 146, 206–216.

DeLong, E.F., Preston, C.M., Mincer, T., Rich, V., Hallam, S.J., Frigaard, N.U., et al., 2006. Community genomics among stratified microbial assemblages in the ocean's interior. Science 311 (5760), 496–503.

DeLong, E.F., 2009. The microbial ocean from genomes to biomes. Nature 459 (7244), 200–206.

Dufresne, A., Salanoubat, M., Partensky, F., Artiguenave, F., Axmann, I.M., Barbe, V., et al., 2003. Genome sequence of the cyanobacterium Prochlorococcus marinus SS120, a nearly minimal oxyphototrophic genome. Proc Natl Acad Sci USA 100 (17), 10020–10025.

Exadactylos, A., 2014. Nutrigenomics in aquaculture research. Fish Aquaculture J 5 (2), 1.

Felix, S., 2010. Marine and Aquaculture Biotechnology. Agrobios (India), Jodhpur.

Fleming, L.E., Broad, K., Clement, A., Dewailly, E., Elmir, S., Knap, A., et al., 2006. Oceans and human health: emerging public health risks in the marine environment. Marine Poll Bull 53 (10), 545–560.

Frias-Lopez, J., Shi, Y., Tyson, G.W., Coleman, M.L., Schuster, S.C., Chisholm, S.W., et al., 2008. Microbial community gene expression in ocean surface waters. Proc Natl Acad Sci USA 105 (10), 3805–3810.

Gilbert, J.A., Dick, G.J., Jenkins, B., Heidelberg, J., Allen, E., Mackey, K.R., et al., 2014. Meeting report: Ocean'omics science, technology and cyberinfrastructure: current challenges and future requirements (August 20–23, 2013). Stand Genomic Sci 9 (3), 1251–1258.

Gilbert, J.A., Dupont, C.L., 2011. Microbial metagenomics: beyond the genome. Ann Rev Mar Sci 3, 347–371.

Gilbert, J.A., Field, D., Huang, Y., Edwards, R., Li, W., Gilna, P., et al., 2008. Detection of large numbers of novel sequences in the metatranscriptomes of complex marine microbial communities. PLoS One 3 (8), e3042.

Giovannoni, S.J., Vergin, K.L., 2012. Seasonality in ocean microbial communities. Science 335 (6069), 671–676.

Hallegraeff, G.M., Bolch, C.J., 1992. Transport of diatom and dinoflagellate resting spores in ships' ballast water: implications for plankton biogeography and aquaculture. J Plankton Res 14 (8), 1067–1084.

Hochstrasser, D.F., Sanchez, J.C., Appel, R.D., 2002. Proteomics and its trends facing nature's complexity. Proteomics 2 (7), 807–812.

Hollywood, K., Brison, D.R., Goodacre, R., 2006. Metabolomics: current technologies and future trends. Proteomics 6 (17), 4716–4723.

Johnson, S.C., Browman, H.I., Hoffmann, G.E., Place, S.P., Dupont, S., Wilson, K., et al., 2007. Introducing genomics, proteomics and metabolomics in marine ecology. Mar Ecol Prog Ser 332, 247–248.

Joyce, A.R., Palsson, B.Ø., 2006. The model organism as a system: integrating 'omics' data sets. Nat Rev Mol Cell Biol 7 (3), 198–210.

Kan, J., Hanson, T.E., Ginter, J.M., Wang, K., Chen, F., 2005. Metaproteomic analysis of Chesapeake Bay microbial communities. Saline Systems 1 (1), 1.

Keeling, P.J., Burki, F., Wilcox, H.M., Allam, B., Allen, E.E., Amaral-Zettler, L.A., et al., 2014. The Marine Microbial Eukaryote Transcriptome Sequencing Project (MMETSP): illuminating the functional diversity of eukaryotic life in the oceans through transcriptome sequencing. PLoS Biol 12 (6), e1001889.

Kristal, B.S., Vigneau-Callahan, K.E., Matson, W.R., 1998. Simultaneous analysis of the majority of low-molecular-weight, redox-active compounds from mitochondria. Anal Biochem 263 (1), 18–25.

Liang, P., Pardee, A.B., 1992. Differential display of eukaryotic messenger RNA by means of the polymerase chain reaction. Science 257 (5072), 967–971.

Luna, G.M., 2015. Biotechnological potential of marine microbes. In: Kim, S.-K. (Ed.), Springer Handbook of Marine Biotechnology. Springer, Berlin, Heidelberg, pp. 651–661.

Margulies, M., Egholm, M., Altman, W.E., Attiya, S., Bader, J.S., Bemben, L.A., et al., 2005. Genome sequencing in microfabricated high-density picolitre reactors. Nature 437 (7057), 376–380.

Maron, P.A., Ranjard, L., Mougel, C., Lemanceau, P., 2007. Metaproteomics: a new approach for studying functional microbial ecology. Microb Ecol 53 (3), 486–493.

Matsunaga, T., Takeyama, H., Okamura, Y., 2008. Marine biotechnology for materials and energy production. J Biotechnol 136.

Metzker, M.L., 2010. Sequencing technologies—the next generation. Nat Rev Genet 11 (1), 31–46.

Ning, M., Lo, E.H., 2010. Opportunities and challenges in Omics. Transl Stroke Res 1 (4), 233–237.

Noaksson, E., Gustavsson, B., Linderoth, M., Zebühr, Y., Broman, D., Balk, L., 2004. Gonad development and plasma steroid profiles by HRGC/HRMS during one reproductive cycle in reference and leachate-exposed female perch (Perca fluviatilis). Toxicol Appl Pharmacol 195 (2), 247–261.

Partensky, F., Hess, W.R., Vaulot, D., 1999. Prochlorococcus, a marine photosynthetic prokaryote of global significance. Microbiol Mol Biol Rev 63 (1), 106–127.

Pedrós-Alió, C., 2010. Genomics and marine microbial ecology. Int Microbiol 9 (3), 191–197.

Pedrós-Alió, C., Simó, R., 2002. Studying marine microorganisms from space. Int Microbiol 5 (4), 195–200.

Poretsky, R.S., Bano, N., Buchan, A., LeCleir, G., Kleikemper, J., Pickering, M., et al., 2005. Analysis of microbial gene transcripts in environmental samples. Appl Environ Microbiol 71 (7), 4121–4126.

Poretsky, R.S., Gifford, S., Rinta-Kanto, J., Vila-Costa, M., Moran, M.A., 2009. Analyzing gene expression from marine microbial communities using environmental transcriptomics. J Vis Exp (24).

Qin, S, W.E.G. Müller, and E.L. Cooper. "Marine Biotechnology." Evidence-based complementary and alternative medicine: eCAM 2011 (2011): Article id: 639140.

Qin, S., Watabe, S., Lin, H., 2012. Omics in marine biotechnology. Chin Sci Bull 57 (25), 3251–3252.

Querellou, J., Børresen, T., Boyen, C., Dobson, A., Höfle, M., Ianora, A., et al., 2010. Marine Biotechnology: A New Vision and Strategy for Europe. Drukkerij De Windroos NV, Beernem, Belgium.

Read, B.A., Kegel, J., Klute, M.J., Kuo, A., Lefebvre, S.C., Maumus, F., et al., 2013. Pan genome of the phytoplankton Emiliania underpins its global distribution. Nature 499 (7457), 209–213.

Rocap, G., Larimer, F.W., Lamerdin, J., Malfatti, S., Chain, P., Ahlgren, N.A., et al., 2003. Genome divergence in two Prochlorococcus ecotypes reflects oceanic niche differentiation. Nature 424 (6952), 1042–1047.

Schweder, T., Markert, S., Hecker, M., 2008. Proteomics of marine bacteria. Electrophoresis 29 (12), 2603–2616.

Siezen, R.J., Van Hijum, S.A., 2010. Genome (re-) annotation and open-source annotation pipelines. Microb Biotechnol 3 (4), 362–369.

Sogin, M.L., Morrison, H.G., Huber, J.A., Welch, D.M., Huse, S.M., Neal, P.R., et al., 2006. Microbial diversity in the deep sea and the underexplored "rare biosphere". Proc Natl Acad Sci USA 103 (32), 12115–12120.

Solanky, K.S., Burton, I.W., MacKinnon, S.L., Walter, J.A., Dacanay, A., 2005. Metabolic changes in Atlantic salmon exposed to Aeromonas salmonicida detected by 1H-nuclear magnetic resonance spectroscopy of plasma. Dis Aquat Organ 65 (2), 107.

Spitz, J., Becquet, V., Rosen, D.A., Trites, A.W., 2015. A nutrigenomic approach to detect nutritional stress from gene expression in blood samples drawn from Steller sea lions. Comp Biochem Physiol A Mol Integr Physiol 187, 214–223.

Stapels, M.D., Cho, J.C., Giovannoni, S.J., Barofsky, D.F., 2004. Proteomic analysis of novel marine bacteria using MALDI and ESI mass spectrometry. J Biomol Tech 15 (3), 191.

Steele, J.H., Thorpe, S.A., Turekian, K.K., 2008. Encyclopedia of Ocean Sciences. Elsevier ScienceDirect, Amsterdam.

Stein, L., 2001. Genome annotation: from sequence to biology. Nat Rev Genet 2 (7), 493–503.

Tacchi L., 2011. Molecular basis of improved feeds for aquaculture: a nutrigenomics approach (Doctoral dissertation, University of Aberdeen).

Titilade, P.R., Olalekan, E.I., 2015. The importance of marine genomics to life. J Ocean Res 3 (1), 1–3.

Venter, J.C., Remington, K., Heidelberg, J.F., Halpern, A.L., Rusch, D., Eisen, J.A., et al., 2004. Environmental genome shotgun sequencing of the Sargasso Sea. Science 304 (5667), 66–74.

Viant, M.R., 2003. Improved methods for the acquisition and interpretation of NMR metabolomic data. Biochem Biophys Res Comm 310, 943–948.

Viant, M.R., Pincetich, C.A., Hinton, D.E., Tjeerdema, R.S., 2006a. Toxic actions of dinoseb in medaka (Oryzias latipes) embryos as determined by in vivo 31 P NMR, HPLC-UV and 1 H NMR metabolomics. Aquat Toxicol 76 (3), 329–342.

Viant, M.R., Pincetich, C.A., Tjeerdema, R.S., 2006b. Metabolic effects of dinoseb, diazinon and esfenvalerate in eyed eggs and alevins of Chinook salmon (Oncorhynchus tshawytscha) determined by 1 H NMR metabolomics. Aquat Toxicol 77 (4), 359–371.

Viant, M.R., Werner, I., Rosenblum, E.S., Gantner, A.S., Tjeerdema, R.S., Johnson, M.L., 2003. Correlation between heat-shock protein induction and reduced metabolic condition in juvenile steelhead trout (Oncorhynchus mykiss) chronically exposed to elevated temperature. Fish Physiol Biochem 29 (2), 159–171.

Walsh, P.J., Smith, S., Fleming, L., Solo-Gabriele, H., Gerwick, W.H. (Eds.), 2011. Oceans and Human Health: Risks and Remedies From the Seas. Academic Press.

Wasser, J.S., Lawler, R.G., Jackson, D.C., 1996. Nuclear magnetic resonance spectroscopy and its applications in comparative physiology. Physiol Zoo 1–34.

Wilmes, P., Bond, P.L., 2006. Metaproteomics: studying functional gene expression in microbial ecosystems. Trend Microbiol 14 (2), 92–97.

Wilmes, P., Bond, P.L., 2006. Towards exposure of elusive metabolic mixed-culture processes: the application of metaproteomic analyses to activated sludge. Water Sci Technol 54 (1), 217–226.

Worden, A.Z., Allen, A.E., 2010. The voyage of the microbial eukaryote. Curr Opin Microbiol 13 (5), 652–660.

Worm, B., Barbier, E.B., Beaumont, N., Duffy, J.E., Folke, C., Halpern, B.S., et al., 2006. Impacts of biodiversity loss on ocean ecosystem services. Science 314 (5800), 787–790.

4

Synthetic Biology: Overview and Applications

Rohini Keshava¹, Rohan Mitra², Mohan L. Gope³ and Rajalakshmi Gope²

¹Bangalore University, Bangalore, India ²NIMHANS, Bangalore, India
³Ajmera Infinity, Bangalore, India

4.1 INTRODUCTION

Synthetic biology is a principal tool for understanding the fundamentals of biological system. It involves design and production of biological gears which would help us to understand living system and would also be beneficial to human kind (Schmidt et al., 2012). It utilizes an assortment of fields such as biology, biotechnology, chemistry, and engineering, thereby frequently interconnecting biochemical and biomedical engineering (Fig. 4.1). In this process it incorporates diverse tactics and protocols from engineering and biotechnology (Gruber et al., 2008). The ultimate goal of synthetic biology is to create living biological forms or systems from existing nonliving matter. By doing so, synthetic biologists are trying to understand the biological foundation of origin of life. Using synthetic chemistry, synthetic DNA/genes or polypeptides can be created from naturally occurring nucleotides and amino acids. These can then be inserted into living cells to assess their biological effects. Synthetic biologists also aim to utilize these technologies to benefit mankind. New synthetic biological parts can be used as platforms to modify or drastically change the naturally existing systems, and these systems would in turn become engineered substitutes. The advancement of synthetic biology depends on technology development, speed, and the cost as well as ease of its utility. Technology to synthesize whole genes or even the entire chromosomes, placing them in heterologous systems and to assess their behavior in a precise manner, is essential in synthetic biology.

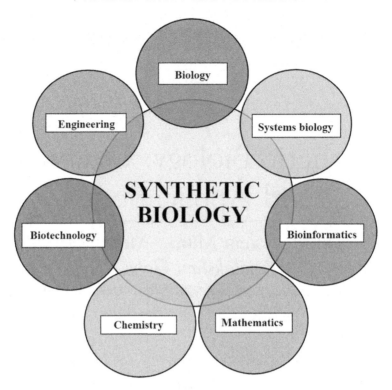

FIGURE 4.1   **Synthetic Biology: A multidisciplinary approach.** Synthetic biology is a multidisciplinary field. It is a combination of several fields such as Biology, Chemistry, Engineering, Biotechnology, Bioinformatics, Molecular Biology and Molecular Genomics.

Ease of synthesis of poly nucleotides and the availability of free genome databases have made gene synthesis cost effective as compared to classical gene cloning. Molecular geneticists have already cloned and assigned functions for a number of genes. These genes can now be made using purely synthetic chemistry and can be used as building blocks of life. Artificial cytoplasm and cell membrane can be constituted by synthetic means, and the synthetic gene can be placed in this artificial cell. Such cell can be directed to perform functions that do not exist in nature, and it can be used for standardizing and understanding biological functions and behavior (Serrano, 2007). Thus synthetic biology requires de novo synthesis of new genes and other cellular components using standard engineering techniques.

The safety and security concerns associated with any new technology and its unknown consequences apply to this emerging field as well. Many groups have raised concerns regarding the rules to govern this new area and to evaluate the impact of new technology, intellectual property right, licensing, and monitoring. Social, legal, and ethical concerns are also being debated in many scientific and social forums (Campos, 2009; Schmidt et al., 2009). The Presidential Commission for the Study of Bioethical Issues has recommended a robust federal monitoring of this evolving technology in the United States.

4.2 HISTORY

The term "synthetic biology" was used in a few publications in the early 1900s. These include Leduc's publication in 1910 entitled "Théorie physico-chimique de la vie et générations spontanées" (Leduc, 1910) and La Biologie Synthétique in 1912 (Leduc, 1912). In the middle 20th century the Polish geneticist Waclaw Szybalski used the term "synthetic biology" which he described as the one with limitless possibilities in biology (Szybalski, 1974). He had also suggested that it can produce better biological control circuits and better synthetic organisms. In 1978 Szybalski indicated that new gene arrangements can be made and studied using restriction enzyme technology, which would have great implications on synthetic biology (Szybalski and Skalka, 1978). In 1980 Barbara Hobom used the title "synthetic biology" to describe genetically engineered bacteria using recombinant DNA technology (Hobom, 1980). This term was again used in 2000 in the annual meeting of the American Chemical Society in San Francisco, where emphasis was given to synthetic organic molecules with biological functions (Rawls, 2000). In this context, it meant to redesigning life from nonliving matters (Szostak et al., 2001; Benner and Sismour, 2005).

4.3 BIOLOGY AND CHEMISTRY

4.3.1 Cell

Synthetic biology could be utilized to build a variance of existing life forms that can be used to get answers that cannot be obtained from natural life forms. In the year 2000, Elowitz and Leibler developed a model termed as "Repressilator," which was used to test gene expression network inside *Escherichia coli*. This repressilator consisted of three artificial transcriptional repressor clocks that do not interfere with the organism's natural biological clock. It also had green fluorescent protein (GFP) as a reporter which would relay the physiological state of the living cell. Synthetic transcriptional repressor DNAs were used to test the hypothesis on how gene expression should work in a living cell. The results showed slight differences between the experimental observations and the expected outcomes. These differences emphasize continued use of such systems to understand the functioning of molecular networks in a living cell (Elowitz and Leibler, 2000). Similar studies together with mathematical analysis to predict the dynamics of biological system would form the basis for constructing a synthetic cell in a logical stepwise manner.

4.3.2 Animal Cell

Individual cells make up the multicellular organisms which are highly complex. A typical animal cell has cell membrane, cytoplasm, and nucleus. The membrane is made up of lipid bilayer and proteins. The cytoplasm consists of many organelles such as mitochondria, endoplasmic reticulum, Golgi complex, vacuole, and lysosomes. The nucleus has the DNA, the genetic material that determines the fate of a cell, organ, and in turn the whole organism. It is possible to divide the components of animal cell into various ascending/descending orders; by using synthetic biology, it is also possible to understand each one of them in a stepwise manner. Molecular biologists used the "top-down" approach to clone and isolate genes and to study their functions. It is also possible to use the "bottom-up" approach to understand the composition and function of each cell type, organ, and finally the organism as a whole. The chemical composition of each cell type would reveal the nature of the corresponding organs and the interaction between the organs within the organism. Such information can be used to constitute a whole synthetic cell that can be used to test various hypothesis regarding the animal cell function, diseases, heritability, and effect of environment and a multitude of other issues (Szostak et al., 2001).

4.3.3 Plant Cell

Most plant species are made up of multiple cell types with definitive functions. The fundamental biochemistry of the animal and plant cells is almost the same but for a few exceptions, such as plastids, chloroplast, and cell wall, which are found only in plant cells. Plants also have a hierarchy in organization from simple to complex forms in their structure and function. Synthetic biology in plant kingdom can be used to test its benefits in economics, medicine, ecology, and environment (Rawls, 2000).

4.3.4 Function of a Cell

The unit of cell is fundamental to life processes. Function of a cell is to carry out multitude of life processes, and each organ and organism comprise varied cell types. Synthetic biology is an extrapolation of biomimetic chemistry where the functional biological units are organically synthesized and they would simulate biological materials such as proteins and enzymes (Breslow, 1972). Thereby synthetic biologist aims to synthesize biological parts that do not currently exist in nature. These parts could have existed in the early life forms, and they could have become extinct, modified, or replaced during evolution. Such studies could provide evidence for Darwinian evolution of biological system (Szostak, 1988). Synthetic biologists also aim to manufacture interchangeable parts, such as promoters, enhancers, enzymes, molecules in signaling cascade, and motifs responsible for weak interactions, which can be tested in biological system (Gibbs, 2004).

4.3.5 Chemistry

The study of chemistry involves analysis of chemicals and their properties. Biological system comprises chemical matters that are assembled to function in a specific manner. The nucleotides, amino acids, carbohydrates, and lipids in the living system are made up of basic elements such as carbon, hydrogen, nitrogen, and oxygen. In the early 20th century chemists started to synthesize new chemicals that created the subject "synthetic chemistry." Many aspects of synthetic biology have overlaps with synthetic chemistry. The chemical approach in synthetic biology is "bottom-up," where the basic molecules that govern life such as nucleic acids, proteins, and carbohydrates can be designed, built, and placed in a living system to test their behavior and effects. In addition, synthetic chemists are examining if novel, biologically, and biochemically functional pathways can be built and tested. Some of these can be tested in already existing living systems and others can be tested in all new synthetic systems. The combination of synthetic chemistry and synthetic biology could formulate and test novel ideas and also could provide evidences to queries concerning the origin of universe and the origin of life. For example, a new synthetic base different than the ones in Watson—Crick base pairs has been shown to function in the living system (Kim et al., 2005). The xDNA containing an unnatural base pair, larger than the 2.4 Å natural base pair, was synthesized and tested for its ability to function like a natural DNA. The naturally existing polymerases read the xDNA that indicates that the synthetic molecule with an unnatural base can replicate in a biological system (Krueger et al., 2011). The rigid DNA polymerase 1, Klenow fragment failed to extend the xDNA, whereas the flexible Y-family polymerase Dpo4 extended the bases in a stepwise manner (Lu et al., 2010). These data indicate that such model systems can teach insight of biology (Benner and Sismour, 2005; Kool and Waters, 2007). Building DNA with bases other than Watson—Crick purine—pyrimidine geometry with more helix stability can be useful in many areas such as biotechnology, biomedicine, and nanotechnology (Winnacker and Kool, 2013)

4.3.6 Chemical Composition of a Cell

The living cell comprises hydrogen, oxygen, carbon, nitrogen, sulfur, and phosphorous. Approximately 90% of the content in the living cell is water. The quantity of protein in a cell always exceeds that of the nucleic acids. All the nonessential lipids are generated from acetyl-CoA. The cell also comprises essential amino acids, unsaturated fatty acids, and other inorganic elements, and these are dietary essentials as they cannot be synthesized within the cell. Great many diseases in the living systems are due to abnormal chemical reactions, imbalance in chemical composition, and altered biochemical pathways which can be studied using synthetic biology (Benner and Sismour, 2005; Cooper and Hausman, 2013).

4.4 CRAFT AND DESIGN

4.4.1 Engineering

Biology is viewed often as technology by physicists, chemists, and engineers and often time they call it as systems biological engineering (Church et al., 2014). Biological engineering is therefore considered as application of ideas and methodologies that are aimed to design and build new biological systems, fabricate materials, and assemble chemicals to form bio parts. The design and fabrication encompasses knowledge of biology as well as understandings of physics, chemistry, mathematics, and computer science as well as structure and processes. The synthetic biological entities thus produced will mimic biological system, and it could help to produce energy, provide food and medicine, maintain safe environment, and to develop cost-effective methods to enhance overall human health (Chopta and Kamma, 2008). One of the differences between synthetic biology and genetic engineering is that the former requires development of fundamental techniques that would make bioengineering easier, reliable, and cost effective. For example, using a redesigned Type III secretion system, the bacterium *Salmonella typhimurium* was directed to secrete spider silk proteins, which is a durable elastic biofiber similar to the natural one but produced with low cost (Widmaier et al., 2009). Another aspect of synthetic biology is to design and manufacture biomedical devices with novel functions that can be used for testing biological functions. One such device is the RNA-based riboregulator which was used as a GFP tracking system to regulate toxic protein synthesis and stress response (Gardner et al., 2000).

The engineering part of synthetic biology has many broad areas such as photocell design, biomolecular engineering, genome engineering, and biomolecular design. The photocell design fabricates components from synthetic materials but has the ability to self-replicate like a biological part. The biomolecular engineering is aimed to create toolkits of functional units that can have novel functions when placed in living cells. Genome engineering includes development of methods to construct synthetic genes and chromosomes for a whole organism or for a minimal life form. The biomolecular design involves designing and producing novel biomolecules and their combination. These engineering technologies converge to produce synthetic part that will mimic or better the existing living system (Channon et al., 2008).

4.4.2 Synthetic Morphology

The task of synthetic morphologists is to introduce inputs different from the natural ones to produce outputs that mimic existing structures or novel structures. Simulated genes and networks are introduced into cells and their functions are manipulated using alternative inputs such as chemicals, light, sound, pressure, or other physical forces. The outputs from such maneuver could be activity of enzymes that can be tracked by GFPs (Davies, 2008). Using this approach, it was verified that simulated genes and networks are able to establish and recover memory depending on specific inputs similar to the Boolean computation model. Synthetic morphology is an extension of synthetic biochemistry, which could alter the shape or behavior of cells in response to selective inputs. For example, the artificial gene network that controls cell adhesion could cause cells to clump together or activate cell mobility. These gene networks can be controlled by various chemical/physical inputs to obtain different outputs such as adhesion and motility. It has been proposed that a limited number of events such as cell proliferation, cell differentiation, status of cell surface receptors, adhesion, fusion, mobility, and chemotaxis can govern the formation of specific cell shape in loose cells and in organs. These events are comparable to that govern morphology in plants and other organisms. It appears that a limited set of combination of input and outputs can be programed into cells to produce artificial designs and shapes, which could form part of morphologically altered tissues. Such studies would lead to designing morphology and anatomy entirely using synthetic biology, and it would also provide a platform for cell, tissue, and organ engineering. Alternatively, it would help in understanding and testing classical embryo morphogenesis pathway where a simple fertilized egg can produce a complex living form consisting of tissues and organs of varying shapes and sizes performing diverse functions. To achieve success in synthetic morphology, we need to create sets of factors that define morphology such as sensors, regulators, and modifiers, which in combination with external input can generate altered cells and tissues that would exhibit the changed morphology (Davies, 2008; Tanaka and Yi, 2009).

4.4.3 Synthetic Chemistry

Chemists believed that there is a clear distinction between organic and inorganic molecules and that inorganic molecules could be made from organic molecules but not vice versa. However, in 1828, the first organic molecule, urea, was synthesized by a German chemist Friedrich Woher from inorganic ammonium cyanate (Wohler, 1828; Jaffe, 1976). In the following few decades knowledge on analytical and synthetic chemistry evolved and combined, which enabled chemists to synthesize hundreds of organic molecules such as acetic acid, ethanol, methanol, and benzene (Asimov, 1965). A complete understanding of chemical structures was not essential for the development of synthetic chemistry. Rather the chemical structures are verified after synthesis based on the proposed structures. The merging of synthetic and analytical approach was key to the development of synthetic chemistry and is repeating itself in synthetic biology. Advanced synthetic procedures led to the total chemical synthesis of oligonucleotides as performed first by Khorana, which helped in the revelation of genetic code (Khorana, 1965). Later, synthesis of polio virus confirmed that live organisms can be created by simple chemical directives, which are placed in proper order and propagated in the cytoplasm (Fig. 4.2) (Cello et al., 2002). Modern experimental technologies such as whole-genome sequencing, micro array, molecular structure determination, analysis of protein isoforms based on posttranslational modification and their functions, advanced microscopy with live sensors, and functional MRI have provided advanced information on the proteome. The proteomic data provide us information on the protein quantity, structure, localization, and weak interactions among protein molecules, and these advancements are comparable to the analytic chemistry in the early 19th century. The history of synthetic chemistry indicates that synthesis of indispensable counterparts of living system is necessary for biologists to truly understand the living systems. Therefore, synthetic chemistry is an essential part of synthetic biology where the various chemical parts are synthesized and reconstituted to form a living cell. This would require synthetic biochemistry and determination and reconstitution of various noncovalently linked chemical forms that are suitable to build a system that can carry out a specific biological function. Serial modifications of such systems to suit specific needs would further the understanding of design and function in a profound manner. The systems thus created would permit us to perform substitutions of synthetic chemical parts to answer questions that the naturally evolved systems fail to provide. For example, we can understand the minimal chemical parts necessary to formulate minimal living being with a particular behavior or mood or feeling. Using such system, we can interchange the chemical composition to evaluate their effects on these parameters. These will be helpful in understanding the logic in biological system, and we could perhaps construct something similar to the periodic table for biology that would unify the networking in living system (Suel et al., 2007; Alon, 2007).

FIGURE 4.2 **de novo synthesis of live polio virus.** Chemical synthesis of a full length, infectious poliovirus cDNA consisting of 7741 bases from its published sequence by chemical and biochemical means using oligonucleotides. Synthetic virus thus produced consisted of its infectious property.

4.4.4 Minimal Cell

A minimal cell is a synthetic vesicle-based system, which consists of minimal sets of genes, proteins, and other biomolecules. In the past decades scientists have used analytical approach to study a living cell. In these studies the various components of the cells such as DNA, RNA, and protein were isolated, purified, and their functions were analyzed in heterologous systems. In recent times the biochemical understanding of the living system has expanded tremendously due to the advancement of technology and the voluminous data generated. In these studies the various components of the live cells were taken apart to understand their characteristics (de Lorenzo and Danchin, 2008). Due to the inherent human curiosity to understand the biological system and to explore the origin of life, the synthetic biological approaches are used to create a minimal living cell. A minimal cell should have the boundary, which is a semipermeable membrane, and the internal parts for it to function, and it should be able to self-produce or undergo autopoiesis. Autopoiesis describes the behavior of all living systems from cell to organelle. This concept was first formulated in 1980, and it describes the requirements of a minimal live cell (Maturana and Varela, 1980). The autopoietic cells will have the ability to construct its own boundary and the internal components, and will have anabolic and catabolic process. It will have the ability to self-organize, self-produce, and will have a self-boundary, homeostasis, and its own identity. Such system does not require a central control unit as all the components work together with a collective dynamics along with the environment. The growth of the autopoietic cell will depend on the environment and the anabolic and catabolic machinery, and it can stay in homeostatic state or die depending on the prevalent conditions within and around it. Autopoiesis seems to be the primary condition of a minimal cell as it is necessary and sufficient for life (de Lorenzo and Danchin, 2008). Later it was suggested that autopoiesis should be combined with cognition and prudent interaction with environment (Bitbol and Luisi, 2004; Dmiano and Luisi, 2010). The first level of cognition is the metabolism that can be influenced by environment; the second level is adaptation of the cell to foreign molecules by changing metabolic process which forms the basis for evolution. Therefore, autopoiesis alone is not sufficient for a minimal cell as it also requires cognition (Bachmann et al., 1992).

To synthesize minimal cells, a boundary forming material is added to the preexisting compartment comprising fatty acids (Fig. 4.3). Fatty acid vesicles are considered as the most likely candidate for a primitive cell and origin of life. Ethyl and methyl esters and anhydrides of fatty acids can be transformed by hydrolysis into precursors to form the boundary of a synthetic compartment. During this interphase the surface of the compartment increases. It is proposed that due to the increase in the surface, the particle becomes structurally unstable and therefore divides giving rise to two identical particles (Bachmann et al., 1992; Stano et al., 2006). The self-reproduction is a by-product of autopoietic mechanism as the growth results in an unstable state. The modern cell has the ability to produce its own DNA, various RNAs, protein, and energy for its survival and has come into existence due to millions of years of evolution. In this regard the synthetic minimal cell may differ from the primitive cell. Therefore, two approaches are possible in synthetic biology, to create a minimal cell and to synthesize a primitive cell model (Stano, 2010; Torino et al., 2013).

4.4.5 Rewriting

Rewriting is defined as changing parts of already existing system or rebuilding the entire natural system from the scratch. This is similar to refactoring that is used in upgradation of computer software or to create new ones. It would provide substitutes or surrogates that are easier to understand and relate to, and they can be used as to obtain answers that the naturally existing systems cannot provide (Stone, 2006; Forster and Church, 2006). Evolution has

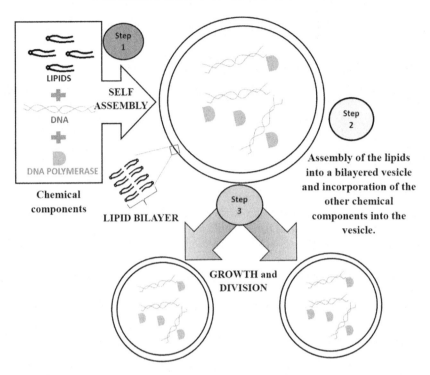

FIGURE 4.3 **Construction of a synthetic minimal cell.** A mixture of minimal number of biological components such as Lipids and DNA along with certain enzymes and low molecular-weight compounds get encapsulated through self-assembly into synthetic lipid-based compartments, such as lipid vesicles (Step1&2). The membrane acts as a boundary to confine the interacting internalized molecules, so that a "unit" is defined. Such an assembled system is capable of producing the components for its own survival and eventually is capable of self reproduction (Step 3).

produced systems that are complicated to understand or maneuver. Using rewriting, it could be possible to generate novel sets of genes, rewrite genetic code, build a whole new organism, or assemble mimics (Bhattacharyya et al., 2006; Chan et al., 2005). Using synthetic "bio bricks," a novel microbe that could produce precursors of antimalarial drug artemisinin has been created (Fig. 4.4) (Ro et al., 2006). Microbes are encoded to produce proteins with unnatural amino acids. Rewriting technology also has produced microbes that could generate hydrogen which is an alternate source of fuel. Yeasts are made to produce ethanol from cellulose, and engineered bacteria are programed to invade cancer cells and kill them (Sedlak and Ho, 2004; Anderson et al., 2006). A microchip-based synthesis and assembly and optimization of all the 21 genes that encode *E. coli* 30S ribosomal subunit provide evidence that it is possible to rewrite the existing system to optimize their functions (Tian et al., 2004).

4.5 TECHNOLOGY

4.5.1 Polypeptides and Proteins

De novo designed peptides and proteins play an important role in synthetic biology as they will be part of a minimal cell or a synthetic primitive cell. In nature, the encoded message in the genes is translated into proteins, and these proteins provide energy and also

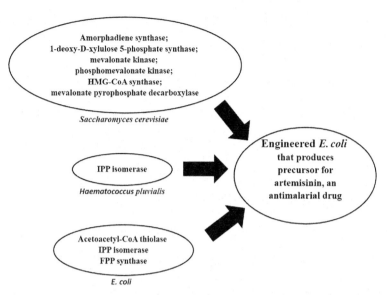

FIGURE 4.4 Engineered *E. coli* that can produce the precursor for an antimalarial drug, artemisinin. The combination of enzymes from different sources (viz., *Saccharomyces cerevisiae, Haematococcus pluvialis, E. coli*) in an Escherichia coli host generated a strain of the bacterium that produced a precursor for artemisinin, an antimalarial drug.

the physiological characteristics to the live cell. All the necessary information for a cell to function is available within the polypeptide molecule. Proteins can be assembled from amino acids by chemical ligation or be synthesized using recombinant DNA technology (Nilsson et al., 2005). Designer proteins can be synthesized either to improve their existing functions or for novel use. Proteins can be folded into a perfect 3D form to execute appropriate functions, and they also can be unfolded and degraded once the function is finished. The unfolded proteins can also be reassembled accurately in a reproducible manner even in in vitro conditions. The folding is done in a stepwise manner starting from a simple α-helix followed by β-sheet and further higher orders. There is a relationship between the amino acid sequence and the final conformation of protein which is governed by rules. However, sometimes the same polypeptide chain can perform variable functions depending on the posttranslational modifications that govern the protein folding. The folded structures take further higher orders to nanoscale to form the functional unit, and they also provide the cell its shape and morphology (Murzin et al., 1995; Taylor, 2002). It is important to understand that some proteins require nonproteinaceous cofactors and chaperones for their proper localization and to perform their functions.

Advanced computer programs and algorithms have been helping to fabricate novel proteins and to assign novel activities to existing proteins. Wealth of information on amino acid sequence of various proteins is procured over the decades, and they are available in various databases such as CATH, SCOP, and PFAM (Murzin et al., 1995; Orengo et al., 1997; Finn et al., 2006). Using these database, functional tectons can be built or synthesized. Tectons are molecules assembled using fundamental units, which can be programed to fold into higher order state to predict final shape of the functional molecules (Bromley et al., 2008; Kajander et al., 2007). Mutations can be inserted into these molecules to provide novel functions (Zahnd et al., 2007). Novel protein designs for globular proteins are generated using Rosetta Design algorithm (Liu and Kuhlman, 2006). Natural proteins such as periplasmic binding protein from *E coli* have been used as a basic frame to introduce novel functions (Dwyer and Hellinga, 2004). A single polypeptide was shown to take two overlapping folded conformations which is a challenge to the notion that each polypeptide can fold into only one form (Ciani et al., 2002; Pandya et al., 2004; Ambroggio and Kuhlman, 2006a). Computational designs have explored the complex structures of transmembrane proteins (Yin et al., 2007). The coiled-coil α-helix is used as a base for creating intricate designs (Lupas, 1996; Mason and Arndt, 2004). Advanced controls and switches are necessary to manage the structure and function of synthetic protein (Ambroggio and Kuhlman, 2006b). In addition, self-replicating peptides are also being developed (Ambroggio and Kuhlman, 2006b; Lee et al., 1996; Yao et al., 1997). There is a need to develop encapsulated part that would allow molecular movements in and out. These are necessary for targeted and higher order functions such as drug delivery and gene expression (Ghosh and Chmielewski, 2004; Carmona-Ribeiro, 2007). Plants and bacteria have cell walls made of polysaccharides, which can be substituted by synthetic analogs with specific permeability (Nomura et al., 2003; Discher and Eisenberg, 2002).

In general, proteins are made up of 20 different types of amino acids. In a reductionist approach, peptides containing only a few amino acids instead of usual 20 were synthesized, and these have biological properties similar to the natural proteins. A synthetic peptide containing 80 amino acids from a random sequence library consisting of predominantly glutamine, leucine, and arginine had stability similar to the natural proteins (Nallani et al., 2007). In experimental condition a polypeptide chain with binary pattern of polar and nonpolar amino acids in a specific order folded into globular α-helical structure (Davidson et al., 1995; Kamtekar et al., 1993). Using a combination of design and selection procedures, a mimic peptide of only nine amino acids was synthesized. This mimic could rescue auxotrophic cells lacking chorismate mutase thus demonstrating its enzymatic property (Walter et al., 2005).

4.5.2 Polynucleotides and DNA and RNA

The first chemically synthesized viral genome of Hepatitis C virus was reported in the year 2000 (Blight et al., 2000). A full-length poliovirus cDNA consisting of 7741 base was synthesized de novo from its published sequence by chemical and biochemical means using oligonucleotides, and it had infectious property. This work took a period of 2 years to get completed (Cello et al., 2002). Craig Venter's group shortened the time period for synthesis and assembly of synthetic gene as they produced infectious Phi X 174 genome of 5386 bp within 14 days (Hamilton et al., 2003). The same team from Craig Venter's institute also synthesized genome of a novel minimal bacterium *Mycoplasma laboratorium* (Gibson et al., 2008). Later Venter's group reported synthesis and assembly of 1.08 Mbp genome of *Mycoplasma mycoides* and transplanted it into a recipient *M. mycoid* cells that are totally controlled by synthetic genes and these cells have the ability to divide continuously (Gibson et al., 2010)

Besides synthetic oligo nucleotides and DNAs, synthetic RNA components are also necessary to construct synthetic biological systems. The versatility and behavior of higher order RNA structure enable them to perform specific functions. This unique property defines the interaction between RNAs with genome and its components and proteins. By varying RNA bases in its sequence, it can be used to attain programmable RNA functions (Issacs et al., 2006).

Riboregulators play an important role in the regulation of RNA function at the posttranscriptional level. Engineered regulatory RNA molecules are tested for their efficacy as riboregulators. Synthetic RNA molecules of specific motif are used successfully as RNA transcriptional activators. Subsequently, these transcriptional activators were modified with aptamers to respond to external factors such as specific small molecules. Engineered ribosomes and mRNAs were tested in the *E. coli* system, and it was found that they recognized and translated only the engineered mRNA and not the native ones. These synthetic ribosome—mRNA pairs are suitable for creating orthogonal cellular pathways which would be a powerful tool to evaluate the translational network. This technology would be applicable in metabolic engineering and designing synthetic bio factory (Issacs et al., 2006; Bayer and Smolke, 2005). Riboswitches also play an important role in gene regulation at the posttranscriptional level. Synthetic riboswitches have been designed and constructed and tested in living cells. Assembling synthetic riboswitches that would work in a genome-wide manner is necessary to produce a functional cell. This would require deeper understanding of the secondary structures of various RNA molecules, so that the synthetic riboswitches can be fabricated to recognize and bind to appropriate regions with high specificity (Issacs et al., 2006). Ribozyme-mediated switches and ligand-mediated control of the same are necessary to fine-tune the gene regulation at the posttranscriptional level (Winkler and Breaker, 2005). Transcriptional activators are necessary for proper initiation and elongation of RNAs. These activators have DNA binding domain and transcription initiation domain. Synthetic transcriptional activators were tested with variable efficiency (Saha et al., 2003).

4.5.3 Restriction Endonucleases

The discovery of restriction enzymes has created a new tool for gene manipulation where anyone can change the arrangement of gene order. Novel genes can be made from existing gene(s) by cut-and-paste mechanism. Gene expression and its regulation can be modified by replacing promoters and regulatory sequences using restriction endonucleases. Thus, tailor-made genes can be assembled and introduced into a synthetic/minimal cell which could be used to produce novel drugs, manage pollution, and to clean up environment (Szybalski and Skalka, 1978).

4.5.4 DNA Amplification and Sequencing

Many important technologies are essential for the advancement of synthetic biology. These are necessary to test the synthetic components and to validate their functions to form a simple functional system that can be upgraded to a complex one by subsequent addition, deletion, or other modifications (Bio FAB Group et al., 2006). Computer aided designs can be helpful to perform this task. Furthermore, it would also be necessary to amplify and sequence the synthetic DNA to create multiples of the synthetic system. Sequence information of large number of naturally existing organisms would provide the consolidated DNA sequence database. This would provide a foundation for constructing synthetic parts. Verification of synthetic DNA would require its amplification and sequencing. Affordable and reliable technologies can enable speedy identification of synthetic system and organisms.

4.5.5 DNA and Protein Transfer

Once the DNA is amplified and sequenced, it can be inserted into plasmid vehicles and transferred into bacterium. The synthetic DNA can interact with the bacterium, which could lead to altered functions of the bacterium. These DNAs could also express and produce novel protein products that could help mankind to tackle health and environmental problems. Thus bacteria can be used as cyborgs because their behavior can be changed by the foreign DNA. Proteins are important biological building blocks and their functions are complicated. Because of the complexity in protein chemistry, synthetic biology in this area is lagging behind nucleic acid and gene regulator aspects. However, synthetic biologists have been toying with the flow of information in the natural signaling pathways, replacing some components involved in these processes. It is important to identify and understand protein and their domains that interact with other proteins as well as cellular molecules. In this process synthetic biologists have swapped modules of protein components and rewired signaling in yeast. These studies are possible with the help of protein transfer techniques where the transferred proteins interfere with intracellular functions. Such studies would help in our understanding of protein complexity to form a well-defined framework where the subunits can be exchanged with predictive outcome. This would help the advancement of engineering of minimal and synthetic cell (Grunberg and Serrano, 2010).

4.5.6 Systems Biology

Systems biology or "systeomics" is a biology-based interdisciplinary field of study that focuses on complex interactions within biological systems. It is an emerging approach applied to biomedical and biological scientific research. One of the aims of systems biology is to model and discover properties of cells, tissues, and organisms functioning as a system. Typically it involves metabolic networks or cell signaling networks. Systems biology makes heavy use of mathematical and computational models. It also includes the study of the interactions between the molecules and networks of biological systems and their effects on behavior and function, for example, the enzymes and metabolites in a metabolic pathway. The systems biology protocols include many steps such as derivation of operational hypothesis, computer and mathematic modeling and experimental validation. A variety of modules such as transcriptomics, metabolomics, proteomics, and high-throughput techniques are used to collect quantitative data for the construction and validation of models (Liu et al., 2014). Systems biology and synthetic biology have many overlaps (Table 4.1). Because systems biology requires the ability to obtain, integrate,

TABLE 4.1 Comparison Between Systems Biology and Synthetic Biology

Sl No.	Systems Biology	Synthetic Biology
1	Systems biology refers broadly to the study of biological interactions and processes at the integrated "systems" level and how different parts of biological systems interact.	Synthetic biology draws on systems biology in seeking to understand and create synthetic biological devices and systems.
2	In systems biology, we use all the available tools of molecular biology, genetics, statistics and modeling to produce accurate and predictive models of existing biological systems.	Synthetic biology draws heavily on engineering principles.
3	The end point of systems biology is understanding biological systems.	The end point of synthetic biology is sometimes referred to as the "industrialization" of biology.

and analyze complex data sets from multiple experimental sources using interdisciplinary tools, some technology platforms involved therein are Phenomics, Genomics, Epigenomics/Epigenetics, Transcriptomics, Interferomics, Translatomics/Proteomics, Metabolomics, Glycomics, Lipidomics, Interactomics, NeuroElectroDynamics, Fluxomics, Biomics, and Semiomics.

Systems biology helps scientists better understand the interaction of microbes with ecosystems, which can be useful to clean up environmental toxins, create biofuels from waste materials, thereby increasing sustainability. Systems medicine will eventually be able to predict when an organ will become diseased or when a perturbation in a biological network could progress to disease. Systems biology has given rise to the "P4 medicine" that is "Predictive, Preventive, Personalized, and Participatory" (Cesario et al., 2014). Significant efforts in computational systems biology have been made to create realistic multiscale in silico models of various brain and nervous system disorders including cancer. The systems biology approach often involves the development of mechanistic models, such as the reconstruction of dynamic systems from the quantitative properties of their elementary building blocks which is necessary for the advancement of synthetic biology (Romualdi et al., 2005; di Bernardo et al., 2005).

4.5.7 Measurements and Calculations

Quantitative measurements and precise calculations are necessary to understand biological system. These data would explain the biological functions that could help in formulation of theories and validation of the same. Differences between the calculated measurements and the actual biological process in the living system can reveal and explain why the synthetic systems differ from the predictions. Technological development such as advanced microscopy and flow cytometry to measure bio process in a time-bound manner is useful in synthetic biology. The living cell comprises many biomolecules in a very crowded and confined environment. Understanding the supramolecular chemistry in such a crowded and confined environment is necessary to create synthetic supramolecular structures and validate them for their functions. This exercise would require precise measurements on inter- and intramolecular affinity and their interactions, localization within the crowded environment and their exchangeability (Foffi et al., 2013). The chemical formula of the *Magnetospirillum gryphiswaldense* was calculated using the elemental mass balance during its growth in bioreactor. Such approach would enhance our understanding of chemical composition of living cell and the chemical logic of origin of life (Naresh et al., 2012). Ensemble modeling methods are used to calculate the folding pattern of disordered protein states, and it suggests that it is easier to calculate the molecular size of the secondary structure but not the tertiary structure which depends on other factors in their environment (Marsh and Forman-Kay, 2012). A low-order constraint-based algorithm was used to assess large-scale gene regulator network, and the data were collected in RegulonDB and tested. The results indicate that this methodology is effective and efficient in reconstructing gene regulatory network from microarray data (Wang et al., 2010). Furthermore, nonnatural DNA sequences showed higher affinity for histone binding and nucleosome formation than the natural ones. Based on this observation, it was concluded that the eukaryotic genome did not evolve towards the highest affinity or nucleosome positioning power (Thastrom et al., 1999). Therefore, to create a synthetic life it is important to estimate the affinities between the synthetic parts that is required for their optimum function within the crowded and confined environment of a synthetic cell.

4.6 APPLICATIONS

4.6.1 Fundamental and Applied Synthetic Biology

The major application of synthetic biology is to design new metabolic pathways in microorganisms to direct them to produce new biofuels, building blocks, and molecules which will be useful for human beings (Fig. 4.5). Such a microbe will utilize renewable materials as stocks thus creating a sustainable bio economy, and it can be further modified to create more effective or even new microbe by changing certain parameters in the metabolic pathways that affect their growth behavior. Micro bioreactors have been built by a few companies that can monitor a variety of parameters of synthetic cell or minimal cell simultaneously. Thus synthetic biology would apply engineering to understand the versatility of biology to solve the challenges of today's world. Redesigning the protein and metabolic pathways, metabolic engineering, and programmed synthesis of therapeutics have already benefited from application of synthetic biology. Robust engineering technology will emerge as this field advances and it should transform biotechnology, pharmaceuticals, and chemical industries. These technologies will provide tools, reagents, and other services for the advancement of synthetic biology (Dalchau et al., 2012).

4.6.2 Artificial Life

Progress in the area of synthetic biology has enabled creation of synthetic cells, in the range of simple protocell to artificial cell to living bacteria, which indicates that creation of live cell from nonliving chemicals is possible. These areas are fueled by the knowledge of primitive membrane, nucleic acid synthesis and replication, synthetic protein, transcription and translation machinery, metabolic enzymes, etc. Replication of synthetic polymer in a

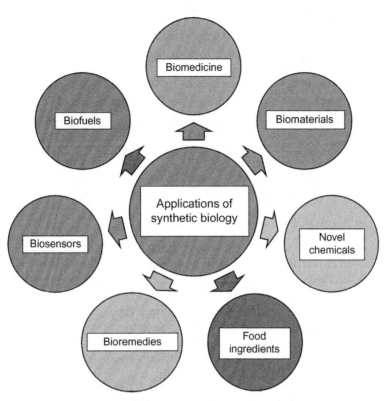

FIGURE 4.5 **Industrial Scale Applications of Synthetic Biology.** The ultimate goal of synthetic biology is to make it useful in an industrial scale. Synthetic biology will have application in the diagnosis and treatment of diseases, production of custom drugs, biofuel, food ingredients, novel chemicals, biomaterials, biosensors etc. It is also applicable in bioremedies, protection of environment and conservation of natural resources.

nonenzymatic manner seems to be the barrier in creating artificial life. One of the problems is strand reannealing and complete template copying of the artificial gene. Another important goal is to coordinate all the process to follow an order in a cell-cycle–dependent manner along with the replication of the boundary membrane. The artificial membrane made with amphiphiles such as fatty acids and related molecules show that it is possible to create artificial life. However, these fatty acid membranes are not compatible with high salt environment but can exist only in fresh water environment. In addition, the requirement of variable temperature for nucleic acid replication and strand separation indicates that the cold fresh water springs locally heated by hot springs such as volcano could be an ideal condition to create artificial life (Mansy and Szostak, 2009).

4.6.3 Biosensor

Biosensors are synthetic cells or organisms (generally bacterium) or minimal cells that can detect environmental phenomenon such as toxins, heavy metals, or other pollutants

(Fig. 4.5). One of the examples of biosensors is the lux operon that has 4-α-helical protein gene monomers that are sufficient for bioluminescence in *Aliivibrio fischeri* (Farid et al., 2013). These genes can be placed under inducible promoters that are regulated by random environmental stimuli. The "critter on a chip" developed by Oak Ridge National Laboratory contains bioluminescent bacteria on a light-sensitive computer chip which is used to detect certain petroleum products that pollute the environment. In presence of the pollutant the bacteria can generate light. In another study, living cells are incorporated into biosensor platform to recognize biotoxins (Park et al., 2013). The use of fluorescence reporter—based biosensors or Fluorescence resonance energy transfer—based sensors allow noninvasive and faster analysis of environmental pollutants, and it also reports minor changes in a time-dependent manner (Constantinou and Polizzi, 2013).

4.6.4 Synthetic Biological Circuits

Synthetic biological circuits are parts when inserted inside a minimal or synthetic cell would perform logical functions just like the natural ones. These can be used to modify cellular functions or direct cells to respond to environmental conditions (Kobayashi et al., 2004). A synthetic analog gene circuit performed sophisticated functions with just three transcription factors and they worked in a logical manner similar to the natural circuits. These can be used in an environment where controlled gene expression is necessary (Daniel et al., 2013). They can also be used in a "plug and play" manner which would help us understand the various bioprocesses within the cell. Many synthetic biological circuits have already been made and tested (Table 4.2).

4.6.5 Industrial Scale Applications

The ultimate goal of synthetic biology is to make it useful in an industrial scale. We are in the verge of developing synthetic biology coupled with systems biology as a viable industry. This will have profound implication on the economy and environment as well as medicine. The applications include diagnosis and treatment of diseases, production of

TABLE 4.2   List of Synthetic Biological Circuits

Sl No.	Synthetic Circuit	Reference
1	Repressilator	Elowitz and Leibler (2000)
2	Toggle-switch	Gardner et al. (2000)
3	Mammalian tunable synthetic oscillator	Tigges et al. (2009)
4	Bacterial tunable synthetic oscillator	Chen et al. (2014)
5	Coupled bacterial oscillator	Siuti et al. (2013)
6	Globally coupled bacterial oscillator	Nielsen et al. (2013)
7	*E coli* with "trigger element" and "memory element" synthetic circuits to monitor and report in a complex, ill-defined environment	Kotula et al. (2014)

custom drugs, understanding aging, biofuel, artificial leaf, and artificial enzymes. Natural disasters like oil spill and escape of harmful radiation can be tackled with synthetic biology, and it can also produce biodegradable packaging. In agriculture it would have effect on seed production, agro fuels, starch synthesis, and optimize food production. Synthetic biology can also influence other industries such as biological computers, logical gates, bio switches, bio fabrication, and nanoparticles (Fig. 4.5) (Khalil and Collins, 2010).

4.7 EXAMPLES

4.7.1 Synthetic Biology Today

In spite of advancements in many areas of synthetic biology, this field is still in its infancy. Nature has created the variety of structures and functions through evolution. Synthetic biology is dependent on the biomaterials to build what molecular evolution took millions of years to create. Thus, application of natural principles and design mimics and variations are made from nonnatural materials. The field of synthetic biology has made considerable advancement within the past two decades in creating functional mimics (Khalil and Collins, 2010). Synthetic biology has given rise to man-made genes such as polio virus, but it has been restricted to only prokaryotic system (Fig. 4.2) (Cello et al., 2002; Blight et al., 2000; Hamilton et al., 2003; Gibson et al., 2008). However, a functional eukaryotic yeast chromosome III of *Saccharomyces cerevisiae*, synIII, was chemically synthesized. The designer synIII with 272,871 bp was synthesized based on the native chromosome III sequence of 316,617 bp (Annaluru et al., 2014). The synIII is expected to form the foundation for the synthesis of other functional eukaryotic parts or even whole organism.

4.7.2 Applications

Application of synthetic biology depends on its advancement in reconstructing more complex biological system with interchangeable parts and understanding and fine-tuning various biological networks. This would be followed by "mix-and-match" of these parts and their behaviors when they are combined. High-throughput analysis of the functions of these "mix-and-match" parts in a time-bound and a real-time manner is essential to understand these networks. Technological advancement such as bio-computing to handle data from complex biological systems should evolve, and it should also incorporate higher order decision-making in the circuits. The "mix-and-match" of various biological and genetic components along with synthetic parts would result in engineered organism with refined automation for biosensing and related applications. Majority of the synthetic biology is currently practiced on microbial system. However, various disease processes are found in complex organisms like human beings. Synthetic biology can be effectively used to manage disease condition, develop next generation diagnostics and therapeutics, and understand differentiation only if we fathom the complexity of human system (Fig. 4.6) (Daniel et al., 2013; Khalil and Collins, 2010; Bryksin et al., 2014).

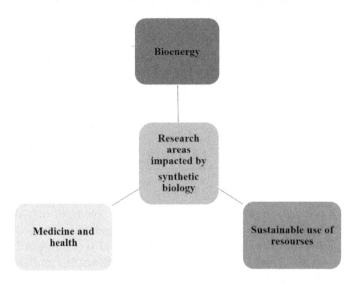

FIGURE 4.6 **Key areas of research impacted by synthetic biology.** Synthetic biology will revolutionize biological research and development in the future. Some of the important areas of research that will significantly benefit from synthetic biology include drug development, bioenergy and gene therapy. Hence synthetic biology holds great promise for the benefit of mankind in the coming future.

4.7.3 iGEM

To spread the research in synthetic biology, the international Genetically Engineered Machine Competition (iGEM) assembles group of biologists, engineers, design experts, mathematicians, etc., to design and develop synthetic mimics using bio-bricks. It also has computer programmers to decode BioBrick grammar and provide theoretical perspective of synthetic biology. iGEM trains students and encourages competition and collaboration in this field. It consists of advisers and undergraduate students. The aim of iGEM is to tackle variety of problems such as malnutrition and one of the iGEM groups is aiming to construct a strain of *S. cerevisiae* that would produce vitamin. Another group is producing antifreeze protein that has wide application in medicine and food preservation technology. One of the groups designed a bio-factory by immobilizing enzymes that would allow the biochemical reaction flow in a linear sequence. One iGEM group is developing Tailings Pond Clean Up kit that would clean toxins from ponds and lakes, and another group has developed nanoparticles that would provide clean drinking water, and these technologies would have tremendous application internationally (Balmer and Bulpin, 2013; Cai et al., 2010).

4.8 ETHICS AND SAFETY

4.8.1 Society, Ethics, and Synthetic Biology

Opposition to synthetic biology was led by the action group on Erosion, Technology, and Concentration (ETC), and they have called for a global moratorium on developments in this field. In 2006, 38 civil society organizations authored an open letter opposing voluntary regulation of the field. In 2008 ETC group released the first critical report on the societal impacts of synthetic biology under the title "Extreme Genetic Engineering." On March 13, 2012, more than 100 environmental and civil society groups, including Friends of the Earth, the International Center for Technology Assessment, and the ETC group issued the statement, "The Principles for the Oversight of Synthetic Biology," which called for a worldwide moratorium on the release and commercial use of synthetic organisms until more robust regulations and rigorous biosafety measures are established.

Discussion of "societal issues" took place at the SYNBIOSAFE forum on issues regarding ethics, safety, security, IPR, governance, and public perception. A report from the Woodrow Wilson Center and the Hastings Center, a prestigious bioethics research institute, found that ethical concerns in synthetic biology have received scant attention. In January 2009, the Alfred P. Sloan Foundation funded the Woodrow Wilson Center, the Hastings Center, and the J. Craig Venter Institute to examine the public perception, ethics, and policy implications of synthetic biology. On July 9–10, 2009, the National Academies Committee of Science, Technology, and Law convened a symposium on "Opportunities and Challenges in the Emerging Field of Synthetic Biology." After a series of meetings in the fall of 2010, the Presidential Commission for the study of Bioethical Issues released a report, on December 16, 2010, titled "New Directions: The Ethics of Synthetic Biology and Emerging Technologies" to the President calling for enhanced federal oversight in the emerging field of synthetic biology (Anderson et al., 2012).

4.8.2 Safety Regulations

Three major risks are identified in synthetic biology. The first and foremost is that the synthetically created organisms can have unintended side effects. Some scientists claim that synthetic organisms could help to solve environmental problems like contamination of soil, which means that the synthetic organisms should be released into the contaminated soil. However, the artificial organism could have unintended or negative side-effects. The second risk is that the synthetic organisms could transfer genes to natural organisms. Finally, there is a fear that the self-replication of synthetic cells or organisms could become out-of-control (Fig. 4.7).

There is a need to develop uniform and standardized screening tools to determine what is dangerous and what is not, especially in the case of synthetic genomics. It is important to create a rationalized list of agents to determine the most dangerous and prioritize screening. Managing risk in this case is complex because the mechanisms of pathogenic agents are not fully understood, and it is currently difficult to identify agents that may be hazardous. A database of risky sequences or experiments is necessary to help stratify and keep track of risk. Synthetic biology can be used by terrorists, and there should be a mechanism to deal with it. The unpredictable functions of designer circuits should be anticipated and safety mechanisms should be in place to contain it. The opinion of public should be taken into consideration, and there should be discussion on the benefit and risks of synthetic biology. Malicious application of synthetic biology could have adverse effect globally. The international regulatory agencies such as United Nations, Chemical Weapons

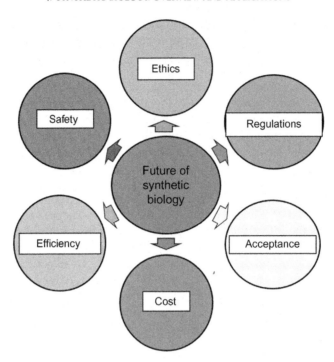

FIGURE 4.7   **Parameters that affect the future of synthetic biology.** Different social aspects can affect the success of future applications based on synthetic biology. The Legal, Ethical, and Safety concerns are of primary importance. Furthermore, the cost, efficiency of the applications and the extent of acceptance of these by the society play an important role in determining the future of synthetic biology.

Conventions, and Biological Weapons Conventions limit the risks involved with new technologies. Specific regulations are in place in different countries to control the use of DNA database (this includes DNA synthesis, DNA sequences, microorganisms, and toxins). There should be control of DNA synthesis, monitoring access to database, and mechanisms to identify suspicious DNA orders that are in place in the United States and France. The awareness about synthetic biology should be spread through education, conferences, and seminars. Some universities have modules dealing with biosecurity and biosafety issues and many more are underway (Rager-Zisman, 2012).

Robust regulations for the safe use of biotechnology and recombinant DNA technology are in place. Synthetic biology is frequently conceived as an extension of these established technologies and therefore requires more appropriate biosafety procedures and regulations. A wide array of existing laws and regulations apply to the emerging field of synthetic biology. Applicable also are local institutional, municipal, and state requirements, many of which focus on safety and security. Prior experience with potentially hazardous materials such as RNA viruses can provide useful lessons for synthetic biology and assist in safety and security efforts. There are documented cases of virus leaks from laboratories, including smallpox in the United Kingdom (1978), SARS in Singapore (2003) and China (2004), and several cases of accidental polio release, including the identification in 2003 of a laboratory strain circulating in the general population. This makes abundantly clear the importance of regulations and guidelines, their enforcement, and the vigilant containment of pathogens (Rager-Zisman, 2012).

The NIH Recombinant DNA Advisory Committee (RAC), which advises the Director of the NIH on rDNA research, was charged with considering the application of the NIH guidelines to synthetic biology. The RAC made a series of proposed modifications including a revised definition of rDNA molecules and an exemption for synthetic nucleic acids that cannot replicate, provided they are not used in human gene transfer. The NIH was noted for its foresight in updating its guidelines to accommodate advances in synthetic biology, and delegates discussed the situation in other countries. In Japan, researchers are required by law to notify the Ministry of Education of any research involving rDNA. In Europe, biosafety is captured within legally binding directives, and regulatory bodies in member states undertake compliance checks. Continuous efforts are being made by the various regulatory bodies to refine the regulatory framework pertaining to synthetic biology and its applications (Rager-Zisman, 2012).

4.8.3 Safe Human Practices

Safety is a critical point that needs to be discussed up front to prevent unintentional risk from pathogens, toxins, and otherwise potentially harmful biological material. It is important to protect humans, animals, and the environment from the accidental release of harmful synthetic biological agents. To maintain biotechnology at its current safety level, or even improve safety measures, we need to know the potential risks. Researchers must decide whether a new synthetic biology technique or application is safe enough for humans, animals, and the environment and its use in restricted and less-restricted settings. For example, it is important to understand if the newly created DNA-based systems and parts are substantially different from existing life forms. It is necessary to assess survival of minimal life forms or a protocell in various environments and develop safeguard against their potential infectious nature (Fig. 4.7) (Marliere, 2009).

An important task of a safety discussion is to explore how synthetic biology itself may contribute towards overcoming existing and possible future biosafety problems by contributing to the design of safer biosystems, for example, design of less competitive organisms by changing metabolic pathways; replacing metabolic pathways with others that have an in-built dependency on external biochemicals; design of evolutionary robust biological circuits; use of biological systems based on an alternative biochemical structure like the XNA (xeno-nucleic acid) molecules to avoid gene flow to and from wild species; and design of protocells that lack key features of living entities, such as growth or replication (Herdewijn and Marlière, 2009).

Biosecurity should also deals with the prevention of misuse through loss, theft, diversion, or intentional release of pathogens, toxins, and other biological materials. Technical advances in DNA synthesis have given rise to a completely new threat potential, namely, the synthesis of harmful viruses without the need for a physical template (Cello et al., 2002). Scientists are also able to generate viruses using DNA from previously infected cadavers and archeological samples. The reconstructed 1918 Spanish influenza pandemic virus is one such example (Tumpey et al., 2005).

Given these possibilities, the issue of biosecurity is getting immediate attention and a suite of solutions has already been developed and implemented. The DNA synthesis companies will need to check and screen orders to avoid production of select agents, such as harmful viruses and bacterial DNA, as well as to further develop and improve the technical means (e.g., software and databases) used to screen DNA orders. Awareness needs to be enhanced through better communication and cooperation between the synthetic biology and biosecurity communities. Issues such as misuse of the life sciences for offensive bioweapon programs, the inadvertent results of security-related research, and the existence and operation of the Biological Weapons should be systematically included in undergraduate and graduate biology curricula. Addressing questions of governance and oversight of biosecurity will require regulatory tools. The involvement of all stakeholders is required to develop useful tools and avoid an oversight system with severe restrictions (Fig. 4.7) (Garfinkel et al., 2007; Kelle, 2009).

In developing synthetic biology tools for human health, we must continually consider the relevant ethical and social issues surrounding this research. Our experience in the development of experimental therapies has underscored the importance of caution in developing human trials (Burrill et al., 2011; Marshall, 1999). Balancing risks with the transformative potential benefits of synthetic biology will be a continual and necessary challenge as synthetic biology enters the clinic (Gabardi et al., 2011).

4.8.4 Technology Transfer

The increasing ease of access to materials and supplies used to generate synthetic agents poses another unique oversight challenge. Gene and oligonucleotide sequences or parts can be commercially obtained with ease, and reagents and automated equipment for synthesizing nucleic acid sequences are available as well. Deviant uses of synthetic biology could therefore, at least theoretically, occur outside of the scope of existing supervision mechanisms. However, efforts are being made to prevent any such misuse of technology. In the United States, the Bureau of Industry and Security within Department of Commerce (DOC) administers the Export Administration Regulations. These regulations govern the export and reexport of dual-use commodities, that is, materials with both commercial applications and military or other defense applications, software, and technology from the United States and apply to any individual or entity seeking to export. Included in this group may be researchers collaborating with overseas colleagues, manufacturers with foreign plants, and gene synthesis providers shipping orders outside of the United States. Particularly relevant to the oversight of synthetic biology are provisions designed to restrict access to materials that have dual use applications. For each controlled item, detailed licensing requirements and policies for screening potential recipients are imposed. Licenses are provided depending on the nature of the threat. The end user receiving the product must also be screened against lists of individuals and organizations (Fig. 4.7).

4.9 CONCLUSIONS AND FUTURE PERSPECTIVES

4.9.1 Promises of Synthetic Biology

Synthetic biology is aimed at creating living, self-replicating cells and organisms using standard parts which would be beneficial to mankind. When it is put into practice, we could have designer drugs, novel food products that would provide all essential vitamin and minerals to lead a healthy life, early diagnostic tools, and technology to deal with large scale pollution such as huge oil spills, chemical spills, and many more. In addition, it would create self-sustainable models that would not deplete natural resources. It also promises brand new industries and job opportunities for millions. New technology and affordable, feasible, effective designs will be created using synthetic biology to do major tasks such as waste management, providing quality drinking water supply, and affordable medical care to the entire human race. Synthetic biology promises to provide means to affordable and sustainable living conditions and to protect the existence of all life forms (Serrano, 2007).

4.9.2 Challenges

Synthetic biology differs from systems biology as it uses engineering and novel functions that do not exist in nature. Systems biology attempts to explore the functions of the existing systems, whereas synthetic biology is trying to assemble or build these functions from the scratch using chemicals that exist in nature. In this regard synthetic biology will use standard parts such as amino acids, nucleotides, and other chemical matters. It also requires standard parts, computer design and software, and knowledge to assemble a system that is operational. However, it is important to understand the complexity of the living system, and it is in multiple order in eukaryotes as compared to prokaryotes. Therefore, there is no guarantee that the same parts will work similarly in both the systems. Moreover, new properties could develop as we start assembling parts together and these will have to be identified and characterized. Our thought process should be more flexible and imaginative to use synthetic biology to benefit all. Synthetic biology requires "out-of-the box" thinking, and it is different from biotechnology. Biological system is capable of withstanding tinkering to certain extent. It is important to understand the difference between standard versus totally synthetic systems such as minimal cell versus synthetic cell. There are many existing challenges in synthetic biology, and many more would emerge as more ideas are tested. Therefore, it is important to evolve safety mechanisms as we proceed further in this exciting area of science (Serrano, 2007).

References

Alon, U., 2007. Network motifs: theory and experimental approaches. Nat. Rev. Genet. 8, 450–461. PMID:17510665.

Anderson, J., Strelkowa, N., Stan, G.B., Douglas, T., Savulescu, J., Barahona, M., et al., 2012. Engineering and ethical perspectives in synthetic biology. Rigorous, robust and predictable designs, public engagement and a modern ethical framework are vital to the continued success of synthetic biology. EMBO Rep 13 (7), 584–590.

Available from: https://doi.org/10.1038/embor.2012.81. PubMed PMID: 22699939; PubMed Central PMCID: PMC3389334.

Ambroggio, X.I., Kuhlman, B., 2006a. Computational design of a single amino acid sequence that can switch between two distinct protein folds. J. Am. Chem. Soc. 128, 1154–1161. PMID:16433531.

Ambroggio, X.I., Kuhlman, B., 2006b. Design of protein conformational switches. Curr. Opin. Struct. Biol. 16, 525–530. PMID:16765587.

Anderson, J.C., Clarke, E.J., Arkin, A.P., Voigt, C.A., 2006. Environmentally controlled invasion of cancer cells by engineered bacteria. J. Mol. Biol. 355, 619–627. PMID:16330045.

Anderson, J., Strelkowa, N., Stan, G.B., Douglas, T., Savulescu, J., Barahona, M., et al., 2012. Engineering and ethical perspectives in synthetic biology. Rigorous, robust and predictable designs, public engagement and a modern ethical framework are vital to the continued success of synthetic biology. EMBO Rep 13 (7), 584–590. Available from: https://doi.org/10.1038/embor.2012.81. PubMed PMID: 22699939; PubMed Central PMCID: PMC3389334.

Annaluru, N., Muller, H., Mitchell, L.A., Ramalingam, S., Stracquadanio, G., Richardson, S.M., et al., 2014. Total synthesis of a functional eukaryotic chromosome. Science 343. Available from: https://doi.org/10.1126/science.1249252.

Asimov, I., 1965. A Short History of Chemistry. Anchor Press Doubleday, New York, NY, ISBN 10: 0385036736 / ISBN 13: 9780385036733.

Bachmann, P.A., Luisi, P.L., Lang, J., 1992. Autocatalytic self-replicating micelles as models for prebiotic structures. Nature 357, 57–59. Available from: https://doi.org/10.1038/357057a0.

Balmer, A.S., Bulpin, K.J., 2013. Left to their own devices: post-ELSI, ethical equipment and the international Genetically Engineered Machine (iGEM) competition. Biosocieties 8, 311–335. PMID:24159360.

Bayer, T.S., Smolke, C.D., 2005. Programmable ligand-controlled riboregulators of eukaryotic gene expression. Nat. Biotechnol. 23, 337–343. PMID:15723047.

Benner, S.A., Sismour, A.M., 2005. Synthetic biology: act natural. Nature 421, 118.

Bhattacharyya, R.P., Remenyi, A., Yeh, B.J., Lim, W.A., 2006. Domains, motifs, and scaffolds: the role of modular interactions in the evolution and wiring of cell signaling circuits. Annu. Rev. Biochem. 75, 655–680. http://dx.doi.org/10.1146/annurev.biochem.75.103004.142710; PMID:16756506.

Bio FAB Group, Baker, D., Church, G., Collins, J., Endy, D., Jacobson, J., et al., 2006. Engineering life: building a fab for biology. Sci. Am. 294, 44–51. PMID:16711359.

Bitbol, M., Luisi, P.L., 2004. Autopoiesis with or without cognition: defining life at its edge. J. R. Soc. Interface 1, 99–107. http://dx.doi.org/10.1098/rsif.2004.0012; PMC1618936.

Blight, K.J., Kolykhalov, A.A., Rice, C.M., 2000. Efficient initiation of HCV RNA replication in cell culture. Science 290, 1972–1974. http://dx.doi.org/10.1126/science.290.5498.1972. PMID:11110665.

Breslow, R., 1972. Centenary lecture. Biomimetic chemistry. Chem. Soc. Rev. 1, 553–580. Available from: https://doi.org/10.1039/CS9720100553.

Bromley, E.H.C., Channon, K., Moutevelis, E., Woolfson, D.N., 2008. Peptide and protein building blocks for synthetic biology: from programming molecules to self-organised biomolecular systems. ACS Chem. Biol. 3, 38–50. Available from: http://dx.doi.org/10.1001/archgenpsychiatry.2011.51. PMID:18205291.

Bryksin, A.V., Brown, A.C., Baksh, M.M., Finn, M.G., Barker, T.H., 2014. Learning from nature—novel synthetic biology approaches for biomaterial design. Acta Biomater. 10, 1761–1769. Available from: http://dx.doi.org/10.1016/j.actbio.2014.01.019. PMID: 24463066.

Burrill, D.R., Boyle, P.M., Silver, P.A., 2011. A new approach to an old problem: synthetic biology tools for human disease and metabolism. Cold Spring Harb. Symp. Quant. Biol. 76, 145–154. Available from: http://dx.doi.org/10.1101/sqb.2011.76.010686. PMID: 22169233.

Cai, Y., Wilson, M.L., Peccoud, J., 2010. GenoCAD for iGEM: a grammatical approach to the design of standard-compliant constructs. Nucleic Acids Res. 38, 2637–2644. Available from: http://dx.doi.org/10.1093/nar/gkq086. PMID: 20167639.

Campos, L., 2009. That was the synthetic biology that was. In: Schmidt, M., Kelle, A., Ganguli-Mitra, A., Vriend, H. (Eds.), Synthetic Biology: The Technoscience and Its Societal Consequences. Springer Academic Publishing, pp. 5–21.

Carmona-Ribeiro, A.M., 2007. Biomimetic particles in drug and vaccine delivery. J. Liposome Res. 17, 165–172. PMID:18027236.

Cello, J., Paul, A.V., Wimmer, E., 2002. Chemical synthesis of poliovirus cDNA: generation of infectious virus in the absence of natural template. Science 297, 1016–1018. PMID:12114528.

Cesario, A., Auffray, C., Russo, P., Hood, L., 2014. P4 medicine needs p4 education. Curr. Pharm. Des. 14, PMID: 24641231.

Chan, L.Y., Kosuri, S., Endy, D., 2005. Refactoring bacteriophage T7. Mol. Systems Biol. 1, 2005.0018. Available from: http://dx.doi.org/10.1038/msb4100025. PMID:16729053.

Channon, K., Bromley, E.H.C., Woolfson, D.N., 2008. Synthetic biology through biomolecular design and engineering. Curr. Opin. Struct. Biol. 18, 491–498. Available from: http://dx.doi.org/10.1016/j.sbi.2008.06.006. PMID 18644449.

Chen, A.Y., Deng, Z., Billings, A.N., Seker, U.O.S., Lu, M.Y., Citorik, R.J., et al., 2014. Synthesis and patterning of tunable multiscale materials with engineered cells. Nat. Mater. Available from: http://dx.doi.org/10.1038/nmat3912. PMID:24658114.

Chopta, P., Kamma, A., 2008. Engineering life through synthetic biology. In Silico Biol. 6, 6–9. PMID:17274769.

Church, G.M., Elowitz, M.B., Smolke, C.D., Voigt, C.A., Weiss, R., 2014. Realizing the potential of synthetic biology. Nat. Rev. Mol. Cell Biol. 15, 289–294. Available from: http://dx.doi.org/10.1038/nrm3767. PMID:24622617.

Ciani, B., Hutchinson, E.G., Sessions, R.B., Woolfson, D.N., 2002. A designed system for assessing how sequence affects alpha to beta conformational transitions in proteins. J. Biol. Chem. 277, 10150–10155. PMID:11751929.

Constantinou, A., Polizzi, K.M., 2013. Opportunities for bioprocess monitoring using FRET biosensors. Biochem. Soc. Trans. 41, 1146–1151. Available from: http://dx.doi.org/10.1042/BST20130103. PMID:24059500.

Cooper, G.M., Hausman, R.E., 2013. The Cell: A Molecular Approach. Sinauer Associates, Inc, Sunderland, CT, USA. Issue No. 978-0-87893-964-0.

Dalchau, N., Smith, M.J., Martin, S., Brown, J.R., Emmott, S., Phillips, A., 2012. Towards a rational design of synthetic cells with prescribed population dynamics. J. R. Soc. Interface 9, 2883–2898. Available from: http://dx.doi.org/10.1098/rsif.2012.0280. PMID:22683525.

Daniel, R., Rubens, J.R., Sarpeshkar, R., Lu, T.K., 2013. Synthetic analog computation in living cells. Nature 497, 619–623. Available from: http://dx.doi.org/10.1038/nature12148. PMID: 23676681.

Davidson, A.R., Lumb, K.J., Sauer, R.T., 1995. Cooperatively folded proteins in random sequence libraries. Nat. Struct. Biol. 2, 856–864. Available from: http://dx.doi.org/10.1038/nsb1095-856. PMID:7552709.

Davies, J.A., 2008. Synthetic morphology: prospects for engineered, self-constructing anatomies. J. Anat. 212, 707–719. PMID 18510501.

de Lorenzo, V., Danchin, A., 2008. Synthetic biology: discovering new worlds and new words. EMBO Rep. 9, 822–827. Available from: http://dx.doi.org/10.1038/embor.2008.159. PMID:18724274.

di Bernardo, D., Thompson, M.J., Gardner, T.S., Chobot, S.E., Eastwood, E.L., Wojtovich, A.P., et al., 2005. Chemogenomic profiling on a genome-wide scale using reverse-engineered gene networks. Nat. Biotechnol. 23, 377–383. PMID: 15765094.

Discher, D.E., Eisenberg, A., 2002. Polymer vesicles. Science 297, 967–973. PMID:12169723.

Dmiano, L., Luisi, P.L., 2010. Towards an autopoietic redefinition of life. Prog. Life Evol. Biosph. 40, 145–149. Available from: http://dx.doi.org/10.1007/s11084-010-9193-2. PMID: 20213162.

Dwyer, M.A., Hellinga, H.W., 2004. Periplasmic binding proteins: a versatile superfamily for protein engineering. Curr. Opin. Struct. Biol. 14, 495–504. PMID:15313245.

Elowitz, M.B., Leibler, S., 2000. A synthetic oscillatory network of transcriptional regulators. Nature 403, 335–338. Available from: http://dx.doi.org/10.1038/35002125. PMID 10659856.

Farid, T.A., Kodali, G., Solomon, L.A., Lichtenstein, B.R., Sheehan, M.M., Fry, B.A., et al., 2013. Elementary tetrahelical protein design for diverse oxidoreductase functions. Nat. Chem. Biol. 9, 826–833. Available from: http://dx.doi.org/10.1038/nchembio.1362. PMID:24121554.

Finn, R.D., Mistry, J., Schuster-Böckler, B., Griffiths-Jones, S., Hollich, V., Lassmann, T., et al., 2006. Pfam: clans, web tools and services. Nucleic Acids Res. 34 (database issue), D247–D251. PMID:16381856.

Foffi, G., Pastore, A., Piazza, F., Temussi, P.A., 2013. Macromolecular crowding: chemistry and physics meet biology (Ascona, Switzerland, 10–14 June 2012). Phys. Biol. 10, 040301. PMID:23912807.

Forster, A.C., Church, G.M., 2006. Towards synthesis of a minimal cell. Mol. Syst. Biol. 2, 45. Available from: http://dx.doi.org/10.1038/msb4100090. PMID:16924266.

Gabardi, S., Halloran, P.F., Friedewald, J., 2011. Managing risk in developing transplant immunosuppressive agents: the new regulatory environment. Am. J. Transplant. 11, 1803–1809. Available from: http://dx.doi.org/10.1111/j.1600-6143.2011.03653.x. PMID: 21827622.

Gardner, T.S., Cantor, C.R., Collins, J.J., 2000. Construction of genetic toggle switch in *Escherichia coli*. Nature 403, 339–342. PMID:10659857.

Garfinkel, M.S., Endy, D., Epstein, G.L., Friedman, R.M., 2007. Synthetic genomics|options for governance. Biosecur. Bioterror. 5, 359–362. PMID: 18081496.

Ghosh, I., Chmielewski, J., 2004. Peptide self-assembly as a model of proteins in the pre-genomic world. Curr. Opin. Chem. Biol. 8, 640–644. PMID:15556409.

Gibbs, W.W., 2004. Synthetic life. Sci. Am. 290, 74–81.

Gibson, D.G., Benders, G.A., Andrews-Pfannkoch, C., Denisova, E.A., Baden-Tillson, H., Zaveri, J., et al., 2008. Complete chemical synthesis, assembly, and cloning of a Mycoplasma genitalium genome. Science 319, 1215–1220. Available from: http://dx.doi.org/10.1126/science.1151721. PMID 18218864.

Gibson, D.G., Glass, J.I., Lartigue, C., Noskov, V.N., Chuang, R.Y., Algire, M.A., et al., 2010. Creation of a bacterial cell controlled by chemically synthesized genome. Science 329, 52–56. Available from: http://dx.doi.org/10.1126/science.1190719. PMID:20488990.

Gruber, A.R., Lorenz, R., Bernhart, S.H., Neubock, R., Hofacker, I.L., 2008. The Vienna RNA websuite. Nucleic Acids Res. 36 (suppl 2), W70–W74. Available from: https://doi.org/10.1093/nar/gkn188.

Grunberg, R., Serrano, L., 2010. Strategies for protein synthetic biology. Nucleic Acids Res. 38, 2663–2675. PMID:20385577.

Hamilton, S.O., Hutchison 3rd, C.A., Pfannkoch, C., Venter, J.C., 2003. Generating a synthetic genome by whole genome assembly: phiX174 bacteriophage from synthetic oligonucleotides. Proc. Natl. Acad. Sci. U.S.A. 100, 15440–15445. Available from: http://dx.doi.org/10.1073/pnas.2237126100. PMID 14657399.

Herdewijn, P., Marlière, P., 2009. Toward safe genetically modified organisms through the chemical diversification of nucleic acids. Chem. Biodivers. 6, 791–808. Available from: http://dx.doi.org/10.1002/cbdv.200900083. PMID: 19554563.

Hobom, B., 1980. Surgery of genes. At the doorstep of synthetic biology. Medizin. Klinik 75, 14–21.

Issacs, F.J., Dwyer, D.J., Collins, J.J., 2006. RNA synthetic biology. Nat. Biotechnol. 24, 545–554. Available from: http://dx.doi.org/10.1038/nbt1208. PMID:16680139.

Jaffe, B., 1976. Crucibles: The Story of Chemistry From Ancient Alchemy to Nuclear Fission, 4th edition Dover Publications, New York. ISBN-13: 9780486233420.

Kajander, T., Cortajarena, A.L., Mochrie, S., Regan, L., 2007. Structure and stability of designed TRP protein superhelices: unusual crystal packing and implications for natural TRP proteins. Acta Crystallogr. D Biol. Crystallogr. 63 (Pt 7), 800–811. PMID:17582171.

Kamtekar, S., Schiffer, J.M., Xiong, H., Babik, J.M., Hecht, M.H., 1993. Protein design by binary patterning of polar and nonpolar amino acids. Science 262, 1680–1685. Available from: http://dx.doi.org/10.1126/science.8259512. PMID:8259512.

Kelle, A., 2009. Synthetic biology and biosecurity. From low levels of awareness to a comprehensive strategy. EMBO Rep. Suppl. 1, S23–S27. Available from: http://dx.doi.org/10.1038/embor.2009.119. PMID: 19636299.

Khalil, A.S., Collins, J.J., 2010. Synthetic biology: applications come of age. Nat. Rev. Genet. 11, 367–379. Available from: http://dx.doi.org/10.1038/nrg2775. PMID:20395970.

Khorana, H.G., 1965. Polynucleotide synthesis and the genetic code. Fed. Proc. 24, 1473–1487. PMID:5322508.

Kim, T.W., Delaney, J.C., Essigmann, J.M., Kool, E.T., 2005. Probing the active site tightness of DNA polymerase in subangstrom increments. Proc. Natl. Acad. Sci. U.S.A. 102, 15803–15808. PMID:16249340.

Kobayashi, H., Kaern, M., Araki, M., Chung, K., Gardner, T.S., Cantor, C.R., et al., 2004. Programmable cells: interfacing natural and engineered gene networks. Proc. Natl. Acad. Sci. U.S.A. 101, 8414–8419. PMID:15159530.

Kool, E.T., Waters, M.L., 2007. The model student: what chemical model systems can teach us about biology. Nat. Chem. Biol. 3, 70–73. Available from: http://dx.doi.org/10.1038/nchembio0207-70. PMID:17235337.

Kotula, J.W., Kerns, S.J., Shaket, L.A., Siraj, L., Collins, J.J., Way, J.C., et al., 2014. Programmable bacteria detect and record an environmental signal in the mammalian gut. Proc. Natl. Acad. Sci. U. S. A. 111, 4838–4843. Available from: http://dx.doi.org/10.1073/pnas.1321321111.

Krueger, A.T., Peterson, L.W., Chelliserry, J., Kleinbaum, D.J., Kool, E.T., 2011. Encoding phenotype in bacteria with an alternative genetic set. J. Am. Chem. Soc. 133, 18447–18451. Available from: http://dx.doi.org/10.1021/ja208025e. PMID:21981660.

Leduc, S., 1910. Théorie physico-chimique de la vie et générations spontanées. A. Poinat, Paris, Internet Archive: thoriephysicoc001eduuoft; open library: OL23348076M.

Leduc, S., 1912. La biologie synthétique, Etude de biophysique. A. Poinat Associated Press, Paris.

Lee, D.H., Granja, J.R., Martinez, J.A., Severin, K., Ghadiri, M.R., 1996. A self-replicating peptide. Nature 382, 525–528. PMID:8700225.

Liu, M., Bienfait, B., Sacher, O., Gasteiger, J., Siezen, R.J., Nauta, A., et al., 2014. Combining chemoinformatics with bioinformatics: in silico prediction of bacterial flavor-forming pathways by a chemical systems biology approach "reverse pathway engineering". PLoS One 9, e84769. Available from: http://dx.doi.org/10.1371/journal.pone.0084769. eCollection2014; PMID:2441682.

Liu, Y., Kuhlman, B., 2006. RosettaDesign server for protein design. Nucleic Acids Res. 34, W235–W238. PMID:16845000.

Lu, H., Krueger, A.T., Gao, J., Liu, H., Kool, E.T., 2010. Toward a designed genetic system with biochemical function: polymerase synthesis of single and multiple size-expanded DNA base pair. Org. Biomol. Chem. 8, 2704–2710. Available from: http://dx.doi.org/10.1039/c002766a. PMID:20407680.

Lupas, A., 1996. Coiled coils: new structures and new functions. Trends Biochem. Sci. 21, 375–382. PMID:8918191.

Mansy, S.S., Szostak, J.W., 2009. Reconstructing the emergence of cellular life through the synthesis of model protocells. Cold Spring Harb. Symp. Quant. Biol. 74, 47–54. Available from: http://dx.doi.org/10.1101/sqb.2009.74.014. PMID:19734203.

Marliere, P., 2009. The farther, the safer: a manifesto for securely navigating synthetic species away from the old living world. Syst. Synth. Biol. 3, 77–84. Available from: http://dx.doi.org/10.1007/s11693-009-9040-9. PMID: 19816802.

Marsh, J.A., Forman-Kay, J.D., 2012. Ensemble modeling of protein disordered states: experimental restraint contributions and validation. Proteins 80, 556–572. Available from: http://dx.doi.org/10.1002/prot.23220. PMID:22095648.

Marshall, E., 1999. Gene therapy death prompts review of adenovirus vector. Science 286, 2244–2245. PMID: 10636774.

Mason, J.M., Arndt, K.M., 2004. Coiled coil domains: stability, specificity, and biological implications. Chembiochem 5, 170–176. PMID:14760737.

Maturana, H.R., Varela, F.J., 1980. Autopoiesis and Cognition: The Realization of the Living. D. Reidel Publishing Company, Dordrecht, Holland.

Murzin, A.G., Brenner, S.E., Hubbard, T., Chothia, C., 1995. SCOP: a structural classification of proteins database for the investigation of sequences and structures. J. Mol. Biol. 247, 536–540. PMID:7723011.

Nallani, M., de Hoog, H.P., Cornelissen, J.J., Palmans, A.R., van Hest, J.C., Nolte, R.J., 2007. Polymersome nanoreactors for enzymatic ring-opening polymerization. Biomacromolecules 8, 3723–3728. PMID:17994700.

Naresh, M., Das, S., Mishra, P., Mittal, P., 2012. The chemical formula of magnetotactic bacterium. Biotechnol. Bioeng. 109, 1206–1216. Available from: http://dx.doi.org/10.1002/bit.24403. PMID:22170293.

Nielsen, A.A., Segall-Shapiro, T.H., Voigt, C.A., 2013. Advances in genetic circuit design: novel biochemistries, deep part mining, and precision gene expression. Curr. Opin. Chem. Biol. 17, 878–892. Available from: http://dx.doi.org/10.1016/j.cbpa.2013.10.003. PMID:24268307.

Nilsson, B.L., Soellner, M.B., Raines, R.T., 2005. Chemical synthesis of proteins. Annu. Rev. Biophys. Biomol. Struct. 34, 91–118. PMID:15869385.

Nomura, S-iM, Tsumoto, K., Hamada, T., Akiyoshi, K., Nakatani, Y, Yoshikawa, K., 2003. Gene expression within cell-sized lipid vesicles. Chembiochem 4, 1172–1175. PMID:14613108.

Orengo, C.A., Michie, A.D., Jones, S., Jones, D.T., Swindells, M.B., Thornton, J.M., 1997. CATH: a hierarchic classification of protein domain structures. Structure 5, 1093–1108. PMID:9309224.

Pandya, M.J., Cerasoli, E., Joseph, A., Stoneman, R.G., Waite, E., Woolfson, D.N., 2004. Sequence and structural duality: designing peptides to adopt two stable conformations. J. Am. Chem. Soc. 126, 17016–17024. PMID:15612740.

Park, M., Tsai, S.L., Chen, W., 2013. Microbial biosensors: engineered microorganisms as the sensing machinery. Sensors (Basel) 13, 5777–5795. Available from: http://dx.doi.org/10.3390/s130505777. PMID:23648649.

Rager-Zisman, B., 2012. Ethical and regulatory challenges posed by synthetic biology. Perspect. Biol. Med. 55, 590–607. Available from: http://dx.doi.org/10.1353/pbm.2012.0043. PMID: 23502567.

Rawls, R.L., 2000. Synthetic biology makes its debut. Chem. Eng. News 78, 49–53.

Ro, D.K., Paradise, E.M., Ouellet, M., Fisher, K.J., Newman, K.L., Ndungu, J.M., et al., 2006. Production of the antimalarial drug precursor artemisinic acid in engineered yeast. Nature 440, 940–943. PMID:16612385.

Romualdi, C., Vitulo, N., Del Favero, M., Lanfranchi, G., 2005. MIDAW: a web tool for statistical analysis of microarray data. Nucleic Acids Res. 33 (Web Server issue), W644–W649. PMID: 15980553.

Saha, S., Ansari, A.Z., Jarrell, K.A., Ptashne, M., 2003. RNA sequences that work as transcriptional activating regions. Nucleic Acids Res. 31, 1565–1570. PMID:12595565.

Schmidt, M., Ganguli-Mitra, A., Torgersen, H., Kelle, A., Deplazes, A., Biller-Andorno, N., 2009. A priority paper for the societal and ethical aspects of synthetic biology. Syst. Synth. Biol. 3, 3–7. Available from: http://dx. doi.org/10.1007/s11693-009-9034-7. PMID: 19816794.

Schmidt, M., Porcar, M., Schachter, V., Danchin, A., Mahmutoglu, I., 2012. Biofuels. In: Schmidt, M. (Ed.), Synthetic Biology: Industrial and Environmental Applications. Wiley-VCH Verlag GmbH & Co. KGaA, Weinheim, Germany. http://dx.doi.org/10.1002/9783527659296.ch1 ISBN 3-527-33183-2.

Sedlak, M., Ho, N.W., 2004. Production of ethanol from cellulosic biomass hydrolysates using genetically engineered Saccharomyces yeast capable of cofermenting glucose and xylose. Appl. Biochem. Biotechol. 114, 403–416. PMID:15054267.

Serrano, L., 2007. Synthetic biology: promises and challenges. Mol. Syst. Biol. 3, 158. http://dx.doi.org/10.1038/msb4100202; PMID: 18091727; PMCID: 2174633.

Siuti, P., Yazbek, J., Lu, T.K., 2013. Synthetic circuits integrating logic and memory in living cells. Nat. Biotechnol. 31, 448–452. http://dx.doi.org/10.1038/nbt.2510; PMID:23396014.

Stano, P., 2010. Synthetic biology of minimal living cells: primitive cell models and semi-synthetic cells. Syst. Synth. Biol. 4, 149–156. PMID: 21886680.

Stano, P., Wehrli, E., Luisi, P.L., 2006. Insights on the oleate vesicles self-reproduction. J. Phys. Condens. Matter. 18, S2231–S2238. Available from: https://doi.org/10.1088/0953-8984/18/33/S37.

Stone, M., 2006. Life Redesigned to suit the engineering crowd. Microbe 1, 566–570.

Suel, G.M., Kulkarni, R.P., Dworkin, J., Garcia-Ojalvo, J., Elowitz, M.B., 2007. Tunability and noise dependence in differentiation dynamics. Science 315, 1716–1719. PMID:17379809.

Szostak, J., 1988. Structure and activity of ribozymes. In: Benner, S.A. (Ed.), Redesigning the Molecules of Life. Springer, Heidelberg, pp. 87–113.

Szostak, J.W., Bartel, D.P., Luisi, P.L., 2001. Synthesizing life. Nature 409, 378–390.

Szybalski, W., 1974. In vivo and in vitro initiation of transcription, pp 405, and Discussion pp. 404–405; Szybalski's Concept of Synthetic Biology, 411–412, 415–417 In: Kohn, A., Shatkay, A. (Eds.), Control of Gene Expression, pp. 23–24. Plenum Press, New York.

Szybalski, W., Skalka, A., 1978. Nobel prizes and restriction enzymes. Gene 4, 181–182. Available from: http://dx.doi.org/10.1016/0378-1119(78)90016-1. PMID:744485.

Tanaka, H., Yi, T.M., 2009. Synthetic morphology using alternative inputs. PLoS One 4 (9), e6946. PMID:19746161.

Taylor, W.R., 2002. A 'periodic table' for protein structures. Nature 416, 657–660. PMID:11948354.

Thastrom, A., Lowary, P.T., Widlund, H.R., Cao, H., Kubista, M., Widom, J., 1999. Sequence motifs and free energies of selected natural and non-natural nucleosome positioning DNA sequences. J. Mol. Biol. 288, 213–229. PMID:10329138.

Tian, J., Gong, H., Sheng, N., Zhou, X., Gulari, E., Gao, X., et al., 2004. Accurate multiplex gene synthesis from programmable DNA microchips. Nature 432, 1050–1054. PMID:15616567.

Tigges, M., Marquez-Lago, T.T., Stelling, J., Fussenegger, M., 2009. A tunable synthetic mammalian oscillator. Nature 457, 309–312. Available from: http://dx.doi.org/10.1038/nature07616. PMID:19148099.

Torino, D., Martini, L., Mansy, S.S., 2013. Piecing together cell-like systems. Curr. Org. Chem. 17, 1751–1757. PMID: 24348089.

Tumpey, T.M., Basler, C.F., Aguilar, P.V., Zeng, H., Solórzano, A., Swayne, D.E., et al., 2005. Characterization of the reconstructed 1918 Spanish influenza pandemic virus. Science 310, 77–80. PMID: 16210530.

Walter, K.U., Vamvaca, K., Hilvert, D., 2005. An active enzyme constructed from a 9-amino acids alphabet. J. Biol. Chem. 280, 37742–37746. PMID:16144843.

Wang, M., Augusto Benedito, V., Zuechun Zhao, P., Udvardi, M., 2010. Inferring large-scale gene regulatory networks using a low-order constraint-based algorithm. Mol. Biosyst. 6, 988–998. Available from: http://dx.doi.org/10.1039/b917571g. PMID:20485743.

Widmaier, D.M., Tullman-Edcek, D., Mirsky, E.A., Hill, R., Givindarajan, S., Minshull, J., et al., 2009. Engineering the Salmonella type III secretion system to export silk monomers. Mol. Syst. Biol. 5, 309. Available from: http://dx.doi.org/10.1038/msb.2009.62. Epub 2009 Sep 15. PMID: 19756048.

Winkler, W.C., Breaker, R.R., 2005. Regulation of bacterial gene expression by riboswitches. Annu. Rev. Microbiol. 59, 487–517. PMID:16153177.

Winnacker, M., Kool, E.T., 2013. Artificial genetic sets composed of size-expanded base pairs. Angew. Chem. Int. Ed. Engl. 52, 12498–12508. Available from: http://dx.doi.org/10.1002/anie.201305267. PMID:24249550.

Wohler, F., 1828. Ueber kunstiliche Bildung des Harnstoffs. Ann. Phys. Chem. 12, 253–256.

Yao, S., Ghosh, I., Zutshi, R., Chmielewski, J., 1997. A pH modulated, self-replicating peptide. J. Am. Chem. Soc. 119, 10559–10560.

Yin, H., Slusky, J.S., Berger, B.W., Walters, R.S., Vilaire, G., Litvinov, R.I., et al., 2007. Computational design of peptides that target transmembrane helices. Science 315, 1817–1822. PMID:17395823.

Zahnd, C., Wyler, E., Schwenk, J.M., Steiner, D., Lawrence, M.C., McKern, N.M., et al., 2007. A designed ankyrin repeat protein evolved to picomolar affinity to Her2. J. Mol. Biol. 369, 1015–1028. PMID:17466328.

Weblinks

http://scop2.mrc-lmb.cam.ac.uk/
http://pfam.sanger.ac.uk/search/sequence
http://www.ung.igem.org/
http://www.syntheticbiolgy.org
http://www.sysbio.med.harvard.edu/
http://www.systemsbiology.org/
http://www.oecd.org/sti/biotechnology/synbio-
http://www.bioethics.gov/documents/synthetic-biology/PCSBI-Synthetic-Biology-Report-12.16.10.Pdf

Reverse Engineering and Its Applications

Attya Bhatti, Nida A. Syed and Peter John

National University of Sciences and Technology (NUST), Islamabad, Pakistan

5.1 INTRODUCTION

Engineering is defined as the science of designing, manufacturing, constructing, and sustaining products, systems, and structures. The first of the two primary methods applied in engineering is forward engineering, which trails a customary methodology starting from the highest level abstractions and logical strategies to the physical application of a system. This, in turn, may be without any technicalities such as drawings, BoM (bills-of-material), or any thermal and electrical features.

The second method, reverse engineering, is often characterized as the reversal of the first method of engineering, initiating with the replication of a prevailing component, sub-assembly, or product itself without the facilitation of drawings, documentation, or computer modeling. It performs comprehensive analyses of components of a given system in the following ways:

1. Identifying a system's components and their interdependence.
2. Creating polymorphic or advanced conjectural illustrations of a system.
3. Creating physical representations of a system

Most engineering platform, whether mechanical, electrical, software, or chemical, has some sort of reverse engineering being applied to it. Competing manufacturers often use the second methodology as a learning curve for either duplication or enhancement.

Reverse engineering may be applied by a chemical manufacturer to outgrow a patent on a rival's manufacturing process. It is highly practiced in car manufacturing as well, where competitors yearn to reengineer a leading production vehicle and introduce their version with improved features. A working source code often works as a benchmark in software engineering. An integral part of the reverse engineering process is the development of a CAD model using the information from the physical model of the product. This is also denoted as the part-to-CAD process. Reverse engineering has emerged as a specifically attractive approach as it greatly reduces the time duration of product manufacture. Rapid product development is the process of developing such technologies and techniques that enable manufacturers and designers in meeting the demands of reduced product development time. By using reverse engineering, a three-dimensional product or model can be quickly imaged in digital form, remodeled, and outsourced for rapid prototyping/ tooling or rapid manufacturing.

In reverse engineering, duplication of an existing component is made possible by attaining its physical dimensions, features, and material features. However, before embarking on such a trail, a carefully planned life-cycle and cost/benefit analysis must be carried out to substantiate the reverse engineering projects. Such a venture can only be cost-effective if the items will be manufactured on a large scale with high financial investment. On the contrary, even if it is not deemed cost-effective, vitality of the part to the functioning of a system is considered to be enough validation for undertaking its reverse engineering.

With technological advances in both Systems Engineering and Systems Biology, reverse engineering has made in-ways into the life sciences as well. This chapter describes its applications in various biological contexts such as bioinformatics, biosystems, software design, gene regulatory networks, medical device design, pharmaceutical drug design, and therapeutic peptide production.

5.2 APPLICATIONS OF REVERSE ENGINEERING

5.2.1 Medical Device Design

This particular application comes under the field of Medical Reverse Engineering (MRE) in which the production methodology is governed by the end use application of the technology. The state-of-the-art end-use applications of MRE include personalized implants for bone reconstruction, dental implants and simulations, surgical tools, medical training, vision science and optometry, orthopedics, ergonomics, orthosis, prosthesis, and tissue engineering. The production methodology of MRE comprises four distinct phases: MRE inputs, data acquisition, MRE data processing, and biomedical application development and research.

MRE input is the most crucial first step as it determines the techniques used for data acquisition, its processing, and analysis as well as medical application and research. It also defines the requisite accuracy level with which 3D models need to be generated for further medical development. However, it is pertinent to mention that the MRE inputs are also dependent on the end use application.

MRE information procurement is basically of two types: contact based and non-contact based. Contact-based strategies make utilization of sensing devices with mechanical arms, Coordinate Measurement Machines, and Computer Numerical Control machines, to digitize a surface. Non-contact-based strategies obtain the 2D cross-sectional pictures and point clouds, illustrative of the geometry of an object, by projecting energy sources (light, sound, or magnetic fields) onto an object and watch the transmitted or the reflected energy. The geometrical information of an object is ascertained by utilizing triangulation, time of flight, wave-obstruction data, and picture preparing calculations. Due to the complex nature of object geometries and shapes of anatomical structures or biomedical objects, contact-based techniques have lesser utility compared to non-contact-based techniques as they often cause distortion of soft objects due to contact pressure exerted by the probe and data collection is rather slow. However, there are some indispensable advantages that are associated with contact-based techniques: its level of precision is unmatched as it can even measure deep slots and pockets and is insensitive to color or transparency, is highly accurate, and bears a low cost. In contrast, non-contact-based techniques also claim the following advantages: there is no physical contact hence faster digitizing of significant volumes, good accuracy and resolution for mainstream applications, ability to detect colors and scan highly detailed objects where mechanical touch probes may be too large to accomplish the task. Potential limitations for data acquisition using non-contact-based techniques include lower accuracy of data obtained from colored or transparent or reflective surfaces.

MRE data processing differs on the basis of the kind of input that is being used for data acquisition, point clouds, or 2D slice images. Point clouds from various scans are combined and aligned to arrange all the point clouds in a series and proper orientation with respect to each other in a common coordinate system. Specialized image processing tools and packages are required for image processing for 3D data reconstruction of the hard and soft tissues or objects of interest.

In both cases, the output is 3D triangle mesh models or 2D contours of the region of interest or anatomical structures, which are finally optimized, manipulated, and controlled or converted into 3D Computer Aided Design models to meet the requirements from the end-use applications.

Despite having well-documented benefits, this technology has not been transferred to health care facilities for large-scale application in diagnosis and treatment. Complex design, lack of interdisciplinary communication, and collaboration and high cost of technology and investment are frequently cited as the reasons for this lack of widespread application.

5.2.2 Pharmaceutical Product Design

In the pharmaceutical industry, Reverse Engineering is commonly used for the characterization of pharmaceutical drugs having unknown formulations or for comparing two drug formulations, often in conjunction with Raman spectroscopy. This is also referred to as Pharmaceutical Deformulation. It often leads to the identification, quantification, and characterization of active pharmaceutical ingredients and all the excipients, both in the core and coating of a drug formulation. This can also be used to develop a reformulated product having superior bioequivalence; which is a prerequisite that needs to be qualified to apply for generic approval under ANDA (abbreviated new drug approval) to FDA.

Apart from Raman Spectroscopy, a wide array of techniques is now being employed in Pharmaceutical Deformulation such as Fourier Transform Infrared Spectroscopy, Thermogravimetric Analyzer, Pyrolysis Gas Chromatography/Mass Spectroscopy (Pyro GC/MS), Flame Ionization Detector (FID), Karl Fisher (an instrument used to identify the amount of water that is within a sample), Scanning Electron Microscope/Energy Dispersive X-Ray Spectroscopy (SEM/EDXA), X-Ray Diffraction (XRD), Liquid Chromatography/Mass Spectrometry (LC/MS), Nuclear Magnetic Resonance, and High-Performance Liquid Chromatography. Similar pharmaceutical analyses can also be performed in the following contexts:

- Troubleshoot product defects through a failure analysis report.
- Compare competing products at a very detailed level.
- Discover the active and inactive ingredients within a product.
- Identify the purity of a raw material.
- Determine the identity of an unknown material or contaminant.

5.2.3 Reverse Engineering in Therapeutic Peptide Production

Therapeutic peptide production greatly relies on the knowledge of biosynthetic pathways that are involved in the indigenous production of these peptides, and reverse engineering of these pathways has been demonstrated in certain studies. So far, using reverse engineering techniques, physicochemical origins of potency and broad spectrum action of antimicrobial peptides have been demonstrated on the basis of structure–activity relationships' study of de novo peptides that are typically analogues of putative peptides. This approach makes it possible to not only characterize the physicochemical characteristics of antimicrobial peptides but also to study the interplay between different biological factors parameters that regulate antimicrobial activity. Investigation of physicochemical characteristics like charge and helicity of a peptide gives insight into the biological activity that it possesses and facilitates the development of synthetic analogues based on amino acid sequence variation (Anantharaman and Sahal, 2010).

Another example of this approach comes from a study that was aimed at reverse engineering the biological synthesis pathway of polyunsaturated fatty acids in transgenic plants, which are typically found in aquatic microorganisms. This approach provides a more reliable and sustainable source of long-chain polyunsaturated fatty acids (LC-PUFA) as the marine stocks are declining rapidly. This comprises identification and characterization of the biosynthetic pathway in an appropriate LC-PUFA-producing organism, followed by the transfer of the genes encoding the primary biosynthetic enzymes of LC-PUFA into a heterologous host such as a transgenic plant (Napier et al., 2004).

5.2.4 Reverse Engineering in Bioinformatics

Genome projects are allowing us to rapidly list the genes and proteins that affect cellular behavior. Moreover, advances in oligonucleotide and cDNA microarray technologies have made it possible, on a genome-wide scale, to quantify gene expression levels (Lockhart and Winzeler, 2000). Study of transcriptional regulation of gene expression has shown that many genes encode regulatory proteins that either activate or suppress other genes, giving rise to complex gene regulatory networks (Dickson et al., 1975; Jacob and Monod, 1961). Therefore, at the genetic level, cellular processes can be considered analogous to complex electrical circuits (Vogelstein et al., 2000). In the context of this analogy, there exists a need to deduce equations that can quantify information within the circuit. One approach that can be used is forward engineering, which involves designing simple circuits and testing them with respect to equations derived based on the underlying biochemistry (Elowitz and Leibler, 2000; Gardner et al., 2000). However, data collected from simultaneous measurement of expression of various genes generate catalogs of patterns of gene activity under several biochemical (DeRisi et al., 1997) and physiological (Wen et al., 1998) circumstances that necessitate reverse engineering of genetic networks. This involves inferring the connectivity of a multifarious genetic network that brings about an enormous set of expression data from experimental measurements.

In this context, research in the field of reverse engineering is primarily focused on extracting the topology of genetic networks, which is, deducing the connectivity of genes. This approach requires an enormous amount of experimental data and computational resources. Typically, a large number of network architectures are consistent with a given set of expression data which increases the complexity of the task. Therefore, studies have targeted small networks using genetic algorithms, neural networks, and Bayesian models (D'Haeseleer et al., 2000; Hartemink et al., 2001; Wahde and Hertz, 2000). Recent studies have been focusing on developing methods to circumvent problems associated with data shortage and computational inefficiency to recover network topology of large-scale networks. Scientists have adopted linear models and used singular value decomposition (Yeung et al., 2002) and have also developed an iterative reverse engineering approach (Tegner et al., 2003). In a recent study, importance of considering network behavior as well as network structure while inferring gene networks was indicated, and an integrated approach was developed that includes a parameter identification procedure and a parameter optimization procedure. The two procedures work iteratively to search sensitive parameters, determine their value ranges, and search for the best fitting solutions to infer networks that satisfy both requirements simultaneously (Hsiao and Lee, 2014).

Generally speaking, in terms of systems biology, reverse engineering addresses situations where an observation of dynamic properties of a biological system has been made but the mechanisms governing these properties are unknown. A comprehensive in silico quest is performed to enlist all possible network topologies leading to the particular dynamic property, and the most compatible network topologies are identified. Individual topology search is a basic reverse engineering procedure in which each network topology is modeled against a collection of random parameter sets, and the robustness of each topology is evaluated (Ma et al., 2006, 2009; Yao et al., 2011). To avoid modeling individual network topologies separately, another approach was recently developed that searches the continuous parameter space and defines an assembly of individual topologies derived from the full network (Fu et al., 2012). There is an overall consistency in the results of the two studies that suggests that the common reverse engineering principle is effective (Mondal et al., 2014).

5.2.5 Reverse Engineering in Biosystems

Systems biology approaches are now being utilized to understand the comprehensive and quantitative details of biological systems. Advances have been made, which allow genome scale models to be built promptly using biological catalogs and system-wide molecular dimensions (Ashworth et al., 2012).

Various approaches can be employed, which generate organism-scale models combining statistical associations, causative abstractions, and known molecular mechanisms to delineate and predict quantitative and complex phenotypes. This is known as top-down reverse engineering approach (Ashworth et al., 2012).

High-throughput technologies have also played a major role in the field of reverse engineering, as their application to -omics data (genomics, proteomics, and metabolomics) may result in new acumens in to biological systems, which is not possible from low-throughput platforms.

Data-driven approaches, also known as data-mining, are being used for enormous volumes of biochemical data at molecular level, to extract patterns from the data to unravel new information (Hunter and Borg, 2003; Kidd et al., 2008; Tweedie et al., 2009). Whereas, design-driven approaches, also referred to as systems modeling, are used to simulate nascent properties of biological systems (Kitano, 2002), including motif distribution (Kaluza et al., 2008), scale-free organization (Jeong et al., 2000), and feedback (Brandman and Meyer, 2008).

Therefore, it can be inferred that combining high-throughput data mining with design-driven systems modeling for biochemical pathways and cells can give rise to a new era of reverse engineering biological systems, which is referred to as "reverse engineering biomolecular systems" (Quo et al., 2012).

With the rapid increase in the quantity of gene expression data, large time series data sets for measurements of expression levels from different tissue types and different organisms are now available, which can be used to determine regulatory interactions between genes. Studies have shown that reverse engineering of genetic regulatory networks is able to present an accurate representation of the parameters of a network, if measurements are in the form of several time series (Wahde and Hertz, 2000).

In scenarios where no strictures are available for in silico modeling, techniques that calculate parameters are useful. Initially, scientists proposed a technique for the dynamic modeling of complicated network structures by combining a genetic algorithm and the S-system (Kikuchi et al., 2003). However, with recent advances, methods emerged that use genetic programming to predict biochemical equations from only time-course information that can instantaneously and precisely predicted the equation topologies and numerical parameters (Sugimoto et al., 2005).

A fundamental problems in functional genomics is the production of functional predictions of metabolic networks and gene regulatory circuits by learning the combination of genome sequence data with gene expression data and previous biological knowledge (DeRisi et al., 1997; Hartemink et al., 2001).

Information about whole-genome gene expression data may in principle help us elucidate the transcriptional level of gene regulation, i.e., gene regulatory networks, in a global manner. The dissection of these complex molecular networks is the ultimate aim of "reverse engineering" in molecular biology. The goal is to develop methods that are able to rank alternative hypotheses about the structure of a regulatory network, given a set of gene expression arrays under a certain experimental design. Reverse engineering algorithms (REAs) enable us to do just that. Results show that it is not necessary to determine all parameters of the genetic network to rank hypotheses. The ranking process is easier the more experimental environmental conditions are used for the data set. During ranking, numerous fixed parameters undergo a concomitant increase with the number of environmental conditions, while some errors in the hypothetical network structure may pass undetected, due to a maintained dynamic behavior (Repsilber et al., 2002).

5.2.6 Reverse Engineering in Software Design

Reverse engineering is the systematic teardown and analysis of what lies under the hood of a product. It typically provides an understanding of how a device works as well as what intellectual property (IP) went into its creation. RE supports strategy and decision-making across the entire IP lifecycle, from concept creation to product retirement. It also enables a company to protect its own IP as well as ensure that it has a rock-solid defense against allegations of patent infringement from competitors. RE is also a proven method to determine if a competitor's comparable product is implementing patented inventions, benchmark one's self against the market, and identify technology trends and key innovation impacting market dynamics.

5.2.7 Reverse Engineering of Software

Software reverse engineering can provide an understanding of behaviors, operations, or functions in products; make it technically viable to enforce IP licensing of system-level and software-based patents; and expand IP licensing opportunities, particularly as the consumer and medical device worlds begin to overlap. Because software has become a bigger part of medical products, it is important to understand some of the factors that differentiate it from hardware, and that can make reverse engineering of the products more complex. Those factors include:

- Software is a hidden part of the product that isn't subject to an evident manufacturing process. Thus disassembly is much more complex than simply removing screws or decapping semiconductor packages to see what is inside.
- Software has a complexity based on its ability to branch, based on different inputs, that makes it more complex and difficult to track than the circuit traces on a board or integrated circuit. Software is reconfigurable, which can be a great advantage when resolving bugs or adding new capabilities via an update, but it can change the product's functionality dramatically based on the new code. This flexibility is a particularly difficult feature in the highly regulated medical device world.
- Although software has begun to be more reusable and standard modules exist, it remains a much more independent design realm than hardware, and relatively small changes in the code can make a big difference in product performance.

Some of the tools available to aid in this effort include firmware extraction, binary code inspection, and static code analysis techniques such as software disassembly and software decompiling. Dynamic analysis methods include custom test application development, I2C/JTAG capture, and code trace and debugging at the application and operating system level.

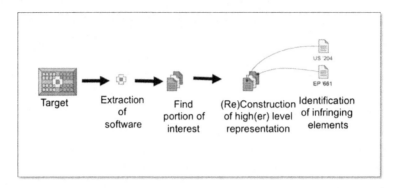

FIGURE 5.1   This flow chart depicts the process of software reverse engineering.

To demonstrate the software reverse engineering process, consider a couple of examples of its use in addressing possible patent infringement. The process flow chart is depicted in Fig. 5.1:

- The first step after acquisition of the target device is the extraction of the software from the device, code update, installer, etc.
- The analyst then assesses the software by identifying the specific code of interest, and then performing a reconstruction of a higher level representation of the software.
- Finally, the potential infringing elements are identified based on this sequence of operations.

Example: An analysis of a smart phone application.

It is an area of interest in medicine due to the rapid increase in health and fitness as well as medical applications. With the rapid proliferation in these software products, the probability of encroaching upon someone else's IP has increased as well.

To understand how the application applies the sensor data to calculate orientation of the displayed image, full software Reverse Engineering was performed. The application was downloaded, and then the app and OS framework were unpacked. The compass and map apps as well as the orientation and image rendering frameworks for analysis were disassembled. A detailed code analysis identified the relevant code segments, strings, and data tables. A decompilation of the disassembled code into native C code was done using an ARM decompiler to create a high-level diagram of relevant functional elements. By linking the functional testing of the app to the code analysis, it will conclusively determine whether the claim elements of the IP mapped to this app. It was also determined how this patent might apply to similar apps (Fig. 5.2).

In summary, medical devices continue to evolve, particularly as more commonly used off-the-shelf hardware is utilized. That tendency alone will drive a further increase in the importance of algorithms and software as key product differentiators. However, the rapid proliferation of smart phones and their attendant apps will further complicate this situation.

Software reverse engineering is a proven method to protect not only your company's IP but also to identify possible licensing opportunities through identification of potential infringement. It should be part of any company's strategic toolbox.

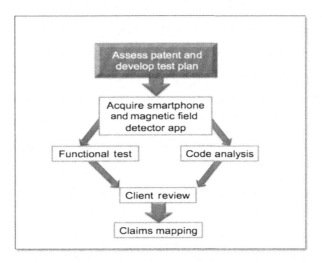

FIGURE 5.2   Outlines the general high-level approach to smart phone app analysis.

5.2.8 Reverse Engineering Human Regulatory Networks

Gene regulatory models are a graphical representation of molecular interactions, whereby nodes represent the individual product of the genes and arcs represent their interactivity. A gene regulatory model typically allows dependent variables to represent the yield of a regulatory process in the cell as a function of independent variables that are reflective of the input of the regulatory process. Examples of dependent variables include expression of the target gene of a transcription factor (TF) or the sustainability of a cellular phenotype, and examples of independent variables come from the gene expression or protein abundance of the TF, as well as small-molecule, RNAi, or environmental perturbations. A good regulatory model has an inherent capability to generate such a hypothesis that can be experimentally validated. Computational models are only relevant if their theoretical predictions can be authenticated in an experimental setting in the biological context. A systems biologist can only be regarded as successful if he possesses a sound understanding of both the computational principles essential to generate key testable hypotheses and the experimental techniques required for their authentication. This emerging paradigm highlights the highly interdisciplinary nature of this field (Lefebvre et al., 2012).

There are different incarnations of gene regulatory models, the simplest being the *noncausal or topological interaction model*. This is a unidirectional graph of physical interactions and nonphysical interactions between proteins. These models constitute an indispensable means for statistical integration of multiple data inputs from different gene products, whose effects converge together to bring about a specific cellular phenotype (Ideker and Sharan, 2008; Costanzo et al., 2010). Second type of gene regulatory network is an *influence map* that represents both direct and indirect fundamental monitoring relationships among gene products in a cell. These are harder to assemble, as causality is relatively difficult to infer from molecular data but gives invaluable insight into cause−effect mechanisms between various physical and nonphysical interactions that are intrinsically causal in nature. *Physical regulatory maps* are yet another type and represent a more complex regulatory model that is restricted to physical interactions that can be experimentally validated, such as binding of a TF to its promoter sequence and activating or repressing its expression. This usually involves predictions made on the basis of an algorithm and validation using an experimental technique (Basso et al., 2005). *ODE (ordinary differential equation) Kinetic models* further add another layer of detail by allowing quantitative modeling of causal interactions undergone by an endogenous variable for example mRNA concentration of a gene with other endogenous and exogenous variables over time (Polynikis et al., 2009). *PDE (partial differential equation) Kinetic models* are less tractable than ODE models and represent the concentration gradient of each important molecular species as a function of time, space, and intracellular or extracellular variables. These models are specifically employed for molecular species that demonstrate to have a concentration gradient rather than a fixed concentration in space (Newman et al., 2008). *Stochastic models* have been developed to take into account specific molecular species that have low availability and therefore cannot be accounted for an unwavering concentration. These models are intensively computational and complex (Gillespie, 2007). These types and their salient features are also summarized in Table 5.1.

Gene regulatory networks governing over gene products have three distinct functional layers: transcriptional, posttranscriptional, and posttranslational. Regulation at transcriptional interactions determines regulation of mRNA transcription by DNA-binding proteins

TABLE 5.1 Types and Features of Gene Regulatory Models

1. Noncausal or Topological Interaction Model	• Simplest model. • Unidirectional graph of physical interactions among various proteins in complexes or nonphysical synthetic lethality interactions.	Ideker and Sharan, (2008); Costanzo et al. (2010)
2. Influence Map	• Represents both direct and indirect underlying regulatory interactions between gene products in a cell. • Gives invaluable insight into cause–effect mechanisms between various physical and nonphysical interactions that are intrinsically causal in nature.	Basso et al. (2005)
3. Physical Regulatory Maps	• Complex model that is limited to physical interactions based on algorithms that can be experimentally validated.	Basso et al. (2005)
4. ODE Kinetic Models	• Detailed in nature. • Employed for modeling causal interactions undergone by an endogenous variable with other endogenous and exogenous variables over time.	Polynikis et al. (2009)
5. PDE Kinetic Models	• Less tractable than ODE models. • Specifically employed for molecular species that possess a concentration gradient rather than a fixed concentration in space.	Newman et al. (2008)
6. Stochastic Models	• Intensively computational and complex models. • Takes into account molecular species that are rare and cannot be represented as a continuous concentration.	Gillespie (2007)

called TFs, their cofactors, and their modulators, which may or may not bind DNA directly. Posttranscriptional interactions regulate mRNA stability and translation, via microRNA and other noncoding RNAs. Posttranslational interactions include both transient protein–protein interactions (PPIs) involved in signal transduction and stable PPIs involved in stable molecular complex formation, such as those in ribosomal subunits. Lately, another nonphysical level has gained significant relevance to the study of synergistic trait regulation and borders on synthetic lethal and synthetic functional interactions (Bandyopadhyay et al., 2010).

These regulatory networks are highly specific to a given cellular context and hence are not universally applicable to all molecular mechanisms.

Reverse engineering of gene regulatory networks depends on the following factors:

1. Indirect rather than direct proof of physical interactions. Here, it is pertinent to mention that simply because a physical interaction between DNA and protein has been reported does not necessitate that it is being regulated as well.
2. Accessibility of numerous measurements over a broad range of concentrations of appropriate biological components and multiple observations of both dependent and independent variables.
3. Availability of genome-wide gene expression profiles.

Reverse Engineering of transcriptional interactions has been reported using yeast and prokaryotes as model organisms with favorable results as a manifestation of the predicted interactions. REAs have also been developed and validated in mammalian cells (Lamb et al., 2006). Although the number of proposed algorithms clearly exceeds those which have been validated, out of the five putative approaches only two have been validated in mammalian cells (Carro et al., 2010). Contrary to this scenario, reverse engineering of signal transduction networks has seen very little progress, specifically in the context of posttranslational modifications like phosphorylation, acetylation, and ubiquitination (Zeitlinger et al., 2003). This is also corroborated by the limitation of the number of experimentally validated algorithms for the analysis of signaling complexes in mammalian cells (Linding et al., 2007).

Posttranscriptional regulation networks mostly emphasize on the regulatory function of short (16–21 bp) RNA molecules denoted as mature microRNAs that originate from the highly specific, DICER-mediated cleavage of short RNA hairpin structures resulting from the DROSHA-mediated processing of longer precursor RNA molecules. This processing is reinforced by RISC-complex–mediated partial hybridization of the mature microRNA molecules with complementary sites in the 3′ UTR region of messenger RNA encoding specific gene transcripts. Such a process, which leads to both reduction in transcript translation efficiency and transcript degradation, ultimately reduces the concentration of the corresponding protein isoforms. In this particular context, REAs rely primarily on available gene expression data and sequence information for the inference of miRNA targets (Lewis et al., 2003).

PPIs in stable complexes of macromolecules are involved in the regulation of virtually every process in the cell. Human disease often ensues when the interactions between these protein complexes and their respective targets are perturbed. Large-scale identification of PPIs typically makes use of experimental techniques such as yeast 2 hybrid (Y2H) technology and tandem affinity purification coupled with mass spectrometry (Volkel et al., 2010, Venkatesan et al., 2009). The main downside associated with these ex vivo techniques is the high rates of false-negatives and false-positives, which also skews the cellular contexts that are being assessed and provide a conceptual insight into the phenotypic interactions that take place between various components of protein—protein complexes. Thus these have low inferential power and rely completely on genome-wide linkage and association studies (Lage et al., 2007).

5.3 LAWS, ECONOMICS, AND ETHICS GOVERNING REVERSE ENGINEERING

Reverse engineering has a long-standing history of being an acceptable practice. Lawyers and economic experts have both endorsed it as an appropriate way of obtaining information about another product, even if the endpoint is to make a new product that is directly competing with the first product and may affect its market value/share. Although largely acceptable, reverse engineering has been target of various legal framework conflicts in the past two decades, which have also seen a concomitant rise in the practice of reverse engineering.

The legislature governing reverse engineering drawn up in the United States defines reverse engineering as "starting with a known product and working backwards to devise the process which was followed for its development or manufacture" and terms it as "always been a lawful way to acquire a trade secret, as long as acquisition of the known product is by fair and honest means, such as purchase of the item on the open market."

The legitimate "right" to figure out a competitive innovation, that is, to dismantle an item to find data about the item's organization and how to make it (that is, connected mechanical expertise) is so settled that neither courts nor analysts have for the most part saw a need to clarify the reason for this regulation. Courts have likewise regarded figuring out as a vital variable in keeping up equalization in licensed innovation law. Government patent law permits pioneers to have up to a quarter century selective rights to make, utilize, and offer the invention, but just in return for divulgence of noteworthy insights about their innovations to people in general.

However, reverse engineering rights do not exist as such in patent law, as patent law demands the specification of all the relevant technical details regarding the design, development, and manufacturing process of a product before it can be patented. As a result, the information is publicly available and can be of utility for reverse engineering but does not indicate infringement of a patent.

Economic impact of reverse engineering is dependent on a couple of dynamics, starting with the intended aims and objectives which are to be achieved, the manufacturing context for which it is undertaken, and the cost and time that will be incurred. It also includes the lead time that the original innovator/developer has before the reverse engineered product is developed and introduced into the market, whether the reverse engineered product is licensed for innovation, and the further utilization of information yielded by the process. In the traditional manufacturing context, it is the most frequent and fiscally significant driving force behind reverse engineering. An economic assessment of the effects of reverse engineering undertaken to develop a contending manufactured commodity must take into account that reverse engineering reflects one of the four staged product development process. A second comer's development progression as a whole will generally be sufficiently time-consuming, difficult, and costly that an innovator will have significant de facto lead time protection within which to recover Research and Development overheads. In this regard, the first stage is the awareness stage where by a second comer will gain awareness and recognition of the product that reflects itself worthy enough of the time, expense, and effort that needs to be put into reverse engineering. Second stage is where the reverse engineering takes place itself, when a second comer attains the innovator's product and starts to take it to pieces and analyze it to gain an understanding of its makings and workings. This entire step may be extravagant in terms of finances, consumes a lot of time and pose a challenge, and depends greatly on the skill and expertise of the engineer performing the reversal task. However, in practical terms, a reverse engineer might just spend lesser time and energy discerning the workings of a product than the actual innovator. This advantage comes from the ability to avoid wasteful experimentation and its associated costs that yield no results just by learning from the mistakes of the innovator. In addition, advancement in technologies also greatly facilitates him, reducing the costs of rediscovery over time. This is followed by the stage of implementation. All the knowledge a reverse engineer has gained will now be utilized and put to use in the design and manufacture of a competitive product. This ranges from making prototypes and testing them during experimentation, reorganizing production facilities, and reiteration of the manufacturing process until it yields an agreeable product. In certain cases, need may arise to go back to the previous stage if it becomes ostensible that a key piece of information or manufacture has remained elusive. Such realizations are typically encountered in the implementation phase. Information obtained may provide valuable insights into novel and innovative means of developing the product, even though it makes significant additions to the financial expenses being incurred.

The fourth and final stage is the introduction of the product into the market. It remains to be seen if the product will reduce the market share for the innovator's product either via competitive pricing or better manufacturing quality because it is under the influence of various market factors.

An economic assessment of reverse engineering should balance four factors:

1. the effects of reverse engineering on incentives to innovate,
2. its effects on prices,
3. its potential for creating incentives to invest in improved products, and
4. its potential for duplicated or other socially wasteful expenditures of resources.

There are considerable economic beneficial effects such as creation of competition in the marketplace, resulting in lower prices and inspires innovators to bring together further novelties into the market. It will almost promote the practice of licensing among innovators, and this licensing should be beneficial for both innovator and potential reverse engineers. Licensing can achieve the same knowledge sharing and market outcomes as reverse engineering without incurring any sort of costs or financial liability (Samuelson and Scotchmer, 2002).

5.4 CONCLUSION

Reverse engineering has emerged as a diverse field that has found several biologically relevant applications due to recent advances in the field of Systems Biology, which have yet to become accessible to the general public as is the case with reverse engineered products in other fields. A major impedance in this regard is the stringent rules and regulations governing any therapeutic approach or product that is to provide amelioration to human health under pathogenic conditions and the clinical trials of such products travailing across years.

5.4 FUTURE DIRECTION

The biggest challenge in reverse engineering to date is the convergence of approaches and technologies from a vast array of fields that have recently undergone an exponential increase. This is greatly facilitated by the digitization and reverse engineering of CAD, which has become the obvious choice for design and manufacturing purposes. However, it is of paramount importance that reverse engineering does not remain a niche market and undergoes necessary evolution to cater the needs of various different kinds of applications. It is also pertinent that reverse engineering undergoes a necessary shift from CAD-driven to knowledge-driven approaches to overcome obstacles faced in current approaches.

References

Anantharaman, A., Sahal, D., 2010. Reverse engineering truncations of an antimicrobial peptide dimer to identify the origins of potency and broad spectrum of action. J. Med. Chem. 53, 6079–6088.

Ashworth, J., Wurtmann, E.J., et al., 2012. Reverse engineering systems models of regulation: discovery, prediction and mechanisms. Curr. Opin. Biotechnol. 23, 598–603.

Bandyopadhyay, S., Chiang, C.Y., Srivastava, J., Gersten, M., White, S., Bell, R., et al., 2010. A human MAP kinase interactome. Nat. Methods 7, 801–805.

Basso, K., Margolin, A.A., Stolovitzky, G., Klein, U., Dalla-Favera, R., Califano, A., 2005. Reverse engineering of regulatory networks in human B cells. Nat. Genet. 37, 382–390.

Brandman, O., Meyer, T., 2008. Feedback loops shape cellular signals in space and time. Science 322, 390–395.

Carro, M.S., Lim, W.K., Alvarez, M.J., Bollo, R.J., Zhao, X., Snyder, E.Y., et al., 2010. The transcriptional network for mesenchymal transformation of brain tumours. Nature 463, 318–325.

Costanzo, M., Baryshnikova, A., Bellay, J., Kim, Y., Spear, E.D., Sevier, C.S., et al., 2010. The genetic landscape of a cell. Science 327, 425–431.

Gillespie, D.T., 2007. Stochastic simulation of chemical kinetics. Annu. Rev. Phys. Chem. 58, 35–55.

D'Haeseleer, P., Liang, S., Somogyi, R., 2000. Genetic network inference: from co-expression clustering to reverse engineering. Bioinformatics 16, 707–726.

DeRisi, J.L., Iyer, V.R., Brown, P.O., 1997. Exploring the metabolic and genetic control of gene expression on a genomic scale. Science 278, 680–686.

Dickson, R.C., Abelson, J., Barnes, W.M., Reznikoff, W.S., 1975. Genetic regulation: the Lac control region. Science 187, 27–35.

Elowitz, M.B., Leibler, S., 2000. A synthetic oscillatory network of transcriptional regulators. Nature 403, 335–338.

Fu, Y., Glaros, T., Zhu, M., Wang, P., Wu, Z., Tyson, J.J., et al., 2012. Network topologies and dynamics leading to endotoxin tolerance and priming in innate immune cells. PLoS Comput. Biol. 8, e1002526.

Gardner, T.S., Cantor, C.R., Collins, J.J., 2000. Construction of a genetic toggle switch in *Escherichia coli*. Nature 403, 339–342.

Hartemink, A.J., Gifford, D.K., Jaakkola, T.S., Young, R.A., 2001. Using graphical models and genomic expression data to statistically validate models of genetic regulatory networks. Pac. Symp. Biocomput. 422–433.

Hsiao, Y.T., Lee, W.P., 2014. Reverse engineering gene regulatory networks: coupling an optimization algorithm with a parameter identification technique. BMC Bioinformatics 15, S8.

Hunter, P.J., Borg, T.K., 2003. Integration from proteins to organs: the Physiome Project. Nat. Rev. Mol. Cell Biol. 4, 237–243.

Ideker, T., Sharan, R., 2008. Protein networks in disease. Genome Res. 18, 644–652.

Jacob, F., Monod, J., 1961. Genetic regulatory mechanisms in the synthesis of proteins. J. Mol. Biol. 3, 318–356.

Jeong, H., Tombor, B., et al., 2000. The large-scale organization of metabolic networks. Nature 407, 651–654.

Kaluza, P., Vingron, M., et al., 2008. Self-correcting networks: function, robustness, and motif distributions in biological signal processing. Chaos 18, 026113.

Kidd, J.M., Cooper, G.M., et al., 2008. Mapping and sequencing of structural variation from eight human genomes. Nature 453, 56–64.

Kikuchi, S., Tominaga, D., et al., 2003. Dynamic modeling of genetic networks using genetic algorithm and S-system. Bioinformatics 19 (5), 643–650.

Kitano, H., 2002. Systems biology: a brief overview. Science 295, 1662–1664.

Lage, K., Karlberg, E.O., Storling, Z.M., Olason, P.I., Pedersen, A.G., Rigina, O., et al., 2007. A human phenome-interactome network of protein complexes implicated in genetic disorders. Nat. Biotechnol. 25, 309–316.

Lamb, J., Crawford, E.D., Peck, D., Modell, J.W., Blat, I.C., Wrobel, M.J., et al., 2006. The connectivity map: using gene-expression signatures to connect small molecules, genes, and disease. Science 313, 1929–1935.

Lefebvre, C., Rieckhof, C., Califano, A., 2012. Reverse-engineering human regulatory networks. Wiley Interdiscip. Rev. Syst. Biol. Med. 4, 311–325.

Lewis, B.P., Shih, I.H., Jones-Rhoades, M.W., Bartel, D.P., Burge, C.B., 2003. Prediction of mammalian microRNA targets. Cell 15, 787–798.

Linding, R., Jensen, L.J., Ostheimer, G.J., van Vugt, M.A., Jorgensen, C., Miron, I.M., et al., 2007. Systematic discovery of in vivo phosphorylation networks. Cell 129, 1415–1426.

Lockhart, D.J., Winzeler, E.A., 2000. Genomics, gene expression and DNA arrays. Nature 405, 827–836.

Ma, W., Lai, L., Ouyang, Q., Tang, C., 2006. Robustness and modular design of the Drosophila segment polarity network. Mol. Syst. Biol. 2, 70.

Ma, W., Trusina, A., El-Samad, H., Lim, W.A., Tang, C., 2009. Defining network topologies that can achieve biochemical adaptation. Cell 138, 760–773.

Mondal, D., Dougherty, E., Mukhopadhyay, A., Carbo, A., Yao, G., Xing, J., 2014. Systematic reverse engineering of network topologies: a case study of resettable bistable cellular responses. PLoS One 9, e105833.

Napier, J.A., Beaudoin, F., Michaelson, L.V., Sayanova, O., 2004. The production of long chain polyunsaturated fatty acids in transgenic plants by reverse-engineering. Biochimie 86, 785–793.

Newman, S.A., Christley, S., Glimm, T., Hentschel, H.G., Kazmierczak, B., Zhang, Y.T., et al., 2008. Multiscale models for vertebrate limb development. Curr. Top. Dev. Biol. 81, 311–340.

Polynikis, A., Hogan, S.J., di Bernardo, M., 2009. Comparing different ODE modelling approaches for gene regulatory networks. J. Theor. Biol. 261, 511–530.

Quo, C.F., Kaddi, C., et al., 2012. Reverse engineering biomolecular systems using -omic data: challenges, progress and opportunities. Brief. Bioinform. 13, 430–445.

Repsilber, D., Liljenstrom, H., et al., 2002. Reverse engineering of regulatory networks: simulation studies on a genetic algorithm approach for ranking hypotheses. Biosystems 66, 31–41.

Samuelson, P., Scotchmer, S., 2002. The law and economics of Reverse Engineering. Yale Law J. 111, 1575–1663.

Sugimoto, M., Kikuchi, S., et al., 2005. Reverse engineering of biochemical equations from time-course data by means of genetic programming. Biosystems 80, 155–164.

Tegner, J., Yeung, M.K., Hasty, J., Collins, J.J., 2003. Reverse engineering gene networks: integrating genetic perturbations with dynamical modeling. Proc. Natl. Acad. Sci. U.S.A. 100, 5944–5949.

Tweedie, S., Ashburner, M., et al., 2009. FlyBase: enhancing Drosophila Gene Ontology annotations. Nucleic Acids Res. 37 (Database issue), D555–559.

Venkatesan, K., Rual, J.F., Vazquez, A., Stelzl, U., Lemmens, I., Hirozane-Kishikawa, T., et al., 2009. An empirical framework for binary interactome mapping. Nat. Methods 6, 83–90.

Vogelstein, B., Lane, D., Levine, A.J., 2000. Surfing the p53 network. Nature 408, 307–310.

Volkel, P., Le Faou, P., Angrand, P.O., 2010. Interaction proteomics: characterization of protein complexes using tandem affinity purification-mass spectrometry. Biochem. Soc. Trans. 38, 883–887.

Wahde, M., Hertz, J., 2000. Coarse-grained reverse engineering of genetic regulatory networks. Biosystems 55, 129–136.

Wen, X., Fuhrman, S., Michaels, G.S., Carr, D.B., Smith, S., Barker, J.L., et al., 1998. Large-scale temporal gene expression mapping of central nervous system development. Proc. Natl. Acad. Sci. U.S.A. 95, 334–339.

Yao, G., Tan, C., West, M., Nevins, J.R., You, L., 2011. Origin of bistability underlying mammalian cell cycle entry. Mol. Syst. Biol. 7, 485.

Yeung, M.S., Tegnér, J., Collins, J.J., 2002. Reverse engineering gene networks using singular value decomposition and robust regression. Proc. Natl. Acad. Sci. U.S.A. 99, 6163–6168.

Zeitlinger, J., Simon, I., Harbison, C.T., Hannett, N.M., Volkert, T.L., Fink, G.R., et al., 2003. Program specific distribution of a transcription factor dependent on partner transcription factor and MAPK signaling. Cell 113, 395–404.

Further Reading

Hieu L.C., Sloten, J.V., Hung, L.T., Khanh, L., Soe, S., Zlatov, N., et al., (2010). Medical reverse engineering applications and methods. In: 2nd International Conference on Innovations, Recent Trends and Challenges in Mechatronics, Mechanical Engineering and New High-Tech Products Development. MECAHITECH'10, Bucharest, September 23–24, 2010.

Omics Approaches in Forensic Biotechnology: Looking for Ancestry to Offence

Syed B. Nizami[1], Sayyada Z. Hassan Kazmi[1], Fatima Abid[2], Mustafeez M. Babar[3], Aneeqa Noor[4], Najam-us-Sahar S. Zaidi[4], Sami U. Khan[5], Humna Hasan[6], Mohsin Ali[4] and Alvina Gul[4]

[1]University of Karachi, Karachi, Pakistan [2]The University of Lahore, Islamabad, Pakistan [3]Shifa College of Pharmaceutical Sciences, Shifa Tameer-e-Millat University, Islamabad, Pakistan [4]National University of Sciences and Technology, Shifa Tameer-e-Millat University, Shifa, Pakistan [5]University of Haripur, Khyber Pakhtunkhwa, Pakistan [6]Quaid-I-Azam University, Islamabad, Pakistan

6.1 INTRODUCTION

Omics approaches have seen a great deal of advancement in the recent years. With more sophisticated tools and techniques now available, forensic analysts have a major advantage when it comes to having a complete picture of the crime scene. The omics approaches namely, Genomics, Transcriptomics, Proteomics, and Metabolomics, are now a crucial part of forensic biotechnology. One of the reasons being the complexities of certain investigational cases that these approaches are able to tackle with ease. Often times in forensics an analyst comes across evidence samples that are very scant or sometimes even degraded. These omics approaches coupled with the newer analytical techniques such as Massive Parallel sequencing, Mass Spectrometry, and 2-Dimensional High-Performance Liquid Chromatography (2D HPLC) have been very successful in overcoming such challenges. There are also certain sensitive cases where techniques such as DNA fingerprinting and profiling do not provide sufficient details which might be demanded. Cases involving sexual assault need more than just DNA profiling and sometimes require more details such as the source of the DNA from the body. The newer omics, i.e. Transcriptomics and Proteomics, have proven to be a major solution for such challenges. Unclear deaths have also posed a challenge to the forensic scientists in the past and it continues to do so. A major breakthrough in solving these unclear deaths is by using the metabolomic approach. A complete picture of all metabolites, endogenous changes, and toxicity is very helpful in these cases. This approach is of particular importance in certain cases that include illicit drug abuse and toxicity. These omics approaches have been very successful in investigations especially methods like DNA fingerprinting. These methods, using a very small amount of evidence, give a complete DNA profile of the suspect present at the crime scene. All of these omics approaches have proven vital for investigations in today's world and hold a lot of promises in the future.

6.2 IMPORTANCE OF OMICS APPROACHES IN STUDYING DNA

Recent achievements and breakthroughs in omics have opened up the horizon for the application of these approaches in studying DNA and have given us new hope for solving the complexities and challenges it is coupled with. Genomics, Proteomics, Transciptomics, and Metabolomics are now essential tools in forensic sciences and hold an extremely promising future. DNA profiling, since its inception, has been used extensively and successfully in solving criminal cases. DNA fingerprints were originally composed of pattern of bands analogous to a barcode assigned to bins. These absence or presence of bands in each bin in comparison with the sample subjects would show a mismatch or match, respectively. Since the 1990s this technique has been taken over by a newer technique "DNA profiling." Many other techniques are used now in DNA analysis, which will be discussed later.

6.2.1 The DNA

To fully understand the role of DNA analysis, one must have a comprehensive knowledge about the structure of DNA. DNA or deoxyribonucleic acid is molecular structure arranged into a double helix. The building blocks of this structure are the nucleotide triphosphate molecules. These molecules are made up of a triphosphate group, a deoxyribose sugar, and a nucleotide base. There are four of these nitrogenous bases. Differentiated by their structure, they can be classified as either purine bases or pyrimidine bases. Adenine and Guanine are the two purine bases, while Thymine and Cytosine are the two pyrimidine bases. These bases bond to their complementary bases which means that Adenine will always bond with Thymine through two hydrogen bonds, whereas Guanine will bind to Cytosine with three hydrogen bonds. A detailed figure of the structure is given below.

It is also very important to know the function of the DNA. This will shed a lot of light on its usefulness in forensic analysis. DNA contains all of the genetic information of an organism composed of over 3 billion base pairs and organized into 23 pair of chromosomes and is present in every cell of every individual. One of the most important functions of DNA is the coding and the regulation of protein synthesis, which is controlled by the genes. Only 15% of the entire genome is responsible for this; 75% of the genome consists of the extragenic DNA, which do not contain any known sequence. Around half of this extragenic DNA is composed of repetitive DNA. This repetitive DNA, in forensic DNA analysis, is the most important element of the genome. This repetitive DNA is then further divided into tandem repeats and interspersed repeats. Tandem repeats are then further divided into (1) satellite DNA, (2) minisatellite DNA, (3) microsatellite DNA, whereas interspersed repeats are subdivided into (1) Short Interspersed Nuclear Elements (SINE), (2) Long Interspersed Nuclear Elements (LINE), (3) Long Terminal Repeats (LTR), and (4) Transposon.

In forensic DNA analytical techniques, the polymorphism in tandem repeats is of prime importance. The number and location of these polymorphs create unique bands and are different in each individual, and the chances of two individuals having the same DNA profile is about 1 in 594 trillion (Romeika and Yan, 2013).

6.2.2 The Genomic Approach

We have studied earlier about how DNA is useful for the forensic analysis. This analysis is of particular importance in more than one way when it comes to forensics. Genomic approach is usually very reliable and gives a very strong evidence about the presence of a person on the crime scene after a match from the crime scene profile and National DNA database is done. There are, however, certain problems that are associated with this approach. Firstly, there can be difficulty and errors in interpreting the results if there is a "mixed sample." A mixed sample contains DNA from various individuals and requires expertise in interpreting the results. The second problem associated with genomic approach is "Partial Profiles." A partial profile is a result obtained from very small fragments of DNA or partially degraded DNA. They are particularly very hard to interpret and are not very reliable when it comes to veracity of the results. Another troublesome feature of this approach is "contamination" of the sample. This is usually a handling error. It occurs when the sample comes into contact with other DNA during the handling by the police or laboratory staff.

The genomic approach is very useful in criminal investigation, being extremely successful in catching numerous criminals. An idea of the rate of success can be made by the statistics of 2005–06 in England and Wales. The crime detection rate there increased from 26% to 40% when there was a DNA evidence present (Whittall, 2008). The DNA database has been growing larger each year, and the number of profiles has doubled in the recent years. This database can also be used for familial searching of a suspect. If the database does not have a match with the DNA sample collected, we can search for the family members who will have a partial match with the sample DNA.

6.2.3 The Transcriptomic Approach

Transcriptomics, also commonly known as expression profiling, is the study of transcriptome of organisms resulting from the expression of genes under specified conditions.

A Transcriptome consists of all RNA molecules that include mRNA, tRNA, rRNA, and noncoding RNA. This transcriptomic approach is of particular importance in case of blood doping. Athletes very often use a variety of substances to improve their sporting ability. One of the methods used that improves the sporting performance of an athlete is the administration of recombinant human erythropoietin (rHumanEPO). This rHumanEPO is prohibited by World Anti-Doping Agency, WADA. Over the years, there are certain approaches used to detect rHumanEPO but they have not been so successful and have had many imperfections. Transcriptomic approach has shown some promising results in the recent studies.

A transcriptome-based longitudinal screening approach, which is currently being developed, identify a molecular signature of rHumanEPO doping. The gene expression of numerous genes is altered by the use of rHumanEPO. The gene transcripts are differentially expressed after the first injection. The further transcripts are then greatly upregulated and then are subsequently downregulated up to 4 weeks of postadministration of the rHumanEPO.

In forensics, the DNA material collected is often very scantly present, and a DNA analysis will tell you about the presence of an individual at a crime scene, but it will not indicate if the DNA came from saliva, vaginal secretions, urine, or any other source. An example of this is the collection of "touch DNA." This touch DNA consists of shed skin which the suspect has transferred to an object or a person. Although DNA profiling is done using this scant piece of evidence but it does not reveal much about the tissue origin of the sample due to the sample size. This lack of nature of the evidence is a hindrance in investigations. A transcriptomic approach in this case will identify highly sensitive and specific mRNA biomarkers for the identification of the skin using only 5–25 pg of total RNA input (Hanson et al., 2012).

Often the biological samples collected from the crime scene are substantially degraded and is scarcely present. This poses a serious challenge in the analysis of the sample. Massively Parallel Sequencing (MPS) could possibly be used to counter this challenge. In a study using MPS, a high-sequencing output was generated even by using low RNA integrity numbers using a modified sample preparation. High-quality sequencing was generated, and body fluid–specific RNA markers were detected from the degraded body fluids that are very commonly encountered during investigations and pose serious problems for the analysts. (Lin et al., 2015).

6.2.4 The Proteomics Approach

Blood and semen protein were once used as identification markers to individualize biological stains but now this technique is replaced by the DNA markers that offer the advantage of being detectable even in scant quantities. Although a major drawback of DNA markers, as we have discussed before, is the lack of identification of the source of the DNA. This is of great importance in cases of sexual assaults. For example, the DNA profile from the skin of the victim matches with the suspect and the suspects claims that the DNA was present due to a casual touch. In this case the DNA will not provide the ample details that are required. The proteomic approach comes in handy in such cases.

One approach is to test the sample for specific proteins present in the sample. The biological fluids that most analysts encounter are blood, saliva, semen, vaginal fluid, urine, feces, and nasal secretions. One approach is to test for the specific proteins present in the biological sample. For example, testing for amylase as indicator of saliva, testing for bilirubin and creatinine for fecal and urine matter, respectively, and testing for hemoglobin as an indicator of blood. Most of these tests are laborious and usually need very large quantities of sample, which are often times not available at the crime scene. Also many of these tests use chemicals which can be very dangerous for the analyst, e.g., the use of mercurial chloride for testing fecal matter.

The recent advancements in the field of proteomics has opened up more sophisticated and advanced options that offer much greater sensitivity when it comes to the scant amounts of samples encountered in forensics. These techniques involve the detection of several different biomarkers present in the different fluids of the body. Saliva is characterized by the presence of alpha-amylase 1 and blood contains hemoglobin. Similarly, vaginal fluid is characterized by the presence of cornulin and cornifin and/or involucrin. Semen contains semenogelin, prostate-specific antigen, and acid phosphatases as biomarkers. Urine can be identified by the presence of uromodulin or Alpha-1-microglobulin/bikunin precursor protein. Plunc protein is usually an identification of nasal secretions. Feces can be identified by the presence of immunoglobulins and the absence of hemoglobin.

Nowadays commercial forensic immunoassay kits have been developed to test for these biomarkers. These test kits use the phenomenon of antigen—antibody reaction. These kits are very helpful in identifying certain markers such as hemoglobin for the identification of blood and prostate-specific antigen for the identification of semen. They are, however, unable to test for other biomarkers because the target proteins still need to be identified and characterized. Other techniques are utilized for the identification of these biomarkers such as Mass Spectrometry. In this technique, the samples are first applied on sterile cotton-tipped swabs. These are then dried and then digested with trypsin to obtain peptides. These are then injected on a mass spectrometer and accurate identification is done (Van Steendam et al., 2013).

Other technique involves the high-resolution fractionation and quantitative mapping of the complete proteomic profile using 2-D HPLC, (Danielson, 2006).

Proteomics can also be used as a tool to figure out which bullet or stab wound caused the death of a person. This can be done by matching the bullet to the wound using organ-specific protein signatures. This enables one to reconstruct the crime scene. This is of particular importance in cases where there has been a shootout and several people are involved in the shooting (Dammeier et al., 2015).

6.2.5 The Metabolomics Approach

Metabolomics, also referred to as metabolic profiling, is the comprehensive study of metabolome. Metabolome is the total number of metabolites present in an organism. It usually involves but not limited to the study of small-sized metabolites, i.e., <1500 Daltons, present in the body. Metabolomic studies also include the molecules obtained from environmental exposure and medications. Metabolomic approach is of particular interest to forensic scientists when it comes to toxicological analysis. It gives a complete picture of the metabolome at any given time. Various methods can be used to determine the concentration of these metabolomes such as Gas Chromatography Mass Spectrometry, Capillary Electrophoresis Mass Spectrometry, Liquid Chromatography Mass Spectrometry, and magnetic resonance spectroscopy (Dinis-Oliveira, 2014).

A particular use of this approach is to study the unclear deaths caused by herbal agents that are used for certain treatments. Herbal agents can be toxic on their own, e.g., *Ricinus communis*, or they could act synergistically with other compounds. Certain herbal drugs also interact with certain pharmaceutical agents and lead to a drug−drug interaction that is often times fatal. Adulteration of herbal medicines with herbicides and heavy metals is also a much known cause of death in individual using herbal medicines.

These herbal medicines are usually not documented and are seldom prescribed, so it is very challenging to evaluate the cause of death in such individuals. Sometimes the herbal drug is documented but the label does not show the ingredient that caused death of an individual.

A metabolomic approach is of great help in such cases. The screening will reveal all the primary and secondary metabolites of a preparation and is able to produce a "finger print" that characterizes the ingredients present and the composition of a particular preparation the individual has been taking (Byard et al., 2015).

Another use of metabolomics is the screening of illicit drugs. Screening of blood for different narcotics is frequently performed in forensic analysis. It can further be used to determine untargeted metabolites and endogenous changes caused by the drug response and its toxicity. In a study on methylenedioxymethamphetamine (MDMA) users, it was seen that several acylcarnitines, adenosine monophosphate, adenosine, inosine, thiomorpholine 3-carboxylate, tryptophan, S-adenosyl-l-homocysteine, and lysophospatidylcholine changed in response to MDMA use (Nielsen et al., 2016).

6.2.6 The Toxicogenetics/Pharmacogenetics Approach

Every individual responds differently to certain drugs, toxic material, environmental chemicals, and allergens. Some individuals respond very well and no serious consequence is seen, whereas in other individuals, the same drug, chemical or toxic material may have dire consequences and might even be fatal. Toxicogenetics or Pharmacogenetics deals with the genetic explanations of such incidences. It involves the study of the genome and genome polymorphism and might be utilized by forensic scientists to determine the cause of death (Shen et al., 2007).

6.3 ANCESTRY AND PHYLOGENY

The genetic variations compared in chimpanzee species are comparatively more than found in human population. Drastic differences in genetic makeup of primate individuals residing in different parts of the world or the continents (Chimpanzee Sequencing and

Analysis Consortium, 2005) are observed as compared to the two samples taken from the humans from different continents (Barbujani and Colonna, 2010). This diversity in human genetics is due to the geographical barriers existing between the sampled volunteers, creating distances and isolation between the individuals (Handley et al., 2007; Ramachandran et al., 2005; Rosenberg, 2005; Jay et al., 2013). The 80% of this variation is explained by individual differences in genetics, whereas the 5% is due to population genetic diversity (Lewontin, 1972). It is however controversial if humans should be considered as a genetically homogenous group or be grouped under continental population (Edwards, 2003; Barbujani, 2005; Risch, 2006). To infer the substructure of population genome, an assumption is made on the existence of individual and population genome mixture to study the current human population to trace ancestry (Pugach et al., 2011). Study of genetic ancestry is hence complex defined at different scales ranging from population to individuals.

In the era of omics, the approach of genomics and transcriptomics is currently widely employed in molecular biology and genetics as a tool of studying forensic or other medical genetics. The Single Nucleotide Polymorphisms (SNPs) and Next Generation Sequencing are currently widely used technological tools for study of genomics and transcriptomics (Ansorge, 2009; Mardis, 2008; Shendure and Ji, 2008). With the introduction of these tools, the sequencing of whole genome has become possible and cost effective (Magi et al., 2010). Information on species' genome and transcriptome nucleotide in gigabytes has become efficiently possible by the introduction of Short Reads Technology in the world of forensic biotechnology. With this, it has become convenient to sequence large amount of RNA—DNA reads, rapidly generating fruitful information on genetic makeup of multiple various species (Roos, 2001; Schuster, 2007; Carvalho and Silva, 2010). For the analysis of individuals' ancestry from forensic DNA, the identification of ancestry markers for genome characterization has a tremendous role (Phillips, 2015). Transcriptome of species is however profiled using high-throughput sequencing under RNA sequencing approach (Nagalakshmi et al., 2010; Haas and Zody, 2010; Lister et al., 2008). The information trapped within the human genome reveals personal features and characteristics of a person including age, physiological and physical physiognomy, and ethnicity (Shriver et al., 1997; Wei et al., 2011; Eiberg et al., 2008). Such advances in omics and molecular biology are likely improving the forensic case work of humans as a result. Consequently, it has become evitable for investigators to find the previously unknown individuals (Kayser and de Knijff, 2011).

Inferring of phylogeny, however, prefers mitochondrial and chloroplast genome for whole-genome analysis of living organisms rather than single gene sequencing. For obtaining a better organism's phylogenic inference, the approach of combining fossil data with genome molecular records is considered (Pyron, 2015). Molecular phylogeny, nevertheless, has helped in developing a deep understanding of relationship existing between organism genes by comparing protein sequences or DNA homologs. This involves nucleotide sequencing of genetic material, that is, RNA and DNA, along with the amino acid sequencing of proteins determined by modern tools of genomics and transcriptomics. As a result, hierarchy of organism' phylogenetic relationship can be revealed by comparing different species' homologous molecules. Organism morphology is a product of its genome, transcriptome, and proteome profiles. Application of these molecular analytical tools in combination with each other can establish development of a strong phylogenetic tree (Patwardhan et al., 2014).

6.4 DNA FINGERPRINTING

6.4.1 History

Being one of the greatest discoveries of the 20th century, DNA fingerprinting has brought a revolution to forensic studies and examinations. In the United Kingdom, heritable and variable patterns were found at the University of Leicester by Alec Jeffrey who is regarded as the Father of DNA fingerprinting (Jeffreys et al., 1985a,b). This marked the beginning of a new era in scientific advancement. The application of this remarkable technique is very diverse, playing a vital role in many biological, clinical, and anthropological studies, studying the conservation and diversity in many species to name a few.

Apart from the biological implications, this extraordinary discovery has political and social dimensions as well. These became transparent when its use in numerous criminal investigations, along with the spectacular precision and very high success rate, was brought to light by different publications. New technologies in DNA fingerprinting are being discovered every year, which has brought a revolution to our judicial system especially criminal justice. Thus the implications of DNA profiling has brought fruitful evidence to court cases and investigations (Jeffreys et al., 1985a,b).

6.4.2 What is DNA Profiling?

DNA fingerprinting/profiling is a series of tests and techniques employed to evaluate and identify the genetic information contained within an individual's cell and in forensics; it is defined as the comparison of genetic information of a person to evidence found at the scene of offence or crime. DNA is the chemical code, which is present in every living and dead cell of a person's body. Factually, the DNA sequencing in almost all human beings is 99.9% the same. Variation is present in only 0.1% of the DNA sequence and only 0.1% is unique to every person. The odds of two people, who are not related by blood, having the exact same DNA fingerprint is about 1 in 594.1 trillion individuals.

However, the DNA analysis and profiling for an individual cannot be done without biological samples. When using this tool in forensic studies and sciences, these biological samples, or let us say biological evidence, for the purpose of DNA profiling, comprise saliva, blood, urine, skin, hair, and semen, keeping in mind that some may be better than the other, i.e., show better and more accurate results. Depending upon the purpose and area of investigation and study, there may be variation in the use of such samples evidences. Some techniques include blood typing, DNA profiling, gender determination through the analysis of chromosome, and lately, DNA phenotyping in forensics (Findlay et al., 1997).

Since the introduction of DNA profiling in the 1980s, it has been effectively employed in investigating criminal cases and has had many implications, for example, in the identification of disaster victim and testing to check the identity of person's biological father, i.e., paternity testing (Findlay et al., 1997). Although this technique has numerous advantages, DNA profiling is not preferably used as the solitary method in criminal studies and investigations and must be done in combination with certain other evidences in the United Kingdom and other countries as well (Findlay et al., 1997).

DNA analysis/fingerprinting in forensics generally involves the following steps (Ley et al., 2012):

1. Collection of evidence from the crime scene and from the suspect.
2. Sample preparation from the collected sample.
3. Extraction of genetic material (DNA) from the evidence obtained at the crime scene.
4. Multiplication and amplification of whole or certain regions of the DNA exponentially, so that the sample becomes detectable.
5. DNA extraction: DNA is isolated from the unknown crime-scene evidence (and/or any bodily fluids from the suspect).
6. Separation of DNA fragmentation and their analysis through spectrophotometer and through other techniques, i.e., DNA quantification.
7. Entering of the DNA fingerprinting performed on the sample obtained at the crime scene to the database to find a match or with the DNA of the suspect, i.e., DNA profile matching.

6.4.3 Sample Collection From the Suspect

Usually sample is collected by the use of touch DNA and cotton swabs. Epithelial cells of the suspect are generally collected from the buccal cavity. Cottons swabs (buccal swabs) are used for attaining sample of the suspect's DNA to compare it with evidence found on the scene of crime. In this, a swab is rubbed along the inside of the buccal cavity (cheek), causing scraping of epithelial cells of the suspect, and these cells are then use for DNA fingerprinting and cross matching the results with the evidence (Findlay et al., 1997). In touch DNA, skin cells of the suspect are used. It is based on the fact that every touch leaves skin cells on the object he/she touches. In the next step, these cells are analyzed for DNA fingerprinting. In contrast with fingerprinting, touch DNA provides better forensic evidence and more accurate results keeping in mind that it does not imply to more individualized IDs (Nunn, 2013).

As mentioned above, an alternative technique for the collection of sample is a sterile cotton swab. There are many approaches to execute this. In one approach, swab is moistened by sterile water and the area of specimen is wiped with it. This is followed by a dry swab for collection of water and cells. Instead of sterile water, solutions like SDS (sodium dodecyl sulfate) can also be used. Detergents like triton X-100 can also replace sterile water. The purpose of using detergents is to increase the DNA yields after DNA profiling (Thomasma and Foran, 2013). The reason for this increased yield is the amphiphilic properties, and the mechanism through which these detergents act is the more rapid attraction of cells to these chemicals. The recommended concentration of SDS is 1%—2%.

In conditions like fire arm extraction of DNA, multiple swabs, although increasing the chance of contamination, are used, and this practice is found to be more appropriate. These swabs are combined for better DNA yield to achieve detectable results.

The material of swab is also a factor, which may affect the DNA yield, and in certain cases cotton swab is preferred over polyester swabs, e.g., sexual assault cases (Hopwood and Elliott, 2012) (Fig. 6.1).

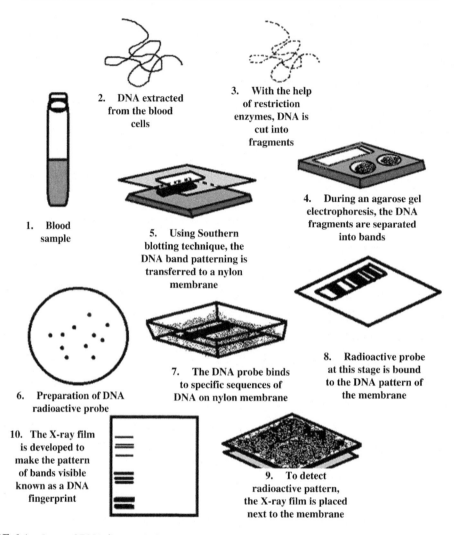

FIGURE 6.1   Steps of DNA fingerprinting.

6.4.4 The Process of DNA Extraction

As discussed earlier, important DNA evidence sources may be semen, blood, hair, and skin debris. Other sources are also present but are less useful such as urine and feces, as the isolation of genetic material can only be obtained from them in theory but practically these are not as much of use, making them very poor sources of evidence due to the presence of very few cells that are still intact. Moreover, urine and feces contain very high level of contaminants that alter the success rate of the analysis.

One crucial and important step that must be carefully performed for all techniques employed in DNA fingerprinting is the extraction of DNA prior to the analysis. The general approach for extraction of genetic material (DNA) is as follows:

1. Lysis of cells in a solution of buffer.
2. Centrifugation for the pelleting of denatured fats and proteins.
3. As the DNA is negatively charged, its lysate is passed through a column containing a medium having positive charge. Due to attraction, it binds to the DNA.
4. Washing of the sample for the removal of contaminants like proteins and salts.
5. Quantification of DNA through spectrophotometer and other techniques after it has been obtained in an appropriate buffer or water. (Findlay et al., 1997).

6.4.5 Various Techniques and Analysis Employed in DNA Profiling

As science has evolved, so has the process and technique of DNA profiling. The original technique by Jeffreys is now obsolete for forensic use and thus has undergone a period of crucial developmental phases. It underwent important developments regarding its basic methodology, moving from radioactive to fluorescent labels, Southern blot to PCR, from slab gels to capillary electrophoresis making the process more accurate, simple, precise, sensitive, and automated, in turn making the statistical treatment more straightforward.

6.4.5.1 Dideoxy Method

This method was developed by Fred Sanger in 1975, hence is termed as Sanger's method. First, the DNA is synthesized through suitable primers that are radiolabelled. The synthesis is halted by the addition of four dideoxynucleotides. Because DNA is now present in fragment, it is separated through the process of gel or capillary electrophoresis. Different band patterns are obtained at the completion of electrophoresis. The band patterns obtained are then converted into DNA sequences through highly specialized software.

Earlier, radioactive tags and labels were used but these posed some serious hazards, and as the process evolved and became more refined, radiolabels were replaced by fluorescent dyes. As a consequence the precision of the technique decreased, because dyes showed erratic behavior and showed varying mobility indicating that nucleotide sequence obtained may not be in the correct order or the bands may not correspond to the actual order of the sequence. Hence, the dyes cannot be relied upon and this technique cannot be used solely in criminal investigations and forensics and is not the ideal choice. Not only the fluorescent dye but also other aspects of this technique make is less proficient that is, ideal conditions are required for proper reaction between DNA templates or the dideoxynucleotides. Alteration in the conditions would lead to false and inaccurate results as the reaction would be unsuccessful.

Another shortcoming of this technique was the faulty computer software that gave overlapping peaks and incorrect order of nucleotides (Findlay et al., 1997).

6.4.5.2 Single Nucleotide Polymorphisms

Human genome contains approximately 10 million SNPs, i.e., one after every 150−300 bp, which act as biological markers. Matching the DNA from the suspect and the crime scene and concluding that the samples are from the same person has been made much convenient and reliable by the variability of the DNA polymorphism, i.e., SNP. SNP constitutes about 90% of DNA variation in human beings and thus is the most simple and common type of genetic variation. Each SNP represents one nucleotide difference and these variations occur during DNA replication during meiosis. An example of this is the replacement of cytosine with thymine in a DNA stretch but the length of the DNA is unaffected.

DNA sequencing techniques may be employed in SNP. The specific base in the SNP in the analysis and testing of SNP which may comprise up to four alleles should be established, and this is done through DNA sequencing.

6.4.5.3 Variable Number Tandem Repeats/Minisatellites

The exon region of DNA contains highly polymorphic sequences called VNTRs (Variable Number Tandems Repeats). These sequences are repetitive at numerous points along the length of chromosomes. No individuals have the same VNTRs as that of their parents because they are inherited from both maternal and paternal sides. The specific lengths of VNTRs are not definite and their length may range from 20 to 100 bp. In case of criminal investigations, VNTRs were one of the first forms of polymorphism to be used.

This is an expensive process, and because minisatellites have greater length than short tandems repeats (STRs), it is a time-consuming process; hence, it is less commonly used (Jeffreys et al., 1985b; Findlay et al., 1997).

6.4.5.4 STRs/Microsatellites

STRs that are used in criminal investigations are mostly penta- and tetra-nucleotides because of their increased robustness. These are the areas of genome repeated up to 17 times and comprise 1–5 bases. STRs are present on all 22 autosomal chromosomes along with both sex chromosomes (X and Y) and they be simple, compound, or complex. Because these show very high polymorphism, they are extensively used in forensics (Jeffreys et al., 1985b; Findlay et al., 1997).

6.4.5.5 Restriction Fragment Length Polymorphisms

In this technique, DNA is digested by endonucleases restriction enzymes, which cuts it into sequences of specific patterns. Due to the difference in location of the cuts made by restriction enzymes, the DNA sequence differs. These are then shifted to a membrane through Southern blot after being separated through electrophoresis. In the next step, the fragments are detected, and the technique used for this is called probe hybridization. Restriction Fragment Length Polymorphism (RFLP) is less commonly used on account of some drawbacks, like the requirement of large DNA samples for detectable results (Jeffreys et al., 1985b; Findlay et al., 1997).

6.4.5.6 Analysis of Degraded or Low Template DNA

Achieving complete STR typing becomes tedious in case of low or degraded DNA and in such cases DNA needs to be amplified. Dropouts of alleles lead to the deficiency of alleles and thus is overcome through PCR, increasing PCR cycles, more PCR products, purification of the product obtained at the end of PCR, leading to an increased sensitivity of the analysis but the increased sensitivity poses the threat of appearance of allele which was not the portion of original DNA, i.e., allele drop-ins (Fattorini et al., 2011; Benschop et al., 2013; Benschop et al., 2012). In such cases, we can use SNPs as an alternative to altering the PCR sensitivities of STRs as they result in lesser allele drop-ins (Børsting et al., 2013).

In case of highly degraded and low template (LT) DNA, whole-genome amplification (WGA) can be employed to increase the sensitivity for the genetic profiling of STRs, and this is then further tested by PCR. Degenerate oligonucleotide primed-PCR (DOP-PCR), Primer Extension Preamplification (PEP), Multiple Displacement Amplification (MDA), blunt-end ligation-mediated whole genome amplification (BL-WGA) and Rolling Circle Amplification (RCA) are some of the variants of WGA. Among these, the most effective is PEP as it gives very good results in case of low copy number or highly degraded DNA (Maciejewska et al., 2013).

6.5 STUDYING PARENTHOOD

Genomics is an important approach to study parenthood. The DNA, as we have studied earlier, contains all the genetic information of an organism. We can define genes as an instruction book that specifies which protein is to be used and when it is to be used. The genetic instruction or the particular set of genes in the DNA is called the genotype. The physical expression of a particular gene is termed as phenotype. This information is passed down from the parent to the offspring as genetic traits, and there are several different methods to test the lineage. This approach is called as lineage-based approach. These lineage-based approach analyzes the Y chromosome of an offspring as it passed on directly and unchanged from the male parent or the mitochondrial DNA which is passed on to the offspring from the mother. In Forensic Sciences this approach can be used to study the parenthood of an individual as well as the ancestry of the individual. Studying the DNA without a suitable match from the DNA database is particularly hard. This can narrow down things with respect to the investigation, and it gets easier if we could narrow down things to ancestry or even with information as little as the color of the hair and skin.

This investigation method is known as forensic phenotyping in which a description of the suspect with respect to external and behavioral features, geographic origin, and may be surname can be composed (Koops and Schellekens, 2008).

6.5.1 Mitochondrial DNA Analysis

With a 16.65-bp length, the shape of this DNA is circular and is inherited from maternal side. This DNA is present in the mitochondria that are responsible for generating cell energy. The sequence of base pair in this DNA is called Andersons sequence. Each mitochondrion contains many copies of the DNA and there are around 100−1000 mitochondria in the human cell. Although most of the DNA is identical among all individuals with very little or no variation, a noncoding region, which is called Control region, is present, which is contained with HV1 and HV2. These HV1 and HV2 are hyper variable regions. The control region is present at the D-loop.

The ability of mitochondrial DNA to repair itself is very poor; thus the chances of mutation are quite high. The variation in the control region of DNA is of SNP type; thus we know that the length of DNA is not affected by this variation.

In the analysis of mitochondrial DNA, for the amplification of HV1 and HV2 regions, PCR is performed but before the DNA is extracted. The amplified HV1 and HV2 are then used to determine the base pair sequence through DNA sequencing. The results are then compared with the Cambridge Reference Sequence and any variation between the two is established and noted. The samples from the crime scene are then compared with this to observe the similarities between the two.

This technique is mostly employed after the ruling out of STR analysis (Jeffreys et al., 1985a,b; Watson and Crick, 1953; Findlay et al. 1997).

6.5.2 Y-Chromosome Analysis

As we discussed earlier in this chapter, the STRs are present in autosomal as well as sex chromosome pairs. Unlike most of the analysis, here we study and evaluate the Amelogenin maker present on the sex chromosomes. This is especially useful for evaluating the Y chromosome that is present in males and not in females. The female sex chromosome pair consists of two X chromosomes. Y chromosome is a small chromosome and shows very little variation and alteration. Mutation, which does not occur frequently, causes its alteration. In theory, the Y chromosome should contain such combination of alleles that are identical between father and son, just like the mitochondrial gene in maternal case. One of the methods employed for the analysis of this chromosomes are Applied Biosystems' Y filter that determines the STRs present on this.

This technique has wide application in cases involving sexual assaults and rape where more than one DNA samples may be found (Jeffreys et al., 1985a,b; Watson and Crick, 1953; Findlay et al. 1997).

6.5.3 Forensic DNA Phenotyping

Phenotypic characteristics are the morphological and apparent characters of the organism, meaning that they are the expression of the genotype in an individual. Hair color, skin, height, and eye color are the examples of phenotypic characters that are specific in different individuals depending upon their genotypes. Analysis of such characters is done through Forensic DNA Phenotyping (FDP).

FDP is performed using SNPs rather than STRs, as they have less odds of mutation, thus they become fixed in a particular population (an example of this may be the fact that many people in a population have the same phenotypic characters of their eye color).

This technique is mostly employed in studying pigmentation genes, e.g., *MC1R* (Melanocortin Receptor Gene). Mutation in this gene is usually associated with red hair.

The allele for this phenotype is recessive, so the likelihood of a person to get red hair is when a person inherits alleles of red hair from maternal and paternal sides. Similarly, the oculocutaneous albinism II (*OCA2*) gene is analyzed for the determination of eye color as it encodes the individual's eye and skin pigmentation.

Another application of this is the determination, identification, and studying of ancestry of an individual, as certain studies of the STRs through this method has proven to be very effective in establishing ancestry.

SNPs study in phenotyping is not enough to provide a clear picture an individual's phenotype, and thus this contributes to the limitation of this technique (Jeffreys et al., 1985a,b; Watson and Crick, 1953; Findlay et al. 1997).

6.6 APPLICATION OF OMICS IN CRIMINOLOGY

As we have studied before, the various ways in which different omics namely genomics, proteomics, transcriptomics, and metabolomics are used in different investigational procedures. All these approaches together cannot give a complete picture of the crime as the samples are present in minimal amounts.

Genomics has been successfully used for quite some time by forensic scientists; the rest of the approaches are much newer techniques and are not as widely applied throughout the world. In England and Wales DNA samples from suspects were compared against around 50,000 samples obtained from crime scenes (Whittall, 2008). A national DNA database is used against which the sample DNA from crime scene can be run. This database usually contains the DNA samples of persons who have been arrested for a recordable offence. Some countries allow greater powers to the police regarding the collection of DNA, for example, England and Wales. In these countries the police have the power to collect DNA samples without the consent of the individuals who have committed a recordable offence. It has also been proposed to allow the police to collect the DNA sample from individuals with nonrecordable offences. Biological and DNA samples can also be collected from the witnesses, victims, and volunteers only by their consent. DNA evidence is extremely powerful in court although the statistical inference offered by it is rather difficult to comprehend by the legal professionals.

The newer omics, transcriptomics and proteomics, both are used for the identification of the source of DNA in complicated cases such as rape cases. Both these omics are not competing techniques but rather complementary with each other. Transcriptomics offers the benefit of amplification using PCR and compatibility with the existing DNA, whereas proteomics offers the advantage of direct identification of fluid without the necessity of amplification (Danielson, 2006).

The application of metabolomics is employed in cases of drug abuse and drug toxicity. It is also an important technique for forensic scientists in cases of deaths related to poison. Metabolomics gives a complete image of the metabolites, active or inactive, present in the body at a certain time and also gives the complete parameters and composition of preparations an individual has taken.

6.6.1 DNA Databases

A combined DNA index system (CODIS) is a DNA index system and thus contains a set of standard DNA markers. CODIS and such other DNA databases contain information of DNA profiles of random individuals especially of those suspects who have been convicted of any offence previously to aid criminal investigations and court cases (Jeffreys et al., 1985a,b; Watson and Crick, 1953; Findlay et al. 1997; Jiang et al., 2013).

6.7 CONCLUSION

Omics approaches, as we have seen, offer a lot to forensic science. Approach such as genomics have been used widely for some time now has incredibly been successful in investigations and numerous criminals have been identified. The recent advancements in genetics have enabled a more in-depth knowledge of the DNA; this, together with improved techniques of DNA profiling, has played a pivotal role in the advancement of this field. This approach is not only used in forensic science to identify people present at a crime scene but also now widely used to study parenthood and even gives a complete account of the ancestry of an individual.

Other techniques such as transcriptomics, proteomics, and metabolomics are relatively new but hold a lot of promises for the future. These techniques, as we have seen, when used with advance analytical techniques are a possible solution to many of the baffling and complicated cases a forensic scientist often encounters.

References

Ansorge, W.J., 2009. Next-generation DNA sequencing techniques. New Biotechnol 25 (4), 195–203.

Barbujani, G., 2005. Human races: classifying people vs understanding diversity. Curr Genomics 6 (4), 215–226.

Barbujani, G., Colonna, V., 2010. Human genome diversity: frequently asked questions. Trend Genet 26 (7), 285–295.

Benschop, C., Haned, H., Sijen, T., 2013. Consensus and pool profiles to assist in the analysis and interpretation of complex low template DNA mixtures. Int J Legal Med 127 (1), 11–23.

Benschop, C.C., Haned, H., de Blaeij, T.J., Meulenbroek, A.J., Sijen, T., 2012. Assessment of mock cases involving complex low template DNA mixtures: a descriptive study. Forensic Sci Int Genet 6 (6), 697–707.

Børsting, C., Mogensen, H.S., Morling, N., 2013. Forensic genetic SNP typing of low-template DNA and highly degraded DNA from crime case samples. Forensic Sci Int Genet 7 (3), 345–352.

Byard, R.W., Musgrave, I., Hoban, C., Bunce, M., 2015. DNA sequencing and metabolomics: new approaches to the forensic assessment of herbal therapeutic agents. Forensic Sci Med Pathol 11 (1), 1–2.

Chimpanzee Sequencing and Analysis Consortium, 2005. Initial sequence of the chimpanzee genome and comparison with the human genome. Nature 437 (7055), 69–87.

Carvalho, M.C., Silva, D.C., 2010. Sequenciamento de DNA de nova geração e suas aplicações na genômica de plantas. Ciência Rural 735–744.

Dammeier, S., Nahnsen, S., Veit, J., Wehner, F., Ueffing, M., Kohlbacher, O., 2015. Mass-spectrometry-based proteomics reveals organ-specific expression patterns to be used as forensic evidence. J Proteome Res 15 (1), 182–192.

Danielson, P.B., 2006. Isolation of Highly Specific Protein Markers for the Identification of Biological Stains: Adapting Comparative Proteomics to Forensics. US Department of Justice, pp. 1–37.

Dinis-Oliveira, R.J., 2014. Metabolomics of drugs of abuse: a more realistic view of the toxicological complexity. Bioanalysis 6 (23), 3155–3159.

Edwards, A.W., 2003. Human genetic diversity: Lewontin's fallacy. BioEssays 25 (8), 798–801.

Eiberg, H., Troelsen, J., Nielsen, M., Mikkelsen, A., Mengel-From, J., Kjaer, K.W., et al., 2008. Blue eye color in humans may be caused by a perfectly associated founder mutation in a regulatory element located within the HERC2 gene inhibiting OCA2 expression. Human Genet 123 (2), 177–187.

Fattorini, P., Marrubini, G., Sorçaburu-Cigliero, S., Pitacco, P., Grignani, P., Previderè, C., 2011. CE analysis and molecular characterisation of depurinated DNA samples. Electrophoresis 32 (21), 3042–3052.

Findlay, I., Taylor, A., Quirke, P., Frazier, R., Urquhart, A., 1997. DNA fingerprinting from single cells. Nature 389 (6651), 555–556.

Haas, B.J., Zody, M.C., 2010. Advancing RNA-seq analysis. Nature Biotechnol 28 (5), 421.

Handley, L.J., Manica, A., Goudet, J., Balloux, F., 2007. Going the distance: human population genetics in a clinical world. Trend Genet 23 (9), 432–439.

Hanson, E., Haas, C., Jucker, R., Ballantyne, J., 2012. Specific and sensitive mRNA biomarkers for the identification of skin in 'touch DNA' evidence. Forensic Sci Int Genet 6 (5), 548–558.

Hopwood, A.J., Elliott, K., 2012. Forensic DNA research: keeping it real. Int J Legal Med 126 (2), 343–344.

Jay, F., Sjödin, P., Jakobsson, M., Blum, M.G., 2013. Anisotropic isolation by distance: the main orientations of human genetic differentiation. Mol Biol Evol 30 (3), 513–525.

Jeffreys, A.J., Brookfield, J.F., Semeonoff, R., 1985a. Positive identification of an immigration test-case using human DNA fingerprints. Nature 317 (6040), 818–819.

Jeffreys, A.J., Wilson, V., Thein, S.L., 1985b. Hypervariable 'minisatellite'regions in human DNA, Nature 314 (6006), 67–73.

Jiang, X., Guo, F., Jia, F., Jin, P., Sun, Z., 2013. Development of a 20-locus fluorescent multiplex system as a valuable tool for national DNA database. Forensic Sci Int Genet 7 (2), 279–289.

Kayser, M., de Knijff, P., 2011. Improving human forensics through advances in genetics, genomics and molecular biology. Nat Rev Genet 12 (3), 179–192.

Koops, B.J., Schellekens, M., 2008. Forensic DNA phenotyping: regulatory issues. Colum Sci Tech L Rev 9, 158–160.

Lewontin, R.C., 1972. The apportionment of human diversity. Evolutionary Biology. Springer, New York, USA, pp. 381–398.

Ley, B.L., Jankowski, N., Brewer, P.R., 2012. Investigating C.S.I.: Portrayals of DNA testing on a forensic crime show and their potential effects. Public Underst Sci 21 (1), 51–67.

Lin, M.H., Jones, D.F., Fleming, R., 2015. Transcriptomic analysis of degraded forensic body fluids. Forensic Sci Int Genet 17, 35–42.

Lister, R., O'Malley, R.C., Tonti-Filippini, J., Gregory, B.D., Berry, C.C., Millar, A.H., et al., 2008. Highly integrated single-base resolution maps of the epigenome in Arabidopsis. Cell 133 (3), 523–536.

Maciejewska, A., Jakubowska, J., Pawłowski, R., 2013. Whole genome amplification of degraded and nondegraded DNA for forensic purposes. Int J Legal Med 127 (2), 309–319.

Magi, A., Benelli, M., Gozzini, A., Girolami, F., Torricelli, F., Brandi, M.L., 2010. Bioinformatics for next generation sequencing data. Genes 1 (2), 294–307.

Mardis, E.R., 2008. Next-generation DNA sequencing methods. Annu Rev Genomics Hum Genet 9, 387–402.

Nagalakshmi, U., Waern, K., Snyder, M., 2010. RNA-Seq: a method for comprehensive transcriptome analysis. Curr Protoc Mol Biol 4–11.

Nielsen, K.L., Telving, R., Andreasen, M.F., Hasselstrøm, J.B., Johannsen, M., 2016. A metabolomics study of retrospective forensic data from whole blood samples of humans exposed to 3, 4-methylenedioxymethamphetamine (MDMA)—a new approach for identifying drug metabolites and changes in metabolism related to drug consumption. J Proteome Res 15 (2), 619–627.

Nunn, S., 2013. Touch DNA collection versus firearm fingerprinting: comparing evidence production and identification outcomes. J Forensic Sci 58 (3), 601–608.

Patwardhan, A., Ray, S., Roy, A., 2014. Molecular markers in phylogenetic studies—a review. J Phylogen Evol Biol 2014 (2), 131–140.

Phillips, C., 2015. Forensic genetic analysis of bio-geographical ancestry. Forensic Science International: Genetics 18, 49–65.

Pugach, I., Matveyev, R., Wollstein, A., Kayser, M., Stoneking, M., 2011. Dating the age of admixture via wavelet transform analysis of genome-wide data. Genome Biol. 12 (2), R19.

Pyron, R.A., 2015. Post-molecular systematics and the future of phylogenetics. Trend Ecol Evol 30 (7), 384−389.

Ramachandran, S., Deshpande, O., Roseman, C.C., Rosenberg, N.A., Feldman, M.W., Cavalli-Sforza, L.L., 2005. Support from the relationship of genetic and geographic distance in human populations for a serial founder effect originating in Africa. Proc Natl Acad Sci USA 102 (44), 15942−15947.

Risch, N., 2006. Dissecting racial and ethnic differences. N Engl J Med. 354, 408−411.

Romeika, J.M., Yan, F., 2013. Recent advances in forensic DNA analysis. J Forensic Res 2014.

Roos, D.S., 2001. Bioinformatics—trying to swim in a sea of data. Science 291 (5507), 1260−1261.

Rosenberg, N.A., 2005. Algorithms for selecting informative marker panels for population assignment. J Comput Biol 12 (9), 1183−1201.

Schuster, S.C., 2007. Next-generation sequencing transforms today's biology. Nature 200 (8), 16−18.

Shen, D.N., Yi, X.F., Chen, X.G., Xu, T.L., Cui, L.J., 2007. [The prospect of application of toxicogenetics/pharmco-genetics theory and methods in forensic practice]. Fa Yi Xue Za Zhi 23 (5), 362−364.

Shendure, J., Ji, H., 2008. Next-generation DNA sequencing. Nat Biotechnol 26 (10), 1135−1145.

Shriver, M.D., Smith, M.W., Jin, L., Marcini, A., Akey, J.M., Deka, R., et al., 1997. Ethnic-affiliation estimation by use of population-specific DNA markers. Am J Hum Genet 60 (4), 957.

Thomasma, S.M., Foran, D.R., 2013. The influence of swabbing solutions on DNA recovery from touch samples. J Forensic Sci 58 (2), 465−469.

Van Steendam, K., De Ceuleneer, M., Dhaenens, M., Van Hoofstat, D., Deforce, D., 2013. Mass spectrometry-based proteomics as a tool to identify biological matrices in forensic science. Int J Legal Med 127 (2), 287−298.

Watson, J.D., Crick, F.H., 1953. Molecular structure of nucleic acids. Nature 171 (4356), 737−738.

Wei, Y.L., Li, C.X., Li, S.B., Liu, Y., Hu, L., 2011. Association study of monoamine oxidase A/B genes and schizophrenia in Han Chinese. Behav Brain Func 7 (1), 1.

Whittall, H., 2008. The forensic use of DNA: scientific success story, ethical minefield. Biotechnol J 3 (3), 303−305.

Further Reading

Alcaraz, N., Friedrich, T., Kötzing, T., Krohmer, A., Müller, J., Pauling, J., et al., 2012. Efficient key pathway mining: combining networks and OMICS data. Integrative Biol 4 (7), 756−764.

Alonso, A., Martín, P., Albarrán, C., García, P., Primorac, D., García, O., et al., 2003. Specific quantification of human genomes from low copy number DNA samples in forensic and ancient DNA studies. Croat Med J 44 (3), 273−280.

Cavalli-Sforza, L.L., 2007. Human evolution and its relevance for genetic epidemiology. Annu Rev Genomics Hum Genet 8, 1−5.

Falk, M., Hausmann, M., Lukasova, E., Biswas, A., Hildenbrand, G., Davidkova, M., et al., 2014. Determining omics spatiotemporal dimensions using exciting new nanoscopy techniques to assess complex cell responses to DNA damage: part A—radiomics. Crit Rev Eukaryot Gene Expr 24 (3).

Hawkins, T.L., Detter, J.C., Richardson, P.M., 2002. Whole genome amplification—applications and advances. Curr Opin Biotechnol 13 (1), 65−67.

Jordan, B., 2010. [Your DNA will talk to us!]. Med Sci 27 (6-7), 667−670.

Liu, G.L., Yin, Y., Kunchakarra, S., Mukherjee, B., Gerion, D., Jett, S.D., et al., 2006. A nanoplasmonic molecular ruler for measuring nuclease activity and DNA footprinting. Nature Nanotechnol 1 (1), 47−52.

Nardone, R.M., MacLeod, R.A., Capes-Davis, A., 2016. Cancer: authenticate new xenograft models. Nature 532 (7599), 313.

Nuijten, R.J., Bosse, M., Crooijmans, R.P., Madsen, O., Schaftenaar, W., Ryder, O.A., et al., 2016. The use of genomics in conservation management of the endangered Visayan warty pig (Sus cebifrons). Int J Genomics 2016.

Ovenden, S.P., Gordon, B.R., Bagas, C.K., Muir, B., Rochfort, S., Bourne, D.J., 2010. A study of the metabolome of Ricinus communis for forensic applications. Aust J Chem 63 (1), 8−21.

Pitsiladis, Y.P., Durussel, J., Rabin, O., 2014. An integrative 'Omics' solution to the detection of recombinant human erythropoietin and blood doping. Br J Sport Med 48 (10), 856−861.

Rocchiccioli, S., Tedeschi, L., Citti, L., Cecchettini, A., 2013. [Proteomics and personalized medicine]. Recenti Progressi in Medicina 104 (5), 189−199.

Roos, C., Zinner, D., 2014. Natural hybridization in primates and what "-omics" contributed to primate taxonomy and systematics. Challenges and Opportunities of the Integrative Taxonomy for Research and Society − Taxonomic

Research in the Age of OMICS Research. National Academy of Sciences, Leopoldina. Online supplement for comment. German Academy of Natural Scientists Leopoldina e.V., National Academy of Sciences, Halle/Saale, pp. 85–89.

Starr, D., 2016. When DNA is lying. Science 351 (6278), 1133–1136.

Vincini, L., 2010. The Characterisation and Identification of Body Fluid Proteins for Forensic Purposes. Department of Pharmacy and Life Sciences.

Waye, J.S., Presley, L.A., Budowle, B., Shutler, G.G., Fourney, R.M., 1989. A simple and sensitive method for quantifying human genomic DNA in forensic specimen extracts. Biotechniques 7 (8), 852–855.

White, P., 2010. Crime Scene to Court: The Essentials of Forensic Science. Royal Society of Chemistry, Cambridge, United Kingdom.

Yang, Y., Xie, B., Yan, J., 2014. Application of next-generation sequencing technology in forensic science. Genomics Proteomics Bioinformatics 12 (5), 190–197.

Biotechnology and Bioengineering in Astrobiology: Towards a New Habitat for Us

Sameen R. Imadi[1], Mustafeez M. Babar[2], Sami U. Khan[3], Humna Hasan[4], Mohsin Ali[1] and Alvina Gul[1]

[1]National University of Sciences and Technology, Islamabad, Pakistan [2]Shifa College of Pharmaceutical Sciences, Shifa Tameer-e-Millat University, Islamabad, Pakistan [3]University of Haripur, Khyber Pakhtunkhwa, Pakistan [4]Quaid-I-Azam University, Islamabad, Pakistan

7.1 INTRODUCTION TO ASTROBIOTECHNOLOGY AND ASTROBIOENGINEERING

Every ecological system on earth hosts life. A majority of living organisms inhabit the biologically favorable conditions that are accepted by a variety of species. Hence, most organisms that are distributed in these conditions live in a temperature range of 5°C−50°C. Moreover, the atmospheric and the overall environmental conditions that support the growth of these organisms are known and are regulated within an optimum range. These conditions, known as benign ambient habitats, provide the habitat for a vast variety of organisms (Seckbach, 2013).

Life emerged on earth over billions of years following the big bang of the initial supermatter. The living organisms originated in the presence of suitable conditions that supported their existence and development. The impact of comets and asteroids followed by the rapid cooling and provision of various elements helped in the origin of life on earth (Wickramasinghe et al., 2015). On similar grounds, the interstellar "space" is defined as a nonhomogenous, dynamic environment, which is strongly influenced by solar activity and is characterized by the presence of radiations of a wide range of energies and particle fluxes that might affect the living organisms (Cano et al., 2011). The presence of living organisms and their behavior in the outer space is studied as the science of astrobiology. The field helps us in deciphering essential questions about the origin of life on earth and the possibility of life on other planets. Hence, astrobiology can be defined as the branch of biology that addresses the origin and distribution of living organisms, study of fossilized records, and destiny or the presence of life in the Universe. Discovery of new planets and even new satellites for search on extraterrestrial life also falls under the umbrella of astrobiology (Briot, 2012). Astrobiology, hence, owes its discoveries to the impact of finding life, whether microbial or complex, beyond earth (Dick, 2014).

It has been observed that environmental conditions in upper atmosphere of Earth resemble a lot with surface conditions of Mars. These conditions include extreme cold, irradiation, hypobaria, and dessication (Smith, 2013). Most extremophiles are microorganisms, which belong to Archaea or bacteria. Among these organisms, methanogens which are a domain of Archaea are intriguing. It is because they can live and grow under extreme environmental conditions, are metabolically versatile, and have enormous diversity. Growth conditions of methanogens are compatible with extraterrestrial conditions throughout the solar system (Taubner et al., 2015). Extreme environments like hot springs, hyper saline soda lakes, and volcanic regions have enhanced arsenic metabolism. Certain extremophile bacteria tend to grow in a high concentration of arsenic and utilize it for fulfilling their metabolic needs, hence, are referred to as arsenotrophs. These observations help in predicting that the microbial life might exist in certain places within the solar system where volcanic eruptions are quite common. Dense brines existing in regolith of Mars or ice layers in Titan, Europa, Enceladus, and Ganymede might be the places where microbes perform arsenotrophy for energy conservation (Stolz et al., 2013).

Research conducted on spore-forming bacteria has helped in predicting that these bacteria have tough endospores that are able to withstand a variety of sterilization processes. Similarly, they can even withstand harsh environments of outer space. Experimentations were conducted in which artificial Mars conditions were produced and spores tagged with Trip to Mars and Stay on Mars were provided with different conditions. Trip to Mars spores were subjected to extraterrestrial solar radiations, space vacuum, and cosmic radiations, whereas the Stay on Mars spores were subjected with simulated environment of Mars which involved higher atmospheric pressure, cosmic radiations, and ultraviolet (UV) radiations. It was seen that both kinds of spores suffered minimum damage and their survival rate was 50% or more which proves that these spores can survive in harsh Martian conditions. Moreover, if these spores can be prevented from direct solar radiations, then their survival rate is significantly high (Horneck et al., 2012).

Bacillus subtilis strain MW01 spores which are very resistant to high UV radiations were exposed to a combination of lower Earth orbital and Martian environmental conditions for a period of 559 days. It was seen that extraterrestrial radiations were the most deleterious factors applied to these spores. Only 8% of spores were able to survive under the lower Earth orbit conditions, whereas survival rate in Martian conditions was observed to be 100%. This helped in indicating that the *Bacillus subtilis* MW01 spores can survive in outer space as well as Martian land (Wassmann et al., 2012). In another study, extremophile bacterial species were exposed to Martian conditions under simulated low pressure. It was observed that *Hyacinthoides hispanica* and *Geobacillus thermantarcticus* were highly resistant to dessication and low pressure hinting, thereby, they can survive on Mars as well

(Mastascusa et al., 2014). Furthermore, under simulated hyper gravity conditions, growth rate of microorganisms showed a robust increase (Deguchi and Horikoshi, 2013).

Bacterial species tend to protect themselves from the environmental threats by adapting a number of protective mechanisms. Extremophillic cyanobacteria, for instance, produce an UV radiation—protective biological molecule known as Scytonemin. Cyanobacteria produce this molecule under stressed terrestrial environments. Scytonemin and related molecules can be used for astrobiological missions to explore and search for extinct and extant life in extraterrestrial space (Varnali and Edwards, 2014). Extreme dry conditions, such as Dry valleys in Antarctica, Atacama Desert in Chile, and Mojave Desert in California, are analogs of Martian conditions. Genus *Chroococcidiopsis* of cyanobacteria is a rock-dwelling genus in these places. These cyanobacteria have an extraordinary ability to tolerate desication, UV radiations, and ionization. Life support systems for Martian environments can be made by mimicking the mechanisms adapted by these cyanobacteria (Billi et al., 2013).

Spores of *Bacillus pumilus* strain SAFR-032 have been isolated from spacecraft-associated surfaces. These spores show high resistance towards UV radiations and peroxide treatment (to generate oxygen-stress environments). The spores when exposed to dark space conditions for a period of 18 months showed up to 40% survival rate. Similarly, exposure to dark Martian conditions showed a survival rate of 85%—100%. One important observation, however, has been that the spores that are exposed to space conditions generated on Earth showed less viability as compared to spores that were exposed to natural space conditions. The space exposed samples were also shown to possess UV resistance in comparison to those studied under simulated laboratory conditions (Vaishampayan et al., 2012).

7.2 BIOTECHNOLOGICAL APPROACHES IN DETECTING LIFE IN SPACE

The progress in the natural sciences and the employment of multidisciplinary approach to study various aspects of biological agents have helped in deciphering many fundamental secrets of biology. Raman spectroscopy is considered to be an effective approach for identification and detection of microorganisms that colonize hostile and extreme environments on Earth. Space missions like ExoMars rover was equipped with a Raman spectrometer to analyze samples from Martian surface (Dartnell et al., 2012). Raman spectroscopy has now been adopted as an essential, nondestructive instrumentation for robotic exploration of Mars. It helps in search of traces of life in space. Miniaturized Raman spectrometers equipped with 532 nm lasers have been found to be excellent for the analysis of powdered specimens. It is due to high sensitivity of these lasers that they can efficiently sense relatively low concentrations of biological agents in bulk powdered specimens. Similarly, the Raman spectrometers equipped with 785 nm laser were observed to be even more sensitive towards Scytonemin (Vítek et al., 2012).

Experiments were also conducted on *Arabidopsis thaliana* under space conditions. In these experiments, biologically replicated DNA microarray and averaged RNA digital transcript profiling was performed. It was observed that more a hundred different genes in seedlings and the cell cultures of the plants were adversely affected by the spaceflight. However, only a few number of genes were induced by more than sevenfolds. Hence, it can be said that survival during spaceflight requires adaptive changes. *Arabidopsis* plants are observed to be dependent on gravity for their growth. Changes in gravity and simulation of microgravity also results in a change in the expression levels in the plant (Paul et al., 2012).

Parabolic flight environment is observed to be suitable for research in molecular biology, which involved transition to microgravity. This has been explained by experiments conducted on *Arabidopsis thaliana*, the model plants. Changes in gene expression due to parabolic flights can be observed in time frames by manipulating replication, proper controls, and analysis. Responses that are generated as a result to parabolic flights are the same responses that are involved in gravity sensing in plants. These responses include metabolic processes which are induction of auxin metabolism and signaling, alteration in calcium-mediated signaling caused by differential expression of associated genes, decrease in disease resistance, and alteration in cell wall biochemistry (Paul et al., 2011).

Vacuum technology can be developed to simulate planetary atmospheric conditions. In this technique, vacuum chambers are made, which are capable of simulating atmospheres and surface temperatures resembling the space environments. These chambers are appropriate to study physical, biological, and chemical changes that are induced in a sample by in situ irradiation or physical parameters in controlled environments. Laboratory simulations of planetary environments are up till now considered as the best ways to create feasible research options for advanced planetary sciences. Among the vacuum chambers is the Planetary Atmosphere and Surfaces Chamber (PASC) that produce versatile planetary environment with computer-controlled gas compositions, temperature, and pressure. These conditions are representative of most of the planets in solar system. With PASC, much research has been conducted on UV absorbing properties of Martian analogs, presence, and stability of saline water on Mars and survival capacity of microorganisms under these conditions. Moreover, the study of increased resistance and survival of lichens (simple eukaryotes) in simulated Mars and likely planetary conditions has also been studied using these chambers (Mateo-Marti, 2014).

Furthermore, microbial cellular processes are largely affected by microgravity. Microgravity is observed to cause alterations in cell growth, gene expression, biotechnological products, and natural pathways. Antimicrobial susceptibility, production of secondary metabolites, enzyme activity, recombinant protein production, and resistance to stresses are all affected by alterations in the gravitational force. Microgravity can be simulated to produce altered secondary metabolites, enzymes, and recombinant proteins from plants and microorganisms which also have significant biopharmaceutical importance (Huangfu et al., 2015).

7.3 INTEGRATION OF BIOTECHNOLOGY AND ASTROBIOLOGY

Palaeo-geological evidence has shown that earlier the Mars and the Earth had similar environmental conditions that could support the growth of Archaea and cyanobacteria. Raman spectroscopy, as mentioned earlier, has been used for space exploration to detect biological agents. Incorporating the instrument in space shuttles and rovers has helped in analyzing the surface and subsurface specimens on Mars. Certain novel analytical systems have also been developed to specifically seek biominerals and biological molecules arising from extinct or extant planetary life (Edwards et al., 2014). Raman spectroscopy, hence, defines a way to isolate and identify organic and inorganic materials. It is generally considered to be a highly effective technique for detection of mineralogical and biological markers for missions on Mars (Vandenabeele and Jehlička, 2014).

Future astrobiology missions heading for Mars are expected to emphasize on rover users with in situ petrologic capabilities. For the purpose, multispectral microscopic imagers would be deployed on the space rovers (Nunez et al., 2014). Research is currently being carried out on the development of "Search for Extra Terrestrial Genomes" instruments that are aimed to isolate, extract, and sequence nucleic acids in situ on Mars. Although it can be said that nucleic acids are not the most general approach to search for life on Mars. These strategies will help in characterization of any life on Mars which might be evolutionarily related to life on Earth (Lui et al., 2011).

On similar grounds, the Space Environment Survivability of Living Organisms experiment measures the long-term survival, germination, and growth responses of living organisms. In one of these experimental procedures, the metabolic activity of *Bacillus subtilis*, which were exposed to micro gravity, ionizing radiation, and heavy ion bombardment of high inclination orbit, was measured (Nicholson et al., 2011). The organisms were found to survive under these variable stress conditions. Similarly, under altered gravitational conditions, vascular smooth muscle cells exhibit different physiological responses. Exposure of cultured rat aortic smooth muscle cells to altered gravitational conditions resulted in reduction of cell surface heparan sulfate proteoglycans (HSPG) and activation of nitric oxide synthase. Glycocalyx, especially HSPG, was established to be a potential gravity sensor because its expression levels were changed in altered conditions (Kang et al., 2013).

Experiments have also been conducted to examine the alterations in chemical behavior of organic molecules originating from plants under the space environment. It was observed that resistance to irradiation is a protective function developed by many plant species to defend themselves from various chemicals and radiations. In plants, the most altered compounds in space environment are dipeptides like aspartic acid and aminobutyric acid. Conversely, the most resistant amino acids are alanine, aminoisobutyric acid, valine, and glycine (Bertrand et al., 2012). Plant seeds were exposed to 1.5 years of solar UV rays, temperature fluctuations, space vacuum, and galactic cosmic radiations. It was observed that out of 2100 exposed seeds of wild-type *A. thaliana* and *Nicotiana tabacum*, around 23% produced viable plants after returning to earth conditions. Survival rate was lower in mutant plants that lacked UV screens. Tobacco seeds, however, had a significantly higher survival rate (Tepfer et al., 2012). *Arabidopsis thaliana* plants were grown in a space flight, and it was observed that these seeds had an 89% germination rate in space flight as compared to 91% rate in ground control. These experimental conditions, therefore, helped in concluding that the seedlings are expected to grow equally in control as well as experimental conditions. However, in flight-grown plants, number of adventitious root from axis of hypocotyls was quite high (Millar et al., 2011). Hence, integrating the biotechnological techniques with the astrobiological methods can help in determining the presence of life in outer space and other planetary conditions.

7.4 ADVANCED INTEGRATED TECHNOLOGIES IN ASTROBIOLOGY FOR FINDING ADEQUATE HABITAT CONDITIONS

As per the biological point of view, all environments in the Universe can be classified in only three forms: the uninhabitable, inhabited habitat, or uninhabited habitat (Cockell, 2014). Color of a planet reveals knowledge about its properties especially structural features and environment of rocky planets. Detectable surface features of planets can be linked to extreme niches in which extremophiles might survive (Hegde and Kaltenegger, 2013). For detecting these properties, a miniature near-infrared point spectrometer, which is based on acousto-optic tunable filter technology, has been developed. It can be used to screen and corroborate analysis of samples which might contain organic biomarkers or mineralogical signatures. This can detect and analyze extant or extinct organic material collected in situ from different planetary surfaces. The equipment works with a laser desorption time of flight (LDTOF) mass spectrometer. LDTOF will prescreen the samples and will gain evidence of volatile or refractory organics before laser desorption (Chanover et al., 2012). The robotic missions that are to be set in space in future require highly capable instruments, spacecraft subsystem, or ground support technologies to study the conditions that can presumably support life (Singleton et al., 2011).

For advanced integrated studies on astrobiology, new eukaryotic models are being developed. For this purpose, UV-resistant yeast, which has been isolated from high-altitude volcanic area, like the Atacama Desert, can be used. *Cryptococcus friedmannii*, *Exophiala* spp., *Holtermanniella watticus*, and *Rhodosporidium toruloides* have been found to be resistant to UV B, UV C, and other environmental UV radiations. They can also grow on saline media and at moderately cold temperatures too. Some species have also been observed to grow in extremely cold temperatures. It is due to their extreme tolerance that these organisms are an interesting tool for astrobiological research (Pulschen et al., 2015).

Space synthetic biology is a branch of biotechnology, which is aimed at engineering of biological systems for exploration of space. Space synthetic biology is expected to design reliable and robust organisms, which may assist the long-duration space missions by providing efficient biological products. Advanced synthetic space biology approaches can be used for transforming biological wastes as well as the in situ destination planet resources for practical products. Synthetic space biology can be used for resource utilization, manufacturing, life support, space medicine, space cybernetics, and terraforming in space (Menezes et al., 2015a,b). Space synthetic biology can also help in developing products that can be utilized for finding ways and means to support the survival of life after they inhabit new planets. A few of the developed products have been discussed in the following sections.

7.4.1 Proteins (Antibodies)-Based Approaches

It has been observed that frozen and vacuum-dried fluorescent tracer antibodies are stable even after excessive exposure to gamma radiations. Gamma radiations are expected during Mars missions. These engineered antibodies can be used to explore planetary system for life, based on the antigen–antibody reactions. They can be generated and stored for space missions (de Diego-Castilla et al., 2011).

7.4.2 Microfluidics

A fully integrated microfluidic system for small space craft has been prepared. This system is fully automated and is capable of in situ measurement of hundreds of microbial genes expression. It can process multiple samples at a time. This miniaturized microfluidic system will be capable of lysis of cell wall of bacteria from cultures which are grown in space, extraction, and purification of RNA from cells, hybridization of RNA on microarrays, and eventually reading the microarray signals. It is a suitable device to be deployed on NASA nano-satellite platforms. This system can also be used with minor modifications for specific microbial applications (Andrew et al., 2012). A full integrated multilayer microfluidic chemical analyzer for automated sample processing and labeling as well as analysis that uses capillary zone electrophoresis has also been developed and characterized. This microfluidic system is fully automated and portable. It is capable of autonomous analysis of diverse compound classes in space (Kim et al., 2013). Furthermore, biosensors have emerged to be a promising tool for the real-time monitoring. They can be used to determine the resistance of bacterial strains like *Chlamydomonas reinhardtii* mutants in space conditions (Cano et al., 2011).

7.4.3 Microarray Technology

Laboratory on a chip system with electrochemical sensing capabilities can be used in space stations because they can provide real-time physiological measurements in space flight environments. These systems are worth using in space because they can easily be miniaturized and, hence, can be integrated with existing space hardware systems. Development of electrochemical sensing system is still in progress for the research on microgravity. In this system, ion-selective electrodes are miniaturized and used as electrochemical sensors (Salim et al., 2013). Biochips are known to be a promising instrument for searching of organic compounds in solar system. Nucleic acid aptamers are considered to be powerful candidates for space biochip development as they have a high affinity and specificity. High irradiations, a common feature of Martian atmosphere, do not affect performance of nucleic acid aptamers. Hence, it can be said that they can be used for space missions aiming towards detection of life on different planets (Baqué et al., 2011).

Life marker chips (LMCs) are a result of research on antibody microarray technology. In this technique, a microfluidic fused silica chip includes the relevant components for analysis. The design of LMCs is based on three parallel fluidic pathways that allow three different classes of assays. These chips have been incorporated in the ExoMars mission that would be sent to Mars in 2018 (Prak et al., 2011). LMCs will be used to detect molecular signatures of life in samples obtained from shallow subsurface of Mars. ExoMars mission level radiation doses do not affect the LMC. It can be concluded that expected radiation environment of Mars does not possess any significant risks to packaged antibodies in the form anticipated for LMC instrument (Derveni et al., 2012).

7.5 SOLAR SYSTEM EXPLORATION

Exploration of solar system for life and the appropriate habitat that supports life is the key concept of astrobiotechnology. A number of tools have been employed for the purpose. Remote chemical analysis of Enceladus, Saturn's moon, is to be performed on board a space craft. The instrumentation employed is meant to look for any evidence for the presence of life in subsurface ocean habitat. It would examine nascent ice grains that are collected by flying the space craft directly through plume or jet of Enceladus (Kirby et al., 2013). Human space exploration and space research can be accelerated by applications of synthetic biology. Synthetic biology and biotechnology can help to the point when permanent human basis on Mars are feasible. At present, we know that all the organisms evolved on planet Earth. For growth of organisms under space conditions, they should survive in metabolically active state. This demands that they have decreased requirements of nutrients and they should produce secondary metabolites and all needed compounds of interest while they only rely on minimal nutrients and exposed environments (Verseux et al., 2015).

Research explains that radiations of stars play a key role in development and evolution of organic compounds in planetary environments and astrophysical atmospheres. Although electromagnetic radiations of stars comprise of as low as 8% of UV radiations, this level is critical for organic compounds because these can be destroyed by UV rays (Cottin et al., 2015). One of the major space exploration targets for Mars missions are lithofacies (lithological subdivisions) suggestive of biotic activities. Although it is known that lithofacies are not a confirmation of the presence of any biological activity, still they can serve as sources for providing samples, which help in further analysis. It has been observed through gas chromatography/mass spectrometry experiments that carbonate samples identified as microbial had a higher concentration of diverse biomarkers as compared to nonmicrobial carbonate samples. Microbial carbonate biomarkers can act as efficient detection technology for selecting samples and analysis on Mars (Marshall and Cestari, 2015).

Space is a unique laboratory, which allows exposure of all of its samples to space parameters. It also causes irradiation of samples under appropriate conditions. Titan's atmosphere might convert simple organic molecules into organic aerosols (Cottin et al., 2015). This fact cannot be denied that any extraterrestrial organism identified on Mars or in space may have been originated on Earth and then transported to planetary neighbors. This can either happen by human intervention or even by phenomenon known as panspermia process (deposition of chemical building bricks through delivery by meteorites, asteroids, or comets) (Edwards et al., 2014). Solar UV radiations can destroy traces of organic materials and life on surface of planets including Mars. It can also destroy organic molecules that are present on meteors and meteorites. Scientists presume that solar radiations can generate reactions incorporating organic molecular chemistry on surface of Titan (Cottin et al., 2013).

Presence of oxygen or ozone in atmosphere of a rocky planet acts as a biosignature. It has been observed that a biological system that synthesizes its constituents from raw materials and energy has an inherent adaptation to become widespread and dominant. Under these kinds of reactions, oxygen acts as an efficient photocatalyst. Hence, it can be said that oxygen appears to be a good biosignature (Léger et al., 2011). Search for organic molecules is now the top most priority of various space missions. These missions have been trying to identify the evolution of organic molecules under simulated environmental conditions. However, one of the most challenging aspect is that the UV spectrum which reaches the surface of Mars needs to be reproduced in these simulated conditions. Filtered extraterrestrial solar electromagnetic radiations actually mimic Mars-like surface UV radiation conditions. In one of the studies, it was noted that 1.5-year-old exposure to Mars conditions can degrade the organic compounds completely (Noblet et al., 2012).

There is a class of exoplanets that reside in the habitable zone in the Universe. Habitable zone is considered to be the zone that contains liquid water on its surface. According to theoretical calculations, these exoplanets have weak planetary magnetic field, which is specific for the case of super earths. Due to the weak magnetic field, there is a high flux of galactic cosmic rays. Hence, energy particles including muons penetrate through them. It is imperative that if radiation doses of these muons are high, chances of sustaining long-term biosphere on the planet are very low (Atri et al., 2013). Mars meteorite sample is found to contain a presumptive fossilized bacterial biofilm that is composed of wide variety of bacteria-like morphologies. It has been reported that these life-like features are present in Martian environment and other meteorites. These reports have not been substantiated up till now because the mineral features of meteorites mimic bacteria and microorganisms. Hence, it is difficult to distinguish between fossilized bacteria and nonbiological artefacts (Wainwright et al., 2014).

Comets are also expected to contain organic molecules that can serve as chemical building blocks of life. There is evidence of the presence of refrozen lakes in comet 67 P/C-G. The presence of these lakes predicts that freeze-dried microorganisms including bacteria and viruses might be present there. There is also probability of signs of life as ice-living frozen microorganisms might get active as their habitats warm up (Wickramasinghe et al., 2015). Space exploration missions of NASA are mentioned in Table 7.1.

TABLE 7.1 Current and Future Astrobiological Missions in Space

S. No	Name	Launch date	Arrival date	Purpose	Mission type and target
1.	Mars Atmosphere and Volatile Evolution (MAVEN)	2013	2014	Studying how atmosphere loss triggered the disappearance of liquid water	Orbiter mission for Mars
2.	Lunar Atmosphere and Dust Environment Explorer (LADEE)	2013	2013	Data about lunar atmosphere and air-borne dust	Orbiter mission for Our Moon

(Continued)

TABLE 7.1   (Continued)

S. No	Name	Launch date	Arrival date	Purpose	Mission type and target
3.	Mars Science Laboratory (MSL)	2011	2012	MSL's curiosity rover is studying whether Mars ever had environment capable of supporting microbial life	Lander/Rover for Mars
4.	Organism/ Organic Exposure to Orbital Stresses (O/OREOS)	2010	–	How living organisms are affected by space environment	Orbiter for space environment
5.	Kepler	2009	–	Search for Earth-sized planets around distant stars	Space telescope for Cygnus Lyra region of our Milky way galaxy
6.	Dawn	2007	2011	New information about early solar system and formation of rocky planets by visiting protoplanets in the main asteroid belt	Flyby for Vesta and Ceres
7.	Stratospheric Observatory for Infrared Astronomy (SOFIA)	2007	–	Making observations of new solar systems, complex molecules in space, and planets in our own solar system	Space telescope for our Universe
8.	Spitzer space telescope	2003	–	Search for habitable world around distant stars, study of astrobiology targets in our solar system	Space telescope for milky way galaxy
9.	Mars Exploration Rovers (MER)	2003	2004	MERs have shaped our understanding of Mars geology and past environments at the surface that may have been suitable for life	Lander/Rover for Mars
10.	Mars Odyssey	2001	2001	Provides detailed maps of Mars that are used to determine evolution of Mars environment and its potential for life	Orbiter for Mars
11.	Cassini-Huygens Saturn Orbiter and Titan Probe	1997	2005	Revealed new details about the potential for life on moons around giant planets	Orbiter/Lander for Saturn, Titan and Enceladus
12.	Hubble Space Telescope	1990	–	Provided observations of everything from distant galaxies to nearby planets in our own solar system	Space telescope for the Universe
13.	2020 Mars Rover	2020	2021	Will provide details about the potential for life on Mars, both past and present	Lander/Rover for Mars
14.	James Webb Space Telescope (JWST)	2018	–	Will be used to hunt for solar systems capable of supporting habitable planets	Space telescope for the Universe

7.6 CONCLUSION AND FUTURE PROSPECTS

NASA defines life as self-sustaining system capable of Darwinian evolution. This definition incorporates molecular genesis with replicative procedures and also avoids several pitfalls of alternative definitions which are based on capabilities of organisms to reproduce and divide (Edwards et al., 2014). It can be said that life on Mars, if it may exist, is related to life on Earth. Study of Mars analog sites is an interesting part of planetary studies and space exploration. Mars like features which exist on Earth include volcanic and erosional systems, glacial and thermokarst systems, dunes, impact craters, and playas. For the Mars exploration missions, first Mars analog sites were studied to search about a proper place for landing of Viking mission (Lui et al., 2011).

We are near the end of the space shuttle era. Future belongs to long space flights and planetary settlement. Settlements on a different planet require chemical as well as nutritional sources which are to be generated in space. These include amino acids, vitamins, and oxygen. With synthetic biotechnology, organisms can be designed and synthesized, which can supply all nutritional needs of humans in space. For this purpose, photosynthetic microbes can be produced as they produce more yield, biomass, and nutritional molecules as compared to green plants. They can be quickly engineered and will have an excellent ability to adapt in environment as well as will reproduce new microbes. A bioreactor that includes genetically engineered nutrient producing microorganisms should be tested for planetary settlement in future (Way et al., 2011).

It really is a small world after all. We have to go further down in Mars research and research on life in space. After then we would be able to find out that the Earth life and Mars life are both rooted in same big tree (Rummel, 2014). Space synthetic biology is expected to answer a lot of questions about life in space and new habitat for human race (Menezes et al., 2015a,b).

References

Andrew, P., Kia, P., Fathi, K., Antonio, R., 2012. Gene Expression Measurement Module (GEMM)—a fully automated, miniaturized instrument for measuring gene expression in space. Aerospace Medicine. American Society for Gravitational and Space Research. Patent Number: ARC-E-DAA-TN5901.

Atri, D., Hariharan, B., Grießmeier, J.M., 2013. Galactic cosmic ray-induced radiation dose on terrestrial exoplanets. Astrobiology 13 (10), 910−919.

Baqué, M., Le Postollec, A., Ravelet, C., Peyrin, E., Coussot, G., Desvignes, I., et al., 2011. Investigation of low-energy proton effects on aptamer performance for astrobiological applications. Astrobiology 11 (3), 207−211.

Bertrand, M., Chabin, A., Brack, A., Cottin, H., Chaput, D., Westall, F., 2012. The PROCESS experiment: exposure of amino acids in the EXPOSE-E experiment on the international space station and in laboratory simulations. Astrobiology 12 (5), 426−435.

Billi, D., Baqué, M., Smith, H.D., McKay, C.P., 2013. Cyanobacteria from extreme deserts to space. Adv. Microbiol. 3 (6A), article ID: 38613.

Briot, D., 2012. A possible first use of the word astrobiology? Astrobiology 12 (12), 1154−1156.

Cano, J.B., Giannini, D., Pezzotti, G., Rea, G., Giardi, M.T., 2011. Space impact and technological transfer of a biosensor facility to earth application for environmental monitoring. Recent Planets Space Tech. 1, 18−25.

Chanover N., Tawalbeh R., Glenar D., Voelz D., Xifeng Xiao, Uckert K., et al. (2012). Rapid assessment of high value samples: an AOTF-LDTOF spectrometer suite for planetary surfaces. In: IEEE Aerospace Conference Proceedings. DOI: 10.1109/AERO.2012.6187060.

Cockell, C.S., 2014. Types of habitat in the Universe. Int. J. Astrobiol. 13 (2), 158–164.

Cottin, H., Saiagh, K., Coll, P., Fray, N., Raulin, F., Stalport, F., et al., 2013. PSS: photochemistry and space station. A low earth orbit laboratory for astrochemistry and astrobiology outside the international space station. EPSC Abstracts 8, 583-1.

Cottin, H., Saiagh, K., Guan, Y.Y., Cloix, M., Khalaf, D., Macari, F., et al., 2015. The AMINO experiment: a laboratory for astrochemistry and astrobiology on the EXPOSE-R facility of the International Space Station. Int. J. Astrobiol. 14 (1), 67–77.

Dartnell, L.R., Page, K., Jorge-Villar, S.E., Wright, G., Munshi, S., Scowen, I.J., et al., 2012. Destruction of Raman biosignatures by ionising radiation and the implications for life detection on Mars. Anal. Bioanal. Chem. 403 (1), 131–144.

de Diego-Castilla, G., Cruz-Gil, P., Mateo-Marti, E., Fernández-Calvo, P., Rivas, L.A., Parro, V., 2011. Assessing antibody microarrays for space missions: effect of long-term storage, gamma radiation, and temperature shifts on printed and fluorescently labeled antibodies. Astrobiology 11 (8), 759–773.

Deguchi, S., Horikoshi, K., 2013. Expanding limits for life to a new dimension: microbial growth at hypergravity. Polyextremophiles 27, 467–481 (Cellular Origin, Life in Extreme Habitats and Astrobiology).

Derveni, M., Hands, A., Allen, M., Sims, M.R., Cullen, D.C., 2012. Effects of simulated space radiation on immunoassay components for life-detection experiments in planetary exploration missions. Astrobiology 12 (8), 718–729.

Dick, S.J., 2014. Analogy and the societal implications of astrobiology. Astropolitics: Int. J. Space Politics Policy 12 (2-3), 210–230.

Edwards, H.G.M., Hutchinson, I.B., Ingley, R., Jehlička, J., 2014. Biomarkers and their Raman spectroscopic signatures: a spectral challenge for analytical astrobiology. Phil. Trans. R. Soc. A 372. Available from: 20140193.

Hegde, S., Kaltenegger, L., 2013. Colors of extreme exo-Earth environments. Astrobiology 13 (1), 47–56.

Horneck, G., Moeller, R., Cadet, J., Douki, T., Mancinelli, R.L., Nicholson, W.L., et al., 2012. Resistance of bacterial endospores to outer space for planetary protection purposes—experiment PROTECT of the EXPOSE-E Mission. Astrobiology 12 (5), 445–456.

Huangfu, J., Zhang, G., Li, J., Li, C., 2015. Advances in engineered microorganisms for improving metabolic conversion via microgravity effects. Bioengineered 6 (4), 251–255.

Kang, H., Liu, M., Fan, Y., Deng, X., 2013. A potential gravity-sensing role of vascular smooth muscle cell glycocalyx in altered gravitational stimulation. Astrobiology 13 (7), 626–636.

Kim, J., Jensen, E.C., Stockton, A.M., Mathies, R.A., 2013. Universal microfluidic automaton for autonomous sample processing: application to the Mars Organic Analyzer. Anal. Chem. 85 (16), 7682–7688.

Kirby, J.P., Price, K., Willis, P., Jones, S., 2013. Remote Chemical Analysis at Enceladus: An Astrobiology Science Instrument Concept. American Geophysical Union, San Francisco, CA, USA.

Léger, A., Fontecave, M., Labeyrie, A., Samuel, B., Demangeon, O., Valencia, D., 2011. Is the presence of oxygen on an exoplanet a reliable biosignature? Astrobiology 11 (4), 335–341.

Lui, C., Carr, C.E., Rowedder, H., Ruvkun, G., Zuber, M., 2011. SETG: an instrument for detection of life on Mars ancestrally related to life on Earth. Aerospace Conference . Available from: http://dx.doi.org/10.1109/AERO.2011.5747299.

Marshall, A.O., Cestari, N.A., 2015. Biomarker analysis of samples visually identified as microbial in the Eocene Green River formation: an analogue for Mars. Astrobiology 15 (9), 770–775.

Mastascusa, V., Romano, I., Di Donato, P., Poli, A., Della Corte, V., Rotundi, A., et al., 2014. Extremophiles Survival to Simulated Space Conditions: An Astrobiology Model Study. Orig. Life Evol. Biosph. 44 (3), 231–237.

Mateo-Marti, E., 2014. Planetary atmosphere and surfaces chamber (PASC): a platform to address various challenges in astrobiology. Challenges 5, 213–223.

Menezes, A.A., Cumbers, J., Hogan, J.A., Arkin, A.P., 2015a. Towards synthetic biological approaches to resource utilization on space missions. J. R. Soc. Interface 12 (102), 20140715. https://doi.org/10.1098/rsif.2014.0715.

Menezes, A.A., Montague, M.G., Cumbers, J., Hogan, J.A., Arkin, A.P., 2015b. Grand challenges in space synthetic biology. J. R. Soc. Interface 12 (113), 20150803. https://doi.org/10.1098/rsif.2015.0803.

Millar, K.D.L., Johnson, C.M., Edelmann, R.E., Kiss, J.Z., 2011. An endogenous growth pattern of roots is revealed in seedlings grown in microgravity. Astrobiology 11 (8), 787–797.

Nicholson, W.L., Ricco, A.J., Agasid, E., Beasley, C., Diaz-Aguado, M., Ehrenfreund, P., et al., 2011. The O/OREOS mission: first science data from the Space Environment Survivability of Living Organisms (SESLO) payload. Astrobiology 11 (10), 951–958.

Noblet, A., Stalport, F., Guan, Y.Y., Poch, O., Coll, P., Szopa, C., et al., 2012. The PROCESS experiment: amino and carboxylic acids under Mars-like surface UV radiation conditions in low-earth orbit. Astrobiology 12 (5), 436–444.

Nunez, J.I., Farmer, J.D., Sellar, R.G., Swayze, G.A., Blaney, D.L., 2014. Science applications of a multispectral microscopic imager for the astrobiological exploration of Mars. Astrobiology 14 (2), 132–169.

Paul, A.-L., Manak, M.S., Mayfield, J.D., Reyes, M.F., Gurley, W.B., Ferl, R.J., 2011. Parabolic flight induces changes in gene expression patterns in *Arabidopsis thaliana*. Astrobiology 11 (8), 743–758.

Paul, A.-L., Zupanska, A.K., Zhang, Y., Sun, Y., Li, J.-L., Shanker, S., et al., 2012. Spaceflight transcriptomes: unique responses to a novel environment. Astrobiology 12 (1), 40–56.

Prak A., Leeuwis H., Heidemann R.G., Leinse A., Borst G., (2011). Integration of optical waveguides and microfluidics in a miniaturized antibody micro-array system for life detection in the NASA/ESA ExoMars mission. Proc. SPIE7928.

Pulschen, A.A., Rodrigues, F., Duarte, R.T., Araujo, G.G., Santiago, I.F., Paulino-Lima, I.G., et al., 2015. UV-resistant yeasts isolated from a high-altitude volcanic area on the Atacama Desert as eukaryotic models for astrobiology. Microbiol. Open 4 (4), 574–588.

Rummel, J.D., 2014. Carl Woese, Dick Young, and the roots of astrobiology. RNA Biol. 11 (3), 207–209.

Salim, A.W., Wan, W., H, Park, J., Ul Haque, A., Marshall Porterfield, D, 2013. Lab-on-a-chip approaches for space-biology research. Recent Patents Space Technol 3 (1), 24–39.

Seckbach, J., 2013. Life on the edge and astrobiology: who is who in the polyextremophiles world? Polyextremophiles 27, 61–79 (Cellular Origin, Life in Extreme Habitats and Astrobiology).

Singleton, J., Kremic, T., Hughes, P., Perry, R.B., Beauchamp, P.M., Clarke, J.T., et al., 2011. Technology development for NASA science missions: challenges and potential opportunities. Aerospace Conference . Available from: http://dx.doi.org/10.1109/AERO.2011.5747646.

Smith, D.J., 2013. Microbes in the upper atmosphere and unique opportunities for astrobiology research. Astrobiology 13 (10), 981–990.

Stolz, J.F., Oremland, R.S., Switzer Blum, J., Hoeft, S.E., Baesman, S.M., et al., 2013. Arsenic, Anaerobes, and Astrobiology. American Geophysical Union, San Francisco, CA, USA.

Taubner, R.-S., Schleper, C., Firneis, M.G., Rittmann SK-MR, 2015. Assessing the ecophysiology of methanogens in the context of recent astrobiological and planetological studies. Life 5 (4), 1652–1686.

Tepfer, D., Zalar, A., Leach, S., 2012. Survival of plant seeds, their UV screens, and nptII DNA for 18 months outside the International Space Station. Astrobiology 12 (5), 517–528.

Vaishampayan, P.A., Rabbow, E., Horneck, G., Venkateswaran, K.J., 2012. Survival of Bacillus pumilus spores for a prolonged period of time in real space conditions. Astrobiology 12 (5), 487–497.

Vandenabeele, P., Jehlička, J., 2014. Mobile Raman spectroscopy in astrobiology research. Phil. Trans. R. Soc. A372. Available from: 20140202.

Varnali, T., Edwards, H.G.M., 2014. Reduced and oxidized Scytonemin: theoretical protocol for Raman spectroscopic identification of potential key biomolecules for astrobiology. Spectrochim. Acta A Mol. Biomol. Spectrosc. 117, 72–77.

Verseux, N., Paulino-Lima, I.G., Baqué, M., Billi, D., Rothschild, L.J., 2015. Synthetic biology for space exploration: promises and societal implications. In: Hagen, K., Engelhard, M., Toepfer., G. (Eds.), Ambivalences of Creating Life, 45. Springer, Berlin, pp. 73–100. (Ethics of Science and Technology Assessment).

Vitek, P., Jehlička, J., Edwards, H.G.M., Hutchinson, I., Ascaso, C., Wierzchos, J., 2012. The miniaturized Raman system and detection of traces of life in halite from the Atacama Desert: some considerations for the search for life signatures on Mars. Astrobiology 12 (12), 1095–1099.

Wainwright, M., Rose, C.E., Omairi, T., Baker, A.J., Wickramasinghe, C., Alshammari, F., 2014. A presumptive fossilized bacterial biofilm occurring in a commercially sourced Mars meteorite. Astrobiol. Outreach. 2 (2).

Wassmann, M., Moeller, R., Rabbow, E., Panitz, C., Horneck, G., Reitz, G., et al., 2012. Survival of spores of the UV-resistant Bacillus subtilis strain MW01 after exposure to low-Earth orbit and simulated Martian conditions: data from the space experiment ADAPT on EXPOSE-E. Astrobiology 12 (5), 498–507.

Way, J.C., Silver, P.A., Howard, R.J., 2011. Sun-driven microbial synthesis of chemicals in space. Int. J. Astrobiol. 10 (4), 359–364.

Wickramasinghe, N.C., Tokoro, G., Wainwright, M., 2015. The transition from earth centered biology to cosmic life. J. Astrobiol. Outreach. 3, 1.

Further Reading

Wan, WAWS., Joon, HP., Aeraj, UI Haque, Porterfield, M., 2013. Lab-on-a-chip approaches for space-biology research. Recent Planets Space Tech. 3 (1), 24–39.

Wickramasinghe, N.C., Wainwright, M., Smith, W.E., Tokoro, G., Al-Mufti, S., Rosetta studies of Comet 67P/Churyumov-Gerasimenko: Prospects for establishing cometary biology. Journal of Astrobiol Outr 3: 126–133.

8

Lab-on-a-Chip Technology and Its Applications

Burak Yılmaz and Fazilet Yılmaz

Turgut Ozal University, Ankara, Turkey

8.1 INTRODUCTION

A microfluidic platform contains set of well-defined fabricated fluidic operational units. Lab-on-a-chip (LOC) platforms provide a solid way for miniaturization, integration, automation, and parallelization of chemical processes (Mark et al., 2010).

The main advantage of LOC devices is cost efficiency because of using fewer reagents in modular fabricated miniaturized devices. Another important advantage is speed; parallelization of reaction chambers in LOC devices speed up the operation in the small area. Moreover, LOC technologies increase throughput and automation of analytical systems. LOC devices can be designed to control fluidics in micro- and nanoscales, and according to this scale difference sometimes they are called as microfluidics and nanofluidics. In LOC technology, micro- or nanochannels allow control of fluids in low quantities to enable biochemical reactions at very small volumes. Integrated circuits and microchannels are designed by using photolithography method, and it allows creating paths and control elements for fluids. LOC also requires integrated pumps, electrodes, valves, electrical fields, and electronics to become complete systems (Casquillas and Timothée, 2015).

Development of integrated circuit technology and wafer fabrication facilities leads the rise of microfluidics and lead to smaller and faster electronic devices. Microfluidic fabrication is based on silicon or glass and polymers. Silicon and glass have well-controlled mechanical and chemical properties, but they need more manufacturing costs and high processing complexity.

However, polymers can easily be fabricated via soft lithography or hot embossing, where a single mold can be used as a template for many devices. This enables production of high-volume disposable devices. By contrast, the mechanical and chemical properties of

Omics Technologies and Bio-engineering: Towards Improving Quality of Life
DOI: https://doi.org/10.1016/B978-0-12-804659-3.00008-7

polymers have reliability problem, and it makes surface modification, which is important for robust device functionality (Neuži et al., 2012).

LOC has been used in amount applications, and molecular biology is the core technology for LOC; applications mainly focus on diagnostic and genomic analysis, and other applications are biochemical analysis, proteomic and cell research, biosensor, and drug development (Table 8.1).

LOC are being applied in lots of molecular biology experiments such as DNA isolation, PCR, qPCR, electrophoresis, and sequencing. LOC has huge potential to build high-speed and more sensitive DNA/RNA amplification on PCR microfluidics systems. During any pandemic, LOC-PCR devices offer rapid detection of viruses and bacteria to overcome infection. Scaling down PCR-based diagnostic devices by LOC technologies enables laboratory-independent genetic diagnostic services such as HIV and HBV infection. These developing LOC devices can also be used for biomarker identification of genetically based disease. Moreover, the food industry can use LOC devices for rapid pathogen detection because it speeds up customs clearance during food transporting (Timothée, 2015a).

DNA sequencing has been used in molecular biology for several decades. The first human genome project was completed in 15 years. Today, using LOC devices to sequence a human genome can be completed within hours, which is thousands times faster. Next-generation sequencing platforms are using LOC technologies to read millions of DNA

TABLE 8.1 LOC Technologies and Applications

Area of LOC	Application of LOC	Products in Market	References
1. Diagnostics	DNA isolation, PCR, qPCR, electrophoresis, sequencing	Fluidigm BiomarkHD, Agilent LabChip, Elvesys system	Timothée, 2015b
			Kaigala et al., 2010
			Beer et al., 2007
2. Genomics	Next-generation sequencing microarray	Illumina Hiseq, Ion torrent, Nanopore, PasificBio	Metzker, 2005
			Wang, 2000
3. Biochemical assays	Immunological assays, glucose monitoring	Dexcom G5	Wang et al., 2003
4. Proteomics	SDS-PAGE, MS	908 Devices Zip Chip	Schasfoort, 2004
			Mouradian, 2002
5. Biosensors	Bioaffinity and biocatalytic devices	IST AG IV4 biosensor	Wang, 2000
6. Cell research	Cell culturing and monitoring, flow cytometers	Fluidigm Helios	Huh et al., 2005

fragments in a single chip, most of them are used in laboratories; Nanopore technologies is one among them that developed the smallest and fastest platform waiting to be commercialized.

Protein analysis includes cell extraction, electrophoresis, blotting, digestion, and mass spectrometry (MS) analysis. LOC integrates all these protein analysis steps in one chip. Integrating steps reduce analysis time into few minutes. Crystallization process can be parallelized and speeded up in LOC systems to study their structure. Additionally, immunoassays are the other LOC applications to shorten the reaction compared to the conventional approach. LOC microchannels are the most suitable platform for cell research; microfluidics enables controlling cell flow, labeled antibody staining and imaging, cell differentiation, cell sorting, and cytometry in cell biology experiments.

8.1.1 Diagnostics

LOCs are being applied in lots of molecular diagnostic steps: DNA extraction and purification, PCR, qPCR, molecular detection, electrophoresis, etc.

8.1.1.1 DNA Extraction and Purification on LOC Devices

Nucleic acid needs to be purified from biological cells for diagnostic assays. DNA extraction basically requires two steps: cell lysis by disrupting cells membrane and nucleus membrane, and DNA purification from unwanted samples such as membrane lipids, proteins, and RNA. LOC cell lysis devices can be based on several types of lysis methods: chemical lysis, thermal lysis, ultrasonic lysis, electrical lysis, mechanical lysis, etc.

Most of the macroscopic DNA purification methods made in extraction columns, using the DNA adsorption on silica beads under certain buffer conditions, are adaptable to microfluidic techniques. In LOC devices, silica beads can thus either be blocked by mechanical obstacles or silica-coated magnetic beads to easily block these particles with a magnet for the washing and elution phases (Timothée, 2015b).

8.1.1.2 PCR, qPCR, and Molecular Detection on LOC Devices

After DNA extraction, the most common analysis is the PCR (Polymerase Chain Reaction). PCR has lots of applications that are directly and indirectly like sequencing techniques. One of the direct PCR applications is obviously the amplification of DNA sequences that helps to make detectable low amounts of DNA (e.g., for pathogen detection, like bacteria or virus).

The importance of PCR in genomic analysis affects the development of numerous LOC devices for PCR. Miniaturization of volume and the high surface to volume ratio leads rapid thermal transfer for rapid and integrated PCR. PCR also requires a post-analysis so that amplicons' size detection carried out by electrophoresis have been made to integrate PCR and electrophoresis on-chip (Timothée, 2015a).

Originally electrophoresis is done by using gels, mainly made using agarose (for longer DNA) and polyacrylamide (for shorter DNA). With the advent of LOCs, DNA electrophoresis was one the first molecular processes that could be integrated *on a chip* (Curtis Saunders et al., 2013). This miniaturization enabled to increase even more the process time

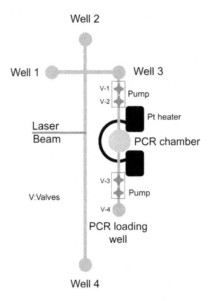

FIGURE 8.1   PCR microfluidic system.

to realize electrophoresis, reduce reagent consumption, and assemble on a chip other DNA process steps as mentioned before (Fig. 8.1).

qPCR is another technique that was adapted to LOC devices that present the advantage to be faster (automated detection during PCR), more sensitive, and sustainable.

As an example, using ultra-fast pressure controller and fluorescence reader and based on the ultra-fast temperature control, ultra-fast qPCR microfluidic system had been developed by *Elvesys system* for the molecular detection of diseases like Anthrax and Ebola in less than 8 minutes with a detection efficiency identical to commercial systems that are 7–15 times slower (Ramalingam et al., 2010).

Another emerged field is digital microfluidics that deals with emulsion and droplets within LOC devices.

Ultra-low amount of DNA can be captured within *droplets*, and limits can be increased with one copy number detection within LOC droplet qPCR (Beer et al., 2007) (Fig. 8.2).

8.1.2 Genomic Application

The field of genomics has had a tremendous growth rate in recent years because of the effects of sequencing projects such as the Human Genome Project. The projects related to sequencing accelerate the sequencing technologies. Sequencing is the analysis determining the nucleotide order of DNA sequences. The first sequencing was developed based on the Sanger method; in this method, DNA fragments with different lengths and different fluorophores corresponding to different nucleotide ends will be obtained. Miniaturization of the standard DNA sequencing, using a slab of polyacrylamide gel, was achieved by replacing the gel slab with a gel-filled small capillary. The small dimensions of the capillaries decreased the required DNA sample volume (Figeys and Pinto, 2000).

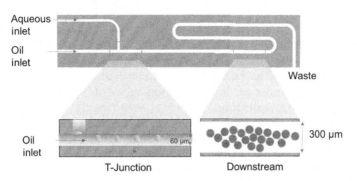

FIGURE 8.2 LOC droplet qPCR.

TABLE 8.2 Comparison of Time and Cost for Next-Generation Sequencing Systems That Use LOC Technologies

	Runtime	Read Length	Output Nucleotide	Cost/MB
Illumina Miseq	27 h	2 × 300 bp	15 GB	$0.15
Illumina Hiseq 2500	1–10 days	2 × 250 bp	3000 GB	$0.05
Ion Torrent	2 h	400 bp	2 GB	$1
Pacific Biosciences	30 min–4 h	10–40 kb	500–1000 MB SMRT cell	$0.13–$0.80
Nanopore Technologies	<48 h	5–200 kb	50–100 GB	$0.003–$0.01

Next-generation sequencing for ultra-fast DNA sequencing currently under development include sequencing-by-hybridization (SBH), nanopore sequencing, and sequencing-by-synthesis (SBS), the latter of which encompasses many different DNA polymerase-dependent strategies. Using LOC approach for these technologies enables high-throughput sequencing to reduce drastically its cost and duration (Table 8.2). Most famous next-generation sequencing systems are Illumina and its sequencing by synthesis technology, Life technologies and its Ion Torrent using semiconductor technologies, Pacific Bio with its single molecule real-time sequencing using zero-mode waveguides, and Roche with its 454 sequencing system using pyrosequencing technology (Metzker, 2005).

8.1.3 Microarray

The analysis of complex DNA samples expression needs integration of multiple biosensors in connection with DNA microarrays. The most important features of these LOCs are miniaturization, speed, and accuracy. This technology offers huge potential for rapid multiplex analysis of nucleic acid samples, including the diagnosis of genetic diseases, detection of infectious agents, measurements of differential gene expression, drug screening, or forensic analysis. Such use of DNA microarrays is thus revolutionizing many aspects of genetic analysis (Wang, 2000).

8.1.4 Biochemical Applications

LOC strategy for coupling enzymatic and immunological assays on a single-channel microfluidic device is also applicable, for example simultaneous glucose and insulin measurements. The availability of an LOC capable of simultaneously monitoring both insulin and glucose holds great promise for improved management of diabetes.

Crucial for the successful realization of such glucose/insulin monitoring is the integration of relevant sample pretreatment/cleanup procedures essential for whole-blood analysis. The new LOC approach can be applicable to the integration of other assays and additional sample handling steps, as is desired for creating miniaturized clinical instruments (Wang et al., 2003).

8.1.5 Proteomics

Proteomics is one of the great scientific challenges in the post-genome era. The most basic form of proteomics is proteome profiling, identifying all the proteins expressed in each sample, which is a demanding task. The proteome has unique analytical challenges, including molecular diversity, wide concentration range, and a tendency to adsorb to solid surfaces.

LOC devices are useful for developing new methods to solve complex analytical problems, such as proteome profiling. LOC devices seem to be progressive in four key areas related to this application: chemical processing, sample preconcentration and cleanup, chemical separations, and interfaces with mass MS (Freire and Wheeler, 2006).

LOC is a miniaturized device suited for separation and detection of proteins which enables less reagent consumption, easy operation, and very fast analysis. Because aim is to obtain the maximum amount of data from each sample, LOC protein devices can generate a huge amount of data points. Another advantage is that the data from the LOC devices can easily be compared to those obtained using 2D-PAGE (Schasfoort, 2004).

LOC protein sodium dodecyl sulfate polyacrylamide gel electrophoresis (SDS-PAGE) separation with fluorescence detection is the most commonly used method to separate and size protein mixtures. Normally, SDS-PAGE is time consuming and labor intensive. LOC devices speed up separation and lay the building blocks toward automation of protein sizing (Mouradian, 2002).

LOC MS protein profiling is one of the applications to quantitate proteins. Mouradian (2002) describes LOC devices that are used for direct infusion into the mass spectrometer, including Capillary Electrophoresis (CE) separation, on-chip sample digestion, and infusion or CE before MS analysis. Researchers use electrospray ionization MS and focus on the interface design between LOC and the mass spectrometer (Mouradian, 2002).

8.1.6 Biosensors

LOC biosensors are small devices in which biological reactions occur for detecting target analytes. Such devices closely link together a biological recognition element (interacting with the target analyte) with a physical transducer converting the biorecognition event into an electrical signal. There are two types of LOC biosensors, which are bioaffinity and biocatalytic devices. Bioaffinity devices are based on selective binding of the target analyte to a surface-enclosed ligand partner (e.g., antibody, oligonucleotide). For example, in hybridization biosensors, there are immobilized single-stranded (ss) DNA probes onto the transducer surface. The duplex formation can be detected following the association of an appropriate hybridization indicator after binding. On the other way, in biocatalytic devices, an immobilized enzyme is used for recognizing the target substrate. For example, sensor strips with immobilized glucose oxidase for monitoring of diabetes (Wang, 2000).

8.1.7 Cell Research

LOC devices enable the researcher to handle small-scale experiments with cells. In many cases, the single cell, when positioned within the microchannel, will also need to be positioned or trapped in a predetermined location, making single-cell handling and manipulation necessary. The microsystem should be able to handle small volumes of fluid containing small quantities of analyte. LOC systems also provide an ability to control the local environment around the cell with precision, using methods that include having control over the physical or chemical nature of the surface to which the cell adheres, the local pH, and temperature. When the cell is to be exposed to a different kind of stimuli, multiple and often complex fluid-handling components can be used. Together, these ideas imply the need for the integration of cell positioning and chemical stimulation. If the cell is then to be analyzed, lysis and analysis should be introduced onto the same platform to avoid sample loss or dilution, which would inevitably occur if multiple devices were used (Sedgwick et al., 2008).

Flow cytometers are one of the successful instruments in the earlier stage and the most useful among LOC devices developed to date. In the past decade, LOC systems of flow cytometers have been developed rapidly by two main factors: (1) the increased need for higher quality and larger quantity of cellular analysis data, and (2) the enhanced sophistication and accessibility of microfabrication technologies. The emerging needs coupled with the new technical capabilities have led to a variety of exciting developments in microfluidics for flow cytometric analysis of cells and particles (Huh et al., 2005).

8.1.8 Drug Development

The industry needs new tools to guide the development of new drugs — especially to help predict the behavior of potential new drugs in humans from performance in animals and cells. Some analytical applications of LOC systems in the production and use of biopharmaceuticals seem straightforward (e.g., analytical systems to monitor and optimize the production of protein drugs such as therapeutic antibodies); others (such as assays based on primary human cells that could predict performance in human clinical trials) are technically more complicated, but also feasible, at least in some instances. In either case, LOC systems in a highly reproducible and easily manipulated format could be used routinely (Whitesides, 2006).

8.2. CONCLUSION AND FUTURE DIRECTIONS

Miniaturization of laboratory devices into chip-sized devices is a new area called LOC. Cost, time, and parallelization are main advantage of this field, and it integrates several operations with better quality of sensitivity.

By looking into commercialized LOC devices such as PCR, DNA sequencing, and glucose monitoring, we can say that LOC will change diagnostic concepts in the near future. And we will see more products in the market developed using LOC technologies. The industry will reach more economical production of LOC devices day by day. LOC development directly affects molecular-based innovations to be used in daily life.

References

Beer, N.R., Hindson, B.J., Wheeler, E.K., Hall, S.B., Rose, K.A., Kennedy, I.M., et al., 2007. On-chip, real-time, single-copy polymerase chain reaction in picoliter droplets. Anal Chem 79 (22), 8471–8475. Available from: https://doi.org/10.1021/ac701809w.

Casquillas, G., Timothée, H., 2015. Introduction to lab-on-a-chip 2015: review, history and future. Retrieved from http://www.elveflow.com/microfluidic-tutorials/microfluidic-reviews-and-tutorials/introduction-to-lab-on-a-chip-2015-review-history-and-future/.

Curtis Saunders, D., Holst, G.L., Phaneuf, C.R., Pak, N., Marchese, M., Sondej, N., et al., 2013. Rapid, quantitative, reverse transcription PCR in a polymer microfluidicchip. Biosens Bioelectron 44, 222–228. Available from: https://doi.org/10.1016/j.bios.2013.01.019.

Figeys, D., Pinto, D., 2000. Lab-on-a-chip: a revolution in biological and medical sciences. Anal Chem 72 (9), 330A–335A. Available from: https://doi.org/10.1021/ac002800y.

Freire, S.L.S., Wheeler, A.R., 2006. Proteome-on-a-chip: mirage, or on the horizon? Lab Chip 6 (11), 1415. Available from: https://doi.org/10.1039/b609871a.

Huh, D., Gu, W., Kamotani, Y., Grotberg, J.B., Takayama, S., 2005. Microfluidics for flow cytometric analysis of cells and particles. Physiol Meas 26 (3), R73–R98. Available from: https://doi.org/10.1088/0967-3334/26/3/R02.

Kaigala, G.V., Behnam, M., Bidulock, A.C.E., Bargen, C., Johnstone, R.W., Elliott, D.G., et al., 2010. A scalable and modular lab-on-a-chip genetic analysis instrument. Analyst 135 (7), 1606–1617. Available from: https://doi.org/10.1039/b925111a.

Mark, D., Haeberle, S., Roth, G., von Stetten, F., Zengerle, R., 2010. Microfluidic lab-on-a-chip platforms: requirements, characteristics and applications. Chem Soc Rev 39 (3), 1153–1182. Available from: https://doi.org/10.1039/b820557b.

Metzker, M.L.M.L.L., 2005. Emerging technologies in DNA sequencing. Genome Res 15 (12), 1767–1776. Available from: https://doi.org/10.1101/gr.3770505.with.

Mouradian, S., 2002. Lab-on-a-chip: applications in proteomics. Curr Opin Chem Biol 6 (1), 51–56. Available from: https://doi.org/10.1016/S1367-5931(01)00280-0.

Neuži, P., Giselbrecht, S., Länge, K., Huang, T.J., Manz, A., 2012. Revisiting lab-on-a-chip technology for drug discovery. Nat Rev Drug Discov 11 (8), 620–632. Available from: https://doi.org/10.1038/nrd3799.

Ramalingam, N., Rui, Z., Liu, H.B., Dai, C.C., Kaushik, R., Ratnaharika, B., et al., 2010. Real-time PCR-based microfluidic array chip for simultaneous detection of multiple waterborne pathogens. Sensors Actuat B Chem 145 (1), 543–552. Available from: https://doi.org/10.1016/j.snb.2009.11.025.

Schasfoort, R.B.M., 2004. Proteomics-on-a-chip: the challenge to couple lab-on-a-chip unit operations. Expert Rev Proteomics 1 (1), 123–132. Available from: https://doi.org/10.1586/14789450.1.1.123.

Sedgwick, H., Caron, F., Monaghan, P.B., Kolch, W., Cooper, J.M., 2008. Lab-on-a-chip technologies for proteomic analysis from isolated cells. J R Soc Interface 5 (Suppl 2), S123–S130. Available from: https://doi.org/10.1098/rsif.2008.0169.focus.

Timothée, H., 2015a. Microfluidic PCR, qPCR, RT-PCR & qRT-PCR. Retrieved from http://www.elveflow.com/microfluidic-tutorials/microfluidic-reviews-and-tutorials/microfluidic-pcr-qpcr-rtpcr/.

Timothée, H., 2015b. Microfluidics for DNA analysis. Retrieved from http://www.elveflow.com/microfluidic-tutorials/microfluidic-reviews-and-tutorials/microfluidics-for-dna-analysis-pcr/.

Wang, J., 2000. From DNA biosensors to gene chips. Nucleic Acid Res 28 (16), 3011–3016. Available from: https://doi.org/10.1093/nar/28.16.3011.

Wang, J., Ibáñez, A., Chatrathi, M.P., 2003. On-chip integration of enzyme and immunoassays: simultaneous measurements of insulin and glucose. J Am Chem Soc 125 (28), 8444–8445. Available from: https://doi.org/10.1021/ja036067e.

Whitesides, G.M., 2006. The origins and the future of microfluidics. Nature 442 (7101), 368–373. Available from: https://doi.org/10.1038/nature05058.

Further Reading

Blazej, R.G., Kumaresan, P., Mathies, R.A., 2006. Microfabricated bioprocessor for integrated nanoliter-scale Sanger DNA sequencing. Proc Natl Acad Sci U S A 103 (19), 7240–7245. Available from: https://doi.org/10.1073/pnas.0602476103.

Robotics and High-Throughput Techniques

Humna Hasan[1], Muhammad Hassan Safdar[1], Sana Zahid[1], Maria Bibi[1] and Alvina Gul[2]

[1]Quaid-I-Azam University, Islamabad, Pakistan [2]National University of Sciences and Technology, Islamabad, Pakistan

9.1 INTRODUCTION

The idea of incorporating a mechanical assembly of hardware called "the robot" in the realm of biology seemed a far cry from being transformed into a reality soon. Highly inspired by the animals' anatomy, the robotic engineers incorporated technologies into mechanical designs. The incorporation of materials and technologies in life ultimately gave rise to a new field known as "Biorobotics." The novel structures being incorporated from nature into machines is the goal of biorobotics. A few applications of biorobotics have been listed in Table 9.1.

The main aim is to use biological inspiration to invent machines that imitate the performance of animals (Trimmer, 2008). Animal systems serve as acting models for robotics with enhanced performance (Taubes, 2000). Novel structures and mechanisms from living organisms are incorporated into the framework of mechanical bodies through biomimetics. Biomimetics, or bionics, has enabled to produce mechanical systems that possess characteristics of living systems (Vogel, 1998). Furthermore, the diverse forms of life with multifarious morphologies present an excellent resource of designs that can be incorporated into the making of biorobotics. The machines produced in turn present a superior performance (Bandyopadhyay et al., 1997). On the contrary, complete adherence to biological systems while designing a mechanical model can rarely produce any practical results (Vogel, 1994). **One of the most important constraints faced while making mechanical systems is the phenomenon of evolution**

TABLE 9.1   Main Areas Where Biorobotics Are Employed

Sr. No.	Applications in	Areas
1.	Biomedical Engineering	• Neural Engineering (generate signals causing movement) • Rehabilitation engineering (speedy recovery from injury or stroke) (Hammel et al., 1989)
2.	Medicine	• Development of artificial sensing skin (can help detect pressures and contacts) • Tissue engineering • BioMEMs (microfluidics) (Fig. 9.1) • Development of surgical robotics (minimally invasive surgery) (Paul et al., 1992)
3.	Industry	• Development of borescopes and snake robots (to help visual inspection and accessibility) • Painting and coating deposition
4.	Biomechanics	• Biotribology (addressing mechanical friction and lubrication) • Micro and nanotechnologies (DNA repair and drug delivery)

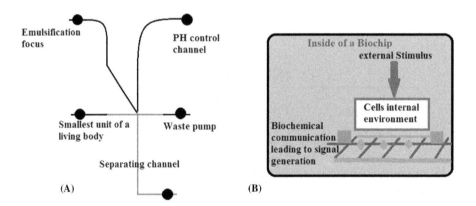

FIGURE 9.1   (A) Model of a plastic biochip. (B) Controlled microenvironment in a biochip.

(Fish and Beneski, 2013). The biotic and abiotic factors are also to be taken care of as they are important in natural selection.

In addition, biorobotics have also introduced us to the term "cyborg"—an organism comprising both artificial and natural parts (Clynes and Kline, 1960). These organisms pose serious questions of free will and empathy on the differences between humans and their mechanically made models. They are also shown to be incorporated with physical and mental abilities better than human counterparts. Examples include prosthetic applications, e.g., C-Leg system.

9.1.1 Technologies in Biorobotics

The biggest challenge faced by biorobotics is to induce in robots such flawless soft technologies that closely replicate or mimic the nature. One of the major soft technologies is *Actuation* that helps to mimic the functionality of muscles in the soft robots. Numerous soft and hard actuators are used for this purpose. Another technology that holds a significant place in the field of soft robotics is *Stiffness Modulation*. Even the softest of the objects require some degree of hardness if they have to perform different functions regarding movement or special tasks, such as tissue sampling. On the basis of particle jamming, a soft gripper serves the function of variable stiffness (Amend et al., 2012, Cheng et al., 2012). There is another technology commonly used in soft robotics known as *Soft Materials* technology. The body made up of high-modulus materials is highly prone to damage. The body must comprise low-modulus materials, such as elastomers, to resist high damaging forces. Silicone rubbers and highly stretchable tough hydrogels have been developed as soft body materials (Sun et al., 2013). Certain dissolvable robots comprising soft materials serve in drug delivery to specific tissues (Tan and Marra, 2010).

The dynamic modeling and kinematics, which are known techniques in robotics, cannot be applied on to soft robots due to their nonlinear deformation and continuum structure. Soft robots have a heterogeneous structure due to which constitutive models formulated for rubber-like materials are inapplicable to them. So, appropriate dynamic modeling of soft robots remains a challenge.

9.1.2 Soft Robotics

Soft robotics has invited researchers from interdisciplinary sciences, including biological sciences, soft materials sciences, organic chemistry, and robotics. The main purpose is to induce bio-inspired capabilities in the robots so that they interact more effectively and flexibly with the humans, as well as unpredictable and real-world environments. In addition to enhancing adaptability, effectiveness, and robustness, this added compliance or softness implements a modern view of intelligence, i.e. embodied intelligence, rendering physical strength to the soft robot body (Pfeifer et al., 2012). Soft robots, inspired by flexible animals like octopuses, are mostly built up of elastic and pliable materials helping them to mold according to the environment. These machines can now twist, squish, stretch and scrunch in altogether new ways, wrapping around objects, transforming in size or shape and touching the subjects more safely.

9.1.2.1 Structure

To substitute the functionality of muscles in the robot body, actuators are required. Soft robots require a number of soft and hard actuators to change the body shape. Currently, there are three types of actuators. First are the Dielectric Elastomeric Actuators, primarily composed of soft materials and actuate via electrostatic forces requiring high voltages (O'Halloran et al., 2008, Carpi and Smela, 2009). Second are the Shape Memory Alloys having temperature-dependent high mass-specific force and suitable for soft actuation (Lin et al., 2011, Laschi et al., 2009, Kim et al., 2009). Third are the actuators that are deformed as a response to compressed air and pressurized fluids, producing high displacements and forces, such as McKibben actuators and other soft orthotic pneumatic actuators (Park et al., 2012).

9.1.2.2 Bio-inspired Soft Robots

A new wave of inspiration released by soft biological materials has traveled across the field of robotics giving rise to a variety of miraculous soft-bodied robots. A variety of aquatic and terrestrial soft robots have been developed by Robotics engineers. Robot models have been built on the basis of neuromechanical strategies used for locomotion by soft-bodied animals (Seok et al., 2010). One of the interesting bio-inspired robots includes worm-like robots, based on their hydrostatic structures governed by various soft and hard actuators. An annelid robot has been made that generates worm-like movements with the help of dielectric elastomers (Jung et al., 2007). Caterpillar-inspired robot, GoQBot, changes its body shape from elongated to circular and depicts ballistic rolling locomotion (Lin et al., 2011). To produce limbed locomotion, numerous octopus-inspired robots have been invented utilizing the concept of compartmentalized deformation (Ilievski et al., 2011, Shepherd et al., 2011). Octopus-like robot arm has been developed by Laschi et al. (2012). Robotics researchers have developed some hard robots with soft capabilities, like snake-like robots. Snake robots and snake-like climbing robots have been built on the basis of the crawling ability and undulatory locomotion of the snakes (Miller, 2008, Wright et al., 2007).

9.1.2.3 Advantages

Soft robots come into play where hard robots fail to perform certain functions due to their hyper-redundancy. Soft robots easily conform to obstacles and depict delicacy in gripping, manipulating, and controlling deformable and fragile objects without causing them any harm, thus, rendering them ideal to be used as personal robots. Their rubbery appearance, adaptable morphology, and delicacy enhance their utilization especially as medical robots in the field of surgery. The major advantage of soft robotics that has particularly increased its importance in the field of robotics is inexpensive development of the soft robots.

9.2 WONDERS IN THE FIELD OF BIOROBOTICS

Biologists have taken the field of robotics to a whole new level by marking their glory in the field of biorobotics. By attaching and detaching some cables with external motors, researchers can make a tentacle wave or curve around a human hand. Swimming fish bots, crawling robotic caterpillars, and undulating artificial jellyfish have already been produced (Katzschmann, 2015, Umedachi et al., 2013, Villanueva et al., 2011). A robotic fabric shows movements in response to the electric current (Yuen et al., 2014). A soft robotic fish has been created that has the capability to emulate escape maneuvers with the help of fluidic elastomer actuators, which depicts the self-containment of soft robots (Andrew et al., 2014). For the nondestructive biological sampling of plants in deep reefs, a soft robotic gripper has recently been created, which has made the benthic fauna sampling an easier task (Kevin et al., 2016). A team of researchers has invented a jumping roly-poly soft robot. The special thing about this untethered robot is that it is driven by combustion (Michael et al., 2015). The next ubiquitous platform is stretchable and flexible ultrathin electronic skin (Nathan et al., 2012). Advanced work is being done on stretchable LEDs and solar cells. Moreover, stretchable batteries and accumulators are being worked upon by most robotics engineers (Kaltenbrunner and Bauer, 2013).

9.2.1 Techniques Used in Biorobotics

Biorobotics is aimed to focus on the new technological and scientific area to increase scientific understanding about the working of biological systems in humans, plants, and animals. For this purpose, living organisms can be analyzed from a biomechatronic viewpoint, and a variety of innovative technologies and methodologies can be developed by exploiting this knowledge and used for the benefit of mankind in various ways (Fig. 9.2). Bio-inspired robots are used by scientists for various purposes including studying of living organisms' behaviors, evaluation, and integration for biological models to develop best possible applications in engineering, validation of hypotheses, and scientific theories (Dario, 2003), studying plant as a system in plant sciences for the formulation of new hypotheses and studying the complex biological processes taking place inside the plants (Mazzolai et al., 2010).

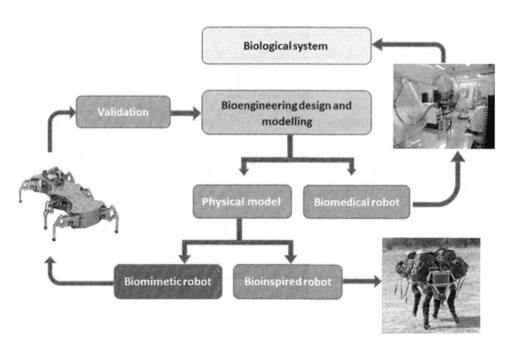

FIGURE 9.2   A schematic representation of a biological system model by experimental validation.

9.2.1.1 Electromyography

Electromyography (EMG) is a technique capable of measuring the intensity of human motion directly. Moreover, this technology can also reflect the muscle activity of the user. Therefore, it is one of the most recurrent techniques used in the control methods of biorobotics applications. EMG-based control methods are very effective, including wheelchairs (Onishi et al., 2008; Felzer and Freisleben, 2002), prosthetics (Shenoy et al., 2008; Pons et al., 2004), and exoskeletons/orthoses (Kiguchi and Hayashi, 2012; Rosen et al., 2001). But, there are also some challenges associated with EMG-based techniques depending upon the user application. For example, EMG-based methods are not useful in case of the user who is unable to generate sufficient muscle signals. For instance, device like exoskeleton cannot be used by a person with totally paralyzed limb due to the inefficient generation of control signals from the muscles of the paralyzed limb.

9.2.1.2 Electroencephalography: Brain—Computer Interfaces or Brain—Machine Interfaces

Brain—computer interfaces (BCI) or brain—machine interfaces (BMI) is of core importance in biorobotics as these are the recent advancements in this area (Fig. 9.3). These technologies can be of great use for directly decoding the user's brain signal and using them to control the biorobotic equipment such as exoskeletons or wheelchairs, and prosthetics. Several methods are available for capturing brain signals, among which electroencephalography (EEG) is a noninvasive method, especially appropriate for practical systems. Various attempts have been made to implement these EEG signal—based interface in prosthetics (Murguialday et al., 2007; Muller-Putz and Pfurtscheller, 2008), wheelchairs (Iturrate et al., 2009; Millan et al., 2009), and exoskeletons/orthoses (Chen et al., 2009; King et al., 2011). However, EEG-signals alone—based applications are not much reliable and are less accurate. Also, these have low data transfer rates and are minimum user

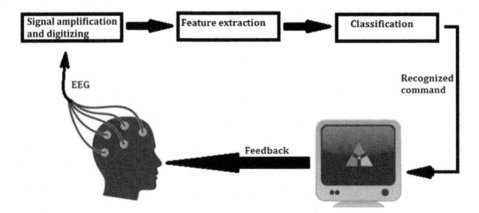

FIGURE 9.3 Schematic diagram of EEG-based BCI. The interface comprises EEG acquisition system, feature extraction through data processing software and pattern classification, and a system capable of transferring the command to external devices, providing feedback to operator.

adaptable, that is why are not yet fully acceptable in biorobotics (Wolpaw et al., 2002; Wolpaw et al., 2000; Vaughan et al., 1996).

9.2.1.3 Hybrid EEG–EMG Control Interface

The efficiency of the techniques used in biorobotics can be improved by using both the EEG- and EMG-based interfaces together within a particular control approach. This technology is called as Hybrid EEG–EMG Control Interface (Fig. 9.4). This fusion of signals can be carried out in various ways and is also dependent upon factors, like abilities of the users and specific application. Hybrid EEG–EMG Control Interface is capable of processing the input signals either sequentially or simultaneously (Lalitharatne et al., 2013).

For people with physical disability, weakness, older age, or any injury, EMG signal inputs devices are recommended, such as exoskeleton or wheelchairs and prosthetics, as such people have residual activities of their muscles. Yet EMG-based approaches have a few drawbacks including muscle fatigue, which is the reason for affecting the EMG spectrum frequency and amplitude, thus disturbing its overall efficiency (Hagberg, 1981; Sadoyama et al., 1983).

The muscle fibers become smaller and weaker with increasing age of individual, thus leading to reduced strength and person suffers from fatigue (Martini, 2000). In old age, the mental and physical condition of individuals keeps changing throughout the day, thus, physical exhaustion may lead to muscles tiredness. So, it is important to design EMG-based control approaches that are capable of dealing with the condition of muscle fatigue, and with a few exceptions (Artemiadis and Kyriakopoulos, 2011), such

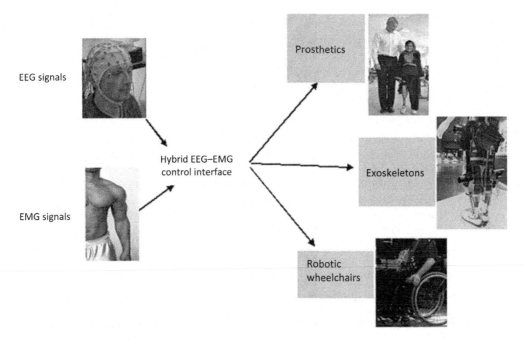

FIGURE 9.4 Hybrid EEG–EMG Control Interface Graphical interpretation.

EMG-based control approaches are not reported yet. To overcome this problem, EMG is used along with the EEG signals to obtain better results by handling this condition effectively. A hybrid BCI system containing the fusion of muscle and brain signals is reported to overcome the challenges faced by single approaches alone (Leeb et al., 2010; Leeb et al., 2011).

9.2.1.4 *Plant Roots–Inspired Robotic Solutions: The PLANTOID Robot*

A totally new approach in robotics is initiated by the development of a plant system that grows at the apical area, representing artificial growth (Sadeghi et al., 2013). This system is composed of two zones, a growing zone and a stationary mature zone. The mature zone is composed of layer-by-layer deposition of hollow, tubular structure that is designed to allow new material from a spool outside the robotic root and power to the growing zone. The growing zone is designed to generate a penetration force into the soil and transfer material from the roots into the head or tip. The first prototype of this system is named as PLANTOID Robot, consisting of two roots capable of generating bending motion and elongation movement with a sensorized tip. The bending roots have sensors for humidity, gravity, touch, and temperature (Mazzolai et al., 2014). Thus, as a whole, the system is composed of two roots connected to a main trunk. One root elongates and penetrates into the soil by the addition of material from outside (Sadeghi et al., 2013), while the other root has a sensory system for humidity, temperature, touch, and gravity along with the capability of bending in three directions (Mazzolai et al., 2014).

Such plant-inspired robot artifacts are useful for the scientists in a number of ways, including studying the anchoring capabilities in plants, in situ monitoring analysis and chemical detections, formulation of new communication strategies and processes, water searching, studying plan behavior by development of physical models, and plants-inspired control algorithms. This biorobot is considered as the largest system to study the complex processes occurring inside the plant, thus helps in the development of "design-based engineering principles" by the use of biorobotics and synthetic biology in combination.

9.2.1.5 *Sperm-driven Micro-Biorobot*

Scientists has contributed to science yet another innovative biorobot, the sperm flagella-driven Micro-Biorobot (MBR) based on the principle of magnet-based motion control, thus is driven by the help of magnetic field. Sperm-driven MBR contains bovine spermatozoon apprehended inside the Ti/Fe nanomembranes and is able to propel with the velocity of $25 \pm 10\,\mu m/s$ by the influence of magnetic field that produces magnetic torque on the dipole of its rolled-up microtube by the use of electromagnetic coils. The future of MBR can enlighten the development of new technologies applicable in microactuation and targeted drug delivery, solving many scientific problems and providing biomedical field with a new platform (Khalil et al., 2014).

9.2.1.6 *Robotic Cell Injection*

Injection approach is yet another significant innovation of biorobotics system. In this technique, a fine needle is used by an automated device to inject substances into a single living cell. In most of such microinjection systems, the needle is precise to insert substances within the accurate position or location, but the injection force is not regulated.

So, due to the uncontrolled injection force, the cell is more prone to damage and may lead to death (Xie et al., 2009). Microinjection in the living cells is a biomanipulation operation that is used in pronuclei DNA injection, intracytoplasmic sperm injection, gene therapy, cloning, etc. (Lacal et al., 1999). For manipulation of the cells with robotic microinjections autonomously, automatic embryo pronuclei DNA injections are carried out through a hybrid visual control scheme. Then these embryos are inoculated into pseudo-pregnant foster female mice, and the reproduced transgenic progeny is used for cancer studies (Sun and Nelson, 2002). Experiments have proved that these robotic cell injections are 100% successful (Yu and Nelson, 2001).

9.2.1.7 Biorobotics in Medicine

Biorobotics itself is one of the most advanced technologies, which is used for various applications in various fields including the field of medicine. A robot-assisted therapy, especially designed for the evaluation of upper limb and its rehabilitation in chronic post-stroke patients is the successful achievement of science in the field of medicine. In this technology, admittance controlled biorobots are designed to improve the movement ability of upper limb in these chronic post-stroke patients by targeting the enhancement of motor outcome. Moreover, clinical quantitative evaluation of the course of therapy can also be assessed by the use of these devices. So, patients' rehabilitation strategies can be developed and modified accordingly to trigger the improvement in the patients' motor outcome. Motor outcome of these patients can be enhanced directly by reducing their disability. Hence, biorobots are yet another success for humankind in the form of robot-aided neurorehabilitation. This provides room for the development of more targeted rehabilitation approaches in future. (Colombo et al., 2005)

The improvement of motor recovery of hemiparetic subjects is also reported by another group of scientists by using simple mechatronic system named "MEchatronic system for MOtor recovery after Stroke" (MEMOS). The experiments were successful and it effectively reduced the level of impairment in the stroke patients (Micera et al., 2005).

9.2.1.8 Natural Orifice Surgery Through Biorobotics

This technology uses an endoluminal mobile robot for the treatment of skin incisions. It is called as natural orifice transgastric endoscopic surgery and is considered more promising and effective over conventional laparoscopy. The technology uses a small mobile in vivo robot to explore the abdominal cavity and is also effective against the reduction of patient trauma. Moreover, this miniature robot can also be used for appendectomy (Rentschler et al., 2007).

9.3 FUTURE PERSPECTIVE

Biorobotics is a promising new field inspiring scientists for novel researches. High-throughput technology provides a feasible option for the automation of experiments in the field of robotics. Furthermore, high throughput has also enabled biologists to meet the great challenge of scientific research, e.g., understand the body metabolism and the underlying basic mechanisms. It has also invaded in interdisciplinary sciences and bringing in better aspects to life.

References

Amend, J.R., Brown, E.M., Rodenberg, N., Jaeger, H.M., Lipson, H., 2012. A positive pressure universal gripper based on the jamming of granular material. IEEE Trans. Robot. 28, 341–350.

Andrew, M.D., Cagdas, O.D., Daniela, R., 2014. Autonomous soft robotic fish capable of escape maneuvers using fluidic elastomer actuators. Soft Robot 1 (1), 75–87. Available from: http://dx.doi.org/10.1089/soro.2013.0009.

Artemiadis, P.K., Kyriakopoulos, K.J., 2011. A switching regime model for the EMG-based control of a robot arm. IEEE Trans Syst Man Cybern B Cybern 41, 53–63.

Bandyopadhyay, P.R., Castano, J.M., Rice, J.Q., Philips, R.B., Nedderman, W.H., Macy, W.K., 1997. Low-speed maneuvering hydrodynamics of fish and small underwater vehicles. Trans ASME 119, 136–144.

Carpi, F., Smela, E., 2009. Biomedical Applications of Electroactive Polymer Actuators. John Wiley and Sons Ltd, Wiley Press, Chichester, UK.

Chen, C.W., Lin, K., Ju, M.S., 2009. Hand orthosis controlled using brain-computer interface. J Med Biol Eng 29, 234–241.

Cheng, N., Lobovsky, M., Keating, S., Setapen, A., Gero, K.I., Hosoi, A., et al., 2012. Design and analysis of a robust, low-cost, highly articulated manipulator enabled by jamming of granular media. IEEE ICRA 4328–4333.

Clynes, M.E., Kline, N.S., 1960. Cyborgs and space, in Gray [1960(1996a)], pp 22–30

Colombo, R., Pisano, F., Micera, S., Mazzone, A., Delconte, C., Carrozza, C.M., et al., 2005. Robotic techniques for upper limb evaluation and rehabilitation of stroke patients. IEEE Trans Neural Syst Rehabil Eng 13 (3), 311–324.

Dario, P., 2003. Biorobotics. IEEE Robot Autom Mag 10, 4–5.

Felzer T., Freisleben B., 2002. HaWCoS: The "Hands-free" Wheelchair Control System: In: Proceedings of 5th International ACM SIGGRAPH Conference on Assistive Technologies; ACM: Edinburgh, Scotland; July 8–10, 2002; 127–134.

Fish F.E., Beneski J.T., 2014. Evolution and Bio-Inspired Design: Natural Limitations. In: Goel A., McAdams D., Stone R. (eds) Biologically Inspired Design. Springer, London.

Hagberg, M., 1981. Electromyographic signs of shoulder muscular fatigue in two elevated arm positions. Am J Phys Med 60, 111–121.

Hammel, J., Hall, K., Lees, D., Leifer, L.J., Van Der Loos, H.F.M., Perkas, I., et al., 1989. Clinical evaluation of a desktop robotic assistant. J Rehabil Res Develop 2 (3), 1–16.

Ilievski, F., Mazzeo, A.D., Shepherd, R.F., Chen, X., Whitesides, G.M., 2011. Soft robotics for chemists. Angew. Chem. Int. Ed. Engl. 50, 1890–1895.

Iturrate, J.M., Antelis, A., Kubler, J., Minguez, A., 2009. Noninvasive brain-actuated wheelchair based on a P300 neurophysiological protocol and automated navigation. IEEE Trans Robot 25, 614–627.

Jung, K., Koo, J.C., Nam, J.D., Lee, Y.K., Choi, H.R., 2007. Artificial annelid robot driven by soft actuators. Bioinspir. Biomim. 2, S42–S49.

Kaltenbrunner, M., Bauer, S., 2013. Power supply, generation, and storage in stretchable electronics. In: Someya, T., Sekitani, T., Someya, T. (Eds.), Stretchable Electronics. Wiley-VCH, Weinheim, Germany.

Katzschmann R.K., Marchese A.D., Rus D., 2015. Hydraulic autonomous soft robotic fish for 3D swimming. In: International Symposium on Experimental Robotics (ISER), Marrakech and Essaouira, Morocco, June 15–18, 2014.

Kevin, C.G., Kaitlyn, P.B., Brennan, P., Jordan, K., Stephen, L., Dan, T., et al., 2016. Soft robotic grippers for biological sampling on deep reefs. Soft Robot. 3 (1), 23–33.

Khalil, I.S., Magdan, V., Sanchez, S., Schmidt, O.G., Misra, S., 2014. Biocompatible, accurate, and fully autonomous: a sperm-driven micro-bio-robot. J Micro-Bio Robot 9 (3-4), 79–86.

Kiguchi, K., Hayashi, Y., 2012. An EMG based control for an upper-limb power-assist exoskeleton robot. IEEE Trans Syst Man Cyber Part B 42, 1064–1071.

Kim, S., Hawkes, E., Cho, K., Joldaz, M., Foleyz, J., Wood, R., 2009. Micro artificial muscle fiber using niti spring for soft robotics. In: IEEE/RSJ International Conference on Intelligent Robots and Systems. IEEE, New York, pp. 2228–2234.

King, C.E., Wang, P.T., Mizuta, M., Reinkensmeyer, D.J., Do, A.H., Moromugi, S., Nenadic, Z., 2011. Noninvasive brain-computer interface driven hand orthosis. In: Proceedings of Annual International Conference of the IEEE Engineering in Medicine and Biology Society 5786–5789.

Lacal, J.C., Perona, R., Feramisco, J., 1999. Microinjection. Birkhauser, Berlin.

Lalitharatne, T.D., Teramoto, K., Hayashi, Y., Kiguchi, K., 2013. Towards hybrid EEG-EMG-based control approaches to be used in bio-robotics applications: current status, challenges and future directions. Paladyn 4 (2), 147−154.

Laschi, C., Mazzolai, B., Mattoli, V., Cianchetti, M., Dario, P., 2009. Design of a biomimetic robotic octopus arm. Bioinspir. Biomim. 4, 015006.

Laschi, C., Cianchetti, M., Mazzolai, B., Margheri, L., Follador, M., Dario, P., 2012. A soft robot arm inspired by the octopus. Adv. Robot. 26, 709−727.

Leeb R., Sagha H., Chavarriaga R., Millan J.d.R., 2010. Multimodal fusion of muscle and brain signals for a hybrid-BCI. In: Proceedings of Annual International Conference of the IEEE Engineering in Medicine and Biology Society; 4343−4346.

Leeb, R., Sagha, H., Chavarriaga, R., Millan, JdR, 2011. A hybrid brain-computer interface based on the fusion of electroencephalographic and electromyographic activities. J Neur Eng 8. Available from: http://dx.doi.org/10.1088/1741-2560/8/2/025011.

Lin, H.T., Leisk, G.G., Trimmer, B., 2011. GoQBot: a caterpillar-inspired soft-bodied rolling robot. Bioinspir. Biomim. 6, 026007.

Martini R., 2000. Aging and the muscular system, Chapter 10: Muscle Tissue, In: Fundamentals of Anatomy and Physiology, 5th Edition, Benjamin-Cummings Publishing Company.

Mazzolai, B., Beccai, L., Mattoli, V., 2014. Plants as model in biomimetics and biorobotics: new perspectives. Front Bioeng Biotechnol 2 (10), 3389.

Mazzolai, B., Laschi, C., Dario, P., Mugnai, S., Mancuso, S., 2010. The plant as a biomechatronic system. Plant Signal Behav 5 (2), 90−93.

Micera, S., Carozza, M.C., Guglielmelli, E., Cappiello, G., Zaccone, F., Freschi, C., et al., 2005. A simple robotic system for neurorehabilitation. Auton. Robot 19 (3), 271−284.

Michael, L., Christoph, S.M., Urs, L.B., Wendelin, S.J., 2015. An untethered, jumping roly-poly soft robot driven by combustion. Soft Robot 2 (1), 33−41. Available from: http://dx.doi.org/10.1089/soro.2014.0021.

Millan, J.d.R., Galan, F., Vanhooydonck, D., Lew, E., Philips, J., Nuttin, M., 2009. Asynchronous non-invasive brain-activated control of an intelligent wheelchair. In: Proceedings of Annual International Conference of The IEEE Engineering in Medicine and Biology Society; 3361−3364.

Miller, G., 2008. Snake robots by Dr. Gavin Miller.

Muller-Putz, G.R., Pfurtscheller, G., 2008. Control of an electrical prosthesis with an SSVEP based BCI. IEEE Trans Biomed Eng 55, 361−364.

Murguialday, A.R., Aggarwal, V., Chatterjee, A., Cho, Y., Rasmussen, R., O'Rourke, B., et al., 2007. Brain-computer interface for a prosthetic hand using local machine control and haptic feedback. In: Proceedings of IEEE 10th International Conference on Rehabilitation Robotics 609−613.

Nathan, A., Ahnood, A., Cole, M.T., Lee, S., Suzuki, Y., Hiralal, P., et al., 2012. Flexible electronics: the nextubiquitous platform. Proceedings of the IEEE 100, 1486−1517.

O'Halloran, A., O'Malley, F., McHugh, P., 2008. A review on dielectric elastomer actuators, technology, applications, and challenges. J. Appl. Phys. 104, 071101.

Onishi, Y., Oh, S., Hori, Y. 2008. New control method for power-assisted wheelchair based on upper extremity movement using surface myoelectric signal. In: Proceedings of IEEE 10th International Workshop on Advanced Motion Control; 498−503.

Paul, H.A., Mitlestadt, B., Musits, B., Taylor, R.H., Kazanzides, P., Zuhars, J., et al., 1992. Development of a surgical robot for cementless total hip arthoplasty. Clin. Orthop. 285, 57−66.

Pfeifer, R., Lungarella, M., Iida, F., 2012. The challenges ahead for bio-inspired "soft"robotics. Commun.ACM 55, 76−87. Available from: http://dx.doi.org/10.1145/2366316.2366335.

Pons, J.L., Rocon, E., Ceres, R., Reynaerts, D., Saro, B., Levin, S., Moorleghem, W.V., 2004. The MANUS-HAND dextrous robotics upper limb prosthesis: mechanical and manipulation aspects. Proceedings of International Conference on Autonomous Robots 143163.

Rentschler, M.E., Dumpert, J., Platt, S.R., Farritor, S.M., Oleynikov, D., 2007. Natural orifice surgery with an endoluminal mobile robot. Surg Endosc. 21 (7), 1212−1215.

Rosen, J., Brand, M., Moshe, B., Arcan, M., 2001. A myosignal based powered exoskeleton system. IEEE Trans Syst Man Cybern Part A: Syst Hum 31, 210−221.

Sadeghi, A., Tonazzini, A., Popova, L., Mazzolai, B., 2013. Robotic mechanism for soil penetration inspired by plant root. In: Proceedings of the 2013 IEEE International Conference on Robotics and Automation, ICRA2013, Karlsruhe, Germany; 3457−3462. https://doi.org/10.1109/ICRA.2013.6631060.

Sadoyama, T., Masuda, T., Miyano, H., 1983. Relationship between muscle fiber conduction velocity and frequency parameters of surface EMG during sustained contraction. Eur J Appl Physiol 51, 247–256.

Seok, S., Onal, C.D., Wood, R., Rus, D., Kim, S., 2010. Peristaltic locomotion with antagonistic actuators in soft robotics. In: 2010 IEEE International Conference on Robotics and Automation 1228–1233.

Shenoy, P., Miller, K.J., Crawford, B., Rao, R.P.N., 2008. Electromyographic control of a robotic prosthesis. IEEE Trans Biomed Eng. 55, 1128–1135.

Shepherd, R.F., Ilievski, F., Choi, W., Morin, S.A., Stokes, A.A., Mazzeo, A.D., et al., 2011. Multigait soft robot. Proc. Natl. Acad. Sci. U.S.A. 108, 20400–20403.

Sun, J., Zhao, X., Illeperuma, W.R.K., Chaudhuri, O., Oh, K.H., Mooney, D.J., et al., 2013. Highly stretchable and tough hydrogels. Nature 489, 133–136.

Sun, Y., Nelson, B.J., 2002. Biological cell injection using an autonomous microrobotic system. Int J Robot Res 21 (10-11), 861–868.

Tan, H., Marra, K.G., 2010. Injectable, biodegradable hydrogels for tissue engineering applications. Materials 3, 1746–1767.

Taubes, G., 2000. Biologists and engineers create a new generation of robots that imitate life. Science. 288, 80–83.

Trimmer, B.A., 2008. New challenges in biorobotics: Incorporating soft tissue into control systems. *Applied Bionics and Biomechanics* 5 (3): 119–126.

Umedachi, T., Vikas, V., Trimmer, B.A., 2013. *Proceedings of the IEEE/RSJ International Conference on Intelligent Robots and Systems*; 4590–4595 (IROS).

Vaughan, T.M., Wolpaw, J.R., Donchin, E., 1996. EEG-based communication: prospects and problems. IEEE Trans Rehabil Eng 4 (4), 425–430.

Villanueva, A., Smith, C., Priya, S., 2011. A biomimetic robotic jellyfish (Robojelly) actuated by shape memory alloy composite actuators. Bioinsp Biomim 6, 036004.

Vogel, S., 1994. Life in Moving Fluids. Princeton University Press, Princeton.

Vogel, S., 1998. Cat's Paws and Catapults. W. W. Norton, New York.

Wehner, M., Park, Y., Walsh, C., Nagpal, R., Wood, R.J., 2012. Experimental characterization of components for active soft orthotics. IEEE International Conference on Biomedical Robotics and Biomechatronics 1586–1592.

Wolpaw, J.R., McFarland, D.J., Vaughan, T.M., 2000. Brain computer interface research at the Wadsworth Center. IEEE Trans Rehabil Eng 8 (2), 222–226.

Wolpaw, J.R., Birbaumer, N., McFarlanda, D.J., Pfurtschellere, G., Vaughan, T.M., 2002. Brain-computer interfaces for communication and control. Clin Neurophysiol 113, 767–791.

Wright, C., Johnson, A., Peck, A., McCord, Z., Naaktgeboren, A., Gianfortoni, P., et al., 2007. Design of a modular snake robot. IROS. IEEE/RSJ International Conference on Intelligent Robots and Systems. IEEE, San Diego, CA, pp. 2609–2614.

Xie, Y., Sun, D., Liu, C., Cheng, S.H., Liu, Y.H., 2009. A force control based cell injection approach in a bio-robotics system. In: Robotics and Automation. ICRA'09 IEEE International Conference; 3443–3448.

Yu, S., Nelson, B.J., 2001. Microrobotic cell injection. In: Robotics and Automation. Proceedings ICRA IEEE International Conference on. 1:620–625.

Yuen, M., Cherian, A., Case, J.C., Seipel, J., Kramer, R.K., 2014. *IEEE/RSJ International Conference on Intelligent Robots and Systems*; 580–586 (IROS).

Further Reading

Fish, F.E., Lauder, G.V., Mittal, R., Techet, A.H., Triantafyllou, M.S., Walker, J.A., et al., 2003. Conceptual Design for the Construction of a Biorobotic AUV Based on Biological Hydrodynamics. In: Proceedings of the 13th International Symposium on Unmanned Untethered Submersible Technology (UUST). Durham, New Hampshire.

3D Printing Technologies and Their Applications in Biomedical Science

Syeda M. Bakhtiar, Hina A. Butt, Shuja Zeb,
Darrak M. Quddusi, Saima Gul and Erum Dilshad

Capital University of Science and Technology, Islamabad, Pakistan

10.1 INTRODUCTION

3D printing is basically a manufacturing technique that includes objects formation by depositing or fusing materials such as liquids, plastic, powders, metal, ceramics, or even living cells in layers to produce a 3D object (Schubert et al., 2014).

It is also denoted as rapid prototyping, solid-free form, computer automated or layered manufacturing depending on the kind of production method used. Rapid prototyping principle is to reconstruct 3D physical model with addition of material layers by using 3D computer models (Ventola, 2014).

Through literature review on 3D printing, it is seen that all sources share the mutual characteristics of providing a definition. Ultimate ideas may vary among sources, but most of them agreed on one idea that 3D printing consists of transferring a special computer file or a design to a printer capable of printing 3D objects through additive process that prints layers of material (Bradshaw et al., 2010; Satyanarayana et al., 2015).

Three fundamental approaches for constructing objects were distinguished by Bradshaw (Satyanarayana et al., 2015). To make an object, the following should be considered.

1. Cutting object out of material
2. Constructing a mold
3. Filling of object
4. Adding shapes together to make an object

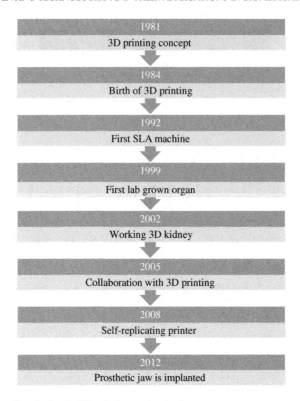

FIGURE 10.1 History of evolution in 3D printing technologies.

3D printing technology plays an essential role in advancing Biomedical Sciences. It increases diagnostic quality, aids further operational patient consultation, provides template for surgical resection, and advances surgical planning. Hence, this emerging technology has great potential to be positively constructive for doctors and patients in terms of patient-specific individualized medicine (Choi and Kim, 2015) (Fig. 10.1).

Advancement in 3D printing applications include creation of scaffolds for tissue engineering, bio cell-printing for 3D development of organs and tissues and actual clinical application for various medical parts.

3D technology has developed the opportunity of the application that has greatly extended especially in medicine. Its usage is increasing day by day because in a short time periods it can deliver individualized product. It suits the objective of personalized medicine where every patient needs a tailored, specific, therapeutic approach. Around two dozen 3D printing processes are used. These processes include varying printer technologies, resolutions, hundreds of materials, and speeds.

Computer-aided design (CAD) defined as 3D printing technology can construct 3D object virtually in any imaginary shape (Ventola, 2014).

In a basic setup the 3D printer first follows the instructions in the CAD file to build the foundation for the object, then moving the print head along the $X-Y$ plane. The printer then continues to follow the instructions and moving the print head along the z-axis to build the object vertically layer by layer. It is important to note that two-dimensional (2D) radiographic images such as X-rays, magnetic resonance imaging (MRI), or computerized tomography (CT) scans can be converted to digital 3D print files, allowing the creation of complex, customized, anatomical, and medical structures (Fig. 10.2). A 3D printer uses instructions in a digital file to create a physical object.

3D printing has been used by the manufacturing industry for decades primarily to produce product prototypes. Many manufacturers use large and fast 3D printers called rapid prototyping machines to create models and molds. A large number of STL files are available for commercial purposes. Many of these printed objects are comparable to traditionally manufactured items (Ventola, 2014) (Fig. 10.3).

Pre-bioprinting involves imaging and digital design in addition to material selection. CT and MRI are considered the two most common imaging technologies for medically applied bioprinting. After medical imaging the tomographic reconstruction is performed

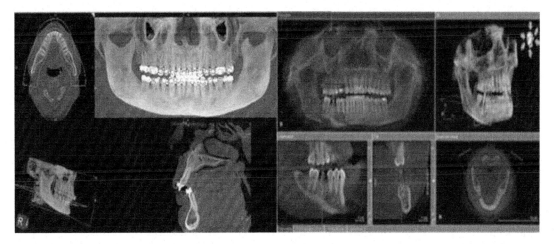

FIGURE 10.2   Normal image conversion into 3D image.

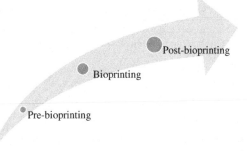

FIGURE 10.3   Process of bioprinting.

to achieve segmental 2D images for the layer-by-layer fabrication process. Finally, by using digital preparations, the 3D representation of the organ, tissue, or target object is created in stereolithography files (STL files) and transferred to the printer. Material selection and another vital part of the process is governed by the printer type used and the requirements of the final product (Mironov et al., 2006).

10.2 TYPES OF PRINTERS

There are many different ways to 3D print an object. But nearly all of them utilize computer-aided design files also known as CAD files. CAD files are digitalized representations of an object. They are used by engineers and manufacturers to turn ideas into computerized models that can be digitally tested, improved, and most recently called as 3D printed. In 3D printing or additive manufacturing, the CAD files must be translated into a language or file type that 3D printing machines can understand. Standard Tessellation Language (STL) is one such file type and is the language most commonly used for Stereolithography. Because additive manufacturing works by adding one layer of material on top of another, the CAD models must be broken up into layers before being printed in 3D STL files cut up CAD models and giving the 3D printing machine the information it needs to print each layer of an object. Table 10.1 summarizes the major properties of various techniques used for 3D printing.

10.2.1 Stereolithography

Stereolithography is one of the several methods used to create 3D printed objects. It is the process by which a uniquely designed 3D printing machine called a stereolithograph

TABLE 10.1   Comparison of the Types of 3D Printers

Technology	Advantages	Disadvantages
STL	• Complex geometries • Smooth finish	• Postfinishing required • Requires support structures
FDM	• Strong parts • Easy to print yourself	• Poorer surface finish and slower • Requires support structures
Selective laser sintering	• No support need • High heat and chemical resistant • High speed	• Precision limited to powder particle size • Rough surface finish
Inkjet printer	• Good precision • Good surface finish • No removal of support material	• Slow build process
LOM	• Lower price • No toxic materials • Quick to make large parts	• Less accurate • Nonhomogenous parts

apparatus that converts liquid plastic into solid objects. The process was patented as a means of rapid prototyping in 1986 by Charles Hull, the co-founder of 3D Systems, Inc., a leader in the 3D printing industry (Gross et al., 2014). Fig. 10.4 shows the comparison between Stereolithographic Apparatus (SLA) and Desktop Printers.

10.2.1.1 *Parts of Stereolithographic Machine*

Following are the parts of the stereo lithographic machine as shown in Fig. 10.2.

In the initial step of the SLA process a thin layer of photopolymer usually between 0.05 and 0.15 mm is exposed above the perforated platform. The UV laser hits the perforated platform painting the pattern of the object being printed. The UV curable liquid hardens instantly when the UV laser touches it forming the first layer of the 3D-printed object. Fig. 10.5 shows the parts of the Stereolithographic Machine.

Once the initial layer of the object has hardened, then the platform is lowered and exposing a new surface layer of liquid polymer. The laser again traces a cross section of the object being printed, which instantly bonds to the hardened section beneath it. This process is repeated again and again until the entire object has been formed and is fully submerged in the tank. The platform is then raised to expose a 3D object. After it is rinsed with a liquid solvent to free it of excess resin, the object is baked in an ultraviolet oven to further cure the plastic.

Objects made using stereolithography generally have smooth surfaces, but the quality of an object depends on the quality of the SLA machine used to print it. The amount of time it takes to create an object with stereolithography also depends on the size of the

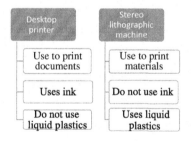

FIGURE 10.4   Comparison of desktop printer and stereolithographic machine.

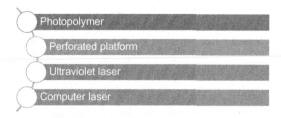

FIGURE 10.5   Parts of the stereolithographic machine.

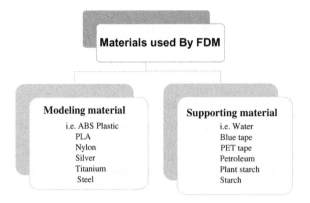

FIGURE 10.6　Materials used by FDM.

machine used to print it. Small objects are usually produced with smaller machines and typically take between 6 and 12 hours to print. Larger objects that can be of several meters in three dimensions take days.

10.2.2 Fused Deposition Modeling

There are several different methods of 3D printing, but the most widely used is a process known as Fused Deposition Modeling (FDM). FDM printers use a thermoplastic filament that is heated to its melting point and then extruded layer by layer to create a 3D object. The technology behind FDM was invented in the 1980s by Scott Crump, the co-founder and chairman of Stratasys Ltd., and a leading manufacturer of 3D printers. Fig. 10.6 shows the materials used by FDM.

Objects created with an FDM printer start out as CAD files. Before an object can be printed, its CAD file must be converted to a format that a 3D printer can usually understand, i.e., STL format. FDM printers use two kinds of materials: a modeling material, which constitutes the finished object, and a support material, which acts as a scaffolding to support the object as it is being printed.

During printing these materials take the form of plastic threads that are unwound from a coil and fed through an extrusion nozzle. The nozzle melts the filaments and extrudes them onto a base referred as build platform. Both the nozzle and the base are controlled by a computer that translates the dimensions of an object into X, Y, and Z coordinates for the nozzle and base to follow during printing.

In a typical FDM system, the extrusion nozzle moves over the build platform horizontally and vertically drawing a cross-section of an object onto the platform. This thin layer of plastic cools and hardens immediately binding to the layer beneath it. Once a layer is completed, the base is lowered usually by about one-sixteenth of an inch to make room for the next layer of plastic. Fig. 10.7 shows the printing time in FDM.

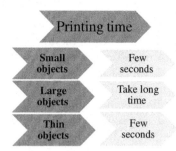

FIGURE 10.7   Printing time of FDM.

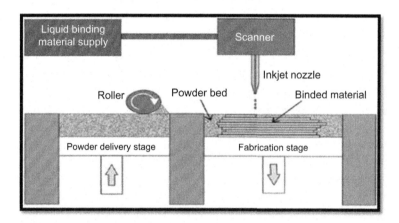

FIGURE 10.8   Schematic inkjet printing apparatus. Source: *http://pubs.acs.org/doi/abs/10.1021/ac403397r*.

10.2.3 Inkjet Printing

The concept of inkjet printing was initially described in 1878 by Lord Rayleigh, and in 1951 Siemens patented the first 2D inkjet type printer called a Rayleigh breakup inkjet device (Strutt, 1878). 3D inkjet printing is mainly a powder-based method where layers of solid particles typically $200\,\mu m$ in height with particle sizes ranging between 50 and $100\,\mu m$ are bound together by a printed liquid material to generate a 3D model.

In this technique a first layer of powder is distributed evenly on the top of a support stage, e.g., by a roller after which an inkjet printer head prints droplets of liquid-binding material onto the powder layer at desired areas of solidification. After the first layer is completed, the stage drops and a second powder layer is distributed and selectively combined with printed binding material. These steps are repeated until a 3D model is generated after which the model is usually heat treated to enhance the binding of the powders at desired regions. Unbound powder serves as support material during the process and is removed after fabrication (Gross et al., 2014). Fig. 10.8 shows the inkjet printing apparatus.

10.2.4 Selective Sintering Printing

Selective Laser Melting is an additive manufacturing technique that can print metal parts in 3D. A laser is used to melt metallic powder in specific places. Selective laser melting uses a laser to melt successive layers of metallic powder.

The laser will heat particles in specified places on a bed of metallic powder until completely melted. The CAD 3D file dictates where melting will occur. Then the machine will successively add another bed of powder above the melted layer until the object is completely finished. The most common applications for this technology are in the aerospace industry as complex parts can be made with additive manufacturing that overcomes the limitations of conventional manufacturing. It can also result in the reduction of parts needed. It can also be used in medical field where prosthetics are created, as this technology allows the model to be customized to the patient's anatomy.

Selective Laser Sintering is a technique that uses laser as power source to form solid 3D objects. This technique was developed by Carl Deckard, a student of Texas University and his Professor Joe Beaman in 1980s. Laser sintering is a 3D printing technique consisting of the fabrication of an object by melting successive layers of powder together to form an object. The process most notably facilitates in the creation of complex and interlocking forms. It is available for Plastic and Alumide.

10.2.5 Laminated Object Manufacturing

Laminated object manufacturing (LOM) is more rapid prototyping system that was developed by the California-based company Helisys, Inc. During the LOM process the layers of adhesive coated paper, plastic, or metal laminates are fused together using heat and pressure and then cut to shape with a computer-controlled laser or knife. Postprocessing of 3D printed parts includes steps as machining and drilling. The LOM process involves the CAD file, transformed to computer format that are usually STL or 3DS. LOM printers use continuous sheet coated with an adhesive, which is laid down across substrate with a heated roller. The heated roller that is passed over the material sheet on substrate melts its adhesive. Then laser or knife traces desired dimensions of the part. Also the laser crosses hatches of any excess material to remove it easily after the printing is done.

After the layer is finished the platform is moved down by about one-sixteenth of an inch. A new sheet of the material is pulled across substrate and adhered to it with a heated roller. The process is repeated until 3D parts are fully printed. When any excess material has been cut, the part can be sanded or sealed with paint. If paper materials were used during printing, then the object would have wood-like properties which means it needs to be protected from moisture. Therefore, covering it with a lacquer or paint might be a very good idea. Fig. 10.9 shows the apparatus of LOM.

10.3 APPLICATION OF 3D PRINTING

3D bioprinting is a flexible automated on demand platform for the free-form fabrication of complex living architectures and is a novel approach for the design and engineering of human organs and tissues.

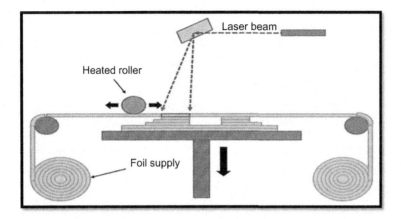

FIGURE 10.9   Schematic LOM printing apparatus. Source: *http://en.topmaxtech.net/content/uploads/types-of-3d-printers-3d-printing-technologies-13.jpg.*

Organ shortage has become more problematic in spite of an increase in willing donors. From July 2000 to July 2001 for example, approximately 80,000 people in the United States awaited an organ transplant with less than one-third receiving it. The solution to this problem as with those to other grand engineering challenges requires long-term solutions by building or manufacturing living organs from a person's own cells. For the past three decades, the tissue engineering has emerged as a multidisciplinary field involving scientists, engineers, and physicians for the purpose of creating biological substitutes mimicking native tissue to replace damaged tissues or restore malfunctioning organs. The traditional tissue-engineering strategy is to seed cells onto scaffolds which can then direct cell proliferation and differentiation into 3D functioning tissues. Both synthetic and natural polymers have been used to engineer various tissue grafts like skin, cartilage, bone, and bladder. To be successfully used for tissue engineering, these materials must be biocompatible and biodegradable with the mechanical strength to support cell attachment, proliferation, and direct cell differentiation toward certain lineages (Visconti et al., 2010).

Although 3D printing technology is evolving, its clinical applications are actually evolving more rapidly. The affordability and convenience of this technology have increased its adoption in a various medical fields. Examples are as follows:

- In pediatrics cardiac surgeons are using 3D printing—based tactile models for analyzing and visualizing complex congenital heart diseases.
- Urologic surgeons are simulating surgery of complex renal cell carcinoma in advance using 3D printed tactile prototype models that include the vessels and parenchyma of the kidney.

Neurosurgeons are using similar approaches for neurosurgery of brain tumors. These kinds of efforts afford the various types of surgeon's involved huge benefits by aiding in the advanced analysis of the patient's specific status.

Organ printing is the biomedical application of very well-established rapid prototyping technology. It is defined as a layer-by-layer additive biofabrication approach using self-assembling tissue spheroids as building blocks. The fundamental biological and biophysical principle of organ printing technology is tissue fusion driven by surface tension forces or the capacity of closely placed tissue spheroids to fuse into a single entity. Fusion is usually defined as melting together. This implies that organ printing is in essence a microfluidic technology. There are many rapidly emerging and competing variants of bioprinting technology.

Characterisics of Unique 3D printing	• Solid scaffold free • Cell Aggregation • Direct Process

The three principal characteristics of organ printing technology, which make it unique are (1) organ printing is not based on using solid scaffolds. It is a solid scaffold–free technology; (2) it is based on using self-assembling tissue spheroids or cell aggregates with maximum possible cell density; and (3) it is a direct processing technology where cells and a hydrogel are dispensed simultaneously without intermediate steps such as creation of a synthetic solid supporting scaffold.

10.3.1 Craniofacial Plastic Surgery

The clinical application of 3D printing in craniofacial surgery are based on more than 500 craniofacial cases conducted using 3D printing tactile prototype models. Of various medical fields, the craniofacial plastic surgery is one of the areas that pioneered the use of the 3D printing concept. The procedure for the fabrication of medical models consists of multiple steps: Acquisition of high-quality volumetric 3D image data of the anatomical structure to be modeled, 3D image processing to extract the region of interest from the surrounding tissues, Mathematical surface modeling of the anatomic surfaces, Formatting of data for rapid prototyping, Model building, and Quality assurance of the model and its dimensional accuracy (Fig. 10.10).

10.3.2 Skull Reconstruction

Calvarial bone reconstruction would be the most pioneering use of Rapid Prototype (RP) models. In 1994 the Mankovich et al. first applied 3D technology for skull reconstruction. Because autogenous bone grafting would be the ideal standard for skull reconstruction, donor bone should be harvested. The ideal curvature should be researched in advance because the bone is so rigid that bending is quite difficult and risky. Experience has shown that tactile prototype models are very helpful in identifying the ideal donor site. For example, the skull reconstruction can be done with a split calvarial bone grafting technique, and the ideal donor area can be determined in advance to match the ideal recipient calvarial bone curvature (Choi and Kim, 2015).

Although many materials can be used for 3D printing, few materials can be permanently inserted into the human body. The titanium is the ideal realistic material for human body use. Therefore, once the recently developed 3D technology starts to provide 3D

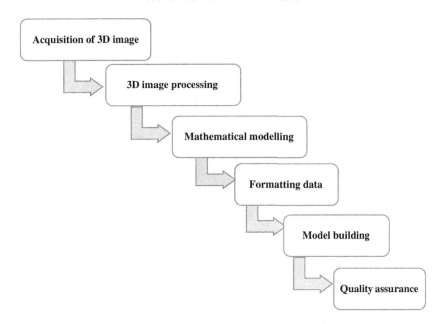

FIGURE 10.10 Procedure for fabrication of organs.

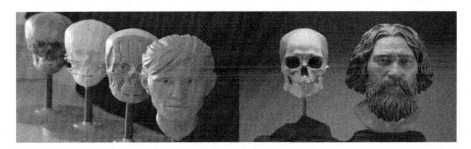

FIGURE 10.11 Skull reconstruction with split calvarial bone grafting.

titanium-based implants, they could be used in the human body. The technique currently appears to be quite successful and without any complications. The implants fit very well onto preexisting defects such as calvarium or maxillary defects (Fig. 10.11).

10.3.3 Cranioplasty for Correction of Syndromic Craniosynostosis

Similar to the skull reconstruction with bone grafting, the extensive bone grafting is needed to correct craniosynostosis. Current 3D printing technology can provide an osteotomy guide that is very useful in the reconstruction process. Moreover, the surgeons can simulate the surgery in advance using the 3D printed tactile model (Fig. 10.12).

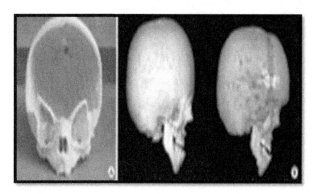

FIGURE 10.12   Syndromic craniosynostosis correction using 3D printing technology.

10.3.4 Facial Bone Fractures

Although, many facial bone fractures can be managed with 3D printing technology, the orbital wall fractures would be the ideal target for these methods. The orbit has such a complex anatomy that ideal reconstruction is not particularly easy. Unless the orbital wall is repaired very precisely, the postoperative enophthalmos or diplopia can occur. Nevertheless, the limited surgical fields during surgery of orbital wall fractures often cause reconstruction in the wrong plane. These kinds of difficulties can be overcome by using 3D printed titanium mesh implants or by preventing the malpositioning of implant material. Medpore or titanium mesh based on the RP model is manufactured from the mirroring technique of CAD–CAM. The contralateral orbit can be a reference. Using the contralateral orbital anatomy, the ideal ipsilateral orbital structures can be simulated on computer software and can be manufactured using 3D printing technology (Fig. 10.13).

3D printing techniques have been most actively used in craniofacial surgery and based on extensive experience with 3D printing in craniofacial reconstruction. First, the computer software used for craniofacial reconstruction should be much more specifically designed. However, the preoperative design of surgery is not especially easy. Because the segmentation process in computer simulations is time consuming, it needs to be more automated. If the various software programs were more suitable and specific for craniofacial reconstruction, the 3D printing technique could be more actively used.

Second, a connection between the preoperative simulations and the real surgery environment should be made. Surgical wafers such as intermediate and final dental splints would be an example in orthognathic surgery. In addition, a navigational system could act as a surgical guide to connect the preoperative simulation and the actual surgery. To apply the 3D printed titanium implant, the surgical cut or osteotomy should be matched precisely with the preoperative planning. Because the 3D printed implant is so solid that it is not easy to cut or bend, planning and surgery should be identical and efforts should be made to ensure that the preoperative planning and intraoperative defects are in agreement. Therefore, a surgical osteotomy guide should be made.

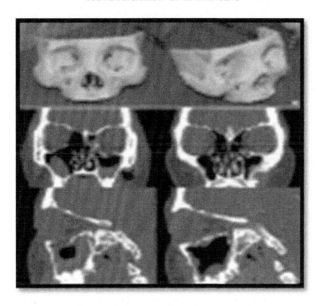

FIGURE 10.13   Orbital wall reconstruction based on 3D printing.

The third issue is accuracy. Although, CT scans are made in very thin slices, the imaging modality can only provide the accumulation of the multiple slices. Error can inevitably occur between the slices. In particular, the orbital wall is too thin to be reconstructed by only a 3D printing technique, and a 3D printed orbit model represents the orbit as vacant fields.

Finally, the artifacts associated with the metal can discourage the use of 3D printing models. For example, the dental models cannot be recreated with CT scanning because of accuracy issues. Dental occlusion requires such delicate precision that the 3D rendering of CT data cannot provide a sufficiently high resolution. Hence, the dental scanners are currently being used and techniques that merge CT and dental scans are common. Despite these obstacles, the 3D printing technology could be a new medical modality. Although this basic technology was initially developed over 20 years ago, further technological advances could enable medical doctors to realize patient-specific individualized medicine in the near future. 3D printing technology enables more effective patient consultations, increases diagnostic quality, improves surgical planning, acts as an orientation aid during surgery, and provides a template for surgical resection. In addition, as Biocell printing technology further evolves, tissues or organs might be made using 3D printing methods in the future. 3D printing technology thus has the potential to be very beneficial to patients and doctors in terms of patient-specific individualized medicine (Choi and Kim, 2015).

10.3.5 Mandibular Reconstruction

Mandibular reconstruction is mostly being performed using fibular osteocutaneous−free flaps. Although the curvature of the original mandible can be reconstructed using the

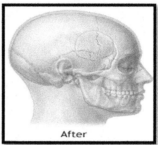

FIGURE 10.14   Mandibular reconstruction.

conventional method, the 3D CAD–CAM technology can provide a more precise reconstruction modality that includes fibular osteotomy and fixation guides. In addition, the 3D printed titanium fixation plates were recently tried and have been shown to be very useful for the ideal reconstruction of the mandible. In the near future, the 3D printed titanium implants that can be inserted into the human body as a whole might be possible (Choi and Kim, 2015) (Fig. 10.14).

10.3.6 Human Skin

Skin is the largest part of the human body, and it plays a vital role in maintaining homeostasis as well as in providing protection from the external environment. The highly complex, hierarchical, and stratified structure of the skin provides a physical barrier to the entry of xenobiotics into the body while regulating the transport of water and small metabolites out of the body (Lee et al., 2014). Wounds originating from physical or chemical trauma can significantly compromise the skin barrier and impair its physiological functions. Instances in which a considerable amount of the skin has been lost to injuries, it becomes critical to replace the impaired skin via grafts to protect water loss from the body and to mitigate the risk posed by opportunistic pathogens.

Skin grafts can also greatly facilitate the wound healing process and can potentially restore the barrier and regulatory functions at the site of the wound. Beyond grafts, the tissue-engineered skin can serve as an extremely valuable in vitro platform to evaluate the permeability as well as the adverse inflammatory responses of topical agents in a high-throughput manner during the preliminary stages of transdermal and topical drug discovery and formulation development. The engineered skin provides several advantages compared with animal skin by better mimicking human skin physiology as well as by alleviating ethical concerns and conforming to emerging regulations on animal use. In addition, the engineered skin models can provide fundamental insights into the etiology of skin diseases and elucidate the pathophysiological mechanisms in skin disease progression and treatment (Fig. 10.15).

Construction was achieved using a layer-by-layer fabrication approach as described in Fig. 10.15. The multilayered collagen structure with embedded cells was constructed on poly-D-lysine-coated glass bottom petri dishes. Before printing was initiated, nebulized

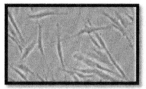

FIGURE 10.15 Reconstruction of skin.

sodium bicarbonate vapor was applied onto the petri dish surface enabling the quick gelation of the first printed collagen layer and thereby increasing its adhesion to the bottom surface. Subsequent to the printing of the first collagen layer, the nebulized $NaHCO_3$ vapor was applied onto the printed collagen layer for gelation. To provide a firm base for the next layer of printing, a time lapse of 1 minute was allowed to facilitate the phase transition of collagen to a gel. The printed skin structure contains eight collagen layers.

10.3.7 Tissue Engineering

The potential of 3D bioprinting for tissue engineering uses human skin as a prototypical example. Keratinocytes and fibroblasts were used as constituent cells to represent the epidermis; dermis and collagen was used to represent the dermal matrix of the skin. Preliminary studies were conducted to optimize printing parameters for maximum cell viability and for the optimization of cell densities in the epidermis and dermis to mimic physiologically relevant attributes of human skin. Printed 3D constructs were cultured in submerged media conditions followed by exposure of the epidermal layer to the air–liquid interface to promote maturation and stratification. Histology and immunofluorescence characterization demonstrated that 3D printed skin tissue was morphologically and biologically representative of in vivo human skin tissue. In comparison with traditional methods for skin engineering, the 3D bioprinting offers several advantages in terms of shape and form retention, flexibility, reproducibility, and high-culture throughput. It has a broad range of applications in transdermal and topical formulation discovery, dermal toxicity studies, and in designing autologous grafts for wound healing (Mironov et al., 2006; Bajaj et al., 2014).

10.3.8 Ears

Maxillofacial prosthodontics gains utmost significance as the process not only tries to develop acceptable esthetics but also tries to enhance the psychological comfort of the patient by restoring the defective or missing craniofacial structures. This science has developed into a more reliable and predictable process due to ever increasing development of materials and equipments used in such procedures, but no single procedure stands out as the most effective method in all the situations.

The fabrication of any extra oral maxillofacial prosthesis presents the prosthodontist with several phenomenal challenges. Foremost among them is the flawless recreation of the same anatomic form that was present preoperatively. The challenge of recreating preoperative anatomic contour is sometimes complicated by the absence of a preoperative record. This is common when the loss is congenital or unexpected as with the loss of human ear. In these conditions, the preoperative bilateral symmetry can be recreated using mirror image modeling.

CT scanning acquires data by recording in slices, which can be used to reconstruct anatomical structures or fabricate prosthesis with accuracy and details. The widespread use of CT scan also casts a shadow on the application as it may expose the patient to high radiation dosage. Techniques to obtain 3D models using data obtained from CT scan and milled using polyurethane blocks to obtain ideal soft tissue models for waxing have been developed in recent times. MRI proves a viable alternative as the exposure to radiation is substantially reduced in this imaging technique. The disadvantage being its application in situations with many metallic restorations leading to scattering and artifact formation. Another disadvantage being the necessity of patient to remain still during the entire length of procedure. 3D laser scanning system has the disadvantage of inability to record the details in the undercut areas as the laser can be directed only vertically. Such problems are corrected with the advent of 3D laser scanning technique. To obtain a scan from cast of existing ear and create a mirror image of the scan to obtain a rapid prototype model of the deficient side using resin material, a 3D laser scanner develops an integrated 3D digital image of the unaffected ear, which is copied and then mirrored. A rapid prototyping machine collects the necessary data to manufacture the definitive resin ear (Thotapalli, 2013; He et al., 2014) (Fig. 10.16).

10.3.9 Cartilage

Orthopedic surgeons commonly face clinical and surgical challenges for which current therapeutic strategies are not able to provide a satisfactory result. One example is young patients with large osteochondral defects due to injury which represents a difficult and frustrating clinical scenario for both the patient and the surgeon. Previous hyaline cartilage damage has been reported to predispose individuals to osteoarthritis, possibly due to the limited capacity of hyaline cartilage to repair itself. The inability to halt degenerative changes in the articular surface in patients with chondral and osteochondral lesions has brought scientists, clinicians, and surgeons together to tackle the difficulties in cartilage tissue engineering.

FIGURE 10.16   3D printing of human ear.

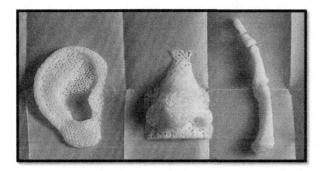

FIGURE 10.17 3D printing of cartilage.

The goal of such collaboration is to produce mature hyaline cartilage that can maintain its physical and functional properties in the long term without accelerated degeneration that may lead to arthritic changes. Tissue engineering has the potential to address the issue of osteoarticular loss and may provide a viable alternative to current treatment modalities. For example, the established in vitro and in vivo tissue engineering techniques have successfully led to the creation of living cartilage and bone.

Without blood vessels, nerves, lymphatics, and with only one type of cells, mature hyaline cartilage appears to be easy to create in laboratory (Di Bella et al., 2015). However, these characteristics also mean that cartilage injuries cannot heal spontaneously and that any type of repair will be characterized by fibrocartilage, which represents a scar type tissue. This tissue lacks the properties that make hyaline cartilage so unique including its resistance to shear, compression, and load thus leading to degenerative changes and arthritis (Fig. 10.17).

Despite its simple appearance, the cartilage is in fact a tissue that shows great heterogeneity and is characterized by a composition that exhibits differences depending on the depth of the tissue.

10.3.10 Heart Valve

Heart valve disease is a serious and growing public health problem for which prosthetic replacement is most commonly indicated. Current prosthetic devices are inadequate for younger adults and growing children. Tissue engineered living aortic valve conduits have potential for remodeling, regeneration, and growth, but fabricating natural anatomical complexity with cellular heterogeneity remains challenging (Visconti et al., 2010). Heart valves ensure critical one-way blood flow through the cardiovascular system. The aortic valve is particularly important as it directs blood through the aorta and coronary arteries. Efficiency of valve performance is controlled by its complex anatomical geometry and heterogeneous tissue biomechanics. Congenital malformations or acquired valve diseases compromise valve shape and tissue mechanics leading to the narrowing of valve orifice area or leaking.

Aortic valve disease is a significant cause of morbidity and mortality. While the aortic valve can sometimes be surgically repaired, prosthetic replacement is the only option for

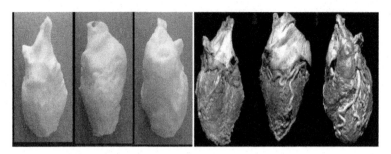

FIGURE 10.18   Stages of printing heart.

the vast majority of patients. Mechanical valves are durable but suffer risks of clot formation on their prosthetic surfaces necessitating lifelong anticoagulant drug therapy (Duan
et al., 2013). Bioprosthetic valves, however, have minimal risk for bleeding events but are
much less durable and unsuitable for pediatric applications.

A common fabrication process is to mold a valve shape from a biodegradable scaffold
material, seed it with cells, and culture it in vitro often under dynamic stimulation, before
implantation. A wide range of synthetic polymers such as polyglycolic acid and poly-4-
hydroxybutyrate and natural biomaterials such as fibrin and collagen have been employed
to generate 3D heart valve scaffolds. 3D bioprinting implements rapid prototyping techniques that follow computer-assisted design or computer-assisted manufacturing blueprints
to build a complex tissue construct (Fig. 10.18).

10.3.11 Tooth and Peridontal Regeneration

A tooth is a major organ consisting of biological viable pulp encased in mineralized den-
tin that may be covered with cementum and enamel ontogenetically in various species. Life
ends in wildlife species after complete tooth loss. In humans the tooth loss can lead to physical and mental suffering that compromises self-esteem and quality of life. Contemporary
dentistry restores missing teeth with dental implants or dentures. Dental implants despite
being the preferred treatment modality can fail and have no ability to remodel with surrounding bone, which undergoes physiologically necessary remodeling throughout life.

Tooth regeneration by cell delivery encounters translational hurdles. The anatomically
shaped human molar scaffolds and rat incisor scaffolds were fabricated by 3D bioprinting
from a hybrid of poly-ε-caprolactone and hydroxyapatite with 200-μm-diameter interconnecting microchannels. Regeneration of tooth-like structures and periodontal integration
by cell homing provide an alternative to cell delivery and may accelerate clinical applications (Hirayama et al., 2015) (Fig. 10.19).

10.4 ONCOLOGY AND 3D PRINTING

Cancer is the major cause of death worldwide and one in every four deaths in the
United States is due to cancer-related diseases. Although cells in normal tissue reside in
defined locations and maintain steady numbers, cancer cells remove these constraints

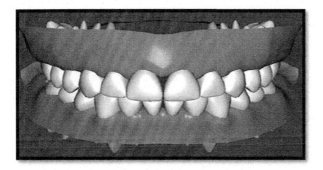

FIGURE 10.19   3D printing of tooth.

through mutations in oncogenes and tumor suppressor gene. Consequently, the cells in the tumor tissues can sustain proliferative signaling, evade growth suppressors, resist cell death, enable replicative immortality, induce angiogenesis, and activate invasion and metastasis. During cancer progression and metastasis, the malignant cells maintain their close interactions with surrounding cells and the stromal extracellular matrices. Numerous stromal cells including endothelial cells of the blood and lymphatic circulation, stromal fibroblasts, and innate and adaptive infiltrating immune cells together comprise the complex tumor microenvironment (Xu et al., 2014). Tumor growth and aggressiveness are influenced by the microenvironment surrounding the tumor mass. The native tumor microenvironment is composed of extracellular matrix (ECM), cell-to-cell contact, and cell—matrix interactions. The ECM components are involved in various cell signaling pathways. These cell-to-cell and cell—matrix interactions regulate tumor growth, angiogenesis, aggression, invasion, and metastasis.

In the early stages of cancer the tumor cells undergo certain alterations, a process called immunoediting, to initiate signaling pathways that inactivate the immune system to prevent their elimination from the body. Such alterations allow cancer cells to avoid the body's immune response and grow abnormally to form a large tumor mass (Asghar et al., 2015). Latest studies show that tumors may have the ability to generate their own capillary network.

This problem can be solved by tissue engineering that can provide a 3D environment mimicking organs or tissues with or without vascular networks. Components and properties of the microenvironment such as ECM, adhesion integrins, tissue architectures, and tissue modulus regulate growth, differentiation, and apoptosis of cells. These properties control cell fate through complex signals that are influenced either by interactions between neighboring cells or by stimulated cell-surface receptors. Regarding different cancers and malignant tumors that have been subject to research in combination with tissue engineering techniques, some of the recent developments are highlighted. Various types of bioreactors have been found to play a crucial role for studying complex 3D interactions of stem or tumor cells under conditions that resemble the clinical situation.

Studies are conducted to observe effects of combining soluble factor signaling and cell—matrix interactions on cell behavior. These studies are important to optimize the cell microenvironment in Tissue Engineering (TE) applications and for further cancer cell research. Various groups have studied cells in the interstice, and there is growing evidence that these cells may play a crucial role in tissue regeneration as well as potentially in cancer

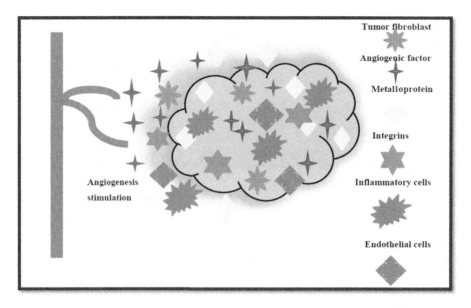

FIGURE 10.20   Tumor cell microenvironment containing different proteins, cells, and angiogenesis.

genesis (Horch et al., 2013). Advances in tissue engineering and microtechnology have enabled researchers to more easily generate in vitro tissue models that mimic the tissue geometry and spatial organization found in vivo. However, the widespread adoption of these models for biological studies has been slow in part due to the lack of direct comparisons between existing 2D and 3D cell culture models and new organotypic models that better replicate tissue structure. Using previously developed vessel and mammary duct models with 3D lumen structures (Bischel et al., 2014) Fig. 10.20 shows the Tumor cell microenvironment and interplay of key proteins involved in angiogenesis.

10.4.1 Cancer Microenvironment Engineering for In Vitro 3D models

Two-dimensional in vitro cancer models and small in vivo animal models are used conventionally for drug testing and screening. However, because of the difficulty in recapitulating the natural tumor microenvironment in 2D culture as well as the cost and issues associated with animal models, both approaches have become less attractive for routine drug testing. New 3D in vitro cancer models have emerged as an alternative approach to conventional methods and have shown the potential to recapitulate the natural microenvironment of tumors in a relatively simple and inexpensive way when compared to conventional methods (Asghar et al., 2015). They can restore cellular morphologies and phenotypes characteristic of in vivo tumor development. Further advancements have led to the fabrication of microfluidic culture models, which replicate native 3D microenvironments in combination with gradient and flow control and thus enable systematic investigation of both physiological and pathological phenomena in vitro (Buchanan et al., 2014a,b).

Various approaches that are currently available for generating 3D tumor models are spheroids, hanging drop, bioprinting, and magnetic levitation. In addition, following are the effects of materials, e.g., basement membrane matrix, hydrogels, and scaffolds; physical parameters, e.g., stiffness, morphology, flow, and shear stress on the growth, invasiveness, differentiation, and regulation of biomarker expression of cancer cells.

Multicellular tumor spheroids (MCTS) consist of cancer cells from established cell lines or disaggregated human tumor fragments. They can be studied in suspension in bioreactors or in 3D matrices and closely resemble cell—cell and cell—matrix interactions in vivo. MCTS consist of actively proliferating cells on the outside of the spheroid with quiescent cells in the inner, nutrient-deprived zone. Therefore, the spheroid size is limited to 400—600 μm as with increasing size the cells inside the spheroid become necrotic. This is likely due to the limited oxygen and nutrient availability, mimicking the natural scenario in a tumor in situ.

Most importantly the 3D structure of MCTS creates a penetration path that needs to be overcome by any agent to be tested. Therefore, MCTS are often used for testing of chemotherapeutics. An alternative approach to mimic the native in vivo tumor microenvironment is to use fibroblast-derived 3D matrices. Fibroblasts harvested from stroma at different stages of tumor progression of an in vivo model do not retain their different characteristics in a 2D model-like growth and ECM secretion such as desmoplastic markers but exhibited them in a 3D model.

Furthermore, the secreted 3D matrices determined the morphology of newly seeded fibroblasts, for example, the advanced tumor 3D matrix—induced desmoplastic characteristics. Adenocarcinoma cell lines showed distinct aggregation, growth, proliferation, and morphology profiles on fibroblast-derived 3D matrices which did not necessarily correlated to 2D behavior and were grouped into different classes depending on the degree of proliferative versus morphological response. A 3D scaffold is a temporary structure that supports cells in a given environment, which may eventually be incorporated into the tissue. Scaffolds have been widely used in tissue engineering for the purpose of wound healing or bone regeneration. They have been adapted as matrices for cell culture and for the investigation of proliferation, growth, and migration of cancer cells. Scaffold properties including composition, configuration, and porosity dictate the extent to which cells can

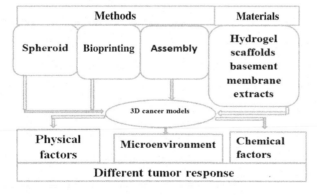

FIGURE 10.21 Methods and materials used to engineer 3D cancer models.

migrate, proliferate, and aggregate. As a 3D structure, a scaffold has the potential of recapitulating the native geometry (Buchanan et al., 2014a,b) (Fig. 10.21).

Various scaffolds and models have been published with regard to liver cancer or brain cancer cell studies on the basis of different biomaterials for long-term culture of cells to study drug toxicity and hepatocyte metabolism in humans and develop a bioartificial liver model. Sphere-templated polymeric scaffolds could offer the potential to serve as an adaptable cell culture substrate for engineering a 3D prostate cancer model. The evolution from 2D to 3D cell cultures and tandem bioreactors to investigate tumor cell growth, tumor-related angiogenesis, attachment, and spreading of tumor cells in TE matrices, as well as the prospect of modulating these phenomena in a nature mimicking environment in combination with supra microsurgical methods of vascularization, opens new horizons for anticancer research (Horch et al., 2013).

10.5 FUTURE DIRECTIONS

Different types of 3D printing are utilized for applications that range from studies of cellular behavior to investigations of tissue pharmacodynamics or toxicological mechanisms. Although 3D printers have high precision and reproducibility, printing organs and functional tissues with entire structures still requires layer-by-layer assembly with "bio-glue." Studies in the near future will likely bring great progress in printing microorgans, such as pancreas islet tissues that function in the absence of the complete pancreas structure which will benefit hundreds of millions of diabetic patients around the globe. Chang et al. (2011) successfully fabricated microlivers that were utilized for testing drug metabolism. As printing technology develops, additional biomimetic, tissue engineered organs will be created.

References

Asghar, W., El Assal, R., Shafiee, H., Pitteri, S., Paulmurugan, R., Demirci, U., 2015. Engineering cancer microenvironments for in vitro 3-D tumor models. Mater. Today 18 (10), 539–553.

Bajaj, P., Schweller, R.M., Khademhosseini, A., West, J.L., Bashir, R., 2014. 3D biofabrication strategies for tissue engineering and regenerative medicine. Annu. Rev. Biomed. Eng. 16, 247–276.

Bischel, L.L., Sung, K.E., Jimenez-Torres, J.A., Mader, B., Keely, P.J., Beebe, D.J., 2014. The importance of being a lumen. FASEB J. 28 (11), 4583–4590.

Bradshaw, S., Bowyer, A., Haufe, P., 2010. The intellectual property implications of low-cost 3D printing. ScriptEd 7 (1), 5–31.

Buchanan, C.F., Verbridge, S.S., Vlachos, P.P., Rylander, M.N., 2014a. Flow shear stress regulates endothelial barrier function and expression of angiogenic factors in a 3D microfluidic tumor vascular model. Cell Adh. Migr. 8 (5), 517–524.

Buchanan, C.F., Voigt, E.E., Szot, C.S., Freeman, J.W., Vlachos, P.P., Rylander, M.N., 2014b. Three-dimensional microfluidic collagen hydrogels for investigating flow-mediated tumor-endothelial signaling and vascular organization. Tissue Eng. Part C Methods 20 (1), 64–75.

Chang, R.C., Emami, K., Jeevarajan, A., Wu, H., Sun, W., 2011. Microprinting of liver micro-organ for drug metabolism study. Methods Mol. Biol. 671, 219–238.

Choi, J.W., Kim, N., 2015. Clinical application of three-dimensional printing technology in craniofacial plastic surgery. Arch. Plast. Surg. 42 (3), 267–277.

Di Bella, C., Fosang, A., Donati, D.M., Wallace, G.G., Choong, P.F.M., 2015. 3D bioprinting of cartilage for orthopedic surgeons: reading between the lines. Front. Surg. 2 (August), 1–7.

Duan, B., Hockaday, L.A., Kang, K.H., Butcher, J.T., 2013. 3D bioprinting of heterogeneous aortic valve conduits with alginate/gelatin hydrogels. J. Biomed. Mater. Res. A 101 (5), 1255–1264.

Gross, B.C., Erkal, J.L., Lockwood, S.Y., Chen, C., Spence, D.M., 2014. Evaluation of 3D printing and its potential impact on biotechnology and the chemical sciences. Anal. Chem. 86 (7), 3240–3253.

He, Y., Xue, G., Fu, J., 2014. Fabrication of low cost soft tissue prostheses with the desktop 3D printer. Sci. Rep. 4, 6973.

Hirayama, M., Tsubota, K., Tsuji, T., 2015. Bioengineered lacrimal gland organ regeneration in vivo. J. Funct. Biomater. 6 (3), 634–649.

Horch, R.E., Boos, A.M., Quan, Y., Bleiziffer, O., Detsch, R., Boccaccini, A.R., et al., 2013. Cancer research by means of tissue engineering—is there a rationale? J. Cell. Mol. Med. 17 (10), 1197–1206.

Lee, V., Singh, G., Trasatti, J.P., Bjornsson, C., Xu, X., Tran, T.N., et al., 2014. Design and fabrication of human skin by three-dimensional bioprinting. Tissue Eng. Part C Methods 20 (6), 473–484.

Mironov, V., Reis, N., Derby, B., 2006. Review: bioprinting: a beginning. Tissue Eng. 12 (4), 631–634.

Strutt, Lord Rayleigh, J.W., 1878. On the instability of jets. Proc. Lond. Math. Soc. 10, 4–13.

Satyanarayana, B., Prakash, K.J., Jyothi, V., 2015. Component replication using 3D printing technology. Procedia Mater. Sci. 10, 263–269.

Schubert, C., van Langeveld, M.C., Donoso, L.A., 2014. Innovations in 3D printing: a 3D overview from optics to organs. Br. J. Ophthalmol. 98 (2), 159–161.

Thotapalli, S., 2013. Fabrication of mirror image prosthetic ears—a short review. Anaplastology 2 (5), 5–7.

Ventola, C.L., 2014. Medical applications for 3D printing: current and projected uses. P T 39 (10), 704–711.

Visconti, R.P., Kasyanov, V., Gentile, C., Zhang, J., Markwald, R.R., Mironov, V., 2010. Towards organ printing: engineering an intra-organ branched vascular tree. Expert Opin. Biol. Ther. 10 (3), 409–420.

Xu, X., Farach-Carson, M.C., Jia, X., 2014. Three-dimensional in vitro tumor models for cancer research and drug evaluation. Biotechnol. Adv. 32 (7), 1256–1268.

11

Next-Generation Sequencing and Data Analysis: Strategies, Tools, Pipelines and Protocols

Pablo H.C.G. de Sá[1], Luis C. Guimarães[1], Diego A. das Graças[1], Adonney A. de Oliveira Veras[1], Debmalya Barh[2,3], Vasco Azevedo[2], Artur L. da Costa da Silva[1] and Rommel T.J. Ramos[1]

[1]Federal University of Pará, Belém, Pará, Brazil [2]Federal University of Minas Gerais, Belo Horizonte, Minas Gerais, Brazil [3]Institute of Integrative Omics and Applied Biotechnology (IIOAB), Purba Medinipur, India

11.1 INTRODUCTION

The emergence of next-generation sequencing (NGS) technologies, starting in 2005, led to major advances in genomics and improvements in contrast to the Sanger method, such as a massively parallel sequencing capacity, high-throughput sequencing, and cost reductions (Liu et al., 2012). Among the new sequencing technologies that have been released over the last 10 years, the most commonly used are the 454 GS FLX (Roche), HiSeq (Illumina), SOLiD (Sequencing by Oligonucleotide Ligation and Detection; Thermo Fisher Scientific), Ion Torrent (Thermo Fisher Scientific), and PacBio (Pacific Biosciences) platforms (Pareek et al., 2011; Liu et al., 2012; Rhoads and Au, 2015). The commercialization of the SOLiD platform was discontinued, and a new version of the Ion Torrent system was launched by Thermo Fisher Scientific in 2015.

The evaluation of the data produced by these platforms remains a major challenge for functional and structural analyses. In this chapter, the chemistry used by the main sequencing platforms, the characteristics of the data produced by each of them, the main tools for de novo and reference genome assembly, the effects of the assembly process on genome annotation will be discussed. The concepts related to RNA-Seq data analysis with the most used software, ChIP-Seq technology, and the protocols and pipelines used in metagenomic approaches will be discussed. The sequencing chemistry of some of these NGS technologies is described below.

11.2 SEQUENCING PLATFORMS

The available sequencing platforms present differences regarding their chemistry, read length, and throughput (Table 11.1).

11.2.1 Illumina Platform

Illumina technology performs clonal amplification and sequencing via synthesis using reversible terminator chemistry with DNA polymerase and fluorophore-labeled terminator nucleotides. The principle of sequencing by synthesis consists of using solid-phase PCR, in which single-stranded DNA fragments ligated to adapters are attached to a flow cell via hybridization. The amplification of DNA begins with the adapter of the free 3' end of the molecule binding to a complementary oligonucleotide, attached to the flow cell surface, forming a bridged structure. Thus, the PCR occurs through the addition of unlabeled nucleotides and other necessary reagents. Clusters of identical molecules are obtained at the end of this stage, and sequencing reactions occur within each cluster, with steps for the incorporation of fluorophore-labeled terminator nucleotides (chemically blocked 3'OH ends), excitation and reading; these steps are repeated for each nucleotide of the sequence. Base reading is performed in each sequencing cycle through sequential analysis of images captured by charge-coupled device cameras (Shendure and Ji, 2008; Mardis, 2008; https://www.illumina.com).

The MiSeq platform is suitable for the sequencing of small genomes such as those of bacteria, as its throughput ranges from 0.3 to 15 Gb, with reading lengths of up to 300 bp, using paired libraries. The HiSeq 2500 platform is the best option for studies that require high-throughput sequencing. Currently, it produces up to 1 Tb, with a read length of

TABLE 11.1   Sequencing Platforms' Features

Technology	Instrument	Read Length	Throughput	Most Frequent Error Type
Illumina	HiSeq 2500	2×125 bp	1 Tb	Single nucleotide substitution error
Illumina	MiSeq	2×300 bp	0.3–15 Gb	Single nucleotide substitution error
Ion Torrent	Ion PGM 318v2	400 bp	2 Gb	Short deletions
Ion Torrent	Ion Proton	200 bp	10 Gb	Short deletions
Pacific Biosciences	PacBio	$\sim$14 kb	1 Gb	CG deletions

250 bp, and allows the use of paired libraries. Illumina has two other versions of this equipment, HiSeq 3000 and 4000, which exhibit an approximate throughput of 125–750 Gb and 125–1500 Gb, respectively, with the same maximum read length for both platforms (150 bp), and allow the use of paired libraries (Van Dijk et al., 2014).

11.2.2 Ion Torrent Platform

The Ion Torrent platform Personal Genome Machine (PGM) uses the sequencing-by-synthesis method, in which one H^+ ion is released upon the incorporation of each nucleotide in a DNA molecule by the polymerase enzyme, changing the pH and allowing the identification of the nucleotide that is added during the synthesis of the DNA strand. This process occurs through emulsion PCR in microwells present on semiconductor chips, which are composed of three layers: a surface layer that contains microwells for the deposition of beads and reagents for the sequencing process, an intermediate layer comprising a semiconductor that allows signal transmission to the bottom layer, and a third layer in which the detection of pH changes occurs. This technology differs from the others in that neither fluorescence or chemiluminescence methods are used, being characterized as a post-light sequencing platform (Rothberg et al., 2011; Liu et al., 2012; Merriman et al., 2012; Yeo et al., 2012). Insertions and deletions during the reading of homopolymer regions are sequencing errors' characteristic of this technology (Faircloth and Glenn, 2012).

The throughput of this platform varies according to the chip used (100 Mb—314 v2, 1 Gb—316 v2, and 2 Gb—318 v2), and the maximum length of the reads produced is 400 bp. The Ion Proton platform with the PI chip, which achieves a throughput of up to 10 Gb, was launched in 2011, allowing the sequencing of one to three human exomes or one to eight human transcriptomes per run (Ekblom and Wolf, 2014; Fox et al., 2014). Due to these characteristics, this equipment is indicated for the sequencing of more complex genomes, such as those of eukaryotes, as well as transcriptomes and exomes. The Thermo Fisher Scientific Company has currently introduced a new version of this platform, Ion S5, which offers three types of chips: Ion 520, Ion 530, and Ion 540, which allow the generation of 2 Gb, 8 Gb, and 15 Gb of data, respectively, with the first two chips generating a read length of up to 400 bp, in comparison to 200 bp for Ion 540 (https://www.thermofisher.com).

11.2.3 PacBio Platform

PacBio technology from Pacific Biosciences is based on Single-Molecule Real-Time (SMRT) sequencing and is considered a next-generation technology. PacBio sequencing identifies the sequence information of the target DNA molecule during the replication process. A closed single-stranded circular DNA model referred to as SMRTbell is initially produced from adapters ligated at both ends of the double-stranded target DNA in hairpin format (Travers et al., 2010; Rhoads and Au, 2015).

The SMRTbell produced is loaded onto a chip referred to as an SMRT cell, which houses a patterned zero-mode waveguide (ZMW) array. Each ZMW contains an immobilized polymerase that can bind to any SMRTbell and initiate replication (Pacific Biosciences). The nucleotides are fluorescently labeled and generate different light emission spectra (Eid et al., 2009). The ZMW exhibits a wide range (50 nm) in which light is not able to propagate through the waveguide, but energy can travel a short distance and excite the fluorophores bound to the nucleotides, which are located near the polymerase at the bottom of the well. A distinct fluorescence pulse is detected in real time after the incorporation of each base (Quail et al., 2012; Rhoads and Au, 2015).

This technology is suitable for de novo assembly studies because it presents a lower sensitivity to GC (guanine and thymine) contents and a greater read length; nevertheless, it exhibits a high error rate, of approximately 11% (Aly et al., 2015; Reuter et al., 2015). The first PacBio RS platform with first-generation chemistry (C1 chemistry) generates a read length of approximately 1500 bp (Brown et al., 2014). The PacBio RS II system with the current C4 chemistry exhibits average read lengths ranging from 10 to 15 kb (Pacific Biosciences). This SMRT approach provides a larger set of reads that can cover more replicates and support genome assembly, thus increasing the efficiency of de novo assembly (Roberts et al., 2013; Rhoads and Au, 2015).

11.3 STRUCTURAL GENOMICS

11.3.1 De Novo Assembly

The de novo genome assembly process consists of grouping reads obtained through sequencing via base pairing to represent the complete genome (Fig. 11.1), without the aid of reference sequences. The biggest problems in this process are repetitions in the genome, especially in eukaryotic organisms, regions with low sequencing coverage due to biases such as GC bias, and sequencing artifacts peculiar to each platform (Chen et al., 2013; Ross et al., 2013). Thus, the development of assembly algorithms to evaluate issues such as these is quite complex. Therefore, different computational approaches are used for genome assembly, such as greedy algorithms, overlap—layout—consensus (OLC), and De Bruijn graphs (Wojcieszek et al., 2014; Ekblom and Wolf, 2014; Miller et al., 2010).

The greedy algorithm approach is characterized by an extensive computational effort because it is based on the analysis of all possible sequence alignments. SSAKE (Warren et al., 2007), SHARCGS (Dohm et al., 2007), and VCAKE (Jeck et al., 2007) are examples of software that utilize this approach. The application of this method in conjunction with De Bruijn graphs and OLC is common (Wojcieszek et al., 2014; Ekblom and Wolf, 2014; Miller et al., 2010).

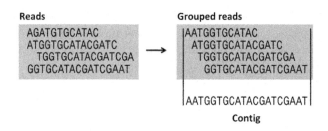

FIGURE 11.1   Grouping of reads derived from sequencing, considering the identity of the bases to generate contigs.

The OLC is an approach that is divided into three steps: the first consists of identifying the possible overlap in the set of reads. Second is followed by the construction of a graph based on the identified overlaps, and the consensus sequence is then finally generated through an algorithm that tours the graph, visiting each node exactly once. Newbler (Reinhardt et al., 2009), Mira (Chevreux et al., 2004), and Edena (Hernandez et al., 2008) are examples of software that adopt this approach (Wojcieszek et al., 2014; Ekblom and Wolf, 2014; Miller et al., 2010).

The approach based on De Bruijn graphs consists of the fragmentation of reads into *k-mer* sizes, in which *k* represents the read length. Subsequently, they are evaluated to find overlaps of *k-1-mers* among the *k-mers* initially generated, considering the identity of the bases (Fig. 11.1). One of the most commonly used software platforms for the assembly of prokaryotic genomes using short reads is Velvet (Zerbino and Birney, 2008). However, SPADES is employed for data obtained through the Ion Torrent platform, whose reads exhibit random lengths (Bankevich et al., 2012). SOAPdenovo (Luo et al., 2012) and ALL-PATHS-LG (Gnerre et al., 2011) software are examples of applications of De Bruijn graphs and are widely used for the assembly of eukaryotic genomes, particularly because of their ability to optimize memory consumption during the process. This is important because these genomes require a high computational effort due to their complexity, with the presence of the major repetitive regions and various chromosomes, and their large size, generally on the order of gigabases, in addition to the small number of complete eukaryotic genome projects (Wojcieszek et al., 2014; Ekblom and Wolf, 2014; Gnerre et al., 2011).

A single consensus sequence representing a genome, or various sequences representing portions of a genome that need to be oriented and arranged in a process known as scaffolding, can be obtained after the assembly process using each one of the aforementioned approaches (Boetzer et al., 2011; Pop, 2004; Kurtz et al., 2004). The scaffolds exhibit regions that were not represented during the assembly process for reasons such as the stringency of the adopted parameters and coverage biases, producing regions referred to gaps, and represented by the letter "N" (Luo et al., 2012; Boetzer and Pirovano, 2012).

The use of different sequencing strategies can be useful in resolving these regions, which often originate from repetitions in the genome. Paired genomic libraries, such as paired-end and mate-pair libraries, are used to represent these areas because they employ long fragments of several kilobases and only sequence the ends, such that the distance between the pairs is known (Schatz et al., 2010; Paszkiewicz and Studholme, 2010).

Some software is applied after the assembly process to generate scaffolds and to resolve gap regions. Thus, the overlap between the ends of contiguous sequences (contigs) is analyzed and can ensure the extension of the sequence, as is the case for SSPACE software (Boetzer et al., 2011). However, there are also tools that use paired libraries to resolve gaps, such as GAPFILLER (Boetzer and Pirovano, 2012).

The evaluation of the assembly process is performed using metrics such as N50, which evaluates the length of the sequences produced, in addition to the average length and number of contigs as well as largest and smallest contigs. The assembled contigs can be evaluated through the mapping of paired reads, when available, to confirm the results (Wences and Schatz, 2015). The core eukaryotic genes can be sought in the results of the assembly for eukaryotic genomes using the Genome Assembly Gold-standard Evaluation (GAGE) tool (Salzberg et al., 2012).

11.3.2 Reference Assembly

The mapping of sequences produced by NGS platforms is one of the most common applications of genomic data. However, this process represents a computational challenge due to the characteristics of the sequences originating from these platforms, mostly short reads. Several obstacles must be overcome to perform this task, such as accurate mapping of the reads to reference sequences, distinguishing between errors that occurred during sequencing and true genetic variations, and the difficulty of handling the large amount of data produced by these platforms (Reinert et al., 2015).

Alignment software platforms can be grouped based on the approaches they use, among which local alignment, which considers only parts of a read and usually omits the bases at the end of the sequence, is deemed to be faster compared with global alignment. In global alignment, the reads are used considering their full length (Fonseca et al., 2012).

Currently, several software are available that can accomplish this task, many of which were developed after the advent of NGS sequencers. The features of these tools include the ability to handle a large amount of data and capacity to work with various fragments or paired genomic libraries (Drucker et al., 2014; Fonseca et al., 2012; Reinert et al., 2015).

Despite a significant number of tools available for mapping, defining the most appropriate tool depends on factors such as the type of analysis to be carried out (DNA, RNA, miRNA). The available software platforms include the following: Bowtie which is very efficient in terms of memory management but has some issues regarding reads that do not exhibit a perfect match, though its parameters can be adjusted (Langmead et al., 2009); BWA which uses the Burrows–Wheeler transformation algorithm to increase mapping speed (Li and Durbin, 2009); SHRiMP which is compatible with data in letter space and color space format produced by the SOLiD platform (Rumble et al., 2009); SOAP2 which was developed to conduct single nucleotide polymorphism studies, whose current version features significant improvements in memory management and alignment speed (Li et al., 2009); TopHat2 which is suitable for data from the Ion Torrent platform (Kim et al., 2013); and mrsFAST which examines all possibilities for mapping to the reference genome, making it useful for variance detection studies (Hach et al., 2010).

The variety of tools suitable for carrying out the sequence alignment makes the process of choosing the most appropriate platform according to the requirements of the work more complex. The main alignment software and input and output file types used are listed in Table 11.2. However, biases related to the sequencing platforms, their throughput, and the low quality at the end of sequences should be considered, as they also impact the quality of alignments (Fonseca et al., 2012). Thus, it is important to carry out pretreatment of data, such as trimming of low-quality bases and the use of quality filters, to achieve more accurate results (Carneiro et al., 2012; Gurevich et al., 2013).

11.3.3 Genome Annotation

Genome annotation consists of describing the function of the product of a predicted gene (through an *in silico* approach). This can be achieved using bioinformatics software with specific features, including (1) signal sensors (e.g., for TATA box, start and stop codon, or poly-A signal detection), (2) content sensors (e.g., for $G + C$ content, codon

TABLE 11.2 List of Available Sequence Alignment Software for Data Obtained From NGS Platforms

Mapper	Input	Output
Bowtie2	Fasta/FastQ	SAM
BWA	Fasta/FastQ	SAM
MapReads	Fasta/FastQ	SAM
ELAND	Fasta	TSV
Mummer	Fasta	TSV
BFAST	Fasta/FastQ	SAM

usage, or dicodon frequency detection), and (3) similarity detection (e.g., between proteins from closely related organisms, mRNA from the same organism, or reference genomes) (Stein, 2001).

However, the method for predicting gene and genome structures (e.g., tRNAs, rRNAs, promoter regions) is associated with the applied assembly strategies and sequencing platforms (Chen et al., 2013).

Genome annotation can be divided into three basic categories. The first is a nucleotide-level annotation, which seeks to identify the physical location of DNA sequences to determine where components such as genes, RNAs, and repetitive elements are located. Sequencing and/or assembly errors at this stage can result in false pseudogenes through indels. The second is a protein-level annotation, which seeks to determine the possible functions of genes, identifying which one a given organism does or does not have. The third is a process-level annotation, which aims to identify the pathways and processes in which different genes interact, assembling an efficient functional annotation. In the last two levels, sequencing and/or assembly errors may compromise the inference of the true gene function because of reduced similarity (Miller et al., 2010; Reeves et al., 2009; Stein, 2001).

11.4 FUNCTIONAL GENOMICS

11.4.1 RNA-Seq: De Novo and Reference-Based Approaches

RNA-Seq is an RNA analysis technique based on NGS that allows the identification, measurement, and comparison of gene expression in a target transcriptome. The data produced through RNA-Seq can be applied to functional studies, such as expression profile analysis, annotation correction, characterization of differentially expressed genes, and gene prediction (Creecy and Conway, 2015; Marchant et al., 2016; Finotello and Di Camillo, 2015; Pinto et al., 2014).

The reads generated in the sequencing process can be subjected to two types of processing. The first is the mapping of reads against an annotated reference genome, to characterize gene expression and subsequently identify differentially expressed genes. The second consists of de novo assembly, in which there is no available annotated reference and the reads are used to represent transcripts in the assembly process (Chopra et al., 2014; Finotello and Di Camillo, 2015; Pinto et al., 2014).

When an annotated reference is available, the reads can be mapped to it using software such as Bioscope (Pinto et al., 2014), which quantifies the expression of each gene. The result is then analyzed using software such as DEGseq (Wang et al., 2010), which runs a statistical analysis of gene expression under the different conditions studied, to identify differentially expressed genes. The TopHat and Cufflinks pipelines may also be employed to identify differentially expressed genes (Trapnell et al., 2012). The TopHat pipeline maps the reads to the annotated reference genes, generating a mapping file in *bam* format. The Cufflinks pipeline then uses this result, together with the reference, to calculate the expression of the genes and identify the genes that are differentially expressed between the analyzed samples (Trapnell et al., 2012).

In the absence of a reference, the reads are used to represent transcripts in the de novo assembly process, thus allowing analyses such as a differential gene expression analysis. For the de novo assembly of RNA-Seq reads, the most widely used software are SOAPdenovo-Trans (Xie et al., 2014), Trans-AByss (Robertson et al., 2010), and Trinity (Grabherr et al., 2011). Trinity software can produce a high-quality assembly of a transcriptome with a low error rate and can identify multiple isoforms. Trans-Abyss generates optimized assemblies, representing a significant number of transcripts with high coverage through the merging of assemblies with different *k-mers*. SOAPdenovo-Trans is a platform that achieves the greatest transcript contiguity with the least amount of redundancy and is also the fastest of the three software platforms (Chopra et al., 2014). The assembled transcripts are used as a reference; thus, it is possible to map the reads to quantify the expression of the transcripts, followed by the identification of differentially expressed transcripts (Pinto et al., 2014; Finotello and Di Camillo, 2015; Chopra et al., 2014).

The correction of genome annotations is also possible through RNA sequencing. One approach to performing such correction is a mapping of the reads to the annotation, thus enabling mapping coverage to be evaluated for all annotated genes and intergenic regions, allowing the identification of potential new transcripts (intergenic regions of high coverage). These regions might be "exons" belonging to the genes they flank or new non-annotated genes. The available software for performing this type of analysis includes Cufflinks and Scripture (Garber et al., 2011). Another use for RNA-Seq data is gene prediction using the obtained reads, thus enabling new genes to be identified and incorporated into an existing annotation. The GeneMark-ET tool (Lomsadze et al., 2014) allows reads derived from RNA-Seq to be used in the training stage of the software, increasing the accuracy of the gene prediction step. The application of GeneMark-ET was shown to enhance the accuracy of the prediction of *Aedes aegypti* genes by approximately 24.5% (Lomsadze et al., 2014).

11.4.2 ChIP-Seq

ChIP-Seq, or ChIP-sequencing, is a combination of the chromatin immunoprecipitation technique with massively parallel sequencing. Chromatin immunoprecipitation allows the identification of specific DNA sequences that are bound to proteins of interest in vivo. This process involves fixation of chromatin with formaldehyde through covalent linkages between DNA-binding proteins and DNA, cell lysis, and subsequent fragmentation of DNA into smaller fragments, wherein specific DNA—protein complexes are isolated through immunoprecipitation with protein-specific antibodies. Then, the DNA isolated from the complexes is amplified by PCR. After this step, the amplicons are sequenced using NGS technologies (ChIP-Seq). The reads resulting from sequencing are then mapped to the genome, and the location of the isolated DNA may be identified (Shin et al., 2013; Solomon et al., 1988).

Thus, the ChIP-Seq technique allows the identification of transcription factors and DNA-bound proteins influencing phenotypes. The identification of proteins that interact with DNA can provide valuable information about the process of gene regulation when combined with transcriptomic profiles or microarray expression, thus enabling a better understanding of various biological processes (Metzker, 2010; Worsley Hunt et al., 2014). Studies employing ChIP-Seq and examining the regulation of gene expression have provided insights into the functional interactions between the binding sites of transcription factors and the promoters of their target genes (Shin et al., 2013).

This technique has been employed in several studies of, for example, the regulation of gene expression in breast and prostate cancer (Li et al., 2014; Zhang et al., 2014); the association of transcription factors such as Oct4, Nanog, or cMyc with cell fate in stem cells (Rahl et al., 2010; Young, 2011); and the architecture of nucleosomes (DNA methylation and modification of histones), which is altered during development and in cases of diseases such as cancer and therefore have a wide effect on chromatin dynamics (O'Geen et al., 2011).

A large number of algorithms have been developed for the analysis of the DNA—protein interaction data generated through ChIP-Seq. These algorithms are based on the detection of genomic regions where a larger number of reads are mapped than expected by chance (also referred to as peak detection). The developed methods include MACS (Zhang et al., 2008), which empirically calculates the change in the coverage of ChIP-Seq reads and uses this measure to improve the resolution of the prediction of the binding sites, and ChIPDiff (Xu et al., 2008) and ODIN (Allhoff et al., 2014), which identify the significant differences in two ChIP-Seq signals under different biological conditions using hidden Markov models.

Table 11.3 points some of the main applications, advantages, and disadvantages of DNA sequencing, RNA-Seq, and Chip-Seq.

11.5 PROTOCOLS AND PIPELINES FOR METAGENOMIC ANALYSIS

Microorganisms are critical for the maintenance of life (Falkowski et al. 2008), and only a fraction of the actual microbial diversity of various terrestrial environments is known (Handelsman, 2004). This situation has spurred interest in understanding how microbial communities interact and modulate the environment. However, there is great difficulty in culturing microorganisms whose physiological characteristics are still unknown. These factors have driven culture-independent microbial analysis, which is defined as the study of

TABLE 11.3   Applications, Advantages, and Disadvantages of DNA Sequencing, RNA-Seq, and Chip-Seq

	Applications	Advantages	Disadvantages
DNA sequencing	Genome assembly	Identifies the whole genome	High coverage needed
	Gene annotation	No reference genome required	Affected by sequencing errors
	Mapping	Fast sequencing in comparison with Sanger method	
	Gene prediction		
RNA-Seq	Transcript assembly	Less expensive than other methods	High coverage needed
	Expression profile	No reference genes required	Affected by sequencing errors
	Gene prediction	Less noise in comparison with other techniques	
Chip-Seq	Protein interactions with DNA	Higher sensitivity and speed compared to ChIP-chip	High cost and availability
		Low amount of ChIP DNA (10–50 ng)	Bias in high GC-rich content in fragment selection
	Gene expression	Coverage limited only by alignability of reads to the genome; repetitive regions can be covered	
	Epigenetic		

genes and microbial genomes obtained directly from environmental samples without the need for culturing.

The most commonly performed types of culture-independent analyses involve environmental DNA extraction and (1) PCR amplification of marker genes, such as 16S rRNA, or (2) whole-genome metagenomics using the shotgun approach, in which genomic fragments are randomly sequenced. The choice of approach depends on the purpose of the study; however, due to the greater wealth of information obtained, the whole-genome metagenomic analysis is the most frequently applied method. The protocols and pipelines used for these two approaches will be addressed in this section.

11.5.1 Analysis of the Microbial Diversity Through PCR Targeting 16S rRNA Genes

The determination of microbial diversity using sequencing data from 16S rRNA genes amplified from environmental DNA most often involves the clustering into operational taxonomic units (OTUs), in which similar 16S rRNA sequences are considered to be from the same taxon (Edgar, 2013). Alpha diversity analysis and the determination of microbial diversity richness can be performed based on the number of OTUs observed and the distance between taxa (Kemp and Aller, 2004). Other metrics may also be implemented when comparing environments regarding diversity (i.e., beta-diversity analysis). Various

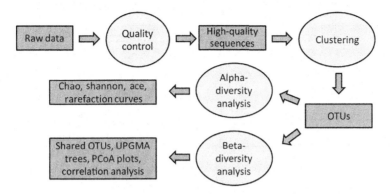

FIGURE 11.2 General steps (balloons) in a workflow applied to evaluate microbial diversity using Qiime and RDP.

software platforms perform such analyses; however, the most efficient, consuming less computational time and effort, are Qiime (Caporaso et al., 2010) and RDP (Cole et al., 2014), although all of these platforms involve general analytical steps (Fig. 11.2).

Analyses involving the sequencing of 16S rRNA genes are necessary for increasing knowledge about diversity and microbial ecology. However, these analyses provide tiny information about the function of such communities and how they interact with the environment and other life forms. Whole-genome metagenomic analyses can be very useful for that purpose.

11.5.2 Whole-Genome Metagenomic Analysis

Unlike gene-specific analysis, metagenomic sequencing aims to sequence the whole genomes (or most of the genome) of the microorganisms present in an environmental sample. The main advantage of this approach is that information on the role of the communities is obtained; i.e., the influence of the microorganisms in the environment is determined. Diversity analyses can also be performed because 16S rRNA fragments can also be sequenced at random, but these analyses are less robust than the standard analyses based on amplicons that were addressed in the previous section (Fig. 11.3).

The data analysis involves annotation and classification of coding DNA sequences, for which MG-RAST (Meyer et al., 2008) and the EBI Metagenomics Portal (Hunter et al., 2014) are the most commonly used platforms. The sequences are processed with these two platforms, including steps from quality control to gene prediction, thus enabling functional inferences about microbial communities.

11.5.3 Obtaining Genomes From Metagenomic Data

Metagenomic studies seek to structurally and functionally characterize the organisms in a given microbial community through sequencing on NGS platforms. The main advantage of these analyses is that they do not rely on a reference sequence to perform

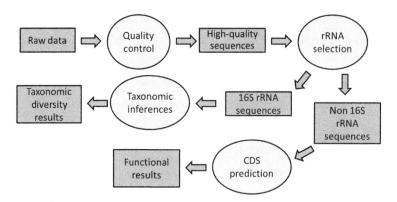

FIGURE 11.3 General steps (balloons) of a workflow used to analyze shotgun metagenome data MG-RAST and EBI metagenomics portal. CDS, coding DNA sequence.

genome assembly for the studied community, instead of employing the de novo approach that is useful for genome studies when a characterized reference is lacking. This is usually the case when a microbial community is being studied through metagenomic experimental analysis (Howe and Chain, 2015; Sharpton, 2014; Ekblom and Wolf, 2014; Wojcieszek et al., 2014).

Metagenome assembly mainly depends on sequencing coverage because this information is relevant for the assemblers to extend the reads obtained in sequencing through the overlaps between the reads and to form contigs that will represent the sequences of each organism within the studied microbial community. Furthermore, the higher the sequencing coverage achieved, the better the results of metagenome assembly, which are measured through metrics such as the number of contigs, contig length, and N50 values (Howe and Chain, 2015; Sharpton, 2014).

Currently, there are specific assemblers available for assembling metagenome data. Most of these assemblers use a De Bruijn graph—based assembly approach. The most common assemblers are MetaVelvet (Namiki et al., 2012), Meta-IBDA (Peng et al., 2011), and PRICE (Ruby et al., 2013). Also, other software platforms perform de novo metagenome assembly in an independent manner, including MetAMOS (Treangen et al., 2013)—a pipeline that carries out the assembly, taxonomic analysis, and functional annotation of the assembled contigs (Howe and Chain, 2015; Sharpton, 2014).

11.6 FUTURE DIRECTIONS

Among the several features of NGS technology, read length, error rate, and high throughput/coverage are in constant improvement. These improvements led to more correct reads, more complete assembly, more exact prediction, and less expensive sequencing. The launch of newer technologies, such as Ion Proton, has been expected as promising in the field of bioinformatics and omics, providing even longer reads, higher coverage, and cheaper base sequencing. It is clear that the progress and improvement of all of these platforms will directly contribute to the deeper understanding of NGS data and analysis. The development of new bioinformatics tools and algorithms will be crucial to handle this vast amount of data and perform analysis with the available computer systems, and overcome the necessities for supercomputers (Aly et al., 2015; Rhoads and Au, 2015; Teeling and Glöckner, 2012).

References

Allhoff, M., Seré, K., Chauvistré, H., Lin, Q., Zenke, M., Costa, I.G., 2014. Detecting differential peaks in ChIP-seq signals with ODIN. Bioinformatics 30, 3467–3475. Available from: https://doi.org/10.1093/bioinformatics/btu722.

Aly, S.M., Mostafa, E.M., Sabri, D.M., 2015. Overviews of "next-generation sequencing". Res. Rep. Forensic Med. Sci. 5, 1–5.

Bankevich, A., Nurk, S., Antipov, D., Gurevich, A.A., et al., 2012. SPAdes: a new genome assembly algorithm and its applications to single-cell sequencing. J. Comput. Biol. 19, 455–477. Available from: https://doi.org/10.1089/cmb.2012.0021.

Boetzer, M., Henkel, C.V., Jansen, H.J., Butler, D., Pirovano, W., 2011. Scaffolding pre-assembled contigs using SSPACE. Bioinformatics 27, 578–579. Available from: https://doi.org/10.1093/bioinformatics/btq683.

Boetzer, M., Pirovano, W., 2012. Toward almost closed genomes with GapFiller. Genome Biol. 13, R56. Available from: https://doi.org/10.1186/gb-2012-13-6-r56.

Brown, S.D., Nagaraju, S., Utturkar, S., De Tissera, S., Segovia, S., Mitchell, W., et al., 2014. Comparison of single-molecule sequencing and hybrid approaches for finishing the genome of *Clostridium autoethanogenum* and analysis of CRISPR systems in industrial relevant Clostridia. Biotechnol. Biofuels 7, 40. Available from: https://doi.org/10.1186/1754-6834-7-40.

Caporaso, J.G., Kuczynski, J., Stombaugh, J., Bittinger, K., Bushman, F.D., Costello, E.K., et al., 2010. correspondence QIIME allows analysis of high-throughput community sequencing data Intensity normalization improves color calling in SOLiD sequencing. Nat. Publ. Gr. 7, 335–336. Available from: https://doi.org/10.1038/nmeth0510-335.

Carneiro, A.R., Ramos, R.T.J., Barbosa, H.P.M., Schneider, M.P.C., Barh, D., Azevedo, V., et al., 2012. Quality of prokaryote genome assembly: indispensable issues of factors affecting prokaryote genome assembly quality. Gene 505, 365–367. Available from: https://doi.org/10.1016/j.gene.2012.06.016.

Chen, Y.-C., Liu, T., Yu, C.-H., Chiang, T.-Y., Hwang, C.-C., 2013. Effects of GC bias in next-generation-sequencing data on de novo genome assembly. PLoS One 8, e62856. Available from: https://doi.org/10.1371/journal.pone.0062856.

Chevreux, B., Pfisterer, T., Drescher, B., 2004. Using the miraEST assembler for reliable and automated mRNA transcript assembly and SNP detection in sequenced ESTs. Genome Res. 1147–1159. Available from: https://doi.org/10.1101/gr.1917404.

Chopra, R., Burow, G., Farmer, A., Mudge, J., Simpson, C.E., Burow, M.D., 2014. Comparisons of de novo transcriptome assemblers in diploid and polyploid species using peanut (Arachis spp.) RNA-seq data. PLoS One 9, e115055. Available from: https://doi.org/10.1371/journal.pone.0115055.

Cole, J.R., Wang, Q., Fish, J.A., Chai, B., McGarrell, D.M., Sun, Y., et al., 2014. Ribosomal database project: data and tools for high throughput rRNA analysis. Nucleic Acids Res. 42, D633–D642. Available from: https://doi.org/10.1093/nar/gkt1244.

Creecy, J.P., Conway, T., 2015. Quantitative bacterial transcriptomics with RNA-seq. Curr. Opin. Microbiol. 23, 133–140. Available from: https://doi.org/10.1016/j.mib.2014.11.011.

Dohm, J.C., Lottaz, C., Borodina, T., Himmelbauer, H., 2007. SHARCGS, a fast and highly accurate short-read assembly algorithm for de novo genomic sequencing. Genome Res. 17, 1697–1706. Available from: https://doi.org/10.1101/gr.6435207.

Drucker, T.M., Johnson, S.H., Murphy, S.J., Cradic, K.W., Therneau, T.M., Vasmatzis, G., 2014. BIMA V3: an aligner customized for mate pair library sequencing. Bioinformatics 30, 1627–1629. Available from: https://doi.org/10.1093/bioinformatics/btu078.

Edgar, R.C., 2013. UPARSE: highly accurate OTU sequences from microbial amplicon reads. Nat. Methods 10, 996−998. Available from: https://doi.org/10.1038/nmeth.2604.

Eid, J., Fehr, A., Gray, J., Luong, K., Lyle, J., Otto, G., et al., 2009. Real-time DNA sequencing from single polymerase molecules. Science 323, 133−138. Available from: https://doi.org/10.1126/science.1162986.

Ekblom, R., Wolf, J.B.W., 2014. A field guide to whole-genome sequencing, assembly and annotation. Evol. Appl. 7, 1026−1042. Available from: https://doi.org/10.1111/eva.12178.

Faircloth, B.C., Glenn, T.C., 2012. Not all sequence tags are created equal: designing and validating sequence identification tags robust to indels. PLoS One 7, e42543. Available from: https://doi.org/10.1371/journal.pone.0042543.

Falkowski, P.G., Fenchel, T., Delong, E.F., 2008. The microbial engines that drive Earth's biogeochemical cycles. Science 320, 1034−1039. Available from: https://doi.org/10.1126/science.1153213.

Finotello, F., Di Camillo, B., 2015. Measuring differential gene expression with RNA-seq: challenges and strategies for data analysis. Brief. Funct. Genomics 14, 130−142. Available from: https://doi.org/10.1093/bfgp/elu035.

Fonseca, N.A., Rung, J., Brazma, A., Marioni, J.C., 2012. Tools for mapping high-throughput sequencing data. Bioinformatics 28, 3169−3177. Available from: https://doi.org/10.1093/bioinformatics/bts605.

Fox, E.J., Reid-Bayliss, K.S., Emond, M.J., Loeb, L.A., 2014. Accuracy of next generation sequencing platforms. Next Gener. Seq. Appl. 1, 1000106. Available from: https://doi.org/10.4172/jngsa.1000106.

Garber, M., Grabherr, M.G., Guttman, M., Trapnell, C., 2011. Computational methods for transcriptome annotation and quantification using RNA-seq. Nat. Methods 8, 469−477. Available from: https://doi.org/10.1038/nmeth.1613.

Gnerre, S., Maccallum, I., Przybylski, D., Ribeiro, F.J., Burton, J.N., Walker, B.J., et al., 2011. High-quality draft assemblies of mammalian genomes from massively parallel sequence data. Proc. Natl. Acad. Sci. U.S.A. 108, 1513−1518. Available from: https://doi.org/10.1073/pnas.1017351108.

Grabherr, M.G., Haas, B.J., Yassour, M., Levin, J.Z., Thompson, D.A., Amit, I., et al., 2011. Full-length transcriptome assembly from RNA-Seq data without a reference genome. Nat. Biotechnol. 29, 644−652. Available from: https://doi.org/10.1038/nbt.1883.

Gurevich, A., Saveliev, V., Vyahhi, N., Tesler, G., 2013. QUAST: quality assessment tool for genome assemblies. Bioinformatics 29, 1072−1075. Available from: https://doi.org/10.1093/bioinformatics/btt086.

Hach, F., Hormozdiari, F., Alkan, C., Hormozdiari, F., Birol, I., Eichler, E.E., et al., 2010. mrsFAST: a cache-oblivious algorithm for short-read mapping. Nat. Methods 7, 576−577. Available from: https://doi.org/10.1038/nmeth0810-576.

Handelsman, J., 2004. Metagenomics: application of genomics to uncultured microorganisms. Am. Soc. Microbiol. 68, 669−685. Available from: https://doi.org/10.1128/MBR.68.4.669-685.

Hernandez, D., Francois, P., Farinelli, L., Osteras, M., Schrenzel, J., 2008. De novo bacterial genome sequencing: millions of very short reads assembled on a desktop computer. Genome Res. 18, 802−809. Available from: https://doi.org/10.1101/gr.072033.107.

Howe, A., Chain, P.S.G., 2015. Challenges and opportunities in understanding microbial communities with metagenome assembly (accompanied by IPython Notebook tutorial). Front. Microbiol. 6, 10−13. Available from: https://doi.org/10.3389/fmicb.2015.00678.

Hunter, S., Corbett, M., Denise, H., Fraser, M., Gonzalez-Beltran, A., Hunter, C., et al., 2014. EBI metagenomics—a new resource for the analysis and archiving of metagenomic data. Nucleic Acids Res. 42, D600−D606. Available from: https://doi.org/10.1093/nar/gkt961.

Jeck, W.R., Reinhardt, J.A., Baltrus, D.A., Hickenbotham, M.T., Magrini, V., Mardis, E.R., et al., 2007. Extending assembly of short DNA sequences to handle error. Bioinformatics 23, 2942−2944. Available from: https://doi.org/10.1093/bioinformatics/btm451.

Kemp, P.F., Aller, J.Y., 2004. Bacterial diversity in aquatic and other environments: what 16S rDNA libraries can tell us. FEMS Microbiol. Ecol. 47, 161−177. Available from: https://doi.org/10.1016/S0168-6496(03)00257-5.

Kim, D., Pertea, G., Trapnell, C., Pimentel, H., Kelley, R., Salzberg, S.L., 2013. TopHat2: accurate alignment of transcriptomes in the presence of insertions, deletions and gene fusions. Genome Biol. 14, R36. Available from: https://doi.org/10.1186/gb-2013-14-4-r36.

Kurtz, S., Phillippy, A., Delcher, A.L., Smoot, M., Shumway, M., Antonescu, C., et al., 2004. Versatile and open software for comparing large genomes. Genome Biol. 5, R12. Available from: https://doi.org/10.1186/gb-2004-5-2-r12.

Langmead, B., Trapnell, C., Pop, M., Salzberg, S.L., 2009. Ultrafast and memory-efficient alignment of short DNA sequences to the human genome. Genome Biol. 10, R25. Available from: https://doi.org/10.1186/gb-2009-10-3-r25.

Li, H., Durbin, R., 2009. Fast and accurate short read alignment with Burrows-Wheeler transform. Bioinformatics 25, 1754–1760. Available from: https://doi.org/10.1093/bioinformatics/btp324.

Li, Q., Wang, H., Yu, L., Zhou, J., Chen, J., Zhang, X., et al., 2014. ChIP-seq predicted estrogen receptor biding sites in human breast cancer cell line MCF7. Tumour Biol. 35, 4779–4784. Available from: https://doi.org/10.1007/s13277-014-1627-4.

Li, R., Yu, C., Li, Y., Lam, T.W., Yiu, S.M., Kristiansen, K., et al., 2009. SOAP2: an improved ultrafast tool for short read alignment. Bioinformatics 25, 1966–1967. Available from: https://doi.org/10.1093/bioinformatics/btp336.

Liu, L., Li, Y., Li, S., Hu, N., He, Y., Pong, R., et al., 2012. Comparison of next-generation sequencing systems. J. Biomed. Biotechnol. 2012, 1–11. Available from: https://doi.org/10.1155/2012/251364.

Lomsadze, A., Burns, P.D., Borodovsky, M., 2014. Integration of mapped RNA-Seq reads into automatic training of eukaryotic gene finding algorithm. Nucleic Acids Res. 42, e119. Available from: https://doi.org/10.1093/nar/gku557.

Luo, R., Liu, B., Xie, Y., Li, Z., Huang, W., Yuan, J., 2012. SOAPdenovo2: an empirically improved memory-efficient short-read de novo assembler. Gigascience 1, 18.

Marchant, A, Mougel, F., Mendonça, V., Quartier, M., Jacquin-Joly, E., da Rosa, A., et al., 2016. Comparing de novo and reference-based transcriptome assembly strategies by applying them to the blood-sucking bug Rhodnius prolixus. Insect Biochem. Mol. Biol. 69, 25–33. Available from: https://doi.org/10.1016/j.ibmb.2015.05.009.

Mardis, E.R., 2008. The impact of next-generation sequencing technology on genetics. Trends Genet. 24, 133–141. Available from: https://doi.org/10.1016/j.tig.2007.12.007.

Merriman, B., Torrent, I., Rothberg, J.M., 2012. Progress in ion torrent semiconductor chip based sequencing. Electrophoresis 33, 3397–3417. Available from: https://doi.org/10.1002/elps.201200424.

Metzker, M.L., 2010. Sequencing technologies—the next generation. Nat. Rev. Genet. 11, 31–46. Available from: https://doi.org/10.1038/nrg2626.

Meyer, F., Paarmann, D., D'Souza, M., Olson, R., Glass, E., Kubal, M., et al., 2008. The metagenomics RAST server—a public resource for the automatic phylogenetic and functional analysis of metagenomes. BMC Bioinformatics 9, 386. Available from: https://doi.org/10.1186/1471-2105-9-386.

Miller, J.R., Koren, S., Sutton, G., 2010. Assembly algorithms for next-generation sequencing data. Genomics 95, 315–327. Available from: https://doi.org/10.1016/j.ygeno.2010.03.001.

Namiki, T., Hachiya, T., Tanaka, H., Sakakibara, Y., 2012. MetaVelvet: an extension of Velvet assembler to de novo metagenome assembly from short sequence reads. Nucleic Acids Res. 40, e155. Available from: https://doi.org/10.1093/nar/gks678.

O'Geen, H., Echipare, L., Farnham, P.J., 2011. Using ChIP-Seq technology to generate high-resolution profiles of histone modifications. Methods Mol. Biol. 791, 265–286. Available from: https://doi.org/10.1007/978-1-61779-316-5_20.

Pareek, C.S., Smoczynski, R., Tretyn, A., 2011. Sequencing technologies and genome sequencing. J. Appl. Genet. 52, 413–435. Available from: https://doi.org/10.1007/s13353-011-0057-x.

Paszkiewicz, K., Studholme, D.J., 2010. De novo assembly of short sequence reads. Brief. Bioinform. 11, 457–472. Available from: https://doi.org/10.1093/bib/bbq020.

Peng, Y., Leung, H.C.M., Yiu, S.M., Chin, F.Y.L., 2011. Meta-IDBA: a de novo assembler for metagenomic data. Bioinformatics 27, i94–i101. Available from: https://doi.org/10.1093/bioinformatics/btr216.

Pinto, A., de Sá, P.H.C.G., Ramos, R.T.J., Barbosa, S., Barbosa, H.P.M., Ribeiro, A., et al., 2014. Differential transcriptional profile of Corynebacterium pseudotuberculosis in response to abiotic stresses. BMC Genomics 15, 14. Available from: https://doi.org/10.1186/1471-2164-15-14.

Pop, M., 2004. Hierarchical scaffolding with bambus. Genome Res. 14, 149–159. Available from: https://doi.org/10.1101/gr.1536204.

Quail, M.A., Smith, M., Coupland, P., Otto, T.D., Harris, S.R., Connor, T.R., et al., 2012. A tale of three next generation sequencing platforms: comparison of Ion Torrent, Pacific Biosciences and Illumina MiSeq sequencers. BMC Genomics 13, 341.

Rahl, P.B., Lin, C.Y., Seila, A.C., Flynn, R.A., McCuine, S., Burge, C.B., et al., 2010. c-Myc regulates transcriptional pause release. Cell 141, 432–445. Available from: https://doi.org/10.1016/j.cell.2010.03.030.

Reeves, G. a, Talavera, D., Thornton, J.M., 2009. Genome and proteome annotation: organization, interpretation and integration. J. R. Soc. Interface 6, 129–147. Available from: https://doi.org/10.1098/rsif.2008.0341.

Reinert, K., Langmead, B., Weese, D., Evers, D.J., 2015. Alignment of next-generation sequencing reads. Annu. Rev. Genomics Hum. Genet. 16, 133–151. Available from: https://doi.org/10.1146/annurev-genom-090413-025358.

Reinhardt, J. A, Baltrus, D. A, Nishimura, M.T., Jeck, W.R., Jones, C.D., Dangl, J.L., 2009. De novo assembly using low-coverage short read sequence data from the rice pathogen *Pseudomonas syringae* pv. oryzae. Genome Res. 19, 294–305. Available from: https://doi.org/10.1101/gr.083311.108.

Reuter, J.A., Spacek, D.V., Snyder, M.P., 2015. High-throughput sequencing technologies. Mol. Cell 58, 586–597. Available from: https://doi.org/10.1016/j.molcel.2015.05.004.

Rhoads, A., Au, K.F., 2015. PacBio sequencing and its applications. Genomics Proteomics Bioinformatics 13, 278–289. Available from: https://doi.org/10.1016/j.gpb.2015.08.002.

Roberts, R.J., Carneiro, M.O., Schatz, M.C., 2013. The advantages of SMRT sequencing. Genome Biol. 14, 405. Available from: https://doi.org/10.1186/gb-2013-14-6-405.

Robertson, G., Schein, J., Chiu, R., Corbett, R., Field, M., Jackman, S.D., et al., 2010. De novo assembly and analysis of RNA-seq data. Nat. Methods 7, 909–912. Available from: https://doi.org/10.1038/nmeth.1517.

Ross, M.G., Russ, C., Costello, M., Hollinger, A., Lennon, N.J., Hegarty, R., et al., 2013. Characterizing and measuring bias in sequence data. Genome Biol. 14, R51. Available from: https://doi.org/10.1186/gb-2013-14-5-r51.

Rothberg, J.M., Hinz, W., Rearick, T.M., Schultz, J., Mileski, W., Davey, M., et al., 2011. An integrated semiconductor device enabling non-optical genome sequencing. Nature 475, 348–352. Available from: https://doi.org/10.1038/nature10242.

Ruby, J.G., Bellare, P., DeRisi, J.L., 2013. PRICE: software for the targeted assembly of components of (meta) genomic sequence data. G3 3, 865–880. Available from: https://doi.org/10.1534/g3.113.005967.

Rumble, S.M., Lacroute, P., Dalca, A.V., Fiume, M., Sidow, A., Brudno, M., 2009. SHRiMP: accurate mapping of short color-space reads. PLoS Comput. Biol. 5. Available from: https://doi.org/10.1371/journal.pcbi.1000386e1000386.

Salzberg, S.L., Phillippy, A.M., Zimin, A., Puiu, D., Magoc, T., Koren, S., et al., 2012. GAGE: a critical evaluation of genome assemblies and assembly algorithms. Genome Res. 22, 557–567. Available from: https://doi.org/10.1101/gr.131383.111.

Schatz, M.C., Delcher, A.L., Salzberg, S.L., 2010. Assembly of large genomes using second-generation sequencing. Genome Res. 20, 1165–1173. Available from: https://doi.org/10.1101/gr.101360.109.

Sharpton, T.J., 2014. An introduction to the analysis of shotgun metagenomic data. Front. Plant Sci. 5, 1–14. Available from: https://doi.org/10.3389/fpls.2014.00209.

Shendure, J., Ji, H., 2008. Next-generation DNA sequencing. Nat. Biotechnol. 26, 1135–1145. Available from: https://doi.org/10.1038/nbt1486.

Shin, H., Liu, T., Duan, X., Zhang, Y., Liu, X.S., 2013. Computational methodology for ChIP-seq analysis. Quant. Biol. 1, 54–70. Available from: https://doi.org/10.1007/s40484-013-0006-2.

Solomon, M.J., Larsen, P.L., Varshavsky, A, 1988. Mapping protein-DNA interactions in vivo with formaldehyde: evidence that histone H4 is retained on a highly transcribed gene. Cell 53, 937–947. Available from: https://doi.org/10.1016/S0092-8674(88)90469-2.

Stein, L., 2001. Genome annotation: from sequence to biology. Nat. Rev. Genet. 2, 493–503. Available from: https://doi.org/10.1038/35080529.

Teeling, H., Glöckner, F., 2012. Current opportunities and challenges in microbial metagenome analysis—a bioinformatic perspective. Brief. Bioinform. 13, 728–742. Available from: https://doi.org/10.1093/bib/bbs039.

Trapnell, C., Roberts, A., Goff, L., Pertea, G., Kim, D., Kelley, D.R., et al., 2012. Differential gene and transcript expression analysis of RNA-seq experiments with TopHat and Cufflinks. Nat. Protoc. 7, 562–578. Available from: https://doi.org/10.1038/nprot.2012.016.

Travers, K.J., Chin, C.-S., Rank, D.R., Eid, J.S., Turner, S.W., 2010. A flexible and efficient template format for circular consensus sequencing and SNP detection. Nucleic Acids Res. 38, e159. Available from: https://doi.org/10.1093/nar/gkq543.

Treangen, T.J., Koren, S., Sommer, D.D., Liu, B., Astrovskaya, I., Ondov, B., et al., 2013. MetAMOS: a modular and open source metagenomic assembly and analysis pipeline. Genome Biol. 14, R2. Available from: https://doi.org/10.1186/gb-2013-14-1-r2.

Van Dijk, E.L., Auger, H., Jaszczyszyn, Y., Thermes, C., 2014. Ten years of next-generation sequencing technology. Trends Genet. 30, 1−9. Available from: https://doi.org/10.1016/j.tig.2014.07.001.

Wang, L., Feng, Z., Wang, X., Wang, X., Zhang, X., 2010. DEGseq: an R package for identifying differentially expressed genes from RNA-seq data. Bioinformatics 26, 136−138. Available from: https://doi.org/10.1093/bioinformatics/btp612.

Warren, R.L., Sutton, G.G., Jones, S.J.M., Holt, R.A., 2007. Assembling millions of short DNA sequences using SSAKE. Bioinformatics 23, 500−501. Available from: https://doi.org/10.1093/bioinformatics/btl629.

Wences, A.H., Schatz, M.C., 2015. Metassembler: merging and optimizing de novo genome assemblies. Genome Biol. 16, 207. Available from: https://doi.org/10.1186/s13059-015-0764-4.

Wojcieszek, M., Pawełkowicz, M., Nowak, R., Przybecki, Z., 2014. Genomes correction and assembling: present methods and tools. SPIE Proc. 9290, 92901X. Available from: https://doi.org/10.1117/12.2075624.

Worsley Hunt, R., Mathelier, A., Del Peso, L., Wasserman, W.W., 2014. Improving analysis of transcription factor binding sites within ChIP-Seq data based on topological motif enrichment. BMC Genomics 15, 472. Available from: https://doi.org/10.1186/1471-2164-15-472.

Xie, Y., Wu, G., Tang, J., Luo, R., Patterson, J., Liu, S., et al., 2014. SOAPdenovo-Trans: de novo transcriptome assembly with short RNA-Seq reads. Bioinformatics 30, 1660−1666. Available from: https://doi.org/10.1093/bioinformatics/btu077.

Xu, H., Wei, C.L., Lin, F., Sung, W.K., 2008. An HMM approach to genome-wide identification of differential histone modification sites from ChIP-seq data. Bioinformatics 24, 2344−2349. Available from: https://doi.org/10.1093/bioinformatics/btn402.

Yeo, Z.X., Chan, M., Yap, Y.S., Ang, P., Rozen, S., Lee, A.S.G., 2012. Improving indel detection specificity of the Ion Torrent PGM benchtop sequencer. PLoS One 7, e45798. Available from: https://doi.org/10.1371/journal.pone.0045798.

Young, R.A., 2011. Control of the embryonic stem cell state. Cell 144, 940−954. Available from: https://doi.org/10.1016/j.cell.2011.01.032.

Zerbino, D.R., Birney, E., 2008. Velvet: algorithms for de novo short read assembly using de Bruijn graphs. Genome Res. 18, 821−829. Available from: https://doi.org/10.1101/gr.074492.107.

Zhang, Y., Liu, T., Meyer, C.A., Eeckhoute, J., Johnson, D.S., Bernstein, B.E., et al., 2008. Model-based analysis of ChIP-Seq (MACS). Genome Biol. 9, R137. Available from: https://doi.org/10.1186/gb-2008-9-9-r137.

Zhang, Y., Huang, Z., Zhu, Z., Liu, J., Zheng, X., Zhang, Y., 2014. Network analysis of ChIP-Seq data reveals key genes in prostate cancer. Eur. J. Med. Res. 19, 47. Available from: https://doi.org/10.1186/s40001-014-0047-7.

12

Computational Techniques in Data Integration and Big Data Handling in Omics

Adonney A. de Oliveira Veras[1], Pablo H.C.G. de Sá[1], Kenny da Costa Pinheiro[1], Debmalya Barh[2,3], Vasco Azevedo[2], Rommel Thiago Jucá Ramos[1] and Artur L. da Costa da Silva[1]

[1]Federal University of Para, Belem, Brazil [2]Federal University of Minas Gerais, Belo Horizonte, Minas Gerais, Brazil [3]Institute of Integrative Omics and Applied Biotechnology (IIOAB), Purba Medinipur, India

12.1 INTRODUCTION

Big Data is defined as a "term that describes large volumes of high velocity, complex and variable data that require advanced techniques and technologies to enable the capture, storage, distribution, management and analysis of the information" (Gandomi and Haider, 2015; TechAmerica Foundation: Federal Big Data Commission, 2012).

It has been adopted by many areas: social networks, astronomy, physics, meteorology, and in the omics sciences.

Since the advent of deep sequencing platforms, which are characterized by large data production, new computational methods have been developed to allow handle and process the data to address complex computational biology problems to areas like computational simulation of proteins, de novo assembly, the genomes, and gene network.

Omics Technologies and Bio-engineering: Towards Improving Quality of Life
DOI: https://doi.org/10.1016/B978-0-12-804659-3.00012-9

These techniques have brought benefits to humanity, such as the development of personalized medicine and new medicines based on information extract from raw data, beyond to increasing research in genomic, transcriptomic, and proteomic analyses.

This chapter presents the concept of Big Data, how these techniques have been applied on computational biology, and what can be developed in the future through Big Data approaches.

12.2 BIG DATA CONCEPT

The term Big Data originated because of an exponential growth in the data that is generated by applications (such as Facebook and YouTube) and astronomical studies. In spite of their different characteristics, these applications share the common feature of producing large amounts of data, which are better defined as Big Data. For example, since the advent of YouTube in 2005, the number of visualized videos per day has reached 4 billion, and every minute, approximately 48 hours of video are uploaded (Sagiroglu and Sinanc, 2013); Facebook stores more than a million photos per second, and the space that is needed to store this amount of data is estimated to be over 20 petabytes (one petabyte corresponds to 1024 terabytes); for these reasons, currently, multimedia data represent 70% of the traffic growth on the Internet backbone (Gandomi and Haider, 2015; Sagiroglu and Sinanc, 2013).

Similarly, the omics sciences (e.g., genomics, transcriptomics, metagenomics, and proteomics), with their myriad lines of research and experiments, have been producing large-scale data, especially since the development of new sequencing platforms, which are usually referred to as NGS (Next Generation Sequencing) or deep sequencing (Fig. 12.1).

Currently, NGS platforms use semiconductor technology (Merriman and Rothberg, 2012) or nanotechnology (Clarke et al., 2009), and throughout the years, such platforms

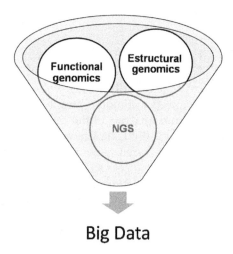

FIGURE 12.1   Fields' knowledge that put together forms the Big Data. Estructural and functional experiments produced by NGS technologies increase the amount of omics data.

have been improving accuracy, reducing costs, and significantly increasing data production. Currently, used NGS platforms include Ion Torrent PGM from Thermo Fisher (http://www.thermofisher.com), with 600 MB–1 GB throughput using Ion Chip 318 v2, and HiSeq 2500 (http://www.illumina.com/), with 180 GB throughput. In 2012, Oxford Nanopore (https://www.nanoporetech.com/) launched two other platforms, MinIon and GridIon, which can both produce long reads of ~100 KB with 10 GB throughput and are also capable of sequencing a single DNA molecule; the reads length is shown on Fig. 12.2.

Another platform that uses Single Molecule Real Time (SMRT) for sequencing is PacBio from Pacific Biosciences (http://www.pacificbiosciences.com/pro-ducts/). Although PacBio presents a ~500 MB throughput, it can produce long reads of ~20 KB. However, the platforms of sequencing is not free of errors; the rates are listed on Table 12.1 (Loman et al., 2012; Stephens et al., 2015; Van Dijk et al., 2014).

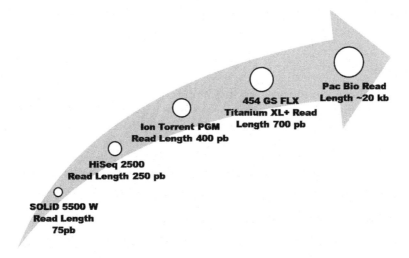

FIGURE 12.2 Reads length on different platforms of sequencing.

TABLE 12.1 Error Rates Related to Different Sequencing Platforms

Technology	Platform	Error Rate (%)
Sanger	3730xl	0.1–1
Illumina	HiSeq 2500	>0.1
SOLiD	SOLiD 5500	>0.06
454	FLX Titanium	1
SMRT	PacBio RS	16
Ion Torrent	PGM	1

These advances in data generation create new challenges for bioinformatics, such as storage and analysis of large amounts of data, which must be overcome. One solution for these challenges is the development and use of Big Data tools. In this chapter, we will discuss the main concepts in Big Data, the applications to Life Sciences for the management and analysis of large volumes of data, and the tools that can be used to manipulate such data.

12.3 THE THREE V's

Big Data can be understood to be the capacity to manage and analyze a large volume of data. The characteristics of Big Data can be subdivided into three key approaches, known as the three V's, which help to understand and characterize data: Volume, Velocity, and Variability (May, 2014). We present these approaches in the following sections.

12.3.1 Volume

The exponential growth in data production has a direct impact on currently available IT (Information Technology) infrastructure storage capacity, which makes data storage an enormous challenge to be overcome. To accomplish this goal, several solutions have been adopted for data storage, including cloud technology (O'Reilly Media, 2012). The volume of genomic data shows similar growth; primarily, this growth is due to a reduction in the sequencing costs. Although sequencing the first human genome took more than a decade to accomplish, currently, research centers can sequence an entire human genome in a few hours. This increased sequencing velocity drove an increase in the number of biological analyses that were performed in the different Omics areas, which is reflected in medicine by changes in medical treatments. Once the genomic profile of a patient is known, it is possible to develop more efficient treatment, which supports the development of personalized medicine (also defined as precision medicine—PM) (Costa, 2012; Servant et al., 2014).

Personalized medicine has been presented as the most recent possible treatment for several diseases, including cancer. However, to implement personalized medicine, many challenges still require resolution, such as answering ethical and legal questions and developing bioinformatics tools that can handle and integrate large amounts of complex data that are generated by this type of analysis. Nevertheless, all of the knowledge that has been obtained on the genomic profile of cancer can be useful for providing researchers with new insights into patients' individual treatments, disease progression, and treatment responses (Servant et al., 2014).

Among the existing methods and techniques for manipulating large volumes of data, it is relevant to cite Big Table, MapReduce, and Hadoop. Big Table has been developed by Google to promote the distributed storage of information, and it has the capacity to manipulate data in the petabyte range. The main characteristics of Big Table are high performance, flexibility, scalability, high availability, and wide applicability. Currently,

Big Table is used by several Google systems, such as Google Analytics, Google Finance, Google Earth, and Personalized Search (Chang et al., 2008).

Although presenting high similarities to database systems, Big Table does not offer support to the relational data model system; instead, it promotes the dynamic control model in a simplified data model that allows format control and data layout (Chang et al., 2008).

MapReduce is a model of distributed programming that was developed by Google, and it has the capacity to manipulate large amounts of data by using a cluster system. The user selects a function map that will be used in the process of generating information pairs, called key and value, which are then submitted to reduce a function that unites all of the values that are associated with a single key (Dean and Ghemawat, 2004).

Hadoop (http://hadoop.apache.org) is a framework that was created from MapReduce and Big Table; it was designed to allow parallel processing in clusters of high latency. In addition, Hadoop implements tolerance requirements for failures and load balancing. Two main components of Hadoop are the massively scalable archival system, which allows for manipulation of data in the petabyte range, and the MapReduce engine, which is used for batch data processing (Hurwitz et al., 2013).

12.3.2 Velocity

Several solutions that are aimed at large-scale data processing have been developed throughout the years, such as supercomputers, clusters, and grids. However, Cloud Computing arises because of its promise to solve two large problems: data storage and processing. As an example of applications that use Cloud Computing technology, we can cite EasyGenomics Cloud at the Beijing Genomics Institute; Embassy Cloud, which results from a cooperation project between various countries: Sweden, Czech Republic, Estonia, Norway, Netherlands, and Denmark (Alyass et al., 2015); and Amazon Web Services— AWS (http://aws.amazon.com/publicdatasets), which is a system that includes many public databases and gathers information on a wide range of disciplines, such as biology, astronomy, chemistry, and economics (Fusaro et al., 2011).

Currently, the large challenge for the use of Cloud resources is the development of tool kits that are suitable for manipulating and processing data from NGS sequencing platforms. Among the available applications, the CGtag tool kit allows the use of genome annotation tools in a Cloud environment (Hiltemann et al., 2014); CloudBLAST (Matsunaga et al., 2008) is a version of BLAST developed by NCBI to be used in the Cloud; and CloudAligner (Nguyen et al., 2011), which is a Cloud application to map NGS data. These new tools have eliminated the need to perform local installations, which in many cases depend directly on advanced computing skills in addition to allowing greater agility in the process of software maintenance; this new software model can be considered Software as a Service (SaaS) (Dai et al., 2012).

12.3.3 Variability in Data Origin

A common Big Data application that uses data from various origins is information mining from social networks (Facebook, Twitter, YouTube, etc.), in which generated data

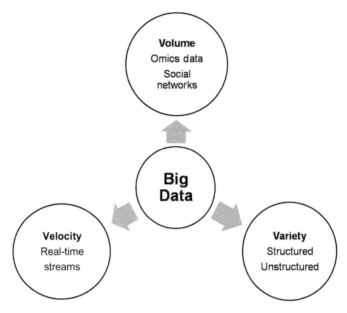

FIGURE 12.3   Three V's in the Big Data.

can be applied to decision support systems in the financial, marketing, social, and business fields, among others (Trifonova et al., 2013).

Currently, sequencing platforms are more accessible to small laboratories, in particular Benchtop platforms, which allow doing functional and structural analyses; these systems originate a large variety of data because each analysis yields a wide array of information, such as tissue (samples), cellular size and form, and methodology or protocol used in the study (Costa, 2012; Loman et al., 2012).

Based on information that was extracted from the Swedish and Canadian Health Systems' Big Data, patterns that regard population health and treatment status were identified, through the association of socioeconomic information (education, age, income, family group, marital status, and gender). From this information, researchers found that low-income patients are usually women with less than 10 years of education, who are usually single, divorced, or widowed, and who suffer from diabetes and heart disease (Alter, 2015; Issa et al., 2014; Stephens et al., 2015). The concept of three V's is shown in Fig. 12.3.

12.4 BIG DATA ON COMPUTATIONAL BIOLOGY APPLICATIONS

Computational Biology is composed of scientific areas that rely on high computational cost tasks, such as protein modeling, de novo assembly, and the mapping of reads (in particular, when using high volumes of reads, which require the development of alignment programs that are capable of manipulating efficiently this amount of data). These tasks create the need to develop algorithms and new knowledge extraction methods in addition to building high-performance platforms (Abuín et al., 2015; Alonso et al., 2015).

Computational biology data must be organized to generate knowledge. For example, neuroscience data combine, in a systematic way, results from high-resolution analytical methods with data from experiments that are designed to understand how the brain works. Currently, large cerebral simulation projects that integrate high-resolution CT scan and light/electronic microscopy data are being developed with the aim of extracting information that is relevant for medical treatment from these data (Issa et al., 2014; Noor et al., 2015; Trifonova et al., 2013).

Similarly, new research that is related to the study of diseases has been performed using a combination of genomic, transcriptomic, and proteomic analyses. To better understand diseases, because each organism has a genetic predisposition and an environmental exposure history, multidimensional analyses are performed with the goal of analyzing normal and disease-specific samples from different organisms, healthy and nonhealthy, to be able to associate the obtained results with survival rates, treatments, responses to treatments, and incidences of other diseases, thereby generating large amounts of information (Costa, 2012; Issa et al., 2014; Noor et al., 2015; Trifonova et al., 2013).

A new trend in Big Data is the integration of Omics data with clinical data, which is defined as personalized medicine, another's examples are list in Table 12.2. This integration will require hardware for processing and new software to extract and associate information that has already been described by the scientific community, which will allow us to choose the best therapies for each patient based on molecular metrics, previous diagnoses, prognosis, collateral effects, and other information that is available in clinical databases (Alonso et al., 2015; Alyass et al., 2015; Costa, 2012; Issa et al., 2014; Merelli et al., 2014; Noor et al., 2015).

NCBI (http://www.ncbi.nlm.nih.gov), the most currently used biological database, is a good example of how data should be available to doctors and scientists who are involved in personalized medicine. Another important aspect to be considered is the privacy of the generated information. Building on these challenges, several companies and institutions are offering services and solutions that are related to the storage, generation, analysis, and visualization of omics and clinical data, such as Appistry (www.appistry.com), BGI (www.genomics.cn/en), CLC Bio (www.clcbio.com), DNAnexus (www.dnanexus.com), Genome International Corporation (www.genome.com), GNS Healthcare (www.gnshealth-care.com), Foundation Medicine (www.foundationmedicine.com), Knome (www.knome.com), and NextBio (www.nextbio.com) (Noor et al., 2015).

TABLE 12.2 Application of Big Data in Biotechnology and Human Benefits

Application of Big Data	Human Benefits
Genomics	Personalized medicine
Drug discovery	Development the new medicines
Drug recycling	Biomedical research

12.5 TOOLS FOR ANALYSIS

Although there is a high availability of computational tools for microarray analysis, the majority of these tools cannot manipulate large data sets, which increases the time that is required to analyze complex data sets (Kashyap et al., 2015).

The tool caCORRECT can remove artifact noise from high-throughput microarray data and can be used to improve the integrity and quality of the microarray data that are available on public databases as well as to provide universal quality value for validation (Stokes et al., 2007).

Another tool used for gene expression analysis is omniBiomarker, which is a web-based application that uses knowledge-driven algorithms to identify differentially expressed genes; omniBiomarker aims at finding biomarkers from high-throughput gene expression data. Additionally, omniBiomarker helps to search and identify stable and reproducible biomarkers (Phan et al., 2012).

Gene expression data sets have grown considerably throughout the years, and for this reason, new tools for gene expression network analysis are needed (Bolouri, 2014). One of these tools is FastGCN (Liang et al., 2015), which is a tool that explores the parallelism that is associated with the GPU architecture to find optimal coexpression networks.

The UCLA Gene Expression Tool (UGET) (Day et al., 2009) performs large-scale coexpression analysis to detect associations between genes and specific diseases. Disease networks have significant gene correlations, and UGET calculates all of the possible correlations between gene pairs. Validation tests with effective results have been performed using Celsius (Day et al., 2009), the largest co-normalized microarray data set of Affymetrix-based gene expression data.

Other technologies that are used include Computational Intelligence tools for the analysis and processing of large volumes of data, such as supervised, unsupervised, and hybrid machine learning approaches, which are among the most frequently used tools for Big Data analysis. In addition, several mathematical techniques have been adopted, which aim to reduce the dimensionality to minimize the problem of large data volumes. Linear mapping methods, such as Principal Component Analysis (PCA) and Singular Value Decomposition, as well as nonlinear mapping methods, such as Sammon's mapping, Kernel PCA, and Laplacian eigenmaps, have been broadly used to reduce dimensionality (Kashyap et al., 2015).

Mathematical optimization has become another important tool for Big Data analysis. Scientific areas such as constraint satisfaction programming, dynamic programming, and heuristics/metaheuristics are present in Computational Intelligence and, especially, in Machine Learning techniques. Other optimization methods exist, such as multiobjective methods, for example, optimization and multimodality methods, in which evolutionary algorithms should be emphasized.

Machine learning techniques address the development of fast and efficient algorithms for data processing and predictive data analysis in real time. Statistical concepts such as expectation–maximization and PCA are frequently used to solve machine learning problems. In addition, some machine learning techniques, such as Probably Approximately Correct learning, are commonly used in applied statistics, which means that statistical and machine learning tools complement each other for Big Data analysis (Kashyap et al., 2015).

Data mining techniques are also used for Big Data analysis. In this case, the challenge is larger than in traditional data mining because the common procedure is to extend data mining algorithms to manipulate large data sets (Shukla and Dubey, 2014). These approaches are known as clustering methods, and well-studied examples of such algorithms are CLARA (Clustering LARge Applications) (Leonard Kaufman, 2005) and BIRCH (Balanced Iterative Reducing using Cluster Hierarchies) (Zhang et al., 1996).

However, the amount of time that is required to perform the analysis of statistical and machine learning data methods is still large when these methods are applied to large-scale data sets, and for this reason, parallel and distributed computing techniques have been used to solve this problem. Therefore, distributed data analysis algorithms have been proposed in the literature (Cohen et al., 2009).

To address problems that are derived from sequence analysis, several tools have been developed on top of the platform Hadoop MapReduce to perform large-scale sequence data analysis (Taylor, 2010). Bigpig (Nordberg et al., 2013) is a sequence analysis tool that scales automatically with the data size and can be ported directly to many Hadoop infrastructures. The Crossbow tool (Langmead et al., 2009b—Searching for SNPs with cloud computing) combines the software Bowtie, an ultrafast and memory-efficient short read aligner, and SoapSNP (Langmead et al., 2009a), an accurate genotyper, to search and identify SNPs using Cloud Computing or a local Hadoop cluster.

The free software revolution and the Big Data phenomenon are strongly associated. Free software such as R (open source programming language and software environment designed for statistical computing and visualization) and Apache Mahout (Scalable machine learning and data mining open source software based mainly in Hadoop) are broadly used in Big Data mining (Fan and Bifet, 2013). There are packages in both R and WGCNA that are specific for coexpression gene analysis that can also be used in an R-Hadoop distributed computing system (Langfelder and Horvath, 2008).

Recent Big Data applications are related to Protein—Protein Interaction (PPI) networks. For this reason, new tools have been developed for PPI networks, such as NeMo (Rivera et al., 2010), MCODE (Bader and Hogue, 2003), and ClusterONE (Nepusz et al., 2012). These three tools can be installed either locally or as a Cytoscape plugin. However, these types of software cannot be used in distributed systems with the goal of achieving better efficiency in large-scale PPI data. Web platforms for the analysis of interaction networks have also been developed, including Path BLAST (Kelley et al., 2004), which is useful for the alignment of these networks.

12.6 FUTURE DIRECTION ABOUT BIG DATA

The use of data generated from Big Data is still considered to be controversial because there are still no methodologies that can define the utility of these data. However, in the future, several fields of study will use information that is generated by Big Data tools. For example, economics will rely on studies that are related to market dynamics and the stock market (Choi and Varian, 2012). Big Data analysis in the life sciences will probably affect personalized medicine and lead to the development of more effective medical treatments, because currently we have broad spectrum treatments that produce better results on certain patients in comparison to others (Chiavegatto Filho, 2015; Taylor et al., 2014).

To reach a stage at which using information derived from Big Data is possible, it is necessary to overcome some of the questions that are still unanswered regarding the development and analysis of Big Data. These unanswered questions hamper the distribution of results from these analyses because Big Data information is unstructured due to the different programming languages that are used to generate the data (Chiavegatto Filho, 2015).

The current trend indicates that more applications and technological improvements will continue to emerge in future years as an effective answer to Big Data challenges (Suciu et al., 2015). Several platforms based on Cloud have been developed and optimized, such as Galaxy (Goecks et al., 2010) and CloudBlast (Matsunaga et al., 2008), to integrate specific tools and give a rapid and comprehensive solution to questions that are associated with the analysis and sharing of large volumes of biological data.

In recent decades, Cloud Computing has become a promising solution to manipulating Big Data, and the applications of Cloud Computing continue to grow together with recent technological advances (Suciu et al., 2015). Because of its capacity to exploit computational resources from a variety of computers and its ability to provide tools for analysis and storage through the Internet, using dynamically allocated virtual resources, Cloud Computing has grown a substantial amount, and it is expected that more and more scientific community users will use these services (Dai et al., 2012). These technologies are an excellent alternative to centers with limited resources because cloud users do not need to finance or maintain hardware (Kashyap et al., 2015; Marx, 2013). Clouds have a large variety of data storage, acquisition, sharing, and analysis services, which can usually be divided into three categories: SaaS, Platform as a Service (PaaS), and Infrastructure as a Service (IaaS) (O'Driscoll et al., 2013).

It is predicted that PaaS and SaaS become viable alternatives to traditional biological data analysis in which data from public sites (e.g., NCBI, Ensembl) are frequently downloaded to allow data to be processed and analyzed locally by preinstalled tools. PaaS technology provides an environment for the development, test, and implementation of specific applications for the analysis of biological Big Data. SaaS refers to the process of taking advantage of available applications using a remote Cloud structure through the web browser, which represents a change in the current pattern of bioinformatics bibliographic research because there would be no need to install software locally in workstations. The user would connect to a desktop through a Virtual Machine in which all of the necessary data analysis software is already installed. IaaS provides a virtual computational infrastructure that has various resources available from the Internet, such as hardware and software (Fig. 12.4; O'Driscoll et al., 2013).

Expectations for new data transfer technologies have also been substantially raised. Big Data transfer solutions are currently under development, with viable methodology proposed using data transfer innovations. The Beijing Genomics Institute (BGI) is currently exploring various technologies and tools to accelerate the process of electronic data transfer. The software *FASP*, developed by Aspera in Emeryville, California, has performed very well on tests carried out by BGI. In one of these tests, the transfer of 24 GB of data occurred in only 30 seconds, which is a velocity much higher than any other transfer protocols that are currently used on the web (Marx, 2013).

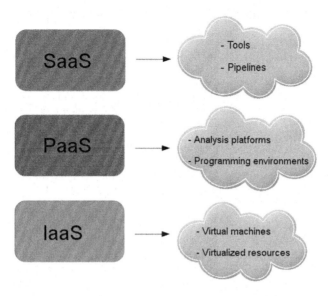

FIGURE 12.4 Cloud services.

In spite of the evolution of Big Data storage and manipulation technologies, current Big Data architecture does not offer solutions for the analysis and association of such a large volume of information (Marx, 2013). To overcome this challenge, machine learning techniques, together with statistical methods, have been used to analyze small- and large-scale data, with techniques such as sampling, feature selection, and distributed computation. Other techniques, such as Cloud Computing, Data Mining, and Computational Intelligence, including evolutionary algorithms, have been optimized to improve the analysis and reduce the processing time (Kashyap et al., 2015). However, current tools are still not fully suitable for Big Data (Shukla and Dubey, 2014).

To overcome these challenges, it is necessary to develop statistical methods that are more robust and appropriate for large volumes of information and that are also capable of performing treatment for noise and weak dependencies (Fan and Bifet, 2013). Clearly, the scientific community is making a noticeable effort to integrate data from various sources to enable us to answer biological questions that are related to cancer, as seen by cancer databases: cancergenome.nih.gov https://www.facs.org/quality%20programs/cancer/ncdb (Sineshaw et al., 2014), and http://cancer.sanger.ac.uk/cosmic (Bamford et al., 2004); disease databases (www.omim.org/ (Amberger et al., 2015), http://www.gideononline.com/ (Edberg, 2005), and http://www.disgenet.org/ (Pinero et al., 2015)); attempts to unravel the interactions between organisms and hosts (http://www.malacards.org (Rappaport et al., 2014), http://www.phi-base.org/ (Winnenburg et al., 2006), and http://www.agbase.msstate.edu/hpi/main.html (Kashyap et al., 2015)); and environment analyses to understand the behavior of an organism in its habitat (www.uniprot.org/help/unimes (Apweiler et al., 2011), https://www.ebi.ac.uk/metagenomics/ (Zhan et al., 2013), and http://omictools.com/rrna-databases-c297-p1.html (Peng et al., 2010); such effort demonstrates the need to overcome the remaining challenges in the Age of Big Data.

References

Abuín, J.M., Pichel, J.C., Pena, T.F., Amigo, J., 2015. BigBWA: approaching the Burrows–Wheeler aligner to Big Data technologies. Bioinformatics 31 (24), 4003–4005. Available from: https://doi.org/10.1093/bioinformatics/btv506.

Alonso, N., Lucas, G., Hysi, P., 2015. Big data challenges in bone research: genome-wide association studies and next-generation sequencing. Bonekey Rep. 4, 635. Available from: https://doi.org/10.1038/bonekey.2015.2.

Alter, D.A., 2015. Merits and pitfalls of using observational "Big Data" to inform our understanding of socioeconomic outcome disparities. J. Am. Coll. Cardiol. 66, 1898–1900. Available from: http://dx.doi.org/10.1016/j.jacc.2015.08.037.

Alyass, A., Turcotte, M., Meyre, D., 2015. From big data analysis to personalized medicine for all: challenges and opportunities. BMC Med. Genomics 8, 33. Available from: https://doi.org/10.1186/s12920-015-0108-y.

Amberger, J.S., Bocchini, C.A., Schiettecatte, F., Scott, A.F., Hamosh, A., 2015. OMIM.org: Online Mendelian Inheritance in Man (OMIM(R)), an online catalog of human genes and genetic disorders. Nucleic Acids Res. 43, D789–D798. Available from: https://doi.org/10.1093/nar/gku1205.

Apweiler, R., Martin, M.J., O'Donovan, C., Magrane, M., Alam-Faruque, Y., Antunes, R., et al., 2011. Ongoing and future developments at the universal protein resource. Nucleic Acids Res. 39, 214–219. Available from: https://doi.org/10.1093/nar/gkq1020.

Bader, G.G., Hogue, C.C., 2003. An automated method for finding molecular complexes in large protein interaction networks. BMC Bioinformatics 4, 2. Available from: https://doi.org/10.1186/1471-2105-4-2.

Bamford, S., Dawson, E., Forbes, S., Clements, J., Pettett, R., Dogan, A., et al., 2004. The COSMIC (Catalogue of Somatic Mutations in Cancer) database and website. Br. J. Cancer 2, 355–358. Available from: https://doi.org/10.1038/sj.bjc.6601894.

Bolouri, H., 2014. Modeling genomic regulatory networks with big data. Trends Genet. 30, 182–191. Available from: https://doi.org/10.1016/j.tig.2014.02.005.

Chang, F., Dean, J., Ghemawat, S., Hsieh, W.C., Wallach, D.A., Burrows, M., et al., 2008. Big table. ACM Trans. Comput. Syst. 26, 1–26. Available from: https://doi.org/10.1145/1365815.1365816.

Chiavegatto Filho, A.D.P., 2015. Uso de big data em saúde no Brasil: perspectivas para um futuro próximo. Epidemiol. e Serviços Saúde 24, 325–332. Available from: https://doi.org/10.5123/S1679-49742015000200015.

Choi, H., Varian, H., 2012. Predicting the present with Google trends. Econ. Rec. 88, 2–9. Available from: https://doi.org/10.1111/j.1475-4932.2012.00809.x.

Clarke, J., Wu, H.-C., Jayasinghe, L., Patel, A., Reid, S., Bayley, H., 2009. Continuous base identification for single-molecule nanopore DNA sequencing. Nat. Nanotechnol. 4, 265–270. Available from: https://doi.org/10.1038/nnano.2009.12.

Cohen, J., Dolan, B., Dunlap, M., Hellerstein, J.M., Welton, C., 2009. MAD skills: new analysis practices for Big Data. In: Proceedings of the VLDB Endowmen; 2; 1481–1492. http://dx.doi.org/10.14778/1687553.1687576.

Costa, F.F., 2012. Big Data in genomics: challenges and solutions. G.I.T. Lab. J. 1–4.

Dai, L., Gao, X., Guo, Y., Xiao, J., Zhang, Z., 2012. Bioinformatics clouds for big data manipulation. Biol. Direct 7, 43. discussion 43. http://dx.doi.org/10.1186/1745-6150-7-43.

Day, A., Dong, J., Funari, V.A., Harry, B., Strom, S.P., Cohn, D.H., et al., 2009. Disease gene characterization through large-scale co-expression analysis. PLoS One 4, e8491. Available from: https://doi.org/10.1371/journal.pone.0008491.

Dean, J., Ghemawat, S., 2004. MapReduce: simplified data processing on large clusters. In: Proceedings of the 6th Symposium on Operating Systems Design and Implementation; 137–149. http://dx.doi.org/10.1145/1327452.1327492.

Edberg, S.C., 2005. Global Infectious Diseases and Epidemiology Network (GIDEON): a world wide web-based program for diagnosis and informatics in infectious diseases. Clin. Infect. Dis. 40, 123–126. Available from: https://doi.org/10.1086/426549.

Fan, W., Bifet, A., 2013. Mining Big Data: current status, and forecast to the future. ACM SIGKDD Explor. Newsl. 14, 1–5. Available from: https://doi.org/10.1145/2481244.2481246.

Fusaro, V.A., Patil, P., Gafni, E., Wall, D.P., Tonellato, P.J., 2011. Biomedical cloud computing with Amazon web services. PLoS Comput. Biol. 7, e1002147. Available from: https://doi.org/10.1371/journal.pcbi.1002147.

Gandomi, A., Haider, M., 2015. Beyond the hype: Big data concepts, methods, and analytics. Int. J. Inf. Manage. 35, 137–144. Available from: https://doi.org/10.1016/j.ijinfomgt.2014.10.007.

Goecks, J., Nekrutenko, A., Taylor, J., 2010. Galaxy: a comprehensive approach for supporting accessible, reproducible, and transparent computational research in the life sciences. Genome Biol. 11, R86. Available from:
https://doi.org/10.1186/gb-2010-11-8-r86.

Hiltemann, S., Mei, H., de Hollander, M., Palli, I., van der Spek, P., Jenster, G., et al., 2014. CGtag: complete genomics toolkit and annotation in a cloud-based Galaxy. Gigascience 3, 1. Available from: https://doi.org/
10.1186/2047-217X-3-1.

Hurwitz, Judith, Nugent, Alan, Halper, Fern, Kaufman, Marcia, 2013. Big Data for Dummies, First. ed. John Wiley
& Sons, Inc, Hoboken, New Jersey.

Issa, N.T., Byers, S.W., Dakshanamurthy, S., 2014. Big data: the next frontier for innovation in therapeutics and
healthcare. Expert Rev. Clin. Pharmacol. 7, 293−298. Available from: https://doi.org/10.1586/
17512433.2014.905201.

Kashyap, H., Ahmed, H.A., Hoque, N., Roy, S., Bhattacharyya, D.K., 2015. Big Data analytics in bioinformatics: a
machine learning perspective. J. Latex Class Files 13, 1−20.

Kelley, B.P., Yuan, B., Lewitter, F., Sharan, R., Stockwell, B.R., Ideker, T., 2004. PathBLAST: a tool for alignment of
protein interaction networks. Nucleic Acids Res. 32, 83−88. Available from: http://dx.doi.org/10.1093/nar/
gkh411.

Langfelder, P., Horvath, S., 2008. WGCNA: an R package for weighted correlation network analysis. BMC
Bioinformatics 9, 559. Available from: https://doi.org/10.1186/1471-2105-9-559.

Langmead, B., Schatz, M.C., Lin, J., Pop, M., Salzberg, S.L., 2009a. Searching for SNPs with cloud computing.
Genome Biol. 10, R134. Available from: https://doi.org/10.1186/gb-2009-10-11-r134.

Langmead, B., Trapnell, C., Pop, M., Salzberg, S.L., 2009b. Ultrafast and memory-efficient alignment of short
DNA sequences to the human genome. Genome Biol. 10, R25. Available from: https://doi.org/10.1186/gb-
2009-10-3-r25.

Leonard Kaufman, P.J.R., 2005. Finding Groups in Data: An Introduction to Cluster Analysis. John Wiley & Sons,
Inc., Hoboken, New Jersey.

Liang, M., Zhang, F., Jin, G., Zhu, J., 2015. FastGCN: a GPU accelerated tool for fast gene co-expression networks.
PLoS One 10, e0116776. Available from: https://doi.org/10.1371/journal.pone.0116776.

Loman, N.J., Constantinidou, C., Chan, J.Z.M., Halachev, M., Sergeant, M., Penn, C.W., et al., 2012. High-
throughput bacterial genome sequencing: an embarrassment of choice, a world of opportunity. Nat. Rev.
Microbiol. 10, 599−606. Available from: https://doi.org/10.1038/nrmicro2850.

Marx, V., 2013. Biology: the big challenges of big data. Nature 498, 255−260. Available from: https://doi.org/
10.1038/498255a.

Matsunaga, A., Tsugawa, M., Fortes, J., 2008. CloudBLAST: combining MapReduce and virtualization on distributed resources for bioinformatics applications. In: 2008 IEEE Fourth International Conference on eScience;
222−229. http://dx.doi.org/10.1109/eScience.2008.62.

May, M., 2014. Life Science Technologies: big biological impacts from big data. Science 344, 1298−1300. Available
from: https://doi.org/10.1126/science.344.6189.1298.

Merelli, I., Pérez-Sánchez, H., Gesing, S., D'Agostino, D., 2014. Managing, analysing and integrating Big Data in
medical bioinformatics: open problems and future perspectives. BioMed Res. Int. 2014 (2014), Article ID
134023, 13 pages. Hindawi.Com 2014. doi:10.1155/2014/134023.

Merriman, B., Rothberg, J.M., 2012. Progress in Ion Torrent semiconductor chip based sequencing. Electrophoresis
33 (23), 3397−3417. Available from: http://dx.doi.org/10.1002/elps.201200424.

Nepusz, T., Yu, H., Paccanaro, A., 2012. Detecting overlapping protein complexes in protein-protein interaction
networks. Nat. Methods 9, 471−472. Available from: https://doi.org/10.1038/nmeth.1938.

Nguyen, T., Shi, W., Ruden, D., 2011. CloudAligner: a fast and full-featured MapReduce based tool for sequence
mapping. BMC Res. Notes 4, 171. Available from: https://doi.org/10.1186/1756-0500-4-171.

Noor, A.M., Holmberg, L., Gillett, C., Grigoriadis, A., 2015. Big Data: the challenge for small research groups in
the era of cancer genomics. Br. J. Cancer 1−8. Available from: https://doi.org/10.1038/bjc.2015.341.

Nordberg, H., Bhatia, K., Wang, K., Wang, Z., 2013. BioPig: a Hadoop-based analytic toolkit for large-scale
sequence data. Bioinformatics 29, 3014−3019. Available from: https://doi.org/10.1093/bioinformatics/btt528.

O'Driscoll, A., Daugelaite, J., Sleator, R.D., 2013. "Big data", Hadoop and cloud computing in genomics. J.
Biomed. Inform. 46, 774−781. Available from: https://doi.org/10.1016/j.jbi.2013.07.001.

O'Reilly Media, 2012. Big Data Now: 2012 Edition. O'Reilly Media, Sebastopol, CA.

Peng, Y., Leung, H.C.M., Yiu, S.M., Chin, F.Y.L., 2010 IDBA—A practical iterative de Bruijn Graph de novo assembler. Research in Computational Molecular Biology. In: Proceedings of the 14th Annual International Conference, RECOMB 2010, Lisbon, Portugal, April 25–28, 2010. 426–440. http://dx.doi.org/10.1007/978-3-642-12683-3_28.

Phan, J., Young, A., Wang, M., 2012. OmniBiomarker: a web-based application for knowledge-driven biomarker identification. IEEE Trans. Biomed. Eng. 60, 3364–3367. Available from: https://doi.org/10.1109/TBME.2012.2212438.

Pinero, J., Queralt-Rosinach, N., Bravo, A., Deu-Pons, J., Bauer-Mehren, A., Baron, M., et al., 2015. DisGeNET: a discovery platform for the dynamical exploration of human diseases and their genes. Database 2015. Available from: http://dx.doi.org/10.1093/database/bav028bav028.

Rappaport, N., Twik, M., Nativ, N., Stelzer, G., Bahir, I., Stein, T.I., et al., 2014. MalaCards: a comprehensive automatically-mined database of human diseases. Curr. Protoc. Bioinforma. 1, 1.24.1–1.24.19. Available from: https://doi.org/10.1002/0471250953.bi0124s47.

Rivera, C.G., Vakil, R., Bader, J.S., 2010. NeMo: network module identification in cytoscape. BMC Bioinformatics 11 (Suppl 1), S61. Available from: https://doi.org/10.1186/1471-2105-11-S1-S61.

Sagiroglu, S., Sinanc, D., 2013. Big data: a review. Int. Conf. Collab. Technol. Syst. 42–47. Available from: https://doi.org/10.1109/CTS.2013.6567202.

Servant, N., Roméjon, J., Gestraud, P., La Rosa, P., Lucotte, G., Lair, S., et al., 2014. Bioinformatics for precision medicine in oncology: Principles and application to the SHIVA clinical trial. Front. Genet. 5, 1–16. Available from: https://doi.org/10.3389/fgene.2014.00152.

Shukla, V., Dubey, P.K., 2014. Big Data: moving forward with emerging technology and challenges. Int. J. Adv. Res. Comput. Sci. Manag. Stud. 2, 187–193.

Sineshaw, H.M., Gaudet, M., Ward, E.M., Flanders, W.D., Desantis, C., Lin, C.C., et al., 2014. Association of race/ethnicity, socioeconomic status, and breast cancer subtypes in the National Cancer Data Base (2010-2011). Breast Cancer Res. Treat. 145, 753–763. Available from: https://doi.org/10.1007/s10549-014-2976-9.

Stephens, Z.D., Lee, S.Y., Faghri, F., Campbell, R.H., Zhai, C., Efron, M.J., et al., 2015. Big Data: astronomical or genomical? PLoS Biol. 13, e1002195. Available from: https://doi.org/10.1371/journal.pbio.1002195.

Stokes, T.H., Moffitt, R. A, Phan, J.H., Wang, M.D., 2007. Chip artifact CORRECTion (caCORRECT): a bioinformatics system for quality assurance of genomics and proteomics array data. Ann. Biomed. Eng. 35, 1068–1080. Available from: https://doi.org/10.1007/s10439-007-9313-y.

Suciu, G., Suciu, V., Martian, A., Craciunescu, R., Vulpe, A., Marcu, I., et al., 2015. Big Data, Internet of Things and Cloud Convergence—an architecture for secure E-health applications. J. Med. Syst. 39, 141. Available from: https://doi.org/10.1007/s10916-015-0327-y.

Taylor, L., Schroeder, R., Meyer, E., 2014. Emerging practices and perspectives on Big Data analysis in economics: bigger and better or more of the same? Big Data Soc. 1, 1–10. Available from: https://doi.org/10.1177/2053951714594877.

Taylor, R.C., 2010. An overview of the Hadoop/MapReduce/HBase framework and its current applications in bioinformatics. BMC Bioinformatics 11, S1. Available from: https://doi.org/10.1186/1471-2105-11-S12-S1.

TechAmerica Foundation: Federal Big Data Commission, 2012. A Practical Guide To Transforming The Business of Government. 1–40.

Trifonova, O.P., Il'in, V.A., Kolker, E. V, Lisitsa, A. V, 2013. Big Data in Biology and Medicine: Based on material from a joint workshop with representatives of the international Data-Enabled Life Science Alliance, July 4, 2013, Moscow, Russia. Acta Naturae 5, 13–16.

Van Dijk, E.L., Auger, H., Jaszczyszyn, Y., Thermes, C., 2014. Ten years of next-generation sequencing technology. Trends Genet. 30. Available from: https://doi.org/10.1016/j.tig.2014.07.001.

Winnenburg, R., Baldwin, T.K., Urban, M., Rawlings, C., Köhler, J., Hammond-Kosack, K.E., 2006. PHI-base: a new database for pathogen host interactions. Nucleic Acids Res. 34, D459–D464. Available from: https://doi.org/10.1093/nar/gkj047.

Zhan, X., Pan, S., Wang, J., Dixon, A., He, J., Muller, M.G., et al., 2013. Peregrine and saker falcon genome sequences provide insights into evolution of a predatory lifestyle. Nat. Genet. 45, 563–566. Available from: https://doi.org/10.1038/ng.2588.

Zhang, T., Ramakrishnan, R., Livny, M., 1996. BIRCH: an efficient data clustering databases method for very large databases. ACM SIGMOD Int. Conf. Manag. Data 1, 103–114. Available from: https://doi.org/10.1145/233269.233324.

CHAPTER

13

Bioinformatics and Systems Biology in Bioengineering

Joseph J. Nalluri[1], Debmalya Barh[2,3], Vasco Azevedo[3] and Preetam Ghosh[1]

[1]Virginia Commonwealth University, Richmond, VA, United States [2]Institute of Integrative Omics and Applied Biotechnology (IIOAB), Purba Medinipur, India [3]Federal University of Minas Gerais, Belo Horizonte, Minas Gerais, Brazil

13.1 BIOINFORMATICS AND MAJOR DATABASES

Bioinformatics has witnessed a surge in the last decades mainly due to the emerging trend of tools and technologies capable of delivering high-throughput biological data. The term "Big Data" is frequently applied to biological and biomedical data. As such, the amount of access to biological data we have today is unprecedented. Omics studies encompass, but not limited to, proteomics, genomics, transcriptomics, and metabolomics. Each technology yields itself to a specific field of data gathering, collection, analysis, and prediction for proteome, genome, and metabolome, respectively. In Chapter 12, Computational Techniques in Data Integration and Big Data Handling in Omics, we observed the techniques handled in data integration of huge data sets of omics data. Bioinformatics is the field of study focused on developing computer-based methods and models to study the various facets of biological data. It consists of development of methods for data curation, data organization, data processing, annotation, and retrieval. The data retrieved are further used for comprehensive analysis, e.g., statistical tests, developing novel models of simulation and prediction, etc. Systems biology has data that are multidimensional in nature and scale. The type and structure of data is extremely crucial in building databases and data repositories for the scientific community. This structure is often called as "schema" in database terminology. Creating new schemas for emerging new biological data and updating previous schemas is an active field in bioinformatics.

Omics Technologies and Bio-engineering: Towards Improving Quality of Life
DOI: https://doi.org/10.1016/B978-0-12-804659-3.00013-0

223

Bioinformatics also focuses on developing large software suites and frameworks for processing the data, step-by-step. To infer and extract knowledge and information from biological data and to test several scientific hypotheses, one must fetch the data and perform several subsequent tasks such as data cleaning, data formatting, tests, analysis, and pruning of the data. Hence, software tools such as network-based tools, statistical tools, and database tools often work in cohesion to answer the questions regarding the topology, structure, and regulation between the entities captured in the biological data.

Of all the types of biological data repositories, most popular are genomic databases containing DNA sequence and sequence analysis. Diverse bioinformatics data repositories featuring varied biological studies and experiments are:

1. The National Center for Biotechnology—a suite of databases and tools to download and analyze the data related to almost various biological experimentation researched and documented (Wheeler et al., 2003). It is available online at http://www.ncbi.nlm.nih.gov/.

   Similar database center is maintained at the European Bioinformatics Institute (EMBL-EBI) that provides free access to comprehensive and integrated molecular data sets (Brooksbank et al., 2014). It is available online at http://www.ebi.ac.uk/.
2. GenBank (Benson et al., 2008) is one of the most widely used databases. It is a repository of all discovered nucleotide sequences, ranging from the human genome, mouse genome, fly genome, etc. It is available online at http://www.ncbi.nlm.nih.gov/genbank/.
3. Protein database (Berman et al., 2000)—a database of protein structures, their structure types, and information. It is available at http://www.rcsb.org/pdb/home/home.do.
4. Kyoto Encyclopedia of Genes and Genomes (Kanehisa and Goto, 2000)—database for storing biological functions and cell signaling pathways. It is available at http://www.genome.jp/kegg/.
5. Pathway Commons (Cerami et al., 2011)—database for biological pathways in multiple organisms. It is available at http://www.pathwaycommons.org/about/.
6. Gene Expression Omnibus (Edgar et al., 2002)—database with extensive collection of microarray and sequence- based data. It is available at http://www.ncbi.nlm.nih.gov/geo/.
7. Online Mendelian Inheritance in Man (Hamosh et al., 2005)—database catalog of human genes and genetic disorders. It is available at http://www.omim.org/
8. Encyclopedia of DNA elements (ENCODE) (Consortium TEP, 2004)—List of all functional elements in DNA sequences. It is available at https://www.genome.gov/10005107/encode-project/.

There are lot more databases specific to biological types and entities. The goal of these databases is to enlarge our understanding of the biological phenomena and help us create larger network models of connected information that represents cellular functions. For example, by combining OMIM data and Gene Bank data, we can try to determine how variations/mutations in genes are associated to diseases by analyzing the gene expression patterns. Also these data serve as prior knowledge in our pursuit of discovering new pathways, subcellular components, and behaviors associated with our functions of interest. The prior knowledge is also inserted in the development of mathematical models.

13.2 SYSTEMS BIOLOGY—A BRIEF OVERVIEW

The above-listed bioinformatics resources and many more provide comprehensive information about these biological entities in isolation. However, the goal of the research community is to try and integrate these pieces of data in an intelligent manner to answer specific biological questions. Systems biology prominently interacts with the field of cell biology because of an underlying fundamental assumption, that studying the cell as a whole entity will yield more discoveries as opposed to studying individual components in isolation. The study of systems biology often tries to bridge the gaps between molecular interactions to cellular interactions to its effect in tissues/organs and finally its result in physiological functions of the human body. This requires researchers to think of the entire process as cellular machinery. Systems biology is therefore the study of molecular reactions and groupings at cellular level, which produces subcellular machinery that acts as functional units that perform operations at the cell, tissue/organ level physiological functions. The crucial difference between biology and systems biology is that the discipline of systems biology focuses prominently on the mechanism of biological components while it is *functioning*. This is further emphasized by Kitano that "to understand biology at systems level, we must examine the structure and dynamics of cellular and organismal function rather than the characteristics of isolated parts of a cell or organism" (Kitano, 2002).

Systems biology rightly finds itself similar to the field of complex systems, which was the term used for these studies prior to the term "systems biology." Fundamentally all approaches in systems biology fall either into the category of top-down approach or bottom-up approach. Top-down approaches essentially try to capture the system's functionality or behavior as a whole and attempt to understand entire system-wide characteristics and capabilities. These include high-throughput data technologies, such as DNA microarrays, image analysis, mass spectrometry, and gene expression microarrays, which are survey-type experiments. In contrast, bottom-up approaches capture the information at the cellular level of genes, proteins, and lipids and develop an understanding of *interdependent* functionality between these components. Bottom-up approaches provide mechanistic understanding of the way in which these entities are assembled, controlled, and function. Top-down approaches include these high-level experiments, also termed as "omics"", are geared towards systems biology-oriented experiments. Fig. 13.1 displays an overview of this methodology.

Few popular data sets that systems biologists often use are of mRNA profiles in GEO, microRNAs in TargetScan (Lewis et al., 2003) (http://www.targetscan.org/), protein characteristics in Swiss-Prot (Bairoch and Boeckmann, 1991) (http://web.expasy.org/docs/swiss-prot_guideline.html), genome-wide association studies in DbGAP (Mailman et al., 2007) (http://www.ncbi.nlm.nih.gov/gap), and disease genes from OMIM and drugs data sets in Pharm GKB (Hewett et al., 2002) (https://www.pharmgkb.org/). By integrating these "omics" data sets from these sources, network models can be built, which can be statistically analyzed for spotting of correlations and associations between these entities (Joyce and Palsson, 2006; Markowetz and Spang, 2007; Sharan and Ideker, 2006). Several key papers (Iyer et al., 1999; Bhalla and Iyengar, 1999; Alon et al., 1999) among many others demonstrate the necessity and effectiveness of combining the different facets of

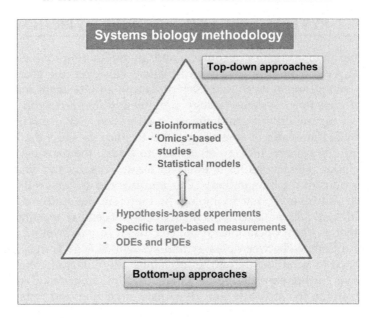

FIGURE 13.1   Systems Biology methodology overview. The merger of these two kinds of approaches provides the best description of a biological process.

biological data to discover novel pathways and characteristics of cell functioning. In the experimental study of Iyer et al. (1999), we can observe the measurement levels of thousands of mRNAs simultaneously, via the usage of microarrays, which show changes in cellular components system-wide in response to a stimulus. Similarly, using computational analysis and simulations, we can observe the manner, in which interactions between cellular components often create a system-wide functional capability like *switching* behavior (Bhalla and Iyengar, 1999). These system-wide functional capabilities are a result of multiple interactions between the cellular components and not a result of a single component. The system-wide behavior such as *robustness* of cellular networks is studied by integrating experimental data and computational analysis. Bacterial chemotaxis displayed insensitivity towards varying levels of concentration of protein components thereby demonstrating a trait of adaptation and robustness of functionality (Alon et al., 1999). Systems biology utilizes the fields of molecular biology and biochemistry of cellular components for the development of an engineering model, to understand the physiological functions at the cell/ tissue/organ level. There are various studies that deploy technologies like yeast-two hybrid and chromatin immunoprecipitation to protein–protein or protein–DND interactions to create large protein interaction networks that demonstrate the cell's information transmission behavior in response to a stimuli (Bar-Joseph et al., 2003; Haugen et al., 2004). Once a particular cell's behavior to stimuli is deciphered, then the stimuli can be engineered and the corresponding pathway behavior can be executed. After a predicted model is simulated and proved accurate based on experimental corroboration, the goal is to test the model prediction or create a technology or product. This involves molecular level manipulations with the help of genetic, biochemical, material, or engineering interventions (Ideker et al., 2006). A few examples of such studies and their modeling strategies are mentioned in the next section.

13.2.1 Mathematical Representations and Modeling

Agents such as hormones and neurotransmitters evoke various kinds of physiological responses. Cell signaling pathways have their effect through multiple layers of organization, e.g., from signaling pathways (G-protein receptor) to biochemical functions (metabolic proteins, transcriptional machinery) to cell/tissue level functions to organismal functions (organismal homeostasis). To regulate the activity of the receptors, it is essential to study quantitatively the manner in which certain amounts of ligands (i.e., hormones and neurotransmitters) regulate the activity of the receptors. To understand this dynamical aspect of a physiological response evoked via the transmitter, various mathematical models are deployed. Ordinary differential equations (ODEs) are a class of differential equations used to model and simulate such a phenomenon. ODEs are used to model the rate of change of an entity with respect to time given the initial concentrations of the entities and rates of reaction.

In case of ligand plus receptor binding reaction (Fig. 13.2) to determine the rate of formation of ligand−receptor complex, ODE-based modeling is deployed in the form of following differential equations,

$$[L] + [R] \xrightarrow{k_{forward}} [LR]$$

$$[L] + [R] \xrightarrow{k_{backward}} [LR]$$

$$\frac{d[L]}{dt} = -k_{forward} \times [L][R] + k_{backward} \times [LR]$$

$$\frac{d[R]}{dt} = -k_{forward} \times [L][R] + k_{backward} \times [LR]$$

$$\frac{d[LR]}{dt} = -k_{forward} \times [L][R] - k_{backward} \times [LR]$$

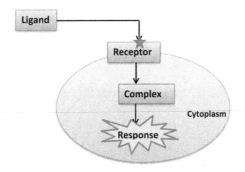

FIGURE 13.2 Cell signaling information flow—from ligand to receptor to ligand−receptor complex to response.

where, L is the ligand, R is the receptor, and LR is the ligand–receptor complex. $k_{forward}$ and $k_{backward}$ are rate constants for forward reaction and backward reaction, respectively. Here, K_D would be the dissociation constant calculated by $K_D = \frac{k_{backward}}{k_{forward}}$. In the above formulation, $\frac{d[LR]}{dt}$ gives the rate of ligand–receptor complex formation over time. Hence, this provides an overall affinity of the receptor. These models are called deterministic models because if the initial concentrations and rates of reaction constants are known, then the formation of the final product can be determined or specified.

For setting up an ODE-based simulation model, every interaction or reactions involved in the processes are identified. These reactions are thereafter parameterized, i.e., the concentrations of reactants, reaction rates are determined. If the exact reaction rates are not available, the parameters are often estimated. In this model it is essential that the simulation model obeys basic thermodynamic principles of mass conservation and microscopic reversibility, i.e., product of all the forward reactions is the same as the product of reverse reactions.

Moreover, behavior of biological network motifs such as positive feedback loops and negative feedback loops is best comprehended through differential equation–based models. Static network models (such as Boolean dynamics) have limited ability to uncover temporal dynamics of a cell system. Often times, the best understanding of a behavioral property in a cell system is obtained by a coupling of network analysis and dynamical modeling.

An ODE-based cell cycle model developed by Tyson (1991) is considered a classic example of the effectiveness of ODE-based modeling to represent complex cellular behavior such as proliferation. The proteins cdc2 and cyclin form a heterodimer (maturation promoting factor; MPF) controlling the major movements of a cell. The interaction between cyclin and cdc2 is shown in Fig. 13.3 (Tyson, 1991).

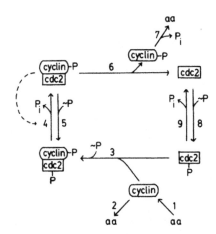

FIGURE 13.3   In Step 1 (bottom right), cyclin is synthesized and combines with cdc2 (Step 3). After forming a heterodimer, the cyclin subunit is phosphorylated and later dephosphorylated to form active MPF (Step 4). After a sufficient quantity of active MPF is activated, a nuclear division is triggered, which destroys the active MPF. This leads to MPF complex releasing phosphorylated cyclin. After subsequent steps the cycle starts again.

Tyson (1991) developed a mathematical model for the interactions of cdc2 and cyclin, and postulated that the cell's control system can operate in any of the following three modes—as a steady state with high MPF activity, as a spontaneous oscillator, or as an excitable switch. The mathematical model consisted of the following kinetic equations:

$$\frac{d[cyclin]}{dt} = k_1[aa] - k_2[cyclin] - k_3[cdc\_P][cyclin]$$

$$\frac{d[MPF]}{dt} = [P\_cyclin\_cdc2\_P]F([MPF]) - k_5[\sim P][MPF]] - k_6[MPF]$$

$$\frac{d[P\_cyclin\_cdc2\_P]}{dt} = k_3[cdc2\_P][cyclin] - [P\_cyclin\_cdc2\_P]F([MPF]) + k_5[\sim P][MPF]$$

$$\frac{d[cyclin\_P]}{dt} = k_6[MPF] - k_7[cyclin\_P]$$

$$\frac{d[cdc2]}{dt} = k_6[MPF] - k_8[\sim P][cdc2] + k_9[cdc2\_P]$$

$$\frac{d[cdc2\_P]}{dt} = -k_3[cdc2\_P][cyclin] + k_8[\sim P][cdc2] - k_9[cdc\_P]$$

Further explanation and details can be studied in Tyson (1991). This model is also implemented in the software, E-cell (Tomita et al., 1999).

In addition to ODEs, partial differentiation equation (PDE)−based modeling can be used to capture the spatial dynamics of cell signaling. Neves et al. (2008) used PDEs to model the flow from the β-adrenergic receptor to MAPK1,2 through the cAMP/PKA/ B-Raf/MAPK1,2 network in neurons using real geometries. Based on their model they predicted that negative regulators control the flow of spatial information to downstream components. These predictions were experimentally verified in the rat hippocampal slices.

Mathematical model is an integral part of system biology. Table 13.1 presents some of the models devised to study various biological phenomena.

13.3 REVERSE ENGINEERING OF NETWORK INTERACTIONS

Interplay of mathematical modeling with experimental data to predict and control biological phenomena is a crucial beginning phase of systems biology. Once the multiscale data are extracted and modeled into a well-informed and comprehensively designed network, reverse engineering methodology is applied. Reverse engineering principles are the same as the principles of biology, which is to *understand* the system and not *design* the system, primarily. In simple terms, the quest is to recreate mathematically an understanding model of the original observed phenomena from the derived data sets, as close to the reality as possible. Reverse engineering studies have primarily focused on the model of microbial cells (Villaverde et al., 2013).

Reverse engineering modeling approaches are broadly classified into three popular categories—interaction based (Sharan and Ideker, 2006), in which dynamics and

TABLE 13.1 Type of Mathematical Representations and Techniques Used to Model Various Biological Phenomena and Their Corresponding Publications

Biological Phenomena	Model	Reference
Cell's emergent property of ultrasensitivity	Goldbeter–Koshland kinetics based on ODE modeling	Villaverde et al. (2013)
Ultrasensitivity of MAP kinase cascading pathway	Numerical solution model implemented in Runge–Kutta-based NDSolve algorithm in Mathematica	Wolkenhauer and Mesarovic (2005)
Detailed molecular representation of human breast cancer	Statistical correlation–based model	Butte et al. (2000)
Computational model of a whole bacterial cell	Different functions split into models such as flux balance analysis for metabolism, Poisson process for protein degradation	Shannon (1948)
Norepinephrine to persistent MAPK activity	ODEs	Farber et al. (1992)
The activation of Ras at different time scales at different locations by EGFR receptor	PDEs	Farber et al. (1992)
Modeling of the shape of cell (neuron) in relation to spatially restricted accumulation of cAMP	PDEs	Neves et al. (2008)
Study of Interferon-β on gene expression pattern in blood cells	Graph theory. Network building and modeling	Liang et al. (1998)
Mapping of hepatitis-C virus infection network by study of protein–protein interaction network	Graph theory. Network building	Margolin et al. (2006)
Linking a gene to phenotype	Statistical/probabilistic model to capture probabilistic associations	Faith et al. (2007)

parameters are not involved; constraint-based, in which there are no dynamics but stoichiometric parameters; and finally, mechanistic models involving the dynamical and kinetic parameters (Tyson, 1991). Regardless of the approach embraced, an accurate description of a biological phenomenon is incomplete without taking into account system dynamics; especially because cell functions are a resulting occurrence of system dynamics (Wolkenhauer and Mesarovic, 2005). Reverse engineering in interaction networks can be stated as *given a list of nodes (biological entities) infer the connections (dependencies) among them using the dataset provided* (Villaverde et al., 2013). This is a simplistic abstraction of the problem because it deals mainly with determination of interactions rather than the characterization of these interactions; hence these models lack dynamical and kinetic parameters. These models rely heavily, if not completely, on statistical inference methods. The main strategies for inferring or recovering interaction structures are correlation-based, information-theoretic, and Bayesian-methods. We shall briefly discuss their theory and

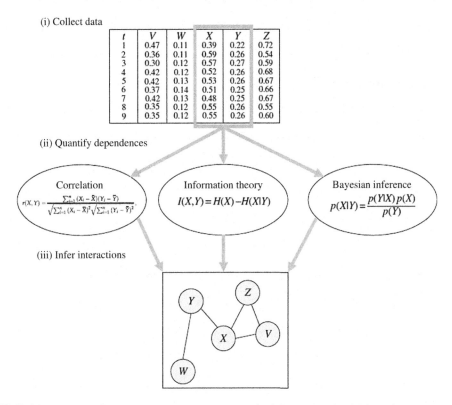

FIGURE 13.4 Overview of steps in reverse engineering methodology to predict biological interactions.

applications. A brief overview of this methodology is displayed in Fig. 13.4, entailing the steps of (1) collecting data, (2) quantifying the dependence metrics, and (3) predicting interactions based on probabilities (Villaverde et al., 2013).

13.3.1 Correlation

The correlation coefficient r, is also popularly termed as Pearson's correlation coefficient, which quantifies the dependence between two random variables X and Y as,

$$r(X, Y) = \frac{\sum_{i=1}^{n}(X_i - \overline{X})(Y_i - \overline{Y})}{\sqrt{\sum_{i=1}^{n}(X_i - \overline{X})^2}\sqrt{\sum_{i=1}^{n}(Y_i - \overline{Y})^2}}$$

where X_i and Y_i are the data points (n) and $\overline{X}$ and $\overline{Y}$ are their respective averages. $r(X, Y)$ is 0 if both the variables are linearly independent, and if a variable is completely dependent upon another variable, $r(X, Y)$ is ± 1. Correlation-based methods have been widely used in determining biological relationships such as for metabolic networks and genetic networks (Butte et al., 2000) in which interdependence and functional ties between the entities can be uncovered.

13.3.2 Information Theory

While Pearson's correlation is helpful in determining linear correlations, its accuracy decreases for nonlinear associations which are true for most of the biological associations. Here, the concept of mutual information from information theory is a suitable option (Shannon). Mutual information is based on the concept of entropy, which essentially represents the uncertainty of a single random variable. The entropy for a discrete random vector X with alphabet Z and probability mass function $p(x)$ is given as,

$$H(X) = - \sum_{x \in Z} p(x) \log p(x)$$

and the conditional entropy $H(X|Y)$ is the entropy of the variable Y conditional based on the knowledge of variable X,

$$H(Y|X) = \sum_x p(x) H(Y|X=x)$$
$$= - \sum_x \sum_y p(x,y) \log p(y|x)$$

Based on the above equations, the general dependencies between the variables, i.e., mutual information is calculated by,

$$I(X, Y) = H(X) - H(X|Y)$$

The mutual information of the two variables measures the amount of information that one variable possesses about another. The application of mutual information to biological networks has been well established (Farber et al., 1992; Liang et al., 1998). Reverse engineering has been extensively applied to the area of gene regulatory network (GRN) inference. The method, ARACNE is a popular tool for inferring GRNs (Margolin et al., 2006). CLR (Faith et al., 2007) and Inferelator (Madar et al., 2009) are other popular mutual information—based algorithms. The annual DREAM (Dialogue for Reverse Engineering Assessment and Methods) challenge aims to bridge the information disciplines of theoretical and experimental methods in the field of network inference. The DREAM5 challenge attracted a wide range of statistical algorithms for GRN inference (Marbach et al., 2012).

13.3.3 Bayesian Inference

Bayesian inference strategy allows for the incorporation of prior knowledge into the statistical inference model. Reverse engineering, a Bayesian network, comprises determining the directed acyclic graph, i.e., a network structure that best represents the data. The fitness score of the model to the data is calculated using the Bayes rule. The Bayes rule for two variables X and Y with probabilities $p(X)$ and $p(Y)$ as

$$p(X|Y) = \frac{p(Y|X)p(X)}{p(Y)}$$

where $p(X)$ and $p(X|Y)$ are prior and posterior probabilities, respectively. Probabilistic graphical models calculate joint distributions as a product of local distributions in which only a few variables are involved (Larrañaga et al., 2005). Bayesian approaches were also used for reverse engineering of genetic networks (Friedman et al., 2000). However, Bayesian models only apply to data that are acyclic, which is very rare in biomolecular interactions.

Although, reverse engineering methodologies have been implemented successfully in the area of network inference, they have not been able to present high accuracy. The compendium of DREAM challenge stated that the "vast majority of the team's predictions were statistically equivalent to random guesses" (Prill et al., 2010). Hence, there is a lot of opportunity for improvement in this field in the coming years with a need for cross-talk among multiple disciplines for a quest of better model.

13.4 BIOENGINEERING AND SYSTEMS BIOLOGY

An embedded computer system is considered as an appropriate architectural model for a cell (Reeves and Hrischuk, 2016). A cell's self-preservation and replication traits can be compared to engineering control systems. DNA is indeed a code for the cell's computing machine language (Haugen et al., 2004; Ideker et al., 2001). As such, many of the simulation and analytical engineering models can be tailored to understand the cellular mechanisms. And hence, apart from bioengineering, the disciplines of computer science, mathematics, genetics, biochemistry, and molecular/cell biology have gained relevance mainly due to the powerful omics technologies that provide data that are quantitative and large scale in nature. Modeling these biochemical phenomena and networks allows us to regulate and tweak cellular functions and biological processes such as metabolism, cell differentiation, and proliferation. Bioengineering modeling encompasses modeling of molecular function to tissue/organ with ramifications to physiological functions of human health and disease condition. An example of this is the modeling of the heart, in which cardiac action potentials were shown to exhibit long-duration plateau phase (Noble, 1960).

13.4.1 Synthetic Biology

Synthetic biology is interpreted as recreating biological entities for novel applications/purposes through an engineering-driven approach. The development of technologies like de novo DNA synthesis (Tian et al., 2004), creation of artificial gene networks (Sprinzak and Elowitz, 2005), and protein engineering have given rise to enormous potential in the ability to program/engineer a biological process or application in the field of biotechnology. A large focus of an engineering approach is on design and fabrication. However, translating those practices in the field of biology requires accurate assessment of the host systems and the nature of fabrication, as discussed below:

1. *Design*: Interpreting design from an engineering discipline is a challenging issue. Understanding and structuring a biological complexity is done with mathematical modeling. Although, we cannot manage or comprehend absolute complexity, we have

been successful with managing small artificial gene networks whose DNA basis can be designed independently from its cellular environment. Another challenge is the continuous information flow through the molecules in the cytoplasm that has many ramifications of cellular interactions and cross-talk that interfere with the intended design functionality of an element. This is difficult to contain because intercellular interactions have multiple functionalities and therefore hard to anticipate and account for, in the design (Heinemann and Panke, 2006).

2. *Fabrication*: The engineering equivalent of fabrication in biology is de novo DNA synthesis, a technology that has proved to be widely popular and prevalent. Biological fabrication requires one to take into account the fact that these created biological entities will degrade and evolve. This property would affect the long-term stability of the component and may require frequent monitoring of the integrity of the component (Heinemann and Panke, 2006).

The tools that have been employed for fabricating synthetic biological parts are:

1. *Cloning*: Conventional cloning technique is considered as the most important fabrication tool in synthetic biology. Herein, using PCR-based technology an existing recombinant DNA fragment of choice is inserted in the host organism and directed to replicate. This method has also been industrialized such that standardized cloning vectors (which meet the established criteria) have been developed and made available on a large scale by many groups such as *Registry of Standard Biological Parts* (iGEM foundation, 2004) for easy experimentation and assembly.

2. *De novo DNA synthesis*: This technology constitutes of assembling overlapping short oligonucleotides that are roughly 25–70 bp in size and chemically synthesized into a lengthier DNA molecule through a PCR-based assembling process (Stemmer et al., 1995). De novo synthesized DNA sequences allow for desirable changes ranging from promoter strength to codon usage directly into a novel sequence which would be assembled within weeks of design. A complete reconstruction of polio virus genome has been achieved through this process (Cello et al., 2002). This process is however expensive and has limits to accuracy (Young and Dong, 2004).

3. *Engineering the biological environment*: This is the final step of placing the novel bioengineered application into a host organism that would act as the *chassis* of the new synthesized process. The novel synthesized DNA segments that encode for desired functionality need to be deployed into a suitable organism. One of the major challenges in this area is the amount of vast biological complexity within an organism that adds to the hindrance of the new synthesized application. Hence, to reduce this complexity an organism with minimal biological complexity or minimal genome is suitable. Theoretically, an organism with least biological complexity would only require the essential genes, about 206 which would code for basic rudimentary cellular functions such as DNA replication, transcription and translation machinery, basic DNA repair mechanisms, protein processing, and degradation (Forster and Church, 2006). *Mesoplasma floru*, a bacterial species is considered a viable chassis environment with optimal cultivating properties and with a genome size of 793 genes (Heinemann and Panke, 2006). Synthetic biology has been successfully deployed for engineering proteins (Blancafort et al., 2004), as well.

13.4.2 Tissue Engineering

Human organs are prone to degrading based on lifestyle choices, diseases, and aging. Subsequently, they can fail to deliver their functionality. Hence, tissue engineering is a viable discipline, which aids towards human health care and therapeutics. Tissue engineering aims to engineer and create replacements for living tissues and organs. It aims to provide biologically created constructs that would recapitulate the original functionalities of the tissue or organ. The biological constructs would closely imitate the native entity. Tissue engineers use natural and nonimmunogenic synthetic materials and cells that are derived from native tissue. The field of tissue engineering has been successfully deployed in the construction of kidney, cornea, cartilage, liver, muscle, vascular graft, and nerve (Rajagopalan et al., 2013). Tissue engineering is a scrupulous process because it involves an architecture that combines proteins, proteoglycans, membranes, and corresponding oxygen specifications for the involved entities. Furthermore, the construction also has to factor in material synthesis and tissue assembly, via the usage of nonimmunogenic material (Mikos et al., 2006). Unintended cellular information exchange via molecular interactions often interferes with the assembly's functionality, structure, and usage. Hence, the functionality of the tissue and its interaction with the existing environment has to be designed in a cohesive manner which is an extensively laborious work.

The application of systems biology approach onto biological tissue engineering has been vital in several studies. Multipotent stromal cells (MSCs) are considered as biomimetic scaffolds to facilitate tissue regeneration. The migration of MSCs involving cell speed and directional persistence was understood and controlled by the usage of data-driven computational modeling with decision-tree modeling for prediction and assertion of optimal cellular responses (Wu et al., 2011). One of the concerns with tissue engineering is the assimilation of the biological construct in vivo. This aspect was addressed in a study (Sun et al., 2012) in which ordinary and stochastic differential equations were used to model the dynamics of balance of bone regeneration and resorption. Based on the equations, a combinatorial approach was used to deliver cytokines from hydrogels for obtaining the desired bone density after implanting a bone graft. The usage of stem cells for bioengineering is an expanding field with high degree of potential outcomes. The characteristic of stem cells to transform/generate new cell types provides a huge opportunity to tackle cancer-affected dysfunctional organs. Hierarchical clustering was used to identify similarities between human embryonic stem cells and embryonal carcinoma, based on which genes can be tested for their possible role of candidate gene that can potentially transform into cancerous counterparts (Sperger et al., 2003). Using detailed gene-expression profiles of MSCs from different sources revealed a core set of gene-expression profile was conserved in all the four cell types. Also, there were similarities found between the gene-expression profiles of MSCs and other fetal organs with respect to regulation of ECM proteins, Wnt signaling, and TGF-β receptor signaling. This provides an opportunity in the future for the usage of MSCs in implants and therapeutical applications (Wang et al., 2011; Tsai et al., 2007). Such systems biology–driven approaches provide efficient predictive computational models of engineered tissues and steer the subsequent novel experimental analysis and strategies.

13.5 FROM BIOLOGICAL NETWORKS TO MODERN THERAPEUTICS

An important goal of systems biology is to understand how the interactions and behavior at the molecular level of genome, proteome, and metabolome give rise to functionalities or dysfunctionalities at the human physiological level of cells, tissues, and organs, i.e., connecting the link from genotype to phenotype. This hugely aids in prognosis and treatment of human diseases and illnesses. A disease state is termed as an undesirable physiological state caused due to an unusual perturbation in the hierarchical network regulatory structure. These perturbations or errors in the cellular information processing pathway are the causes for cancer, autoimmunity, and diabetes (Klipp et al., 2010). Clinical researchers collect multivariate data sets of patients—genetic, transcriptional, multimodal images, proteomic data apart from the conventional clinical data pertaining to patients belonging to a specific disease group, and develop models that can uncover novel insights regarding disease progression, mechanics, and biological markers for therapeutics. Interactions at the cellular/molecular level evoke a corresponding tissue response; however, different organs comprising different tissues further interact to evoke a physiological response. Characterizing a disease condition is a challenging task mainly because the interacting components are at a multilevel hierarchical network structure and respond at multiple spatiotemporal scales, resulting in a certain physiological state (Somvanshi and Venkatesh, 2014). Furthermore, information obtained at the biomolecular level is not sufficient enough to directly make a linkage to the physiological state.

Systems biology addresses these issues by treating disease as a "system state" rather than a single defect in a gene or molecular component, thereby applying system design and analysis principles on the whole system. Designing a disease state is primarily designing an interaction network geared towards a healthy genome state (nondisease, in this case) comprising the signal input to the transcription factor that triggers a specific gene expression event which is the source of the phenotypic response. Upon developing a detailed model of the interacting network, it can be tested for perturbations that mimic mutations in certain genes and environmental abnormalities that disrupt the network behavior and lead to a disease state (del Sol et al., 2010). Monogenic diseases are a resulting occurrence of single gene, while most of the diseases are multigenic diseases, which are caused by mutation in multiple genes. Comprehensive understanding of the regulatory network structure is crucial in deciphering the cause of disease state. In case of multigenic diseases, information of protein—protein interaction network is extremely vital to uncover the cause and effect mechanisms (Zhu et al., 2007). Although, network capturing and graph-based analysis allow us to understand the interrelations and correlations between various diseases and help in the identification of disease candidates, they fail to provide us with dynamic nature of disease progression in these networks (Kann, 2007).

13.5.1 Disease Modeling

The extension of mathematical modeling/representation of a disease state from a molecular to physiological level is termed as disease modeling (Tegnér et al., 2009). Many cell signaling pathways have been modeled using Michalis—Menten and Hill equations. The dynamics of disease pathways have been successfully coded using stochastic modeling approaches such as Monte-Carlo simulations, Fokker−Plank formulations, and rule-based approaches (Somvanshi and Venkatesh, 2014). Efficient disease modeling involves detailed hierarchical network model development comprising information from protein−protein interaction network and metabolic networks, upon which a hierarchical regulation analysis can be performed to understand transcriptional, signaling, and metabolic inputs leading to a disease phenotype (Ideker and Sharan, 2008).

By conducting perturbation analysis, various possibilities of the disease dynamics of the network can be explored (Ma'ayan, 2011). Perturbation analysis involves varying system parameters, protein concentration, node sizes, edge sizes, strengths of interactions, and nodal knockouts to monitor the system-wide response. This kind of analysis helps in building a matrix of sensitive parameters that define the healthy and disease phenotype.

13.5.2 Drug Modeling

Upon derivation of sensitive parameters, a drug-based discovery analysis can be performed. After the essential and prominent interactions and participants in the network are revealed, we can optimize the network performance by tweaking the essential parameters and identify the specific target. Such kind of flux-control analysis has been applied in proteome of *Mycobacterium tuberculosis* for identification of potential drug targets (Verkhedkar et al., 2007). Based on the flux parameters, the devised drug mimics the experimental perturbation to the system and steers the state of the network in the desired direction and state. Several clinical trials are required to study the effectiveness of the drug at various levels of progression of disease. Furthermore, the output of clinical trials of the drug performance can be fed into the original network model and can be further enhanced to accurately mimic the experimental observation. This area of research where the computational efforts are geared more prominently towards the treatment of human diseases is known as *computational medicine* and is a rapidly increasing field of study.

13.6 BIOINFORMATICS TOOLS AND RESOURCES

There are various softwares that have in-built capabilities and functionalities for modeling and simulating of cellular systems described in this chapter. The methods consisting of mathematical models and numerical solutions (mentioned in Table 13.1) are built into bioinformatics softwares and resources. Some of the bioinformatics resources are mentioned in Table 13.2.

13.7 FUTURE ADVANCEMENTS

Currently much of the process for understanding and decoding of complex biological systems is an iterative and isolated process in which several software tools have to be applied at each step of the discovery process. These tools are part of a computational

TABLE 13.2   A List of Bioinformatics Resources for Systems Biology

Software	Developed By	URL
Virtual cell (Loew et al., 2010)—web application for modeling cell biological systems	Center for Cell Analysis and Modeling—University of Connecticut	http://vcell.org/
Modules available: Mathematical models, computer simulations, image-based geometries, diffusion on surfaces, and stiff spatial solver		
COPASI (Hoops et al., 2006)—Software application for simulation and analysis of biochemical networks and their dynamics	Biocomplexity Institute of Virginia Tech, the University of Heidelberg, and the University of Manchester	http://copasi.org/
Modules available: Models in SMBL standard, ODEs, Gillespie's stochastic simulation, and arbitrary discrete events		
Matlab—A numerical computing environment	Mathworks	https://www.mathworks.com/
Modules available: ODEs, PDEs, and supports multiple programming languages		
E-cell (Tomita et al., 1999)—Software for modeling, simulation, and analysis of multiscale systems like cells	Laboratory for Bioinformatics at Keio University, Japan	http://www.e-cell.org/
Modules available: Discrete and continuous models, Gillespie algorithm, ODEs, PDEs, stochastic simulator, flux rate equations, Python API, and prebuilt cell cycles, and cellular models such as Drosophila circadian clock, and dual phosphorylation cycle		
PhysioDesigner (Asai et al., 2014) 1.2—Open platform for supporting multilevel modeling of physiological system	The Systems Biology Institute	http://www.physiodesigner.org/
Modules available: Combining of mathematical models of biological and physiological functions, integration of morphometric data on model, PDEs, and multilevel modeling of physiological systems		
Cytoskape—Modeling and analysis of networks	The Cytoscape Consortium	http://www.cytoscape.org/
Modules available: Visualization of molecular interaction networks, pathways, annotation of networks, and gene expression profiling		

(Continued)

TABLE 13.2   (Continued)

Software	Developed By	URL
Garuda 1.1—An integrated, community-driven platform with several tools embedded for applications and services in drug discovery and biomedical research. A service-oriented software	The Systems Biology Institute	www.garuda-alliance.org
Modules available: Prebuilt several network drawing, algorithms and analyses tools, smart virtual assistant to guide in analysis of the data, modeling, simulation and visualization, and a host of community-provided tools		

pipeline of workflow that deal with data curation, data cleaning and standardizing, application of suitable algorithms, subsequent modeling/analysis, and finally interpretation of results. These individual steps are handled by separate bioinformatics resources for the most part. This disjointed computational workflow has been mitigated largely due to the advent of several integration platforms. However, due to the ever-increasing high-throughput next-generation data sequencing capabilities, gathering and curation of vast amounts of data, which is inherently complex in nature, poses a great computational challenge. Hence, leveraging and maximizing of computational and bioinformatics resources are imperative.

Moreover, biological and biomedical data are heavily unstructured. To understand these systems, every discipline of study has to be tapped into. The vast heterogeneity of data formats, lack of universal standardized formats for data exchange and storage make it essentially impossible to develop common interfaces around data management integration framework. Hence, the future systems would have to be built in keeping with these upcoming challenges. Several efforts for standardizing data representation and annotation such as SMBL (Hucka et al., 2003), SBGN (Le Novère et al., 2009), and MIRIAM (Le Novère et al., 2005) have already paved a way for solving these issues. Hence, an integrated software platform largely aids in maintaining and achieving a cohesive workflow in which formats, tools, and algorithms are not disparate in their approaches. Garuda-alliance developed by *The Systems Biology Institute* is an example of such a platform of community-provided tools integrated into one platform in the form of gadgets.

Another step towards solving this challenge is the implementation of crowd-sourcing methodologies. Crowd-sourcing is a collaborative and community-driven effort to solve complex problems and sharing of solutions. Designing a comprehensive in-depth model of a biological system is extremely difficult for a single group alone and hence via formulation of problems in a crowd-sourcing challenge alleviates that burden. Several such challenges, such as DREAM (Stolovitsky and Friend, 2016), Disease module identification, sbvIMPROVER (Meyer et al., 2011), and CASP (The Critical Assessment of Protein Structure Prediction) (Moult, 2005) have already gained traction and have yielded results.

13.8 CONCLUSION

Systems biology has enormous impact and applications in the fields of engineering, medicine, and drug discovery. It is a truly interdisciplinary area that thrives on the convergence of many disciplines working in cohesion. This chapter briefly deals with the manifold aspects of biological data, complexity, engineering, modeling, and therapeutics, and its applications in each arena. It also demonstrates the necessity and the effectiveness of cross-talk between several disciplines and collaborations. SMBL (Hucka et al., 2003), BioModels Database (Li et al., 2010), and DREAM challenges are successful collaborative and interdisciplinary initiatives, which have produced remarkable strides in the field of systems biology.

References

Alon, U., Surette, M.G., Barkai, N., Leibler, S., 1999. Robustness in bacterial chemotaxis. Nature. 397 (6715), 168–171.

Asai, Y., Abe, T., Oka, H., et al., 2014. A versatile platform for multilevel modeling of physiological systems: SBML-PHML hybrid modeling and simulation. Adv. Biomed. Eng. 3, 50–58. Available from: https://doi.org/10.14326/abe.3.50.

Bairoch, A., Boeckmann, B., 1991. The SWISS-PROT protein sequence data bank. Nucleic Acids Res. 19 (Suppl), 2247–2249. http://www.ncbi.nlm.nih.gov/pubmed/2041811. Accessed April 12, 2016.

Bar-Joseph, Z., Gerber, G.K., Lee, T.I., et al., 2003. Computational discovery of gene modules and regulatory networks. Nat. Biotechnol. 21 (11), 1337–1342. Available from: https://doi.org/10.1038/nbt890.

Benson, D.A., Karsch-Mizrachi, I., Lipman, D.J., Ostell, J., Wheeler, D.L., 2008. GenBank. Nucleic Acids Res. 36 (suppl 1), D25–D30.

Berman, H.M., Westbrook, J., Feng, Z., et al., 2000. The protein data bank. Nucleic Acids Res. 28 (1), 235–242. Available from: https://doi.org/10.1093/nar/28.1.235.

Bhalla, U.S., Iyengar, R., 1999. Emergent properties of networks of biological signaling pathways. Science 283 (5400), 381–387.

Blancafort, P., Segal, D.J., Barbas, C.F., 2004. Designing transcription factor architectures for drug discovery. Mol. Pharmacol. 66 (6), 1361–1371. Available from: https://doi.org/10.1124/mol.104.002758.

Brooksbank, C., Bergman, M.T., Apweiler, R., Birney, E., Thornton, J., 2014. The European bioinformatics institute's data resources 2014. Nucleic Acids Res. 42 (D1), D18–D25.

Butte, A.J., Tamayo, P., Slonim, D., Golub, T.R., Kohane, I.S., 2000. Discovering functional relationships between RNA expression and chemotherapeutic susceptibility using relevance networks. Proc. Natl. Acad. Sci. U S A. 97 (22), 12182–12186. Available from: https://doi.org/10.1073/pnas.220392197.

Cello, J., Paul, A.V., Wimmer, E., 2002. Chemical synthesis of poliovirus cDNA: generation of infectious virus in the absence of natural template. Science 297 (5583), 1016–1018. Available from: https://doi.org/10.1126/science.1072266.

Cerami, E.G., Gross, B.E., Demir, E., et al., 2011. Pathway Commons, a web resource for biological pathway data. Nucleic Acids Res. 39 (Database issue), D685–D690. Available from: https://doi.org/10.1093/nar/gkq1039.

Consortium TEP, 2004. The ENCODE (ENCyclopedia Of DNA Elements) project. Science 306 (5696), 636–640. Available from: https://doi.org/10.1126/science.1105136.

del Sol, A., Balling, R., Hood, L., Galas, D., 2010. Diseases as network perturbations. Curr. Opin. Biotechnol. 21 (4), 566–571. Available from: https://doi.org/10.1016/j.copbio.2010.07.010.

Edgar, R., Domrachev, M., Lash, A.E., 2002. Gene Expression Omnibus: NCBI gene expression and hybridization array data repository. Nucleic Acids Res. 30 (1), 207–210. Available from: https://doi.org/10.1093/nar/30.1.207.

Faith, J.J., Hayete, B., Thaden, J.T., et al., 2007. Large-scale mapping and validation of Escherichia coli transcriptional regulation from a compendium of expression profiles. PLoS Biol. 5 (1), e8.

Farber, R., Lapedes, A., Sirotkin, K., 1992. Determination of eukaryotic protein coding regions using neural networks and information theory. J. Mol. Biol. 226 (2), 471−479. http://www.ncbi.nlm.nih.gov/pubmed/1640461. Accessed April 13, 2016.

Forster, A.C., Church, G.M., 2006. Towards synthesis of a minimal cell. Mol. Syst. Biol. 2, 45. Available from: https://doi.org/10.1038/msb4100090.

Friedman, N., Linial, M., Nachman, I., Pe'er, D., 2000. Using Bayesian networks to analyze expression data. J. Comput. Biol. 7 (3−4), 601−620. Available from: https://doi.org/10.1089/106652700750050961.

Hamosh, A., Scott, A.F., Amberger, J.S., Bocchini, C.A., McKusick, V.A., 2005. Online Mendelian Inheritance in Man (OMIM), a knowledgebase of human genes and genetic disorders. Nucleic Acids Res. 33 (Database issue), D514−D517. Available from: https://doi.org/10.1093/nar/gki033.

Haugen, A.C., Kelley, R., Collins, J.B., et al., 2004. Integrating phenotypic and expression profiles to map arsenic-response networks. Genome Biol. 5 (12), R95. Available from: https://doi.org/10.1186/gb-2004-5-12-r95.

Heinemann, M., Panke, S., 2006. Synthetic biology—putting engineering into biology. Bioinformatics. 22 (22), 2790−2799. Available from: https://doi.org/10.1093/bioinformatics/btl469.

Hewett, M., Oliver, D.E., Rubin, D.L., et al., 2002. PharmGKB: the Pharmacogenetics Knowledge Base. Nucleic Acids Res. 30 (1), 163−165. Available from: https://doi.org/10.1093/nar/30.1.163.

Hoops, S., Gauges, R., Lee, C., et al., 2006. COPASI—a COmplex PAthway SImulator. Bioinformatics. 22 (24), 3067−3074. Available from: https://doi.org/10.1093/bioinformatics/btl485.

Hucka, M., Finney, A., Sauro, H.M., et al., 2003. The systems biology markup language (SBML): a medium for representation and exchange of biochemical network models. Bioinformatics 19 (4), 524−531. Available from: https://doi.org/10.1093/bioinformatics/btg015.

Ideker, T., Sharan, R., 2008. Protein networks in disease. Genome Res. 18 (4), 644−652. Available from: https://doi.org/10.1101/gr.071852.107.

Ideker, T., Galitski, T., Hood, L., 2001. A new approach to decoding life: systems biology. Annu. Rev. Genomics Hum. Genet. 2 (1), 343−372. Available from: https://doi.org/10.1146/annurev.genom.2.1.343.

Ideker, T., Winslow, L.R., Lauffenburger, A.D., 2006. Bioengineering and systems biology. Ann. Biomed. Eng. 34 (2), 257−264. Available from: https://doi.org/10.1007/s10439-005-9047-7.

iGEM Foundation. Registry of standard biological parts. http://parts.igem.org/. 2004.

Iyer, V.R., Eisen, M.B., Ross, D.T., et al., 1999. The transcriptional program in the response of human fibroblasts to serum. Science 283 (5398), 83−87.

Joyce, A.R., Palsson, B.Ø., 2006. The model organism as a system: integrating "omics" data sets. Nat. Rev. Mol. Cell. Biol. 7 (3), 198−210. Available from: https://doi.org/10.1038/nrm1857.

Kanehisa, M., Goto, S., 2000. KEGG: Kyoto Encyclopedia of Genes and Genomes. Nucleic Acids Res. 28 (1), 27−30. Available from: https://doi.org/10.1093/nar/28.1.27.

Kann, M.G., 2007. Protein interactions and disease: computational approaches to uncover the etiology of diseases. Brief Bioinform. 8 (5), 333−346. Available from: https://doi.org/10.1093/bib/bbm031.

Kitano, H., 2002. Computational systems biology. Nature. 420 (6912), 206−210. Available from: https://doi.org/10.1038/nature01254.

Klipp, E., Wade, R.C., Kummer, U., 2010. Biochemical network-based drug-target prediction. Curr. Opin. Biotechnol. 21 (4), 511−516. Available from: https://doi.org/10.1016/j.copbio.2010.05.004.

Larrañaga, P., Inza, I., Flores, J.L., 2005. A guide to the literature on inferring genetic networks by probabilistic graphical models. Data Analysis and Visualization in Genomics and Proteomics. John Wiley & Sons, Ltd, Chichester, UK, pp. 215−238, < https://doi.org/10.1002/0470094419.ch13 > .

Le Novère, N., Finney, A., Hucka, M., et al., 2005. Minimum information requested in the annotation of biochemical models (MIRIAM). Nat. Biotechnol. 23 (12), 1509−1515. Available from: https://doi.org/10.1038/nbt1156.

Le Novère, N., Hucka, M., Mi, H., et al., 2009. The systems biology graphical notation. Nat. Biotechnol. 27 (8), 735−741. Available from: https://doi.org/10.1038/nbt.1558.

Lewis, B.P., Shih, I., Jones-Rhoades, M.W., Bartel, D.P., Burge, C.B., 2003. Prediction of mammalian microRNA targets. Cell 115 (7), 787−798. http://www.ncbi.nlm.nih.gov/pubmed/14697198. Accessed April 12, 2016.

Li, C., Donizelli, M., Rodriguez, N., et al., 2010. BioModels Database: an enhanced, curated and annotated resource for published quantitative kinetic models. BMC Syst. Biol. 4 (1), 92. Available from: https://doi.org/10.1186/1752-0509-4-92.

Liang, S., Fuhrman, S., Somogyi, R., 1998. Reveal, a general reverse engineering algorithm for inference of genetic network architectures. Pac. Symp. Biocomput. 18−29. http://www.ncbi.nlm.nih.gov/pubmed/9697168. Accessed April 13, 2016.

Loew, L.M., Schaff, J.C., Slepchenko, B.M., Moraru, I.I., 2010. The virtual cell project. Systems Biomedicine. Academic Press, pp. 273–288. Available from: http://dx.doi.org/10.1016/B978-0-12-372550-9.00011-0.

Ma'ayan, A., 2011. Introduction to network analysis in systems biology. Sci. Signal. 4 (190).

Madar, A., Greenfield, A., Ostrer, H., Vanden-Eijnden, E., Bonneau, R., 2009. The Inferelator 2.0: a scalable framework for reconstruction of dynamic regulatory network models. Conf. Proc. IEEE Eng. Med. Biol. Soc. 2009, 5448–5451. Available from: https://doi.org/10.1109/IEMBS.2009.5334018.

Mailman, M.D., Feolo, M., Jin, Y., et al., 2007. The NCBI dbGaP database of genotypes and phenotypes. Nat. Genet. 39 (10), 1181. Available from: https://doi.org/10.1038/ng1007-1181.

Marbach, D., Costello, J.C., Küffner, R., et al., 2012. Wisdom of crowds for robust gene network inference. Nat. Methods. 9 (8), 796–804. Available from: https://doi.org/10.1038/nmeth.2016.

Margolin, A.A., Wang, K., Lim, W.K., Kustagi, M., Nemenman, I., Califano, A., 2006. Reverse engineering cellular networks. Nat. Protoc. 1 (2), 662–671. Available from: https://doi.org/10.1038/nprot.2006.106.

Markowetz, F., Spang, R., 2007. Inferring cellular networks—a review. BMC Bioinformatics. 8 (Suppl 6), S5. Available from: https://doi.org/10.1186/1471-2105-8-S6-S5.

Meyer, P., Alexopoulos, L.G., Bonk, T., et al., 2011. Verification of systems biology research in the age of collaborative competition. Nat. Biotechnol. 29 (9), 811–815. Available from: https://doi.org/10.1038/nbt.1968.

Mikos, A.G., Herring, S.W., Ochareon, P., et al., 2006. Engineering complex tissues. Tissue Eng. 12 (12), 3307–3339. Available from: https://doi.org/10.1089/ten.2006.12.3307.

Moult, J., 2005. A decade of CASP: progress, bottlenecks and prognosis in protein structure prediction. Curr. Opin. Struct. Biol. 15 (3 SPEC. ISS.), 285–289. Available from: https://doi.org/10.1016/j.sbi.2005.05.011.

Neves, S.R., Tsokas, P., Sarkar, A., et al., 2008. Cell shape and negative links in regulatory motifs together control spatial information flow in signaling networks. Cell. 133 (4), 666–680. Available from: https://doi.org/10.1016/j.cell.2008.04.025.

Noble, D., 1960. Cardiac action and pacemaker potentials based on the Hodgkin-Huxley equations. Nature. 188 (4749), 495–497. Available from: https://doi.org/10.1038/188495b0.

Prill, R.J., Marbach, D., Saez-Rodriguez, J., et al. Towards a rigorous assessment of systems biology models: the DREAM3 Challenges. Isalan M, ed. *PLoS One*. 2010;5(2):e9202. https://doi.org/10.1371/journal.pone.0009202.

Rajagopalan, P., Kasif, S., Murali, T.M.M., 2013. Systems biology characterization of engineered tissues. Annu. Rev. Biomed. Eng. 15 (1), 55–70. Available from: https://doi.org/10.1146/annurev-bioeng-071811-150120.

Reeves, G.T., Hrischuk, C.E., 2016. Survey of engineering models for systems biology. Comput. Biol. J. 2016, 1–12. Available from: https://doi.org/10.1155/2016/4106329.

Shannon, C.E., 1948. A mathematical theory of communication. Bell Labs Tech. J. 27, 379–423.

Sharan, R., Ideker, T., 2006. Modeling cellular machinery through biological network comparison. Nat. Biotechnol. 24 (4), 427–433. Available from: https://doi.org/10.1038/nbt1196.

Somvanshi, P.R., Venkatesh, K.V., 2014. A conceptual review on systems biology in health and diseases: from biological networks to modern therapeutics. Syst. Synth. Biol. 8 (1), 99–116. Available from: https://doi.org/10.1007/s11693-013-9125-3.

Sperger, J.M., Chen, X., Draper, J.S., et al., 2003. Gene expression patterns in human embryonic stem cells and human pluripotent germ cell tumors. Proc. Natl. Acad. Sci. U S A. 100 (23), 13350–13355. Available from: https://doi.org/10.1073/pnas.2235735100.

Sprinzak, D., Elowitz, M.B., 2005. Reconstruction of genetic circuits. Nature. 438 (7067), 443–448. Available from: https://doi.org/10.1038/nature04335.

Stemmer, W.P., Crameri, A., Ha, K.D., Brennan, T.M., Heyneker, H.L., 1995. Single-step assembly of a gene and entire plasmid from large numbers of oligodeoxyribonucleotides. Gene 164 (1), 49–53. http://www.ncbi.nlm.nih.gov/pubmed/7590320. Accessed April 12, 2016.

Stolovitzky G, Friend S. DREAM4—in silico network challenge. https://www.synapse.org/#!Synapse:syn3049712/wiki/. Published 2009. Accessed January 1, 2016.

Sun, X., Su, J., Bao, J., et al., 2012. Cytokine combination therapy prediction for bone remodeling in tissue engineering based on the intracellular signaling pathway. Biomaterials 33 (33), 8265–8276. Available from: https://doi.org/10.1016/j.biomaterials.2012.07.041.

Tegnér, J.N., Compte, A., Auffray, C., et al., 2009. Computational disease modeling—fact or fiction? BMC Syst. Biol. 3 (1), 56. Available from: https://doi.org/10.1186/1752-0509-3-56.

Tian, J., Gong, H., Sheng, N., et al., 2004. Accurate multiplex gene synthesis from programmable DNA microchips. Nature 432 (7020), 1050–1054. Available from: https://doi.org/10.1038/nature03151.

Tomita, M., Hashimoto, K., Takahashi, K., et al., 1999. E-CELL: Software environment for whole-cell simulation. Bioinformatics 15 (1), 72−84. Available from: https://doi.org/10.1093/bioinformatics/15.1.72.

Tsai, M.-S., Hwang, S.-M., Chen, K.-D., et al., 2007. Functional network analysis of the transcriptomes of mesenchymal stem cells derived from amniotic fluid, amniotic membrane, cord blood, and bone marrow. Stem Cells 25 (10), 2511−2523. Available from: https://doi.org/10.1634/stemcells.2007-0023.

Tyson, J.J., 1991. Modeling the cell division cycle: cdc2 and cyclin interactions. Proc. Natl. Acad. Sci. U S A 88 (16), 7328−7332. Available from: https://doi.org/10.1073/pnas.88.16.7328.

Verkhedkar, K.D., Raman, K., Chandra, N.R., Vishveshwara, S., 2007. Metabolome based reaction graphs of M. tuberculosis and M. leprae: a comparative network analysis. Dermitzakis E, ed. PLoS One 2 (9), e881. Available from: https://doi.org/10.1371/journal.pone.0000881.

Villaverde, A.F., Banga, J.R., Mesarović, M., et al., 2013. Reverse engineering and identification in systems biology: strategies, perspectives and challenges. J. R. Soc. Interface. 11 (91), 20130505-20130505. https://doi.org/10.1098/rsif.2013.0505.

Wang, T.-H., Lee, Y.-S., Hwang, S.-M., 2011. Transcriptome analysis of common gene expression in human mesenchymal stem cells derived from four different origins. Methods Mol. Biol. 698, 405−417. Available from: https://doi.org/10.1007/978-1-60761-999-4_29.

Wheeler, D.L., Church, D.M., Federhen, S., et al., 2003. Database resources of the National Center for Biotechnology. Nucleic Acids Res. 31 (1), 28−33.

Wolkenhauer, O., Mesarovic, M., 2005. Feedback dynamics and cell function: why systems biology is called Systems Biology. Mol. Biosyst. 1 (1), 14−16. Available from: https://doi.org/10.1039/b502088n.

Wu, S., Wells, A., Griffith, L.G., Lauffenburger, D.A., 2011. Controlling multipotent stromal cell migration by integrating "course-graining" materials and "fine-tuning" small molecules via decision tree signal-response modeling. Biomaterials. 32 (30), 7524−7531. Available from: https://doi.org/10.1016/j.biomaterials.2011.06.050.

Young, L., Dong, Q., 2004. Two-step total gene synthesis method. Nucleic Acids Res. 32 (7), e59. Available from: https://doi.org/10.1093/nar/gnh058.

Zhu, X., Gerstein, M., Snyder, M., 2007. Getting connected: analysis and principles of biological networks. Genes Dev. 21 (9), 1010−1024. Available from: https://doi.org/10.1101/gad.1528707.

PART II

ANIMAL AND MEDICAL BT

Techniques for Nucleic Acid Engineering: The Foundation of Gene Manipulation

Şükrü Tüzmen[1,2], Yasemin Baskın[3], Ayşe Feyda Nursal[4], Serpil Eraslan[5], Yağmur Esemen[1], Gizem Çalıbaşı[3], Ayşe Banu Demir[3], Duygu Abbasoğlu[6] and Candan Hızel[7]

[1]Eastern Mediterranean University (EMU), Mersin, Türkiye [2]Arizona State University (ASU), Phoenix, AZ, United States [3]Dokuz Eylul University, İzmir, Türkiye [4]Hitit University, Çorum, Türkiye [5]Bogazici University, İstanbul, Türkiye [6]Anadolu University, Eskişehir, Türkiye [7]Opti-Thera Inc., Montreal, QC, Canada

14.1 NUCLEIC ACID ISOLATION TECHNIQUES

14.1.1 Introduction

DNAs and RNAs carry the coding system of living organisms. The initial studies of Dr Friedrich Miescher in 1869 revealed a novel substance with the acidic nature, later named as nucleic acid. However, science of genetics and the innovative speed of technology took an evolutionary turn after Watson and Crick's accomplishment on revealing the structure of DNA. Nucleic acid analysis, which requires high-quality DNA and RNA as a primary material, has advanced from manual methods to a user-friendly, harmless, and economic high-throughput technologies. However, the basics of irrespective of the method or the material used have four major steps. The first step is *cell disruption (cell lysis)* mainly by osmotic imbalance; the second step is the *removal of lipids and proteins*, with the help of detergents and enzymes; in the third step, it has to be harvested from the mixture by *precipitation* using ethanol; and the final product has to be *washed and resuspended* for a long-term storage. Chelaters such as ethylenediaminetetraacetic acid (EDTA) are used to sequester divalent cations, such as Mg^{2+} and Ca^{2+}, thereby inhibiting the deoxyribonuclease (DNase), ribonuclease (RNase), and restriction enzyme activity, and preventing the degradation of DNA or RNA. The techniques are applied manually or with automation. There are three technical categories: inorganic methods, organic methods, and solid-based technology. The advancement of the latter enabled the implementation of automated high-throughput applications. The success, quality, and the quantity of an extraction depend on choosing the appropriate technique. The cell type (eukaryote and prokaryote), nucleic acid type [genomic DNA (gDNA), messenger RNA (mRNA), ribosomal RNA (rRNA), transfer RNA (tRNA), microRNA (miRNA), mitochondrial DNA (mtDNA), plasmid, virus, etc.], and material (whole blood, serum, stool, urine, hair, bone, plant, tissue, paraffin block, etc.) form the basis of important selection criteria. The "quality check" is an important step, which involves a single or combined application of agarose gel analysis, spectrophotometry, and microfluidics-based technology. These procedures will help to confirm the material of interest, the efficiency of recovery, and the purity results. It is almost impossible to get a good quality pass without following the guidelines for good laboratory practice (GLP).

14.1.2 Basics of Nucleic Acid Isolation: A Brief Introduction

The initial encounter of the scientific world with nucleic acids was after the discovery made by a Swiss doctor Friedrich Miescher, while working on the chemical composition of leukocytes in Felix Hoppe-Seyler's laboratory at Tubingen, Germany, in 1869. The research of Miescher revealed a precipitated substance with an acidic nature, containing large amounts of phosphorous and did not show any characteristics of lipids and proteins. The historical development underlies a procedure that, as basic as it may look, is the most crucial step of molecular biology research. This novel substance was called nuclein because it was discovered from the nuclei of the cells and was later called *nucleic acid*.

It is a well-known fact that nucleic acids are macromolecules that are the building blocks of living organisms. They constitute monomers called *nucleotides* that line up in a specific order to produce this polymer, and we are also familiar that nucleotides consist of three parts: a nitrogen containing base, a five carbon sugar—pentose sugar, and a phosphate group.

The sugar component lacking the hydroxyl (−OH) group in C2' is called deoxyribose and is the determinant of the name DNA. There is a hydrogen atom in the relevant position, making the macromolecule less reactive and, therefore, more stable under alkaline conditions. DNA provides a guideline of essential genetic information that is transferred from one generation to the next. This information is deciphered into proteins, which are responsible for functional and structural integrity of an organism, by another nucleic acid called RNA. All RNA molecules with different functions (mRNA, tRNA, and rRNA) constitute a ribose in their sugar—phosphate backbones, and unlike double-stranded DNA (dsDNA), they are found in single-stranded form. These properties contribute to the unstable characteristic of RNA molecules. The acidic nature of both DNA and RNA is a common feature due to the presence of negatively charged phosphate groups.

All constituents of a nucleotide define individual chemical properties of different types of nucleic acid molecules and form the basis of extraction processes as well as being the means of distinct protocols developed for DNA or RNA isolation.

A great deal of knowledge has been put forward about the chemical properties of nucleic acids, especially DNA before the world was enlightened about its structure by James D. Watson and Francis Crick in 1953. This accomplishment was a breakthrough for science of genetics boosting the progression of the technology behind it tremendously.

A better understanding of nucleic acid chemistry and structure enabled scientists to discover and implement countless different techniques, all of which are used in numerous different disciplines such as clinical genetics, forensics, environmental sciences, and systems biology. All these alternate disciplines use DNA and RNA as a primary material. The experimental procedures require the isolation of a satisfactory amount of quality material. The source of nucleic acids has a large spectrum, i.e., bacteria, plant, animal cell and tissue, human whole blood, urine, stool, bone, spinal fluid, paraffin-embedded tissues, hair, saliva, serum, and many more. The developments have led to improving the isolation procedures, so that it would be possible to work with user-friendly, harmless, and economic high-throughput technologies.

14.1.3 Components of Nucleic Acid Isolation

As has been mentioned in the Introduction section, the nucleic acid of interest and the material used for isolation can be various; however, the basic steps of isolation are irrespective of all these details. The components of extraction and purification involve four major steps: (1) disruption of the cell/tissue to expose the molecule to the outer environment; (2) removal of proteins and lipids; (3) fishing out the molecule by precipitation; and (4) washing and drying the sample before; finally, dissolving the sample under appropriate conditions. Now, let us have a look at each step in detail.

14.1.3.1 Cell Disruption (Cell Lysis)

It is conspicuous that the cell has to be disrupted so that the material within the cell can be exposed.

The natural cell lysis can occur by lytic enzymes produced during viral infections and osmotic imbalance that mainly represents itself in two different ways.

1. Cytolysis: Hypotonic environments may cause excess water to move into the cells. The cells with only cell membranes, such as animal cells, eventually swell and burst. However, cytolysis cannot occur in plant cells, bacterium, and yeast due to their strong cell walls.
2. Plasmolysis: Hypertonic environment or hot/dry weather conditions may cause the cells, with a cell wall, to lose water. This process eventually induces the cell membrane to collapse inside the cell wall resulting in gaps between the cell wall and cell membrane and lysis occurs as the cell shrivels and dies.

The procedure of cell lysis in laboratory conditions involves the use of buffers that help to mimic the natural conditions. The structure of the cell is an important aspect.

Prokaryotic and eukaryotic cells have different cell packages. The cell wall contained by all of prokaryotic organisms (bacteria), plants, and some eukaryotic single cells such as yeast, gives the cell additional strength. Therefore, it has to be broken down using stringent circumstances. During cell lysis, the main process is to break down the cell and compartment wall/membrane without damaging the nucleic acid material. There are two ways to disrupt the cell package.

14.1.3.1.1 CHEMICAL LYSIS

The major factors effective are pH, ionic strength, osmotic strength, surfactants/detergents, chaotropes, and enzymes (nucleases and proteases).

Osmotic imbalance is almost always a part of all procedures. Lysis buffers contain salts (i.e., Tris-HCl, KCl, and NaCl) to control the pH of the lysate and create an osmotic pressure, while the detergent additives (sodium dodecyl sulfate (SDS), cetyltrimethylammonium bromide (CTAB), and Triton X-100) help to destabilize the cell membrane. Chemical lysis of cells is facilitated with proteases like proteinase K, zymolase, lyticase, lysozyme, and cellulase. The characteristics of the most used chemicals are summarized in Table 14.1.

14.1.3.1.2 MECHANICAL LYSIS: METHODS INVOLVE GRINDING, SHEARING, BEATING, AND SHOCK

This method involves the use of equipment such as sonicator, homogenizer, cyro-grinders, blender, vortex, and bead disruption to facilitate cell disruption. In each case, the method of choice for the cell lysis has to be dependent on the characteristics of the cell package. Mechanical lysis is often preferred for the cells with a cell wall. In the case of plants, due to their high cellulose contents in their cell walls, more stringent force is applied by freezing the tissue in liquid nitrogen and breaking down the cell by a grinder.

The mixture after the cell lysis procedures is referred to as *lysate*; a "bimolecular soup" of the cell content. The buffer content, the choice of enzyme, and the approach to lysis differ according to the organism, material, or nucleic acid of interest.

14.1.3.2 Removal of Artifacts

In the case of nucleic acid isolation, obviously the aim is to acquire an intact group of RNAs or DNAs and discard the rest in the lysate. Monomers such as nucleotides, amino acids, metabolic, and intermediate products and their derivatives gradually clear out during the procedure. However, peptides, proteins, and lipids need special treatment to be discarded.

Lipid removal: Lipids in the cell/compartment membranes and/or cell walls are removed by adding detergents. A detergent is partly hydrophilic and partly hydrophobic surfactant that enables the interfusion with hydrophobic compounds such as lipids. The surface active property of detergents helps the compound to disintegrate the membrane and thus intracellular contents are exposed to the extracellular environment. SDS ($CH_3(CH_2)_{11}OSO_3Na$) is a highly negatively charged organic anionic surfactant, which is extensively used in molecular biology laboratories during nucleic acid extraction and protein unfolding. As much as it denatures proteins, it also removes oils from any surface that it encounters. There are many detergents used in nucleic acid isolation such as Triton X-100, NP-40, and Tween 20.

TABLE 14.1 The Names, Categories, Specific Function, and Usage of Detergents, Chaotropes, and Enzymes Used During DNA or RNA Extractions

Chemical Name	Category	Specific Function	Usage
SDS	Anionic detergent	Disrupts cell membrane denature proteins	General nucleic acid isolation
CTAB	Cationic detergent	Separates DNA from plant carbohydrates	Plant DNA isolation
Triton X-100	Nonionic detergent	Solubilizing membrane proteins in alkaline pH	General nucleic acid isolation
Sodium iodide, guanidine salts, urea	Chaotropes	Disrupts the weak interactions between proteins	General nucleic acid isolation Spin column-based nucleic acid purification
Guanidine isothiocyanate, guanidine-HCl	Chaotropes	Disrupts the weak interactions between proteins	Mostly used for RNA extraction in manual protocols
		pH-dependent partition of nucleic acids	Successive separation of RNA, DNA, and proteins
EDTA	Chelator	Chelates divalent cations required by metalloproteases	General nucleic acid isolation
2-Mercaptoethanol	Reducing agent	Linearizing proteins by disrupting tertiary and quaternary structure by breaking the S–S bonds	General nucleic acid isolation Mostly used in RNA extraction
Proteinase K	Protease	Digestion of proteins after hydrophobic amino acids inactivates nucleases	General nucleic acid isolation
Lysozyme	Protease	Splits polysaccharide chains	Bacteria (especially Gram-positive bacteria)
Cellulase	Protease	Cleaves cellulose	Plants
Chitinase	Protease	Cleaves chitinase	Filamentous fungi
Zymolyase, lyticase, and glusulase	Protease	Endoglucanase activity	Yeast and fungi
RNase inhibitor (RI)	Nuclease	Catalyzes degradation of RNA. Used during DNA extraction for pure DNA	General DNA extraction
DNase inhibitor (DI)	Nuclease	Catalyzes degradation of DNA. Used during RNA extraction for pure RNA	General RNA extraction

Deproteinization: Cellular and histone proteins are removed by adding proteases to the lysis mix. The conformation of the protein is greatly disrupted by the addition of SDS. Therefore, the protein unfolds and becomes more liable to degradation. Proteolytic enzymes such as proteinase K, lyticase, Zymolyase, glusulase catabolize proteins by peptide bond hydrolyzes. There are different types of proteases in all living organisms. Each protease has a different catalytic mechanism (Table 14.1). High salt concentrations are also a way to precipitate and thereby discard proteins.

Nuclease treatment: Although considered to be an optional step in many isolation protocols, nucleases added to the lysate prevent unwanted nucleic acid contamination in the mixture. For example, adding DNase during RNA isolation degrades all contaminant DNA molecules, leaving a pure and high-quality RNA sample and vice versa; RNase can be added during DNA isolation for a better result.

The lysis method will require a centrifugation step to be able to discard what is called "cell *debris,*" which consists parts of cell membrane/wall, insoluble molecules, precipitated peptides or proteins, and large impurities. Therefore, centrifugation is considered to be the first step toward removing the unwanted fraction.

14.1.3.3 Precipitation

Water has two hydrogen and one oxygen atom covalently bonded with an uneven distribution of electron density. The "electronegativity" of oxygen (δ^-) makes hydrogen slightly positive (δ^+). Therefore, charged compounds have the ability to dissolve readily in water. For example, when sodium chloride crystals are put in water, the electric forces are weakened and the ions form a hydration shell. The phosphate group of DNA and RNA gives them their negatively charged polar, hydrophilic characteristic. Simple chemical properties enable these molecules to be readily recovered from the lysate.

- The addition of salt (sodium chloride [NaCl], ammonium acetate [$CH_3CO_2NH_4$], sodium acetate [CH_3COONa], and potassium acetate [CH_3COOK]) increases the ionic strength of a given solution.
- DNA and RNA are insoluble in alcohol, such as ethanol and isopropanol.
- Optimized amount of salt and alcohol in the lysate leads the water molecules, surrounding the nucleic acids, to form a shell around the salt molecules and DNA or RNA immediately precipitated in the solution.
- In manual DNA isolation, the DNA precipitate is collected by the aid of a pipette or a loop. This process is called "fishing-out." If the sample is not visible, the sample is recovered by full speed centrifugation.

14.1.3.4 Washing and Resuspension

Washing usually involves submerging the precipitated sample in dilute alcohols (usually 70% ethanol in water). The water content of this solution is optimum to dissolve and discard the salt and water-soluble impurities but inadequate to dissolve the DNA or RNA. The samples are collected by centrifugation, the supernatant is discarded, and excess alcohol should be removed by air or vacuum drying. The precipitate is resuspended in water or EDTA-containing buffers to prepare them for laboratory use and storage.

All buffers prepared for nucleic acid extraction have a chelating agent, such as EDTA. The chelators sequester divalent cations such as Mg^{2+} and Ca^{2+}. These cations act as a cofactor of the enzymes such as DNase and RNase. Thus, the addition of such an agent helps prevent enzyme activity that in turn prevents degradation of DNA or RNA.

14.1.4 Principles of Methods of Extraction

There are numerous methods for nucleic acid isolation, all in which the major components, as discussed in Section 14.2, remain the same. The conventional methods involve the use of organic or inorganic substances, and they have been utilized, modified, and optimized over the years. Nucleic acid material has become the object of many disciplines, and so the materials used have also became numerous, various, and challenged the scientists in a way that the methods have become less and less handmade and more automated over the years.

14.1.4.1 Organic Methods

14.1.4.1.1 PHENOL–CHLOROFORM METHOD

Phenol (carbolic acid, C_6H_5OH), at room temperature, is an immiscible, slightly acidic, and highly hazardous crystalline substance that is less polar because its aromatic ring balances the electronegative concentration around oxygen of the hydroxyl (−OH) group. It is soluble in chloroform but almost insoluble in water. Phenol crystals are either liquefied at 45°C–50°C in oven or dissolved in chloroform or aqueous buffers.

Chloroform (trichloromethane; $CHCl_3$) is a clear colorless volatile liquid miscible with most organic solvents. It is only slightly soluble in water. It has a wide range of industrial, medical, and pharmaceutical use.

Both phenol and chloroform can form phases.

Proteins are more soluble in phenol and chloroform mixture than DNA and RNA. Addition of chloroform also helps the removal of lipids, and isoamyl alcohol is added to the phenol and chloroform to prevent foaming. The ratio of the mixture is 25:24:1, v/v. The lysate and Phenol-Chloroform Isoamyl Alcohol (PCI) mixed together in 1:1 ratio, and protein partition will remain in the organic phase, which is denser than water and stays in the lower phase. Nucleic acids, however, remain in the upper aqueous phase. The interphase is the layer in between the upper and lower phases, containing precipitated artifacts composed of proteins and a minute amount of nucleic acids mostly.

Phenol–chloroform extraction is a pH-dependent liquid–liquid organic extraction method. The phase partitioning happens according to the pH of the solution. If the aim is to extract RNA from any biological material, the pH of the solution should be slightly acidic. At pH lower than 7 (pH 4−6.5) DNA will be retained in the organic phase and RNA will stay in the aqueous phase. The addition of chloroform prevents the formation of insoluble RNA–protein complexes, reduces the cleavage of RNA by reducing the nuclease activity by precipitation, and increases the recovery of good quality RNA. High salt concentrations also help reduce endogenous RNAse activity. However, more recent in-house techniques use RIs during RNA isolation that drastically reduced the need of using salt in the extraction solution. During DNA isolation, however, the pH of the solution should be basic, increased above pH 7 (pH ~7−9), so that the partition of DNA into the upper aqueous phase is possible. The contaminants such as salts and sugars can be removed by washing the pellet with dilute ethanol after precipitation of nucleic acids using isopropanol and ethanol.

14.1.4.1.2 GUANIDINIUM THIOCYANATE−PHENOL−CHLOROFORM METHOD (COMMERCIAL NAMES: TRI, TRIZOL)

The addition of a chaotropic agent, guanidinium thiocyanate−phenol−chloroform, is often preferred in RNA extraction. Its ability to denature proteins including RNases and ribosomal proteins is much more efficient than phenol-only procedures.

14.1.4.1.3 CETYLTRIMETHYLAMMONIUM BROMIDE METHOD

This method is usually suitable for extracting nucleic acid from high polysaccharide−producing organisms such as plants and some Gram-negative bacteria. CTAB is a nonionic detergent. The procedure involves mixing the lysate with a CTAB containing high ionic strength solution in which CTAB cannot precipitate nucleic acids and form complexes with proteins. The centrifugation step allows collecting the supernatant containing the sample of interest, and the sample is precipitated with alcohol.

14.1.4.2 *Inorganic Methods*

14.1.4.2.1 SALTING-OUT METHOD

Weak chemical bonds, hydrophilic and hydrophobic interactions govern the conformation of the molecule. Hydrophobic parts do not interact with the surrounding water, therefore, they tend to fold and aggregate together, whereas the hydrophilic parts are composed of charged amino acids that interact with water. Ionic strength of the solution determines the number of water molecules surrounding both the ions and proteins. High salt concentrations tend to increase the ionic strength of the solution, and water molecules no longer surround proteins and large organic molecules because they are the least soluble in water. This process is called salting out. This technique involves the precipitation of proteins using high concentrations of salt such as sodium chloride or ammonium acetate. The centrifugation removes precipitated organic artifact. Because this procedure also prompts the precipitation of minute amounts of nucleic acids, there tends to be a material loss and sometimes the proteins cannot be totally eliminated Nasiri et al., (2005).

14.1.4.2.2 CESIUM CHLORIDE DENSITY GRADIENT METHOD

Cesium chloride (CsCl) is an inorganic, colorless, hygroscopic crystalline compound. It has a large mass and is highly soluble in water (1865 g/L). Due to its hygroscopic characteristic, when put in water, it forms a dense solute that is not very viscous. Therefore, it is a good material for equilibrium gradient differential centrifugation where the separation of the particles is size and density dependent. The method is used to separate cellular compartments (organelles), different types of DNA (viral, plasmid, mitochondrial, etc.), or RNA types (mRNA, tRNA, and rRNA).

The separation of different types of DNA usually involves the centrifugation of the DNA resuspension layered on top of CsCl gradient, at a certain speed for a decided length of time. Addition of ethidium bromide (EtBr) enhances the density difference between linear and circular/coiled DNA and also makes the bands formed after centrifugation visible under ultraviolet (UV) light.

Centrifugation forms a gradient of CsCl by forcing the sedimenting cesium ions moving to the bottom of the tube, away from the rotor. However, concentration of solvated CsCl forms a gradient and diffusion toward the center occurs. Once stabilized, density decreases from the bottom of the tube to the top. Typically the separation of nucleic acids depends not on its length (size) but mostly its molecular weight.

14.1.4.3 Solid-Based Extraction Method

The basics of this technique are based on the immobilization of nucleic acids or sample on a solid support. The polymer used as a support can be beads or columns. Silica is the most used solid-phase materials (Wolfe et al., 2002).

14.1.4.3.1 SILICA-BASED PURIFICATION

The lysate is applied onto the column using buffers containing denaturing chaotropic agents (i.e., guanidine hydrochloride and sodium iodide); detergents (i.e., Triton X-100) and isopropanol at a certain alkaline pH ($\sim$8.0); and high salt concentration enabling nucleic acid to be adsorbed to the silica solid phase. The solid phase could be silica column, silica membrane, or magnetic silica—based mini columns. The solid phase is washed with phosphate buffer and ethanol. Elution buffers have relatively low-ionic strength (lower pH $\sim$6−7).

14.1.4.3.2 MAGNETIC SEPARATION

Bead particles are surrounded with a specific nucleic acid binding functional group. NA in the lysate binds to the beads that are then placed on a magnetic area separating them from the media. The beads are collected, washed, and sample is recovered from the beads with a special buffer.

14.1.4.3.3 ANION EXCHANGE PURIFICATION

Specific positively charged substrate binds negatively charged nucleic acids in the presence of high salt, detergents, and enzymatic digestion. Elution is carried out with higher salt concentration.

14.1.4.3.4 FTA TECHNOLOGY (TRADEMARK OF GENERAL ELECTRIC COMPANY): A FAST TECHNOLOGY FOR ANALYSIS OF NUCLEIC ACIDS

It was developed for a long-term storage of sample contents from degradation. The technique applies biological samples, such as blood and saliva, containing the nucleic acid onto a cellulose-based filter paper treated with weak base, chelating agent, anionic surfactant or detergent, and urea salt. Purification solution is applied at high temperatures (95°C), and source of contamination is washed while nucleic acid remains bound to the filter paper. This technique is important for forensic analysis, identification studies and any application for that has to be carried out with a minute amount of material.

14.1.5 Automation and High-Throughput Technology

Many of the manual nucleic acid purification methods involve the use of hazardous material as well as being time-consuming, labor-intensive, and require relatively large quantities of material.

As has been mentioned several times, nucleic acid preparation is an important step for all downstream applications in molecular biology. In the forensics or anthropological genetics, for example, the sample quantities might be as little as a drop on a small piece of cloth or an ancient bone material, whereas in diagnostic genetics, for example, the number samples per day might be hundreds in various sample types. The development of solid base extraction kits had enabled scientists and technicians to do the applications more easily and in a standardized manner for even small sample quantities. However, as the knowledge on genetics increased and the chemistry behind the nucleic acid extraction methods improved, the technology also needed to be improved for a less hands-on, non-hazardous, more standardized, reproducible, and user-friendly high-throughput systems.

Automation of nucleic acid extraction mainly uses robotics with enhanced solid-phase extraction systems. These are "hands-free" systems where you prepare the equipment, the disposables, the kit units, and the samples as described in the user manual. They allow standardized material production, decrease the margin of error, and able to work with up to 96 samples per run.

14.1.6 Choosing the Method

14.1.6.1 Genomic DNA Isolation

DNA is a widely used material for most molecular biology applications such as Southern blotting, polymerase chain reaction (PCR), fingerprinting, restriction digestion, and sequencing. The source of material can vary immensely because DNA is a stable material even under most unsuitable conditions. Section 14.3 gives an overview to the methods applicable for gDNA extraction. However, each method has their advantages and disadvantages that should be considered before any application (Table 14.2).

14.1.6.2 RNA Isolation

High-quality RNA is crucial in the studies involving real-time PCR (reverse transcriptase–quantitative polymerase chain reaction, RT-qPCR), digital PCR, microarray and next-generation sequencing, Northern blot analysis, and complementary DNA (cDNA) library construction. RNA is not a stable molecule like DNA. Therefore, it is always necessary to reverse transcribe the material to cDNA. It is also a condition and time-dependent dynamic molecule. Quantitative analyses of RNA require the samples to be captured at a certain experimental stage at a given time. Cryogenic applications such as using liquid nitrogen or RNA Stabilization Solutions are used to preserve RNA within the cell/tissue. The choice of lysis buffer depends on the cell type. The extraction procedures involve organic extraction methods using strong denaturants like guanidine salts, SDS; spin column methods using glass fiber, silica, or ion exchange membranes; and magnetic beads (0.5–1 μm) with a paramagnetic core and surrounding shell binding the RNA. Table 14.3 shows the list of processes and their comparison for total RNA isolation.

TABLE 14.2 Advantages and Disadvantages of Different Techniques for GDNA Isolation[a]

Technique	Principle	Advantages	Disadvantages
Salting-out procedures	Requires buffers with high ionic strength	• Cost-effective • Modified procedures are time effective	• Time-consuming, not suitable for high throughput • Requires the use of high salt concentrations
Phenol and chaotropic procedures	Chaotropic agents denature proteins Phenol solubilizes and extracts proteins and lipids into an organic phase and DNA in the aqueous phase	• Cost-effective	• Time-consuming, not suitable for high throughput • Requires the use of caustic/toxic organic solvents
CTAB procedure	Cells are lysed enzymatically and the cellular components are separated based on their solubility differences	• Cost-effective • Does not require the use of hazardous phenol/chloroform	• Time-consuming Requires handling and disposal of caustic/toxic organic solvents • Not amenable to high throughput
Silica-based and magnetic bead	Utilization of chaotropes proteolytic digestion in combination with silica purification	• User-friendly • Eluted DNA is ready to use • High throughput	• Poor yield • Irreversibly binding if underloaded or over dried
Anion exchange chromatography	Utilization of detergent and enzymatic digestion in combination with anion exchange purification	• Scalable	• The eluted product has to be desalted or precipitated due to very high salt concentration
Whatman FTA technology	Spot sampling onto a chemically treated, dry filter paper DNA elution is simple, contaminants are washed away and gDNA remains on the matrix	• Suitable for a long-term storage • Storage conditions do not require expensive freezers • DNA can directly be used in PCR	• Spotted samples must have been prepared and thoroughly dried before use

[a]Specific DNA isolation approaches are mentioned in Sections 14.5.3–14.5.7.
In reference to and slightly modified from Nucleic Acid Sample Preparation for Downstream Analyses; Principles and Methods Handbook. GE Healthcare Life Sciences.

14.1.6.3 Plasmid DNA Isolation

Bacterium consists of circular double-stranded extra chromosomal DNA that can replicate independently. These DNA units are called "plasmid" varying between 1 and 1000 kbp. The general principles of isolation are based on its covalently closed circular (supercoiled) conformation, which is very stable under stringent alkali conditions (pH 12.0–12.5). High pH denatures protein and gDNA leaving the plasmid DNA intact. The volume and the growth phase of bacterial cultures are determined by the desired amount of plasmid to be isolated. The cells are harvested by centrifugation and resuspended in EDTA containing buffer to prevent nuclease activity. The sample is incubated

TABLE 14.3   Advantages and Disadvantages of Different Techniques for Total RNA Isolation

Technique	Principle	Advantages	Disadvantages
Guanidinium— phenol/chloroform	Chaotropes, i.e., guanidium salts, lyse cells and inactivate RNases. Proteins are denatured and removed by phenol/chloroform extraction	• Cost effective • Enhanced protection against RNases • Eluted DNA is ready to use	• Time-consuming • Not suitable for high throughput • Requires the use of caustic/ toxic organic solvents
Guanidinium density gradient centrifugation	Chaotropes, i.e., guanidium salts, lyse cells and inactivate RNases. RNA moves in gradient-dependent manner in CsCl and separated from proteins and DNA	• High-purity • No caustic/toxic organic solvents • No precipitation with alcohol	• Requires expensive high-speed centrifuge and overnight centrifugation
Silica purification	Chaotropes, i.e., guanidium salts, lyse cells, and inactivate RNases. Nucleic acids bind to silica and RNA is eluted with low salt concentration buffers	• Easy to work with and fast • No caustic/toxic organic solvents • Reproducible/high throughput	• Yield is low • Requires the use of nucleases (DNase)
Magnetic purification	Nucleic acids bind to magnetic beads at low pH and eluted at high pH	• Easy to work with and fast • No caustic/toxic organic solvents • Reproducible/high throughput	• Requires the use of nucleases (DNase)
Phenol and chaotrope procedures	Chaotropic agents denature proteins Phenol solubilizes and extracts proteins and lipids into an organic phase and DNA in the aqueous phase	• Cost-effective	• Time-consuming, not suitable for high throughput • Requires the use of caustic/ toxic organic solvents
Salting-out procedures	Requires buffers with high ionic strength	• Cost-effective • Modified procedures are time-effective	• Time-consuming, not suitable for high throughput • Requires the use of high salt concentrations

In reference to and slightly modified from Nucleic Acid Sample Preparation for Downstream Analyses; Principles and Methods Handbook. GE Healthcare Life Sciences.

in the lysis buffer containing a high concentration of SDS adjusted to high pH. The material is neutralized, chromosomal DNA and protein are salted out, and supernatant phase containing the plasmids are subjected to ethanol precipitation.

14.1.6.4 Mitochondrial DNA Isolation

Many commercial kits are available for quick isolation of mtDNA. However, conventional methods involve mechanical and/or chemical cell disruption followed by organic extraction.

14.1.6.5 Viral DNA/RNA Isolation

Viruses can only replicate when they are inside a living cell of a plant, animal, human, or bacteria. Their genetic material is either DNA or RNA. They are usually abandoned in plasma, serum, cell-free body fluids (urine, cerebrospinal fluid, spinal fluid, etc.) of human and animal, plant tissues, and many other biological materials. Viral DNA and RNA extraction involve very sensitive methods because the circulating nucleic acids might have a low copy number. Conventional techniques using salt or phenol—chloroform are not preferred because kit technology and automation have made it possible to work with small amounts of material with low viral load.

14.1.6.6 Plant DNA/RNA and Chloroplast DNA Isolation

Cell lysis is the most important part of this application because polysaccharides that hinder restriction and DNA modifying enzyme activity might contaminate the plant DNA. If possible the tissue from a young plant should be preferred. The homogenization of this tissue can be done mechanically using, for example, homogenizers and liquid nitrogen. Among many other methods, CTAB is the most used. Plant chloroplast DNA has the same basics of mtDNA isolation.

14.1.6.7 DNA/RNA Isolation From Other Materials

14.1.6.7.1 PARAFFIN BLOCKS

Xylene is added to remove paraffin. It is washed down with ethanol and the tissue sample is collected, incubated, and lysed in enzyme containing buffer followed by the phenol—chloroform method.

14.1.6.7.2 ANCIENT BONE DNA ISOLATION

Initial method is to grind and powder the bone and incubate it with EDTA, sarcosyl, and enzyme-containing buffers and continue with a phenol—chloroform extraction with ethanol precipitation. Glass-milk, silica suspension, and boiling the powder in Chelex suspension could be alternative procedures.

14.1.7 Quality Control

The succession of any molecular biology application depends immensely on the utilization of sufficient amount of quality material. The primary focus for efficient extraction yielding sufficient amount of good quality material will depend largely on choosing the appropriate methodology for the material of interest. Once this initial necessity has been fulfilled, following step becomes crucial to check the recovery efficiency and the integrity of the material before starting any downstream application.

14.1.7.1 Agarose Gel Electrophoresis Control

Probably, this is the oldest and the still most used method to check the relative quantity and integrity of the nucleic acids. The agarose concentration and the content of the gel depend on the material to be analyzed.

14.1.7.1.1 CHECKING GDNA

The genome sizes could vary from about hundreds of thousands to trillions of base pairs (106–1011 bp) depending on the organism. Therefore, the concentration of the agarose gels should be efficient for the mobility of large molecules. Usually, 0.5%–1% agarose with EtBr is widely used to check the quality of the gDNA sample. The DNA bands should be sharp and clear without any smears. A smear is the sign of degradation.

14.1.7.1.2 CHECKING TOTAL RNA

Because the molecule sizes are much smaller, usually more concentrated gels (1.5%–2%) are prepared. RNA is a single-stranded molecule and form secondary structures. Denaturing gels stained with EtBr help to unravel the molecule enabling migration according to their size. Ribosomal RNA bands are the most abundant RNA species. The electrophoresis of total RNA is expected to give two sharp bands of rRNAs without any smearing. The detectable rRNAs are 28S (5070 nt) and 18S (1869 nt) rRNAs in eukaryotes and 23S (2906 nt) and 16S rRNAs (1542 nt) in prokaryotes.

The size markers during DNA or RNA agarose runs are always used to ensure that the electrophoresis was done properly. Agarose gels are the most economical way for the visual inspection of integrity. Theoretically the amount of DNA or RNA can also be calculated against a control sample of known concentration. However, this method will not give an accurate result. Another drawback is the amount of sample loaded on EtBr-stained gels. The big size and double-stranded nature of gDNA makes them easier to visualize, even with 1 μL loading volume. However, the visualization of RNA requires larger volumes. Pyronin Y, SYBR Gold, and SYBRï Green II RNA gel stain (Molecular Probes, Inc., trademark of Life Sciences) are less hazardous and very effective for little amounts (>1 ng) of RNA samples.

14.1.7.2 Spectrophotometry

Spectrophotometric measurement requires the calculation of the amount of light absorbed by the sample by measuring its intensity. It is a well-known fact that nucleic acids best absorb UV light at 260 nm. Optical density (OD) at 260 nm equals to 1 for a 1-cm path length for 50 μg/mL of dsDNA, 33 μg/mL of ssDNA, and 40 μg/mL of RNA. The concentration of the sample is calculated by:

$$6 = \begin{array}{l} \text{DNA concentration}(\mu g/mL) = (OD\ 260) \times (\text{dilution factor}) \times (50\ \mu g\ DNA/mL)/(1\ OD\ 260\ unit) \\ \text{RNA concentration}(\mu g/mL) = (OD\ 260) \times (\text{dilution factor}) \times (40\ \mu g\ RNA/mL)/(1\ O\ D\ 260\ unit) \end{array}$$

The protein best absorb UV light at 280 nm. The ratio of calculated OD for 260 and 280 nm (OD260/OD280) gives the value for possible protein contamination. The value for pure DNA is ~1.8 and pure RNA is ~2.0, and values lower than these may indicate protein contamination. There are of course other contaminants. Phenol absorbs at 270 nm, which may overlap with the 260 nm reading at times. Organic compounds such as phenolate ion and thiocyanates absorb at 230 nm. Each spectrophotometric measurement should be read against a blank which usually is the solution used for diluting the nucleic acid (water, Tris-EDTA buffer, etc.). Standard cuvette spectrophotometers require larger volumes of diluted samples (10−50 μL sample diluted in 500−1000 μL) and the use of expensive quartz cuvette.

New devices have been developed to hold a sample in place using surface tension. These devices enable the measurements with only 1−2 μL of sample without the need for dilution.

14.1.7.3 *Microfluidics-Based High-Throughput Technology*

This technology is an alternative to agarose gel and spectrophotometer technology. There are microchannels on the chip on which the external standard (ladder), gel polymer mixed with a fluorescent dye, and samples are loaded. Electrodes that are designed to fit each well are connected to their substantive electrodes and once the system begins to circuit the molecules are separated according to the basics of conventional gel electrophoresis. The dye molecules are detected by the system detected by laser-induced fluorescence, and the data are transferred to bands and peaks. The systems have both DNA and RNA chips, and it usually requires a small amount of sample (>1 μL).

Both DNA and RNA concentrations are calculated by comparing the peak areas of sample with the peak areas of the ladder with a known concentration.

RNA integrity number estimation is determined by the whole electrophoretic pattern of the total RNA together with any trace of degradation and including the ratio of the rRNA bands. The integrity numbers from 1 to 10, 1 marking completely degraded samples and 10 marking for a perfect sample, were determined by the RNA states of an RNA integrity database.

If the sample image or data result is not good enough, then this is an indication of poor sample integrity or quantity. In that case, a checklist should be examined before repeating the procedure with a new material (section 7.1, Pg 131).

14.1.8 GLP for Nucleic Acid Isolation

GLP as defined by Organization for Economic Co-operation and Development (OECD) in the report named "Principles on Good Laboratory Practice" is: "Good Laboratory Practice (GLP) is a quality system concerned with the organizational process and the conditions under which nonclinical health and environmental safety studies are planned, performed, monitored, recorded, archived and reported."

The implementation of a set of rules involves meeting the needs of physical requirements; management and personnel skills and responsibilities; system facilities, apparatus, material, and reagents; sample receipt, handling, sampling and storage; procedures; internal external quality control systems; and recording, reporting, and storage of data and material.

14.1.8.1 *A Room of Its Own*

A molecular biology laboratory set-up should have an appropriate independent space for nucleic acid isolation. This is especially important to prevent contamination of crude sample with other nucleic acid material, amplicon, nucleases, and other contaminants.

This separate space can be a clean room, and if viral studies are carried out, a laminar flow should also be in the room. Protective lab-ware should always be worn before the entrance to the room. All disposables, buffers, kits, and consumables should be stocked within this room. All equipment necessary for the extraction process should be accommodated in the room, and their maintenance procedures and calibration set-ups should be followed according to their manual.

14.1.8.2 Sampling

As we have already mentioned, samples can vary immensely especially for DNA analysis. It is crucial that:

- recording to be done carefully to avoid sample mix-up;
- all samples should be considered hazardous and a potential source of infection and handled with care;
- the fresh samples (cultured cells, tissue, blood, etc.) should be collected into an appropriate tube and medium and kept at their optimum temperature conditions prior to extraction;
- RNA sampling may require immediate temperature shock (liquid nitrogen, ice treatment, and cold chain transport)

14.1.8.3 Extraction

Disposables should be preferred to avoid contamination. All tubes, tips, buffers, enzymes, beads, columns, etc., used during the procedures should be DNase/RNase-free. Prior to each procedure, the working space should be prepared. The extraction procedures should be followed according to the protocols. Labeling should be handled with care.

- DNA: Clean the area and always use sterile buffers.
- RNA: All benches, instruments, and pipettes should be cleaned with special solutions to remove RNase activity. Diethylpyrocarbonate (DEPC) is the most used protective solution. Water-treated DEPC (usually 0.1% v/v) is autoclaved and used to wash all surfaces, utensils, and prepare buffers. It is an unstable compound, and by-products are capable of inhibiting downstream processes. There are commercially available RIs containing detergents. These solutions are easy to use, stable, and do not interfere with reactions after isolation.
- Infectious material: Viruses, bacteria, fungi, etc., are potential infectious samples. The working space should always be cleaned with appropriate solutions (bleach, sterilization solutions, etc.), and UV light should be turned on for at least half an hour after the procedures. All tips, tubes, gloves, liquids, etc., should be disposed inside a sterilization solution contained in the disposal bins. The room or laminar flow should meet the safety requirements according to the material being handled.
- Low sample volume: In Section 14.1.4.3.4, we have discussed an efficient, stable, and easy to store FTA Technology. Techniques involving low sample volume need to immobilize the sample onto a pretreated surface. Cheek-swab materials are also a good example.

14.1.8.4 Post Processes

After extraction the quality check should be carried out as soon as possible. If DNA material is going to be used the same or the next day, it is appropriate to keep it in 2°C–8°C conditions. For monthly storages −20°C and for longer storages −80°C should be preferred. RNA should always be handled with care. Separate storage space should be preferred whenever possible. Frequent freeze–thaw should be avoided if possible.

14.1.9 Conclusion and Future Perspectives

Nucleic acid isolation is the most crucial step of all molecular biology studies. It is the first step in providing the material suitable to carry out the rest of the experimental procedures. Therefore, each check point in DNA/RNA isolation should be handled with care. Some manual procedures are still in use due to their economic and relatively less hazardous features. Salting-out method, for example, is still a method of choice in the laboratories where the material is suitable. However, most laboratories, whether involved in research or routine analyses, chose automated systems. These systems are suitable for increased throughput. They are less laborious and safer to work with, and the sample quality is high due to kit-dependent operation systems.

The development of new robotic systems has enhanced the nucleic acid extraction procedures. However, as to every innovation, these systems also need improvement.

- All these robotic systems need a pre-preparation step for the setup of all the buffers, disposables, and the pretreated samples.
- The basic models are of modest size, but the high-throughput models take up a large space.
- In laboratories where there is a large number of multisample flow, there is a great need to do the extractions simultaneously.

Overcoming these obstacles will help to improve the robotic extraction and purification systems, and in the future it will be more thinking, less hand-on work involved in the scientific arena.

14.2 RESTRICTION ENZYME TECHNIQUES

14.2.1 The Discovery of Restriction Enzymes (Endonucleases)

The Nobel Prize for Physiology or Medicine was bestowed upon Werner Arbor, Dan Nathans, and Hamilton Smith for the year 1978 for their roles in the discovery of restriction enzymes. In reality, the discovery of restriction enzymes was an accident. The Nobel laureates were performing experiments to investigate why certain strains of bacteria could not be infected by bacteriophages. They were working with phages, which could not infect certain strains of bacteria. During their studies, they noticed that the bacteriophage DNA was cut into small pieces by some sort of an enzymatic action. They then chose to name this process as host-controlled restriction (Dussoix and Arber, 1962; Lederberg and Meselson, 1964; Meselson and Yuan, 1968; Arber and Linn, 1969).

14.2.2 Where Are They Found?

Restriction enzymes (endonucleases) are "Molecular scissors" that are found in and harvested from bacteria and archaea, which cut DNA strands at predetermined locations on DNA. The importance of these enzymes resides in the fact that they do not cut a strand of DNA arbitrarily. Preferably, they recognize a unique pattern of bases and then excise only at that specific site. Bacteria and archaea use restriction enzymes as part of their defense mechanisms to cut up the invading DNA of bacteriophages (Arber and Linn, 1969; Krüger and Bickle, 1983). Bacteriophages can be described as viruses that infect bacteria and use bacterial DNA to reproduce their own. As part of bacterial defense system, restriction enzymes cut (digest) any foreign DNA they encounter, by binding at their unique recognition sites. DNA that has been cut with these enzymes becomes inoperative (Ausubel et al., 1998; New England Biolabs Inc., 2014). Meanwhile, the host DNA is safeguarded by a modification enzyme (a methylase), which alter the bacterial DNA, and prevent it from being cleaved. Considering that the nucleotide sequences recognized by the restriction enzymes are very short, usually varying between 4 and 8 nucleotides, the bacterium itself undoubtedly has plenty of these nucleotide sequences existing in its own DNA. Consequently, to stop destruction of its own DNA by its own restriction enzymes, the bacterium label its own DNA by attaching methyl groups (CH_3) to its cytosine or adenine nucleotides (Capuano et al., 2014). This alteration should not obstruct the DNA base-pairing, for that reason, generally only a few specific nucleotides are modified on each DNA strand (Kobayashi, 2001). There exists over 3000 restriction enzymes that are studied in detail thus far, and more than 600 of these are commercially available (Roberts et al., 2007). Each enzyme is named after the bacterium from which the enzyme is isolated. The number that comes after the letters represents the order in which the enzyme is isolated. For example, EcoR1 restriction endonuclease was the first restriction enzyme to be isolated from the bacteria *Escherichia coli* (Table 14.4). Similarly, HindIII is the third enzyme isolated from *Haemophilus influenza* bacteria (Smith and Nathans, 1973; Ausubel et al., 1998).

14.2.3 Mechanism of Their Action

Restriction enzymes (endonucleases), with their three-dimensional structure, fit perfectly into the groove formed by the two antiparallel DNA strands. They attach themselves to the DNA and move along the double helix. Restriction enzymes glide along the DNA

TABLE 14.4   Derivation of EcoRI Name (Smith and Nathans, 1973)

Abbreviation	Meaning	Description
E	*Escherichia*	Genus
co	*coli*	Specific epithet
R	RY13	Strain
I	First identified	Order of identification
		In the bacterium

until they reach their recognition sites. These specific sequences of base pairs prompt the restriction enzyme to stop sliding. To excise DNA, all restriction endonucleases make two nicks, one at each sugar—phosphate backbone, namely at each strand of the DNA double helix (Roberts, 1976; Kessler and Manta, 1990; Pingoud et al., 1993). After excision, the enzymes chemically detach themselves from the DNA base pairs at that designated recognition site (Ausubel et al., 1998).

The recognition sequences usually differ between 4 and 8 nucleotides, and many of them are palindromic, namely the nucleotide sequences read the same forwards and backwards (Pingoud and Jeltsch, 2001). Theoretically, there seems to be two types of palindromic sequences that are possible in DNA: the inverted repeat palindromes and mirror-like palindromes. The former is a sequence, which reads the same forwards and backwards, where these sequences are found in cDNA strands, such as the ones found in a dsDNA (i.e., as in CTATAG being complementary to GATATC). Inverted repeat palindromes are found more frequently than the mirror-like palindromes and tend to have more significance in biological terms than mirror-like palindromes. On the other hand, the mirror-like palindromes are identical to those found in ordinary text, where the sequence reads exactly the same as forwards and backwards on an ssDNA as in CGAAGC (Clark, 2005).

14.2.4 Classification of Restriction Enzyme Types

Naturally existing restriction enzymes are frequently grouped into four types, Types I, II, III, and IV, based on their configuration and cofactor needs, their target sequences, and the location of their DNA excision sites in connection with the target sequence (Bickle and Krüger, 1993; Boyer, 1971; Yuan, 1981). These vary in their structure and specific recognition sites (restriction sites) they identify and cleave on their DNA substrate.

Type I enzymes (http://enzyme.expasy.org/EC/3.1.21.3) cut at sites remote from recognition site; require both ATP and S-adenosyl-L-methionine to function; are multifunctional proteins with both restriction and methylase activities.

Type II enzymes (http://enzyme.expasy.org/EC/3.1.21.4) cut within or at short specific distances from recognition site; mostly require magnesium; are single function (restriction) enzymes independent of methylase.

Type III enzymes (http://enzyme.expasy.org/EC/3.1.21.5) cut at sites a short distance from recognition site; require ATP (but do not hydrolyze it); are S-adenosyl-L-methionine stimulates reaction but is not required; exist as part of a complex with a modification methylase.

Type IV enzymes select modified DNA, e.g., methylated, hydroxymethylated and glucosyl—hydroxymethylated DNA.

Type I restriction enzymes bind to the particular recognition site on the DNA, excising arbitrarily along the length of the DNA molecule. Type II restriction enzymes bind at a recognition site, where they cut the DNA molecule by snipping the DNA backbones at certain predetermined regions within these specifically recognized sequence of nucleotides (http://www.accessexcellence.org/AE/AEC/CC/enzyme_glossary.php). Each restriction enzyme has a single, specific recognition sequence and cuts the DNA molecule at a particular site. Thus, treatment of specific DNA molecules with a particular restriction enzyme will always produce the same set of DNA fragments. Recombinant DNA (rDNA) technology benefits extensively form Type II restriction enzymes utilization. Type III restriction enzymes identify two separate nonpalindromic DNA sequences, which are inversely positioned, and cut DNA nucleotides about 20–30 bp following the recognition site (Dryden et al., 2001). These enzymes contain more than one subunit and need S-Adenosyl methionine (AdoMet), a common cosubstrate involved in methyl group transfers, and ATP cofactors for their function in DNA methylation and restriction, respectively (Meisel et al., 1992). *Type IV* restriction enzymes recognize modified, particularly methylated DNA, and are exemplified by the McrBC (McrBC is an endonuclease that cuts DNA harboring methylcytosine on one or both strands) systems of *E. coli*. (http://www.neb.com/products/restriction-endonucleases/restriction-endonucleases/types-of-restriction-endonucleases).

Different restriction enzymes, which recognize the same sequence but cut in different positions in the sequence, are referred to as neoschizomers. Whereas, different enzymes that recognize and cleave in the same location of the same sequence are known as isoschizomers.

If the cut produces overhanging pieces of ssDNA (sticky ends), these enzymes are called sticky-end cutters, yielding differences in sequence, length, and the orientation of the strand (5′ end or 3′ end) of a sticky-end "overhang" of an enzyme restriction site (Goodsell, 2002). These overhanging pieces are capable of base pairing with any DNA molecule that contains complementary sticky ends. These two sticky ends can be joined back together again in vitro by ligation performed by DNA ligase enzymes. Otherwise, if the cut produces non-overhanging pieces of dsDNA (blunt ends), then these enzymes are referred to as blunt-end cutters. These blunt-ended pieces are capable of base pairing with any DNA molecule regardless of the necessity for complementary ends. These two blunt ends then can be joined back together again in vitro by ligation performed by DNA ligase enzymes.

14.2.4.1 *Infidelity Among the Restriction Enzymes*

While most restriction endonucleases cut their unique recognition sequences with great fidelity, some are infamous for their tendency to excise at secondary sites, which are closely related to their recognition sites. One of the earliest examinations of this indiscriminate behavior was observed for EcoRI (cognate recognition site: GAATTC), which was found to cut at sites differing from this "proper" site by one base (Polisky et al., 1975). It was observed that under certain buffer conditions, such as low ionic strength, low Mg^{2+} or the presence of organic solvents, sites such as NAATTC, GNATTC, or GANTTC could be excised, though at reduced efficiency. This undesirable cleavage became known as *star activity* and has caused burden to the researchers looking for faithful cleavage ever since. Additionally, a number of other enzymes also have been shown to exhibit star activity. These include BamHI, PvuII, and EcoRV (Robinson and Sligar, 1995). The conditions leading to this additional cleavage are now well documented. A brief look at most catalogs containing restriction endonucleases indicate that a rather large number of enzymes display this unwanted response. Luckily, under optimal buffer conditions, which can change considerably from one enzyme to another, star activity can be eliminated or at least greatly minimized in many instances (Tables 14.5 and 14.6).

14.2.5 Examples of Restriction Enzymes

14.2.5.1 Sticky-End (Cohesive) Cutters
See Table 14.5.

14.2.5.2 Blunt-End Cutters
See Table 14.6.

TABLE 14.5   Restriction Enzymes That Create Sticky (Cohesive) Ends

Restriction Enzyme	Source	Recognition Sequence	Cut Site
BamHI	*Bacillus amyloliquefaciens*	5′...G↓GATCC...3′	5′...G GATCC...3′
		3′...CCTAG↑G...5′	3′...CCTAG G...5′
HindIII	*Haemophilus influenzae*	5′...A↓AGCTT...3′	5′...A AGCTT...3′
		3′...TTCGA↑A...5′	3′...TTCGA A...5′
EcoRI	*Escherichia coli*	5′...G↓AATTC...3′	5′...G AATTC...3′
		3′...CTTAA↑G...5′	3′...CTTAA G...5′
HpaII	*Haemophilus parainfluenzae*	5′...C↓CGG...3′	5′...C CGG...3′
		3′...GGC↑C...5′	3′...GGC C...5′

TABLE 14.6   Restriction Enzymes That Create Blunt Ends

Restriction Enzyme	Source	Recognition Sequence	Cut Site
HaeIII	*Haemophilus aegyptius*	5′...GG↓CC...3′	5′...GG CC...3′
		3′...CC↑GG...5′	3′...CC GG...5′
AluI	*Arthrobacter luteus*	5′...AG↓CT...3′	5′...AG CT...3′
		3′...TC↑GA...5′	3′...TC GA...5′
SmaI	*Serratia marcescens*	5′...CCC↓GGG...3′	5′...CAG CTG...3′
		3′...GGG↑CCC...5′	3′...GTC GAC...5′
PvuII	Proteus vulgaris	5′...CAG↓CTG...3′	5′...CAG CTG...3′
		3′...GTC↑GAC...5′	3′...GTC GAC...5′

14.2.5.3 *Utilization of Restriction Enzymes in Biotechnology*

The use of restriction enzymes allowed scientists to produce rDNA molecules. The ability of these enzymes revolutionized the study of genetics and unveiled a new field of biotechnology. Everything from PCR-based cloning to DNA fingerprinting benefit from restriction enzymes as either a primary or secondary methodology of confirmation and processing (Geerlof, 2008; Russell and Sambrook, 2001).

14.2.5.4 *Detecting SNPs (PCR Restriction Fragment Length Polymorphism)*

Studying single nucleotide polymorphisms (SNPs) are essential for understanding and diagnosing malignancies and also for improving forensic science methodologies. Many SNP genotyping methods are available; however, they are mostly expensive. Using restriction enzymes for SNP genotyping has been found to be a cost-effective method. PCR-restriction fragment length polymorphism (PCR-RFLP) is a technique in which the analysis of the patterns derived from cleavage of the PCR-amplified DNA pieces enables scientists to distinguish between individual SNPs. If a SNP is found inside the restriction enzyme recognition sites (4−8 bp in length), this would affect the enzyme's ability to digest the DNA efficiently. Analyzing the cut fragments via gel electrophoresis enables scientists to detect the presence of SNPs located in the restriction enzyme recognition sites (Chuang et al., 2008; Zhang et al., 2005).

14.2.5.5 *Gene Cloning*

Restriction endonucleases can be employed to engineer DNA for different scientific approaches. They can be used to facilitate insertion of genes into vectors (i.e., plasmids, bacterial artificial chromosome (BAC)) for gene cloning and protein expression studies. For optimal utilization, plasmids, which are commonly used for gene cloning, are modified to include a short polylinker sequences (i.e., multiple cloning sites, or MCS) rich in restriction enzyme recognition sequences. Nowadays, these cloning vectors are readily available from the commercial vendors (i.e., Life Technologies, Clontech Laboratories Inc., Origene Technologies). This enables flexibility when gene fragments are inserted into the plasmid vector. Restriction enzyme sites that are naturally found within genes affect the choice of endonucleases for digesting the DNA during cloning experiments, because it is necessary to avoid restriction of wanted DNA while intentionally cutting the ends of the DNA. To clone a gene fragment into a vector, both gene insert and plasmid DNA are typically cut with the same restriction enzymes, and then joined together with the assistance of an enzyme known as a DNA ligase (Geerlof, 2008; Russell and Sambrook, 2001). Digesting DNA with a restriction enzyme creates a linear piece of DNA with sticky or blunt ends. Thus, DNA containing a specific restriction enzyme recognition site from any source can be excised and ligated in vitro to another DNA molecule, which has already been cut with the same restriction enzyme. This is possible because the free ends created after using the same restriction enzyme are complementary to each other. This new cloned (recombinant) DNA molecule can be transformed/transfected/trunsducted into a host cell for further investigation (Miles and Wolf, 1989).

14.2.5.6 DNA Footprinting

Restriction enzymes come in handy in protocols such as DNA footprinting as they allow the digestion of the DNA, where proteins of interest are not bound to facilitate the identification of the location of a protein, where the proteins bind. DNA footprinting is a technique that aids in investigation of the sequence specificity of DNA-binding proteins. This method facilitates the study of protein–DNA interactions both within and outside cells. The transcriptional regulation has been studied extensively, and yet there is still so much to be researched. Transcription factors and their associated proteins, which bind promoters, enhancers, and silencers to activate or silence transcription, are the bases of understanding the unique regulation of individual genes within the genome. Methods like DNA footprinting will help explain as to which proteins bind to these regions of DNA and resolve the complexities of transcriptional control (Galas and Schmitz, 1978). (http://www.scientificlib.com/en/Biology/Molecular/DNAFootprinting.html) (Fig. 14.1).

14.2.5.7 Artificial Restriction Enzymes

These restriction enzymes are generated by linking the *Fok*I DNA cleavage domain with an assortment of DNA binding proteins, designated as zinc finger nucleases (ZFN). Artificial restriction enzymes are a powerful tool for host genome editing, owing to their enhanced sequence specificity. ZFN work in pairs. Their dimerization is facilitated in situ through the *Fok*I domain. Each zinc finger array is capable of identifying 9–12 bp, making 18–24 for the pair. A 5–7 bp spacer between the recognition sites further increases the specificity of ZFN. This enables them to become a safe and more precise tool, which can be applied in humans. A recent Phase I clinical trial of ZFN for the targeted disruption of the CCR5 co-receptor for HIV-1 has been tested (Tebas et al., 2014).

14.2.5.8 Restriction Endonuclease Utilization in Diagnostics

Given the variety of restriction endonucleases and the distinct sites they recognize, restriction digests have become one of the methods scientists utilize to pinpoint gene mutations and identify genetic disorders. A diagnostic restriction enzyme digest takes advantage of the fact that restriction enzymes cleave DNA at certain discrete sequences, where a site for a restriction enzyme is created or abolished by the presence of a mutation. Often, the region of the DNA to be studied for diagnostic purposes is amplified via PCR, the size of the PCR-amplified DNA are known, and thus this technique can be quickly used to verify the genetic mutation such as Sickle Cell Disease (Clarke and Higgins, 2000; Tuzmen and Schechter, 2001). The goal of a diagnostic digest is to cut your DNA into specific sized pieces and analyze those resulting fragments by gel electrophoresis. The pattern of the fragments on the gel can indicate if the DNA of interest contains the expected known mutation. In Fig. 14.2, a diagnostic identification of Sickle Cell Disease is illustrated utilizing the DdeI restriction endonuclease (Tuzmen and Schechter, 2001). DdeI enzyme recognition sequence is given as 5'-C^TNAG-3' and 3'-GANT^C-5', where N stands for any nucleotide (A, G, C, T) and ^ stands for DdeI enzyme recognition site.

Restriction endonuclease analysis approach was the original method of diagnosis of beta-thalassemia gene, which was performed indirectly by polymorphism analysis. The method of choice now in most diagnostic laboratories for detecting beta-globin gene

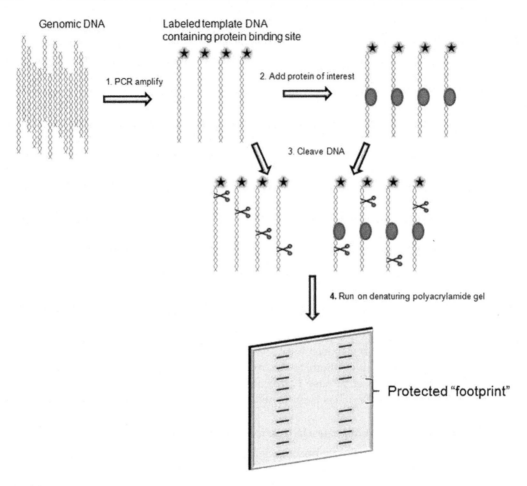

FIGURE 14.1   DNA footprinting workflow. *Adapted from Hampshire et al. (2007).*

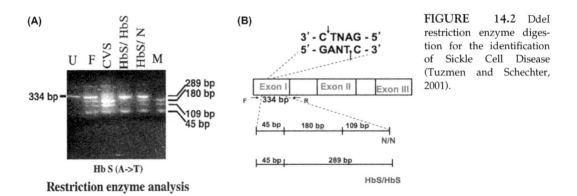

FIGURE 14.2 DdeI restriction enzyme digestion for the identification of Sickle Cell Disease (Tuzmen and Schechter, 2001).

alterations such as the sickle cell disease mutation is restriction enzyme analysis of the HbS gene itself (Pirastu et al., 1989). The loss of DdeI restriction enzyme site at codon 6 of the beta-globin gene ($GAG \rightarrow GTG$) is diagnostic for the presence of the Sickle Cell mutation. Fig. 14.2A is a 3% NuSieve gel picture illustrating DdeI digestion of DNA for detection of sickle cell mutation. Lane 1 shows an undigested 334 bp PCR product. Lane 2 is DNA from a father heterozygous for the sickle cell disease (45 bp, 109 bp, and 180 bp fragments are created from the digestion of the normal allele, 334 bp fragment, and the 289 bp is obtained from the mutant allele, where a DdeI site is abolished and addition of 180 bp and 109 bp fragments gave rise to 289 bp fragment). Lane 3 is from the proband (CVS), who has a normal genotype (N/N). Lane 4 shows a homozygous control for sickle cell disease (HbS/HbS) (only 45 bp and 289 bp fragment are visible). Lane 5 is a heterozygous control (HbS/N), and Lane 6 is from the mother, who has a normal beta-globin gene. Fig. 14.2B illustrates the formation of the banding patterns form an amplified region of beta-globin gene, utilizing (F) and (R) primers shown on the illustration in the presence and absence of sickle cell mutation (Tuzmen and Schechter, 2001).

14.2.6 Restriction Enzyme Digestion Protocol

Select a restriction endonuclease of interest to digest your DNA (i.e., BamHI) (Fig. 14.3). http://www.addgene.org/plasmid_protocols/restriction_digest/.

Note: To find out which restriction endonuclease will cleave your DNA sequence (and where on your sequence they will cut), utilize a sequence analysis program that is available to you (i.e., DNAstar, etc.).

Choose an appropriate reaction buffer (or the buffer supplied by the manufacturer with the enzyme) by reading the instructions for your enzyme of interest.

Note: If you are preparing a double digest (digesting with two restriction enzymes simultaneously in the same reaction), you will need to select the best-optimized buffer,

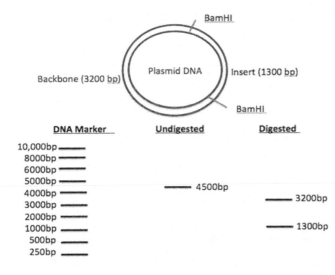

FIGURE 14.3 BamHI restriction endonuclease digestion of a plasmid DNA.

which works for both of your enzymes. Most companies will have a compatibility table, such as the double digestion chart from New England Biolabs. If you cannot find a buffer that is appropriate for both of your enzymes, you will need to digest with one restriction enzyme first in the specific buffer for Enzyme 1, purify the cut DNA, and then perform the second digest in the specific buffer for the restriction Enzyme 2.

In a 1.5 mL tube combine the following:

- DNA
- Buffer
- BSA (if recommended by manufacturer)
- dH$_2$O (distilled H$_2$O) up to total volume
- Restriction enzyme(s)

Note: The concentration of DNA (i.e., microgram, nanogram, etc.) that you digest depends on your application. Diagnostic digests typically involve ∼500 ng of DNA, while molecular cloning often requires 1−3 μg of DNA. The total reaction volume usually varies from 10−50 μL depending on application and is largely determined by the volume of DNA to be digested.

Note: A typical restriction digestion reaction could look like this:

1 μg DNA

1 μL Restriction Enzyme

3 μL 10 × Buffer 3 μL 10 × BSA (if recommended) 22 × μL dH$_2$O (to bring total volume to 30 μL)

Mix gently by pipetting.

Incubate the reaction tube at appropriate temperature (usually 37°C) for 1 hour. Always follow the manufacturer's instructions.

Note: Depending on the application and the amount of DNA in the reaction, incubation time can range from 45 minutes to overnight. For diagnostic digests, 1−2 hours is sufficient. For digests with >1 μg of DNA used for cloning, it is recommended to digest for at least 4 hours.

Note: If you will be using the digested DNA for another application (i.e., digestion with another restriction enzyme in a different buffer) but will not be gel purifying it, you may need to inactivate the enzyme(s) following the restriction digestion reaction. This may involve incubating the reaction at 70°C for 15 minutes or purifying the DNA via a purification kit (i.e., Qiagen DNA cleanup kit). See the enzyme manufacturer's instructions for more details.

To visualize the results of your digest, perform gel electrophoresis (Fig. 14.3).

14.2.7 Summary

Restriction enzymes, also referred to as restriction endonuclease, are proteins synthesized by bacteria, which cut DNA at designated sites on the DNA molecule. In the bacterial cell, restriction enzymes are utilized as a defense mechanism to cleave foreign DNA, thus eliminating infectious bacteriophages. Restriction enzymes can be isolated from bacterial cells and used in the laboratory settings to engineer fragments of DNA, for cloning purposes; for this reason they are indispensible tools of rDNA technology or biotechnology.

14.3 PCR TECHNIQUES

14.3.1 PCR Chronicle

PCR was invented by Kary Mullis in 1993 (Bartlett and Stirling, 2003; US4683195). Today, PCR is a technique frequently utilized in medical and biological research laboratories for a variety of applications (Saiki et al., 1985; Saiki et al., 1988). These applications may include sequencing, DNA cloning, DNA-based phylogeny, functional gene expression analysis, diagnosis of hereditary genetic disorders, identification of genetic fingerprints (i.e., utilized in paternity testing and forensic science), and diagnosis and detection of infectious diseases. In 1993, Kary Mullis along with Michael Smith was awarded the Nobel Prize in Chemistry for his work related to PCR (Kary, 1993). The technique of PCR depends on thermal cycling reactions, comprising cycles of repeated heating and cooling. During these cycles, the template DNA is repeatedly denatured and enzymatically replicated. Synthetically produced short DNA fragments, called primers, containing complementary sequences to the target DNA, reaction buffer, nucleotides (deoxynucleotide triphosphates; dNTPs), together with heat stable DNA polymerase, facilitate selective and repeated amplification. As the cycles in a PCR reaction progress, the copies of DNA generated are themselves used as templates for replication, creating a chain reaction in which the template DNA is copied exponentially. Nowadays, PCR can be modified to facilitate the utilization of a wide array of genetic/genomic manipulations.

14.3.2 PCR Reaction Components

Each PCR reaction contains template DNA, PCR buffer, forward (sense) and reverse (antisense) primers, dNTPs (dATP, dCTP, dTTP, and dGTP), DNA polymerase, and water. Water source need to be nuclease-free to prevent nucleic acid degradation and also should not contain specific ions, organics, or bacteria. PCR buffers contain important divalent ions such as Mg^{+2}, and Mn^{+2} necessary for optimal polymerase activity (Fig. 14.4). Concentration of each component needs to be calculated to ensure successful PCR amplification. Thermo Scientific suggests optimal concentrations for each component as follows (David, 2000; Skerra, 1992).

Template: 0.01−1 ng for both plasmid and phage DNA, 0.01−1 µg range for gDNA
Primers: 0.1−1 µM each forward and reverse
Each dNTP: 0.2 mM
Fig. 14.4 illustrates an example of a typical conventional PCR reaction setup.

14.3.2.1 PCR Primer Design

Primers are designed to flank both upstream and downstream regions of the target sequence. C, G, T, and A contents of forward (sense) and reverse (antisense) primers are carefully chosen to ensure an optimum melting or annealing temperature (Tm) that can work for both primer pairs. Online programs such as Tm Calculator on the NEB (New England Biolabs, Inc.) website can be used to calculate primer Tms. Tm is the temperature at which 50% of the DNA molecules are double-stranded, and the rest of the 50% of the

Component	1 Rxn	Final Conc.	Rxns
10X PCR Buffer Minus Mg^{2+}	5.0 µl	1X	
50 mM MgCl$_2$	1.5 µl	1.5 mM	
10mM dNTP Mix	1.0 µl	0.2 mM each	
10µM sense primer (F) ( )	1.0 µl	0.2µM	
10µM antisense primer (R) ( )	1.0 µl	0.2µM	
Taq DNA Pol (5U/µl)	0.4 µl	1.0 U	
Template DNA ( )	1.0 µl		
RNase/DNase free water	39.1 µl		
Final volume	50 µl		

FIGURE 14.4 An example of a conventional PCR reaction components. Rxn, Reaction; F, Forward primer, sense primer; R, Reverse primer, antisense primer.

PCR conditions (MJ Research Tetrad PCR machine)

94 °C → 5 min

94 °C → 30 sec ⎫
55 °C → 20 sec ⎬ 35 cycles
68 °C → 1 min ⎭

72 °C → 7 min
25 °C → HOLD

DNA molecules are single-stranded. This is a commonly used metric to reflect the stability of a DNA duplex, while more stable duplexes will have higher melting temperatures.

Generally, longer DNA strands have higher melting temperatures, so do sequences with higher C and G content. A comprehensible formula named "Itakura's empirical rule" has been developed to facilitate a quick and dirty calculation of the melting temperature of an oligonucleotide:

$$Tm = 4(G + C) + 2(A + T).$$

Each primer sequence needs to be analyzed for secondary priming sites and inter- and intra-primer complementation. OLIGO Primer Analysis Software (http://www.oligo.net/demo-downl.html) is another helpful tool for primer design and analysis.

14.3.2.2 PCR Steps

PCR consists of three repeated temperature changes (cycles) designated as *denaturation*, *annealing*, and *extension*, which are compiled into a cycle loop of 25–40 cycles depending on the need. At the end of each cycle the amount of DNA target amplified is doubled, leading to exponential amplification of the target DNA fragment.

Denaturation: Denaturation is the first step of a regular PCR cycle, and it requires heating the reaction to 94°C–98°C for 20–30 seconds. High temperature disrupts the hydrogen bonds between the strands of dsDNA. At the end of this step, two complementary ssDNA molecules are produced.

Annealing: During annealing step, the reaction temperature is lowered to 50°C−65°C for 20−40 seconds to allow the primer pairs to anneal to ssDNA molecules. Exact temperature for this step needs to be arranged to provide the best environment for hydrogen bonding between primers and the complementary template DNA bases. Usually annealing temperature is chosen to be 3°C−5°C below the Tm of the primers.

Extension/Elongation: During the extension step, DNA polymerase binds to the primer−template and begins forming the new DNA molecule. The temperature for this step depends on the polymerase used because every polymerase enzyme has its own optimum working temperature. DNA polymerase synthesizes a new DNA molecule that is complementary to the template in 5′ to 3′ direction by using free dNTPs in the environment. Extension time depends both on the enzyme used and the length of the region that is amplified. Rough estimation can be done by assuming that DNA polymerase synthesizes a 1000 bp per minute (Fig. 14.5).

14.3.2.3 PCR Optimization

14.3.2.3.1 PCR BUFFER CONCENTRATION

Reaction buffer is responsible for providing the appropriate pH and ionic concentration for optimal DNA polymerase activity. B different polymerases have different optimal

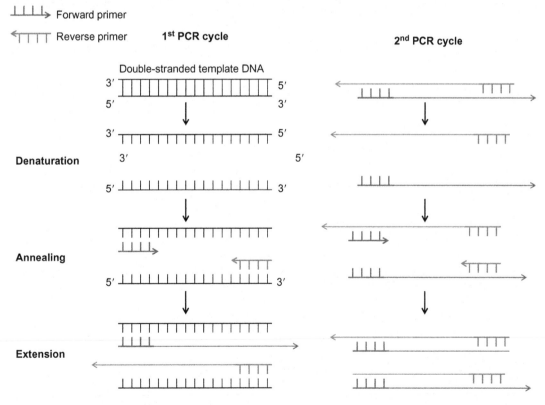

FIGURE 14.5   Steps of PCR.

conditions, their buffers are usually provided with them in high concentrations. Mg^{2+} ions are important for stabilizing primer–template complexes. Mg^{2+} acts as a cofactor for the thermostable DNA polymerases, and thus its concentration can affect the PCR yield. Having too low of Mg^{2+} concentration can result in reduced PCR yield due to reduced polymerase activity. However, too much Mg^{2+} can cause nonspecific PCR products due to reduced fidelity of the polymerase (ThermoScientific, 2012).

14.3.2.3.2 DNA POLYMERASE ENZYME OF CHOICE

Different DNA polymerase enzymes can be chosen for multiple reasons, including increasing the fidelity, minimizing error rates in amplification of long sequences, and improving speed (Parker, 2014). Enzyme selection is an important part of a successful PCR setup. DNA polymerases may vary in their thermostability, elongation rate, processivity, and fidelity. Additionally, there are proofreading and non-proofreading enzymes, according to the needs of a study, commercially available. Nowadays, thermostable DNA polymerases (i.e., Taq polymerase) are the polymerase enzyme of choice for conventional PCR reactions. Some of the commercially available DNA polymerases include; Tfl DNA Polymerases (Promega Corporation), Phusion DNA Polymerases (New England BioLabs, Inc.), Vent DNA Polymerases (New England BioLabs, Inc.), Therminator DNA Polymerases (New England BioLabs, Inc.), Pfu DNA Polymerases (Thermo Scientific), and Q5 High-Fidelity DNA Polymerases (New England BioLabs, Inc.).

14.3.3 PCR Instrumentation

There are selection of complete line of thermal cyclers from different vendors, which may include the following.

14.3.3.1 *Applied Biosystems*

Applied Biosystems thermal cyclers that combines the reliability and performance with the flexible configuration and control features that fit today's research. Dual block configurations allow maximization of throughput (Fig. 14.6).

14.3.3.2 *Bio-Rad*

Bio-Rad's thermal cyclers is a modular thermal cycler platform, with C1000 Touch thermal cycler chassis, and choice of dual 48/48 fast reaction module, 96-well fast and deep well format, 384-well reaction module, with a USB flash drive (Fig. 14.7).

14.3.3.3 *Eppendorf*

Eppendorf's thermal cyclers is for the daily routine of PCR. It is easy to use, does not need much space or energy and sends you an e-mail when it is done (Fig. 14.8).

FIGURE 14.6 The ProFlex PCR System. *Available from Applied Biosystems. http://www.lifetechnologies.com/uk/en/home/life-science/pcr/thermal-cyclers-realtime-instruments/thermal-cyclers/pro-flex-pcr-system.html.*

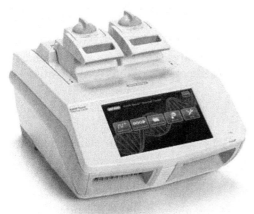

FIGURE 14.7 C1000 Touch Thermal Cycler. *Available from Bio-Rad. http://www.bio-rad.com/en-uk/category/thermal-cyclers?pcp_loc = catprod.*

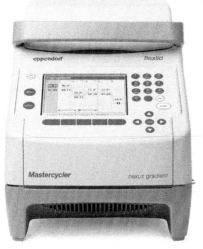

FIGURE 14.8 Eppendorf Mastercycler nexus. *Available from Eppendorf. http://www.eppendorf.com/int/index.php?sitemap = 2.1&pb = 48d 64803d6cfced0&action = products&contentid = 1&catalognode = 88879.*

14.3.4 Recent Advances in PCR Technology and Its Applications

14.3.4.1 Hot Start PCR

This method includes withholding the polymerase using physical barriers such as wax beads or specific inhibitors or antibodies until higher temperatures are attained. It aims to reduce nonspecific amplification that is usually observed by nonspecific primer annealing at low temperatures (Parker, 2014).

14.3.4.2 Reverse Transcriptase PCR

Regular PCR technique is adapted to reverse transcribe cDNA molecules using RNA molecules as a template and reverse transcriptase as an enzyme. RT-quantitative PCR can be used to study gene expression, disease diagnosis, and insert eukaryotic genes into prokaryotes (Parker, 2014).

14.3.4.3 Droplet Digital PCR

Droplet Digital PCR (ddPCR) gives a direct quantitative measurement of absolute DNA concentration without the need for a standard curve. This is a single-molecule counting method, where a single sample is split into many fractions, and all these fractions are analyzed by a standard PCR method. This method allows analysis of DNA that is found in low levels against background noise. The results are recorded digitally as positive or negative (Huggett et al., 2013) (Fig. 14.9).

14.3.4.4 Long PCR

During long PCR, blend of thermostable DNA polymerases are used to allow amplification of fragments that are longer than 5 kbs (Parker, 2014).

14.3.4.5 Multiplex PCR

Multiplex PCR is used to amplify several targets in a single PCR reaction. This is accomplished by using multiple sets of primers. This method is useful for pathogen identifications, high-throughput SNP genotyping, and forensic applications for human identification studies (Parker, 2014).

FIGURE 14.9 QX200 Droplet Digital PCR System (BioRad Inc.) Mainly used for quantification of target DNA/RNA molecules.

14.3.4.6 Colony PCR

Colony PCR is used to screen colonies for the desired plasmid after transformation has been carried out. The length of PCR steps and temperatures are adjusted to release DNA from the small amount of cells added to the PCR mix. The DNA released by this process can be then used for amplification (Parker, 2014).

14.3.4.7 Nested PCR

Following conventional PCR, a second set of primers that are specific to a DNA sequence within the initial PCR amplicon are added to the reaction. Second amplification step with "nested" primers reduces the nonspecific binding and increases the amount of amplicon produced (Parker, 2014).

14.3.4.8 Quantitative Real-Time PCR

To date, many methodologies have been described as tools to investigate gene expression profiling of functional analyses. Standard methodologies for studying transcriptional analysis include Northern blotting (Alberts et al., 2008; Kevil et al., 1997), in situ hybridization (Jin and Lloyd, 1997), semiquantitative RT-PCR (Freeman et al., 1999), nuclease (RNAse) protection assay (Eyler, 2013), competitive RT-PCR, microarray analysis (Pena et al., 2014), and quantitative real-time PCR (qPCR) (Bustin, 2002; Livak and Schmittgen, 2001; Sukru, 2007). qPCR has notable advantages when compared to the rest of the gene expression quantification techniques. These advantages include enhanced sensitivity and accuracy, a large dynamic range, high-throughput capability, ability to perform multiplex reactions, faster and more reliable amplification, and lack of post-PCR manipulations. This powerful technique has become the method of choice for rapid and quantitative monitoring of gene expression studies. Commercial availability of a wide range of gene expression assays for qPCR analysis also make this technology attractive for endpoint evaluation. The analysis of gene expression can be determined either by using relative quantification or absolute quantification methods (Bustin, 2002; Livak and Schmittgen, 2001; Sukru, 2007; Bustin, 2000) (Fig. 14.10).

This technique initially requires extraction of total RNA (Sukru, 2007) from which the gene expression profiling will be determined. There are many commercially available RNA extraction kits (i.e., RNAeasy RNA extraction kit from Qiagen Inc.) [http://www.qiagen.com/products/catalog/sample-technologies/rna-sample-technologies]. For further details, please refer to the manufacturer's recommendations. Following RNA extraction, an aliquot of 10 ng–1 μg total RNA is subjected to a reverse transcription reaction (Sukru, 2007). There are many commercially available kits for cDNA synthesis (i.e., iScript cDNA Synthesis kit from Bio-Rad, Hercules, CA). Perform the reactions according to the manufacturer's protocol. CDNA is then subjected to 40 cycles of qPCR (Sukru, 2007). There exists many commercially available qPCR instruments (i.e., ABI 7900HT Sequence Detection System (ABI)). Perform the reactions according to the manufacturer's instructions. For qPCR, various detection chemistries can be utilized (i.e., TaqMan Gene Expression Assays) (Sukru, 2007) for determining the expression levels of the genes of interest as well as for the reference genes for normalization purposes during data analysis (Sukru, 2007). QPCR is then performed following standard procedures (Sukru, 2007). In brief, qPCR reactions includes 0.2–0.4 mmol (determined empirically) of gene-specific primers, and 2× of the TaqMan Universal

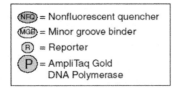

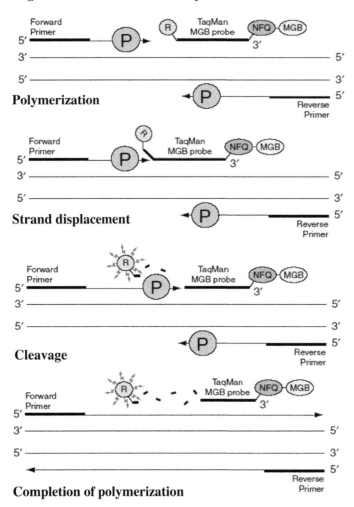

FIGURE 14.10 qPCR detection chemistry. The 5′ nuclease assay illustrating the mechanism of probe cleavage generating a fluorescent signal (Sukru, 2007).

Mastermix (without Amp Erase UNG), and the template cDNA in 10−20 μL reaction volume. The analysis of the data could then be achieved via standard bioinformatics tools (Sukru, 2007). In Fig. 14.11, an example of qPCR utilization as a golden standard for validation of gene expression, following small intervening RNA (siRNA) treatment is illustrated.

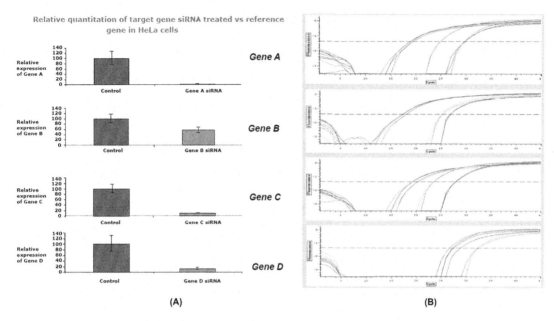

FIGURE 14.11   Validation of gene expression following siRNA treatment by qPCR, utilizing relative quantification method (Sukru, 2007).

Here, $\Delta\Delta CT$ method was employed for facilitating the calculation of relative quantification of knock-downs (Livak and Schmittgen, 2001; Sukru, 2007).

14.3.4.9 PCR Site-Directed Mutagenesis

PCR site-directed mutagenesis, also known as site-specific mutagenesis or oligonucleotide-directed mutagenesis, is one of the most important laboratory techniques for introducing mutations or creating unique restriction enzyme recognition sites for cloning purpose in a DNA sequence (Braman, 2002; Carrigan et al., 2011). Several strategies to this method have been published, which utilize either ssDNA (Kunkel, 1985; Sugimoto et al., 1989; Taylor et al., 1985) or dsDNA as a template (Vandeyar et al., 1988). PCR-driven overlap extension is a frequently used and an inexpensive method for creation of desired mutations/restriction enzyme recognition sites along DNA sequences (i.e., MCS) (Fig. 14.12). The overlap extension method facilitates the amplification of the desired gene/fragment from gDNA or cDNA or any synthetic MCS without a requirement for cloning the gene/region of interest into a plasmid vector (Heckman and Pease, 2007). Fragments of the target gene/MCS are amplified from template DNA/synthetic DNA sequences using two flanking master primers (AccI and PflmI) that label the 5′ ends of both strands, which have the AccI and PflmI restriction enzyme sites synthetically integrated into the 5′ ends of both master primer sequences. Eventually, the PCR products will harbor the AccI and PflmI restriction enzyme recognition flanking sites (Fig. 14.12: PCR II and PCR IV). Two sets of internal primers (SacII (F and R), and MluI (F and R)) together with the two flanking master primers

PCR Site-Directed Mutagenesis

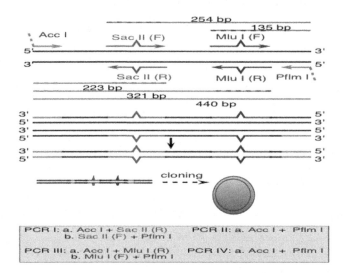

FIGURE 14.12   Schematic of PCR site-directed mutagenesis.

(AccI and PflmI) that eventually introduce AccI, PflmI, SacII, and MluI restriction endonuclease recognition sites via PCR will aid in creating unique restriction enzyme sites (Fig. 14.12: PCR I, PCR II, PCR III, and PCR IV). Both the internal complementary primer pairs (SacII (F and R), and MluI (F and R)) contain the desired restriction enzyme recognition sites for SacII and MluI, respectively. Internal primers must not only contain the desired restriction enzyme recognition sites but also create overlapping nucleotide sequences with their complementary restriction enzyme recognition pairs. Consequently, via mixing the amplified products from PCR I, PCR II, PCR III, and PCR IV, segments will be rejoined during the final PCR reactions (PCRII and PCR IV) utilizing the flanking master primers (AccI and PflmI). The primary limitation to maximal accuracy and the size of amplified product between flanking master primers (AccI and PflmI) is the efficiency of the DNA polymerase used for PCR. Eventually, the desired final PCR product should be sequenced for the accuracy of the mutations/restriction enzyme recognition sites created (Fig. 14.13).

14.3.4.10 Other PCR Techniques

Other PCR Techniques that are worthwhile mentioning can be listed as allele-specific PCR, ARMS (Amplification Refractory Mutation System), Assembly PCR or Polymerase Cycling Assembly (PCA), Asymmetric PCR, Dial-out PCR, Helicase-dependent amplification, Intersequence-specific PCR (ISSR), Inverse PCR, Ligation-mediated PCR, Methylation-specific PCR (MSP), Miniprimer PCR, Multiplex Ligation-dependent Probe Amplification (MLPA), Nested PCR, Overlap-extension PCR or Splicing by overlap extension (SOEing), Solid Phase PCR, Thermal asymmetric interlaced PCR (TAIL-PCR), Touchdown PCR (Step-down PCR), and Universal Fast Walking, In Silico PCR (digital

Confirmation of PCR Site Directed Mutagenesis for SacII and MluI Sites by Restriction Enzyme Digestion and Sequencing

FIGURE 14.13   Schematic of confirmation for PCR site-directed mutagenesis by sequencing.

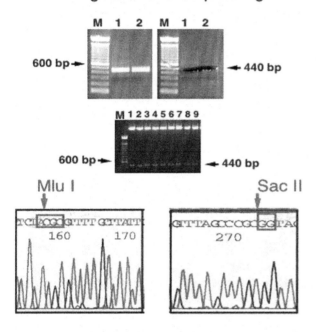

PCR, virtual PCR, electronic PCR, e-PCR). Basically, depending on the specific need for amplification, PCR technology has a versatile application.

14.4 BLOTTING TECHNIQUES

Identifying mutated/abnormal genes is the backbone of clinical research and genetic counseling today. Most of the time, diagnostic medicine relies on our abilities to measure specific proteins in blood and identify unique protein or nucleic acid sequences that a patient may possess. To accomplish these, scientists developed highly specific and sensitive blotting techniques that can be used both in molecular biology and clinical research (Hayes et al., 1989; Nicolas and Nelson, 2013).

14.4.1 General Principle

According to Hayes et al. (1989) and Nicolas and Nelson (2013) the general work principle of blotting techniques are summarized in four steps:

1. Separating protein or nucleic acid fragments by electrophoresis
2. Transferring to and immobilizing the fragments on paper (membrane) support

TABLE 14.7   Blotting Techniques (Nicolas and Nelson, 2013).

Name	Target	Probes Used	Newer Alternative techniques
Southern	DNA	Complementary (antisense) Sequence of DNA/RNA	PCR, real-time PCR (qPCR), FISH
Northern	RNA	Complementary sequence of DNA/RNA	RT-PCR
Western	Protein	Monoclonal antibody	ELISA, immunohistochemistry, immunofluorescence, flow cytometry

3. Using an analytical probe that binds to the target on the paper
4. Detecting and visualizing the probe bound to the fragment.

The blotting techniques can be listed under three main blotting categories, which include Southern Blotting, Northern Blotting, and Western Blotting (Hayes et al., 1989; Nicolas and Nelson, 2013) (Table 14.7).

14.4.1.1 Southern Blotting

Southern blot analysis exhibits information related to DNA identity, molecular weight, and abundance. It is a conventional method that facilitates separating DNA fragments based on their size, utilizing gel electrophoresis, transferring them to a membrane (nitrocellulose or nylon), hybridization with a labeled sequence-specific probe (radioactive or non-radioactive), series of washing, and eventually detection of labeled DNA fragment(s) (Southern, 1975). There are commercial companies, i.e., Life Technologies, which offer comprehensive portfolios of products for Southern blot analysis. Southern Blotting technique has facilitated the detection of RFLP and variable number of tandem repeat polymorphism. The latter being the basis of DNA fingerprinting (Tamaki Jeffreys, 2005; Jeffreys, 2005).

14.4.1.1.1 THE ORDER OF SEQUENCE OF SOUTHERN BLOT ANALYSIS
- Step 1. Restriction enzymes are utilized to digest high-molecular weight DNA strands into smaller fragments (Fig. 14.14).
- Step 2. Digested DNA fragments are subjected to electrophoresis, on an agarose gel (0.5%–1.5%), to separate the fragments according to their sizes (Fig. 14.14).
  Note: (1) Should one expect to get DNA fragments larger than 15 Kb as a result of enzyme digestion, then it is recommended that prior to blotting, the gel on which the DNA strands were run be treated with dilute HCl. This allows the DNA fragments to break into smaller pieces, hence permitting more effective transfer of DNA bands from the gel to the membrane. (2) If alkaline transfer methods are utilized, the DNA gel is placed into an alkaline solution (i.e., containing sodium hydroxide) to denature the dsDNA. The alkaline condition ameliorates binding of the thymine residues of DNA (negatively charged) to amino groups of membrane (positively charged). This causes the DNA strands to become single stranded, helping for later hybridization of the

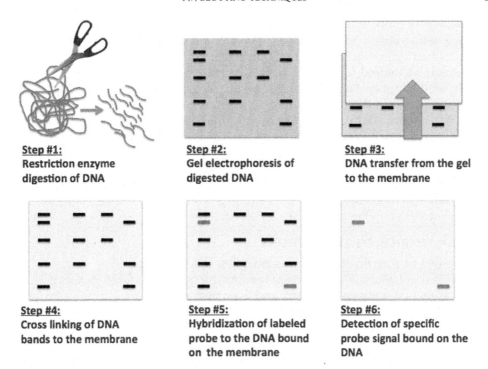

Step #1:
Restriction enzyme
digestion of DNA

Step #2:
Gel electrophoresis of
digested DNA

Step #3:
DNA transfer from the gel
to the membrane

Step #4:
Cross linking of DNA
bands to the membrane

Step #5:
Hybridization of labeled
probe to the DNA bound
on the membrane

Step #6:
Detection of specific
probe signal bound on the
DNA

FIGURE 14.14   Southern blot analysis steps.

ssDNA to the probe, and destroying any residual RNA that may still be present with the DNA.
- Step 3. A sheet of nylon or nitrocellulose membrane is placed on top of (or below, depending on the direction of the transfer) the gel in a buffer solution. A uniform pressure is applied to the gel (either using suction, or by placing a stack of paper towels and a weight on top of the membrane and gel), to make certain that there is a good and even contact between gel and membrane. If transferring by suction, the buffer of choice is $20 \times$ SSC, to ensure a seal, and prevent drying of the gel. The capillary action property is utilized during buffer transfer, which takes place from a region of high water concentration to a region of low water concentration. This process usually uses filter paper and paper tissues to move the DNA from the gel on to the membrane. Due to the negative charge of the DNA and positive charge of the membrane, ion exchange interactions bind the DNA to the membrane (Fig. 14.14).
- Step 4. Following Step 3, the membrane (nitrocellulose or nylon membrane) is then either baked in a vacuum oven or a regular oven at 80°C for 2 hours, or exposed to UV radiation in the case of a nylon membrane, to permanently fix the transferred DNA to the membrane (Fig. 14.14).
- Step 5. The membrane is then exposed to a hybridization probe (a single-stranded synthetic oligonucleotide fragment), with a specific complementary sequence, which will eventually bind the target DNA of interest. The single-stranded synthetic

oligonucleotide probe is labeled (usually by incorporating radioisotopes or tagging the molecule with a fluorescent dye). The most common hybridization methods utilize salmon or herring sperm DNA to block the membrane surface and target DNA. Additionally, deionized formamide and detergents (i.e., SDS) are used to reduce nonspecific binding of the probes and to secure the specificity of the binding of the probe to the sample DNA (Fig. 14.14).

- Step 6. After hybridization protocol in Step 5, utilizing $1 \times$ SSC buffer, excess probe is washed from the membrane, and the pattern of hybridization is visualized on X-ray film by autoradiography. If a chromogenic detection method is used via utilizing a fluorescent probe, then development of a color on the membrane is observed, where a hybridization between the DNA fragments and the fluorescent probes are achieved (Fig. 14.14).

14.4.1.1.2 SOUTHERN BLOT APPLICATIONS

- Identification of a particular DNA fragment in a DNA sample
- Isolation of a necessary DNA fragment for construction of an rDNA molecule
- Identification of mutations, deletions, and gene rearrangements
- Utilization for prediction of cancer and for prenatal diagnosis of genetic diseases
- Utilization in RFLP
- Utilization in phylogenetic analysis
- Diagnosis of infectious disease, i.e., HIV
- Utilization in DNA fingerprinting
  - Maternity and Paternity testing
  - Forensics
  - Personal identification

14.4.1.2 *Northern Blotting*

This is a methodology utilized in molecular biology research to study gene expression regulation via detecting RNA molecules in a sample (Alberts et al., 2008; Kevil et al., 1997; Streit, 2009). Northern blotting facilitates observation of cellular control over structure and function by determining the particular gene expression levels during cellular differentiation, morphogenesis not only in wild type but also in aberrant conditions (Schlamp, et al., 2008). Northern blotting first requires the utilization of electrophoresis to separate RNA molecules by size. Then, the separated RNA molecules are identified via using labeled probes to hybridize with the complementary target sequences. The term "Northern blot" refers particularly to the capillary transfer of RNA molecules from the gel to the blotting membrane (Schlamp, et al., 2008). However, the whole method is commonly denoted as Northern blotting (Trayhurn, 1996). The Northern blot method was developed in 1977 by James Alwine, David Kemp, and George Stark at Stanford University (Alwine, et al, 1977). Northern blotting takes its name from its similarity to the first blotting methodology, the Southern blot, which is named after the scientist Edwin Southern (Alberts et al., 2008). The main difference is that RNA molecules, rather than DNA molecules are analyzed in the Northern blot (Bor, et al., 2006; Streit, 2009).

RNA Ladder   Sample 1   Sample 2

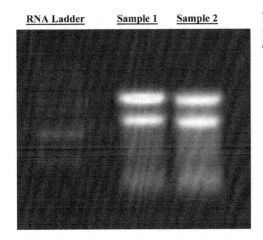

FIGURE 14.15 Formaldehyde (1%) with RNA samples run at 100 V for 1 hour in 1× MOPS buffer. *Adapted from http://creativecommons.org/licenses/by-sa/3.0/.*

14.4.1.2.1 NORTHERN BLOT PROTOCOL

The Northern blot procedure is quite similar to that of Southern blot, except that the nucleic acid utilized is RNA rather than DNA. Northern blots facilitate measuring the molecular weight and relative amounts of RNA/mRNAs present in individual samples.

14.4.1.2.1.1 RNA GELS RNA is usually run on a formaldehyde agarose gel to highlight the rRNA subunits 28S (upper bright band) and 18S (lower bright band) (Fig. 14.15) (Streit, 2009).

RNA samples are frequently separated on agarose gels that are prepared with formaldehyde as a denaturing agent for the RNA. This chemical serves to limit secondary structure formation (Streit, 2009; Yamanaka, 1997). The gels can be stained using EtBr and visualized under UV light to check the quantity and quality of RNA prior to blotting (Strei, 2009).

RNA separation can also be achieved by PAGE (polyacrylamide gel electrophoresis) containing urea, but oftentimes it is utilized for miRNAs and fragmented RNA (Valoczi, 2004). Usually, an RNA ladder is run next to the samples on an electrophoresis gel to observe the RNA fragment size obtained. However, in total RNA samples, the ribosomal subunits can serve as RNA size markers (Strei, 2009). Because the large ribosomal subunit is 28S (roughly 5 kb), and the small ribosomal subunit is 18S (roughly 2 kb) two noticeable bands will be clearly visible on the gel, the larger band being close to twice the intensity of the smaller one (Streit, 2009; Gortner, 1996).

14.4.1.2.2 THE ORDER OF SEQUENCE OF NORTHERN BLOT ANALYSIS

Step 1. RNA is isolated from several biological samples (e.g., various tissues, various cell lines, etc.)

Note (1): RNA is more susceptible to degradation than DNA. This is due to the fact that at the position of 2′ carbon atom of ribonucleotides, RNA molecules possess an -OH group rather than -H group, which openly exposes these molecules to nucleophilic attack, which then results in degradation of the RNA molecules. (2) The decision on whether to isolate total RNA or mRNA usually depends on the type of scientific question put forward. Isolated mRNA molecules comprise 2%−5% of total RNA, and its isolation requires a further purification step using a poly A tail hybridization. If the presence of total RNA needs to be totally avoided, then the utilization of the poly A tail hybridization may be necessary.

Step 2: The RNA samples are separated according to their size on an agarose gel.

Step 3: The gel is then blotted on nylon or nitrocellulose membrane.

Step 4: The membrane is then placed in a container comprising hybridization buffer with a labeled probe (radioactively or chemically labeled). The probe is specifically designed for the sequence of interest. Thus, it will hybridize to the RNA on the blot that corresponds to the sequence of interest.

Step 5: The membrane is washed to remove unbound probe.

Step 6: The labeled probe is detected via autoradiography (if a radioactive probe is used) or via a chemiluminescence reaction (if a chemically labeled probe is used) (Streit, S, 2009). In both cases this results in the formation of a dark band on an X-ray film.

We can compare expression patterns of the sequence of interest in the different samples. Enhanced sensitivity of this technique can be accomplished by utilizing high specific activity antisense RNA probes (Step 4), optimized hybridization buffers (Step 4), and positively charged nylon membranes (Step 3). [http://teachline.ls.huji.ac.il/72320/methods-tutorial/northern.html#procedure] (Fig. 14.16).

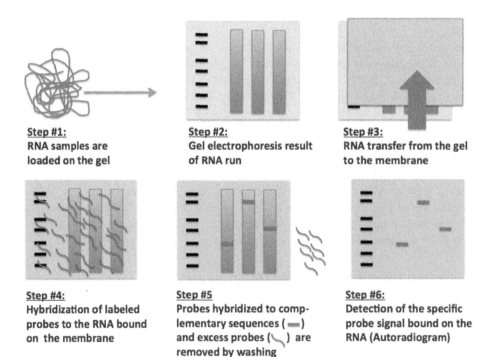

Step #1:
RNA samples are loaded on the gel

Step #2:
Gel electrophoresis result of RNA run

Step #3:
RNA transfer from the gel to the membrane

Step #4:
Hybridization of labeled probes to the RNA bound on the membrane

Step #5
Probes hybridized to complementary sequences (▬) and excess probes (⌇) are removed by washing

Step #6:
Detection of the specific probe signal bound on the RNA (Autoradiogram)

FIGURE 14.16   Northern blot analysis steps.

14.4.1.2.3 APPLICATIONS OF NORTHERN BLOTS

Northern blotting facilitates the observation of an individual gene's expression profile among pathogen infection, environmental stress levels, developmental stages, tissues, organs, and during the period of a particular treatment (Mori, 1991; Liang, 1995; Baldwin, et al., 1999). Northern blotting technique has been described for its use in detecting upregulation of oncogenes and downregulation of tumor-suppressor genes in neoplasia, when compared to "normal" tissues (Streit, 2009). If the outcome of a Northern blot procedure reveals an upregulation of a particular gene, observed as an abundance of mRNA on the Northern blot, the sample can then be further investigated by sequencing to verify whether the gene is known to researchers or if it is a previously undescribed finding (Utans, 1994). The gene expression profile obtained under established conditions can provide understanding into the function of the gene in question. Additionally, because the RNA is first resolved by its molecular weight, if only one type of a probe is utilized, variance in the level of each band on the membrane can provide insight into the size of the product, demonstrating alternatively spliced species of the same gene or repetitive sequence motifs (Durand and Zukin, 1993; Gortner, et al., 1996). Furthermore, the difference in fragment size of a gene product can also demonstrate deletions or aberrations in transcript processing. The missing RNA region can be determined by altering the probe target used along the known sequence of the RNA molecule (Alberts et al., 2008). Nowadays, Northern blots are published on online databases such as *BlotBase*. This database has more than 700 published Northern blots of human and mouse samples, in more than 650 genes across more than 25 different tissue types (Schlamp, 2008). By way of using a blot ID, paper reference, gene identifier, or by tissue type, Northern blots can be searched thoroughly (Schlamp, 2008). The results of a search provide the blot ID, species, tissue, gene, expression level, blot image (if available), and links to the publication that the work initiated from (Schlamp, 2008). BlotBase provides sharing of information between members of the science community, which was not seen in the past for Northern blotting protocols, as it was in sequence analysis, genome determination, protein structure, etc.

14.4.1.3 *Western Blotting*

Western blotting (protein immunoblotting) is an analytical technique used to identify and locate specific proteins in a sample of tissue homogenate or extract, based on their ability to bind to specific antibodies. Western blot analysis can detect protein of interest from a mixture of a vast number of proteins. Western blotting can provide valuable information about the size of individual proteins (can be detected via a comparison to a molecular weight marker in kilodalton), and further yield information on protein expression usually in correlation with a control (i.e., untreated sample or another tissue or a cell type) (Hirano, 2012)

Gel electrophoresis is employed to separate native proteins by three-dimensional structure or denatured proteins by the length of the polypeptide. The proteins are then transferred to a membrane (usually nitrocellulose or polyvinylidene difluoride (PVDF)), where they are treated with antibodies specific to the target protein (Towbin, 1979; Renart, 1979). The gel electrophoresis stage in Western blot analysis facilitates to resolve the issue of the cross-reactivity of antibodies. However, an improved immunoblot methodology, *Zestern analysis* (Zhang, 2012), is able to address this issue without the electrophoresis step, thus notably enhancing the efficacy of protein analysis.

Nowadays, there are many specialized reagent companies, which commercially provide antibodies against various proteins ("Western blot antibody," exactantigen.com. Retrieved 29 January, 2009). Commercial antibodies can be costly, even though the unbound antibodies can be reused between different experiments. This method could be deployed in the fields of biochemistry, molecular biology, immunology, and other fields of biology in general. If the identified protein is a previously undescribed one, then a new antibody should be produced for the detection of the novel protein, either by the research group that discovered the particular protein or its production could be serviced out to a company. For this purpose, at least a small amount of the novel protein is necessary, obtained either from purified cell extracts or made as a recombinant protein via in vitro or in a recombinant protein expression system. Antibodies distinct to recognizing a novel protein are essential to Western blotting. They are able to identify particularly the protein of interest instead of random recognition from the cocktail of proteins on Western blot. Other associated methodologies may include dot blot analysis, Zestern analysis, immunohistochemistry, where antibodies are utilized to identify proteins in tissues and cells by immunostaining, and enzyme-linked immunosorbent assay (ELISA).

14.4.1.3.1 THE ORDER OF SEQUENCE OF WESTERN BLOT ANALYSIS

Step 1. Sample preparation and PAGE
- Note: Samples can be prepared from whole tissue or from cell culture. Solid tissues are first broken down mechanically by utilizing a blender (larger sample volumes), by a homogenizer (smaller sample volumes), or by sonication (sound waves). Cells can also be lysed by one of the above methods. However, virus or environmental samples can also be the source of proteins, and thus Western blotting is not restricted to cellular studies only. Detergents, buffers, and salts may be utilized to break down cells and solubilize proteins. Generally, protease and phosphatase inhibitors are added to hinder the digestion of the samples by their own enzymes. Tissue preparation is often done at cold temperatures to avoid denaturation and degradation of proteins (Nicolas and Nelson, 2013). Integration of biochemical and mechanical techniques, which consist of various types of filtration and centrifugation methods, can be used to separate different cell compartments and organelles.
- Denaturation of proteins denotes "unfolding" their naturally existing three-dimensional (3D) or tertiary structures to a specifically or completely linearized form. The denaturing process also interferes with most protein–protein interactions. Frequently, SDS, a detergent, is employed to disrupt proteins in this setting. SDS also contributes a stronger negative charge, which enables protein gel electrophoresis. The protein samples are then loaded onto a polyacrylamide gel, where an electrical current is applied, facilitating the movement of negatively charged molecules (due to SDS treatment) to separate, based on their size and charge from negative to positive poles of the electrophoresis tank, with smaller and more charged molecules running through the gel quicker and thus moving ahead of large molecules in the same period of time. A "protein marker," consisting of standardized and known sizes of

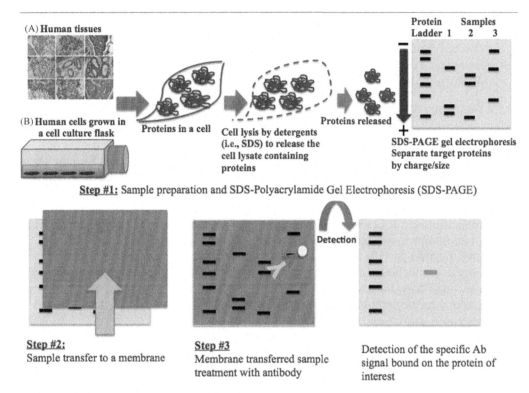

FIGURE 14.17 Western blot analysis steps.

proteins, is run side by side with the samples, which facilitates to estimate the size of each unknown protein (Fig. 14.17) (Nicolas and Nelson, 2013).

Step 2. Sample transfer to a membrane
- The proteins are transferred to a membrane (usually nitrocellulose or PVDF), utilizing an electrical gradient, which supports the vertical migration facilitated by gel electrophoresis. This step is very similar to producing a photocopy of the molecules that are enabled to transfer from the gel to the membrane (Fig. 14.17) (Nicolas and Nelson, 2013). The consistency and total success of transfer of proteins from the gel to the membrane can be determined by Coomassie Blue and Ponceau S dye staining of the SDS-PAGE gel and membrane respectively (Corley, 2005).

Step 3. Membrane transferred sample treatment with labeled antibody
- Lastly, the membrane from Step 2 is treated with an antibody, which binds solely to the particular protein being investigated (Fig. 14.17). The nature of the antibodies varies based on the target protein. A monoclonal/polyclonal antibody (the primary antibody) binds to the particular protein of interest, followed by a labeled second antibody (the secondary antibody), which recognizes the "stem" of the Y shape of the primary antibody, permitting precise detection of the protein of interest. This "two-step" technique is utilized for expense control as labeling each specific

monoclonal antibody would be quite costly. Owing to the utilization of antibodies, the Western blot method is occasionally described as immunoblotting. Irrespective of the antibody utilized, the label may be a radioactive isotope or a fluorescent or chromogenic dye. All of which permit detection of the protein under specific conditions (Fig. 14.17) (Nicolas and Nelson, 2013).

14.4.1.3.2 WESTERN BLOT APPLICATIONS

- To detect the presence of biological probes (circulating antibodies) to a single protein or group of proteins.
- To be used as the confirmatory HIV test to identify anti-HIV antibody in human serum samples. Proteins from known HIV-infected samples are isolated and blotted on a membrane. Next, the serum to be tested is added in the primary antibody incubation step. Following the washing steps, free antibody is washed away, and a secondary antihuman antibody attached to an enzyme signal is administered. Consequently, the stained bands represent the proteins from the patient's serum, which would be considered positive to HIV infection.
- To employ as a conclusive test for Bovine spongiform encephalopathy, commonly known as "mad cow disease."
- To utilize as a test for some forms of Lyme disease.
- To employ as a validation test for Hepatitis B infection.
- To utilize as a validation tool for FIV + status in cats, in veterinary medicine.

14.4.2 Western Blot Instrumentation

14.4.2.1 *Bio-Rad*

Bio-Rad's V3 Western Workflow is a five-step approach to streamlining Western blotting protocol. The V3 Western Workflow enables speed and validation at each step of the process, from running gels to quantifying proteins (Fig. 14.18).

14.4.2.2 *ProteinSimple*

Add the samples and primary antibody to the pre-filled assay plate. Pop in Wes' capillary cartridge and load the plate. Push start and in just 3 hours get quantitated size-based separation data on up to 25 samples. After the run, toss the used cartridge and plate and then the process is complete (Fig. 14.19).

14.4.2.3 *Life Technologies*

Western blots are considered to be a gold standard for examining protein expression. They are still used routinely in research (Kidwai et al., 2013; Wang et al., 2013).

14.4.2.4 *Additional Blotting Techniques*

14.4.2.4.1 DOT BLOT

Fig. 14.20 is an illustration of a dot blot analysis (radioactive) (Pirastu et al., 1989; Tuzmen and Schechter, 2001). This method distinguishes between homologous DNA species bound to a nylon membrane by hybridization with a radiolabeled synthetic

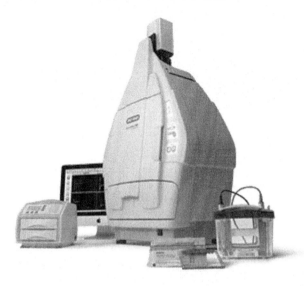

FIGURE 14.18 Bio-Rad's V3 Western Workflow. *http://www.bio-rad.com/en-uk/product/v3-western-workflow? pcp_loc = catprod.*

FIGURE 14.19 ProteinSimple, Wes.http://www.proteinsimple.com/wes.html?gclid = CKOM44bx-b8CFam WtAodCk8AjA.

oligonucleotide probe. Single base changes are easily recognized by using a pair of allele-specific oligonucleotide probes, which will differ in only one nucleotide. The normal probe, as the name states, is complementary to the normal and the mutant probe is complementary to the mutant gene sequences. Two membranes dotted with homologous DNA species are treated with mutant and normal probes separately and labeled as mutant (Mt) and normal (N) membranes so as to be able to interpret the results on the autoradiogram. Fig. 14.20 shows an example of this, where IVS-I-110 probe is used. Numbers 1, 2, 3, and 4 on the autoradiogram illustrates IVS-I-110 heterozygote individuals, and Number 5 shows

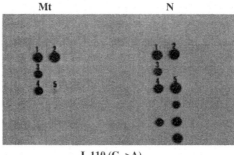

$_\beta$I–110 (G–>A)

FIGURE 14.20   Dot blot analysis of β-thalassemia mutation IVS-I-110 (Tuzmen and Schechter, 2001).

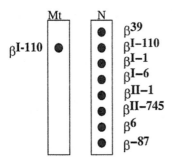

FIGURE 14.21   Reverse dot blot analysis of β-thalassemia mutation IVS-I-110 (Tuzmen and Schechter, 2001).

DNA from a normal individual for IVS-I-110 mutation. Dot blot technique (radioactive) is extremely practical in screening large number of DNA samples for the analysis of molecular and population genetics of β-thalassemia in certain populations at risk (Tadmouri, G.O., et al., 1998; Tuzmen and Schechter, 2001).

14.4.2.4.2 REVERSE DOT BLOT

Reverse dot blot analysis (nonradioactive) (Maggio, A., 1993; Sutcharitchan, P., 1995; Tuzmen and Schechter, 2001) is based on the similar concept as the dot blot analysis (radioactive) with a small difference, where the allele-specific oligonucleotide probes are membrane-bound rather than the individual DNA samples. In this configuration, multiple pairs of mutant and normal allele-specific oligonucleotides are dotted on strips of nylon membranes.

For each diagnostic test, a pair of homologously dotted strips of membranes are used, one being designated as normal and the other mutant. Fig. 14.21 demonstrates an example of a reverse dot blot analysis (nonradioactive), where an individual is tested for the presence of the Mediterranean-specific mutations, which turned out to be the IVS-I-110 carrier state. The use of this method is much valuable in prenatal diagnosis laboratories for two main reasons: one being the application of several population-specific probes to individual DNA samples and the other being the nonradioactive nature of the process (Tuzmen and Schechter, 2001).

14.4.2.4.2.1 POWER OF BLOTTING

- Blotting enables specific and sensitive detection of a protein via Western blot technique or specific DNA/RNA sequence utilizing Southern blot and Northern blot systems respectively within a large number of samples.
- Targets are first separated by size/charge via gel electrophoresis and then picked out utilizing a sensitive complementary probe.
- Alterations in these techniques can detect post-translational modifications and DNA-bound proteins.
- Western blotting may also be used to identify a circulating antibody in a patient sample or confirm an antibody's specificity (Nicolas and Nelson, 2013).

14.4.2.4.2.2 LIMITATIONS OF BLOTTING

- Blotting is more time- and labor-intensive than newer techniques and probably not as sensitive.
- The technique and reagents selected can determine the specificity and sensitivity of blotting.
- Verification of the quantity of molecules present is not as precise as with newer techniques.
- In the case of Western blotting, the tertiary structure of the protein is destroyed; hence the relevant epitope recognized by the primary antibody may not be recognized (Nicolas and Nelson, 2013).

14.5 RECOMBINANT DNA TECHNIQUES

Recombinant DNA technology was made possible through the discovery, and application of restriction enzymes for which Werner Arber, Daniel Nathans, and Hamilton Smith received the 1978 Nobel Prize in Medicine [Physiology or Medicine 1978—Press Release].

With the recent advances in molecular biology, it is now possible to mix genetic material from multiple organisms together (molecular cloning) to create DNA sequences that are otherwise not found naturally in biological organisms. Techniques like molecular cloning are used to create these rDNA molecules in the laboratories. Vectors are used to transfer and express these foreign rDNA fragments in suitable host organisms such as bacteria. R-DNA technology facilitated a whole new world in scientific research. R-DNA technology employs palindromic sequences, and results in the creation of blunt and sticky (staggered) ends (Fig. 14.22). Since its development, many organisms and food products have been genetically modified. Utilizing rDNA technology and synthetic DNA molecules, literally any DNA sequence can be created and inserted into any of a very broad range of living organisms. Thus, it is not surprising that this topic also has many controversies attached to it.

14.5.1 Creation of Recombinant (Artificial) DNA

Recombinant plasmid formation involves construction of rDNA, in which a foreign DNA fragment is inserted into a plasmid vector. The gene indicated by white color in

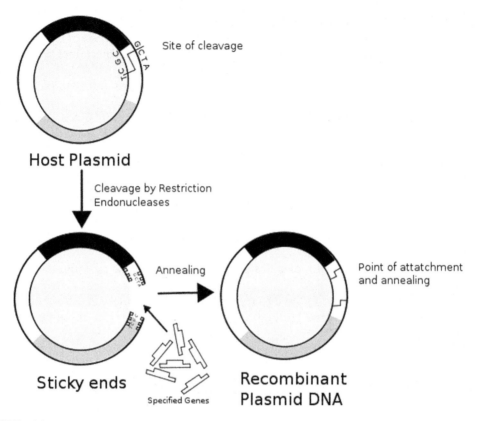

FIGURE 14.22   Recombinant plasmid formation. *Adapted from http://en.wikipedia.org/wiki/Recombinant_DNA.*

Fig. 14.22 is inactivated upon insertion of the foreign DNA fragment illustrated by jigsaw pieces (Fig. 14.22).

14.5.2 Chimeric/rDNA

Recombinant DNA molecules are occasionally referred to as chimeric DNA, because they are usually constructed using materials from two different species. The term "molecular cloning" is used to indicate the laboratory process utilized to make rDNA (Campbell and Reece, 2002; Walter et al., 2008; Berg et al., 2010; Watson, 2007) (Fig. 14.23).

14.5.2.1 Steps of Cloning DNA Fragments (Gene Cloning) to Create rDNA

Step 1. Small circular DNA molecules (plasmids) are removed from bacteria. These plasmids serve as vectors (molecules to carry genes of interest).

Step 2. DNA containing the gene of interest to be cloned is isolated from particular cells/tissues.

Step 3. Unique single site recognizing restriction endonucleases are utilized, which recognize specific restriction site(s) (short sequences of 4–8 bp long).

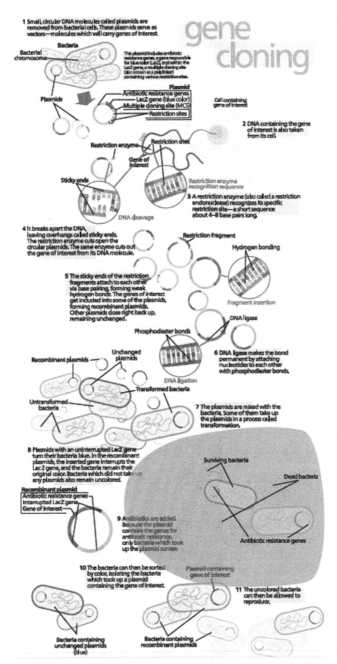

FIGURE 14.23 Gene cloning. *Adapted from http://en.wikipedia.org/wiki/Recombinant_DNA.*

Step 4. These unique restriction endonuclease are used to cut both the plasmid DNA (vector) and the DNA to be cloned (insert) into the plasmid vector, creating either overhangs called sticky ends or filled ends called blunt ends.

Step 5. Following the restriction enzyme, digestion of both the vector and the insert, then the digested vector and insert fragments, whether with sticky ends or blunt ends, are joined together via complementary base pairing, via a DNA ligase enzyme. The gene of interest gets included into some of the plasmids forming recombinant plasmids. Other plasmids close right back up, remaining unchanged (without an insert).

Step 6. DNA ligase enzyme makes the bonds permanent between complementarily paired bases, by attaching nucleotides to each other with phosphodiester bonds.

Step 7. The next experiment is to mix the plasmids from Step 6 with competent bacteria. This process is called transformation. During transformation some of the bacteria take up the plasmids. Here bacteria are utilized to clone (multiply in number) the rDNA.

Step 8. In some transformation experiments, a color-processing gene such as LacZ gene is utilized for confirmation of the molecular cloning (inserting a DNA fragment of interest into a plasmid vector). Plasmids with an uninterrupted LacZ gene turn their bacteria blue. In the recombinant plasmids, the inserted gene interrupts the LacZ gene, and the bacteria remain their original color (white). The bacteria that did not take up any plasmid DNA also remain uncolored/white.

Step 9. For selection against positive bacterial clones (clones that have taken recombinant plasmids), antibiotics are added to the growth media, where the bacteria are grown. As the plasmids contain the genes for antibiotic resistance, only bacteria, which took up the plasmid, survives.

Step 10. The bacteria can then be distinguished by color. The color difference helps to isolate the bacteria, which took up plasmid DNA containing the gene of interest. White-colored bacteria are expected to carry the rDNA plasmids.

Step 11. The white-colored bacteria can then be set aside at 37°C incubator to reproduce, enabling the recombinant plasmids within to multiply in number.

14.5.3 Expression of rDNA

Following transfection (introduction of the rDNA plasmid into the host cell), the foreign DNA contained within the rDNA construct may or may not be expressed. In other words, the DNA may simply be cloned without expression or it may be transcribed and translated so that a recombinant protein is produced. In general, expression of a foreign gene requires restructuring the gene to include sequences that are required for producing an mRNA mole-cule, which can be utilized by the host's translational system (i.e., promoter, translational initial signal, and transcriptional terminator signal) (Hannig and Makrides, 1998).

14.5.4 Applications of rDNA Technology

R-DNA is extensively utilized in medicine, research, and biotechnology. Nowadays, recombinant proteins and other related products, which are derived from the applications of rDNA technology, are found in essentially every doctor's/veterinarian's office, medical testing laboratory, pharmacy, and biological research laboratory. Furthermore, organisms, which have been altered utilizing rDNA technology, as well as products obtained from those organisms, have been introduced into many farms, supermarkets, homes, and even pet shops, such as the ones that sell GloFish, and other genetically altered animals.

The most common utilization of rDNA is in basic research. The technology is also significant in most current research in the biomedical and biological sciences (Brown, 2006). R-DNA technology not only facilitates identification, mapping and sequencing genes, but also it helps to determine their function. R-DNA probes are utilized in characterizing gene expression in individual cells, and the tissues of complete organisms. Recombinant proteins are extensively utilized as reagents in laboratory experiments and to generate antibody probes for analyzing protein synthesis in cells and in organisms (Peter et al., 2008).

Many other functional applications of rDNA are found in industry, food production, human/veterinary medicine, agriculture, and bioengineering (Peter et al., 2008).

The following include some of the examples of the applications for rDNA technology:

- Recombinant human insulin (Gualandi-Signorini and Giorgi, 2001).
- Recombinant human growth hormone (somatotropin) (Von Fange et al., 2008),
- Recombinant blood clotting factor VIII (Manco-Johnson, 2010),
- Golden rice (Paine et al., 2005),
- Insect-resistant crops (Paine et al., 2005).

14.5.5 Controversy of rDNA

The potential for undesirable or hazardous properties of organisms containing rDNA was originally realized by scientists, who were involved with the initial development of rDNA techniques. Nowadays, rDNA molecules and recombinant proteins are usually not considered as dangerous. Nevertheless, concerns still exist regarding some organisms, which express rDNA, especially when they are introduced into the environment or food chain. Some of these concerns include contamination of the nongenetically modified food supply, effects of genetically modified organisms (GMOs) on the environment and nature, the hardship of the regulatory process, and more strict control of the food supply in companies that make and sell GMOs.

14.5.6 Genetically Modified Organisms

A GMO is one whose genetic material has been modified utilizing genetic engineering methodologies. Organisms, which have been genetically altered, include microorganisms such as bacteria and yeast, insects, plants, fish, and mammals. GMOs are the origin of genetically modified foods and are also widely utilized in scientific research.

Genetic alterations involve mutations including, insertion, or deletion of genes. Inserted genes usually originate from different species. In nature, this can take place when exogenous DNA penetrates the plasma membrane for any reason. To achieve this artificially one may require:

- linking the genes to a virus
- physically injecting the foreign DNA into the nucleus of the host

FIGURE 14.24 GloFish is the first genetically modified animal to be sold as a pet. *Adapted from http://en.wiki-pedia.org/wiki/Genetically_modified_organism.*

- with the aid of electroporation (introducing DNA from one organism into the cell of another via an electric pulse)
- with very small particles fired from a gene gun (Sanford et al., 1987; Klein et al., 1987).
- other methods utilize natural forms of gene transfer, including the ability of *Agrobacterium* to transfer genetic material to plants (Lee and Gelvin, 2008) or the ability of lentiviruses to transfer genes to animal cells (Park, 2007) (Fig. 14.24).

14.5.7 Genetically Modified Food

In agriculture, currently marketed genetically modified crops have characteristics such as pest resistance, herbicide resistance, increased nutritional value, or production of beneficial goods such as drugs. Products under development include crops, which are capable of thriving in harsh environmental conditions (i.e., drought or salt resistance).

Since the first commercial cultivation of genetically altered plants in 1996, they have been modified to be resistant to the herbicides, to virus damage, and to produce the Bt toxin, an insecticide, which is documented as nontoxic to mammals [http://www.agf.gov.bc.ca/pesticides/infosheets/bt.pdf]. Plants, including algae, maize, and poplars (Hope, 2013) have been genetically modified for use in producing fuel, known as biofuel.

Second- and third-generation genetically modified crops are available and under development with enhanced nutrition profiles and enhanced yields or potential to grow in harsh environments (http://dtma.cimmyt.org). Genetically modified oilseed crops available today provide enhanced oil profiles for processing or healthier edible oils (Canadian Food Inspection Agency. DD2009-76).

Other examples include:

- Golden rice, developed by the International Rice Research Institute (IRRI) and has been indicated as a possible cure for Vitamin A deficiency.
- A vitamin-enriched corn derived from South African white corn variety.
- *Camelina sativa*, which accumulates high levels of oils similar to fish oils ["Crop plants—'green factories' for fish oils" (2013).

14.6 DNA SEQUENCING TECHNOLOGIES

14.6.1 Brief Introduction

Explanation of nucleic acids, as being essential components of living organisms, is an important era in life sciences by providing into new insights. Understanding the chemical structure of nucleic acids led to the emergence of sequencing technology, which subsequently had a major impact on many areas, especially biochemistry, biology, biotechnology, and medicine (AbouAlaiwi et al., 2012; Alison Van).

Maxam and Gilbert introduced that DNA can be sequenced chemically and he sequenced 24 bp of *E. coli* lac operator in 1973 (Alberts et al., 2008). Early sequencing technology was more time-consuming and laboring. Frederick Sanger introduced a faster sequencing technique called "plus and minus method," also referred as the chain termination method, in which DNA is sequenced with chain-terminating inhibitors, and he sequenced the genome of bacteriophage ϕX174, which was the first genome to be sequenced in 1977 (Anguiano et al., 2012). Sanger's technique was applies to ssDNA. Essentially, with their pioneer discovery of nucleic acid sequencing, Gilbert and Sanger shared the Nobel Prize in Chemistry in 1980 with Paul Berg, due to his fundamental work with biochemistry of nucleic acids.

After Gilbert and Sanger, nucleic acid sequencing techniques improved with alternative separation techniques and visualization strategies and being able to run several parallel samples at a time. Automation in DNA sequencing led to the progress of high-throughput methods in DNA sequencing.

The studies to understand the basics of health and diseases have gathered scientists under the Human Genome Project in 1990. Human genetic structure has been understood by means of intensive works of Human Genome Consortium that is organized by various university and research institute foundations (Arber and Linn, 1969). Human genome project, which aimed at sequencing DNA of human genome, provided the data for 3 billion nucleotides of human (http://www.nobelprize.org/nobel_prizes/chemistry/laureates/1980/). Genomic differences related to diseases and health conditions in humans take shape with structural alterations such as substitution, deletion, insertion of bases [adenine (A), thymine (T), guanine (G), cytosine (C)] that are building blocks of DNA or insertion of DNA fragments that are not in reference genome (Arizona, 1996). In this time, the reveled data updated the project results (Arber and Linn, 1969).

Through sequence analysis of human DNA, the opportunity of the earlier diagnosis and therapy will be achieved by getting information about the reason of every kind of diseases, the possibility of appropriate therapy regimes will be constituted from beginning by predicting the prognosis and the course of disease, and even the generation of disease will be able to be prevented by identifying populations with genetic predisposition against such diseases (Ausubel et al., 1998; Bartlett and Stirling, 2003).

DNA sequencing technology is clearly important in the production of knowledge in biological sciences. Therefore, rapid developments in DNA sequence technologies are progressed to faster, cheaper, accurate, and the most appropriate methods (Berg et al., 2010).

14.6.2 First-Generation Sequencing Techniques

The notion of DNA sequencing was suggested in 1970s. The first form of nucleotide sequencing was RNA sequencing due to easier technical methods. The first step in the RNA sequencing has been taken with the studies performed on Bacteriophage MS2 by Walter et al. Bacteriophage MS2 has been the first known gene in 1972 and the first sequenced genome in 1976 (Bergallo et al., 2006; Bickle and Krüger, 1993; Biss et al., 2014).

The DNA sequencing studies by Frederick Sanger and Allan Maxam and Walter Gilbert (Maxam—Gilbert) majorly contributed to the development of current DNA sequencing methods and made a tremendous impression in 1977.

14.6.2.1 Maxam's and Gilbert's Chemical Method

The sequencing of the DNA bases by degradation of DNA with chemical reaction was developed by Maxam and Gilbert in 1973. First of all, they sequenced the 24 bases of the RNA transcript of E. coli lac operator by their technique based on the restriction of the terminally labeled DNA at each repetition base (Alberts et al., 2008). They performed was to isolate approximately 1000 bp long dsDNA fragments of E. coli, including the lactose promoter operator. The lac repressor binds to cellulose nitrate and when it is mixed with the sonicated DNA fragments from phage that carries lac genes, only fragments bound to the lac operator are filtered. Isopropyl β-D-1-thiogalactopyranoside (IPTG), a synthetic inducer of the lac operon, elutes the repressor bound DNA fragments. The fragments they obtained were approximately 27 bp long, dsDNA. Then they had synthesized RNA molecules from those DNA fragments. They resolved RNA into single molecular species for the synthesis of specific regions. Exploiting the facts that the NTP concentration less than 5 μM represses the triphosphate initiation of RNA polymerase and on denatured DNA, the polymerase would initiate and incorporate into the beginning of RNA chain if it is presented with a di- or poly-nucleotide complementary to the template; they started synthesizing one strand of the operator DNA with ragged 3' ends at a specific point, which resolved them into series of bands during electrophoresis on polyacrylamide gel. GpGp-ApApU oligonucleotide primed RNA synthesis on one strand while dinucleotide UpA and the pentanucleotide ApUpCpCpG on the other strand. GpU primed on both strands, and incorporation of these synthesis led to the 24 bp nucleic acid sequencing of the *E. coli* lac operator (Alberts et al., 2008).

Chemical pathway of reactions was summarized in articles of Maxam—Gilbert in 1977. In this method, DNA of which nucleotide sequence will primarily be detected is marked with 32P or a fluorescent stain from 5′ end. Subsequently, DNA molecules are separated into four tubes. Chemical reactions for a chemical change and degradation of a specific base (A, C, G, or T) are carried out in each of these four individual tubes. These chemical reactions consist of basically two steps. The first step is that glycosidic bond between ribose glucose in nucleotide triphosphates and base is cleaved off with dimethylsulfate in purines and with hydrazine in pyrimidines. The second step is cleavage phosphodiester bonds with piperidine. The first one from four chemical reaction mediums at issue in Maxam—Gilbert sequencing is the medium constituted on cleavage of G nucleotides by dimethylsulfate and piperidine. While dimethylsulfate and piperidine in formic acid generate the second medium cleaving off both G and A nucleotides, similarly, hydrazine and piperidine form both T and C nucleotides (the third medium), and hydrazine and piperidine in 2M NaCl make up the other mediums cleaving C nucleotides only (the fourth medium). Following selective DNA sequencing reactions, ssDNA, of which 5′ end is radioactively marked, is obtained. After reactions, the fragments are loaded into high-percentage polyacrylamide, and electrophoretic separation is done. Because the fragments in gels are radioactively marked, they are enabled visible by using autography technique, and base sequence is identified.

The marked DNA fragments are sequenced by chemical destruction reactions specific for bases. While the method can be used for both ssDNA and dsDNA, DNA polymerase enzyme is not necessary. The fragments of radioactively marked four bases are base-specifically degraded, and four bases are electrophoretically separated in poly-acrylamide planar (slab) gel separation lines (Bishop, 2010).

14.6.2.2 Sanger Sequencing

The modern DNA sequencing method has started after Frederick Sanger and Alan R. Coulson have published their studies for the determination of DNA base sequence in 1975 (Anguiano et al., 2012; Boyd, 2013). Sanger introduced a less complex method that is more likely to be scaled up, which is called "the chain termination method," also referred as "plus and minus sequencing." The discovery of DNA sequencing in biology in those days has become a tool used to decode genes, even all genomes (Boyer, 1971). Although this method has made an overwhelming impression in the scientific world, it has brought the Nobel Prize to Frederick Sanger, its creator, in 1980.

Developments in studies of genotyping compete with Sanger sequencing (accepted as chain termination method, enzymatic method, or dideoxy method in the literature), which is gold standard in this field. Sanger DNA sequencing technique is based on the ground of synthesizing new DNA fragments by joining the reaction of normal deoxynucleotides (dNTP) and dideoxynucleotides (ddNTPs) causing the termination of the new synthesizing DNA chain. In this method, four types of dideoxynucleotides (ddATP, ddGTP, ddCTP, and ddTTP) that are specific chain terminators join in synthesis reaction in very controlled amounts and durations. Due to the use of four kinds of ddNTP, a preparation for four individual reaction mediums is in question. In each of the reaction medium, four kinds of ddNTP, DNA polymerase I (Klenow fragment), four kinds of dNTP (dATP, dGTP, dCTP, and dTTP—these nucleotides are required for DNA polymerization), DNA template of which base sequence will be detected, and primers that are complementary to DNA bases at the 3′ end of this and marked with 32P at the 5′ ends are found (Kunkel and 2002, 1985).

The lack of oxygen in Position 3′ and the presence of only H in deoxyriboses, which are found in the structure of ddNTPs, stop polymerization during synthesis process in the event that these ddNTPs incorporate chain. So ddNTPs cannot form fosfodiester bond with other molecules of the lack of OH groups. At the end of all procedures, the different DNA segments that have occurred in four kinds of reaction medium are individually electrophoresed and become separated according to their lengths. Then, these become distinct black bands with autoradiography, and their base sequences are tried to be detected comparing each other. Sanger sequencing yields reads approximately around 800−1000 bases in length (Boyd, 2013; Bridge, 2008; Brown, 2006; Bustin, 2002; Bustin, 2000).

14.6.3 Automation in DNA Sequencing

In first-generation sequencing techniques, radioactive isotopes were used to label the DNA. To reduce the use of radioactive materials in DNA sequencing, which cause health threats for people working in the laboratory, fluorescent tracers were used in automated DNA sequencing. Around 1980s, most of the automated DNA sequencing methods combined PCR with fluorescence labeling (Canadian Food Inspection Agency; Campbell and Reece, 2002; Catalina et al., 2007; Capuano et al., 2014).

Use of fluorescent-marked primers and ddNTPs: Standard slab PAGE and autoradiography technique equipments were being used to separate sequencing reaction products in the original DNA sequencing systems. PAGE is a separation method according to molecule dimension. PAGE is obtained from constitution of long chains with acrylamide monomers by polymerizing. Gels that have pores in different dimensions are formed with changes in the amount of polyacrylamide placed into the gel. DNA molecules are also decomposed with their capture to these pores. Because the greater DNA molecules would be captured to pores, they would move slower throughout gel. That will also cause that great DNA fragments remain at the upper part of gel in proportion small fragments. Conversely, the smaller fragments move faster in the gel and bands are observed at the lower sheets of the gel.

Autoradiography technique used for detection was being performed by marking DNA fragments with radioactive substances such as 32P and 35S. Gels used were being exposed to X-rays, and detection was being actualized from the diagram obtained. In chain termination method performed by marking DNA fragments with radioactive phosphor, several detection-based changes in the sequence analysis have been made. One of these is the 5′ end of the primer to be labeled using fluorescent dye. However, this detection-based change has not removed the necessity of four individual reaction mediums. The approach of enabling automation with easy, fast, and economic analysis by optic system has developed dye terminator sequencing for marking DNA fragments with dyes. The next development is high automated throughput DNA sequencing with fluorescent-marked primer and ddNTPs, developed by Hood and coworkers (Carrigan et al., 2011; Clark, 2005).

Dye Terminator Sequencing: Dye terminator sequencing, by marking with chain terminators, is an alternative method for the sequencing with marked primers. The greatest advantage of this method in comparison with sequencing performed using the marked primer is the procedure of sequencing to be realized in a single reaction medium (four reaction mediums are used in the marked primary medium). In dye terminator sequencing, each of four dideoxynucleotide chain terminators is marked with a different fluorescent dye and each fluorescent has a different wavelength. The restriction of this method is the dye efficacy that has a voice in the formation of inequalities emerging in length and shape of peaks obtained as a result of capillary electrophoresis. In a majority of sequencing projects, the method of dye terminator sequencing is used along with automated high-throughput DNA sequence analyzers. Thus, fast, low-priced sequencing procedures are performed (Clarke and Higgins, 2000).

Using capillary electrophoresis: Not long after introduction of sequencing systems by using slab gel, capillary electrophoresis (CE)-based sequencing systems have been developed. CE-based sequencing systems, which have more sample capacity than slab gel systems, are bringing quick detection, ease of use, and more developed accuracy (Chuang et al., 2008; Dahm, 2008; Dahm, 2005).

Difficulties and limitations of autoradiography technique have been prevented by means of detection actualizing with the help of dyes that are mounted at DNA fragments in CE technique. CE separation has various advantages as compared with slab gel–based sequencing. The first one of these is capillary systems to have a moveable sheet to enable matrix replacement among separations. In separations with slab gel, the gel is poured between glass sheets and is polymerized. It is difficult to pour gels without creating balloon, and its preparation takes time. However, this has been prevented in capillary-based automated systems. The second advantage is the redundancy of the number of sample worked at a sitting as a consequence of integrating capillary systems with microtiter plates. In the wake of integrating CE-based separation with 384-well plates, separation at the same time up to 384 samples has been enabled. Also the use of dyes attached at DNA fragments instead of radioactive substances in detection to prevent the damage of radioactive substances to employee's health is the third advantage (Kunkel, 1985; Braman, 2002; Burden, 2012).

Analysis of 384 samples can be carried out in a single work in automated DNA sequencing devices modernized with combination of several technologies. In DNA sequencing done with DNA dimension-based separations with CE, fluorescent peak chromatograms, which are raw data, are obtained from the detection made with the record of fluorescent dyes (Kunkel, 2002, 1985; Burden, 2012).

14.6.4 Developments and High-Throughput Methods in DNA Sequencing

Sanger sequencing is doubtlessly a unique method in the identification of polymorphism/mutations previously described. Nevertheless, it is a frequently applied method in the verification of a polymorphism found via other genotyping systems or in the cases that no result is obtained from again such systems. However, apart from being the gold standard, a pursuit of new genotyping methods is in question due to intensive labor and the presence of a long-standing work flow (David, 2000; Downie et al., 1997).

Although Sanger sequencing is the most used base sequencing method, it might be unable to distinguish mutations at very small rates among large wild-type sequences. This situation causes a problem in working with samples incorporating different cell groups in different type such as solid tumors. Solid tumors have a heterogeneous tissue structure, because of neoplastic and non-neoplastic cell groups. Therefore, the identification of gain of function mutations in oncogenes becomes difficult because not only normal cells but also neoplastic cells have two alleles, and one of this allele sets in neoplastic cells generally is wild type. In such cases, methods that are sensitive, cost-effective, and user-friendly become a necessity (Dryden et al., 2001; Dussoix and Arber, 1962).

With increases in sequencing-based researches, the pursuits for sequencing methods from which low-priced and quick results are obtained have caused the emergence of high-throughput sequencing technologies with which thousands, even millions of sequence analysis can be made at a time. Second-generation sequencing tools emerged commercially around 2005 due to the low throughput of first-generation methods. The most fascinating property of these second-generation sequencing tools was their ability to automatically run more than one sample in parallel in shorter time periods. In the second-generation sequencing, the "wash and scan" cycle processes took place. In these cycles, the labeled nucleotides in flooding reagent were incorporated into newly synthesized DNA strands; the incorporation reaction then stopped and washed for the removal of excessive reagents and scanning for the incorporated bases steps. The cycle is repeated until the reaction cannot be determined by scanning (Bridge, 2008).

14.6.4.1 *Pyrosequencing Method*

Pyrosequencing is a sequence-by-synthesis based method, which uses bioluminometric determination in solution (Edwards et al., 1960). In this method, pyrophosphate (PPI) evolved as proportional to the amount of nucleotide joining DNA chain that is newly being synthesized is converted into an apparent light via enzymatic reactions series, and a bioluminometric measure is done. The enzymes controlling chemical reactions in this process are DNA polymerase I (Klenow fragment), ATP sulfurylase, luciferase, and apyrase. First reaction, if the nucleotide added to the reaction base pairs with the sequencing template, DNA polymerization occurs. The inorganic pyrophosphate is released by Klenow DNA polymerase and is used as a substrate for ATP sulfurylase leading to ATP formation as a second reaction. Then ATP is converted to Luciferase and light is detected. Apyrase then removes excess (unincorporated) nucleotides and ATP between different base additions, which are required for synchronized DNA synthesis and precise observation of light upon correct nucleotide addition (Edwards et al., 1960). By using 4-enzyme-based pyrosequencing, approximately 1–100 bp can be sequenced. The read length can be increased up to 300 bases with a 3-enzyme-based pyrosequencing system; however, then the automation of the system fails upon requirement of a watching step in between and the DNA template loss is a limiting factor for this type of pyrosequencing (Eyler, 2013). Pyrosequencing has two important restrictions. While the first one is the necessity of specialized equipment and education about this matter, the second one is that it does not allow to genome sequencing arising from the scantiness of base reading length. Despite these restrictions, however, the excess of analysis options such as working with a large number of samples through 96- or 384-well plates, allelic imbalance, methylation, and determination of the number of gene copy increase its preferability (Fan, 2004).

14.6.4.2 *The Genome Sequencer 454 FLX System*

In 2004, a Roche group (454 Life Technologies) introduced the FLX next-generation sequencing system to the market, which is also based on the pyrosequencing technique (Ford and Hamerton, 1956). Using emulsion PCR amplification (ePCR) in this method, template DNA is synthesized into water drops inside the fat solution (Freeman et al., 1999). Template DNA, which is bound to the primer that is united with bead, is present in each drop. Inside this drop, a clonal colony is generated with PCR procedure. Each well in the sequencing machine including wells that have a picoliter volume contains a single bead and sequencing enzymes. From small to large samples, a wide scale of sequencing can be performed on gDNA, PCR products, BACs, and cDNA. Larger fragments are needed to be fractioned into smaller pieces (300–800 bp) by nebulization before sequencing, while for small samples, no fragmentation required. Short A and B adaptors are added to fragments, which are required for purification, amplification, and sequencing. Single-strand DNA fragments (ssDNA library) with adaptors are mixed with Sepharose beads carrying complementary oligonucleotides, and this mixture is emulsified with amplification reagents in water-in-oil mixture for the ePCR procedure. Each bead is captured where its own clonal amplification of ssDNA fragment occurs, which results in bead-immobilized, clonally amplified DNA fragments. ssDNA library beads are then added into DNA Bead Incubation Mix and layered with Enzyme Beads (to ensure the position of the DNA beads during sequencing) to the 454 PicoTiterPlate device for the sequencing reaction. One library bead fits into each well of the plate. PicoTiterPlate is then loaded into the Genome Sequencer FLXTM Instrument, and while the fluidics flows across the plate, ssDNA molecules are sequenced upon the CCD camera detection of light that is generated due to the incorporation of the flowing complementary nucleotide into the DNA. Approximately 400–600 Mb of sequence can be read per run with the FLX system (Frenzilli et al., 2009; Galas and Schmitz, 1978).

14.6.4.3 *Illumina/Solexa Genome Analyzer*

As another example from new generation high-throughput sequence analysis devices, Illumina/Solexa Genome Analyzer, can be given. It is a sequencing technology developed based on reversible dye terminators. The terminator is then cleaved for the incorporation of the next nucleotide. All four terminator-bound dNTPs are present in each sequencing cycle, ensuring more precise sequencing. A camera in the medium gets images of fluorescent-marked nucleotides, and then, the fluorescent dye at the blocker end at the 3′ end is moved away from DNA with chemical degradation. Illumina can generate more than 200–300 gigabases of data per run. The raw base accuracy for this method is 99.5% and is the most widely used sequencing technique (Galas and Schmitz, 1978; Goodsell, 2002).

14.6.5 Transition Sequencing Techniques

Third-generation sequencing techniques eliminate the reaction pause step that occurs in each base incorporation during second-generation sequencing techniques (Bridge, 2008). This development in sequencing field increases the sequencing rates and sequencing lengths as well as lowering the sample preparation complexity. Second-molecule sequencing technologies (sequencing single DNA/RNA molecule) are the promising third-generation sequencing techniques. However, there exist some technologies that neither belongs to the third generation nor the second generation. These technologies clearly show the developmental period of third-generation sequencing techniques and make us understand that the third-generation sequencing technologies better.

14.6.5.1 Ion-Torrent's Semiconductor Sequencing

In Ion-Torrent's semiconductor sequencing technology, by the use of semiconductor technology, high-density array of microwells were produced and the sequencing is carried out by the sensation of hydrogen ions that are released upon base incorporation process (Bridge, 2008). The technique simplifies the sequencing process and eliminates the need for light to scan the process as well as lowers the cost. However, the technology still uses the second-generation sequencing system, "wash and scan," and uses PCR in each well for DNA template amplification. Therefore, the time required for sequencing and the throughput is limited with the second-generation sequencing systems (Bridge, 2008), which makes this technique stand somewhere between second- and third-generation sequencing methods.

14.6.5.2 Helico's Genetic Analysis Platform

Helico's genetic analysis platform is actually the first commercial system that carries out second-molecule sequencing technology (Geerlof, 2008; Gualandi-Signorini and Giorgi, 2001; Hampshire et al., 2007; Hannig and Makrides, 1998). This platform uses the defined primers, a modified polymerase, and fluorescently labeled nucleotide analogues and visualizes the DNA molecules as they are attached to a planar surface while being synthesized. Fluorescently labeled nucleotide analogues used in this platform, called Virtual Terminator nucleotides, allow step-wise sequencing due to attachment of the dye to the nucleotide by a chemically cleavable group (Gualandi-Signorini and Giorgi, 2001). Similar to second-generation sequencing technology, the sequencing length is approximately 32 nucleotides long and the time to sequence a single nucleotide is long. On the other hand, unlike second-generation sequencing techniques, PCR is not required in this platform. As in all second-molecule sequencing technologies, the reading error rates is more than 5% unlike its high fold coverage and 99% read accuracy (Bridge, 2008).

14.6.6 Third-Generation Sequencing Techniques

Third-generation sequencing techniques (second-molecule sequencing), which are mainly under development, can be summed up in three basic categories as (1) sequence-by-synthesis technologies in which single DNA polymerase molecules are visualized as they synthesize the single DNA molecule, (2) nanopore sequencing technologies in which individual bases are detected as they pass through the nanopore, and (3) direct individual DNA molecule imaging with advanced microscopic techniques. Each category is used for specific applications. In this section, we will try to explain these various third-generation techniques briefly.

In the first category, second molecule sequencing by synthesis, single molecule real-time sequencing (by Pacific Biosciences), which is the first third-generation sequencing technique to observe DNA polymerase in real time, and real-time DNA sequencing with fluorescence resonance energy transfer (FRET) (by Visigen Biotechnologies/ Life Technologies) are the two methods used for sequencing. In real-time DNA sequencing with FRET, the method relies on the emission of an FRET signal by a fluorophore-tagged DNA polymerase when brought in close proximity to an acceptor fluorophore-tagged nucleotide. After the nucleotide incorporation, the released fluorophore label can be detected (Bridge, 2008).

In the second category, direct DNA imaging with advanced microscopic techniques, direct imaging of DNA by transmission electron microscopy (TEM) and by scanning tunneling microscopy (STM) are the two methods, respectively. Direct imaging of DNA sequences by TEM (by Halcyon Molecular) detects nonperiodic atoms on a planar surface and uses the dark-field imaging in scanning TEM (Heckman and Pease, 2007). In direct DNA sequence imaging by tunneling microscopy (by Reveo), the nucleotides are detected electronically from a conductive surface in which the DNA is attached, by using STM (Henderson et al., 1972).

In the third category, DNA sequencing with nanopores, the methods use small amount of single unmodified DNA molecules and relies on the detection of bases as they pass through the nanopore due to their electric current effects or optical signals, and several techniques are proposed for nanopore sequencing (Bridge, 2008).

14.7 CONCLUSION

DNA sequencing technology has a wide application impact. Foundation of DNA as a hereditary material and observation of some disease conditions in several members of the same family increased the demand for identification of the nucleotides and their relation to certain diseases. By the discovery of sequencing technology, DNA sequencing has become one of the milestones of biological sciences. Sequencing of the human genome led to the identification of molecular basis of several disorders, while many still are on hold to be discovered. DNA sequencing is being used for many purposes ranging from whole-genome sequencing of many organisms, DNA/RNA sequences, SNP analysis, methylation analysis to protein—nucleic acid interactions, and gene regulation analysis. Today, the personalized therapies for many diseases are performed with the help of sequencing technologies.

The Sanger sequencing that is a quite inconvenient method in the years when it has been started to be used have become automated by the addition of fluorescent-marked primer-ddNTPs and capillary gel electrophoresis into the method and has enabled scientific researches to get easy and to increase. The rise of information about the method has enabled to increase its applicability. Thus, the growth of the interest in DNA sequencing has come with an increase in the number of publication intended for this method.

The studies to increase its applicability and decrease its cost due to advantages of DNA sequencing methods become crucial. New generation high-throughput sequencing devices achieved with these studies have enabled scientific researches to get easy and to become cheap. With the research outcomes obtained, what the functional effect of the change at the genetic level (SNP, structural or fragmental copies) is will be characterized, and this information will also be used for "personal medicine" or "personal therapy." The cognition of genetic differences between humans through these ways will pave the way for personalized therapies and increase the benefit that physicians expected from clinical practices. Adequately getting cheap of sequencing technologies will involve the prescanning of DNAs of individuals in terms of possible genetic predispositions and to take precautions for this (Henderson et al., 1972; Hope, 2013; Houldsworth and Chaganti, 1994; Hsu, 1952; Huggett et al., 2013; Hughes, 1952). The developments in the direction of personalized therapy, which is a new approach in the medical field, indicate that new generation DNA sequencing methods in particular will be one of the basic laboratory methods in the protective and therapeutic medical applications.

References

AbouAlaiwi, W.A., Rodriguez, I., Nauli, S.M., 2012. Spectral karyotyping to study chromosome abnormalities in humans and mice with polycystic kidney disease. J. Vis. Exp. 60, pii: 3887.

Alberts, B., Johnson, A., Lewis, J., Raff, M., Roberts, K., Walter, P., 2008. Molecular Biology of the Cell, fifth ed. Garland Science, Taylor & Francis Group, New York, pp. 538–539.

Alison, Van E. DNA-based technologies. <http://animalscience.ucdavis.edu/animalbiotech/My_Laboratory/Publications/NBCEC-SireSelectionManualChapter.pdf.>.

Alwine, J.C., Kemp, D.J., Stark, G.R., 1977. Method for detection of specific RNAs in agarose gels by transfer to diazobenzyloxymethyl-paper and hybridization with DNA probes. Proc. Natl. Acad. Sci. U.S.A. 74 (12), 5350–5354.

Anguiano, A., Wang, B.T., Wang, S.R., Boyar, F.Z., Mahon, L.W., El Naggar, M.M., et al., 2012. Spectral karyotyping for identification of constitutional chromosomal abnormalities at a national reference laboratory. Mol. Cytogenet. 5, 3.

Arber, W., Linn, S., 1969. DNA modification and restriction. Annu. Rev. Biochem. 38, 467–500.

Arizona, B., 1996. The Biology Project: Molecular Biology. The University of Arizona, Arizona.

Ausubel, F.M., Moore, D., Struhl, K., Seidman, J.G., Moore, D.D., 1998. Current Protocols in Molecular Biology, first ed. John Wiley, John and Sons Inc., New York.

Baldwin, D., Crane, V., Rice, D., 1999. A comparison of gel-based, nylon filter and microarray techniques to detect differential RNA expression in plants. Curr. Opin. Plant Biol. 2, 96–103.

Bartlett, J.M.S., Stirling, D., 2003. A Short History of the Polymerase Chain Reaction. PCR Protocols 226, pp. 3–6. Patent: US4683195.

Berg, J.M., Tymoczko, J.L., Stryer, L., 2010. Biochemistry (Biochemistry (Berg)), seventh ed. W.H. Freeman and Company, New York.

Bergallo, M., Costa, C., Gribaudo, G., Tarallo, S., Baro, S., Negro Ponzi, A., et al., 2006. Evaluation of six methods for extraction and purification of viral DNA from urine and serum samples. New Microbiol. 29, 111–119.

Bickle, T.A., Krüger, D., 1993. Biology of DNA restriction. Microbiol. Rev. 57, 434–450.

Bishop, R., 2010. Applications of fluorescence in situ hybridization (FISH) in detecting genetic aberrations of medical significance. Biosci. Horizons 3, 1.

Biss, M., Hanna, M.D., Xiao, W., 2014. Isolation of yeast nucleic acids. Meth. Mol. Biol 1163, 15−21.

Bor, Y.C., Swartz, J., Li, Y., Coyle, J., Rekosh, D., Hammarskjold, M.L., 2006. Northern blot analysis of mRNA from mammalian polyribosomes. Nat. Protoc . Available from: https://doi.org/10.1038/nprot.216.

Boyd, S.D., 2013. Diagnostic applications of high-throughput DNA sequencing. Annu. Rev. Pathol. 8, 381−410.

Boyer, H.W., 1971. DNA restriction and modification mechanisms in bacteria. Annu. Rev. Microbiol. 25, 153−176.

Braman, J. (Ed.), 2002. In Vitro Mutagenesis Protocols. Methods in Molecular Biology, vol. 182. second ed. Humana Press, Totowa, NJ.

Bridge, J.A., 2008. Advantages and limitations of cytogenetic, molecular cytogenetic, and molecular diagnostic testing in mesenchymal neoplasms. J. Orthop. Sci. 13, 273−282.

Brown, T., 2006. Gene Cloning and DNA Analysis: An Introduction. Blackwell Pub, Cambridge, MA.

Burden D.W., 2012. Guide to the disruption of biological samples—2012, Version 1.1. Random Primers (12): 1−25 (updated June 4, 2012).

Bustin, S.A., 2000. Absolute quantification of mRNA using real-time reverse transcription polymerase reaction assays. Mol. Endocrinol. 25 (2), 169−193.

Bustin, S.A., 2002. Quantification of mRNA using real-time reverse transcription PCR (RT-PCR): trends and problems. J. Mol. Endocrinol. 29, 23−39.

Campbell, N.A., Reece, J.B, 2002. Biology, sixth ed. Addison Wesley, San Francisco, pp. 375−401.

Canadian Food Inspection Agency. DD2009-76: Determination of the Safety of Pioneer Hi-Bred Production Ltd.'s Soybean (*Glycine max* (L.) Merr.) Event 305423 Issued: 2009-04.

Capuano, F., Muelleder, M., Kok, R.M., Blom, H.J., Ralser, M., 2014. Cytosine DNA methylation is found in *Drosophila melanogaster* but absent in Saccharomyces cerevisiae, Schizosaccharomyces pombe and other yeast specise. Anal. Chem. 86, 3697−3702.

Carrigan, P.E., Ballar, P., Tuzmen, S., 2011. Site-directed mutagenesis. In: DiStefano, J.K. (Ed.), Disease Gene Identification, Part 4, Methods in Molecular Biology (Methods and Protocols), 1, vol. 700: Springer, New York, pp. 107−124.

Catalina, P., Cobo, F., Cortés, J.L., Nieto, A.I., Cabrera, C., Montes, R., et al., 2007. Conventional and molecular cytogenetic diagnostic methods in stem cell research: a concise review. Cell. Biol. Int. 31, 861−869.

Chuang, L.-Y., Yang, C.H., Tsui, K.H., Yu-Huei, C., Chang, P.l., Wen, C.H., et al., 2008. Restriction enzyme mining for SNPs in genomes. Anticancer Res. 28, 2001−2008.

Clark, D.P., 2005. Molecular Biology. Elsevier Academic Press, Amsterdam.

Clarke, G.M., Higgins, T.N., 2000. Laboratory investigation of hemoglobinopathies and thalassemias: review and update. Clin. Chem. 46 (8 Pt 2), 1284−1290.

Corley, R.B., 2005. A Guide to Methods in the Biomedical Sciences. Springer, New York 11.

Dahm, R., 2005. Friedrich Miescher and the discovery of DNA. Dev. Biol. 278, 274−288.

Dahm, R., 2008. Discovering DNA: Friedrich Miescher and the early years of nucleic acid research. Hum. Genet. 122, 565−581.

David, H., 2000. Modern Analytical Chemistry. McGraw-Hill, p. 315.

Downie, S.E., Flaherty, S.P., Matthews, C.D., 1997. Detection of chromosomes and estimation of aneuploidy in human spermatozoa using fluorescence in-situ hybridization. Mol. Hum. Reprod. 3, 585−598.

Dryden, D.T., Murray, N.E., Rao, D.N., 2001. Nucleotide triphosphate-dependent restriction enzymes. Nucleic Acids Res. 29, 3728−3741.

Durand, G.M., Zukin, R.S., 1993. Developmental regulation of mRNAs encoding rat brain kainate/AMPA receptors: a northern analysis study. J. Neurochem. 61 (6), 2239−2246.

Dussoix, D., Arber, W., 1962. Host specificity of DNA produced by *Escherichia coli*. II. Control over acceptance of DNA from infecting phage lambda. J. Mol. Biol. 5, 37−49.

Edwards, J.H., Harnden, D.G., Cameron, A.H., Crosse, V.M., Wolff, O.H., 1960. A new trisomic syndrome. Lancet 1, 787−790.

Eyler, E., 2013. Explanatory chapter: nuclease protection assays. Methods Enzymol 530, 89−97.

Fan, Y.S., 2004. Molecular Cytogenetics, Protocols and Applications. Humana Press, Totowa, NJ.

Ford, C.E., Hamerton, J.H., 1956. The chromosomes of man. Nature 178, 1020.

Freeman, W.M., Walker, S.J., Vrana, K.E., 1999. Quantitative RT-PCR: pitfalls and potential. Biotechniques 26, 112−122. 124−5.

Frenzilli, G., Nigro, M., Lyons, B.P., 2009. The Comet assay for the evaluation of genotoxic impact in aquatic environments. Mutat Res. 681, 80−92.

Galas, D., Schmitz, A., 1978. DNAse footprinting: a simple method for the detection of protein-DNA binding specificity. Nucleic Acids Res 5, 3157−3170.

Geerlof, A., 2008. Cloning using restriction enzymes. European Molecular Biology Laboratory, Hamburg.

Goodsell, D.S., 2002. The molecular perspective: restriction endonucleases. Stem Cells 20, 190−191.

Gortner, G., Pfenninger, M., Kahl, G., Weising, K., 1996. Northern blot analysis of simple repetitive sequence transcription in plants. Electrophoresis 17 (7), 1183−1189.

Gualandi-Signorini, A., Giorgi, G., 2001. Insulin formulations—a review. Eur. Rev. Med. Pharmacol. Sci. 5, 73−83.

Hampshire, A., Rusling, D., Broughton-Head, V., Fox, K., 2007. Footprinting: a method for determining the sequence selectivity, affinity and kinetics of DNA-binding ligands. Methods 42, 128−140.

Hannig, G., Makrides, S., 1998. Strategies for optimizing heterologous protein expression in *Escherichia coli*. Trends Biotechnol. 16 (2), 54−60.

Hayes, P.C., Wolf, R.C., Hayes, J.D., 1989. Blotting techniques for the study of DNA, RNA, and proteins. BMJ 299, 965−968.

Heckman, K.L., Pease, L.R., 2007. Gene splicing and mutagenesis by PCR-driven overlap extension. Nat. Protoc. 2, 924−932.

Henderson, A.S., Warburton, D., Atwood, K.C., 1972. Location of rDNA in the human chromosome complement. Proc. Natl Acad. Sci. U.S.A. 69, 3394−3398.

Hirano, S., 2012. Western blot analysis. Methods Mol. Biol. 926, 87−97.

Hope, A., 2013. News in brief: The Bio Safety Council... Flanders Today, p. 2.

Houldsworth, J., Chaganti, R.S., 1994. Comparative genomic hybridization: an overview. Am. J. Pathol. 145, 1253−1260.

Hsu, T.C., 1952. Mammalian chromosomes in vitro I, The karyotype of man. J. Hered. 43, 167−172.

Huggett, J.F., Foy, C.A., Benes, V., Emslie, K., Garson, J.A., Haynes, R., et al., 2013. The digital MIQE guidelines: minimum information for publication of quantitative digital PCR experiments. Clin. Chem 59, 892−902.

Hughes, A., 1952. Some effects of abnormal tonicity on dividing cells in chick tissue cultures. Quar. J. Micro. Sci. 93, 207−220.

Jeffreys, A.J., 2005. Genetic fingerprinting. Nat. Med. 11 (10), 1035−1039.

Jin, L., Lloyd, R.V., 1997. In situ hybridization: methods and applications. J. Clin. Lab. Anal. 11 (1), 2−9.

Kary, M. Nobel Lecture, December 8, 1993.

Kessler, C., Manta, V., 1990. Specificity of restriction endonucleases and DNA modification methyltransferases a review (Edition 3). Gene 92, 1−248.

Kevil, C.G., Walsh, L., Laroux, F.S., Kalogeris, T., Grisham, M.B., Alexander, J.S., 1997. An improved, rapid Northern protocol. Biochem. Biophys. Res. Comm. 238, 277−279.

Kidwai, F.K., Liu, H., Toh, W.S., Fu, X., Jokhun, D.S., Movahednia, M.M., et al., 2013. Differentiation of human embryonic stem cells into clinically amenable keratinocytes in an autogenic environment. J. Investig. Dermatol. 133, 618−628.

Klein, T.M., et al., 1987. High-velocity microprojectiles for delivering nucleic acids into living cells. Nature 327, 70−73.

Kobayashi, I., 2001. Behavior of restriction-modification systems as selfish mobile elements and their impact on genome evolution. Nucleic Acids Res 29, 3742−3756.

Krüger, D.H., Bickle, T.A., 1983. Bacteriophage survival: multiple mechanisms for avoiding the deoxyribonucleic acid restriction systems of their host. Microbiol. Rev. 47, 345−360.

Kunkel, T.A., 1985. The mutational specificity of DNA polymerase-beta during in vitro DNA synthesis. Production of frameshift, base substitution, and deletion mutations. J. Biol. Chem. 260, 5787−5796.

Lederberg, S., Meselson, M., 1964. Degradtion of non-replicating bacteriophage DNA in non-accepting cells. J. Mol. Biol. 8, 623−628.

Lee, L.Y., Gelvin, S.B., 2008. T-DNA binary vectors and systems. Plant Physiol. 146, 325−332.

Liang, P., Pardee, A.B., 1995. Recent advances in differential display. Curr. Opin. Immunol. 7, 274−280.

Livak, K.J., Schmittgen, T.D., 2001. Analysis of relative gene expression data using real-time quantitative PCR and the 2-(Delta Delta CT). Method. Methods 25, 402–408.

Maggio, A., Giambona, A., Cai, S.P., Wall, J., Kan, Y.W., Chehab, F.F., 1993. Rapid and simultaneous typing of hemoglobin S, hemoglobin C, and seven Mediterranean beta-thalassemia mutations by covalent reverse dot-blot analysis: application to prenatal diagnosis in Sicily. Blood 81, 239–242.

Manco-Johnson, M.J., 2010. Advances in the care and treatment of children with hemophilia. Adv. Pediatr. 57 (1), 287–294.

Meisel, A., Bickle, T.A., Krüger, D.H., Schroeder, C., 1992. Type III restriction enzymes need two inversely oriented recognition sites for DNA cleavage. Nature 355, 467–469.

Meselson, M., Yuan, R., 1968. DNA restriction enzyme from *E. coli*. Nature 217, 1110–1114.

Miles, J.S., Wolf, R.C., 1989. Principles of DNA cloning, BMJ, 299. Scientific Tools in Medicine, pp. 1019–1022.

Mori, H., Takeda-Yoshikawa, Y., Hara-Nishimura, I., Nishimura, M., 1991. Pumpkin malate synthase cloning and sequencing of the cDNA and Northern blot analysis. Eur. J. Biochem. 197 (2), 331–336.

Nasiri, H., Forouzandeh, M., Rasaee, M.J., Rahbarizadeh, F., 2005. Modified salting-out method: high-yield, high-quality genomic DNA extraction from whole blood using laundry detergent. J. Clin. Lab. Anal. 19, 229–232.

Nicolas, M.W., Nelson, K., 2013. North, South, or East? blotting techniques. J. Invest. Dermatol. 133, e10.

New England Biolabs Inc., 2014. Restriction Endonucleases. <https://www.neb.com/products/restriction-endonucleases/restriction-endonucleases>.

c. Jpn Acad. Ser. B Phys. Biol. Sci. 2010. 86, 103–116.

Paine, J.A., Shipton, C.A., Chaggar, S., Howells, R.M., Kennedy, M.J., Vernon, G., et al., 2005. Improving the nutritional value of Golden Rice through increased pro-vitamin content. Nat. Biotechnol. 23, 482–487.

Park, F., 2007. Lentiviral vectors: are they the future of animal transgenesis? Physiol. Genomics 31 (2), 159–173.

Parker, K., 2014. Retrieved from SelectScience: <http://www.selectscience.net/pcr_buying_guide.aspx#section1>.

Pena, R.N., Quintanilla, R., Manunza, A., Gallardo, D., Casellas, J., Amills, M., 2014. Application of the microarray technology to the transcriptional analysis of muscle phenotypes in pigs. Anim. Genet. 45 (3), 311–321.

Peter Walter; Alberts, Bruce; Johnson, Alexander S.; Lewis, Julian; Raff, Martin C.; Roberts, Keith (2008). Molecular Biology of the Cell (5th edition, Extended version). New York: Garland Science. ISBN 0-8153-4111-3. Fourth edition is available online through the NCBI Bookshelf.

Pingoud, A., Jeltsch, A., 2001. Structure and function of type II restriction endonucleases. Nucleic Acids Res 29, 3705–3727.

Pingoud, A., Alves, J., Geiger, R., 1993. Chapter 8: restriction enzymes. In: Burrell, M. (Ed.), Enzymes of Molecular Biology. Methods of Molecular Biology, vol. 16. Humana Press, Totowa, NJ, pp. 107–200.

Pirastu, M., Ristaldi, M.S., Cao, A., 1989. Prenatal diagnosis of beta-thalassaemia based on restriction endonuclease analysis of amplified fetal DNA. J. Med. Genet. 26, 363–367.

Polisky, B., Greene, P., Garfin, D.E., McCarthy, B.J., Goodman, H.M., Boyer, H.W., 1975. Specificity of substrate recognition by the EcoRI restriction endonuclease. Proc. Natl. Acad. Sci. U.S.A. 72, 3310–3314.

Renart, J., Reiser, J., Stark, G.R., 1979. Transfer of proteins from gels to diazobenzyloxymethyl-paper and detection with antisera: a method for studying antibody specificity and antigen structure. Proc. Natl. Acad. Sci. U.S.A. 76 (7), 3116–3120.

Roberts, R.J., 1976. Restriction endonucleases. CRC Crit. Rev. Biochem. 4, 123–164.

Roberts, R.J., Vincze, T., Posfai, J., Macelis, D., 2007. REBASE-enzymes and genes for DNA restriction and modification. Nucleic Acids Res. 35 (Database issue).

Robinson, C.R., Sligar, S.G., 1995. Heterogeneity in molecular recognition by restriction endonucleases:osmotic and hydrostatic pressure effects on BamHI, PvuII, and EcoRV specificity. Proc. Natl Acad. Sci. U.S.A. 11 (92), 3444–3448.

Russell, D.W., Sambrook, J., 2001. Molecular Cloning: A Laboratory Manual. Cold Spring Harbor Laboratory, Cold Spring Harbor, NY.

Saiki, R., Scharf, S., Faloona, F., Mullis, K., Horn, G., Erlich, H., et al., 1985. Enzymatic amplification of beta-globin genomic sequences and restriction site analysis for diagnosis of sickle cell anemia. Science 230, 1350–1354.

Saiki, R., Gelfand, D., Stoffel, S., Scharf, S., Higuchi, R., Horn, G., et al., 1988. Primer-directed enzymatic amplification of DNA with a thermostable DNA polymerase. Science 239, 487–491.

Sanford, J.C., et al., 1987. Delivery of substances into cells and tissues using a particle bombardment process. J. Particul. Sci. Technol. 5, 27–37.

Schlamp, K., Weinmann, A., Krupp, M., Maass, T., Galle, P., Teufel, A., 2008. BlotBase: a northern blot database. Gene 427 (1–2), 47–50.

Skerra, A., 1992. Phosphorothioate primers improve the amplification of DNA sequences by DNA polymerases with proofreading activity. Nucleic Acids Res. 20, 3551–3554.

Smith, H.O., Nathans, D., 1973. Letter: a suggested nomenclature for bacterial host modification and restriction systems and their enzymes. J. Mol. Biol. 81, 419–423.

Smith, L.M., Sanders, J.Z., Kaiser, R.J., Hughes, P., Dodd, C., Connell, C.R., et al., 1986. Fluorescence detection in automated DNA sequence analysis. Nature 321 (6071), 674–679.

Southern, E.M., 1975. Detection of specific sequences among DNA fragments separated by gel electrophoresis. J. Mol. Biol. 98 (3), 503–517.

Streit, S., Michalski, C.W., Erkan, M., Kleef, J., Friess, H., 2009. Northern blot analysis for detection of RNA in pancreatic cancer cells and tissues. Nat. Protoc. 4 (1), 37–43.

Sugimoto, M., Esaki, N., Tanaka, H., Soda, K., 1989. A simple and efficient method for the oligonucleotide-directed mutagenesis using plasmid DNA template and phosphorothioate- modified nucleotide. Anal. Biochem. 179, 309–331.

Sukru, T., 2007. Quantitative real-time polymerase chain reaction in cancer drug target identification and validation. Drug Discov. Genomics (Touch Briefings 2007) 1, 27–28.

Sutcharitchan, P., Saiki, R., Huisman, T.H., Kutlar, A., McKie, V., Erlich, H., et al., 1995. Reverse dot-blot detection of the African-American beta-thalassemia mutations. Blood 86, 1580–1585.

Tadmouri, G.O., Tuzmen, S., Ozcelik, H., Ozer, A., Baig, S.M., Senga, E.B., et al., 1998. Molecular and population genetic analyses of beta thalassemia in Turkey. Am. J. Hematol. 57, 215–220.

Tamaki, K., Jeffreys, A.J., 2005. Human tandem repeat sequences in forensic DNA typing. Leg. Med. (Tokyo) 7 (4), 244–250.

Taylor, J.W., Ott, J., Eckstein, F., 1985. The rapid generation of oligonucleotide-directed mutations at high frequency using phosphorothioatemodified DNA. Nucleic Acids Res. 13, 8765–8785.

Tebas, P., Stein, D., Tang, W.W., Frank, I., Wang, S.Q., Lee, G., et al., 2014. Gene editing of CCR5 in autologous CD4 T cells of persons infected with HIV. N. Engl. J. Med. 370, 901–910.

ThermoScientific., 2012. Components of the reaction mixture protocol. Retrieved from <http://www.thermo-scientificbio.com/uploadedFiles/Resources/components-reaction-mixture.pdf>.

Towbin, H., Staehelin, T., Gordon, J., 1979. Electrophoretic transfer of proteins from polyacrylamide gels to nitrocellulose sheets: procedure and some applications. Proc. Natl. Acad. Sci. U.S.A. 76 (9), 4350–4354.

Trayhurn, P., 1996. Northern blotting. Proc. Nutr. Soc. 55 (1B), 583–589.

Tuzmen, S., Schechter, A.N., 2001. Genetic diseases of hemoglobin: diagnostic methods for elucidating beta-thalassemia mutations. Blood Rev. 15 (1), 19–29.

Utans, U., Liang, P., Wyner, L.R., Karnovsky, M.J., Russel, M.E., 1994. Chronic cardiac rejection: identification of five upregulated genes in transplanted hearts by differential mRNA display. Proc. Natl. Acad. Sci. U.S.A. 91 (14), 6463–6467.

Valoczi, A., Hornyik, C., Varga, N., Burgyan, J., Kauppinen, S., Havelda, Z., 2004. Sensitive and specific detection of microRNAs by northern blot analysis using LNA-modified oligonucleotide probes. Nucleic Acids Res. 32, e175.

Vandeyar, M.A., Weiner, M.P., Hutton, C.J., Batt, C.A., 1988. A simple and rapid method for the selection of oligodeoxynucleotide-directed mutants. Gene 65, 129–133.

Von Fange, T., McDiarmid, T., MacKler, L., Zolotor, A., 2008. Clinical inquiries: can recombinant growth hormone effectively treat idiopathic short stature? J. Fam. Pract. 57 (9), 611–612.

Walter, P., Alberts, B., Alexander, J.S., Lewis, J., Raff, M.C., Roberts, K., 2008. Molecular Biology of the Cell, fifth ed, extended version. Garland Science, New York.

Wang, C.Q.F., Akalu, Y.T., Farines, M.S., Gonzalez, J., Mitsui, H., Lowes, M.A., et al., 2013. IL-17 and TNA synergistically modulate cytokine epression while suppressing melanogenesis: potential relevance to psoriasis. J. Investig. Dermatol. 133 (12), 2741–2752.

Watson, J.D., 2007. Recombinant DNA: Genes and Genomes: A Short Course. W.H. Freeman, San Francisco.

Wolfe, K.A., Breadmore, M.C., Ferrance, J.P., Power, M.E., Conroy, J.F., Norris, P.M., et al., 2002. Toward a microchip-based solid-phase extraction method for isolation of nucleic acids. Electrophoresis 23, 727–733.

Yamanaka, S., Poksay, K.S., Arnold, K.S., Innerarity, T.L., 1997. A novel translational repressor mRNA is edited extensively in livers containing tumors caused by the transgene expression of the apoB mRNA-editing enzyme. Genes Dev. 11 (3), 321–333.

Yuan, R., 1981. Structure and mechanism of multifunctional restriction endonucleases. Annu. Rev. Biochem. 50, 285–319.

Zhang, R., Zhu, Z., Zhu, H., Nguyen, T., Yao, F., Xia, K., et al., 2005. SNP cutter: a comprehensive tool for SNP PCR-RFLP assay design. Nucleic Acids Res. 33 (Web Server issue), W489–W492.

Techniques for Protein Analysis

Gülay Büyükköroğlu[1], Devrim Demir Dora[2], Filiz Özdemir[1] and Candan Hızel[3]

[1]Anadolu University, Eskisehir, Türkiye [2]Akdeniz University, Antalya, Türkiye
[3]OPTI-THERA Inc., Montreal, QC, Canada

15.1 PROTEIN IDENTIFICATION

Proteins are one of the most important macromolecules in nature. Many functions inside the cell are provided with the aid of different types of protein molecules. Despite these different functions, protein structures are similar. Because of their significant roles, it is important to identify the details about their structures, synthesis, functions, and regulations through the lens of molecular biology. The first step in identification of the structure of a protein is to determine its amino acid composition (Murray et al., 2012; Rao et al., 2014; Berg et al., 2002).

15.1.1 Sequencing

The characteristic of each protein depends on its unique amino acid sequence (Garrett and Grisham, 2013). The first protein sequencing was achieved by Frederic Sanger in 1953 bovine insulin for which he was awarded the Nobel Prize in 1958. Protein sequencing is used to identify the amino acid sequence and its conformation. The identification of the structure and function of proteins is important to understand cellular processes.

There are several applications of protein sequencing;

1. Identification of the protein family to which a particular protein belongs and finding the evolutionary history of that protein.
2. Prediction of the cellular localization of the protein based on its target sequence.
3. Prediction of the sequence of the gene encoding the particular protein.

Omics Technologies and Bio-engineering: Towards Improving Quality of Life
DOI: https://doi.org/10.1016/B978-0-12-804659-3.00015-4

317

4. Discovering the structure and function of a protein through various computational methods and experimental methods.

There are two major direct methods for protein sequencing: Edman degradation (Harvey and Ferrier, 2011; Bauer et al., 1997) and mass spectrometry (MS) (Sahukar et al., 2016). A protein of interest is digested by a proteolytic enzyme and subsequently, the enzyme digest is injected into the first mass spectrometer, which separates the oligopeptides according to their mass-to-charge ratios (m/z).

15.1.1.1 Determining Amino Acid Composition With Hydrolysis

Knowledge of the amino acid sequence of proteins is crucial in order to facilitate the discovery of errors during the process of biological information and to distinguish some ambiguous results regarding the process of protein synthesis. To be able to define the number of amino acids in a protein structure, hydrolysis can be used as a general method to assist determination of the number of amino acids that form the protein structure.

The first step is to identify the amino acid composition (Aloisi et al., 2016; Darragh and Moughan, 2005). The peptide bonds, which link amino acids to each other, are broken by hydrolysis. Because peptide bonds are stable at neutral pH, they are hydrolyzed using strong acids and bases; enzymatic catalysis is not effective for a complete hydrolysis. However, it is not possible to hydrolyze proteins without having a partial loss from some amino acid residues. In a typical method for analyzing the amino acid composition of proteins and polypeptides, a protein sample is hydrolyzed in a glass tube at 110°C using 6N HCl. Under these conditions, all tryptophan and most of the cysteine amino acids are disrupted. If there are metals in the structure, methionine and tyrosine are partially disrupted as well. The glutamine and asparagine are delaminated to form glutamate and aspartate, respectively. The serine and threonine contain an OH group in their structure; they are broken down more slowly than the other amino acids. Finally, only 50% of the bonds between neutral residues (Val–Val, Ile–Ile, Val–Ile, Ile–Val) are hydrolyzed after 20 hours. Typically, folded samples are hydrolyzed in 24, 48, 72, and 96-hour time periods. The serine and threonine data are marked on the semilogarithmic paper and are extrapolated backwards through time 0. The valine and isoleucine are calculated from 96-hour data. Dicarboxylic acids and their amides are identified, and they are reported together as either "Glx" or "Asx." Before hydrolysis, cysteine and cystine are converted into a stronger derivative (e.g., cysteic acid). The hydrolysis, which degrades serine, threonine, arginine, and cysteine, is catalyzed with alkalies; it is also used for tryptophan analysis. After hydrolysis, the amino acid composition is determined by either automatic ion-exchange chromatography or by high-performance liquid chromatography (HPLC) (Rombouts et al., 2009; Murray et al., 2012; Garrett and Grisham, 2013).

15.1.1.2 Quantitative Analysis

There are many methods used to determine the amount of protein in a sample; the most important one is the Kjeldahl (Lynch and Barbano, 1999), Warburg (UV absorption method), Coomassie-blue (Bradford method), biuret, and Folin–Lowry methods (Sapan et al., 1999). However, MS can also be used for a precise quantitative analysis of proteins, as it is an important and promising method for protein characterization (refer to Section 15.1.5 for more detail).

15.1.2 Edman Degradation

15.1.2.1 The Edman Degradation Reaction

This reaction is a chemical method for sequencing a whole polypeptide protein from their N-terminus, was developed by Pehr Edman in the early 1950s. This method labels and removes an amino acid residue from the whole sequence without disturbing peptide bonds (Hunkapiller, 1988; Liu et al., 2016).

Edman degradation is a series of chemical reactions that remove amino acids on the N-termini of the proteins. One of the most important reagents in sequence analysis is phenyl isothiocyanate, which was developed by Edman. Moreover, this reagent reacts with N-termini residue of both peptides and proteins. After the reaction with phenyl isothiocyanate, the protein sample is incubated with an anhydrous acid (e.g., trifluoroacetic acid) that breaks the peptide bond between the first and second amino acids of the protein. Thus, the amino acids at the N-termini residue become free as a thiazolinone derivative. This thiazolinone is then extracted using an organic solvent, and dried. Finally, it is converted to a more stable phenylthiohydantoin (PTH) derivative. The PTH derivatives are separated using HPLC and are determined by their order and elution location. In 1967, an automatic amino acid separation and identification method began to be used. Although the automatic Edman method is faster than Sanger method, it is slower and more difficult than other DNA sequencing methods. Because proteins are composed of polypeptides that are bound by noncovalent interactions and disulfide bridges, the first step is to take off these polypeptide structures apart and to make them an individual polypeptide chain. Some denaturation chemicals (urea and guanidine hydrochloride) are used for that purpose. Some oxidative and reductive compounds disrupt the disulfide bonds; the polypeptides are then separated by chromatography (Speicher et al., 2001).

Edman Reactive and Sequencing Methods by Edman Degradation: The automatic sequencing method uses phenyl isothiocyanate (Edman reagent). It involves a series of reactions that end with degradation of the N-termini residue in the form of its PTH derivative (Edman reaction, Fig. 15.1). The instrument consists of a cup-shaped reaction chamber that spins around its axis; the reaction occurs in a thin solution layer on the wall of this cup-shaped chamber making solvent extraction and removal easier. Fully automatic instruments can analyze 30−40 residues, in some cases up to 60 or 80 residues, continuously. This instrument is programmed to the sequential Edman fragmentations on a N-treminal increment of a polypeptide. After the amino acid at N-termini residue is removed, the Edman derivative of the residue is formed. PTH derivatives are separated using HPLC and are defined by their order and elution locations (Nelson and Cox, 2005; Berg et al., 2002; Shively, 2000).

15.1.2.2 Limitations of the Edman Degradation

Edman technique particularly affects proteins at their N-terminus. Accordingly, if this protein is chemically modified, or has stored these proteins in the case of large peptides, this technique does not work as requested to obtain the reliable output (Speicher et al., 2001).

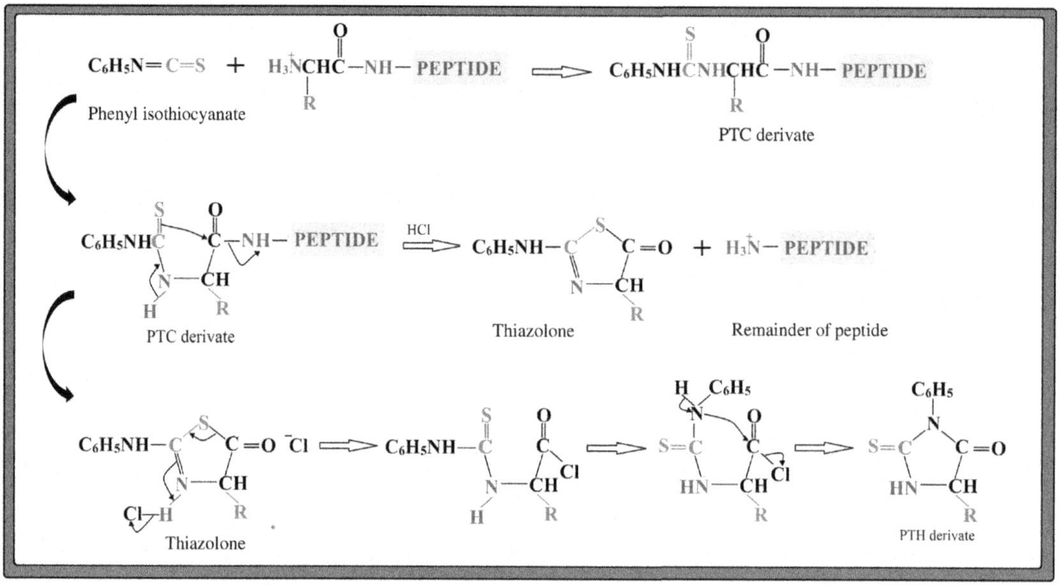

FIGURE 15.1   Schematic illustration of Edman degradation.

Although the main advantage of the Edman technique is to be more easily accessible in many laboratories and so to facilitate the determination of protein sequence and also to improve identification rate in recently developed of gas chromatography and thin layer chromatography method, it is still difficult to realize reliable determination of all the amino acids in single step. Therefore, for discernible results, separate procedure could be required in order to determine the positions of disulfide bridges, and peptide concentrations more than 1 pM (Miyashita et al., 2001).

15.1.3 Gel Electrophoresis

15.1.3.1 *Polyacrylamide Gel Electrophoresis*

Polyacrylamide gels are based on the free radical polymerization principle of acrylamide and cross-linking N,N'-methylene-bis-acrylamide. This material is physically very stable and strong. It is especially used for the electrophoretic separation of small or medium sized (up to about 1×10^6 Da) proteins. Its interaction with the migrating molecules is at the lowest level. The separating power of this gel depends on the dimension of the molecule to be separated as well as the concentration of acrylamide and bis-acrylamide. Low concentration acrylamide and bis-acrylamide polymerization is preferred to prepare gels with larger pores for high molecular weight samples. The difference of polyacrylamide gel electrophoresis (PAGE) from gel permeation chromatography is that small molecules move faster in polyacrylamide gels in comparison with larger molecules (Sadeghi et al., 2006). A standard gel used for the separation of proteins generally contains about 7.5% polyacrylamide.

Pore size in polyacrylamide gels is determined with the values of %*T* [total polyacrylamide percentage (w/v)] and %C_{bis} [the ratio of bis to monomer (w/w)] using the following formulae:

$$\% \, T = \frac{\text{Acrylamide (g)} + \text{Bis (g)}}{\text{Volume (mL)}} \times 100$$

$$\% \, C_{bis} = \frac{\text{Bis (g)}}{\text{Acrylamide (g)} + \text{Bis (g)}} \times 100$$

PAGE is given different names according to the type of gel (tube and slab gel electrophoresis or continuous and discontinuous gel electrophoresis), the position of the gel (vertical and horizontal gel electrophoresis), the chemical composition of the gel (native and sodiumdodecylsulfate (SDS), gel electrophoresis), and the pore distribution of the gel (homogeneous and gradient gel electrophoresis) (Wenk and Fernandis, 2007).

In tube PAGE, glass tubes (10 cm × 6 mm) are used and the gel material is filled into these tubes and polymerization is attained. Gel tube is placed vertically between two different buffer stocks. Cathode is generally located in the upper stock, whereas anode is located in the lower stock. Since most of the biologic materials are negatively charged, they move toward the anode. Hence, the sample to be analyzed is applied on the upper section of the gel with a tracking dye and electrical current is passed through the system. Since the tracking dye moves faster than the compounds in the sample, the current is stopped when it reaches the end of the gel, the gel is taken out of the tube and dyed.

Slab gels are used more in comparison with the columns since they enable the analysis of many samples in the same support environment (at the same conditions). Polyacrylamide gel is prepared between two glass plates (Fig. 15.2). A plastic comb placed on the top of the gel during polymerization enables the formation of small wells in the gel. The comb is removed after polymerization, the wells are washed with the buffer in order to remove the salts and unpolymerized acrylamide. The gel cassette is placed between two buffers; samples are placed inside the wells and current is passed. The gel is painted accordingly at the end of the electrophoresis for imaging.

SDS anionic detergent is used in SDS-PAGE in order to denaturate the proteins surrounding the main skeleton of the polypeptides as well as to give a negative charge to the molecules (Černý et al., 2013). The movement rate of the polypeptides in this method depends on molecular weight as well as internal electrical loads. Thus, it is a method that is frequently used for determining the molecular weights of the applied samples. A protein sample with unknown molecular weight is applied side by side with a protein of known molecular weight on the same gel and then is separated electrophoretically in different lines. The comparison of the bands on the gel after painting gives an idea about the molecular weight of the protein. In addition, it is also possible to determine the molecular weight by evaluating the results of this separation mathematically (Righetti et al., 2001).

Two different buffer systems, *continuous* and *discontinuous*, can be used in electrophoresis. There is only one separating gel in continuous system; the same buffer is used in the tanks and the gel. Whereas in the discontinuous system two-sided gel preparation with different buffers is used. The "stacking" gel with its large porous structure is located at the upper side of the gel providing the order of the applied sample in terms of its size. Whereas the

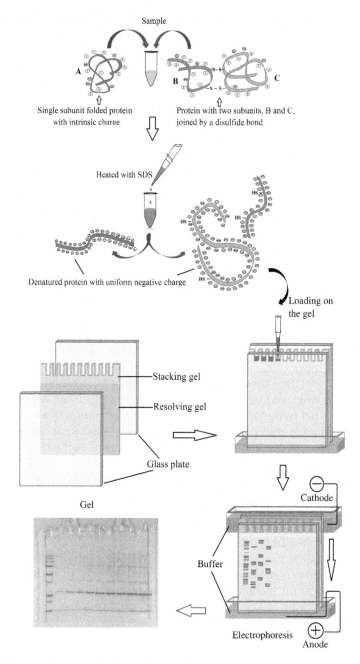

FIGURE 15.2　Schematic diagram of polyacrylamide gel electrophoresis.

"separating" gel with small pores is located at the lower side of the gel and so provides a more sensitive separation of the sample. The buffers used in the preparation of the gels are different from each other. In addition, the tank buffers are prepared differently than the gel buffers so that a better separation can be attained. *Gradient gels* are also used to provide this feature. Acrylamide concentration is gradually increased in gradient gels as we go down from the upper section of the gel. Thus, it is ensured that the pores in the gel decrease in size gradually as we move down. Polypeptides with similar molecular weights are separated more efficiently in this type of gel and form sharper bands (Bolt and Mahoney, 1997).

15.1.3.2 Isoelectric Focusing

Isoelectric focusing (IEF) is an efficient method developed for the electrophoretic analysis of proteins (Mathy and Sluse, 2008). Since the net charge of the proteins depends on pH, electrophoretic is attained by pH changes in this method. The net charge of the proteins is the sum of the negative and positive charges at the amino acid chain regions. This value changes according to pH, and the isoelectric point (pI) is where the net charge is 0. Proteins are charged positively under their isoelectric pH's (pI) and migrate toward the negative charged electrode (cathode) in a medium with fixed pH. Whereas the protein loses a proton at pH values above the isoelectric points therefore becoming negatively charged and migrating toward the positively charged electrode (anode). If the pH of the electrophoresis environment is equal to its pI, the electrophoretic migration stops at its own pI since the net charge of the protein will be 0 (Fig. 15.3). It is theoretically possible to

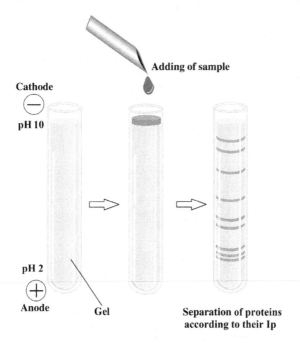

FIGURE 15.3   Representation of isoelectric focusing method.

separate the proteins and to determine the pI value of the protein by monitoring the electrophoretic movements in different experiment series with different environment pH values. For this, the three-dimensional structure of the protein should first be disrupted with chaotropic agents such as urea and the charges should be exposed in the environment (Černý et al., 2013). An acid (generally phosphoric acid) is placed in the anode and a base, such as triethanolamine, is placed in the cathode. The gel medium between the electrodes is adjusted before or during electrophoresis such that the pH will be between 2 and 10. The pI value of the protein is reached by determining the pH value at the point where the protein is focused. The IEF method that enables the separation of polypeptides using the property, that they have different pIs, can be used by itself for analytical or preparative purposes, while it can also form the first dimension of the two-dimensional gel electrophoresis (2DGel) (Curreem et al., 2012). Thus, it can be carried out in both horizontal and vertical (in a capillary tube or column) systems.

15.1.3.3 Two-Dimensional Gel Electrophoresis

Two-dimensional gel electrophoresis (2DGel) is a successful method used for the detection and analysis of proteins. It has been designed as a combination of the 2DGel, IEF and SDS-PAGE methods, and is used in the analysis of complex protein mixtures. In the first step, protein is separated into its charges with IEF, whereas in the second step, the protein is separated according to its mass. The separated protein on the gel with IEF is negatively charged by treatment with SDS, and the electrophoresis is performed by inserting the gel horizontally into the SDS-PAGE gel. (Fig. 15.4). Thus, the proteins that are focused on the pI are separated according to their molecular weights. Generally 20×20 cm large gels are used in SDS-PAGE setup and more than 10,000 proteins can be separated. If the protein amount is around 10 ng, Coomassie dye is used and if the protein amount is around 0.5 ng, silver or fluorescent total-protein satins can be used for detection. Using the system

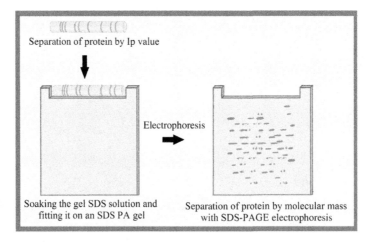

FIGURE 15.4   Steps of two-dimensional gel electrophoresis (2DGel).

known as "ISO-DALT," both IEF and SDS-PAGE can be carried out simultaneously (Chen et al., 2015; Brunelle and Green, 2014; Hanash et al., 1991; Magdeldin et al., 2014).

15.1.4 Isotope Labeling

Quantitative proteomics, which has been performed by 2DGel for over 25 years, is used for analyzing proteins in a cell, tissue, or an organism (Ong et al., 2002; Jungblut, 2014) to obtain quantitative information about all proteins in a sample. Recently, in many research areas, quantitative proteomics methods for relative and absolute quantitation of peptide and proteins are based on MS techniques. Most methods for relative quantification by MS-based techniques use labeling of peptides or proteins with an isotope, but label-free quantification methods are also available (Craft et al., 2013). Labeling of peptides or proteins with an isotope can be done by enzymatic, in vitro (chemical) or in vivo (metabolic) techniques. Labeling stage is an important difference between them; while metabolic labeling requires living cells, chemical labeling can be done on any proteome. Some of enzymatic, chemical (Isotope-Coded Affinity Tag, ICAT) and metabolic (Stable-Isotope Labeling with Amino acids in Cell culture, SILAC) labeling techniques are described later (Craft et al., 2013; Ong et al., 2002).

15.1.4.1 Enzymatic Labeling

Enzymatic labeling, which is achieved by using deuterium and ^{18}O, is one of the most important protein labeling techniques which incorporates functional groups into proteins at specific sites via enzymatic reactions. Enzymatic labeling can be performed during or after the proteolytic digestion with the protease. Isotope labels are introduced into peptides during protein digestion by trypsin or Glu-C-catalyzed incorporation of ^{18}O. Each digested peptide is enzymatically labeled at the C-terminus by ^{18}O which incorporates two heavy oxygen atoms from $H_2{}^{18}O$. Samples are digested with trypsin and either ^{18}O water or ^{16}O water, and then mixed together for MS analysis. Primary amino groups of digested peptides are labeled enzymatically by deuterated (^2H) acylating agents such as N-acetoxysuccinimide (NAS) (Bantscheff et al., 2007).

15.1.4.2 Isotope-Coded Affinity Tag

Isotope-coded affinity tag (ICAT) is the first chemical in vitro labeling method that uses a biotin tag to label proteins containing cysteine residues (Craft et al., 2013). Cysteine-containing proteins extracted from the control cells and diseased cells are labeled in vitro by commercial ICAT reagents containing C12 and C13, respectively. After purification of mixed protein samples digested with trypsin by ion-exchange chromatography, cysteine-containing peptides are isolated by affinity chromatography with avidin. Cysteine-containing peptides with C12 and C13 isotope labels are identified and quantified by LC—MS/MS. In chemical labeling, proteomes to be compared have to be purified and fractionated by exactly the same experimental conditions. Protein profiling of two different cell lines can be achieved by this method (Gygi et al., 1999; Yi et al., 2005).

15.1.4.3 *Stable-Isotope Labeling in Cell Culture*

Stable-isotope labeling in cell culture (SILAC), is a more commonly used metabolic labeling technique for mass spectroscopy-based quantitative proteomics. Nonradioactive isotopic labeling is used for detection of differences in protein abundance between the samples that will be analyzed. SILAC depends on metabolic incorporation of stable isotopes, that is "light" or "heavy" form of the amino acids, into the proteins during normal cell growth and division through the growth medium. Complete substitution of the natural amino acid with a stable isotopic nuclei (e.g., deuterium, 13C, 15N) is needed for successful experimental results. Differently labeled samples can be mixed and analyzed together. In an experiment, different organisms or cell lines are grown in identical culture media, generally containing lysine or arginine with 13C and 15N atoms, except one of them containing a "light" and the other a "heavy" form of a particular amino acid. All natural amino acids in proteins, synthesized after a labeled amino acid has been introduced to a cell, will be replaced by their isotope labeled analog. Isotopically labeled samples are quantified by calculating the ratios of integrated signal intensities for labeled peptides to the signal intensities of the corresponding unlabeled peptides in a chromatogram by mass spectrometer. SILAC data could be analyzed effectively by several software packages like MaxQuant, Mascot Distiller, Xpress, and ASAPratio (Ong and Mann, 2006, 2007; Boumediene et al., 2010; Amanchy et al., 2005; Harsha et al., 2008).

15.1.5 Mass Spectrometry

Mass spectrometry (MS) is a high-throughput analytical detection technique used to get information about the molecular weights and chemical structures of the peptides, proteins, carbohydrates, oligonucleotides, natural products, and drug metabolites (Biemann, 2014).

MS provides some advantages including small amount of sample requirement, label-free detection, fast analysis, capability of defining chemical structures with fragmentation, high sensitivity, and simultaneous detection of multiple analytes (Zhu and Fang, 2013; Glish and Vachet, 2003).

MS is widely used for various molecular biology analysis purposes either alone or combined with other structural proteomics techniques because of its advantages (Pi and Sael, 2013; Wasinger et al., 2013).

Examples of the analysis include molecular weight characterization, posttranslational modifications in proteins, identification of vibrational components in proteins, analysis of protein conformation and dynamics, noncovalent interactions, protein and peptide sequencing, DNA sequencing, protein folding, in vitro drug analysis, and drug discovery (Glish and Vachet, 2003; Benesch and Ruotolo, 2011; Steendam et al., 2013).

15.1.5.1 *Principle and Instrumentation*

The principle of the spectrometer depends on the separation of the molecules based on their mass-to-charge (m/z) ratio by ionization of the molecules with high energy electrons to break a molecule into fragments. Fragmentation pattern and m/z values provide information about the molecular weights and chemical structures of the peptides and proteins. Each peptide has a specific molecular weight. Molecules from solution or solid phase should be transferred into gaseous phase, because MS measurements have to be done on ionized molecules in the gaseous phase (Fenn et al., 1989; Ong and Mann, 2005).

15.1.5.2 Components of the Instrument

A spectrometer basically consists of the following components (Dobson, 2003; Oudenhove and Devreese, 2013):

1. Device for sample input into the machine
2. Molecular ionization source
3. Mass analyzer
4. Detector
5. Vacuum system
6. Computer-based data obtaining and processing system

15.1.5.2.1 DEVICE FOR SAMPLE INPUT INTO THE MACHINE

Introduction of a sample into the spectrometer is a changeable process according to sample specifications which can be a solid, liquid, or vapor and the methods of ionization such as direct insertion with a probe or plate with matrix-assisted laser desorption/ionization (MALDI)-MS and direct infusion or injection into the ionization source with electrospray ionization (ESI)-MS (Kang, 2012).

15.1.5.2.2 MOLECULAR IONIZATION SOURCE

Detection by MS requires ionization and transfer of proteins into gaseous phase. Proteins and peptides are usually ionized via protonation in a spectrometer because of their NH_2 groups that accept a H+ ion. Different ionization sources are given in Fig. 15.5 (Kang, 2012; Hoffmann and Stroobant, 2007).

In proteomics analysis, the most commonly used two devices for ionization are MALDI and ESI. Both of them are soft ionization techniques, and ions undergo little fragmentation (Guerrera and Kleiner, 2005).

- *Matrix-assisted laser desorption/ionization (MALDI):*
  In MALDI, samples to be analyzed are mixed with crystal matrix material in a solvent and dried on a metal plate. After cocrystallization and absorption of samples on a matrix, the matrix is exposed to the pulse of a nitrogen laser beam, which vaporizes the samples for ionizing the molecules inside the source of the mass spectrometer. MALDI causes not only positive ionization for peptide and protein molecules, but also negative ionization for oligonucleotides and carbohydrate molecules. MALDI is more sensitive and more universal than other laser ionization techniques (Lagarrigue et al., 2012; Gessel et al., 2014).
- *Electrospray ionization (ESI):*
  In ESI, molecules are directly ionized from the solution by applying a high-voltage electric field under atmospheric pressure, to a liquid passing through a capillary tube. The high electrical field creates a charge on the surface of the droplets which become much smaller through vaporization of the solvent. ESI produces multiple charged ionized molecules, and is extremely useful for accurate mass measurement, especially for thermally unstable, high molecular mass substances such as proteins, oligonucleotides, and synthetic polymers (Ho et al., 2003).

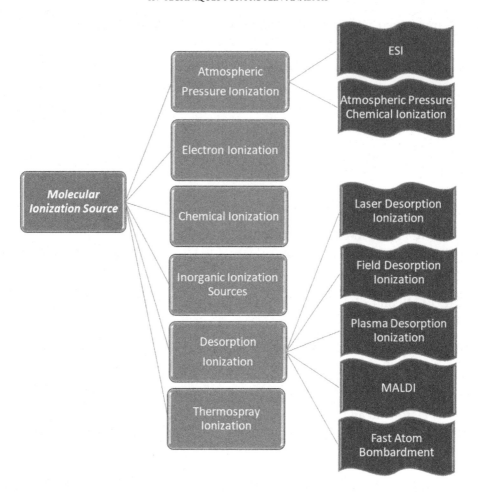

FIGURE 15.5   Types of Ionization.

15.1.5.2.3 MASS ANALYZER

After ionization, the ionized molecules of peptides or proteins enter the mass analyzer section of spectrometer. In the mass analyzer, molecules are separated based on their mass-to-charge ratio by electric and/or magnetic fields or by measuring the time an ion takes to reach a fixed distance from the point of ionization to the detector (Römpp and Spengler, 2013).

For the separation of ionized molecules, different kinds of mass analyzers are available such as quadrupoles, time-of-flight (TOF), magnetic sectors, Fourier transform, and quadrupole ion traps. In proteomics, quadrupole and the TOF analyzers are mostly used. MS with the ESI device usually carries a quadrupole analyzer (El-Aneed et al., 2009).

15.1.5.2.4 DETECTOR

The detector is the final component of a mass spectrometer and is used for monitoring and recording the presence of separated ions coming from the mass analyzer. Depending on the analytical applications and design of the instrument, different detectors can be used such as electron multiplier, Faraday cup, negative-ion detection, postacceleration detector, channel electron multiplier array, photomultiplier conversion dynode, the daly detector, and array detector (Kang, 2012). After detection of ions, the signals are recorded on a graph by plotting the amount of signal versus m/z ratio. MS graphs not only show the presence and abundance of different molecular size peptides and proteins, but also the energy level of the molecules of a particular kind.

15.1.5.2.5 VACUUM SYSTEM AND COMPUTER-BASED DATA OBTAINING AND PROCESSING SYSTEM

MS vacuum pumps are used for free movement of ions within the spectrometer. Computer-Based Data Obtaining and Processing System. Computer-based system is needed for data obtaining and processing.

15.1.5.3 Liquid Chromatography–Mass Spectrometry

Liquid chromatography–Mass spectrometry (LC–MS) technique combines separation and analysis of samples with LC and MS, respectively. LC–MS is a bioanalytical method for quantitative analysis of proteins which has several application areas such as biopharmaceutical drug development, drug metabolism and toxicology studies, quantification of drugs in biological fluids (plasma, urine, tissue, etc.), pharmacokinetic studies, bioavailability studies, doping control, quantification of biogenic amines, and therapeutic drug monitoring. Bioanalytical determination of protein-based biopharmaceuticals in biological matrices can be successfully achieved by liquid chromatography coupled to tandem mass spectrometry (LC–MS/MS) (Kang, 2012; Irene van den Broek et al., 2013).

15.1.5.4 Matrix-Assisted Laser Desorption/Ionization-Time-of-Flight Mass Spectrometry

MALDI-TOF mass spectrometer is used in conjunction with the MALDI device as an ionization source and TOF as a mass analyzer that is described in Section 15.1.6. It is a simple method with high sensitivity which can be coupled by high resolution analyzers and has different application areas such as detection of cancer biomarkers in various cancers (Rodrigoa et al., 2014), characterization of microorganisms including bacteria, fungi, and viruses (Croxatto et al., 2012), and analysis of glycoproteins, oligonucleotides, carbohydrates, and small biomolecules (Susnea et al., 2013).

15.1.5.5 Tandem Mass Spectrometry

A tandem mass spectrometry (TANDEM MS), also named as MS/MS, is a two-step technique used to analyze a sample either by using two or more mass spectrometers connected to each other or a single mass spectrometer by several analyzers arranged one after another. TANDEM MS (MS/MS) contains two or three quadrupoles and a TOF analyzer

(Broek et al., 2013). The mass analyzer coupled to the MALDI source determines the type of MS/MS analysis that can be carried out (Flatley et al., 2014).

MS/MS is especially useful for analyzing complex mixtures and involves two stages of MS. In the first stage of MS/MS, a predetermined set of m/z ions are isolated from the rest of the ions coming from the ion source and fragmented by a chemical reaction. In the second stage, mass spectra are produced for the fragments. TANDEM MS is generally used for bioanalysis of drugs. Identification and determination of phase I and phase II drug metabolites are achieved by TANDEM MS coupled with HPLC (Glish and Vachet, 2003; Holčapek et al., 2008).

15.1.6 Enzyme-Linked Immunosorbent Assay

The enzyme-linked immunosorbent assay (ELISA) is a rapid, high-throughput, quantitative immunoassay for the selective detection of target antigens. This technique is used as diagnostic tools and as quality control measures in biomedical research or various industries for the detection and quantification of specific antigens or antibodies in a given sample. The general principle behind an ELISA is antibody-mediated capture and detection of an antigen with a measurable substrate. The antigens such as proteins, peptides, hormones, or antibody immobilized on a solid surface and then complexed with an antibody that is linked to an enzyme. Alkaline phosphatase (AP) and horseradish peroxidase (HRP) enzymes are commonly used in ELISA applications. The size of HRP (40 kDa) is smaller than AP (140 kDa) and small size allows more molecules to be coupled to antibodies or avidin. This can boost signal generation. For this reason, HRP can be used with a variety of substrates, most of which are more sensitive than AP equivalents. The addition of substrate giving colored, fluorescent, or luminescent reaction products makes it possible to determine the concentrations of the reactants at very low levels. Colorimetric, chemifluorescent, and chemiluminescent substrates are available for both HRP and AP. Detection is accomplished by assessing the conjugated enzyme activity via incubation with a substrate to produce a measureable product. Colorimetric and chemifluorescent substrates are typically able to detect low- to mid-picogram levels of antigen and chemiluminescent substrates are the most sensitive, with antigen detection possible in the subpicogram. Colorimetric substrates are measured using a standard plate reader with the appropriate filters. Chemifluorescence is measured using a fluorometer with the appropriate excitation and emission filters (Fang and Ramasamy, 2015; Watabea et al., 2016; Akama et al., 2016; Ma et al., 2011).

Various types of ELISAs have been employed with modification to the basic steps described earlier.

15.1.6.1 Indirect ELISA

For indirect detection, a sample that must be analyzed for a specific antigen is adhered to on the solid surface. In order to block any areas of this surface that is not coated with the antigen, a solution of nonreacting protein such as bovine serum albumin is added and incubated. First an unlabeled primary antibody, which is specific for the antigen, is applied. Next, an enzyme-linked secondary antibody which is reactive against the primary

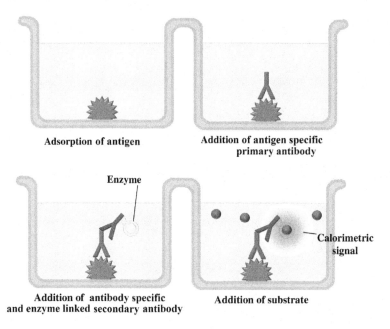

FIGURE 15.6 Schematic illustration of indirect ELISA.

antibody is added. For enzymatic detection and production of a calorimetric signal, the appropriate enzyme substrate is added and this signal is detected with appropriate equipment (Fig. 15.6) (Iannone, 2015).

15.1.6.2 Sandwich ELISA

Sandwich ELISA is also an indirect type of ELISA. The only difference in this ELISA principle is that, just like a sandwich in between two antibodies an antigen is present just a seen in the Fig. 15.7. The sandwich technique is used to identify a specific sample antigen. For this purpose, the well surface is covered with a known quantity of bound antibody (known as capture antibody) to capture the targeted antigen. Nonspecific binding sites of this antibody should be blocked using bovine serum albumin before adding the sample which is including antigen to the plate. A specific primary antibody is then added that "sandwiches" the antigen. Enzyme-linked secondary antibodies (known as detection antibody) are applied that bind to the primary antibody. Unbound antibody–enzyme conjugates are washed off. Substrate is added and the standard colorimetric detection method (described earlier) is used to detect and quantify analyte in the sample (Fig. 15.7) (Goldsby et al., 2000; Iannone, 2015).

15.1.6.3 Competitive ELISA

Competitive ELISA also known as inhibition ELISA is often used to indicate the presence of the antigen in the sample such as blood serum. Targeted antigen-specific antibody is immobilized on the solid surface. The enzyme-labeled antigen and the clinical sample are

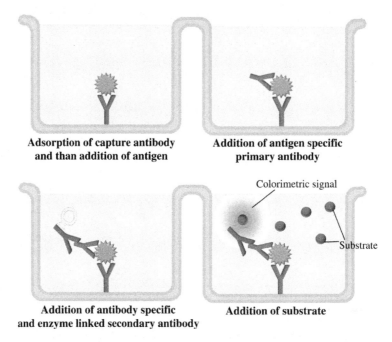

FIGURE 15.7 Schematic illustration of sandwich ELISA.

added at the same time into well and incubated in order to compete for binding with the immobilized antibodies. After an incubation period, any unbound antibody is removed by washing the plate and substrates are added (Aydin, 2015). The more antigen in the sample, the less antibody will be able to bind to the antigen in the well, hence "competition." After that, a standard colorimetric detection method that is related with substrates is used to detect and quantify analyte in the sample (Fig. 15.8). The hydrolyzed substrate amounts are inversely proportional to the amount of antigen in the sample. For this reason, the presence of antigen in the sample does not change the color (Ma et al., 2011).

15.1.6.4 *Reverse ELISA*

The solid phase of the Reverse ELISA consists of polystyrene pins protruding rod with 8−12 protruding ogives. The solid phase is immersed into primary antibody solution and incubated. After washing procedure of these pins, bovine serum albumin is used for blocking step. After each washing, pins were dried on absorbent paper and immersed directly in the collected sample (tube) or the sample dispensed in a microplate wells. After an incubation period, pins are removed and washed. The pins are then immersed into the wells prefilled with specific conjugate(s) antibodies and incubated. After the color development step the pins are removed from the wells and the plate can be immediately analyzed by a microplate reader if necessary (Fig. 15.9). Further developments of the test could foresee the simultaneous detection of an increased number of targets since each pin

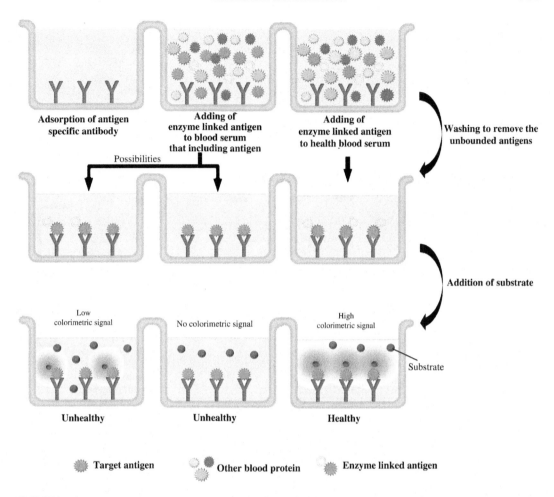

FIGURE 15.8   Representation of competitive ELISA.

could be dedicated to a specific target sensitizing the pin with an appropriate monoclonal antibody (Wilson et al., 2015; Folloni et al., 2011).

15.1.7 Immunohistochemistry

Protein analysis by immunohistochemistry (IHC) is based on the detection of antigens in organ or tissue sections by binding of specific antibodies and visualization of antigen—antibody complexes microscopically after immunohistochemical staining via an appropriate detection system (Ramos-Vara, 2005). Fluorescent dyes, enzymes, colloidal gold particles, and radioactive elements are the markers used for detection. Respectively, fixing and embedding the specimen, sectioning, deparaffinizing and rehydrating the section, antigen retrieval, immunohistochemical staining and visualization under the microscope are the basic steps of the IHC (Renshaw, 2017).

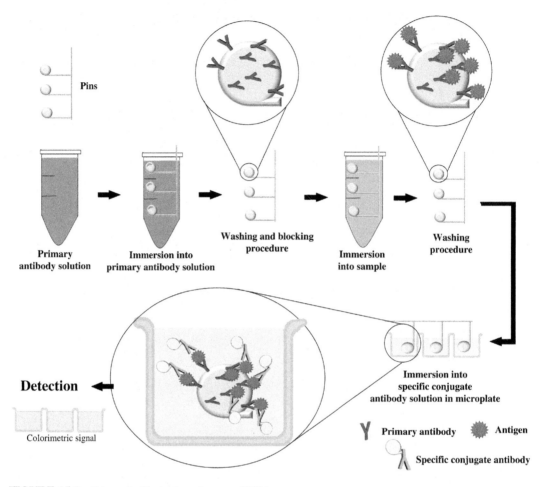

FIGURE 15.9   Schematic illustration of reverse ELISA.

15.1.7.1 Sample Preparation

Sample preparation is a critical step to maintain cellular components and special feature of antigenicity for IHC analysis. The specimen must be properly collected, fixed, embedded, and sectioned in the shortest time to prevent damage on the cell or tissue.

Fixation of the specimen is an important parameter for preserving and detecting cellular morphology and antigenic properties. Depending on the size and type of the specimen and fixative agent, the time for ideal fixation changes. Specimen fixation can be done by air drying, chemical fixation with cross-linking fixatives, and coagulating fixatives (Buchwalow and Böcker, 2010). Cross-linking fixatives stop enzymatic degradation more quickly than coagulating fixatives. Aldehydes such as formalin, paraformaldehyde, and glutaraldehyde are the most common used cross-linking fixatives for specimens that should be embedded in paraffin. Acetone and alcohols are used as coagulating fixatives for cryosections, cell smears, and cell monolayers.

After fixation, specimens are either frozen in liquid nitrogen or dehydrated and embedded in paraffin wax or synthetic resin prior to cutting by microtome to the desired thickness and mounted onto slides.

Prior to immunohistochemical staining, antigenicity of tissue sections must be recovered by antigen retrieval (Hyatt, 2002). Conformational changes of protein macromolecules affected by formaldehyde fixation and paraffin embedding can be reestablished by antigen retrieval, which can be achieved by either heat treatment (90–110°C) or enzymatic (trypsin, pepsin, and proteinase K) digestion. After antigen retrieval, samples are treated with antibodies for detection of antigenic properties of the tissue by antigen–antibody complex reaction (Renshaw, 2017).

15.1.7.2 Sample Labeling

Antigen–antibody immunoreaction can be detected by labeling the antibody before visualization by a microscope. Immunolabeling methods can be classified as direct or indirect according to the kind of procedure. Direct labeling is one step method in which the antigen directly binds to its specific labeled primary antibody in contrast to indirect labeling which could be two, three, or multi step and involves the labeled antibody reacting with the unlabeled primary antibody that will bind to the antigen. Although indirect labeling is a time consuming and complex method and has a potential for cross reactivity, it is more sensitive than the other. Direct labeling has lower signal, higher cost, and less flexibility. Polyclonal antibodies, obtained from many species such as rabbit, goat, pig, sheep, guinea pig, and horse, and monoclonal antibodies, generally obtained from mouse or rabbit hybridoma cells, are used for immunolabeling. Incubation time and antibody titers are important parameters for the antibodies affinity to the antigen (Miller, 2001).

Labeling the antibody by immunofluorescence, immunoenzymological (with HRP), calf intestinal AP, glucose oxidase, and beta-galactosidase) and immunoaffinity methods (with Avidin–Biotin complex) are done by using enzymes, biotin, fluorophore, and colloidal gold particles as labeling agents (Giorno, 1984; Hsu et al., 1981).

15.1.7.3 Sample Visualization

Sample visualization can be achieved by microscopic techniques such as light, fluorescent, and electron microscopes. In order to detect antigen–antibody immunoreaction by the microscope, antibodies should be labeled as described earlier. If the enzyme label is used for antibody labeling, immunoreaction could be visualized by light microscope using the brightfield illumination modus which is the most widely used observation mode in optical microscopy. Fluorophore labeling agents can be visualized in a fluorescent microscope. Electron microscope is needed for visualizing the colloidal gold particles. Depending on the labels, the Avidin–Biotin complex system can be visualized by light, fluorescence, and electron microscopy (Miller, 2001).

15.1.7.4 Applications

IHC has different application areas such as drug development, molecular biology, and diagnosis. Major changes in the expression pattern of antigen, specific cell, or tissue expression pattern of antigen, tissue or cellular localization of antigen can be determined by IHC. IHC is also used for diagnosis of diseases by specific tumor markers that can determine the origin of tumor and grade of tumor cells, identify the cell type, classify the tumor type as malign or benign, and determine the localization. Membrane antigens, antigens localized in the nucleus and structural antigens in the cytoplasm, could be identified by IHC. Quantitative analysis of IHC can be achieved by computer-based programs designed for IHC such as BLISS, ACIS, iVision, GenoMx, ScanScope, LSC, and AQUA (Cregger et al., 2006).

15.2 PROTEIN STRUCTURAL ANALYSIS

Proteins are complex macromolecules that are composed of amino acid residues covalently bonded together by peptide bonds. Four levels of protein structure, such as primary, secondary, tertiary, and quaternary structure, are defined. Primary structure is the amino acid sequence of the specific protein. In a protein chain, the number, chemical structure, and order of amino acid sequences determine the structure and chemical behavior of the protein. Secondary structure is the regularly repeated local structure, stabilized by hydrogen bonds. Alpha helix and beta sheets are two main types of secondary structure. Tertiary structure, the three-dimensional structure of a protein molecule, is the intramolecular arrangement of the secondary structure. The alpha-helixes and beta sheets are folded into a compact structure by the nonspecific hydrophobic interactions. Three-dimensional structure analysis is important for understanding the functions of proteins at molecular level. Quaternary structure is the three-dimensional structure of a single protein complex which is formed by several protein molecules, such as dimers, trimers, tetramers, or even high order aggregates of identical polypeptide chains. Protein structures can be analyzed by some methods such as circular dichroism (CD), nuclear magnetic resonance (NMR) spectroscopy, X-ray crystallography, and electron microscopy which are discussed later (Kamp et al., 1997; Jiskoot and Crommelin, 2005).

15.2.1 Circular Dichroism

Circular Dichroism (CD), which is the difference in the absorption of left-handed circularly polarized light and right-handed circularly polarized light that arise due to structural asymmetry, is the technique used for analyzing secondary and tertiary structures and folding properties of proteins in solutions, which could be changed due to its environmental changes such as temperature or pH.

Folding properties of proteins, characterization of either secondary structure or tertiary structure in the far-UV and near-UV, respectively, comparing the structures of proteins obtained from different sources, determining thermal stability of the proteins, comparing thermal stability of proteins after changes in manufacturing processes or formulations, conformational stability of proteins under different environmental conditions, and kinetics of conformational changes could be done by CD analysis.

Molecules should contain one or more chiral molecules (light-absorbing groups, chromophores) for CD analysis. CD is measured with a CD spectropolarimeter that measures in the far-UV spectral region at 190–250 nm and near-UV spectral region at 250–350 nm.

Chromophores are the peptide bonds at 190–250 nm wavelengths and the aromatic amino acids and disulfide bonds at 250–350 nm wavelengths, which have specific CD signals.

15.2.2 Nuclear Magnetic Resonance Spectroscopy

Three-dimensional structure and conformational dynamics of the macromolecules affect the biological activity. Nuclear Magnetic Resonance (NMR) spectroscopy is generally used for analyzing small-to-medium sized flexible proteins, with molecular weights up to approximately 30 kDa, which could not be crystallized, and detailed information can be obtained about topology, dynamics, and three-dimensional structure of molecules in solutions and the solid state (Dötsch and Wagner, 1998; Arora and Tamm, 2001; Castellani et al., 2002, Loquet et al., 2008). The principle of the method is based on the magnetic properties of the nuclei of certain atoms. When the nuclei of certain atoms are immersed in a static magnetic field and exposed to a second magnetic field, NMR occurs between these nuclei through bond (scalar coupling) or through space (dipolar coupling) interactions (Dyson and Wright, 1996). The nuclei of many isotopes with odd electron numbers such as ^{1}H, ^{13}C, ^{15}N, ^{19}F, and ^{31}P carry magnetic dipoles, and NMR measures the energy levels of magnetic atoms that are orientated differently and have a different energy in a magnetic field (Kwan et al., 2011).

An image of a protein cannot be obtained directly by NMR. Protein structure is calculated from the NMR spectra as a result of interactions between pairs of atoms by extensive data analysis and computer calculations (Wider, 2000).

15.2.3 X-Ray Crystallography

X-ray crystallography is a method used for various materials in the crystallized state to determine the arrangement of atoms within a crystal. Three-dimensional structure and function of many biological molecules, including proteins and nucleic acids, can be discovered by this method. The crystals of a pure protein are exposed to X-ray beam and X-ray is diffracted by atoms present in a protein crystal. Depending on the organization of atoms within a crystal and the number of electrons in the atoms, an X-ray beam is diffracted into many specific directions. From the angles and intensities of these diffracted beams, diffraction pattern is obtained and the electron density map is produced by a crystallographer. From this electron density map, the mean positions of the atoms in a crystallized protein and three-dimensional structure of the protein can be determined (Drent, 1994).

15.2.4 Electron Microscopy

Electron microscopy, in combination with image analysis, is used to determine the shape and three-dimensional structures of large proteins and large macromolecular complexes, with molecular weights greater than 150 kDa, that could not be investigated by conventional X-ray crystallography or NMR methods because of their large size or heterogeneous structure. Direct images of the molecules in their physiological environment can be obtained with the help of electron beams however, resolution of the images is low

(5−15A°) and it usually requires additional information from X-ray crystallography and/ or NMR (Topf and Sali, 2005).

Transmission electron microscopy (TEM) is the original form of electron microscopy and produces two-dimensional, black and white images. Unlike the light microscopes that use glass lenses, electromagnetic and/or electrostatic lenses are used in all electron microscopes to control the path of electrons. TEM requires a high-voltage electron beam which is formed by electromagnetic lenses. The structure of the sample is determined by the electron beam that has been partially transmitted through the sample.

In the scanning electron microscope, the electron beam is scanned across the surface of the sample and image is detected by mapping the detected signals with detectors (Zhou et al., 2006).

15.3 PROTEIN PURIFICATION

The protein source to be purified can be plant, animal (organs and blood samples of animals such as rabbit, cow, pig), human (blood and placenta samples), and microorganism based (bacteria, fungus, yeasts and organisms that produce recombinant protein). The existence of proteins other than the one that is targeted as well as contaminations such as bacteria, virus, and nucleic acids and progens in the extracts obtained from these sources may lead to problems in purification (Tan and Yiap, 2009). Hence, all molecules and structures apart from the proteins should be removed on by one during the first stage of purification. Precipitation of the DNA molecules by adding chemical substances (streptomycin sulfate) to the raw extract, dissolving of the dissoluble proteins in a proper environment, and extraction from other proteins by way of centrifuging may be given as examples to these means of separation.

A purification method is selected afterward taking into account the characteristic properties such as the size and shape of the proteins, the total charge, the hydrophobic groups on the surface as well as the binding capacity with the stable phases used. The most important points when applying these methods are environment temperature and pH. Working temperature should generally be 4°C since changes in temperature and pH are effective in the denaturation and inactivation of proteins and the pH should be kept at the desired value (Takeda and Moriyama, 2015; Nelson and Cox, 2005).

15.3.1 Chromatography

Chromatography is the separation and purification of the substances in a mixture using a two phase system comprised of a stationary phase and a mobile phase that do not mix. There is a stationary phase (matrix) and a mobile phase in all chromatographic applications. The separation of proteins via chromatographic methods is based on driving the protein over the matrix via the mobile phase. Since the migration speeds of different proteins will be different, it is possible that they will group on the stationary phase and separate in the mixture. The factors that lead to this separation are the adsorption, partition, ion-exchange, and affinity properties of molecules or the differences between their molecular weights (Ly and Wasinger, 2011). Various chromatographic methods have been explained later based on these properties.

15.3.1.1 Column Chromatography

The separation of proteins with this method depends on the differential separation of proteins between the mobile phase and the stationary phase (chromatographic medium or adsorbent) and it is one of the most effective methods. In general, the stationary phase is packed in vertical glass, plastic, or stainless steel columns and the mobile phase is pumped to the column. Protein extract is placed on the top part of the column and it is ensured that it moves downward with the mobile phase. The proteins that make up the mixture move at varying speeds in the column according to the differences between their adsorption or dispersion properties. Different proteins can be separated by collecting in fractions the liquid phase that comes out of the column (Carta and Jungbauer, 2010).

The primary materials that are used as stationary phase in column chromatography are classified as inorganic (porous silica, controlled pore glass, and hydroxyapatite), synthetic organic polymers (polyacrylamide, polymethacrylate, and polystyrene), and polysaccharides (cellulose, dextran, and agarose). These materials can be used to pack the column by themselves or in combination. The particle size in the restraining material is expressed in Mesh size (Janson and Jönsson, 2011).

15.3.1.2 Size-Exclusion (Gel-Filtration) Chromatography

Size-exclusion chromatography (SEC) method is also known as gel-filtration chromatography and ensures the separation of biological molecules according to their molecular size. When compared with other chromatographic methods, SEC is not an adsorption method. The column is packed with a solid-phase matrix made up of beads with pores of $100-250 \, \mu m$ and the mobile phase is passed through this structure. The mobile phase fills both the pores of the beads as well as their exterior. A matrix of high porosity generally covers more than 95% of the total liquid in the column. The protein extract applied to the top part of the column moves down in the column with the aid of the mobile phase passing around the beads as well as flowing through the pores of the beads. Since proteins with sizes greater than the pore size cannot go inside the pores, they move around the beads with the help of the mobile phase and reach the lower side of the column thus easily separating. Whereas the movement of the small molecules that can spread inside the pores is slower. Since the small molecules inside the pores can separate more easily than the beads, they are sorted according to their sizes inside the column (Lodish et al., 2000).

The pore size of the matrix used, the physical and chemical stability, its inertness and hydrophilic properties as well as its chemical interactions with the proteins are important. The most commonly used ones are dextran, dextran/bis-acrylamide, agarose, agarose/dextran, polyacrylamide, methacrylate, acrylamide, acrylamide/agarose, and cellulose. Dextran is a natural polysaccharide and is the first gel type that has been developed. Dextran-based gels are suited for the separation of molecules with molecular weights ranging between 1000 and 600,000 Da. The addition of bis-acrylamide can reach up to 107 Da. Whereas the separation limits of polyacrylamide gels range between 100 and 100,000 Da. The agarose gels have the highest separation limit with (1000−5,000,000 Da) (Phillips, 1992).

15.3.1.3 Ion-Exchange Chromatography

This method depends on the principle of the adsorption of charged proteins due to electrostatic interaction with oppositely charged proteins in the column as well as the matrix known as ion exchanger. The adsorption of the proteins to the matrix by way of electrostatic interaction is a reversible process. While the charge loads on the protein arise from different amino acids in the protein, the net charge of the protein depends on the combination of negatively and positively charged amino acids. The hydrogen ion concentration (acidity) of the mobile phase, that is its pH, causes variations to occur in the charges of amino acids. A highly acidic mobile phase will cause many groups to become positively charged thereby making the protein charge positive, whereas high amounts of alkaline mobile phase will form negatively charged proteins. Since the proteins are positively charged at pH values below their isoelectric points and negatively charged at pH values above their isoelectric points, it is important for this method to know their isoelectric points beforehand (Fanali et al., 2017).

The column is packed with the matrix in the first stage. Ion exchangers consist of a matrix with either acidic or basic groups. If the ion exchanger contains positive groups, it is called anion exchanger and if it contains negative groups, it is called as cation exchanger. These matrixes can be nonporous synthetic hydrophilic polymers that have been designed so as to not diffuse inside (Zou et al., 2001) or hydrophobic polystyrene based or partially hydrophobic polymethacrylate-based various polymers or hydrophilic and macroporous synthetic or natural polymers such as polyacrylamide, cellulose, dextran, and agarose (Černý et al., 2013; Ly and Wasinger, 2011).

In the second stage; the mobile phase is added to the positively or negatively charged matrix inside the column and it is ensured that the matrix is surrounded with ions in the buffer that are oppositely charged (Luqman and Inamuddin, 2012).

Whereas, in the third stage, negatively, positively charged or neutral proteins separated are packed in the column. It is ensured that proteins with charges opposite to that of the matrix are bound tightly to the stationary phase. Neutral molecules and molecules that have the same charge with the matrix have either no affinity to the stationary phase or the affinity is very low and thus they move together with the mobile phase and are removed from the column. Whereas the molecules that have been electrostatically bound to the matrix can be taken back from the column using another mobile phase with increased ionic force or pH. The increase in the ionic force of the mobile phase results in the separation of bound molecules; whereas the increase in pH results in the reduction of the charge of the molecule or matrix and the decrease of the electrostatic interaction power (Zou et al., 2001).

15.3.1.4 Affinity Chromatography

The aforementioned chromatographic separations are based on nonspecific physicochemical interactions between the matrix and molecules in the column. Molecular properties of proteins such as charge, size, and polarity do not provide a high selectivity in separation, purification, and isolation. On the other hand, the affinity chromatography is a method based on biological interactions which enables strong-specific separations. The interactions between the protein and the matrix depend not on general properties such as isoelectric point or hydrophobicity, but on selective properties such as the interactions between antigen and antibody, enzyme and substrate analog, nucleic acid and binding proteins as well as hormone and receptors (Fanali et al., 2017).

The biological functions that are carried out by macromolecules such as proteins in biological systems are the results of the interaction with specific molecules known as ligands. These ligands can bind strongly with their protein of interest. That is why ligands are used in affinity chromatography. Ligands bond covalently with the water-insoluble matrix and are immobilized. The matrix is packed to the column and the column is packed with the mobile phase. Afterward, the extract of the protein that will be separated is added and it is ensured that it passes through the column with the mobile phase. There is a slowing down in the movement of the macromolecules that know and bind to the ligand during the passage through the column. Unbound molecules are removed from the column via washing. Whereas the protein bound to the matrix via ligands is recycled after the complex it forms with the ligand is decomposed via various methods.

This method is used in the isolation and purification of almost all biological macromolecules. However, there are various points that should be taken into account; if the interaction between the ligand and the protein is low, there is no adsorption and if the interaction is high, it is difficult to remove the protein from the ligand. The selection of detergents or other chemical substances to strengthen pH, salt concentration, and interaction are important to provide the proper environment for protein–ligand interaction. The most important criteria is that the substrates used for the binding of protein to the ligand or the removal of the protein from the ligand following chromatography will not harm the protein structure in any way (Bailon et al., 2000; Hage et al., 2012; Lodish et al., 2000).

15.3.1.5 *Reverse Phase High-Performance Liquid Chromatography*

This method takes advantage of the large-scale protein separations. Reverse phase high-performance liquid chromatography (RP-HPLC) is a more sensitive, relatively rapid, and accurate method for the purification of peptides, small polypeptides, and related compounds of pharmaceutical interest have not been replicated to the same extent for larger polypeptides and globular proteins. Additionally, this method has been applied on the nano, micro, and analytical scale, and has also been scaled up for preparative purifications, to large industrial scale for proteins.

RP-HPLC has a nonpolar stationary phase and an aqueous, moderately polar mobile phase. The separation of proteins depends on the hydrophobic binding of the solute molecule from the mobile phase to the immobilized hydrophobic ligands attached to the stationary phase. Adding more water to the mobile phase can increase retention times, thereby the affinity of the hydrophobic properties of proteins for the hydrophobic stationary phase gets stronger relative to the more hydrophilic mobile phase. The proteins are, therefore, eluted in order of increasing molecular hydrophobicity. Acetonitrile containing an ionic modifier such as trifluoroacetic acid is used as a common solvent, although other organic solvents such as ethanol also may be used. This acidic solvent has increased solubility at pH values further removed from isoelectric point of protein. This *technique has advantage* for excellent separation of complex mixtures of peptides and proteins with easy experimentation via changes in mobile phase characteristics, temperature. However, these changes can cause the irreversible denaturation of protein samples thereby reducing the potential recovery of material in a biologically active form .

The most commonly employed experimental procedure for the RP-HPLC analysis of peptides and proteins generally involves the use of the more hydrophobic C18 ligands. Peptides and proteins behave as hydrophobic molecules because of their size and most often bind very strongly to C18 ligands. Whereas C4 (*n*-butyl) and C8 (*n*-octyl), phenyl and cyanopropyl ligands can provide different selectivities for peptides and proteins. Increases in column length can increase the resolution of small peptides and proteins. Thus, column lengths between 15 and 25 cm and id of 4.6 mm are generally preferred for applications such as tryptic mapping. However, for larger proteins, low mass recovery and loss of biological activity such as irreversible binding and denaturation may result with these columns. For this reason, shorter columns of between 2 and 20 cm in length are preferred (Aguilar, 2004; Sundaram et al., 2009; Bird, 1989).

15.4 PROTEIN QUANTITATION WITH WESTERN BLOTTING

15.4.1 Tissue Preparation

The original location of the protein inside the cell should be known before the protocol is set. If a tissue sample is used in the study, the most commonly used method is the SDS lysis method, but for cellular studies, ultrasonication is used. Thus, either cells in the tissue structure or cell cultures are homogenized. At the end of the homogenization, the undissolved material in the homogenate is discarded by centrifugation. The protein of interest stays in the liquid phase (supernatant). The cytoplasmic proteins and nuclear proteins are both obtained by disrupting the cell membrane: there are thousands of proteins in the whole cell extract (Kurien et al., 2015).

Because protein-degrading protease enzymes appear when the cell is disrupted, proteins should be protected from these proteases. Therefore protease inhibitors such as diisopropylfluorophosphate, aprotinin, leupeptin, sodium orthovanadate, or phenyl methyl sulfonyl fluoride are used. Thereby, the degradation of proteins is blocked by the inhibition of protease activity. Because proteins are affected by physical conditions, such as pH and temperature, these conditions need to be tightly controlled (Xiong and Gendelman, 2013; Iannone, 2015).

15.4.2 Gel Electrophoresis

This section was provided in detail in Section 15.1.3.

15.4.3 Transfer Methods

A nitrocellulose membrane is the first choice for protein blotting. However, in recent years, other membrane types have also been developed for different protein sizes. The physical features and physical characteristics of the membrane should be selected according to different transfer conditions. Membranes that have pores between 0.45 and 0.2 μm are commonly used. The membrane with 0.2-μm pore size is usually used for small proteins (<15,000 kDa). The nitrocellulose and supported nitrocellulose membranes can warm up easily, making protein transfer easier. Protein binding to nitrocellulose is instantaneous, nearly irreversible, and quantitative up to $80-100\,\mu g/cm^2$. Supported nitrocellulose is an inert support structure with nitrocellulose applied to it. This support structure makes the membrane more flexible. It warms up more quickly than a nitrocellulose membrane, but its protein binding capacity is higher, and it can be autoclaved (121°C) (Mahmood and Yang, 2012).

Polyvinylidene difluoride (PVDF) membrane is an ideal support for N-terminal sequencing, amino acid analysis, and immunoassay of blotted proteins. The nitrocellulose membrane protects proteins that are exposed to acidic–basic conditions or organic solvents. It is effective for detecting low amounts of proteins and also provides an opportunity to determine sequencing manipulations of high molecular weight proteins (Young and Hongbao, 2006; Shively, 2000).

In addition, these membranes can bind proteins even in the presence of the SDS that comes from the transfer buffer. Of note, 1–2 minutes before using a PVDF membrane, it should be wetted with 100% methanol and then incubated in ice-cold transfer buffer. The transfer of proteins from gel to membrane can be done in two ways: wet and semidry transfer. The semidry transfer is usually quicker than wet transfer, but wet transfer keeps membrane from drying out. For both kinds of transfer, the membrane is placed next to the gel. The two are sandwiched between blotting filter paper, and the sandwich is clamped between solid supports to maintain tight contact between the gel and membrane. Blotting filter paper, which is made of 100% cotton fiber, provides a uniform flow of buffer through the gel (Liu et al., 2014).

In wet transfer, the gel and membrane are sandwiched between sponge and paper (sponge/paper/gel/membrane/paper/sponge) and all are clamped tightly together after ensuring that no air bubbles have been formed between the gel and membrane. The sandwich is submerged in transfer buffer to which an electrical field is applied. The negatively charged proteins travel toward the positively charged electrode, but the membrane stops them, binds them, and prevents them from continuing on (Iannone, 2015).

15.4.4 Blocking Buffers

After proteins are transferred onto the membrane, it is blocked with the primary antibody and is washed. After this step, the membrane is incubated with the secondary antibody and is washed once more. Before using antibodies to detect proteins that have been transferred to a membrane, the remaining binding surface must be blocked to prevent the nonspecific binding of the antibodies. Otherwise, the antibodies or other detection reagents will bind to any remaining sites that initially served to immobilize the proteins of interest. A variety of blocking buffers ranging from milk or normal serum to highly purified proteins have been used to block free sites on a membrane. The blocking buffer should improve the sensitivity of the assay by reducing background interference and improving the signal-to-noise ratio.

The primary antibody is specific for the protein of interest, and the secondary antibody enables its detection (Fig. 4.8). The secondary antibody can usually be radiolabeled, labeled with a fluorescent compound, or conjugated to an enzyme-like AP or HRP.

Available detection methods now include colorimetric, chemiluminescent, radioactive, and fluorescent detection (Young and Hongbao, 2006; Thieman and Palladino, 2004; Ghosh et al., 2014; Taylor and Posch, 2014; Taylor et al., 2013).

15.4.5 Detection

After proteins have been transferred to a membrane, they can be visualized using a variety of specialized detection reagents. Total-protein stains allow visualization of the protein pattern on the blot and immunological detection methods, employing antibody or ligand conjugates, and allow visualization of specific proteins of interest. This chapter will provide a brief summary of the immunological detections.

15.4.5.1 Colorimetric Detection

Because chromogenic or precipitating substrates are inexpensive and easily detectable, they are commonly used as a detection method. When these substrates bind to appropriate substrates (e.g., AP and HRP), they become colorful and insoluble products that precipitate onto the membrane. There is no need for an instrument for the detection of this colorful precipitation band. Colorimetric detection is typically considered a medium-sensitivity method, compared to radioactive or chemiluminescent detection (Iannone, 2015).

15.4.5.2 Chemiluminescent Detection

Chemiluminescence is a chemical reaction in which a chemical substrate is catalyzed by an enzyme, such as AP or HRP, and produces detectable light. The light signal can be captured on X-ray film, or by a charge-coupled device (CCD) imager. Faster and more precise measurements can be made with CCD than with other systems such as colorimetric systems and radioisotopic detection. The detection sensitivity is dependent on the affinity of the protein, primary antibody, and secondary antibody, and can vary from one sample to another. In addition, this method does not have negative outcomes such as radioactive exposure, environmental concerns, and high cost (Burgess and Deutscher, 2009).

15.4.5.3 Radioactive Detection

Radioactive detection system does not require enzyme substrates. The antibody itself can be labeled with a radioactive marker such as ^{125}I or ^{35}S. These labeled antibodies are placed in direct contact with X-ray film. After exposure of the membrane to the film for a suitable period, the film is developed and a photographic negative is made of the location of radioactivity on the membrane and dark regions are created which correspond to the protein bands of interest.

The importance of radioactive detections methods is declining, because it is very expensive, health and safety risks are high. Additionally, the signal produced by ^{125}I and ^{35}S is unaffected by enzyme, metal salts, pH, or temperature. Therefore, using the other detection methods, it has become more common in recent years (Walker, 2002).

15.4.5.4 *Fluorescent Detection*

Fluorescent detection is different from the methods explained previously in this chapter (those that work with an enzyme-substrate system). In this system, the secondary antibody is labeled with a fluorophore such as fluorescein (FITC), Texas Red, rhodamine (TRITC), or R-phycoerythrin. Fluorescent detection methods involve detection of light emitted transiently by a fluorescent molecule after it has absorbed light (excitation) and then releases photons (emission) as it returns to its normal state. Fluorescent western blot detection can allow multiplexing and provide better linearity and better quantitation within the detection limits. Application of fluorophores in different channels on the same blot can detect many target proteins at the same time (Walker, 2002).

15.4.6 Protein Microarray

Microarrays constitute a new platform which allows the discovery and characterization of proteins. This technology is also known as protein chips that are miniaturized and parallel assay systems containing small amounts of purified proteins (Hall et al., 2007). These properties allow the simultaneous analysis of thousands of parameters and simultaneously perform high-throughput studies of thousands of proteins within a single experiment. Protein microarrays are typically prepared by immobilizing proteins onto a microscope slide using a standard contact spotter or noncontact microarrayer. Different type slide surfaces can be used such as aldehyde and epoxy-derivatized glass surfaces, nitrocellulose, or gel-coated slides, and nickel-coated slides for different types of protein. After proteins are immobilized on the slides, they can be recognized for a variety of functions/activities. Finally, the resulting signals are usually measured by detecting fluorescent or radioisotope labels.

Three types of protein microarrays are currently used to study the biochemical activities of proteins: analytical microarrays, functional microarrays, and reverse phase microarrays. Additionally, microspheres bead-based systems were noted with different size or color beads as a support of the capture agent to analyze the sample (Sahukar et al., 2016).

15.4.6.1 *Analytical Microarray*

Analytical microarrays are typically used to determine parameters such as the binding affinity, specificities, and protein expression levels of the proteins in the mixture. In this method, antigens or antibodies are immobilized on a glass microscope slide and the test sample, which including proteins dropped on the slide to interact with them. These types of microarrays can be used to monitor differential expression profiles and for clinical diagnostics. Moreover, only selected target proteins can be analyzed by antibody microarrays.

15.4.6.2 *Functional Protein Microarray*

Functional protein microarrays have recently been applied to many aspects of discovery-based biology, including protein−protein, protein−DNA, protein−RNA, protein−phospholipid, protein−drug, and protein−peptide interactions. This technology holds great potential for basic molecular biology research, disease marker identification, toxicological response profiling, and pharmaceutical target screening. Productivities of functional protein microarray are related with creation of a detailed expression clone library, high-throughput protein expression, isolation and purification, adaptation of DNA microarray technology to accommodate protein substrates, and ensuring the stability of arrayed proteins. These allow studying on different areas such as cell-free expression systems (*Escherichia coli*), wheat germ extracts, vaccine development, early detection of biomarkers biochemical activity protein—protein interaction studies.

15.4.6.3 *Reverse Phase Protein Array*

Reverse phase protein array (RPPA): In this technique, target proteins such as cellular or tissue lysate or serum samples are immobilized onto a nitrocellulose slide using a contact pin microarrayer and interacted with an antibody. Fluorochrome-conjugated secondary antibody is added to the first one to achieve a higher fluorescent signal to be detected with chemiluminescent, fluorescent, or colorimetric assays. The fluorescent signal intensity is related with the binding affinity, the specificity, and the steric accessibility of the antibody against the target protein. Reference peptides are printed on the slides to allow for protein quantification of the sample lysates (Krishnan and Davidovitch, 2015; Hall et al., 2007).

15.5 CONCLUSION

Proteins play crucial roles in nearly all biological processes. Moreover, they emerge as an important tool in scientific studies and pharmaceutical industry. Protein analysis aims to explore how amino acid sequences are specified in the structure of proteins and how these proteins bind to substrates and perform their functions. The true diagnosis of some diseases, the determination of where to start to identify the type of mutation and inheritance, the development of animal models for gene therapy, and the production and purification of drugs in peptide/protein structure are within the scope of protein analysis purposes. These analyses should be regarded as indispensable steps in determining the activity of proteins depending on the understanding of their structures.

References

Aguilar, M.I., 2004. HPLC of peptides and proteins: methods and protocols. In: Aguilar, M.-I. (Ed.), Methods in Molecular Biology, vol. 251. © Humana Press Inc., Totowa, NJ.

Akama, K., Shirai, K., Suzuki, S., 2016. A droplet-free digital enzyme-linked immunosorbent assay based on a tyramide signal amplification system. Anal. Chem. 88, 7123—7129.

Aloisi, I., Parrotta, L., Ruiz, K.B., Landi, C., Bini, L., Cai, G., et al., 2016. New insight into quinoa seed quality under salinity: changes in proteomic and amino acid profiles, phenolic content, and antioxidant activity of protein extracts. Front. Plant. Sci. 7, 656.

Amanchy, R., Kalume, D.E., Pandey, A., 2005. Stable isotope labeling with amino acids in cell culture (SILAC) for studying dynamics of protein abundance and posttranslational modifications. Sci. STKE. 2005 (16), l2.

Arora, A., Tamm, L.K., 2001. Biophysical approaches to membrane protein structure determination. Curr. Opin. Struct. Biol. 11, 540—547.

Aydin, S., 2015. A short history, principles, and types of ELISA, and our laboratory experience with peptide/protein analyses using ELISA. Peptides 72, 4—15.

Bailon,, P., Ehrlich,, G.K., Fung,, W.J., Berthold,, W., 2000. An overview of affinity chromatography. In: Bailon, P., Ehrlich, G.K., Fung, W.-J., Berthold, W. (Eds.), Affinity Chromatography: Methods and Protocols. Humana Press, Totowa, NJ, pp. 1–6.

Bantscheff, M., Schirle, M., Sweetman, G., Rick, J., Kuster, B., 2007. Quantitative mass spectrometry in proteomics: a critical review. Anal. Bioanal. Chem. 389, 1017–1031.

Bauer, D.M., Sun, Y., Wang, F., 1997. Nano-electrospray mass spectrometry and Edman sequencing of peptides and proteins collected from capillary electrophoresis. Techniques in Protein Chemistry, vol. VIII. Academic Press, London.

Benesch, J.L.P., Ruotolo, B.T., 2011. Mass spectrometry: come of age for structural and dynamical biology. Curr. Opin. Struct. Biol. 21, 641–649.

Berg, J.M., Tymoczko, J.L., Stryer, L., 2002. Biochemistry, fifth ed. W H Freeman, New York.

Biemann, K., 2014. Laying the groundwork for proteomics: mass spectrometry from 1958 to 1988. J. Proteomics 107, 62–70.

Bird, I.M., 1989. High performance liquid chromatography: principles and clinical applications. BMJ 299 (6702), 783–787.

Bolt, M.W., Mahoney, P.A., 1997. High-efficiency blotting of proteins of diverse sizes following sodium dodecyl sulfate—polyacrylamide gel electrophoresis. Anal. Biochem. 247, 185–192.

Boumediene, S., Kumar, C., Gnad, F., Mann, M., Mijakovic, I., Macek, B., 2010. Stable isotope labeling by amino acids in cell culture (SILAC) applied to quantitative proteomics of Bacillus subtilis. J. Proteome. Res. 9, 3638–3646.

Brunelle, J.L., Green, R., 2014. One-dimensional SDS-polyacrylamide gel electrophoresis (1D SDS-PAGE). Methods Enzymol. 541, 151–160.

Buchwalow, I.B., Böcker, W., 2010. Probes processing in immunohistochemistry. In: Buchwalow, I.B., Bocker, W. (Eds.), Immunohistochemistry: Basics and Methods. Springer, Berlin, pp. 19–29.

Guide to protein purification. In: Burgess, R.R., Deutscher, M.P. (Eds.), Methods in Enzymology, vol. 463. Academic Press, San Diego, CA, pp. 1–820.

Castellani, F., van Rossum, B., Diehl, A., Schubert, M., Rehbein, K., Oschkinat, H., 2002. Structure of a protein determined by solid-state magic-angle-spinning NMR spectroscopy. Nature 420, 98–102.

Carta, G., Jungbauer, A., 2010. Protein Chromatography: Process Development and Scale-Up. WILEY-VCH, Weinheim.

Černý, M., Skalák, J., Cerna, H., Brzobohatý, B., 2013. Advances in purification and separation of posttranslationally modified proteins. J. Proteomics 2–27.

Chen, Z.T., Liang, Z.G., Zhu, X.D., 2015. A review: proteomics in nasopharyngeal carcinoma. Int. J. Mol. Sci. 16, 15497–15530.

Craft, E.G., Chen, A., Nairn, A.C., 2013. Recent advances in quantitative neuroproteomics. Methods 61, 186–218.

Cregger, M., Berger, A.J., Rimm, D.L., 2006. Immunohistochemistry and quantitative analysis of protein expression. Arch. Pathol. Lab. Med. 130 (7), 1026–1030.

Croxatto, A., Prod'hom, G., Greub, G., 2012. Applications of MALDI-TOF mass spectrometry in clinical diagnostic microbiology. FEMS Microbiol. Rev. 36, 380–407.

Curreem, S.O., Watt, R.M., Lau, S.K., Woo, P.C., 2012. Two-dimensional gel electrophoresis in bacterial proteomics. Protein Cell 3 (5), 346–363.

Darragh, A.J., Moughan, P.J., 2005. The effect of hydrolysis time on amino acid analysis. J. AOAC. Int. 88 (3), 888–893.

Dobson, C.M., 2003. Protein folding and misfolding. Nature 426, 884–890.

Dötsch, V., Wagner, G., 1998. New approaches to structure determination by NMR spectroscopy. Curr. Opin. Struct. Biol. 8, 619–623.

Drenth, J., 1994. Principles of Protein X-ray Crystallography. Springer Science + Business Media, LLC, New York.

Dyson, H.J., Wright, P.E., 1996. Insights into protein folding from NMR. Annu. Rev. Phys. Chem. 47, 369–395.

El-Aneed, A., Cohen, A., Banoub, J., 2009. Mass spectrometry, review of the basics: electrospray, MALDI, and commonly used mass analyzers. Appl. Spectrosc. Rev. 44, 210–230.

Fang, Y., Ramasamy, R.P., 2015. Current and prospective methods for plant disease detection. Biosensors. 4, 537–561.

Fanali, S., Haddad, P.R., Poole, C., Riekkola, M.L., 2017. Liquid Chromatography: Applications, second ed.. Elsevier, Netherlands.

Fenn, J.B., Mann, M., Meng, C.K., Wong, S.F., Whitehouse, C.M., 1989. Electrospray ionization for mass spectrometry of large biomolecules. Science 246, 64–71.

Flatley, B., Malone, P., Cramer, R., 2014. MALDI mass spectrometry in prostate cancer biomarker discovery. Biochim. Biophys. Acta 1844, 940–949.

Folloni, S., Bellocchi, G., Kagkli, D.M., Pastor-Benito, S., Aguilera, M., Mazzeo, A., et al., 2011. Development of an ELISA reverse-based assay to assess the presence of mycotoxins in cereal flour. Food Anal. Methods 4, 221–227.

Garrett, R.H., Grisham, C.M., 2013. Biochemistry. Brooks/Cole, Cengage Learning, © Belmont, CA.

Gessel, M.M., Norris, J.L., Caprioli, M.R., 2014. MALDI imaging mass spectrometry: spatial molecular analysis to enable a new age of discovery. J. Proteomics 107, 71–82.

Ghosh, R., Gilda, J.E., Gomes, A.V., 2014. The necessity of and strategies for improving confidence in the accuracy of western blots. Expert. Rev. Proteomics 11, 549–560.

Giorno, R., 1984. A comparison of two immunoperoxidase staining methods based on the avidin-biotin interaction. Diagn. Immunol. 2, 161–166.

Glish, G.L., Vachet, R.W., 2003. The basics of mass spectrometry in the twenty-first century. Nat. Rev. Drug. Discov. 2, 140–150.

Goldsby, R.A., Kindt, T.J., Osborne, B.A., 2000. Kuby Immunology, fourth ed. W. H. Freeman and Company, p. 162.

Guerrera, I.C., Kleiner, O., 2005. Application of mass spectrometry in proteomics. Biosci. Rep. 25, 71–93.

Gygi, S.P., Rist, B., Gerber, S.A., Turecek, F., Gelb, M.H., Aebersold, R., 1999. Quantitative analysis of complex protein mixtures using isotope-coded affinity tags. Nat. Biotechnol. 17 (10), 994–999.

Hage, D.S., Anguizola, J.A., Bi, C., Li, R., Matsuda, R., Papastavros, E., et al., 2012. Pharmaceutical and biomedical applications of affinity chromatography: recent trends and developments. J. Pharm. Biomed. Anal. 69, 93–105.

Hall, D.A., Ptacek, J., Snyder, M., 2007. Protein microarray technology. Mech. Ageing Dev. 128, 161–167.

Hanash, S.M., Strahler, J.R., Neel, J.V., 1991. Highly resolving two-dimensional gels for protein sequencing. Proc. Natl. Acad. Sci. 88, 5709–5713.

Harsha, H.C., Molina, H., Pandey, A., 2008. Quantitative proteomics using stable isotope labeling with amino acids in cell culture. Nat. Protoc. 3 (3), 505–521.

Harvey,, R.H.,,, Ferrier,, D.R., 2011. In: McLaughlin, H.R. (Ed.), Biochemistry, fifth ed. Lippincott William and Wilkins, Philadelphia, p. 482.

Ho, C.S., Lam, C.W.K., Chan, M.H.M., Cheung, R.C.K., Law, L.K., Lit, L.C.W., et al., 2003. Electrospray ionisation mass spectrometry: principles and clinical applications. Clin. Biochem. Rev. 24 (1), 3–12.

Hoffmann, E., Stroobant, V., 2007. Mass Spectrometry Principles and Applications. John Wiley & Sons Ltd, England.

Holčapek, M., Kolářová, L., Nobilis, M., 2008. High-performance liquid chromatography–tandem mass spectrometry in the identification and determination of phase I and phase II drug metabolites. Anal. Bioanal. Chem. 391, 59–78.

Hsu, S.M., Raine, L., Fanger, H., 1981. Use of avidin-biotin-peroxidase complex (ABC) in immunoperoxidase techniques: a comparison between ABC and unlabeled antibody (PAP) procedures. J. Histochem. Cytochem. 29 (4), 577–580.

Hunkapiller, M.W., 1988. Proteins: Structure and Function. Kluwer Academic Publishers Group, Dordrecht Netherlands.

Hyatt, M.A., 2002. Microscopy, Immunohistochemistry, and Antigen Retrieval Methods: For Light and Electron Microscopy. Kluwer Academic/Plenum Publishers, Plenum, New York.

Iannone, E., 2015. Labs on Chip: Principles, Design and Technology. CRC Press, Taylor & Francis Group, Boca Raton.

Janson, J.C., Jönsson, J.A., 2011. Protein Purification: Principles, High Resolution Methods, and Applications, 54. John Wiley & Sons Ltd, New Jersey.

Jiskoot, W., Crommelin, D.J.A., 2005. Methods for Structural Analysis of Protein Pharmaceuticals. American Association of Pharmaceutical Scientists, Arlington.

Jungblut, P.R., 2014. The proteomics quantification dilemma. J. Proteomics 107, 98–102.

Kamp, R.M., Papadopoulou, T.C., Liebold, B.W., 1997. Protein Structure Analysis: Preparation, Characterization, and Microsequencing. Springer, New York.

Kang, J.S., 2012. Principles and applications of LC-MS/MS for the quantitative bioanalysis of analytes in various biological samples. In: Prasain, J.K. (Ed.), Tandem Mass Spectrometry—Applications and Principles. © The Author(s), pp. 441–492.

Krishnan, V., Davidovitch, Z., 2015. Biological Mechanisms of Tooth Movement, second ed. Wiley Blackwell, Oxford.

Kurien, B.T., Scofield, R.H., 2015. Western Blotting: Methods and Protocols. Humana Press, New York.

Kwan, A.H., Mobli, M., Gooley, P.R., King, G.F., Mackay, J.P., 2011. Macromolecular NMR spectroscopy for the non-spectroscopist. FEBS J. 278, 687–703.

Lagarrigue, M., Lavigne, R., Gue'vel, B., Com, E., Chaurand, P., Pineau, C., 2012. Matrix-assisted laser desorption/ionization imaging mass spectrometry: a promising technique for reproductive research. Biol. Reprod. 86 (3), 74 1–11.

Liu, Z.Q., Mahmood, T., Yang, P.C., 2014. Western blot: technique, theory and trouble shooting. N. Am. J. Med. Sci. 6, 160.

Liu, Y., Wang, Z., Zhang, H., Lang, L., Ma, Y., He, Q., et al., 2016. A photothermally responsive nanoprobe for bioimaging based on Edman degradation. Nanoscale 8, 10553–10557.

Lodish, H., Berk, A., Zipursky, S.L., Matsudaira, P., Baltimore, D., Darnell, J., 2000. Molecular Cell Biology, 4th edition Section 3.5, Purifying, Detecting, and Characterizing Proteins. W. H. Freeman, New York.

Loquet, A., Bardiaux, B., Gardiennet, C., Blancher, C., Baldus, M., Nilges, M., et al., 2008. 3D structure determination of the Crh protein from highly ambiguous solid-state NMR restraints. J. Am. Chem. Soc. 130 (11), 3579–3589.

Luqman, M., Inamuddin, M., 2012. In: Inamuddin, I., Luqman, M. (Eds.), Ion Exchange Technology II. Springer, Netherlands.

Ly, L., Wasinger, V., 2011. Protein and peptide fractionation, enrichment and depletion: tools for the complex proteome. Proteomics 11, 513–534.

Lynch, J.M., Barbano, D.M., 1999. Kjeldahl nitrogen analysis as a reference method for protein determination in dairy products. J. AOAC Int. 82 (6), 1389–1398.

Ma, L., Zhang, J., Chen, H., Zhou, J., Ding, Y., Liu, Y., 2011. An overview on ELISA techniques for FMD. Virol. J. 8, 419–428.

Magdeldin, S., Enany, S., Yoshida, Y., Xu, B., Zhang, Y., Zureena, Z., et al., 2014. Basics and recent advances of two dimensional-polyacrylamide gel electrophoresis. Clin. Proteomics 11 (1), 16.

Mahmood, T., Yang, P.C., 2012. Western blot: technique, theory, and trouble shooting. N. Am. J. Med. Sci. 4 (9), 429–434.

Mathy, G., Sluse, F.E., 2008. Mitochondrial comparative proteomics: strengths and pitfalls. Biochim. Biophys. Acta 1777 (7–8), 1072–1079.

Miller R.T., 2001. Technical Immunohistochemistry. Achieving reliability and reproducibility of immunostains. In: Society for Applied Immunohistochemistry Meeting 2001. pp. 1–53.

Miyashita, M., Presley, J.M., Buchholz, B.A., Lam, K.S., Lee, Y.M., Vogel, J.S., et al., 2001. Attomole level protein sequencing by Edman degradation coupled with accelerator mass spectrometry. PNAS 98 (8), 4403–4408.

Murray, R., Bender, D., Botham, K.M., 2012. Harpers Illustrated Biochemistry 29th Edition (Lange BasicScience). © McGraw-Hill Companies, Bridgeton (NJ).

Nelson, D., Cox, M., 2005. Lehninger Principles of Biochemistry, fourth ed. W.H. Freeman and Company, New York.

Ong, S.E., Mann, M., 2005. Mass spectrometry-based proteomics turns quantitative. Nat. Chem. Biol. 1, 252–262.

Ong, S.E., Mann, M., 2006. A practical recipe for stable isotope labeling by amino acids in cell culture (SILAC). Nat. Protoc. 1, 2650–2660.

Ong, S.E., Mann, M., 2007. Stable isotope labeling by amino acids in cell culture for quantitative proteomics. Methods Mol. Biol. 359, 37–52.

Ong, S.E., Blagoev, B., Kratchmarova, I., Kristensen, D.B., Steen, H., Pandey, A., et al., 2002. Stable isotope labeling by amino acids in cell culture, SILAC, as a simple and accurate approach to expression proteomics. Mol. Cell. Proteomics 1 (5), 376–386.

Oudenhove, L.V., Devreese, B., 2013. A review on recent developments in mass spectrometry instrumentation and quantitative tools advancing bacterial proteomics. Appl. Microbiol. Biotechnol. 97, 4749–4762.

Phillips, T.M., 1992. Size Exclusion Chromatography, Analytical Techniques in Immunochemistry. Marcel Dekker, New York.

Pi, J., Sael, L., 2013. Mass spectrometry coupled experiments and protein structure modeling methods. Int. J. Mol. Sci. 14, 20635–20657.

Ramos-Vara, J.A., 2005. Technical aspects of immunohistochemistry. Vet. Pathol. 42, 405–426.

Rao, V.S., Srinivas, K., Sujini, G.N., Sunand Kumar, G.N., 2014. Protein-protein interaction detection: methods and analysis. Int. J. Proteomics 2014, 12.

Renshaw, S., Immunohistochemistry and Immunocytochemistry: Essential Methods, Wiley Blackwell, 2017, Oxford.

Righetti, P.G., Stoyanov, A.V. and Zhukov, M.Y., 2001. Sodium dodecyl sulphate polyacrylamide gel electrophoresis (SDS-PAGE), Journal of Chromatography Library. The Proteome Revisited Theory and Practice of all Relevant Electrophoretic Steps. Elsevier, Amsterdam, 295–302.

Rodrigoa, M.A., Zitkaa, O., Krizkova, S., Moulicka, A., Adama, V., Kizeka, R., 2014. MALDI-TOF MS as evolving cancer diagnostic tool. J. Pharm. Biomed. Anal. 95, 245–5595.

Rombouts, I., Lamberts, L., Celus, I., Lagrain, B., Brijs, K., Delcour, J., 2009. Wheat gluten amino acid composition analysis by high-performance anion-exchange chromatography with integrated pulsed amperometric detection. J. Chromatogr. 1216, 5557–5562.

Römpp, A., Spengler, B., 2013. Mass spectrometry imaging with high resolution in mass and space. Histochem. Cell. Biol. 139, 759–783.

Sadeghi, A.A., Nikkhah, A., Shawrang, P., Shahrebabak, M.M., 2006. Protein degradation kinetics of untreated and treated soybean meal using SDS-PAGE. Anim. Feed Sci. Tech. 126 (1–2), 121–133.

Sahukar, B., Vinodhkumar, P., Selvamani, M., 2016. Proteomics: a new perspective for cancer. Adv. Biomed. Res. 5, 67.

Sapan, C.V., Lundblad, R.L., Price, N.C., 1999. Colorimetric protein assay techniques. Biotech. Appl. Biochem. 29 (2), 99–108.

Shively, J.E., 2000. The chemistry of protein sequence analysis. EXS 88, 99–117.

Speicher, K.D., Gorman, N., Speicher, D.W., 2001. N-terminal sequence analysis of proteins and peptides. Curr. Protoc. Protein Sci. 11 (10), 1–41.

Steendam, K.V., Ceuleneer, M.D., Dhaenens, M., Hoofstat, D.V., Deforce, D., 2013. Mass spectrometry-based proteomics as a tool to identify biological matrices in forensic science. Int. J. Legal. Med. 127, 287–298.

Sundaram, H., Vijayalakshmi, N., Srilatha, K.P., 2009. High performance liquid chromatography and its role in identification of Mycobacteriae: an overview. NTI Bull. 45, 1–4.

Susnea, I., Bernevic, B., Wicke, M., Ma, L., Liu, S., Schellander, K., et al., 2013. Application of MALDI-TOF-mass spectrometry to proteome analysis using stain-free gel electrophoresis. Top. Curr. Chem. 331, 37–54.

Tan, S.C., Yiap, B.C., 2009. DNA, RNA, and Protein Extraction: The Past and The Present. J. Biomed. Biotechnol. 2009, 1–10.

Takeda, K., Moriyama, Y., 2015. Kinetic aspects of surfactant-induced structural changes of proteins-unsolved problems of two-state model for protein denaturation. J. Oleo. Sci. 64 (11), 1143–1158.

Taylor, S.C., Posch, A., 2014. The design of a quantitative western blot experiment. Biomed. Res. Int. 2014, 8 pp.

Taylor, S.C., Berkelman, T., Yadav, G., Hammond, M.A., 2013. Defined methodology for reliable quantification of western blot data. Mol. Biotechnol. 55 (3), 217–226.

Thieman, W.J., Palladino, M.A., 2004. Introduction to Biotechnology, second. ed. Pearson/Benjamin Cummings, San Francisco.

Topf, M., Sali, A., 2005. Combining electron microscopy and comparative protein structure modeling. Curr. Opin. Struct. Biol. 15, 578–585.

van den Broek, I., Niessen, W.M.A., van Dongen, W.D., 2013. Bioanalytical LC–MS/MS of protein-based biopharmaceuticals. J. Chromatogr. B 929, 161–179.

Walker, J.M., 2002. The Protein Protocols Handbook, second ed. Springer.

Wasinger, V.C., Zeng, M., Yau, Y., 2013. Current status and advances in quantitative proteomic mass spectrometry. Int. J. Proteomics 2013, 12 pp.

Watabea, S., Morikawab, M., Kaneda, M., Nakaishi, K., Nakatsuma, A., Ninomiya, M., et al., 2016. Ultrasensitive detection of proteins and sugars at single-cell level. Commun. Integr. Biol. 9 (1), e1124201 (8 pp.).

Wenk, M.R., Fernandis, Z.A., 2007. Manual for Biochemistry Protocols. World Scientific, River Edge, NJ, pp. 21–27.

Wider, G., 2000. Structure determination of biological macromolecules in solution using NMR spectroscopy. Biotechniques 29, 1278–1294.

Wilson, W.C., Daniels, P., Ostlund, E.N., Johnson, D.E., Oberst, R.D., Hairgrove, T.B., et al., 2015. Diagnostic tools for bluetongue and epizootic hemorrhagic disease viruses applicable to north American veterinary diagnosticians. Vector. Borne. Zoonotic. Dis. 15 (6), 364–373.

Xiong, H., Gendelman, H.E. (Eds.), 2013. Current Laboratory Methods in Neuroscience Research. Springer, New York, Heidelberg, Dordrecht, London.

Yi, E.C., Li, X.J., Cooke, K., Lee, H., Raught, B., Page, A., et al., 2005. Increased quantitative proteome coverage with (13)C/(12)C-based, acid-cleavable isotope-coded affinity tag reagent and modified data acquisition scheme. Proteomics 5 (2), 380–387.

Young, J., Hongbao, M., 2006. Western blotting method. J. Am. Sci. 2 (2), 23–27.

Zhu, Y., Fang, Q., 2013. Analytical detection techniques for droplet microfluidics. Anal. Chim. Acta 787, 24–35.

Zhou, W., Apkarian, R., Wang, Z.L., Joy, D., 2006. Fundamentals of Scanning Electron Microscopy (SEM), Scanning Microscopy for Nanotechnology. Springer Science + Business Media, LLC, New York.

Zou, H., Luo, Q., Zhou, D., 2001. Affinity membrane chromatography for the analysis and purification of proteins. J. Biochem. Biophys. Methods 49, 199–240.

Engineering Monoclonal Antibodies: Production and Applications

Gülay Büyükköroğlu and Behiye Şenel

Anadolu University, Eskisehir, Türkiye

16.1 INTRODUCTION

The immune system protects the body against bacteria, viruses, and extracellular proteins (Chaplin, 2010). Antibodies are glycoprotein structures present in tissues, the circulating system, and mucosal areas, and are responsible for either humoral or antibody-dependent immunity (Janeway et al., 2001a; Lee and Jeong, 2015). They also known as immunoglobulins (Ig), are Y-shaped proteins produced by B cells (Merlo and Mandik-Nayak, 2013). These are the only molecules that can recognize a large repertoire of antigenic structures; they can also bond strongly with and distinguish different antigenic structures (Chaplin, 2010). The immune system can produce many antibodies with different specificities: in humans, specific to more than 10^{11} antigens. These antibodies show structural similarities (Janeway et al., 2005).

Antibodies that are specifically reactive to antigens are the most important element of the immune system. There are two types of immunity: *innate immunity* and *adaptive immunity*. Innate immunity is the primary immune response system that occurs when the body encounters a pathogen. In this system, there are three main cell groups: phagocytic cells, T-helper cells, and natural killer cells. The innate immune system gives way to a more complex adaptive immune system in cases where a possible advanced infection develops (Parkin and Cohen, 2001; Turvey and Broide, 2010).

Another immunity type, adaptive immunity, is generated to form a more specialized and effective immune response. B and T lymphocytes, the most important cells of this system, play an important role in developing two types of immunity: *cellular immunity* and *humoral immunity*. T lymphocyte cells are responsible for cellular immunity, and in this type of immunity, bacteria are first determined by sensitivity and then are directly cleared

353

up by phagocytosis. By contrast, B lymphocytes are responsible for humoral immunity, in which most of the pathogens and bacteria are cleared up by means of antibodies that are synthesized by these cells (Bonilla and Oettgen, 2010; Turvey and Broide, 2010).

The antibodies are the most effective branch of the humoral response. Structural features, functions, production techniques, and clinical usage of the antibodies are reviewed in this chapter.

16.2 STRUCTURES AND FUNCTIONS OF ANTIBODIES

Antibodies are produced against antigens and are responsible for antibody-dependent immunity. Basically, antibodies have two roles; the first is to bind pathogens, the second is to disrupt the biological structure of the antigen. Antibodies label foreign molecules that they bind to, and make these molecules recognizable by immune-defense cells (Alberts et al., 2002; Merlo and Mandik-Nayak, 2013).

Antibodies show structurally similar features (Janeway et al., 2001b). In most higher mammals, there are five immunoglobulin classes: IgG, IgM, IgA, IgD, and IgE (Fig. 16.1). These immunoglobulins differ in size, charge, carbohydrate contents, and amino acid composition. There is some heterogeneity in these classes (Kindt et al., 2007). Due to the difference between heavy chains, this heterogeneity is mostly seen on IgGs. There are four IgG subtypes in both human (IgG1, IgG2, IgG3, and IgG4) and mice (IgG1, IgG2a, IgG2b, and IgG3) (Cruse and Lewis, 2010; Williams et al., 2012). From a biotechnological perspective,

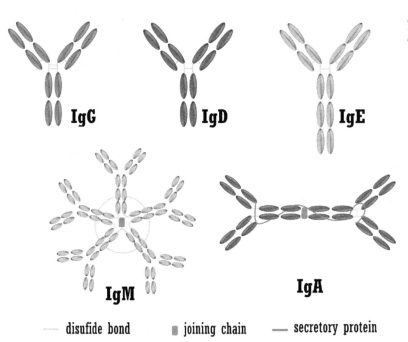

FIGURE 16.1 Classes of immunoglobulin.

disufide bond   joining chain   secretory protein

IgGs are 80% of the immunoglobulins in the serum, and only IgG can cross the placenta. IgGs are the main antibodies that cover and phagocyte the antigens. IgM is the first antibody that is synthesized just after the infection; only IgMs are synthesized by immature B lymphocytes, but the other lymphocytes are synthesized by mature cells (Kindt et al., 2007). IgMs are the largest Ig structures as a result of their pentameric structure. Because they have 10 antigen-binding regions, they are quite effective agglutinators (Kindt et al., 2007; Merlo and Mandik-Nayak, 2013).

A typical antibody is formed from four polypeptides: two identical polypeptide chains that are 50 kDa in size, the heavy (H) chains, and two identical polypeptide chains that are about 25 kDa in size, the light (L) chains (Flaherty, 2012; Hermanson, 2013). Heavy and light chain structures are linked by a disulfide bond. Each heavy chain is also linked by a disulfide bond; this structure forms the Ig skeleton (Janeway et al., 2001b; Hermanson, 2013). The glycoprotein structure of the Igs ranges between 82% and 92%, and around 4%−18% of the structure is made up of carbohydrates. These carbohydrates can be in the form of simple or complex sugars, but are mostly in the form of oligosaccharides. The oligosaccharides bind to asparagine amino acid on the polypeptide chain by their glucosamine ends. Oligosaccharide chains end with galactose, and N-acetylneuraminic acid binds to this galactose. When L chains are isolated from different patient samples, significant variability is found on the amino acid sequence of these L chains. The variations on the amino acid structures start from the N-terminal region of the polypeptide chain; they are seen between 1st and 110th amino acids. In many different amino acid sequences, this variation is not seen after 110th position. Thus the region between 1st and 110th amino acids is called the variable region (Janeway et al., 2001b; Kindt et al., 2007). This variable region is located on the arms of the Y-shaped molecule. The variable region formed by the heavy and light chains is the region that binds to antigens (Hermanson, 2013). Like the L chain, an H chain also comprises variable and constant regions. The variable region on the H chain is as long as the variable region on the L chain, whereas the constant region is three times longer than on the L chain. L and H chains have three segments (Chaplin, 2010). The segment that shows more variabilities with respect to the other regions is termed the hypervariable region. These hypervariable regions are also called complementarity-determining regions, because antibody specificity is provided by these regions (Kindt et al., 2007; Chaplin, 2010). They are 5 or 10 amino acids in length and from 5% to 10% of the amino acid sequence of the antibody. This region determines the specificity of the antibodies, and the remaining 90%−95% of the sequence has similar structures in different antibodies (Janeway et al., 2001b; Kindt et al., 2007; Gupta, 2008).

There are some following common characteristics among amino acid sequences:

1. The L chain on the variable region (VL) and L chain on the constant region (VH) have similar amino acid sequences.
2. There is a constant region on the C-termini of the heavy chain (CH) and it comprises three regions that have similar sequences and are the same length (CH1, CH2, CH3). The constant region has different types that change due to immunoglobulin type, such as γ, α, ε, μ, and δ; however, this difference does not affect specificity.

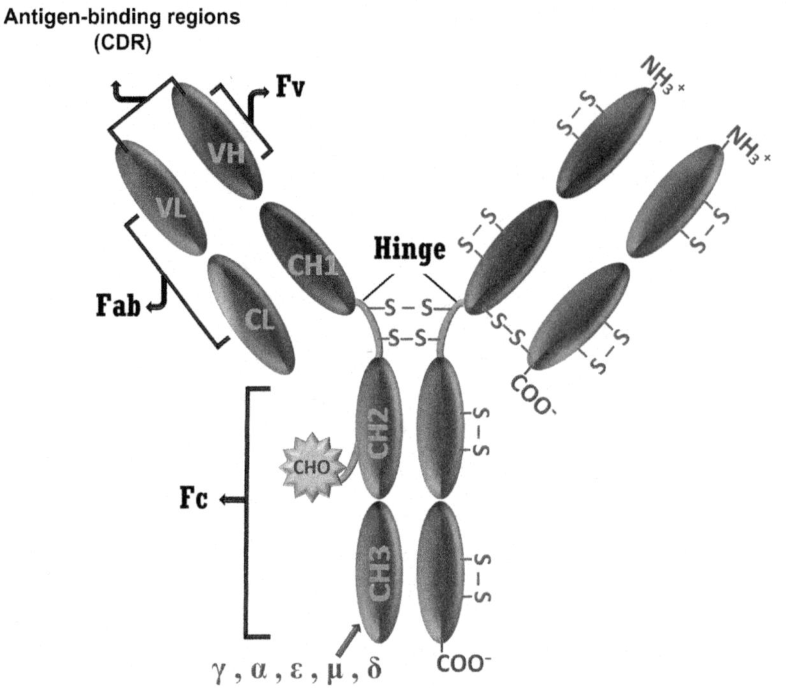

Antigen-binding regions
(CDR)

FIGURE 16.2 Schematic diagram of IgG structure.

3. The L chain of the constant region (CL) is the closest region to the H chain of the constant region.
4. For both L and H chains, inner-chain disulfide bonds are seen in the similar places (Fig. 16.2) (Janeway et al., 2001b; Alberts et al., 2002; Kindt et al., 2007; Liu and Kay, 2012).

The Fab fragment is formed from four globular subunits: VL, VH, CL, and CH; it has a tetrahedral conformation and contains antigen-binding regions. The Fc fragment is formed from two CH2 and CH3 domains and can easily crystalize (Kindt et al., 2007; Dubey, 2014). It does not bind to antigens, but all biological activities occur in this region. It has some roles such as fixation of the complements and inducing the mast cell to synthesize histamine. The antigen—antibody complex binds to an appropriate Fc receptor on the macrophage surface, and it is phagocyted by the macrophage (Chaplin, 2010). The antigens that contain multiple binding sites interact with Fc-bound antibodies; this interaction starts phagocytosis. This complex is endocytosed and digested by the lysosome. Fc fragments are also responsible for antibody transport to fetus by means of the placenta (Kindt et al., 2007). The region where fragments meet is called the "hinge region." The hinge region holds Fab and Fc fragments together and provides flexibility to the molecule. It also allows two Fab arms to move freely. Just as each heavy chain can interact with a light chain, two heavy chains can also interact with each other (Flaherty, 2012; Dubey, 2014).

16.2.1 Polyclonal Antibodies

An antigen has many antigenic regions called "epitopes." Each epitope is recognized by a different antibody. For each epitope, a group of B lymphocyte cells begins to proliferate and produces specific antibodies; at the end of the process, B lymphocyte colonies are formed. The antibodies that are produced against these antigens are called "polyclonal antibodies" (PoAbs) (Stills, 2012; Khan, 2014). PoAbs have different affinities and also microspecificities against different antigens (Khan, 2014). In general, they are produced to provide passive immunity to various pathogens and their toxins (for instance, botulinum antitoxin, tetanus antitoxin, diphtheria antitoxin) (Walsh, 2002).

In many studies done in 1890s, it was observed that when antitoxin serum taken from an animal that had been immunized against diphtheria is given to an unimmunized animal, this unimmunized animal is protected from the disease. After this discovery, hyperimmunized animal-derived PoAbs were used to treat several diseases (pneumonia, meningococcal meningitis, scarlet fever, diphtheria, and measles) (Newcombe and Newcombe, 2007; Saylor et al., 2009).

Modern hyperimmune PoAb therapeutics used for the treatment of acute diseases and medical emergencies are typically derived from either human donors or animals with elevated serum levels of specific PoAbs (Rasmussen et al., 2007). Animal-derived PoAbs are also termed "hyperimmune polyclonal therapeutics". For the production of these antibodies (Fig. 16.3), healthy animals should be selected carefully; they need to be immunized with antigens. Then, hyperimmunized serums are collected from these immunized animals, and the PoAbs are purified using affinity chromatography and centrifugation techniques.

These PoAbs can also show cross-activity because they are more tolerant to small changes on their epitopes and they can recognize and bind to antigens that have some small alterations in structure. In addition, because the stability of monovalent Fab (1) and Fab (2) fragments has higher whole antibody stability, the Fab region is digested by some proteases such as papain and pepsin (Fig. 16.4). After the digestion occurs, purification is done by chromatographic methods; they are then lyophilized and released (Newcombe and Newcombe, 2007).

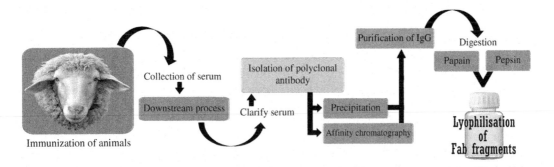

FIGURE 16.3 Schematic illustration of productions of hyperimmune polyclonal therapeutics.

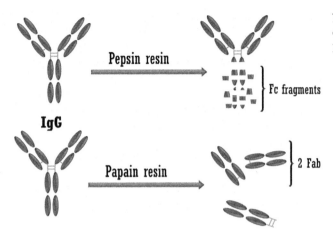

FIGURE 16.4 Schematic representation of IgG fragment generated by pepsin and papain.

There is no need for expert personnel or sophisticated instruments for the production of these animal-derived PoAbs; however, serum disease is the most important side effect of PoAbs (Khan, 2014). In addition, the animals are not protected against infections and other serums. Because maintaining the survival of the animals is expensive, these antibodies are also very expensive.

Alternatively, human blood can also be used for the production of PoAbs. These therapeutics are also termed "pooled PoAbs," and they are produced using special methods such as fractionation of the plasma products. The Igs that are applied intravenously (*immunoglobulins* intraveneous, IgIV) are obtained by purification of human-derived blood samples (Buchacher and Iberer, 2006; Newcombe and Newcombe, 2007). Typical plasma pools range between 4000 and 50,000 L. The World Health Organization (WHO) original guidelines require more than 1000 donors per lot. The IgG amount in the final product should be between 50 and 100 mg/mL. However, these ratios can change in different lots as a result of different production methods. Cold ethanol centrifugation is usually used for purification. The most important step for the production of PoAbs is purification. Therefore new purification techniques have been studied in recent years (Martin, 2006).

IgIV has been used to treat genetic abnormalities, genetic disorders, and chemotherapy-sourced antibody deficiencies for several years. IgIV showed its effect in primary immunodeficiency syndrome for the first time (Buchacher and Iberer, 2006).

In addition, some therapeutic cocktails can be produced in a single drug format by mixing PoAbs with several monoclonal antibodies (mAbs) having the desired specificities. The aim is to reduce the cost, to develop safe human-derived PoAbs, and to avoid the limitations of mAb treatments. These antibodies have been used and evaluated in in vivo models. Human rabies virus-neutralizing mAb cocktails and radiolabeled mAb mixtures for the treatment of carcinoma can be cited as applications of these antibodies (Prosniak et al., 2003).

In the production of these cocktails, natural heterogeneous cell cultures cause differences in the final product. These differences occur due to different cell growth and expression ratios, and to varying genetic stability between different cell lines (Newcombe and Newcombe, 2007).

Antigen-specific recombinant PoAbs are termed "next-generation PoAbs," and for their large-scale production, new technologies have been developed by using mammalian cell cultures (Symphogen A/S, Ballerup, Denmark) (Haurum and Bregenholt, 2005). For the production of recombinant human PoAbs, a modified mammalian cell expression system is used. Only a single antibody-expressing plasmid is placed into the host genomic DNA. This technique is different from traditional mammalian transfection and expression systems by virtue of the site-specific integration of the plasmid into the host genomic DNA. This system blocks the localization of the plasmid in different regions and allows limited expression of the plasmid. Thus modified cell lines that express desired mAbs termed "symphobodies" can determine the desired level of the recombinant human PoAb mixture (Wright, 2006; Newcombe and Newcombe, 2007).

Human antibodies that are produced using transgenic technologies have great potential for the large-scale production of the human PoAbs. Some transgenic strategies are used to produce functional human antibodies in cattle. For this process, researchers transferred sequences containing heavy and light chain loci of human immunoglobulins into fetal cattle fibroblasts, thereby constructing human artificial chromosomes (HACs). In addition to antibody-expressing genes, the HAC also contains different genes. The HAC that was produced in a Chinese hamster ovary (CHO) cell line is transferred into fetal bovine fibroblasts. Eventually, more than 80% of the bovine cells contain HAC; they are called transchromosomic bovines (Robl et al., 2003). Thus PoAbs are synthesized by means of antibody-producing immune cells. Then, human antibodies are collected from bovines and purified. A vast number of studies are being carried out to explore the use of these antibodies for treatments of infections and cancer (Robl et al., 2003; Kuroiwa et al., 2002).

16.2.2 Monoclonal Antibodies

mAbs are produced in large quantities against a specific antigen in a laboratory environment and interact with their specific antigens with a high binding capacity. All produced mAbs have similar characteristics. mAbs are commonly used in diagnosis and treatment of disease and can also be used for disease prevention (passive immunity), antigen purification, diagnostic kit preparation, scientific research, and vaccine development (Breedveld, 2000; Dimitrov, 2010).

16.2.2.1 Production of Monoclonal Antibodies

mAbs are produced from a specific type of B cells. To find and purify, a specific B-cell is quite difficult because B cells have a short life span (Kindt et al., 2007). Because they are not used as a continuous antibody source, they are immortalized by hybridoma technology. This technology, developed in the last quarter of the 20th century, has a great importance for mAb production. Moreover, a recombinant phage display can be produced both in transgenic plants and transgenic animals (Breedveld, 2000).

16.2.2.1.1 MONOCLONAL ANTIBODY PRODUCTION BY HYBRIDOMA TECHNIQUE

In 1975, a technique was developed by G. Köhler and C. Milstein at University of Cambridge. This technique allows a large amount of antibody production that is specific

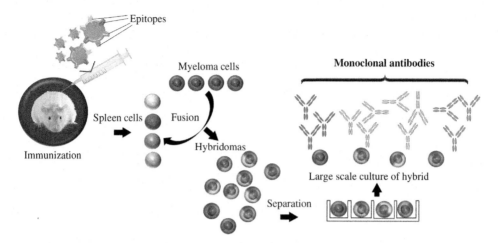

FIGURE 16.5   Representation of hybridoma technique.

to only one epitope; it was termed the "hybridoma" technology. As a result of this discovery, these researchers were awarded with the Nobel Prize in Physiology-Medicine (Bussel et al., 2007; Nagy, 2014).

In this technology, immortal tumor cells were merged with antibody-producing mammalian cells; the hybridoma cells obtained had the capacity to produce an unlimited amount of antibodies (Fig. 16.5). Because these cells are produced from a single type of cells, they are termed *monoclonal cells*, and the antibodies that are produced from these cells are called *mAbs* (Khan, 2008; Pandey, 2010).

Therapeutic immunoglobulins are usually produced in mammalian host cells such as NS0 murine myeloma cells, PER.C6 human cells, and CHO cells. CHO cells are used for the production of 70% of the recombinant proteins (Li et al., 2010). The main reasons for using CHO cells for mAb production are:

- They were proved as safe during the last 20 years, and therefore make it easier to get an approval from some institutions like the Food and Drug Administration (FDA).
- CHO cells are found to be compatible with gene amplification techniques such as dihydrofolate reductase (DHFR) and glutamine synthetase (GS).
- CHO cells have an appropriate molecular repertoire for natural and posttranslational modifications that are in a form comparable with that in humans.
- CHO cells grow well in a suspension environment without serum. For their large-scale production, stainless steel and disposable bioreactors can be used (Jayapal et al., 2007; Li et al., 2010).

Hybridoma technology can be summarized in seven stages (Pandey, 2010; Gorny, 2012; http-1):

Stage 1: Immunization of the mice

For the production of a mAb that is specific against an antigen, an experimental animal model is immunized with this antigen via injection. The immunization activates B cells and causes B lymphocyte synthesis in the spleen (Pandey, 2010; Gorny, 2012; http-1).

Stage 2: Preparation of the splenocytes

The spleen is dissected and splenocytes prepared in the second step. To be able to obtain a single cell, the sample is incubated with Red Blood Cell (RBC) lysis buffer and then washed (Pandey, 2010; Gorny, 2012; http-1).

Stage 3: Mixing the cells

In this step, splenocytes and myeloma cells are mixed in a defined ratio. It is preferred that the selected myeloma cell line should not produce antibodies. Some agents such as polyethylene glycol or sendai virus are used to mix the cell suspension (Pandey, 2010; Gorny, 2012; http-1).

Stage 4: Selection of the hybrid cells

Mixtures of hybridoma cells are taken into an environment that contains hypoxanthine, aminopterin, and thymidine. Aminopterin acts as a folate antagonist and blocks de novo nucleotide biosynthesis. Thus hybridoma cells provide purine and pyrimidine synthesis via a "salvage synthase" pathway using hypoxanthine and thymidine in the environment. Eventually, β cells die due to their short life span, and nonhybridized myeloma cells also die because of the blocking of their de novo production pathway. The only living cells are those that are hybridized with tumor cells (Pandey, 2010; Gorny, 2012; http-1).

Stage 5: Clonal selection

After observing mixed hybridomas, the most important thing is to select optimal hybrid cells that produce the desired antibody rather than selecting living cells. For this experiment, a 96-well plate is used, and the researcher attempts to place only one cell in each well. After incubation, cells proliferate. With this increasing number of cells, antibody production starts at some level, and a specific hybrid cell is selected by examining culture supernatants using various methods (Pandey, 2010; Gorny, 2012; http-1).

Stage 6: Clonal expansion

After the determination of the specific clone, clones are expanded by in vivo and in vitro methods. In the in vitro method, cells are transported to larger plates (e.g., 6−12−24−48−96-well plates). Then, the cells grown in these plates are further transported to tissue culture flasks. After growing in these bottles, they are frozen for future experiments (Pandey, 2010; Gorny, 2012; http-1).

In the in vivo method, antibody-producing cellular clones can be injected into the peritoneal cavity of the mice. These cells then show a tumor-like growth and antibodies begin to be produced from hybridoma cells in the peritoneal cavity in an acidic liquid (Pandey, 2010; Gorny, 2012; http-1).

Stage 7: Purification

mAbs should not always be purified after they are produced; however, they need to be purified under some critical conditions. These critical conditions are:

- the antibody amount is low;
- there is any contamination by other compounds or acidic fluid;
- the antibody is to be used for a specific treatment;
- the antibody is to be used for diagnosis;
- the synthesized antibody is to be further processed (conjugation, alkaline phosphatase, fluorescence staining, and other processes).

In these situations, affinity column chromatography is usually used for the purification of the mAbs. There are several guidelines for antibody purification (Pandey, 2010; Gorny, 2012; http-1).

16.2.2.1.2 MONOCLONAL ANTIBODY PRODUCTION BY PHAGE-DISPLAY TECHNIQUE

The genes that encode different proteins, either in prokaryotic or eukaryotic organisms, are successfully cloned and expressed with the assistance of several vectors (phage, plasmid, cosmid, phagemid, virus, bacterium) in microorganisms (usually in bacteria and bacilli) and/or eukaryotic cells (Azzazy and Highsmith, 2002; Hammers and Stanley, 2014). There are some Ig genes among them. Because there is not just one DNA sequence in an Ig gene, they can be formed by the reorganization of the separate genes. Because it is not possible to obtain these separate genes at the same time, a functional Ig gene has not as yet been cloned. However, either the Fc portion of the heavy chain or only specific part of the antibody (heavy of light chain) were cloned separately.

Antibody molecules can be produced by means of recombinant DNA technology. In the most general sense, phage-display technology can be summarized as presentation of recombinant protein and peptide structures that are able to recognize a specific target molecule on the surface of a filamentous phage (Fig. 16.6). Usually M13, f1 or fd phage vectors are selected (Bazan et al., 2012). All of these vectors are bacteriophages that have an external protein coat. Antibody genes (fragments) are ligated to the phage DNA, which allows expression on the external protein coat of the phage. mRNA of the antigen-recognizing VH and VL regions is isolated. mRNA is reverse transcribed into cDNA and amplified by a polymerase chain reaction. These amplified VH and VL genes are linked by short, synthetic DNA linkers. A linker is formed by approximately 15 amino acids, and the interval in the 3.5 nm length is filled between the 5′-amino terminal end and 3′-carboxy terminal end of the VH. Eventually, the stability and affinity of the prepared molecule is similar to that of the natural antibody molecule. These formed VH and VL genes are termed *single-chain variable fragments* (scFVs). An scFV is placed between pIII protein genes, which are the major coat proteins of either M13 or fd. Then, these phages are transformed into *Escherichia coli*, and replication of the phage inside the *E. coli* is made possible. The fusion gene product is expressed on the surface coat proteins of newly formed phage particles. A few g3p protein-bound antibodies are expressed on the surface of the phage particles that carry the scFv-gene sequence (Azzazy and Highsmith,

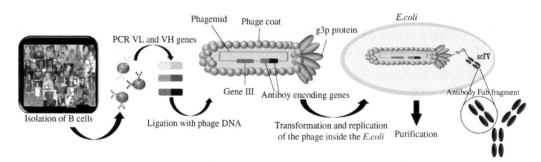

FIGURE 16.6 Schematic illustration of phage display technique.

2002; Carmen and Jermutus, 2002). The presence of the antibody on the end of the phages that leave the cell is checked using immunologic methods (enzyme-dependent immunosorbent analyses, ELISA, for example). The unbound phage and antibodies are removed by washing. The phage that carries the antibody in the end and the *E. coli* TG1 is infected once again, and they are seeded onto a solid environment. Pure colonies are produced, and extra amounts of antibody-carrying phage particles are obtained. These antibody segments are then purified by extraction from the *E. coli* and phage elements (Azzazy and Highsmith, 2002; Kretzschmar and Rüden, 2002; Kindt et al., 2007).

16.2.2.1.3 MONOCLONAL ANTIBODY PRODUCTION USING TRANSGENIC ANIMALS

The ability to make genetic alterations at the molecular level has been confirmed as a revolution in biology. Transgenic technology is found in the intersection of embryology, cellular biology, and molecular genetics techniques. The transgenic production of the Igs is dependent on the transgenic animal production. The first Ig transgene expressed in a transgenic mouse encoded a murine κ isotype light chain. The spleen was used for the transgene expression, and the expressed protein was detected in the serum of the transgenic animal. Not only the transgene was expressed in B and T cells of transgenic mice, but also it combined with the endogenous light chain of the host to produce functional IgM. Thus, when the transgene that encodes either the light of heavy chain of the Ig is combined with an appropriate endogenous molecule, a fully functional Ig can be produced. Disease-specific Ig and humanized Ig can be produced using transgenic animals (Masih et al., 2014).

The microinjection of the large segment of DNA (Translocus), which carries human Ig locus, is another approach. Thus a large number of functional human Ig expression gene loci are transmitted to a transgenic mouse. When this animal is immunized with a specific antigen, it produces antigen-specific human Igs in vivo. In recent years, the focus has been on the production of "transgenic cattle" (Gavin et al., 2014; Masih et al., 2014). Some information about transgenic cattle was given in Section 16.2.1.

To use large Ig loci, allow hypermutation, class alterations, and good expression levels, the human Ig loci have been entirely cloned, initially using phage and cosmid vectors, and more recently bacterial artificial chromosomes and yeast artificial chromosome. This has allowed the determination of the sequences of probably all variable (V), and certainly all diversity (D), joining (J), and constant (C) region genes (Brüggemann et al., 2015).

To be able to reproduce transgenic animals, fertilized mice eggs that are obtained after superovulation are injected into the male pronucleus by microinjection. Thus a transgenic animal that carries multiple copies of human Igs is obtained. In these animals, multiple copies are better expressed, and it is also possible to inject a mixture of two structures. The IgH locus, which is ∼100 kb in length, is reorganized by means of the tail integration of two plasmids; this combination is constructed by segments in each cosmids, is then expressed (Masih et al., 2014).

Obtaining proteins from animal milk is another promising approach for large-scale protein production. The success of this approach was proven by obtaining more than 20 proteins. In particular, cow, goat, pig, rabbits, and mice are promising animal models for mAb production (Echelard et al., 2006).

Although recombinant protein production can be provided by different tissues, the milk is the most common source because milk is a product that can be easily collected and purified. There is approximately 5 g of mAb-like recombinant protein per liter of milk. When we further calculate, the protein amount can yield up to 4 kg per year. For this reason, even a goat flock can produce a few hundred kilograms of antibody in a year at a low cost (Pollock et al., 1999; Echelard et al., 2006).

Transgenic egg production is another approach: large amounts of protein production can be obtained from the contents of an egg. Moreover, harvesting the protein is even easier using this method (Masih et al., 2014).

16.2.2.1.4 PRODUCTION OF ANTIBODIES IN TRANSGENIC PLANTS

The production of the recombinant proteins in transgenic plants was proposed as a good alternative for traditional expression systems. After Ig-encoding genes were expressed in tobacco, the possibility of production of active heterologous proteins became a hot research topic. It is possible to use plants for large-scale production of antibodies with an acceptable cost. The antibodies obtained from plants are termed "plantibodies" (Rani and Usha, 2013). In 1980, plantibodies were first produced by two German graduate students (Hood and Howard, 2002). The *Agrobacterium tumefaciens* bacterium carries Ti-plasmid. This plasmid can integrate into the plant genome causing tumor development between the root and stem of the dicotile plants by changing the hormonally controlled cellular proliferation. Using this method, the production of mAbs by integration of antibody-encoding genes into Ti-plasmid and infection of the plant with *Agrobacterium* strains can be performed; it was designated *Agrobacterium-mediated transformation.* In addition, this plasmid was transferred into the plant cells using electroporation, particle bombing (with a gene gun), and sonication (Stoger et al., 2002; Gelvin, 2003; Rani and Usha, 2013). With these methods, captured genes provide ways to produce the target antibody in the cellular gaps of a plant cell (apoplasts). These antibodies are full-sized IgG and IgA, chimeric IgG and IgA, secretory IgG (sIgG) and IgA (sIgA), single-chain Fv fragments (scFv), Fab fragments, and heavy-chain variable domains (Hood and Howard, 2002). In addition, plant systems are eligible for the production of mammalian secretory antibody-like functional dimeric antibodies. Functional scFv molecules are expressed in the leaves, and especially in the storage organs, of a plant. Most of the scFv molecules also accumulate either in the endoplasmic reticulum or in the apoplast more than in cytosol (Stoger et al., 2002).

There is no consensus on the best plant type for commercial antibody production. For this purpose, usually tobacco, soybean, alfalfa, and rice plants are used because mAbs can remain stable inside the seeds and tubers of these plants. By means of scFv molecules, mAbs can be stored inside rice for 6 months, and they also retain 50% of their functionality during 18 months in potato tubers. In addition, scFv and IgG lose a small amount of mAb when are grown in dried tobacco and alfalfa plants over 5–7 days (Daniell et al., 2001; Stoger et al., 2002).

The most interesting Igs produced from plants are sIgAs. The sIgA protects against *Streptococcus mutans* bacteria, which causes tooth decay; it is used for the treatment of that condition. This IgA antibody is applied topically to teeth, where it blocks colonization of the *S. mutans*. The second example is the humanized anti–herpes-simplex virus (HSV) antibody produced in soybeans. This antibody is effective on HSV-2 transmission in mice. The third example is an antibody against carcinoembryonic antigen that is produced as a surface antigen in rice. It is clear that plant-originated antibody production will be more common for human immune therapeutics and other applications in the future (Daniell et al., 2001).

As a protein source, transgenic plants that are produced by genetic engineering have several advantages when compared with those produced in animal serum/tissues, cell lines that are transfected with recombinant microbes, and transgenic animals. A few of the possibilities are production of raw material on an agricultural scale at low cost with the accompanying advantages of low cost with respect to fermentation; rapid scale-up of production; and correct eukaryotic assembly of multimeric proteins such as antibodies. In addition, plants are safer because HIV, prions, hepatitis viruses, and other harmful molecules pathogenic for humans are not harbored in plants. Moreover, plantibodies do not give an Human anti-mouse antibody (HAMA) response; however, the antibody has some side effects in plants, and it is difficult to develop a process with plants (Twyman, 2008).

16.3 CLINICAL USAGE OF THE ANTIBODIES

16.3.1 Antibodies in Diagnosis

To be able to determine antibodies and the interactions between specific antigen–antibody interactions, several methods are used in the laboratory. Immunoprecipitation, ELISA, quantitative immunofluorescence, immunocytochemistry, immunohistochemistry (IHC), flow cytometry (FC), and routine Western blot (WB) are the widely used experimental tools. Using these immune tests, the antibody's response against infection can be monitored and the source of the infection can be determined. For example, antibodies that are produced against virus antigens can be detected in the serum of HIV-positive individuals (Bordeaux et al., 2010).

16.3.1.1 ELISA

ELISA is the abbreviation of enzyme-bound immunosorbent analysis. In this experimental method, an immunosorbent (either an antigen or antibody that is bound to a solid surface) and an enzyme-linked immune-reactant are used. For example, ELISA is used to detect an unknown concentration using competitive binding between an unlabeled unknown and a labeled reactant (Paulie et al., 2006).

ELISA can be performed by four different techniques: direct, indirect, sandwich, and competitive ELISA. These techniques will be described in detail in Chapter 15: Techniques for Protein Analysis, section 15.1.6. Although they have different names, these techniques work using the same principle (Çırak, 1999).

In an experiment, either an antigen or a specific antibody compound is bound to a solid surface; these experimental compounds have a selective interaction. During the experiment, the molecules that do not interact with the molecule that is adsorbed on the solid surface are removed by washing. To detect the interaction between the compounds, either labeling or an enzymatic conjugation is performed by another antibody conjugation. In these experiments, an enzymatic substrate should be added to make colorless products produce a colorimetric or fluorescent signal (Wakabayashi, 2010).

TABLE 16.1   Advantages and Disadvantages of the ELISA Technique

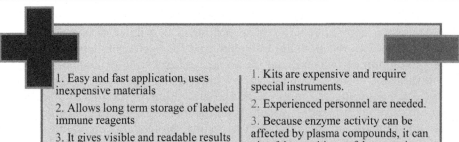

1. Easy and fast application, uses inexpensive materials	1. Kits are expensive and require special instruments.
2. Allows long term storage of labeled immune reagents	2. Experienced personnel are needed.
3. It gives visible and readable results using enzyme-labeled chromogenic substrates.	3. Because enzyme activity can be affected by plasma compounds, it can give false-positive or false-negative results.
4. High sensitivity, specificity, and reliablilty	4. If the blocking agent is insufficient, negative controls can yield positive results (to prevent this, a secondary antibody or a different antigen can be used).
5. Unknown or low concentrations of antigens can be detected using specific antibodies	5. Because the enzyne-substrate reaction has a short-term effect, wells should be read quickly.

Next, optic densities on the plates are read by a spectrophotometer that is set to a specific absorbance. Producers provide some computer programs that calculate positive and negative control titers (Çırak, 1999; Aras, 2011).

In ELISA assays, either polystyrene, polyvinyl, dextran, or polyacrylamide tubes or 96-well plates can be used. Antigens in soluble form are converted to an insoluble form by adsorption on the solid surface (Ananthanarayan and Paniker, 2005). There a number of advantages and disadvantages of ELISA method, some of which are listed in Table 16.1.

16.3.1.2 Western Blotting

This technique was designed by Towbin et al. (1979) for protein detection and analysis making use of their ability to bind to specific antibodies; it became one of the most commonly used laboratory techniques.

Before blotting, one-dimensional (1D) sodium dodecyl sulfate-polyacrylamide gel electrophoresis (SDS-PAGE) is used for the separation of the proteins in the protein mixture according to their molecular weights.

The SDS is used for control of protein purity and determination of molecular weights. This is the first step of *Western blotting*. In this technique, proteins are first denatured by converting their 3D structure to a linear structure; they are then separated in the polyacrylamide gel. Proteins are separated using an electric current according to their size in the gel matrix that is formed by the polymerization of the acrylamide and *bis*-acrylamide monomers. Ammonium sulfate and the tetramethylethylenediamine (TEMED) system are the starter for the polymerization/catalyzer. Unlike other systems, some SDS is added into sample preparation buffer. SDS allows a negative layer to be formed on the protein molecule. Thus separation occurs according to molecular weight of the protein, independent of the individual charge of the proteins. When an electric current is applied to the gel matrix, the migration rate of the proteins changes according to pore size. According to this principle, small proteins move faster and large proteins move slowly in the gel (Wenk and Fernandis, 2007).

The blotting procedure is used for the detection of a specific protein in a mixture of proteins that migrate in the gel according to their molecular size. These separated proteins are transferred onto a membrane. Either nitrocellulose or polyvinylidene fluoride can be used as the membrane.

The transfer process can be performed either with wet blotting or semidry blotting. In wet transfer, the gel and membrane are put on top of each other, and a sandwich structure is prepared by using filter papers. This sandwich is put into an electrophoresis tank, which is filled with transfer buffer.

In semidry transfer, the gel and membrane are prepared like a sandwich and placed between two electrodes, then an electric current is applied horizontally to these electrodes. Of note, although wet transfer is a more effective method for transfer, it takes a longer time than semidry transfer (Arya, 2007).

The aim of the second step of the WB is filling the gaps on the membrane. By means of this process, protein-structured antibodies in the blotting solution bind to the gaps on the membrane; this situation increases the antibody-binding possibility to the band of interest. For this purpose, different blotting solutions/buffers can be used according to the type of membrane and protocol. The most common ones are bovine serum albumin, nonfat milk powder, and gelatin (which is not preferred). Some nonionic detergents such as Tween-20, NP-40, and polyvinylpyrrolidone (PVP-40) can also be used.

After this step, primary PoAbs or mAbs are incubated with the proteins that are transferred onto the membrane. Then, radioactive isotope-labeled secondary antibodies or enzymes [alkaline phosphatase or *horseradish* peroxidase (HRP)] that can bind to primary antibodies are sent to the membrane, and the target protein is labeled (Wenk and Fernandis, 2007).

In some cases, nondenatured conditions are used to separate native proteins. This method is efficient when either the primary antibody recognizes only denatured proteins or the biological activity of the protein is necessary on the membrane.

16.3.1.3 *Flow Cytometric Analyses*

Flow Cytometry (FC) is the simultaneous measurement and analysis of the physical aspects of a structure, such as a cell, in a flowing liquid by means of a light beam. The basic principle behind this method is the reaction between antigens with fluorescence-labeled antibodies.

Fluorescence-activated cell sorting (FACS) is a specialized type of flow cytometry and is used for the evaluation of peptides and DNA in addition to membranes and intracellular proteins. The particles, which are approximately 0.2–150 μm in length, are appropriate for the analysis of the cell suspension (Shapiro, 2003).

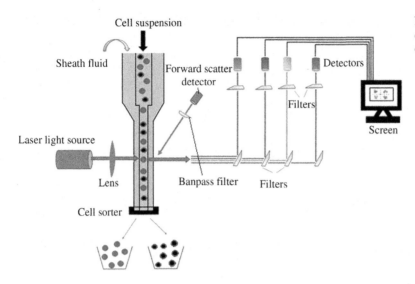

FIGURE 16.7 Schematic overview of a typical flow cytometer setup.

The FC instrument is formed by a liquid system, a light source (laser beam, filters, and signal detectors), and an electronic mechanism (computer and software programs and cell sorting mechanism) components (Sobti and Krishan, 2003; http-2).

In FC, particles are transported to the laser detector in a liquid flow. Simultaneously, a cellular suspension is formed by mixing antigens with antigen-specific antibodies. The cells are forced to follow a straight path (hydrodynamic focusing) by spraying this cellular suspension into a sheath liquid. The cells pass through this light beam one by one. During the passage, some alterations are observed on the light beam according to the cytoplasmic or nucleic structure of the cell. The angle of each alteration is recorded by different detectors that are placed around the light in different angles. Alternatively, if the cells have the antigens, which are the targets of the fluorescent-labeled compounds, the light is first absorbed and then transmitted to the environment at a different wavelength. Finally, this transmission is detected by fluorescent detectors (Sobti and Krishan, 2003; http-2) (Fig. 16.7).

Physical and fluorescent features that are obtained from the cell suspension given to the instrument are computer processed. The results are presented in the form of numbers and histograms. When these data collection and analysis are complete, the amount of an antigen found in a cell is measured.

The measurements that are done at different wavelengths can give qualitative and quantitative information about intracellular molecules such as fluorochrome-labeled cell-surface receptors, DNA, or cytokines (Shapiro, 2003). The advantages and disadvantages of the FC was summarised at Table 16.2.

To detect projected light, different fluorescent channels (FL−) are used in FC. The number of detectors can be changed according to the instrument and producer. These detectors are in the form of silicon photodiode or photomultiplier tubes (PMT). The silicon photodiodes are usually used to measure the scatter, but PMTs are more sensitive instruments and are ideal for scatter and fluorescent readings.

TABLE 16.2 Advantages and Disadvantages of the Flow Cytometric Analyses

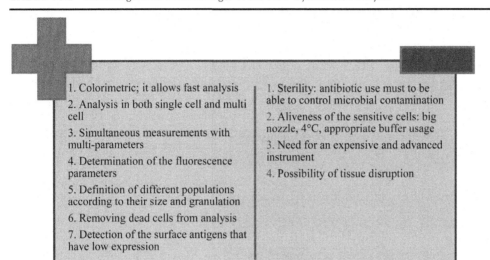

Advantages	Disadvantages
1. Colorimetric; it allows fast analysis	1. Sterility: antibiotic use must to be able to control microbial contamination
2. Analysis in both single cell and multi cell	2. Aliveness of the sensitive cells: big nozzle, 4°C, appropriate buffer usage
3. Simultaneous measurements with multi-parameters	3. Need for an expensive and advanced instrument
4. Determination of the fluorescence parameters	4. Possibility of tissue disruption
5. Definition of different populations according to their size and granulation	
6. Removing dead cells from analysis	
7. Detection of the surface antigens that have low expression	

There are dozens of fluorescent molecules (fluorochromes) for FC applications. Researchers categorize these molecules under two groups:

Single dyes: One of the most important example of a single dye is FITC, which has been in use for 30 years. However, some dyes, which have higher photostability and fluorescence, are currently preferred (Shapiro, 2003).

Tandem dyes: These fluorochrome dyes are the products of cutting-edge technology that is able to detect surface or intracellular epitopes. A small fluorochromal and a large fluorochromal molecules are put together in a tandem dye. Here, the first stimulated dye transmits all its energy to another dye that is close to the first dye. This activation causes the production of the second fluorescent emission. This process is called as FRET (fluorescent resonance energy transfer). Most of the tandem dyes are produced for the 488-nm laser in a multiflow cytometer. Tandem dyes work better, especially in combination with a single dye for multicolored fluorescent studies. For example, Alexa Fluor 488, Phycoerythrin, PerCP-Cy5.5, and PE-Cy7 produce green, yellow, purple, and infrared emissions, respectively, using detectors at 488 nm (Rahman, 2006).

16.3.1.4 *Immunhistochemistry*

IHC is used to detect proteins or other antigens in tissue sections. For this purpose, tissue sections are incubated with labeled antibodies that are specific to epitopes on the target protein. Thus the target region can be visualized by a fluorescent dye, enzyme, radioactive tracer, or a colloidal gold reagent.

The application of these antibodies can be in two forms: either with a direct method (e.g., a tracer-conjugated antibody is bound to the target) or with an indirect method (a labeled secondary antibody bound to a primary antibody) (http-3).

This method is produced by two steps:

1. Preparation of the slides (tissue fixation and processing) and reaction [antigen retrieval (AR), blocking the nonspecific binding, blocking endogen peroxidase, antibody binding, labeling, and storage].

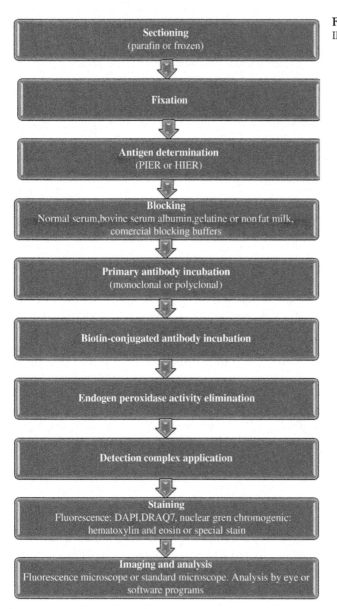

FIGURE 16.8   Schematic overview of the IHC technique.

2. Measurement and evaluation of the obtained expression (Miller, 2001).

This method is briefly represented in Fig. 16.8 (Miller, 2001; Hayat, 2002; de Matos et al., 2010).

Immunohistochemical methods are used in different situations in a research laboratory or a pathologic anatomy laboratory. For example, it is used for histogenetic detection of morphologically undefined neoplasia; subclassification of the neoplasia (e.g., lymphomas); characterization of the primary region of the malign neoplasia; determination of the prognostic factors and therapeutic indications; malign–benign discrimination in some cell proliferations; and definition of the secreted compounds, organisms, and structures (de Matos et al., 2010).

16.3.1.4.1 TISSUE PREPARATION

It is important to keep tissue viable, and because of bacterial and fungal growth, it is difficult to block autolysis. Therefore some fixatives are used such as paraformaldehyde-lysine-periodate or formalin. The most common fixative is 4%−10% formaldehyde. Tissues that are negatively affected by formaldehyde are filled with liquid nitrogen and then cut into sections. The tissues in fixative solution are first dehydrated, then incubated either in alcohol or isopropanol and buried in a solid matrix; they are then cut into sections using a microtome. For sensitive tissues, vibratome microtomes are used to decrease the pressure on the tissue (Hayat, 2002).

16.3.1.4.2 ANTIGEN RETRIEVAL

This step discerns antigen epitopes and is also important for antigen–antibody reactions that will occur in the upcoming steps of the experiment. Two techniques are used for AR: "heat-induced epitope retrieval" (HIER) or "proteolytic-induced epitope retrieval" (PIER). To apply these techniques to tissue sections, samples should be cleared of paraffin and can also be rehydrated (Lin and Shi, 2015; http-4, http-5).

HIER works with the retrieval solution incubation. This preheated buffer usually has different ingredients such as Tris−HCl and citrate at different pH values. The obtained samples are heated (usually 10−60 min) for different time intervals and then are cooled slowly. This process can be performed using a water bath or pressured pot. *PIER* works with the proteinase K, trypsin, pepsin, or other digestive proteinase incubations (Lin and Shi, 2015; http-4, http-5).

There are some cases where HIER and PIER are applied together. However, since the advent of HIER techniques, proteases play a much smaller role in most IHC laboratories (Lin and Shi, 2015; http-4, http-5).

Fig. 16.9 shows some additional physical and chemical methods for AR (D'Amico et al., 2009).

16.3.1.4.3 DETECTION METHODS

The selection of detection methods depends on the expertise of the technician, the type, and the number of the antigens to be tested, and the power of the AR method. Direct and indirect methods are used.

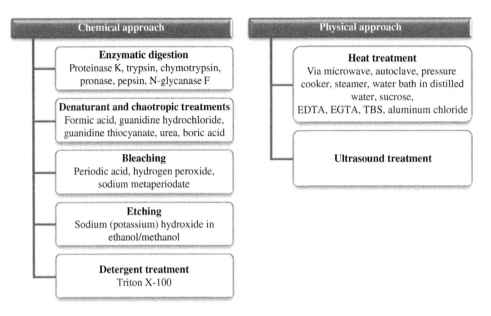

FIGURE 16.9   Schematic overview of alternative physical and chemical techniques.

In the *direct method*, a labeled antibody (e.g., a substrate-chromogen) directly reacts with antigen in the tissue. The advantage is that only one antibody is needed; however, because the antigen is bound to a specific epitope, the signal density is low (Petersen and Pedersen, 2016).

In the *indirect method*, there are two- and three-step methods, the peroxidase-antiperoxidase (PAP) method, streptavidin—biotin complex (ABC) method, and polymeric methods. In the *two-step method*, after the primary antibody binds to an antigen, a labeled secondary antibody also binds to the primary antibody. Because more than one secondary antibody can bind to a primary antibody, the signal is amplified. In the *three-step method*, an additional antibody can be bound to a secondary antibody for signal amplification. This method leads to the formation of a tertiary antibody layer. This application is good for a limited number of antigen labeling cases. The *PAP method* is rarely used. Peroxidase binds to an unconjugated secondary antibody-binding tertiary layer (de Matos et al., 2010). The *ABC method* is the most common method for staining. This method is performed by the help of the high affinity between glycoprotein-containing avidin (chicken egg) and streptavidin (*Streptomyces avidinii*). The biotin usually binds to a secondary antibody to create a colorful reaction. In the *polymeric methods*, antibodies bind to polymers that have the capability to bind many molecules. On average, 10 antibody and 70 enzyme molecules can be bound to these polymers. This step provides high amplification and less nonspecific binding. By means of this system, the amount of background signal is decreased. It also allows the staining of two different antigens at the same time. However, this method is expensive (Petersen and Pedersen, 2016).

To increase the sensitivity of the reaction, more than one variation has been added to the preceding methods: for instance, the vitamin biotin and immune rolling circle amplification methods. The most commonly used enzymes are peroxidase and alkaline phosphatase.

16.3.1.5 *Immunoelectrophoresis*

Immunoelectrophoresis is the rapid version of bi-directional gel diffusion; it involves the cooperation of electrophoresis and immunodiffusion. This method is used for the detection of high antibody concentrations such as albumin and transferrin; it is also used to check the presence, homogeneity, and specificity of antibodies. This technique is usually used for the detection of three immunoglobulin levels in blood: IgM, IgG, and IgA. In this system, a thin layer of agar is poured on a slide. After digging small holes in the agar, antigens are added into these holes and the slide is subjected to an electric current. The antigen molecules migrate at different rates through the electric current because of differing electric charges at a specific pH value (electrophoresis).

After electrophoresis, some bands are cut away from the agar parallel to the migration track, and the cut places are filled with antiserum. When a specific antigen encounters the antibody, a white precipitin band is formed on a black background (immunodiffusion). This precipitin band represents the presence of an antibody for a specific antigen. If the antibody is homogenous, there is only one precipitin band observed. The presence of more than one precipitin band represents antibody heterogeneity. On the other hand, if there is no precipitin band seen, that means there is no antibody specific to the target antigen.

There are four types of immunoelectrophoresis used, such as electroimmunoassay (EIA) or rocket/Laurell rocket electroimmunoassay, classical immunoelectrophoresis, immunofixation electrophoresis (IFE), and capillary electrophoresis (Levinson, 2009).

Rocket immunoelectrophoresis is also known as electro-immunodiffusion; it is a simple and rapid method to quantify the amount of either a protein or a protein mixture. Unknown samples can be analyzed on a single plate. The reference solutions should also be included on the plate (Laine, 1992).

IFE is preferred by comparison with the other methods because of its specificity and characteristic ease of analysis. This technique is commonly used in medical and clinical laboratories for detection of the abnormal mAbs related to lymphocytic diseases such as multiple myeloma (Levinson, 2009).

16.3.1.6 *Immunodiffusion*

Antigens that are separated by electrophoresis are immunodiffused with the addition of antiserum in a well that is cut from agarose gel. Through diffusion, an antigen and antibody density gradient is formed at the equivalent zone (the zone in which antigen and antibody are present at an optimal ratio) and an opaque zone is observed in a shape similar to a "smile" because of the precipitation of the antigen—antibody complex. The precipitation line shows the presence of the specific antibody. If the antibody is homogenous, there is only one precipitation line observed. Not observing a precipitation line confirms the absence of the antibody. On the other hand, more than one line denotes heterogeneity (Ananthanarayan and Paniker, 2005).

There are four different types of immunodiffusion techniques (Ananthanarayan and Paniker, 2005).

1. *One-dimensional single diffusion (Oudis procedure)*:
    This technique depends on the formation of a layer of an antibody solution through an agar column that contains an antiserum gel. During incubation, a precipitin band is formed. Each band represents the presence of a different antigen (Ananthanarayan and Paniker, 2005; Nagoba and Vedpathak, 2008).
2. *One-dimensional double diffusion (Oakley Fulthorpe procedure)*:
    This test is performed in a tube that contains both an antiserum-containing melted agar layer and plain agar. The antigen solution is added slowly and incubated. Through the meeting of the antigens that diffuse and rise through the plain agar and antibodies, a precipitation layer is formed in the plain agar (Ananthanarayan and Paniker, 2005; Nagoba and Vedpathak, 2008).
3. *One-dimensional double diffusion (Mancini procedure)*:
    An antiserum-containing agar is poured on a plate and holes are made. Antigen solution is put inside of these holes, and it is incubated under humidity. Antigens disperse through the holes and a precipitation circle is formed around the holes (Ananthanarayan and Paniker, 2005; Nagoba and Vedpathak, 2008).
4. *Two-dimensional double diffusion (Ouchterlony procedure)*:
    Soluble antigens and antibodies are put in two holes on an agar plate. Antigen and antibody disperse through each other in the agar. When they reach an optimal concentration, a precipitin line is formed. This technique is useful for detecting nonidentical cross-reacting antigens (Ananthanarayan and Paniker, 2005; Bhoosreddy and Wadher, 2010).

16.3.2 Antibodies in Treatment

mAbs are important therapeutic agents for autoimmune diseases and cancer. Antibodies modulate tumor-related signaling, and they can also induce antitumor responses. In addition, antibodies can show some immunomodulatory effects by means of direct activator and inhibitor molecules (Weiner et al., 2010).

16.3.2.1 Hematology/Oncology

mAbs are preferred because of their target-specificity and their having less nonspecific binding affinity (Gümüş and Sunguroğlu, 2001; Schmidt and Wood, 2003).

mAbs show their tumoricidal effects in three different ways: receptor-related signalization, antibody-dependent cellular cytotoxicity (ADCC), and complement-dependent cytotoxicity (CDC). mAbs can activate cell-surface receptors by stimulating these receptors. This signaling may have a different quality in different cases. For example, when an anti-CD-20 mAb induces apoptosis, an epidermal growth factor receptor (EGFR)-specific antibody blocks ligand binding (Van den Eynde and Scott, 1998).

There are two types of mAbs: *conjugated* and *unconjugated* (Scott et al., 2012).

Unconjugated mAbs are developed to block cancer cell–specific antigens; these mAbs lead to death of the tumor cell by activating CDC and ADCC (Gümüş and Sunguroğlu, 2001; Scott et al., 2012).

Conjugated mAbs are conjugated with some structures such as chemotherapeutic agents, radioisotopes, and enzymes. These antibodies target cancerous areas and carry their cargo to these areas. For example, a radioisotope-conjugated antibody exposes the area that carries tumor-specific antigens to radiotherapy. Using this approach, healthy cells are not damaged by radiotherapy (Gümüş and Sunguroğlu, 2001; Scott et al., 2012).

For the treatment of several hematologic diseases, many mAbs have been selected. For example, B cells have many cell-surface antigens; therefore, they are the main target of mAb therapy. Rituximab, the first FDA-approved chimeric (human/mice) monoclonal CD-20 antibody, is used for treatment of metastatic/refractory B-cell non-Hodgkin lymphoma (NHL). Although the CD-20 antigen is 90% more expressed than in B-cell NHL, it is not expressed in stem cells, pro-B cells, normal plasma cells, or normal tissues (Ross, 2013).

Ofatumumab is a human-sourced IgG1-kappa mAb that is specific to the CD-20 antigen. It is effective against B-cell lymphoma/chronic lymphocytic leukemia cells. It has been studied as a possible therapy alternative for NHL, rheumatoid arthritis (RA), and multiple sclerosis (Lemery et al., 2010).

Trastuzumab is a recombinant DNA-sourced, humanized mAb that specifically targets the extracellular region of the epidermal growth factor receptor-2 (HER-2). Because it binds to the HER-2/neu (erbB2) receptor, it is mainly used for HER-2/neu receptor-positive breast cancer cells (Ross, 2013; http-6).

Bevacizumab is used for primary treatment of metastatic colorectal cancer patients. It is a mAb that binds to VEGF. It blocks new vessel formation in tumor cells (Ross, 2013).

Alemtuzumab was developed for treatment of lymphoid malignancies such as NHL, chronic lymphocytic leukemia, and T-cell lymphoma. It is a humanized mAb that is not effectively conjugated against CD-52 (Dumont, 2002; http-6).

Cetuximab (IMC-C225) is a recombinant, chimeric (human/mice) mAb that binds specifically to the nonextracellular region of the human EGFR. Cetuximab was tried for head—neck and colon cancer treatment, and its usage was approved for irinotecan-resistant metastatic colon cancer treatment (Vincenzi et al., 2010).

16.3.2.2 Transplantation

Clinically, transplantation means transfer of organs and islands of Langerhans. Knowledge about transplantation immunology is required to overcome host-versus-graft (HVG) immune response for blocking regeneration. However, in this case, not only the HVG immune response, such as in hematopoietic stem cell transplantation, is important, but also the graft-versus-host (GVH) immune response must be taken into account. Bone marrow and peripheral blood cells that are prepared for transplantation contain some mature T cells; therefore, there is a GVH risk for the recipient after transplantation (Maloney et al., 2002).

Muromonab-CD3 is a murine-originated mAb that is used parenterally to treat acute allograft rejection in patients who have kidney, heart, or liver transplantation surgery. Muromonab recognizes and reacts with T3 (CD3) antigens on the T lymphocyte membrane. Thus it inhibits all T lymphocyte activities, thus, all cytokines are released outside the organ. This drug is effective on T cells in peripheral blood and body tissues, but it does not affect other hematopoietic factors or tissues. This is a reversible reaction (http-7).

Daclizumab is a chimeric mAb that is produced by recombinant DNA technology; it is used to inhibit organ rejection in patients who have renal allograft surgery. It binds to the alpha subunit of the interleukin-2 receptor (p55, CD25, or Tac subunit). IL-2-R is expressed only on the surface of the T cells, and it is important for the clonal expansion of the T cells. Daclizumab inhibits IL-2 binding to IL-Rα; thus, it blocks T-cell activation signals. When it is used in combination with corticosteroids and cyclosporine, the combination decreases the risk of organ rejection (Li et al., 2009).

Like daclizumab, basiliximab is a recombinant, chimeric (murine/human) mAb; it is produced against CD25 antigen (interleukin-2 receptor alpha chain). The effect of basiliximab is similar to that of daclizumab. Basiliximab was determined to be an orphan drug for solid organ transplant rejection prophylaxis (Salis et al., 2008).

16.3.2.3 Cardiology

In cardiovascular medicine, there are a few studies that include mAbs, such as visualization of cardiac necrosis with myosin-specific mAbs and intoxication of reverse digitalis with digoxin-specific antibodies. In clinical cardiology, four main antibodies are used: Digibind, OKT3, Myoscint, and 7E3. Each of these has great potentials for antibody treatments in clinical cardiology (Azrin, 1992).

To decrease ischemic complications, the clinical effectiveness of the antiplatelet agents in atherothrombosis and plaque rupture has been studied. When a percutaneous intervention is done, it has been shown that mAb application is quite effective against platelet membrane receptor glycoprotein IIb/IIIa. Abciximab is an FDA-approved mAb that inhibits thrombocyte aggregation. It is the first member of a group of drugs, called thrombocyte-receptor glycoprotein inhibitors, which are used to treat thrombotic arterial diseases. Abciximab is the chimeric (murine/human), monoclonal Fab fragment of the 7E3 antibody (Acharya et al., 2011). In this chimeric form, murine IgGs substitute for human IgGs. It is also used to block acute cardiac ischemic complications in patients who either have a high embolus risk or who have percutaneous transluminal coronary angioplasty or atherectomy treatment (Lefkovits et al., 1996).

16.3.2.4 Infection

Currently, palivizumab is the only antiinfective mAb on the market. It is a humanized, IgG-sourced antibody that was produced by recombinant DNA technology (95% human, 5% murine). It binds to the antigenic epitope region of the F protein in a respirational syncytial virus (Rodney, 2013). Palivizumab is used for the inhibition of RSV-virus—sourced lower respiratory tract infections in newborns; it can also be used for children who have either bronchopulmonary dysplasia or premature birth histories. Respiratory syncytial virus is a RNA-virus that is a member of the paramyxovirus family. It is the leading factor for respiratory diseases in children. Some patient groups are under RSV infection risk, such as preterm babies, those with bronchopulmonary disease, or congenital heart failure. Palivizumab shows neutralizing and fusion inhibitor effects against RSV virus (Rogovik et al., 2010).

16.3.2.5 Rheumatology

There are two basic targets of antibodies that are used in rheumatology, in particular in RA: tumor necrosis factor-alpha (TNF-α) and interleukine-6 (IL-6). The role of TNF-α in RA pathogenesis was discovered in the 1980s. TNF-α is the key mediator of inflammation-induced joint disruption. mAbs that bind to TNF provide a decrease in the TNF-induced immune responses—such as cytokine production, matrix metalloproteinase production, neutrophil activity, dendritic cell functions, and osteoclast differentiation (Bluml et al., 2012).

To have available an increasing amount of knowledge about immunopathogenesis of rheumatism diseases leads to detection of therapeutic targets at the inflammation step. These targets include cytokines and molecules that are important for initiating immune response (especially B and T cells). All defects in cytokines (TNF-α, IL-1, and IL-6) are very well defined. Biotechnological developments in recent years have led to isolation of these molecules, which provides some therapeutic alternatives for several diseases including RA, psoriatic arthritis (PSA), ankylosing spondylitis (AS), autoimmune syndromes, antineutrophil cytoplasmic antibody-related (ANCA) vasculitis (AAV), and systemic lupus erythematosus (SLE) (Yuvienco and Schwartz, 2011).

In particular, TNF-α mAbs and anticytokine therapy suppress inflammation and prevent symptomatic recurrence. Antilymphocyte mAbs have been found to suppress the disease for an extended time in animals with RA. However, it was disappointing that sufficient amount of synovial penetration into joints was not prevented using antilymphocyte mAb therapy. Eventually, it discontinued for RA treatment; however, it is not known whether or not it has a long-term recurrence effect (Choy et al., 1998).

There are many mAbs that target several cell-surface molecules or costimulators and pro-inflammatory cytokines or their receptors (e.g., infliximab, adalimumab, tocilizumab, belimumab, HuMax−IL-15) that are either in clinical trial or have been clinically approved (Bayry et al., 2007).

Infliximab is a chimeric IgG1 mAb that is used in combination with methotrexate. It was approved by FDA/EMA in 2001 for treatment of individuals who have moderate-to-severe RA (http-3).

Adalimumab is a recombinant IgG1 derivative. It does not contain any murine component: it was produced by phage-change technology. It is used in moderate-to-severe stage RA treatment either as a monotherapy or in combination with other drugs (Nixon et al., 2014).

Golimumab is one of the anti-TNF-α antibody that is produced from human-sourced IgG1. Its structure is similar to that of infliximab; the only difference is that golimumab is human-sourced and does not contain any murine protein (Thorlund et al., 2015).

Certolizumab pegol is obtained from the fusion of human-sourced Fab fragment (without Fc) with 40 kDa polyethylene glycol. By PEGylation, the goal is to improve pharmacological aspects such as bioavailability and the possible localization of the antibody in inflammatory tissues. The absence of the Fc region minimizes Fc-linked CDC and antibody-dependent cytotoxicity (Goel and Stephens, 2010).

Tocilizumab is a humanized IgG1 antibody that binds to both membrane-bound and water-soluble forms of the IL-receptor. In 2010, FDA/EMA approved use of this drug for moderate-to-severe rheumatic diseases and/or RA cases that do not respond to anti-TNFs (http-6).

16.3.2.6 Gastroenterology

In the last 10 years, anti-TNF-α treatment has become a gold standard for the treatment of autoimmune diseases. The results of clinical studies showed that infliximab, adalimumab, and certolizumab pegol have some therapeutic benefits against inflammatory intestinal diseases such as Crohn's disease and ulcerative colitis.

These mAbs help to define new targets for future drug therapy studies by inducing molecular and physiological changes. Through increasing experience with TNF-α usage, these antibodies have begun to be used even for pregnant women (Lee and Fedorak, 2010).

Adalimumab is a human-derived, anti-TNF mAb. Because it is introduced subcutaneously, it does not have to be applied by infusion like other antibodies. In addition, since it is completely originated from human sources, its immunogenicity is decreased. In individuals who have moderate-to-severe Crohn's disease, better therapy outcomes were observed with certolizumab induction with respect to a placebo group (Kawalec et al., 2013).

Natalizumab is an integrin α-4 antagonist (95% human, 5% murine) mAb. It inhibits migration to the intestinal mucosa and leukocyte adhesion to endothelial cells by binding to its α-4 chain. Patients with irritable bowel syndrome, α-4-dependent leukocyte adhesion were observed. The severity of Crohn's disease and inflammation decreases with the inhibition of this pathway. Natalizumab is given by IV infusion (Rodney, 2013; http-6).

16.3.2.7 FDA-Approved Monoclonal Antibodies

In Europe, the United States, and other markets, therapeutic antibodies are approved regularly. In Table 16.3, some information about different MoAbs is shown for those approved by the FDA either in Europe or in the United States since September 2017 (Reichert, 2012; http-6).

16.4 CONCLUSION

Several physicochemical and biochemical properties of mAb-based therapeutics have been associated with specific preclinical and clinical endpoints, and their safety and efficacy parameters have been elucidated. The determination of the pharmacological and toxicological properties of mAbs has enabled the development of new platforms to overcome certain limitations in first-generation biotherapeutics developed to treat various diseases. Therapeutic antibodies are widely used for treating various diseases and disease states, including cardiovascular diseases, autoimmune disorders, malignant diseases, and infections. The binding of this antibody to at least two molecular targets on the cell surface may initiate this therapeutic effect. The treatment starts simultaneously with this binding by inhibiting the surface receptor, blocking the two ligands, and inducing the T cells to get closer to the antibody-bound cells. It has been elucidated that mAbs increase the antitumor responses in the body via various mechanisms. Moreover, the current antibody production techniques have led to the development of certain cancer and chronic disorder therapies mentioned in this chapter.

TABLE 16.3 FDA-Approved Therapeutic Monoclonal Antibodies

Name	Trade name	Target	Origin	Route	Indication	Clinical domain	Company	Approval year Last revision date
Muromonab-CD3	Orthoclone Okt3	CD3	Murine	iv infusion	Reversal of kidney transplant rejection	Immunology	Ortho Biotech	September 14, 1992 Not
Abciximab	Reopro	Anti-GPIIb/IIIa	Chimeric	iv infusion	Prevention of blood clots in angioplasty	Cardiology	Centocor Inc	December .16, 1993 November25, 2013
Arcitumomab	Cea-Scan	Cancer cell receptor	Murine	iv infusion	Diagnostic imagining of colon and rectal cancer	Radiology	Immunomedics	June 28, 1996 Not
Imciromab pentetate	MyoScint	Cardiac myosin heavy chain	Murine	iv infusion	Acute myocardial infarction	Cardiology	Centocor Inc	July 03, 1996 Not
Nofetumomab	Verluma	40 kDa glycoprotin antigen on cancer cell surface	Murine	iv injection	Diagnostic imagining of variety of cancer	Radiology	Boehringer Ingelheim	August 20, 1996 Not
Capromab Pendetide	ProstaScint	Prostate specific membrane antigen (PSMA)	Murine	iv injection	Imagining of prostate cancer	Radiology	Cytogen	October 28, 1996 June 26, 2012
Rituximab	MabThera, Rituxan	CD20	Chimeric	iv infusion	Non-Hodgkin's lymphoma, CLL, rheumatoid arthritis	Oncology Rheumatology	Genentech	November 26, 1997 September 24, 2013
Daclizumab	Zenapax	IL2R	Humanized	iv infusion	Prevention of kidney transplant rejection	Immunology	Hoffman-La Roche	December10, 1997 September 15, 2005
Basiliximab	Simulect	IL2R-alpha	Chimeric	iv infusion	Prevention of kidney transplant rejection	Immunology	Novartis	May 12, 1998 November14, 2003
Palivizumab	Synagis	Respiratory syncytial virus protein	Humanized	im injection	Prevention of respiratory syncytial virus infection	Immunology	Medimmune	June 19, 1998 March 21, 2014
Infliximab	Remicade	TNF	Chimeric	iv infusion	Crohn disease	Gastroenterology	Centocor Inc	August 24, 1998 November 06, 2013

(Continued)

TABLE 16.3 (Continued)

Name	Trade name	Target	Origin	Route	Indication	Clinical domain	Company	Approval year Last revision date
Trastuzumab	Herceptin	Her-2	Humanized	iv infusion	Breast cancer	Oncology	Genentech	September 25, 1998 March07, 2014
Gemtuzumab ozogamicin	Mylotarg	CD-33	Humanized	iv infusion	Acute myeloid leukemia (AML)	Haematology Oncology	Wyeth Pharms Inc/Pfizer Inc	May 17, 2000 January 23, 2006/ September 01, 2017
Alemtuzumab	MabCampath, Lemtrada	CD-52	Humanized	iv infusion	Chronic myeloid leukemia (CML), non-Hodgkin lymphoma	Haematology Oncology	ILEX Pharma-ceuticals	May 07, 2001 September 19, 2007
Ibritumomab tiuxetan	Zevalin	CD-20	Murine	iv injection	Non-Hodgkin's lymphoma	Oncology	Spectrum Pharms	February 19, 2002 August 30, 2013
Adalimumab	Humira	TNF-alpha	Human	sc injection	Rheumatoid arthritis-Crohn's disease	Gastroenterology Rheumatology	Abbvie Inc	December 31, 2002 May 30, 2014
Omalizumab	Xolair	IgE	Humanized	sc injection	Allergic asthma, chronic idiopathic urticaria	Immunology	Genentech	June 20, 2003 March 21, 2014
Tositumomab-I131	Bexxar	CD-20	Murine	iv infusion	Non-Hodgkin lymphoma	Radiology Oncology	Smithkline Beecham	June 27, 2003 August 15, 2012
Efalizumab	Raptiva	CD-11a	Humanized	sc injection	Psoriasis	Rheumatology	Genentech	October 27, 2003 March 13, 2009
Cetuximab	Erbitux	EGFR	Chimeric	iv infusion	Colorectal cancer	Oncology	Imclone	February 12, 2004 March 04, 2013
Bevacizumab	Avastin	VEGF	Humanized	iv injection	Colorectal cancer	Oncology	Genentech	February 26, 2004 December 16, 2013
Natalizumab	Tysabri	α4-Integrin	Humanized	iv infusion	Multiple sclerosis, Crohn's disease	Neurology Gastroenterology	Biogen Idec	November 23, 2004 December 15, 2013
Natalizumab	Tysabri	α4-Integrin	Humanized	iv injection	Multiple sclerosis, Crohn's disease	Neurology/ gastroenterology	Biogen Idec	November 23, 2004 December 15, 2013

(Continued)

TABLE 16.3 (Continued)

Name	Trade name	Target	Origin	Route	Indication	Clinical domain	Company	Approval year Last revision date
Ranibizumab	Lucentis	VEGF	Humanized	intravitreal injection	Macular degeneration	Ophthalmic Disease	Genentech	June 30, 2006 February 27, 2014
Panitumumab	Vectibix	EGFR	Human	iv infusion	Colorectal cancer	Oncology	Amgen	September 27, 2006 May 23, 2014
Eculizumab	Soliris	Compleman protein C5	Humanized	iv infusion	Paroxysmal nocturnal hemoglobinuria, atypical hemolytic uremic syndrome	Immunology	Alexion Pharm	March 16, 2007 April 30, 2014
Certolizumab pegol	Cimzia	TNF blocker	Humanized Fab	sc injection	Crohn disease-rheumatoid arthritis	Gastroenterology Rheumatology	UCB Inc	April 22, 2008-May 13, 2009 October 17, 2013
Golimumab	Simponi	TNF-alpha	Human	sc injection	Rheumatoid and psoriatic arthritis, ankylosing spondylitis, ulcerative colitis	Gastroenterology Rheumatology	Centocor Ortho Biotech Inc	April 24, 2009 January 24, 2014
Canakinumab	Ilaris	IL1beta	Human	sc injection	Cryopyrin-associated periodic syndrome (CAPS)	Rheumatology	Novartis Pharms	June 17, 2009 May 09, 2013
Ustekinumab	Stelara	IL12 and IL23	Human	sc injection	Psoriasis	Rheumatology	Centocor Ortho Biotech Inc	September 25, 2009 March 04, 2014
Ofatumumab	Arzerra	CD20	Human	iv infusion	Chronic lymphocytic leukemia (CLL)	Hematology Oncology	Glaxo GRP Ltd	October 26, 2009 April 17, 2014 January 19, 2016
Tocilizumab	Actemra	IL6 receptor	Humanized	iv infusion or infusion or sc	Rheumatoid arthritis	Rheumatology	Genentech	January 08, 2010 October 21, 2013
Denosumab	PROLIA	RANK-L	Human	sc injection	Cell tumor of bone	Orthopedic Oncology	Amgen	June 01, 2010 June 04, 2014

(Continued)

TABLE 16.3 (Continued)

Name	Trade name	Target	Origin	Route	Indication	Clinical domain	Company	Approval year Last revision date
Denosumab	Xgeva	RANK-L	Human	sc injection	Skeletal-related events in patients with bone metastases from solid tumors.	Orthopedic Oncology	Amgen	June 01, 2010 June 04, 2014
Belimumab	Benlysta	B lymphocyte stimulator	Human	iv infusion	Active systemic lupus erythematosus	Rheumatology	Human Genome Sciences Inc.	March 09, 2011 April 03, 2014
Ipilimumab	Yervoy	Cytotoxic T lymphocyte antigen 4 (CTLA-4)	Human	iv infusion	Metastatic melanoma	Oncology	Bristol Myers Squibb	March 25, 2011 December 05, 2013 October 28, 2015
Brentuximab vedotin	Adcetris	CD30	Chimeric	v infusion	Hodgkin lymphoma	Oncology	Seattle Genetics	August 19, 2011 August 19, 2013
Pertuzumab	Perjeta	HER2.Neu	Humanized	iv infusion	HER2-positive breast cancer	Oncology	Genentech	June 08, 2012 September 30, 2013
Raxibacumab	Raxibacumab	*B. anthracis* protective antigen	Human	iv injection	Anthrax infection	Infectious Diseases	Human Genome Sciences Inc	December 14, 2012 Not
Ado-Trastuzumab emtansine	Kadcyla	HER2	Humanized	iv infusion	HER2-positive metastatic breast cancer	Oncology	Genentech	February 22, 2013 August 29, 2013
Golimumab	Simponi Aria	TNF-alpha	Human	iv injection	Active rheumatoid arthritis	Rheumatology	Janssen Biotech	July 18, 2013 February 12, 2014
Obinutuzumab	Gazyva	CD20	Humanized	iv infusion	Chronic lymphocytic leukemia	Hematology Oncology	Genentech	November 01, 2013 February 26, 2016
Ramucirumab	Cyramza	VEGFR2	Human	iv infusion	Gastric cancer	Gastroenterology	ELI LILLY and Co.	April 21, 2014 April 24, 2015
Siltuximab	Sylvant	IL-6	Chimeric	iv injection or infusion	Multicentric Castleman's disease	Immunology	Janssen Biotech	April 22, 2014 Not

(Continued)

TABLE 16.3 (Continued)

Name	Trade name	Target	Origin	Route	Indication	Clinical domain	Company	Approval year Last revision date
Blinatumomab	Blincyto	CD19	Humanized	iv injection	B-cell precursor acute lymphoblastic leukemia (ALL)	Hematology Oncology	AMGEN INC	May 19, 2014 Not July 11, 2017
Vedolizumab	Entyvio	Alpha-4-beta-7 integrin	Humanized	iv infusion	Ulcerative colitis, Crohn's disease	Gastroenterology	Takeda Pharms USA	March 04, 2015 September 22, 2017
Nivolumab	Opdivo	Anti-PD-L1	Humanized	iv injection	Melanoma or lung cancer	Oncology	Bristol Myers Squibb	March 10, 2015 Not
Dinutuximab	Unituxin	GD2	Chimeric	iv injection	Neuroblastoma	Oncology	Silver Spring	July 24, 2015 Not
Alirocumab	Praluent	PCSK9 inhibitor	Human	sc injection	High cholesterol	Cardiology	SANOFI and REG. Pharm., Inc	August 27, 2015
Evolocumab	Repatha	PCSK9 inhibitor	Human	sc injection	Hyperlipidemia	Cardiology	Amgen	October 02, 2015 September 22, 2017
Pembrolizumab	Keytruda	Anti-PD-L1	Humanized	iv injection	Melanoma or lung cancer	Oncology	Merck & Co.	October 19, 2015 October 16, 2015
Idarucizumab	Praxbind	Anticoagulant effect	Humanized	iv injection	Dabigatran	Hematology	Boehringer Ingelheim	November 04, 2015 Not
Mepolizumab	Nucala	IL-5	Humanized	sc injection	Asthma	Pulmonary Disease	Glaxo Smithkline	November 16, 2015 November 21, 2016
Daratumumab	Darzalex	CD-38	Human	iv infusion	Multiple myeloma	Oncology	Janssen Biotech	November 24, 2015 Not
Necitumumab	Portrazza	EGFR	Human	iv injection	Lung cancer	Oncology	Eli Lilly and Co.	November 30, 2015 Not
Elotuzumab	Empliciti	SLAMF7	Humanized	iv injection	Multiple myeloma	Oncology	Bristol-Myers Squibb	March 21, 2016 Not
Obiltoxaximab	Anthim	*B. anthracis* toxin	Chimeric	iv injection	Anthrax infection	Pulmonary disease	Elusys Therapeutics	March 22, 2016 Not
Ixekizumab	Taltz	IL-17A	Humanized	sc injection	Plaque psoriasis	Autoimmune diseases	Eli Lilly and Co.	

(Continued)

TABLE 16.3 (Continued)

Name	Trade name	Target	Origin	Route	Indication	Clinical domain	Company	Approval year Last revision date
Reslizumab	Cinqair	IL-5	Humanized	iv infusion	Asthma	Pulmonary disease	Teva Pharma-Ceuticals	March 23, 2016 Not
Atezolizumab	Tecentriq	Anti-PD-L1	Humanized	iv injection	Metastatic urothelial carcinoma	Oncology	Genentech	May 18, 2016 Not
Olaratumab	Lartruvo	Platelet-derived growth factor receptor (PDGFR-α)	Humanized	iv injection	Soft tissue sarcoma (STS)	Oncology	Eli Lilly and Co.	November 19, 2016 Not
Durvalumab	Imfinzi	Anti-PD-L1	Human	iv injection	Advanced or metastatic urothelial carcinoma	Oncology	Astrazeneca UK Ltd.	January 05, 2017 Not
Avelumab	Bavencio	Anti-PD-L1	Human	iv injection	Merkel cell carcinoma	Oncology	EMD Serono, Inc	March 23, 2017 Not

The most notable disadvantages of mAbs are genetic mutations associated with some chronic diseases and malignancies, target accessibility, and weak immunological effector responses. Several newly developed technologies, such as transgenic mice and phage display, have led to the development of small therapeutically effective antibody formats, such as single-chain variable fragment and diabody. They have the ability to overcome some of the previously mentioned barriers. The current strategies used for producing therapeutic antibodies are expensive and far beyond the reach of many patients. Hence, new production strategies, such as the use of symphobodies to produce therapeutic antibodies, have been used to overcome these barriers. The development of therapeutic antibody-conjugated nanoparticles using nanotechnology is also emerging as a new approach. The emergence of new-generation therapeutic antibodies with excellent properties and affordable prices, compared with those currently in use, is expected in the next decade. These are significant advances from earlier applications in therapy and diagnostics.

References

Acharya, S., Shukla, S., Mahajan, S.N., Diwan, S.K., 2011. The charisma of "Magic Bullets"—monoclonal antibodies (mAB/moAB) in clinical medicine. J. Indian Acad. Clin. Med. 12 (4), 283–289.

Alberts, B., Johnson, A., Lewis, J., Raff, M., Roberts, K., Walter, P., 2002. Chapter 24: The adaptive immune system, Molecular Biology of the Cell, fourth edition Garland Science; Taylor and Francis Group, New York.

Ananthanarayan, R., Paniker, C.K.J., 2005. Part II, Chapter 13: Antigen-antibody reactions. In: Paniker, C.K.J. (Ed.), Ananthanarayan and Paniker's Text Book of Microbiology. Orient Blackswan, Hyderabad, pp. 92–109.

Aras, Z., 2011. Rapid diagnostic methods used in microbiology, Turk Hij Den Biyol. 68 (2), 97–104.

Arya, S.L., 2007. Immunotechnology. Global Media, Delhi, pp. 109–111.

Azrin, M.A., 1992. The use of antibodies in clinical cardiology. Am. Heart. J. 124 (3), 753–768.

Azzazy, H.M.E., Highsmith, W.E., 2002. Phage display technology: clinical applications and recent innovations. Clin. Biochem. 35 (6), 425–445.

Bayry, J., Lacroix-Desmazes, S., Kazatchkine, M.D., Kaveri, S.V., 2007. Monoclonal antibody and intravenous immunoglobulin therapy for rheumatic diseases: rationale and mechanisms of action. Nat. Rev. Rheumatol. 3, 262–272.

Bazan, J., Całkosiński, I., Gamian, A., 2012. Phage display—a powerful technique for immunotherapy 1. Introduction and potential of therapeutic applications. Hum.Vaccine Immunother. 8 (12), 1817–1828.

Bhoosreddy, G.L., Wadher, B.J., 2010. Basic İmmunology. Himalaya Publishing House, Mumbai, pp. 57–60.

Bluml, S., Schenecker, C., Smolen, J., Redlich, K., 2012. Targeting TNF receptors in rheumatoid arthritis. Int. Immunol. 24, 275–281.

Bonilla, F.A., Oettgen, H.C., 2010. Adaptive immunity. J. Allergy Clin. Immunol. 125 (2), 33–40.

Bordeaux, J., Welsh, A.W., Agarwal, S., Killiam, E., Baquero, M.T., Hanna, J.A., et al., 2010. Antibody validation. Biotechniques 48, 197–209.

Breedveld, F.C., 2000. Therapeutic monoclonal antibodies. Lancet 355, 735–740.

Brüggemann, M., Osborn, M.J., Ma, B., Hayre, J., Avis, S., Lundstrom, B., et al., 2015. Human antibody production in transgenic animals. Arch. Immunol. Ther. Exp. (Warsz). 63 (2), 101–108.

Buchacher, A., Iberer, G., 2006. Purification of intravenous immunoglobulin G from human plasma—aspects of yield and virus safety. Biotechnology 1, 148–163.

Bussel, J.B., Giulino, L., Lee, S., Patel, V.L., Sandborg, C., Stiehm, E.R., 2007. Update on therapeutic monoclonal antibodies. Curr. Probl. Pediatr. Adolesc. Health Care 37, 118–135.

Carmen, S., Jermutus, L., 2002. Concepts in antibody phage display. Brief Funct. Genom. Proteomic 1 (2), 189–203.

Chaplin, D.D., 2010. Overview of the immune response. J. Allergy Clin. Immunol. 125 (2), 1–41.

Choy, E.H.S., Kingsley, G.H., Panayı, G.S., 1998. Monoclonal antibody therapy in rheumatoid arthritis. Br. J. Rheumatol. 37, 484–490.

Çırak, M.Y., 1999. Enzyme linked immunosorbent assay (ELISA) Sistemleri. T Klin Tıp Bilim. 19, 242−248.

Cruse, J.M., Lewis, R.E., 2010. Chapter 7: Immunoglobulin synthesis, properties, structure and function, Atlas of Immunology, third ed. CRC Press, Taylor and Francis, Boca Raton, FL.

D'Amico, F., Skarmoutsou, E., Stivala, F., 2009. State of the art in antigen retrieval for immunohistochemistry. J. Immunol. Methods 341 (1−2), 1−18.

Daniell, H., Streatfield, S.J., Wycoff, K., 2001. Medical molecular farming: production of antibodies, biopharmaceuticals and edible vaccines in plants. Trends Plant. Sci. 6 (5), 219−225.

de Matos, L.L., Trufelli, D.C., Matos, M.G.L., Pinhal, A.S., 2010. Immunohistochemistry as an important tool in biomarkers detection and clinical practice. Biomark. Insights 5, 9−20.

Dimitrov, S.D., 2010. Therapeutic antibodies, vaccines and antibodyomes. MAbs 2 (3), 347−356.

Dubey, R.C., 2014. Chapter 16: Immunology. Advanced Biotechnology. S. Chand & Company PVT. LTD, New Delhi, pp. 715−766.

Dumont, F.J., 2002. CAMPATH (alemtuzumab) for the treatment of chronic lymphocytic leukemia and beyond. Expert. Rev. Anticancer Ther. 2, 23−35.

Echelard, Y., Ziomek, C.A., Meade, H.M., 2006. Procduction of recombinant therapeutic proteins in the milk of transgenic animals. BioPharm. Int. 19, 36−46.

Flaherty, D., 2012. Chapter 9: Antibodies. Immunology for Pharmacy. Elsevier, China, pp. 70−79.

Gavin, W., Chen, L.H., Schofield, M., Masiello, N., Meade, H., Echelard, Y., 2014. Chapter 26: Transgenic cloned goats and the production of recombinant therapeutic proteins. In: Cibelli, et al., (Eds.), Principles of Cloning, second ed. Academic Press, Waltham, MA, pp. 329−342.

Gelvin, S.B., 2003. Agrobacterium-mediated plant transformation: the biology behind the "Gene-Jockeying" tool. Microbiol. Mol. Biol. Rev. 67 (1), 16−37.

Goel, N., Stephens, S., 2010. Certolizumab pegol. MAbs 2 (2), 137−147.

Gorny, M.K., 2012. Human hybridoma technology. Antibody Technol. J. 2, 1−5.

Gümüş, G., Sunguroğlu, A., 2001. Monoclonal Antibodies and Cancer Therapy, Turk J. Biochem. 26 (3), 111−118.

Gupta, P.K., 2008. Part II: Chapter 23: Immunotechnology: Molecular Biology and Fenetic Engineering. Capital Offset Press, New Delhi, pp. 378−392.

Hammers, C.M., Stanley, J.R., 2014. Antibody phage display: technique and applications. J. Invest. Dermatol. 134 (2), 1−5.

Haurum, J., Bregenholt, S., 2005. Recombinant polyclonal antibodies: therapeutic antibody technologies come full circle. IDrugs 8 (5), 404−409.

Hayat, M.A, 2002. Chapter 3: Fixation and embeding. Microscopy, Immunohistochemistry, and Antigen Retrieval Methods. Kluwer Academic Publications, New York, pp. 53−69.

Hermanson, G.T., 2013. Chapter 20: Antibody Modification and Conjugation, Bioconjugate Techniques, Third ed. Academic Press, Elsevier, Waltham, MA, pp. 867−920.

Hood, E.E., Howard, J.A., 2002. Plants as Factories for Protein Production. Kluwer Academic Publisher, The Netherlands, pp. 85−86.

Janeway, C.A., Travers, P., Walport, M., Shlomchik, M.J., 2001a. Chapter 9. The humoral immune response: the distribution and functions of immunoglobulin isotypes, Immunobiology: The Immune System in Health and Disease, fifth ed. Garland Science, Taylor and Francis Group, New York.

Janeway, C.A., Travers, P., Walport, M., Schlomchik, M.J., 2001b. Chapter 3: Antigen recognition by B-cell and T-cell receptors: the structure of a typical antibody molecule, Immunobiology: The Immune System in Health and Disease, fifth ed. Garland Science, Taylor and Francis Group, New York.

Janeway, C.A., Travers, P., Walport, M., Schlomchik, M.J., 2005. Immunobiology: The Immune System in Health and Disease, sixth ed. Garland Science, New York, p. 136.

Jayapal, K.P., Wlaschin, K.F., Hu, W.S., Yap, M.G.S., 2007. Recombinant protein therapeutics from CHO cells-20 years and counting. Chem. Eng. Prog. 103, 40−47.

Kawalec, P., Mikrut, A., Wiśniewska, N., Pilc, A., 2013. Tumor necrosis factor-α antibodies (infliximab, adalimumab and certolizumab) in Crohn's disease: systematic review and meta-analysis. Arch. Med. Sci. 9 (5), 765−779.

Khan, F.H., 2014. Chapter 25: Antibodies and their applications. In: Verma, A., Singh, A. (Eds.), Animal Biotechnology: Models in Discovery and Translation. Elsevier, Kidlington, pp. 473−490.

Khan, M.M., 2008. Chapter 5: Monoclonal antibodies as therapeutic agents. Immunopharmacology. Springer, New York, pp. 107−121.

Kindt, T.J., Goldsby, R.A., Osborne, B.A., 2007. Cahpter 4: Antigens and antibodies, Kuby Immunology, sixth ed. W.H. Freeman and Company, New York, pp. 76−106.

Kretzschmar, T., Rüden, T., 2002. Antibody discovery: phage display. Curr. Opin. Biotechnol. 13, 598−602.

Kuroiwa, Y., Kasinathan, P., Choi, Y.J., Naeem, R., Tomizuka, K., Sullivan, E.J, et al., 2002. Cloned transchromosomic calves producing human immunoglobulin. Nat. Biotechnol. 9, 889.

Laine, A., 1992. Rocket immunoelectrophoresis technique or electroimmunodiffusion. Methods Mol. Biol. 80, 201−205.

Lee, T.W., Fedorak, R.N., 2010. Tumor necrosis factor-α monoclonal antibodies in the treatment of inflammatory bowel disease: clinical practice pharmacology. Gastroenterol. Clin. North Am. 39 (3), 543−557.

Lee, Y.J., Jeong, K.J., 2015. Challenges to production of antibodies in bacteria and yeast. J. Biosci. Bioeng. 120 (5), 483−490.

Lefkovits, J., Ivanhoe, R.J., Califf, R.M., Bergelson, B.A., Anderson, K.M., Stoner, G.L., et al., 1996. Effects of platelet glycoprotein IIb/IIIa receptor blockade by a chimeric monoclonal antibody (abciximab) on acute and six-month outcomes after percutaneous transluminal coronary angioplasty for acute myocardial infarction. EPIC investigators. Am. J. Cardiol. 77 (12), 1045−1051.

Lemery, S.J., Zhang, J., Rothmann, M.D., Yang, J., Earp, J., Zhao, H., et al., 2010. U.S. Food and Drug Administration approval: Ofatumumab for the treatment of patients with chronic lymphocytic leukemia refractory to Fludarabine and Alemtuzumab. Clin. Cancer Res. 16 (17), 4331−4338.

Levinson, S.S., 2009. Immunoelectrophoresis. Wiley, New York. Available from: http://dx.doi.org/10.1002/9780470015902.a0001136.pub2.

Li, F., Vijayasankaran, N., Shen, A.(Y), Kiss, R., Amanullah, A., 2010. Cell culture processes for monoclonal antibody production. MAbs 2 (5), 466−477.

Li, J., Li, X., Tan, M., Lin, B., Hou, S., Qian, W., et al., 2009. Two doses of humanized anti-CD25 antibody in renal transplantation A preliminary comparative study. mAbs 1, 49−55.

Lin, F., Shi, J., 2015. Chapter 2: Standardization of diagnostic immunochemistry. In: Lin, F., Prichard, J. (Eds.), Handbook of Practical Immunohistochemistry: Frequently Asked Questions. Springer, London, pp. 17−30.

Liu, H., Kay, M., 2012. Disulfide bond structures of IgG molecules. MAbs 4 (1), 17−23.

Maloney, D.G., Sandmaier, B.M., Mackinnon, S., Shizuru, A.J., 2002. Non-myeloablative transplantation. ASH Educat. Book 2002 (1), 392−421.

Martin, T.D., 2006. IGIV: Contents, properties, and methods of industrial production—evolving closer to a more physiologic product. Int. Immunopharmacol. 6, 517−522.

Masih, S., Jain, P., El Baz, R., Khan, Z.K., 2014. Chapter 22: Transgenic animals and their applications. Animal Biotechnology. Elsevier Science Publications, Waltham, MA, pp. 407−423.

Merlo, L.M.F., Mandik-Nayak, L., 2013. Chapter 3: Adaptive immunity: B cells and antibodies. Cancer Immunotherapy. Elsevier, Lankenau Institute for Medical Research, Wynnewood, PA, pp. 25−40.

Miller, R.T., 2001. Technical immunohistochemistry: achieving reliability and reproducibility of immunostains. Soc. App. Immunohist. Ann. Meeting, pp. 1−53.

Nagoba, B.S., Vedpathak, D.V., 2008. Chapter 5: Antigen-antibody reactions. Immunology. BI Publications Pvt Ltd, New Delhi, pp. 46−77.

Nagy, Z.A., 2014. Chapter 4: Monoclonal antibodies: the final proof for clonal selection. A History of Modern Immunology: The Path Toward Understanding. Elsevier, Kidlington, pp. 33−40.

Newcombe, C., Newcombe, A.R., 2007. Antibody production: polyclonal-derived biotherapeutics. J. Chromatogr. 848, 2−7.

Nixon, A.E., Sexton, D.J., Ladner, R.C., 2014. Drugs derived from phage display. MAbs 6 (1), 73−85.

Pandey, S., 2010. Hybridoma technology for production of monoclonal antibodies. Int. J. Pharm. Sci. Rev. Res. 1 (2), 88−94.

Parkin, J., Cohen, B., 2001. An overview of the immune system. Lancet 357, 1777−1789.

Paulie, S., Perlmann, P., Perlmann, H., 2006. Chapter 6: Cell line authentication, Cell Biology, third ed. Elsevier, Inc., Kidlington, pp. 533−538.

Petersen, K., Pedersen, H.C., 2016. Part I: Staining protocols, Chapter 6: Detection methods. In: Taylor, C. (Ed.), Immunohistochemical Staining Methods Education Guide. Dako Agilent Technologies, Santa Clara, CA.

Pollock, D.P., Kutzko, Williams, J.L., Echelard, Y., Meade, H.M., 1999. Transgenic milk as a method of production of recombinant antibodies. J. Immunol. Methods 231 (1−2), 147−157.

Prosniak, M., Faber, M., Hanlon, C.A., Rupprecht, C.E., Hooper, D.C., Dietzschold, B., 2003. Development of a cocktail of recombinant-expressed human rabies virus-neutralizing monoclonal antibodies for postexposure prophylaxis of rabies. J. Infect. Dis. 187, 53–56.

Rahman, M., 2006. Introduction to Flow Cytometry. Serotec Ltd., Oxford, pp. 1–36, Published by Serotec Ltd.

Rani, S.J., Usha, R., 2013. Transgenic plants: types, benefits, public concerns and future. J. Pharm. Res. 6, 879–883.

Rasmussen, S.K., Rasmussen, L.K., Weilguny, D., Tolstrup, A., 2007. Manufacture of recombinant polyclonal antibodies. Biotechnol. Lett. 29 (6), 845–852.

Reichert, J.M., 2012. Marketed therapeutic antibodies compendium. MAbs 4 (3), 413–415.

Robl, J.M., Kasinathan, P., Sullivian, E., Kuroiwa, Y., Tomizuka, K., Ishida, I., 2003. Artificial chromosome vector and expression of complex protein in transgenic animals. Theriogenology 59, 107–113.

Rodney, J.Y., 2013. Part II: Therapeutic and clinical applications of biopharmaceuticals. Biotechnology and Biopharmaceuticals: Transforming Proteins and Genes into Drugs. Wiley, Toronto.

Rogovik, A.L., Carleton, B., Solimano, A., Goldman, R., 2010. Palivizumab for the prevention of respiratory syncytial virus infection. Can. Fam. Physic. 56 (8), 769–772.

Ross, J.J., 2013. Chapter 69: Diagnostic-therapeutic combinations, Genomic and Personalized Medicine, second ed. Elsevier, Inc., Oxford, pp. 798–819.

Salis, P., Caccamo, C., Verzaro, R., Gruttadauria, S., Artero, M., 2008. The role of basiliximab in the evolving renal transplantation immunosuppression protocol. Biologics 2 (2), 175–188.

Saylor, C., Dadachova, E., Casadevall, A., 2009. Monoclonal antibody-based therapies for microbial diseases. Vaccine 27 (6), 38–46.

Schmidt, K.V., Wood, B.A., 2003. Trends in cancer therapy: role of monoclonal antibodies. Semin. Oncol. Nurs. 19 (3), 169–179.

Scott, A.M., Allison, J.P., Wolchok, J.D., 2012. Monoclonal antibodies in cancer therapy. Cancer Immun. 12, 1–8.

Shapiro, H.M., 2003. Practical Flow Cytometry. Wiley, Hoboken, NJ.

Sobti, R.C., Krishan, A., 2003. Advanced Flow Cytometry: Applications in Biological Research. Springer Scince Business Media Dortrecht, New York.

Stills, H.F., 2012. Polyclonal antibody production. In: Suckhow, M.A., Stevens, K.A., Wilson, R.P. (Eds.), The Laboratory Rabbit, Guinea Pig, Hamster and Other Rodents. Elsevier Inc, Oxford, pp. 259–274.

Stoger, E., Sack, M., Fischer, R., Christou, P., 2002. Plantibodies: applications, advantages and bottlenecks. Curr. Opin. Biotechnol. 13, 161–166.

Thorlund, K., Druyts, E., Toor, K., Mills, E.J., 2015. Comparative efficacy of golimumab, infliximab, and adalimumab for moderately to severely active ulcerative colitis: a network meta-analysis accounting for differences in trial designs. Expert. Rev. Gastroenterol. Hepatol. 9 (5), 693–700.

Towbin, H., Staehelin, T., Gordon, J., 1979. Electrophoretic transfer of proteins from polyacrylamide gels to nitrocellulose sheets: procedure and some applications. Proc. Natl Acad. Sci. 76 (9), 4350–4354.

Turvey, S.E., Broide, D.H., 2010. Chapter 2: Innate Immunity. J. Allergy Clin. Immunol. 125 (2), 24–32.

Twyman, R.M., 2008. Large-Scale Protein Production in Plants: Host Plants, Systems and Expression. Protein Science Encyclopedia. Wiley-VCH Verlag GmbH & Co. KGaA, Weinheim.

Van den Eynde, B.J., Scott, A.M., 1998. Tumor antigens. Encyclop. Immunol. 1998, 2424–2431.

Vincenzi, B., Zoccoli, A., Pantano, F., Venditti, O, Galluzzo, S., 2010. Cetuximab: from bench to bedside. Curr. Cancer Drug. Targets 10 (1), 80–95.

Wakabayashi, K., 2010. ELISA-A to Z ……from introduction to practice. Technical Consultant. Shibayagi, Co., Ltd, Ishihara, pp. 1–60.

Walsh, G., 2002. Chapter 6: Therapeutic Antibodies and Enzymes, Proteins: Biochemistry and Biotechnology. Wiley, London, pp. 251–278.

Weiner, L.M., Surana, R., Wang, S., 2010. Monoclonal antibodies: versatile platforms for cancer immunotherapy. Nat. Rev. Immunol. 10, 317–327.

Wenk, M.R., Fernandis, Z.A., 2007. Manual for Biochemistry Protocols. World Scientific, River Edge, NJ, pp. 21–27.

Williams, J.W., Tjota, M.Y., Sperling, A.I., 2012. The contribution of allergen-specific IgG to the development of Th2-mediated airway inflammation. J. Allergy 2012, 1–9.

Wright, T., 2006. Biopharmaceuticals a high-cost gamble. Pharma Diag Innovat. 4 (11), 2.

Yuvienco, C., Schwartz, S., 2011. Monoclonal antibodies in rheumatic diseases. Med. Health 94 (11), 320–324.

Weblinks

http-1: http://www.nap.edu/read/9450/chapter/4

http-2: http://www.d.umn.edu/~biomed/flowcytometry/introflowcytometry.pdf

http-3: http://www.antibodies-online.com/resources/17/1216/Immunohistochemistry + IHC/

http-4: https://www.rndsystems.com/resources/protocols/antigen-retrieval-methods

http-5: http://www.ihcworld.com/epitope_retrieval.htm

http-6: http://www.fda.gov/

http-7: http://home.intekom.com/pharm/janssen/orthocln.html

Cell and Tissue Culture: The Base of Biotechnology

Onur Uysal, Tugba Sevimli, Murat Sevimli, Sibel Gunes and Ayla Eker Sariboyaci

Eskisehir Osmangazi University, Eskisehir, Turkey

17.1 INTRODUCTION

Cell culture is basically reproduction and survival of cells in an artificial environment. There are very common areas of use in cell cultures. The areas in which the cell culture is used can be counted as monoclonal antibody, viral and insect vaccine, enzyme, hormone, interleukin, and growth factor productions. Today advancements in the cell culture studies have led to the development of regenerative medicine concept. Here differentiation is an important factor. Cell differentiation is a common situation that is provided by internal cellular program and environmental conditions. Knowing the characteristics and factors of cellular differentiation enables the cells that are studied in the cell culture to be easily manipulated.

Cell culture studies need special working conditions. A cell culture laboratory must be a sterile environment. Contamination is the most common problem in the cell culture studies. When it cannot be resolved, it might cause serious losses. A good cell culture practice happens when necessary appropriate conditions are met and standard contamination factors are removed. The culture environment must necessarily be arranged in a way that is suitable to the characteristics that the cells need. Follow-up of the culture environment via macroscopic and microscopic ways is important. There are two significant factors that limit the reproduction of the cells in culture. The one is the free space in the culture vessel and the other is decrease in the nutrition support that the medium provides. In this condition cell passaging is required. Criteria of the passage needs can be put in order as cell concentration, pH of the environment, period that has passed since the last passaging, and special conditions that the study brings. In cell culture studies sometimes some reasons reveal the need for the cells to be frozen and stored. Also, cell viability tests need to be carried out after a traumatic procedure such as cell freezing, thawing, and subculturing.

Omics Technologies and Bio-engineering: Towards Improving Quality of Life
DOI: https://doi.org/10.1016/B978-0-12-804659-3.00017-8

Although the cells continue to proliferate in culture, at the end of a certain amount of time, the cells lose the ability to proliferate and survive as the result of the changes that occur physiologically or pathologically. Basically, this process is similar to the events in the organisms such as cellular aging and apoptosis. Apoptosis is programmed cell death occurs as a physiological or pathological process. Therefore, in cell culture studies it is important to know the ways of determining apoptosis and measuring cell viability.

17.2 CELL CULTURE LABORATORY

17.2.1 Safety

Safety means the condition of being protected from danger, risk, or injury. From this perspective, compared with other people, scientists are facing more safety risks in working environment. These risks can be divided into two main groups: risks related with the workspace and working materials (Hartung et al., 2002).

Cell culture laboratories are special and distinct areas differing from many other working places. The special techniques used in these laboratories need many different equipment, solvents, or reagents. There are a lot of corrosive, toxic, flammable or inflammable, and mutagenic chemicals. Additionally some of the equipment has complex structures and mechanisms. The insufficient information about the properties and usage of this equipment may cause injuries. Also misuse of them can cause electrical or fire hazard. Some chemicals used in the laboratory can be dangerous for human or environment. Inhalation, contact with skin or mucosal membranes, splash onto skin, explosions or fires, and faults during disposal of waste materials are some of the common safety problems (Freshney, 2015; Mather, 2013).

The risks in the second group stem from the characteristics of biological materials used in the cell culture studies. Basically, tissues, fluids, and cells obtained from human and animals take part in the cell culture studies. Researchers might be infected by various bacterial and viral agents, which these materials might carry according to the type of obtained organism, in the situations where not enough measures are taken. In a study, particularly, the use of cell strain infected or transformed by latent viruses requires to be treated more carefully about safety procedures to be implemented. However, some cell strains carry oncogenic potential spontaneously or as a result of intervention and carry the risk of cancer development upon access to the body (Sewell, 1995; Chosewood and Wilson, 2009).

The risks in the cell culture laboratory might affect both human health and the environment. With the measures to be taken, it must be targeted that these risks be removed completely or decreased. First, study that needs to be carried out with this aim is to determine potential risk factors in the cell culture laboratory. Work and security protocols and current legal regulations that will be formed after the risk factors have been determined must be practiced meticulously by all the employees. Conversely, in relation to failures and problems that might occur, procedures to be implemented should be designated. Also, information and education of the staff on these precautions and procedures carry great importance. In microbiological and biomedical laboratories prepared by National Institutes of Health (NIH), World Health Organization (WHO), and Center for Disease

Control and Prevention (CDC), biosafety guidance is an important document about designating the risks and safety precautions. According to this guidance, four basic biosafety level (BSL) and safety precautions that need to be taken in these have been specified. In the light of basic principles in this and other similar guides, each laboratory must determine ways to follow in relation to safe working principles and problems considering their own working materials, working density, and the equipment that it possesses. For instance, the way to be followed during the transport of inspiratory harmful chemical substance, during an accident that might occur, in the meantime what will be done and by whom it should be determined are considered. It is important to form guidelines (Tables 17.1 and 17.2) that include this and similar scenarios. In addition, staff should be educated in accordance with these guidelines (Caputo, 1988).

TABLE 17.1 Summary of BSL Requirements (Prepared According to WHO, 2004, 2015 updated, BSL Criteria)

	BSL			
	1	2	3	4
Isolation[a] of laboratory	No	No	Yes	Yes
Room sealable for decontamination	No	No	Yes	Yes
Ventilation				
• Inward airflow	No	Desirable	Yes	Yes
• Controlled ventilating system	No	Desirable	Yes	Yes
• HEPA-filtered air exhaust	No	No	Yes/No[b]	Yes
Double-door entry	No	No	Yes	Yes
Airlock	No	No		Yes
Airlock with shower	No	No		Yes
Anteroom	No	No	Yes	–
Anteroom with shower	No	No	Yes/No[c]	No
Effluent treatment	No	No	Yes/No[c]	Yes
Autoclave				
• On site	No	Desirable	Yes	Yes
• In the laboratory room	No	No	Desirable	Yes
• Double-ended	No	No	Desirable	Yes
BSCs	No	Desirable	Yes	Yes
Personnel safety monitoring capability[d]	No	No	Desirable	Yes

[a]Environmental and functional isolation from general traffic.
[b]Dependent on location of exhaust.
[c]Dependent on agent(s) used in the laboratory.
[d]For example, window, closed-circuit television, two-way communication.

TABLE 17.2   Summary of Recommended BSLs for Infectious Agents (Chosewood and Wilson, 2009)

BSL	Agents	Practices	Primary Barriers and SE	Facilities (Secondary Barriers)
1	Not known to consistently cause diseases in healthy adults	Standard microbiological practices	No primary barriers required. PPE (personnel protective equipment): laboratory coats and gloves; eye, face protection, as needed	Laboratory bench and sink required
2	Agents associated with human disease Routes of transmission include percutaneous injury, ingestion, and mucous membrane exposure	BSL-1 practice plus: Limited access Biohazard warning signs "Sharps" precautions Biosafety manual defining any needed waste decontamination or medical surveillance policies	Primary barriers: BSCs or other physical containment devices used for all manipulations of agents that cause splashes or aerosols of infectious materials PPE: Laboratory coats, gloves, face, and eye protection, as needed	BSL-1 plus: Autoclave available
3	Indigenous or exotic agents that may cause serious or potentially lethal disease through the inhalation route of exposure	BSL-2 practice plus: Controlled access Decontamination of all waste Decontamination of laboratory clothing before laundering	Primary barriers: BSCs or other physical containment devices used for all open manipulations of agents PPE: Protective laboratory clothing, gloves, face, eye, and respiratory protection, as needed	BSL-2 plus: Physical separation from access corridors Self-closing, double-door access Exhausted air not recirculated Negative airflow into laboratory Entry through airlock or anteroom Hand washing sink near laboratory exit
4	Dangerous/exotic agents that post high individual risk of aerosol-transmitted laboratory infections that are frequently fatal, for which there are no vaccines or treatments Agents with a close or identical antigenic relationship to an agent requiring BSL-4 until data are available to redesignate the level Related agents with unknown risk of transmission	BSL-3 practices plus: Clothing change before entering Shower on exit All material decontaminated on exit from facility	Primary barriers: All procedures conducted in Class III BSCs or Class I or II BSCs in combination with full-body, air-supplied, positive pressure suit	BSL-3 plus: Separate building or isolated zone Dedicated supply and exhaust, vacuum, and decontamination systems Other requirements outlined in the text

The classification of infectious microorganisms according to risk groups—NIH and WHO—is as follows:

Risk group 1: Agents not associated with disease in healthy adult humans (NIH). No or low individual and community risk: A microorganism that is unlikely to cause human or animal disease (WHO).

Risk group 2: Agents associated with human disease that is rarely serious and for which preventive or therapeutic interventions are often available (NIH). Moderate individual risk, low community risk: A pathogen that can cause human or animal disease but is unlikely to be a serious hazard to laboratory workers, the community, livestock, or the environment. Laboratory exposures may cause serious infection, but effective treatment and preventive measures are available and the risk of spread of infection is limited (WHO).

Risk group 3: Agents associated with serious or lethal human disease for which preventive or therapeutic interventions may be available (high individual risk but low community risk) (NIH). High individual risk, low community risk: A pathogen that usually causes serious human or animal disease but does not ordinarily spread from one infected individual to another. Effective treatment and preventive measures are available (WHO).

Risk group 4: Agents likely to cause serious or lethal human disease for which preventive or therapeutic interventions are not usually available (high individual risk and high community risk) (NIH). High individual and community risk: A pathogen that usually causes serious human or animal disease and that can be readily transmitted from one individual to another, directly or indirectly. Effective treatment and preventive measures are not usually available (WHO).

The relationship between risk groups and BSLs, laboratory type (LT), laboratory practices (LPs), and safety equipment (SE) as follows:

Risk group 1: Basic BSL 1, basic teaching, research (LT), good microbiological techniques (LP), and open bench work (SE).

Risk group 2: Basic BSL 2, primary health services; diagnostic services, research (LT), good microbiological techniques plus protective clothing, biohazard sign (LP), and open bench plus biological safety cabinet (BSC) for potential aerosols (SE).

Risk group 3: Containment BSL 3, special diagnostic services, research (LT), as Level 2 plus special clothing, controlled access, directional airflow (LP), and BSC and/or other primary devices for all activities (SE).

Risk group 4: Maximum containment BSL 4, dangerous pathogen units (LT), as Level 3 plus airlock entry, shower exit, special waste disposal (LP), and Class III BSC, or positive pressure suits in conjunction with Class II BSCs, double-ended autoclave (through the wall), filtered air (SE) (WHO; 2004, 2015 updated, BSL Criteria).

Biosafety equipment and features are as follows:

In Class I, II, and III BSCs, aerosol and spatter hazards must be corrected. Safety features: Class I: Minimum inward airflow (face velocity) at work access opening, adequate filtration of exhaust air, and does not provide product protection. Class II:

Minimum inward airflow (face velocity) at work access opening, adequate filtration of exhaust air, and provides product protection. Class III: Maximum containment, and provides product protection if laminar flow air is included.

In pipetting aids, hazards from pipetting by mouth, e.g., ingestion of pathogens, inhalation of aerosols produced by mouth suction on pipette, blowing out of liquid or dripping from pipette, and contamination of suction end of pipette must be corrected. Safety features: Ease of use, controls contamination of suction end of pipette, protecting pipetting aid, user and vacuum line, can be sterilized, and controls leakage from pipette tip.

In sharps disposal containers, puncture wounds hazards must be corrected. Safety features: Autoclavable, robust, and puncture-proof.

In transport containers, between laboratories, institutions release of microorganisms' hazards must be corrected. Safety features: Robust, watertight; primary and secondary containers to contain spills and absorbent material to contain spills.

In manual or automatic autoclaves, infectious material (made safe for disposal or reuse) hazards must be corrected. Safety features: Approved design and effective heat sterilization (WHO, 2004, 2015 updated, BSL Criteria).

Beside all these preliminaries, Safety Data Sheets that includes the characteristics of the materials used should be examined carefully before use and necessary precautions must be taken. During the studies, considering the problems that might be met, the determination and the use of basic SE such as biosafety cabin, closed carriers, and personal SE such as gloves, glasses, and aprons carry great importance (Freshney, 2015).

17.2.2 Cell Culture Equipment and Laboratory Design

The special techniques used in cell culture laboratories need many different types of equipment.

The basic equipments in the cell culture laboratory are as follows:

- Laminar air flow cabin
- Incubator controlled O_2
- Fridge (4°C)
- Freezers (-20°C, -80°C, -150°C)
- Liquid nitrogen tank
- Light microscope
- Inverted microscope
- Fluorescence microscope
- Stereomicroscope
- Pipet aid
- Micropipettors
- ELISA microplate reader
- qReal-time PCR instruments
- Flow cytometer
- Centrifuge

- Hemocytometer
- Air conditioner
- Sterilization system
- Mini chamber/incubator for microscope
- Water bath (37°C)
- pH meter
- Water purifier
- Vortex mixture
- Orbital shaker

Basic materials needed for cell culture:

- Cells
- Medias and reagents
- Cell culture dishes
- Cell culture flasks
- Cell culture multiwell plates
- Cell strainers (40, 70, and 100 μm)
- Syringe filters (0.22 and 0.45 μm)
- Culture/chamber slides (multiwell)
- Cell chamber slides and coverslips
- Cell counting equipment
- Controlled-rate cell freezing container
- Cryo storage racks
- Cryo vials
- Aluminum canes (liquid nitrogen)
- Cryo Vial Color Coders
- Various pipettes and micropipettes
- Waste containers

Laminar air flow cabin is the cabin that is used for providing a sterile environment in the cell culture studies and protects both the samples and the staff. It can be horizontal or vertical. Cabins have been diversified as Class 1, 2, and 3 according to the research and clinical needs. Class I: Anterior is open, air flow is inward, outward air flow is from HEPA (high-efficiency particulate air) filter and only personnel protective. Class II: Anterior is open, vertical laminar air flow, inward and outward air HEPA filtration, and personnel and product protective. Class III: Cabins that are used for studying known human pathogens and other BSL-4 substances and provide protection at the utmost level for both the personnel and the environment. The air in the cell culture must be at the standard of Class II. The air entering the laboratory passes through the HEPA filters and is given to the environment with positive pressure (Wilson et al., 1998; Freshney, 2015).

Selection of a BSC, by type of protection needed (WHO, 2004, 2015 updated, BSL Criteria):

- Personnel protection, microorganisms in Risk Groups 1–3: Class I, Class II, and Class III.
- Personnel protection, microorganisms in Risk Group 4, glove-box laboratory: Class III.

- Personnel protection, microorganisms in Risk Group 4, suit laboratory: Class I and Class II.
- Product protection: Class II and Class III only, if laminar flow included.
- Volatile radionuclide/chemical protection, minute amounts: Class IIB1, Class IIA2 vented to the outside.
- Volatile radionuclide/chemical protection: Class I, Class IIB2, and Class III.

Incubator The intended use of the incubators is to provide suitable environmental conditions for cells to grow. It must be big enough in accordance with the laboratory requirements of everybody. Stainless steel incubators are especially resistant to abrasion and have the feature of easy cleaning. There are two types of incubators which are dry and humid CO_2. Dry incubators are economical but might cause evaporation. Putting water into a petri plate might provide humid but does not meet the atmospheric conditions in the incubator. Humid CO_2 incubators are expensive but provide culture conditions at the highest level. They provide forming a buffer system with the incubators enhanced by CO_2 and bicarbonate found in the culture medium. CO_2 level ensures that culture pH remain 6.9−7 (Butler, 2003).

Fridges and freezers In a cell culture laboratory there must be storage areas for fluids such as media and reagent, chemicals such as drug and antibiotics, tissues, and cells. Media, reagent, and chemicals should be preserved in the fridge or freezer according to their data sheets. Some media, reagent, and chemicals are photosensitive. Therefore, they should be preserved in the dark or wrapped with aluminum folio. Fridge temperature should be 2°C−8°C. Most cell reagents are preserved at −5°C to −20°C. If necessary, −80°C freezer can be used as well.

Fluid nitrogen storage Cells can be kept at minus degrees for a long time. This enables to store cells without applying to the primer tissue. Cell stocking enables protection against any genetic change or contamination situation.

Centrifugation is necessary for the harvest of the centrifugal culture cells at low speed; 5−10 minutes is sufficient to separate cells from the culture environment at 150−200 G. High speed or long time cause cells to be harmed. When the centrifugation has been completed, supernatant must be emptied carefully. Due to safety reasons, centrifugal rotor and individual buckets must be closeable.

Hemocytometer is necessary for the growth kinetics in the cell culture laboratory. It is a bench-top tool designed to evaluate accurate and precise cell count and vitality (alive, dead, and total cells) in less time.

Inverted and fluorescent microscopes are the essential elements in cell culture studies. It has been used to examine viable cells in cell and tissue cultures.

Plasticware/glassware Although being expensive, plastic materials are preferred more than the glass materials. Glass materials need to be washed and autoclaved. Plastic flasks are sterilized with gamma irradiation and be produced from polystyrene for the cell to connect and grow. Plastic or glass materials that will be used in cell culture must be absolutely tested in terms of not developing toxicity or other problems in the cell.

The following is a list of laboratory systems and components that may be included in a commissioning plan for functional testing, depending on the containment level of the facility being renovated or constructed. The list is not exhaustive. Obviously, the actual commissioning plan will reflect the complexity of the laboratory being planned (WHO, 2004, 2015 updated).

1. Building automation systems including links to remote monitoring and control sites
2. Electronic surveillance and detection systems
3. Electronic security locks and proximity device readers
4. Heating, ventilating (supply and exhaust) and air-conditioning (HVAC) systems
5. HEPA filtration systems
6. HEPA decontamination systems
7. HVAC and exhaust air system controls and control interlocks
8. Airtight isolation dampers
9. Laboratory refrigeration systems
10. Boilers and steam systems
11. Fire detection, suppression, and alarm systems
12. Domestic water backflow prevention devices
13. Processed water systems (i.e., reverse osmosis, distilled water)
14. Liquid effluent treatment and neutralization systems
15. Plumbing drain primer systems
16. Chemical decontaminant systems
17. Medical laboratory gas systems
18. Breathing air systems
19. Service and instrument air systems
20. Cascading pressure differential verification of laboratories and support areas
21. Local area network and computer data systems
22. Normal power systems
23. Emergency power systems
24. Uninterruptible power systems
25. Emergency lighting systems
26. Lighting fixture penetration seals
27. Electrical and mechanical penetration seals
28. Telephone systems
29. Airlock door control interlocks
30. Airtight door seals
31. Window and vision-panel penetration seals
32. Barrier pass-through penetration
33. Structural integrity verification: concrete floors, walls, and ceilings
34. Barrier coating verification: floors, walls, and ceilings
35. BSL-4 containment envelope pressurization and isolation functions
36. BSCs
37. Autoclaves
38. Liquid nitrogen system and alarms
39. Water detection systems (e.g., in case of flooding inside containment zone)
40. Decontamination shower and chemical additive systems
41. Cage-wash and neutralization systems
42. Waste management.

A cell culture laboratory must be a sterile environment in which procedures are carried out with contamination minimally. To set up a completely sterile laboratory without contaminated microorganisms is not practically possible but contamination level must be reduced to a minimum. The most important feature of the laboratory is the separation of sterile working area from other areas. Cell culture areas take place in confined spaces where the personnel are less. Air filtration is one of the most important issues. There must be Class II laminar flow cabin in the room. With this, after taking clean air from outside and sterilizing, it is given to the environment with positive pressure. Thus, the number of particle in the environment is minimal. Laminar flow hood, microscope, and incubator must be close to each other, so that physical contact of the cultures is reduced to a minimum. Areas where the nonsterile processes are done such as washing in the laboratory must take place in the other side of the sterile area. Air cooling is needed due to the heat formed by electronic devices in the laboratory. Such precautions will reduce contaminants in the laboratory environment to minimum. Only the hands of the person who works must have access to the laminar flow cabin that is a sterile area where the culture processes are made. Thus, contamination risk that might result from both laboratory employees and the environment are decreased. Other areas of the laboratory should not be messy and should be cleaned regularly with antiseptic cleaning agents (Freshney, 2015; Zhou and Kantardjieff, 2014).

One of the most important subjects, which needs to be planned before laboratory settlement, is the arrangement and set up of gas systems that feed the laminar air flow cabin and incubator. Tubes that are linked to gas systems must regularly be changed. Therefore, tubes being in the next room are vital both in providing aseptic conditions and in the laboratory layout. Another important subject is that all electrical appliances are requiring to be connected to uninterrupted power supplies. There must be computer systems in which data belonging to the processes carried out in the laboratory can be stored. All the processes should be carried out in accordance with the national/international standards (Marks, 2003).

17.2.3 Aseptic Technique

One of the vital subjects in cell culture is to be able to keep cells away from the contamination. Dirty working area, nonsterile devices, and particles in the air full with microorganism, unclean incubators are the sources of contamination. Aseptic technique is a technique designed to set a barrier between microorganisms in the environment and sterile cell culture. So, how can we prevent contamination (Coté, 2001; Phelan and May, 2007)?

- Before and after the culture, hands should be washed with scentless soap. Latex gloves should be worn.
- SE should be worn to be protected from dangerous materials; in this way, contamination resulting from the clothing is prevented as well.
- The person who works must definitely wear bonnet and mask.
- In terms of meeting the aseptic conditions, the laboratory should not be tidy and crowded.
- Access must be restricted while the experiments continue in the laboratory.
- Working areas must be cleaned with 70% of ethanol.

- Incubators must be opened in a way that only hands can enter, so as to keep the changes in the environment at a minimum during the cell culture studies.
- For all the processes, sterile laminar flow cabin must be used. Cabin should be in a place away from the door and the traffic.
- Working area must be tidy and should not be used as storage.
- Used materials and containers must be appropriately labeled and controlled before starting to work.
- Glass materials that will be used before working must be sterilized.
- Medium preparation and modification operations must certainly be carried out in laminar flow cabin.
- While carrying out operations with the fluids in the cabin, aerosol formation must be prevented.
- Solutions that are sterile in the working area and cultures should not contact with each other.
- The neck portion of medium bottles and culture flasks should never be touched.
- Clean and contaminated materials must be kept at separate areas.
- On a regular basis, maintenance and disinfection of tools and devices must be done.
- After the study has been completed, waste materials must be moved away from the laboratory and sterilization must be ensured with ultraviolet lamp.

17.2.4 Cross Contamination

Cross contamination is the naming of a cell line differently with the wrong definition. With the isoenzyme studies conducted in the 1960s, it has been shown that 18 out of 20 human cells are the same with HeLa cells. The reasons of the cross contaminations are as follows: wrong labeling of culture bottles as a result of carelessness while working with the cell lines, contaminating a cell line to another cell line's medium bottle, carelessness in the use of fluid nitrogen stores, problems that are lived through while transferring cells from one laboratory to another (Nelson-Rees et al., 1981; Ryan, 1994; Zhou and Kantardjieff, 2014; Ulrich and Negraes, 2016).

To prevent cross contamination:

- before the cell lines are frozen and thawed, accuracy with the labels must be determined;
- in laminar flow cabin, each cell line must be studied by itself. Before moving on to other cell line, all the steps of the process must be completed for that line;
- before and after the culture, photographs of the cell should be compared with the standard ones;
- for each cell line, separate medium and materials should be used;
- the same pipette should not be used for different cell lines;
- the pipette that is inserted into the culture flask should not be used for the medium; and
- pipette stopper must be used.

The prominent international cell banks control cell lines in their stocks by isoenzyme check, karyotyping, and short tandem repeat analysis. They announce these results. So, it is important to provide cell lines from these cell banks.

17.2.5 Biological Contamination

We can say that contamination is the most common problem in the cell culture studies. When it cannot be resolved, it might cause serious losses. We can divide cell culture contaminants into two groups:

- Biological contaminants: bacteria, fungus, mycoplasma, parasite, virion, and cross contamination that stems from other cell lines.
- Chemical contaminants: pollution in the water and medium, detergent residues.

Bacteria are the largest group of single cell microorganisms. They can be a few micrometer sized and in various shapes. As it can easily adapt to any environment, it is one of the most common biological contaminations in the cell culture (Fig. 17.1A and B). Bacteria contamination can be detected with phase contrast microscope. Infected cultures are often blurry, and sometimes a thin film layer is determined visually on the surface with eyes. pH of the culture environment drops suddenly. When examined with a microscope, bacteria can be easily spotted as well. Preventing and treating bacterial infections are relatively easier.

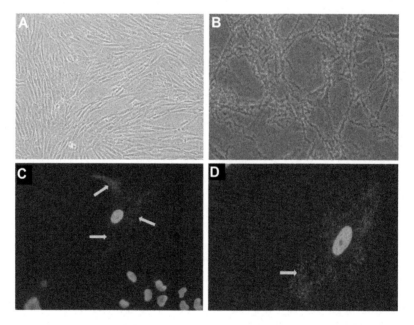

FIGURE 17.1   Phase contrast images are showing adherent human periodontal ligament—derived mesenchymal stem cells and dental pulp—derived mesenchymal stem cells (A and B, respectively) contaminated with bacteria. The spaces between the adherent cells show tiny, shimmering granules under low power microscopy, but the individual bacteria are typically rod-shaped and not easily distinguishable. (C and D): Fluorescence microscopic images are showing adherent human synovial fluid—derived mesenchymal stem cells contaminated with mycoplasmas (arrow). DAPI stains both the nuclei and the mycoplasma in the cell juxtanuclear cytoplasm (original magnification A: x20, B: x40, C: x20 and D: x40). Source: *Courtesy Dr Sariboyaci.*

Mycoplasmas are organisms without cell walls. Compared with bacteria and fungus infections, mycoplasma causes more problems in terms of its incidence, fixation, prevention, and annihilation. As they are very small organisms, it is rather difficult to determine them without reaching a certain density. In cell cultures, mostly bovine mycoplasmas (*Mycoplasma arginini, M. pirum, M. bovis*, and *M. bovoculi*), then human mycoplasmas (*M. orale, M. hominis, M. fermentans*, and *M. salivarum*), pig mycoplasmas (*M. hyorhinis* and *Acholeplasma oculi*), which cause illness in poultry and cats (*M. arthritidis, M. pulmonis*, and *M. canis*) are seen. Some mycoplasmas that grow slowly might deteriorate the physiology of the cell without causing the death of the cell. Chronic mycoplasma infections show themselves with decrease in the cell proliferation, decreased saturation density and agglutination in suspension and adherent cultures. However, the only assured way of detecting mycoplasma contamination is by testing the cultures periodically using fluorescent staining (e.g., Hoechst 33258, DAPI) (Fig. 17.1C and D), enzyme-linked immunosorbent assay (ELISA), polymerase chain reaction (PCR), immunostaining, autoradiography, or microbiological assays.

Viruses Viral contamination is a situation that wants considerable attention. It is quite difficult to detect them without the cytopathic effect or because of the fact that they infect secretly. Host viruses generally do not have any negative impact on the cell cultures but can pose a threat for laboratory personnel. Viral infections in the cell cultures can be detected via ELISA, PCR, electron microscope, and immunohistochemistry methods.

Parasites Contamination with parasites is a rare and quite unexpected situation. Generally it is observed in primer cells that are made of fresh tissues with parasite infection and tissue cultures. The most well-known parasites about the laboratory infections are *Trypanosoma cruzi, Leishmania* sp., *Cryptosporidium parvum, Plasmodium* sp., etc.

Fungi are unicellular eukaryotic organisms. Like bacterial contamination, it creates blurry vision in the culture. In the first phase of the contamination, generally there is a little change in pH. As their number increases, pH starts to increase too. When examined with microscope, they can be seen as oval, globular, or gemmiferous.

Prions Prions are the unique examples of the pathogen proteins that multiply themselves. Although they are protein molecules, they cause serious diseases. Unlike bacteria, virus, and other known pathogens, prions invading the host cells and multiplying in them do not happen with nucleic acids (DNA or RNA). Many cell lines are resistant to prion infection. However, unlike many infectious agents, it is rather difficult to inactive them. This must be kept in mind when using bovine-originated medium or serum.

Molds are hyphae microorganisms from Fungus family. Hyphae nets are called as micelle. Like in fungus contamination, in the first phase of the contamination, there is usually a little change in pH. Afterwards, there occurs a rapid increase and culture becomes blurry. When examined with microscope structures similar to hyphae and sometimes spore clusters can be observed.

A good cell culture practice happens when necessary appropriate conditions are met, and standard contamination originating from bacteria, fungus, mycoplasma, and cross contamination are removed (Ryan, 1994; Langdon, 2004; Ulrich and Negraes, 2016).

17.3 CELL CULTURE

17.3.1 Cell Culture System

What is expressed as cell culture system basically is keeping the cells taken out of a living in an artificial environment alive. In cell culture systems, it is aimed that the cells that are in three-dimensional environment in tissues to be cultured and examined by providing appropriate circumstances under in vitro conditions. These systems are an important research tool because of enabling supervision of the conditions that the cells are in. Especially the examination of normal cellular processes' cell culture systems serves too many fields such as the toxicology studies, development of new drugs, and treatment methods. Besides, nowadays especially in parallel with the improvements that are experienced in stem cell and the field of genetics, cell culture systems have come to an important place in tissue engineering and regenerative medicine field. With all these advantages it provides, these systems carry some restrictions such as special working conditions, special equipment, and labor force requirement that has received education in this field. What is more important than these is that no in vitro condition is sufficient about imitating complex mechanisms in the organism. Despite different definitions and the concepts, it is possible to gather cell culture systems under three headings as primary, secondary, and continuous cell culture (Sinha, 2008; Przyborski, 2016).

Primary cell culture is the removal of the pieces/biopsy (dimension of about $1 \times 1 \times 1$ cm) from tissue or organs in aseptic conditions and then obtaining cells via mechanic (tissue explant culture Fig. 17.2A and B), chemical, or enzymatic digestion method from this biopsy. Although the obtained cells display a mostly heterogeneous population and despite low proliferation rate and the hardship of the techniques used, primary cultures represent the cells' closest form in the tissues. Cells that are replicated with the primary culture cannot maintain their vitality for a long time (Davis, 1994). Because proliferating cells cause the required substances to consume in a short time and the rapid increase of the metabolites and the decrease of their proliferating areas in the culture environment. In this situation, if the cells that are obtained with the primary culture are transferred to a new culture environment in appropriate conditions, they can continue to proliferate. Cultures that are obtained in this way are called secondary culture. Although the cells continue to proliferate with the secondary culture, at the end of a certain amount of time, the cells lose the ability to proliferate and survive as the result of the changes that occur physiologically or pathologically. Basically, this process is similar to the events in the physiologic processes in the organisms such as cellular aging and apoptosis. This situation cannot be observed in the continuous cell culture systems because the cells are divided by a number that accepted as endless. These cells that have a rather high growth speed are less sensitive to negative changes that will decrease the proliferation in culture conditions. However, in relation to division skills that last for a long time and high growth speed, they carry problems such as viability to chromosomal anomalies, morphological deformity, and loss of tissue markers (Freshney, 2015; Fauza and Bani, 2016).

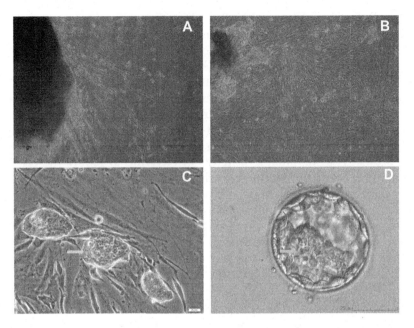

FIGURE 17.2   (A and B) Isolation of mesenchymal stem cells (MSCs) from foreskin (A) and placental villi (B) tissues via tissue explant culture method. Phase contrast images are showing MSCs during the early days of explant culture in passage 0 (P0) on Day 6 (A) and 12 (B). (C) Undifferentiated mouse embryonic stem cells (ESCs). Undifferentiated ESCs formed a compact colony and showed a high nuclear to cytoplasmic ratio (arrow). (D) Mouse blastocyst. ICM (arrow) and trophoblast (arrow head). Original magnification A and B: ×4. Scale bar: C: 20 μm and D: 25 μm. Source: *Courtesy Dr Sariboyaci.*

17.3.2 Cell Line and Culture Monitoring

While the cells are being cultured in an environment where the appropriate conditions are provided, regular control of some parameters is quite important. Proliferation of cells in the cell culture happens in a few phases. Prophases in which no cell increase is observed and growth factors are synthesized and increased are the lag phases. In the following growth phase, it is observed that the cells go into the cell cycle and increase their number by dividing. Then comes the stationary phase in which the cell division rate is balanced with cell death rate. Last of all, there is a phase that with increasing cell death the number of cells decreases. In different cell types and culture systems, these cycles are completed in different period of times. It is important to know well and follow regularly these characteristics belonging to the cells and the growth requirement of the cells that are used to achieve successful results in cell culture studies. Of all the tools that are used with this aim, the most important one is the cell counting. Considering the division rate of the used cell type, the cell counting that are carried out at certain intervals is an important tool to monitor the cell culture. Data gathered from the cell counting that is carried out periodically give us significant information about the anomalies in the culture system. Another way to monitor the culture system is the detection of the intracellular compounds such as DNA or proteins or the changes in the amount of various matters that the cell uses. For example, total protein amount may help the estimation of the cell number or the determination of some proteins can inform about the cell proliferation. Similarly, DNA measurement gives significant information on this subject as well. Another important monitoring tool is the vitality tests of the cells in the culture environment. Here the cells can be evaluated in terms of their metabolic conditions, and information about the cells growth conditions can be provided. The cells are present at the phase which basically realizes DNA synthesis in the cell cycle or mitotic division stage. The fixation of cell cycle stage in which the cells are present is also valuable in terms of ensuring to follow the cell culture. For instance, in a culture environment at growth phase, cells are at the phase that is often DNA synthesis stage or it might be seen that at the stationary phase the cell number of mitosis at the G1 stage increases. In this field, especially the evaluation of the cells through flow cytometry provides important data. Frequent use of cell lines in the cell culture studies reveals the necessity of developing some special monitoring methods. Cell lines are sources that are constantly divided and therefore quite open to genetically and morphological changes. In this respect, cell lines that are used should be searched with different methods being karyotype and antibody analysis, isoenzyme, and DNA studies in the first place. Thanks to this, in the following passages, the changes that might occur in the cells can be detected. This situation, also, allows for the fixation of the cross contaminations (Sinha, 2008; Yadav, 2008; Mather, 2013; Przyborski, 2016).

17.3.3 Primary Culture

Primary culture is the cell culture system that is formed by culture cells directly obtained from tissue. A primary culture starts with the biopsy (~ 1 cm^3) from tissue or organ via dissection. In tissue organization, cells have intercellular and cell basal membrane or cell matrix connections. For cells to be cultured, first of all, they need to get rid of these connections (single cell suspension). In the separation of cells, there are enzymatic or chemical digestion methods in which various proteolytic enzymes are used such as trypsin or collagenase and mechanic separation methods like splitting the tissue (mincing of the tissue) with surgical knives. Thus obtained cell suspensions are purified with serial dilution and centrifugation process and transferred into culture vessels. Then, many cells that are freed from cellular connections stick to the culture vessel. Primary culture process is completed once nonsticky cells and their residual are removed from the cell environment. However, except the cells that are initially wanted to be obtained in primary cultures, it is observed that different cell types especially fibroblasts also hold on to the culture vessel. Then, in secondary cultures that are obtained by passaging, adding to the growth factors that are peculiar to wanted cell type, it can be ensured that a certain cell type proliferate and others are pressed (Ng and Schantz, 2010; Fauza and Bani, 2016).

Cells in the primary culture are quite significant in terms of being the closest forms of the state of the cells that they represent in normal tissues. However, sometimes this situation might bring disadvantages together. Foremost among these come limited life span. Therefore, these cells should be reproduced with serial passages and stored. And in some other cases the number of cells that we are interested in might be less than other cells and that's why their reproduction might be pressed. In such a situation, intervention with special mediums or growth factors might be required. Another important problem that has been encountered in primary cultures is that when cells are taken out of the in vivo conditions and transferred into the culture environment, they might lose their structural and functional characteristics. In this respect, cells that have completely different morphology at the tissue level in the living can display similar morphologies in the culture environment.

17.3.4 Cell Isolation

Cell isolation is the leading of the basic practices in the cell culture studies. Although there are various techniques in this field that have been developed to this day, the method that will be chosen changes depending on tissue type, type of the living from which the tissue is obtained, and its age. Considering these conditions, the choice of the appropriate medium, the choice of the enzymes that will be used, and the arrangement of their amount and application periods are of great importance. No matter which method is chosen, the result that is planned to reach is to obtain more functional and living cells with the minimal cellular damage (Mather, 2013; Ulrich and Negraes, 2016).

The choice of a successful isolation method can be possible with knowing the characteristics that the tissues have better. Because basic structures that are targeted in cell isolation are extracellular matrix, cell—extracellular matrix connections, and intercellular connections. Different tissue types have extracellular matrices and cellular connections with different characteristics. For example, epithelial tissue is a tissue in which the cells hold on to each other tightly with various intercellular connections and in basal it clings to a structure formed by a connective tissue. Here desmosomes, which are one of the connections among epithelial cells, are addicted to calcium and the choice of appropriate enzyme; to remove the connections, the use of calcium-free medium is required. At the top of the tools used as the basis in cell isolation among the tissues come enzymes such as collagenase, trypsin, elastase, hyaluronidase, DNA, and protease. Different tissue organizations necessitate the use of different enzymes. Beside this, another discussion topic is the effect of the damages the enzymes might cause in the cells on the studies (Ross and Pawlina, 2014; Freshney, 2015; Przyborski, 2016).

17.3.5 Culture Environment

When the cells taken out of the living tissue are transferred into the culture environment, this situation leads to cellular stress. That is why the culture environment must necessarily be arranged in a way that is suitable to the characteristics that the cells need. Two fundamental components of the cell culture are physiological environment and physicochemical environment. With physico-chemical environment, factors such as temperature of the environment and cell atmosphere, concentration of oxygen or carbon dioxide, and pH are meant. And with the physiological environment, rather factors such as growth factors and the nutrients that cells physiologically need to grow and reproduce are meant and basically it is arranged through the mediums.

Surface features that the culture vessel possesses are one of the physico-chemical environments. For example, while the surfaces such as plastic and glass form an appropriate environment for the cells that need to cling to growing, it is not suitable for suspension cultures. Hydrogen ion concentration in the culture environment is an important factor that affects the cellular survival and reproduction directly. One of the best indicators of this situation is the pH value of the environment. pH value that each cell type needs can differ but generally pH value for mammalian cells are accepted as 7.2–7.4. Environment pH mostly undergoes change because of metabolic wastes of the cells. Therefore, considering the division rate of the cells, necessary buffer systems must be added to the culture environment or medium change should be planned. One of the factors that affect environment pH is the CO_2 concentration. Because bicarbonate buffer system is in the medium functions with CO_2 that is present in the environment. In many culture environments, frequently a CO_2 concentration of 4%–10% ensures the optimum value. Temperature of the culture environment is important, too. Here basically it should imitate the body temperature of the organism that the cells are taken and the changes depending on the allocation of the tissue taken. For many mammalian cells, the ambient temperature of 37°C is seen as sufficient. Considering the fact that cells are especially more sensitive to temperature rise than temperature drop, optimum value of the ambient temperature should be determined. Besides all these fundamental environmental conditions, moisture, osmotic pressure, surface tension, viscosity, sterility, and osmolarity are other important psycho-chemical factors (Sinha, 2008; Zhou and Kantardjieff, 2014; Fauza and Bani, 2016).

Culture mediums are the most important tools in the arrangement of the physiological conditions in the culture environment. Mediums feature nutrients are necessary for the cells, growth factors, and hormones. A typical medium consists of carbohydrate, amino acid, salts, bicarbonate, vitamins, and hormones. Actually this formula defines a typical basal medium. Except basal medium, two more classes can be defined as enriched mediums and serum-free mediums. Although basal medium meets the basic needs of the cells, the content of it needs to be enriched most of the time. In this sense, the most important content that needs to be added to the basal medium is the serum. Ten percent added serum to the medium supports cell reproduction and growth for the most part. Among different serum types, the most commonly used is the fetal calf serum that includes embryonic growth factors of large quantity. Serum is an important factor for adhesion factors, hormones, lipids, and minerals as well as growth factors. However, at the top of the problems that are met in the use of serum comes the standardization problem. Serums that are obtained in different times can vary as to the content. In addition to this, serums are significant sources for transmission of especially viruses and the prions. Another content that might be added to the basal medium is the antibiotics that are used to decrease the risk of contamination. Often instead of only one antibiotic, nontoxic doses of a few antibiotics to the cells are preferred. This combination frequently comprises penicillin G, streptomycin, amphotericin B, which is an antifungal agent. In some studies carried out with some cell lines, the use of serum-free medium might be mandatory. In such a situation, basal medium needs to be supported through appropriate nutrient and hormonal factors (Salzman, 1961; Jayme, 2001).

17.3.6 Cell Morphology

Follow-up of the culture environment via macroscopic and microscopic ways is very important. Regular follow-ups enable the early indications of the changes that have occurred during the study to be detected and intervened at the proper time. Among different follow-up criteria, one of the most important is the investigation of cell morphologies because the cell morphology is one of the most significant indicators of vitality and health conditions. Control and follow-up of the morphology which the cells show in the culture environment enable the vitality status of the cell as much as the follow-up of contamination status, toxic effects, and cellular aging. Especially vacuolar structures seen in nucleus and cytoplasm, changes that occurred at the nuclear and cell membrane, and distortions in the shape of the cell are significant changes that need to be taken into account (Tuschl and Mueller, 2006).

Mostly the cells lose some of their characteristics and morphologies when taken out of their own microenvironments and transferred to the culture environment. The cells in the culture environment are observed morphologically in three types being epithelioid and fibroblastic (monolayer adherent culture) and lymphoblastic (suspension culture). The cells in lymphoblastic morphology are spherical shaped and do not need to hold on to the surface and therefore are mostly reproduced in suspension cultures (Fig. 17.3A1 and A2). Cells in the epithelioid morphology have a settlement form similar to epithelial layer but do not display a complete layer characteristic. Mostly they grow in separate groups sticking to a surface. They are cells that have round lines with more regular polygonal shapes. Some sources mention cells that have epithelial type morphology unlike epithelioid type. Visually these cells display similar features to epithelioid type, but dissimilarity in these cells reproducing tends to form a continuous layer (Fig. 17.3A3). Cells in fibroblastic morphology usually draw attention with their irregular shapes. These cells are thin, elongated bipolar or multipolar cells that are in contact via various connection complexes. Also, these cells show the need to hold on to a surface to reproduce (Fig. 17.3A4). Except these, neurons and some special cell lines might have a peculiar morphology (Sinha, 2008; Turksen, 2013; Zhou and Kantardjieff, 2014).

17.3.7 Cells

When the cells that are isolated from human or animal tissue or organs are cultured in appropriate conditions, behavior of the cells like an independent organism can be watched. These cells display the characteristic of the tissue which they are originated from. Cells belonging to each tissue have morphological and functional characteristics. Fundamentally, origin of these cells can be counted as epithelial tissue, connective tissue, muscle tissue, and nerve tissue (Butler, 2003).

Epithelial tissue shows characteristics that surrounds the organs and lines the cavities. Epithelial cells can show single layer or an array that is multilayered. In both cases, they connect with each other via intercellular connections. Epithelial tissue clings to the connective tissue that is under again with special cellular connections. Epithelial tissue has merely intercellular space. However, connective tissue cells have a distinct intercellular space and the extracellular matrix that fills this space. In different connective tissue types,

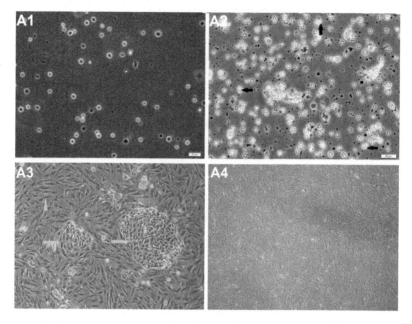

FIGURE 17.3 Lymphoblastic (suspension culture), epithelioid and fibroblastic (monolayer adherent culture) cells in culture. (A1 and A2) Lymphoblastic morphology of T cells in culture. (A1) Inactivated T cells isolated from rat spleen are showing by phase-contrast microscopy. T cells in lymphoblastic morphology are spherical shaped in suspension cultures. A2: Underactivated by concanavalin A for 24 h for maintenance culture of T cells, the morphologies and shapes of the cells changed, and they presented extensions (red arrows). Blue arrows are showing us proliferated T-cells. A3: Epithelioid and fibroblastic morphology of glandular and stromal cells in culture. Phase-contrast microscopy is showing epithelioid glandular (arrow) and fibroblastic stromal (arrow head) cells isolated from human endometrium. Endometrial glandular cells have round lines with more regular polygonal shapes. A4: Phase-contrast microscopy is showing fibroblastic morphology of human dental pulp−derived mesenchymal stem cells (hDP-MSCs) in culture. These cells are thin, elongated bipolar or multipolar. Scale bars: A1: 20 μm, A2: 50 μm, A3: original magnification × 40 and A4: × 40. *Source: Courtesy Dr Sariboyaci.*

cell types both show changes and the structure and characteristics of extracellular matrix changes. Blood tissue, cartilage, adipose tissue, bone tissue, hematopoietic tissue, and lymphoid tissue are specialized forms of connective tissue. Unlike cells in other tissues, muscle tissue cells have the characteristics of contractibility with special proteins and arrangement it possesses. These cells have a different morphologic structure by force of their special functions. Cells that extend to each other in parallel mostly seem to be elongated and form bundles. Accordingly, their nucleus appears to be elongated as well. Nerve tissue basically forms of nerve cells that are called neurons and support cells. Neurons are specialized cells that can constitute and transmit electrical stimuli. Besides neurons form a special connection called synapse with other neurons. Support cells that take place in the nerve system are called as neuroglia and conduct rather important functions that ensure the continuity of the system (Ross and Pawlina, 2014).

The cells that have been obtained at the end of the primary culture and passages are the fundamental sources of the cells that are used in the cell culture. These cells need to have some characteristics. For instance, the cells that we will use should not carry any chromosomal anomaly, must have a limited life span, and should not carry the potential of forming cancer. Cells that are obtained with cell lines are the cells that are formed as suitable to our purposes and have the ability to be continuously divided. Sometimes isolated from a cancer tissue, sometimes intervened through genetically changes, these cells display different characteristics in this regard (Sinha, 2008; Ulrich and Negraes, 2016).

17.3.8 Stem Cells

Zygote that is formed with the fertilization of an oogonium by sperm is the first form of the organism. While, on the one hand, it ensures the numerical increase with the cell divisions which the zygote passes through, on the other hand, these cells differentiate bring about specialized cell types. Nevertheless, need of the cells to reproduce and differentiate lasts not only just during the embryonic development but for a lifetime of an adult organism. With the aim of being able to continue its vitality, the organism constantly replaces new ones instead of the cells that are lost for different reasons such as cellular aging, damage, or injury. Stem cells can be defined as the cells that embody the characteristics of renewing itself and differentiation and that can be colonized. Here with the characteristic of self-renewal, it is stated that at least one of the cells that are brought into being after the stem cell has been divided carry the characteristics of the stem cell. Self-renewal feature explains both as a result of mitosis the ability to reproduce in a certain number and the ability to protect stem cell pool with asymmetrical division. Differentiation is the ability to form different cell types of the stem cells and is defined with the concept of potency. For example, except placental structures, embryonic stem cells (Fig. 17.2C) can differentiate to all the cells that form the organism and therefore are called as pluripotent. Stem cells that are settled in the organs and differentiate to only one or two cell types are named as unipotent and multipotent stem cells, respectively. Colony features are defined as the capacity of stem cells to form new stem cells. Features that can form a colony are the cells that can be cultured successfully (Lanza et al., 2009; Yilmaz et al., 2016).

In mammals, the fertilized oocyte, zygote, 2-cell, 4-cell, 8-cell, and morula resulting from cleavage of the early embryo are examples of totipotent cells (ability to form a complete organism). The inner cell mass (ICM) of the 5- to 6-day-old human blastocyst (Fig. 17.2D) is the source of pluripotent human embryonic stem cells. During embryonic development, the ICM develops into two distinct cell layers, the epiblast and the hypoblast. The hypoblast forms the yolk sac that later becomes redundant in the human, and the epiblast differentiates into the three primordial germ layers (ectoderm, mesoderm, and endoderm). Pluripotent embryonic stem cells can give rise to many cell types in vitro, including cells specific to endodermal tissues. Advances in the understanding as to how ES cells differentiate should provide answers for reprogramming of stem cells from adult tissues.

Classification of stem cells according to different features is possible. But none of these can provide a complete classification. On the other hand, generally, it is an appropriate approach to classify stem cells according to the source they are taken as being embryo-derived stem cells and adult stem cells and according to the potency features as pluripotent, multipotent, and unipotent. Just as embryo-derived stem cells can be pluripotent stem cells that are derived from ICM of blastocyst (Fig. 17.2D) or epiblast, they can be multipotent stem cells that are trophoblast or neural crest sourced. While recently the only known source of pluripotent stem cells is trophoblast, ICM and somatic cells that have been fully differentiated recently have been reprogrammed with the influence of some of the transcription factors and produced pluripotent cells called as induced pluripotent stem cells. Pluripotent stem cells have great importance in terms of regenerative medicine with high differentiation capacity they have. Stem cells that are in different organ and tissues of an organism that has completed its development at a certain level are named as adult stem cells, and mostly these cells have multipotent or unipotent character with lessened potency. Among these come especially mesenchymal stem cells. Nearly all tissues such as epithelial tissue, digestive system, musculoskeletal system, nerve system, reproductive system, and heart and vascular system have stem cells liable to ensure their renewal. Whereas the feature of a cell's high potency also brings the possibility of transforming into a cancer cell, mesenchymal stem cells do not carry this risk (Bongso and Lee, 2011; Przyborski, 2016).

Nowadays the most frequently used cells in the culture of the specialized cells are the stem cells. Stem cell having very special settlements in the organism reveals the necessity of these cells' culture conditions being very special as well because stem cells are found inside of a rather special microenvironment called niche mostly in a silent manner and the signals they receive from here are quite important in terms of determining the fate of the cells. Getting out of the silent state with signals that come from the microenvironment, a stem cell can go to the cell division again; these cells that are divided as a result of the interaction with this microenvironment can differentiate. In this respect, to know well the conditions that must be provided for the cell that is cultured in the stem cell culture is quite significant. For instance, in mouse embryonic stem cell culture, to prevent these cells from differentiating, to plant/cultivate these cells on mouse embryonic fibroblasts, and to add Leukemia Inhibitory Factor (LIF) into the culture environment are required. However, in a suspension culture where this is not provided and grasping feature is prevented, mouse embryonic stem cells will differentiate in a short time and form embryoid body structures that are seen equal to a 6-day-old embryo (Freshney et al., 2007; Fauza and Bani, 2016).

17.4 METHODS

17.4.1 Growth and Maintenance of Cells in Culture

A successful cell culture study can be possible with providing some preconditions and aseptic working conditions. In this section, the methods that will be handled should start with providing aseptic conditions and continue with preparing the cell growth mediums and monitoring of the cells in the culture environment and feeding them.

Providing aseptic conditions The purpose is to provide equipment, material, technique, and environment that will prevent microorganisms such as various bacteria, fungus, and cross contamination which might occur with other cells (Freshney, 2015).

- Before the studies, cabin should be sterilized with 70% alcohol.
- Sterile gloves should be used. After wearing the gloves if possible using 70% alcohol again sterilization should be done before starting to work in the cabin.
- Unnecessary material should not be found in the cabin, it should be noted that materials that have been placed do not prevent air flow.
- All the materials that will be used for the study should be taken after their exterior surface being sterilized with 70% alcohol a short time ago before the study.
- Hands must be held within the cabin during the study; on situations when they need to be taken out it must be very quick; it must be avoided to touch any nonsterile material. Hands must absolutely be disinfected with 70% alcohol before are put in the cabin again.
- During the studies, any kind of material such as medium bottle, box of pipette tips should be wiped with 70% alcohol before taking in the cabin.
- After the studies in the cabin are over, unused equipment, wastes, and materials should be moved away. Afterwards, cabin surfaces must be sprayed with 70% alcohol and dried with the help of a sterile swab.
- Regular disinfection of the cabin and the working environment must be provided at regular intervals. With this aim, the use of disinfectants developed specially with the features that will be used in the cell culture laboratory is appropriate.

Preparation of the mediums Owing to technological advancements, mediums that are needed by different cell types can be acquired easily as ready commercial forms and when there is a need they can be manipulated with adding of different substances (Pollard and Walker, 1997; Mather, 2013).

The culture medium is the most important component of the culture environment, because it provides the necessary nutrients, growth factors, and hormones for cell growth, as well as regulating the pH and the osmotic pressure of the culture. Although initial cell culture experiments were performed using natural media obtained from tissue extracts and body fluids, the need for standardization and media quality, as well as an increased demand led to the development of chemically defined media. The three basic classes of media are basal media, reduced-serum media, and serum-free media (SFM), which differ in their requirement for supplementation with serum.

Serum is vitally important as a source of growth and adhesion factors, hormones, lipids, and minerals for the culture of cells in basal media. In addition, serum also regulates cell membrane permeability and serves as a carrier for lipids, enzymes, micronutrients, and trace elements into the cell. However, using serum in media has a number of disadvantages including high cost, problems with standardization, specificity, and variability, and unwanted effects such as stimulation or inhibition of growth and/or cellular function on certain cell cultures. If the serum is not obtained from reputable source, contamination can also pose a serious threat to successful cell culture experiments.

Basal media The majority of cell lines grow well in basal media, which contain amino acids, vitamins, inorganic salts, and a carbon source such as glucose, but these basal media formulations must be further supplemented with serum.

Reduced-serum media: Another strategy to reduce the undesired effects of serum in cell culture experiments is to use reduced-serum media. Reduced-serum media are basal media formulations enriched with nutrients and animal-derived or recombinant factors, which reduce the amount of serum that is needed.

SFM SFM circumvents issues with animal sera by replacing the serum with appropriate nutritional and hormonal formulations. SFM formulations exist for many primary cultures and cell lines, including recombinant protein producing lines of Chinese Hamster Ovary, various hybridoma cell lines, the insect lines Sf9 and Sf21 (*Spodoptera frugiperda*), and for cell lines that act as hosts for viral production, such as 293, VERO, MDCK, MDBK, and others. One of the major advantages of using SFM is the ability to make the medium selective for specific cell types by choosing the appropriate combination of growth factors (Zhou and Kantardjieff, 2014).

Monitoring the cells in culture and feeding: After the cells are derived from primary culture and a cell line, they must be monitored daily through eye and with the help of an inverted microscope. During these observations, evaluating color change that points out to the change in the pH of the environment, the cells' state of clinging onto the culture container, the cover area of cells and cell morphology it must be given a decision about the course of the culture (for example, necessary interventions such as medium change, passaging, or ending study because of contamination). In the medium change that we can also name feeding of the cells, medium is pulled with the help of a sterile pipette and emptied in a waste container. Evenly fresh medium being pulled with the help of a sterile pipette adds to the culture vessel.

Cell count and viability detection: Especially during the passaging, before the freezing and after the thawing, the count of the cells that we have cultured and the detection of the living cells are quite significant. With this aim, cells in a cell suspension that are of quite a small volume are counted with the help of a microscope and the result out of this proportioning of the number of cell in the culture environment is acquired. A dye such as trypan blue that is used during the count helps us to detect the rate of the living cells (Fig. 17.4A and B). Normally living cells do not take trypan blue into the cell and appear in bright yellow. However, nonliving cells pass the trypan blue to their cytoplasm and are colored in blue. During this count, a kind of slide named as hemocytometer is used. There are different hemocytometer types and out of the living cell number counted in the unit area, the cell number in the culture environment is acquired from these with the help of different formulas. For example; the cell count is done with Thoma slide (Fig. 17.4C) in this way:

- Cells that are lifted with trypsin are washed, centrifuged, after the supernatant is dropped, and 1 mL medium is added to the remaining pellet (These steps are explained in the subculture section 17.4.2).
- Cell suspension is derived with pippetage. 100 µL is drawn from this suspension and equal volume of trypan blue is added.
- On the other hand, Thoma slide is placed to a flat ground and lamella is placed in a way to cover the counting area.
- 50 µL of the mixture that we acquire added to slowly in the counting area.
- In the meantime, it must be drawn attention not to form air bubble and the content to spread to the counting area completely.

- Thoma slide is placed to the microscope. First of all, counting area is found in the 10× lens, the image is made clear, and the counting area is accurately centralized. In the image here, 16 large squares and 25 small squares in each square are observed.
- The volume of this counting area is 0.1 mm^3.
- Living cells in this area are counted in the 16 large squares (different sources count the cells in a few counting areas and take their average).
- Total cell number = counted cell number × dilution rate × standard deviation (10.000).

17.4.2 Subculturing Adherent Cells

Adherent cells proliferate by dividing until the cells cover the whole surface of the culture vessel. There are two significant factors that limit the reproduction of the cells. The one is the free space in the culture vessel and the second is the decrease in the nutrition support that the medium provides. Usually adherent cells filling 70%–80% of the culture vessel show that time has come for the passaging (Fig. 17.3A4). However, time that has passed since the last passaging and the necessity of the other procedures to be implemented are the criteria that are taken into consideration.

Basically passaging of the adherent cells include the incidences of moving away of the old medium, lifting the cells and after the count of the cells transferring to the new culture vessels that contain fresh medium in appropriate rates.

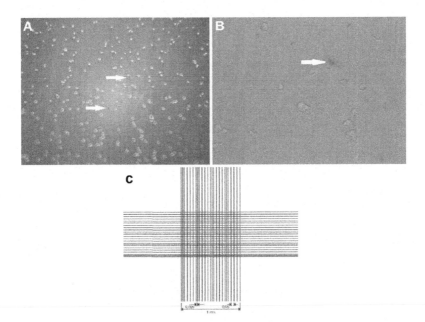

FIGURE 17.4 (A and B) Phase-contrast microscopy is showing cell count with trypan blue on Thoma slide. Normally living cells appear in bright yellow. Nonliving cells are colored in blue (arrow). Original magnification A: ×10; B: ×20. C: Illustration is showing cell counting chamber Thoma pattern. *Photo by courtesy of Dr Sariboyaci.*

- Medium is moved away by withdrawing with the help of a sterile pipette.
- Minimum 120 µL/cm^2 amount of Ca/Mg-free PBS is added to the culture vessel, then moving away with the pipette so washing is made. This process is repeated two times.
- Minimum 120 µL/cm^2 amount of 0.25% g/mL Trypsin EDTA solution is added. After being sure that the surface has been completely covered, lifting to the incubator, 3 minutes are waited (waiting period might show difference between the cell types).
- Whether the cells stand (Fig. 17.5B) or not (Fig. 17.5A) is controlled in the microscope. In case of not standing, incubation period might be repeated at short intervals.
- After the cells stand, it is added to minimum 120 µL/cm^2 culture medium (in ratio of 1:1, Trypsin EDTA:culture medium), pippetage is done and reaction is terminated.
- The content of the culture container is transferred in to a Falcon tube, centrifuged, 1 mL of medium is added to the mixture, and homogenized cell suspension is obtained with pippetage.
- It is counted with the trypan blue of the living cells and total cell number is achieved (this method is explained in the growth and maintenance of cells).
- As for in which culture vessel the study will continue, in number of the cells is determined according to the surface area. Cells also is cultivated with the medium that is in the amount of culture container requires (for example; T25 flask surface area being 2500 mm^2, the cultivate of 1.7×10^5 mesenchymal stem cells with 3.5 mL medium will be appropriate) (Eker Sariboyaci et al., 2014).

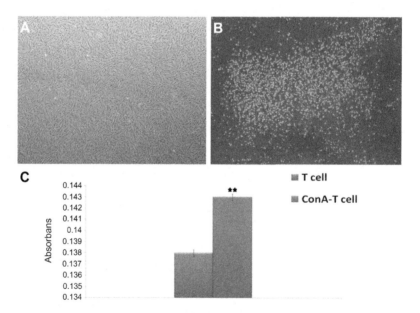

FIGURE 17.5   (A and B) Phase-contrast microscopy is showing whether cells stand (B, after trypsinization) or not (A, before trypsinization) (original magnification ×10). (C) WST-1 analyses of T-cells. Increased levels of proliferation of activated T cells by Concanavalin A (Con A-T cells) was determined by WST-1 ($n = 3$, mean ± SD, ** $P < 0.01$). *Photo by courtesy of Dr. Sariboyaci.*

17.4.3 Subculturing Suspension Cells

Passaging of the cells that is cultured in suspension varies. However; similarly the criteria of the passage needs can be put in order as cell concentration, pH of the environment, period that has passed since the last passaging, and special conditions that the study brings. Here as the equivalent of the 80% confluent concept in the adherent cultures, concentration of 10^6 cells/mL is accepted. In a suspension culture (Fig. 17.3A1 and 2), cells can be found separately or in clusters. For the suspension cultures in which the cells are found in clusters to be passaged, firstly these piles being distributed need to be changed into single cell suspension (Mather, 2013; Ulrich and Negraes, 2016).

- Cell suspension is mixed, if there are cell clusters scattering with the pippetage single cell suspension is acquired.
- From this suspension, cell count is made.
- Cell in appropriate amount is diluted into a new culture vessel with sufficient amount of medium.

17.4.4 Viability and Proliferation Assay of Cultured Cells

Cell viability tests are carried out after a traumatic procedure such as cell freezing, thawing, and cell separation or with the aim of searching the cytotoxic influences of an agent that is implemented. These tests have been developed in relation to different cellular mechanisms.

In tests that are measured by the vitality of cells not taking the dye inside of the cell, basic mechanism is that living cells do not take some dyes into the cell; however, nonliving cells whose membrane is disintegrated is permeable against the dyes. Among these dyes the most frequently used ones are trypan blue (Fig. 17.4A and B) and propidium iodide.

- First of all, in compliance with the culture type, a cell suspension is prepared (1×10^6 cells/mL).
- 1 unit cell suspension is mixed with 1 unit dye solution (1:1).
- The mixture that is loaded to the hemocytometer is soaked about several seconds.
- As living cells do not take the color in the optical microscope, they are observed in bright yellow; dead cells are observed in the color of the dye they take in (Fig. 17.4A and B).
- Here as described in the cell count, the number of the living cells can be counted and proportioning can be made about the vitality of the cells in the culture.

The basic mechanism in the tests that is measured by the cells taking the dye in the cell is based on the fact that living cells let diacetyl fluorescein in and hydrolyze. Fluorescein that is formed as a result of this reaction cannot go outside of the cell membrane. Therefore, while the living cells are dyed with fluorescein green, dead cells are not dyed. Here, for the dead cells to be dyed, propidium iodide can be added as a second dye. In this case, dead cells color red fluorescent. This method is only appropriate to the use with fluorescent microscope (Mather, 2013; Przyborski, 2016).

- Cell suspension is prepared through a medium that do not contain phenol red and suitably to the culture type.
- Diacetyl fluorescein dye at the rate of 1:10 is added to the cell suspension in a way that total volume will be 1 μL.
- Cells are incubated for 10 minutes at 37°C.
- Putting a drop of mixture on it, slide is closed with lamella and examined with the light that is appropriate in wave range through fluorescent microscope.
- The rate of the living cells is calculated.

Metabolic cytotoxicity and proliferation tests are based on the fact that living cells metabolize substances such as MTT, XTT, and WST1 (Fig. 17.5C) in their mitochondria with dehydrogenases and form the colored components named as formazan crystals. Measuring the color intensity with spectrophotometer, the rate of the living cells can be calculated. In this method the first molecule being used is MTT and formazan crystals that are formed here does not have a water-soluble characteristic and requires additional procedures. On the other hand, in later on developed XTT method, crystals that are formed have a water-soluble characteristic and are more reliable compounds. However, WST1 test is the most suitable and reliable method among these (Cetinalp Demircan et al., 2011; Eker Sariboyaci et al., 2014). Today these tests' easily applicable kits are available and they need different protocols. For this reason, with the purpose of understanding the general mechanism of this practice, manual MTT application will be explained as an example.

- The cells are lifted according to the techniques previously told. Then they are centrifuged and thrown supernatant. A 100 μL medium is added to the acquired pellet and pippetage is carried out. This cell suspension will be used in the experiments.
- Preparation of MTT stock solution: 0.5 mg/mL MTT is thawed in the study medium.
- Dimethyl sulfoxide (DMSO)/ethanol solution is prepared at the rate of 1:1 (these compounds are necessary for the formazan crystals to become soluble).
- According to the characteristics of the study, a design is given to the multiwell microplates. Here basically saline, medium, cell, and control wells should take place.
- Saline wells are filled with saline.
- 100 μL studied medium is put to the medium wells.
- 100 μL cell suspensions are put to the cell wells. If the toxic effects of an agent are to be searched, cell wells that contain different drug doses can be formed. (before the study that will be carried out, the cells that will be put into the cell and control wells are removed according to the methods that are explained previously, centrifuged, and then cell suspension is obtained).
- Only 100 μL supernatant is put to the control well.
- 20 μL MTT solution is transferred to the control and cell wells. It should be protected from light after this stage. Then it is held in the incubator for 3—4 hours.
- At the end of this period, all the wells are aspired and 100 μL DMSO/ethanol solution is added; it is held at the room temperature for 3—4 hours.
- Measurement is carried out with spectrophotometer. On interpreting the values that are read between the wells, proliferation states of the cells are interpreted.

17.4.5 Freezing Cells (Cryopreservation)

Cellular aging and genetic changes are observed in relation to the increasing passage numbers in cell culture studies. Also in some situations, carcinogenic transformations and phenotypic changes might happen. In addition to this, contamination risk with different agents possesses an important threat. However, with both economical and scientific reasons, these cells might be needed again in different studies. All these reasons reveal the need for the cells to be frozen and stored (cryopreservation).

Cells that have reached approximately 3rd to 50th passage can be frozen if they display a sufficient growth cycle. The most important point in the freezing process is the crystallization that intracellular water generates during the freezing. That is why freezing process must be slow so that sufficient time must be provided for the water to leave the cell. Also water must be separated with a hydrophilic cryoprotectant, and cells must be saved at the lowest temperature possible so as to prevent protein denaturation (Freshney, 2015). On the other hand, during the freezing cell concentration (optimum 3×10^6 cells/mL) being high will increase the rates of success. As the freezing medium, it is appropriate to use mediums that contain DMSO at the rates of 5%. Except these, different cryoprotectant agents (often used Glycerol) can be used. Freezing conditions in which the rate of cooling is about 1°C per minute until -80°C has high rates of success (Karaoz et al., 2010).

- Cells are prepared to obtain cell suspension according to the culture method. In the milliliter of this cell suspension, there must be about $2-5 \times -10^6$ cells. 35% fetal bovine serum (FBS) (350 μL), 60% basal medium (600 μL), and 5% DMSO (50 μL) should be used to freeze the cells in the final rate of freezing medium (1 mL for a cryo vial).
- 35% FBS (350 μL) that will be added in the 60% basal medium (600 μL; DMEM, RPMI, and MEM) is ready for use and stored at $+4$°C until use.
- Cell pellet (3×10^6 cells) is quickly added with mixture of FBS and basal medium (950 μL) with pipetting. 950 μL of this mixture (cells, FBS, and basal medium) is taken into a cryo vial. Immediately 5% DMSO (50 μL) is added drop by drop without shaking and without pipetting. All steps must be acted quickly because staying in room temperature for a long time with DMSO will cause very toxic impacts for the cells.
- Cryo vial should be keeping in -80°C overnight cooling so as to lose 1°C per minute. There are some containers or methods providing cooling about 1°C per minute at -80°C.
- Cryo vial that cools down until -80°C must be taken in liquid nitrogen tank or special cooler cabin (-150°C) after 80 minutes the next morning.

17.4.6 Thawing Frozen Cells

Cells that are frozen and saved can again be planted/cultivated and used after thawing if required (Helgason and Miller, 2013).

- Firstly, necessary materials and water bath at 37°C are prepared. Then cryo vial is taken out and put into the water bath.
- After the thawing process has been carried out, the tube that is wiped with 70% alcohol is taken into the cabin and tube content is emptied in a culture vessel with the help of a pipette.

- Then about 30–50 mL medium is added slowly onto the cell suspension. Thus, both cells and the cryoprotectant agent are diluted. In many sources, it is emphasized that cryoprotectant agent needs to be moved away from the environment with centrifuge. In this situation, cell suspension is transferred to the centrifuge tube, about 30–50 mL medium is added slowly on it. With the supernatant that is thrown after the centrifuge, cryoprotectant agent is moved away, too. Then medium is added on the pellet, and planting/cultivating is performed obtaining cell suspension.

17.4.7 Transplantation of Cultured Cells

In the face of damage there might be two situations in the tissues and the organs. In the former, the damage is repaired with a cellular matrix and connective tissue. This situation provides a quick repair process, but as it causes deterioration in the structure, it brings the function loss as well. Secondly, in the tissues a repair process might happen according to the structure of the organ called as regeneration. Regenerative medicine concept comes from here as well (Gurtner et al., 2007).

Today advancements in the cell culture studies have led to the development of regenerative medicine concept. Especially tissue engineering, stem cells, and cloning studies have become important application fields of the regenerative medicine. Tissue engineering studies include the processes of the reproduction of autologous cells that are taken from the damaged organ in the culture environment and later on again the transplantation to the patient. In some cases, transplantation can be done with the tissue scaffolds that will comply with the structural integrity of the organs. In some other cases, to obtain cell from the patient's own organ might not be possible. In this situation, different stem cell types are used as an alternative source for the transplantation (Atala, 2006).

First cell transplantation from human to human was performed through blood transfusion in 1818. Then, with the researches and advancements in especially the cell culture field, cornea transplantation that is the first tissue transplantation was achieved by Zirm in 1905. After finding the aseptic techniques and implementing on cell culture studies, first successful bone marrow transplant was performed in 1956 (Orlando et al., 2011). In this respect, regenerative medicine can also be considered as the evolution of organ transplantation. One of its fundamental purposes is to bring back the deteriorated functions. Cell transplantation has great importance in the regenerative medicine beside the tools such as developmental biology, genetics, molecular biology, tissue engineering, and stem cell biology (Daar and Greenwood, 2007; Fauza and Bani, 2016).

17.4.8 Differentiation of Cells

The process of the cells that are formed out of a common progenitor cell being specialized and gaining different structure and function is called as differentiation. Briefly, the differentiation can be defined as gaining the phenotypic characteristics that an adult cell has functionally. While differentiation can sometimes happen irreversibly, sometimes gained characteristics might be lost. Whereas dedifferentiation states the loss of the characteristics that has been gained with differentiation in the culture environment or processes such as becoming cancerous, terminal differentiation defines the situation in which a cell displays all the characteristics in the related phenotype (Sinha, 2008).

Differentiation can happen through various ways in in vivo conditions. Stem cells that take place in the structures such as mucosa and blood tissue form the initial cells primarily. These become terminal differentiated, mature, and no longer dividing cells. However, in the cells that do not have rapid turnover in the tissues, very little differentiation is observed. Generally, during the differentiation cell division stop. Some cancer cells can continue dividing while differentiating on the one hand.

Markers that are expressed by the cells at the early period and continue their expression in the ongoing development steps are named as lineage markers. On the other hand, the markers that are the indicators of terminal differentiation and mature phenotype are mostly special cellular enzymes or products. These usually begin to be produced at the late stages. While from this point of view identity of the cell is determined by lineage-specific markers; differentiation states are detected by the definition of one or a few special cellular markers (Fig. 17.6). For example; for epithelial cells, cytokeratin is a lineage marker. But transglutaminase that is detected in the keratinocyte which is a differentiated epithelial cell is the indicator of the differentiation (Freshney, 2015).

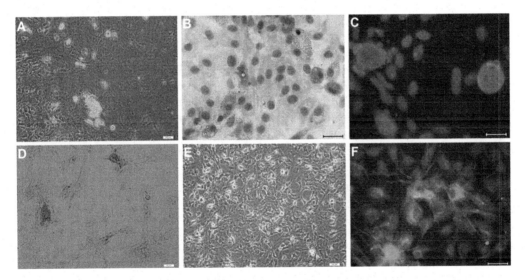

FIGURE 17.6 Photomicrographs of the in vitro differentiation of rat pancreatic islet–derived stem cells (rPI-SCs) cultured in differentiation-inducing media. (A) The phase-contrast microscope appearance of rPI-SC differentiated into an adipogenic lineage after 20 days of incubation. (B) Phase-contrast microscopy showing adipogenic differentiation was marked visually by an accumulation of neutral lipid vacuoles in cultures after 25 days (Oil Red O; ORO staining). (C) Fluorescence imaging of cells ORO-staining (red) and vimentin-labeled (green). (D) Osteogenic differentiation of rPI-SC 34 days after osteogenic induction. Mineral nodules were stained positively with Alizarin red S staining. (E) Differentiation of rPI-SC into neuronal lineages on 2 days. Under phase-contrast microscopy, SC-derived neuron-like cells displayed distinct morphologies, ranging from extensively simple bipolar to large, branched multipolar cells. (F) In vitro differentiation of rPI-SC into neuron-like cells was shown by β-tubulin (red) and β3-tubulin (green). Scale bars: (A and E) 100 μm; (D) 50 μm and (B, C, and F) 30 μm. *Source: Courtesy Dr Sariboyaci.*

Knowing the characteristics and factors of cellular differentiation enables the cells that are studied in the cell culture to be easily manipulated. However, firstly it must be decided that the cultured cells wanted to be differentiated or only proliferated without differentiation. If differentiation is to be wanted, these general approaches must be practiced; right cell type is chosen and cultured, cells are reproduced until to a certain cellular intensity. With this aim, to prevent for the cells to differentiate at this stage, it is necessary to provide suitable adherence surfaces and growth factors. Later, the cells begin to be cultured not with normal medium but the differentiated medium. These mediums need to be arranged so that they contain basic factors that cells need to differentiate. Also cytokines, hormones, and chemicals must be added to these mediums. For instance; after mouse embryonic stem cells have been cultured (Fig. 17.2C), if it is wanted to generate endodermal cells, activin must be added to the culture environment. However; for further differentiation, for example, so as to form hepatic progenitor from these endodermal cells, molecules such as FGF and RA are required. But if we want these mouse embryonic stem cells to reproduce without differentiating, this time there should be embryonic fibroblasts in the culture environment as well, and LIF must be added to the culture (Saitama, 2007; Lanza et al., 2009).

Cell differentiation is a common situation that is provided by internal cellular program and environmental conditions. At the top of the environmental conditions come connection complexes that play a part in the intercellular communication and paracrine hormones, cell–matrix interaction, cell shape, and polarization and physico-chemical factors which the environment provides. In some cases, deteriorations that occur under the control of factors which are both internal and provided by the external environment might cause the development of some pathology such as cancer. Many tumors are observed to form cells which are very little differentiated. In this respect, cancer can also be defined as the deterioration of the cells' normal differentiation processes. There is an inverse relationship between displaying cancer-related characteristics and displaying characteristics belonging to the differentiation (Bosch, 2008; Bongso and Lee, 2011).

17.4.9 Characterization of Cells

One of the most important purposes in characterizing the cells with various methods is to be able to detect whether there is cross contamination or not. Wrong labeling or describing the used cell lines or the possibility of being cross contamination with another cell line should always be considered. Such a case possesses the risk of obtaining wrong results in the studies in which especially molecular mechanisms are searched. In addition to this, whether there are malignant changes as a result of genetic changes in the cells which are being used or because of different reasons should be followed with cell characterization studies as well. Also, determining where the isolated cells are in reality with the detection of the characteristics they carry is of great importance. Beside, characterization studies are needed for the detection of various viral and mycoplasma contaminations.

The methods that are used in cell characterization are mostly determined by the features of the study. For example, in a study that uses molecular biology techniques, cell characterization is mostly done with DNA profile or gene expression analyses. However, in cell characterization only one method is not enough. Use of various methods that deal with a few different characteristics and verification of the information that is gathered is preferred. The first step in the cell characterization is to examine the cells with the help of an inverted, phase contrast, or confocal microscopes and to determine the morphology. Another basic characterization tool is the karyotyping study. Karyotyping study is the best criteria for the type separation. Genetic stabilities of embryonic stem cells must be routinely followed with karyotype analyses. Also, karyotype analyses are necessary to be able to eliminate cross contamination. DNA analyses studies are important in terms of gathering cell type—specific information. In addition to this, isoenzyme studies can be conducted as well (Freshney, 2015). Cells belonging to a certain linage can be proved by the determination of surface antigen, intermediate filaments, special proteins, and enzymes (Fig. 17.7) (Karaoz et al., 2009; Turksen, 2013).

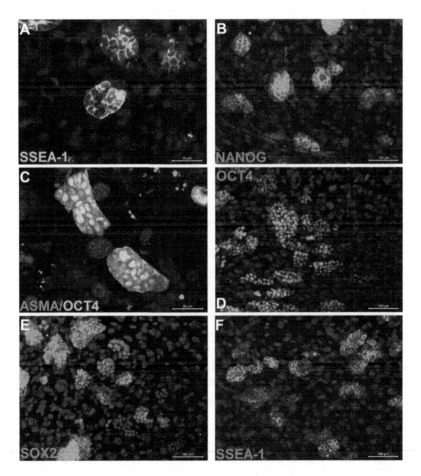

FIGURE 17.7   Characterization of undifferentiated mouse ES cells in culture. Fluorescence microscopy analysis of the expression of pluripotent markers immune-labeling against SSEA-1, NANOG, ASMA, OCT4, and SOX2 was positive. ASMA and OCT4 antibodies were merged in C. Nuclei were stained with DAPI (blue). Scale bars: (B, D, E, and F) 100 μm and (A and C) 50 μm. *Photo by courtesy of Dr. Sariboyaci.*

17.4.10 Apoptosis, Necrosis, Senescence, and Quiescence

Apoptosis is defined as the death that occurs during the normal development in the cells. It is a system that completes its duty and in which cells that are not needed are eliminated and arranged with mechanisms that are controlled genetically. Apoptosis is different in many ways from necrosis which is a classic form of death. Necrosis is not a physiological death. On the other hand, apoptosis happens in both physiological and pathological conditions. In necrosis, the cell swells, cell membrane loses its integrity, intracellular material goes outside of the cell, the cell undergoes lysis, and inflammation is stimulated. In the apoptosis, the cell gets smaller, the structure of membrane protects, the cell is disintegrated into small bodies, but inflammation does not happen (Hengartner, 2000; Sinha, 2008; Przyborski, 2016).

The methods that are used in determining the apoptosis are as follows:

- Morphological imaging methods: It can be examined in optical microscope with hematoxylin staining. As the chromatin is dyed, it is decided according to the nucleus changes. In the fluorescent microscope, because fluorescent dyes (Hoechst dye, propidium iodide etc.) are connected to DNA, nucleus of the cell becomes visible. Although Hoechst dye stains all the cells whether they are living or dead, propidium iodide only stains the dead cells. With this method, the viability of the cells is determined. However, for necrosis and apoptosis separation, nucleus morphology should be examined. With phase contrast microscope, buds that are formed on the apoptotic cells can be observed. Evaluation with the electron microscope is the most valuable method. Morphological changes can be detected easily and distinctly.
- Histochemical methods: Annexin V is the cell membrane that goes to the apoptosis; it is observed that phosphatidylserines that settle normally in the inner surface are observed to go into the outer surface. These molecules become visible using Annexin V that is marked with a fluorescent substance such as FITC. Thus, apoptotic cells are determined. TUNEL method: 3 O'H tips of the broken DNA pieces are in situ marked with dUTP using terminal deoxynucleotidyl transferase enzyme. Apoptotic cells are stained and become visible. M30 method: It is the staining of antigenic area in the apoptotic cells with immunohistochemical methods. Caspase 3 method: Whether the tissue expresses caspase 3 or whether the agent that leads to apoptosis break caspase 3 needs to be known (Fig. 17.8). However, if they are known, apoptotic cells can be detected.
- Biochemical methods: DNA fragments can be shown with agarose gel electrophoresis. This is characteristic for apoptosis. This situation does not apply for the necrosis. With western blot method, it can be detected that whether some proteins belonging to the apoptosis are expressed or fragmented or not. With flow cytometry method using antibodies marked with fluorescent matter, determining any cell surface protein known to be expressed in the apoptosis is possible.
- Immunological methods: With ELISA method, DNA fragmentation can be detected. Also, the levels of M30 can be measured. Fluorimetric method is a method that is used in appointing the caspase activity in the culture cells.
- Molecular biological methods: With microarray method, gene expressions can be searched but it is an expensive method. Comet method: It is a micro-electrophoresis method that is used aimed at showing DNA damage.

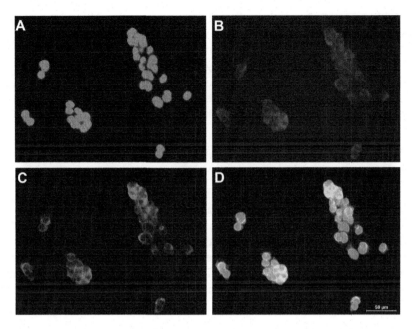

FIGURE 17.8 Apoptosis of T cells activated by concanavalin A detected by active caspase 3 immuno-labeling on Day 4 in transwells by when the cells were co-cultured with rat pancreatic islet−derived stem cells. Fluorescence microscopy analysis of the expression of T cell and apoptotic markers immune-labeling against CD5 (B: green) and active caspase 3 (C: red), respectively, was positive. CD5 (green) and active caspase 3 (red) antibodies were merged in (D). Nuclei were stained with DAPI (A: blue). Scale bars: 50 μm. *Photo by courtesy of Dr. Sariboyaci.*

17.4.11 Senescence

Senescence is the sum of the mistakes of the cell in the biosynthesis mechanisms. Senescence is generally an irreversible phenomenon. Eukaryotic cell cycle consists of countless repetitions and various processes. While in the majority of our lives the capacity of the cells to renew themselves in tissues such as blood and skin is high, this situation is rare in the nerve tissue. Dividing and dying of the cells are within balance. In the aged cells, increased fraction decreases to zero and increase is not observed. Leonard Hayflick showed that the normal human fibroblasts lose their development and division skills at the end of the 50th cell division (Campisi and di Fagagna, 2007).

The aged cells do not make any more cells while staying active metabolically (at the end of the proliferation or "The Limit of Hayflick") then eventually die. The cells that are taken from the young humans show more division than the old humans in the culture environment. While the human embryo cells begin to age after dividing about 60−80 times, the cells that are taken from the middle aged humans begin to age after about 10−20 division. It has been shown that this situation also might be related to the average life span of the living being and division number in the species whose life span is long and might be longer than the short-lived species. For example; while the cells taken from a mouse are divided 10−15 times, the cells taken from a turtle can undergo more than 100 divisions. The findings that are acquired as a result of all these researches reveal the effects of telomere, genetic clock, and destructive damages in the senescence theory. It is adopted that genetic clock was established depending on the replication number, and destructive damages are the damages formed by free radicals. In the cell culture, shortening of the telomeres at the end of each cell cycle is still at the top of the subject that is searched in the senescence mechanisms (Shay and Wright, 2000; Collado et al., 2007; Mather, 2013).

17.4.12 Quiescence

Some cells can stop their reproduction skill but does not lose their division skill. This state is called quiescence. It is a reversible situation. It is seen that the cells go into the cell cycle again and reproduce as a result of the change or manipulations around their environment. When how the cells go into a silent phase with quiescence is searched, it is seen that this situation is enabled with a complex mechanism. Although it is known that a pause occurs especially in the cell cycle of G1 phase, what are the mechanisms that start the reproduction from the quiescence state is not known (Coller et al., 2006; Fauza and Bani, 2016).

17.4.13 Production From Cell Culture

The aim of the cell culture is to keep a group of cell alive, reproduce for further study, to use when necessary freeze and save. There are very common areas of use in cell cultures. It is possible to encounter with these kinds of studies in nearly all the fields. The areas in which the cell culture is used can be counted as the production of monoclonal antibody, viral and insect vaccine, enzyme, hormone, interleukin, and growth factor production (Abbott, 2003).

Many health products are produced through fraction from blood or extraction from tissue. But this situation carries a lot of contamination risks. This risk can be decreased with animal cell cultures that are characterized well. For example, Factor VIII, which was used in the hemophilia treatment in the beginning of the 1980s, was distilled from the human plasma; however, when it was determined that these samples had been contaminated with HIV, this implementation was stopped. Nowadays, in large-scale studies in which mammal cell lines are used, product safety should be enabled carefully. In cell culture, synthesis and overexpression of glycoprotein products from mammal cells can be provided with genetic engineering. On the contrary, protein synthesis with genetic engineering from bacteria cells is difficult because synthesized protein is gathered in the shape of nonsoluble aggregate. Before obtaining biological product in the cell culture studies, which product will be obtained, in which purity and how much it will be produced must be determined correctly. Later on, deciding on the scale of production type must be determined. The last product that is obtained must be refined. All the biochemical and genetically manipulations that are implemented on the cells are a process that is designed to transform them into more effective chemical catalysts (Aunins, 2000; Ozturk and Hu, 2005; Li et al., 2010).

17.5 CONCLUSION

Cell culture system is a prominent field in biotechnology. The first cell culture study is accepted as the study performed by Roux in 1880. Nowadays, we can isolate, reproduce, and manipulate nearly every kind of cell in culture conditions. Moreover, we can obtain some products from these cultured cells or we can transplant them to treat some diseases. Researchers require that cell-based assays achieve high efficiency and high throughput; therefore, the cell culture consumables used need to meet exacting standards when it comes to performance, quality, precision, and safety. For many years it has been assumed that mammalian cells could be grown in vitro in Petri dish or flask just as bacterial cells do, as long as temperature and adequate growth medium were provided. This situation progressively evolved when it appears that culture condition dramatically modifies the cell biology of cultured cells. Cell culture conditions do not act only to support cell growth but are now considered as "instructive." But aseptic techniques are still key point. A simple biological contaminant can easily ruin all these perfect studies. Cell culture holds great potential for groundbreaking applications in cell therapy and regenerative medicine and to advance our understanding of human biology. Successful cell cultures require both optimized protocols and high-quality reagents. Increasing the chemical definition of cell culture media, along with high-quality, cGMP-grade reagents will improve consistency and validity of experimental results and ultimately enable the reproducible generation of quality-assured cells for the development of breakthrough clinical uses.

References

Abbott, A., 2003. Cell culture: biology's new dimension. Nature 424 (6951), 870–872.

Atala, A., 2006. Recent developments in tissue engineering and regenerative medicine. Curr. Opin. Pediatr. 18 (2), 167–171.

Aunins, J.G., 2000. Viral vaccine production in cell culture. In: Spier, R.E. (Ed.), Encyclopedia of Cell Technology. Wiley, New York.

Bongso, A., Lee, E.H., 2011. Stem Cells: From Bench to Bedside. World Scientific, Singapore.

Bosch, T.C.G., 2008. Stem Cells: From Hydra to Man. Springer, the Netherlands.

Butler, M., 2003. Animal Cell Culture and Technology. Taylor & Francis, London and New York.

Campisi, J., di Fagagna, F.d.A., 2007. Cellular senescence: when bad things happen to good cells. Nat. Rev. Mol. Cell. Biol. 8 (9), 729–740.

Caputo, J.L., 1988. Biosafety procedures in cell culture. J. Tissue Culture Method. 11 (4), 223–227.

Cetinalp Demircan, P., Eker Sariboyaci, A., Unal, Z.S., Gacar, G., Subasi, C., Karaoz, E., 2011. Immunoregulatory effects of human dental pulp-derived stem cells on T cells: comparison of transwell co-culture and mixed lymphocyte reaction systems. Cytotherapy 13 (10), 1205–1220. Available from: https://doi.org/10.3109/14653249.2011.605351.

Collado, M., Blasco, M.A., Serrano, M., 2007. Cellular senescence in cancer and aging. Cell 130 (2), 223–233.

Coller, H.A., Sang, L., Roberts, J.M., 2006. A new description of cellular quiescence. PLoS Biol. 4 (3), e83.

Coté, R.J., 2001. Aseptic technique for cell culture. Current Protoc. Cell Biol. 1.3.1–1.3.10.

Daar, A.S., Greenwood, H.L., 2007. A proposed definition of regenerative medicine. J. Tissue. Eng. Regen. Med. 1 (3), 179–184.

Eker Sariboyaci, A., Cetinalp Demircan, P., Gacar, G., Unal, Z., Erman, G., Karaoz, E., 2014. Immunomodulatory properties of pancreatic islet-derived stem cells co-cultured with t cells: does it contribute to the pathogenesis of type 1 diabetes?. Exp. Clin. Endocrinol. Diabetes 122 (03), 179–189. Available from: https://doi.org/10.1055/s-0034-1367004.

Fauza, D.O., Bani, M., 2016. Fetal Stem Cells in Regenerative Medicine: Principles and Translational Strategies. Springer Verlag, New York.

Freshney, R.I., 2015. Culture of Animal Cells: A Manual of Basic Technique and Specialized Applications. Wiley-Blackwell, New York.

Freshney, R.I., Stacey, G.N., Auerbach, J.M., 2007. Culture of Human Stem Cells. Wiley-Blackwell, New York.

Gurtner, G.C., Callaghan, M.J., Longaker, M.T., 2007. Progress and potential for regenerative medicine. Annu. Rev. Med. 58, 299–312.

Hartung, T., Balls, M., Bardouille, C., Blanck, O., Coecke, S., Gstraunthaler, G., et al., 2002. Good cell culture practice. Altern. Lab. Anim. 30, 407–414.

Helgason, C.D., Miller, C.L., 2013. Basic Cell Culture Protocols. Humana Press, New York.

Hengartner, M.O., 2000. The biochemistry of apoptosis. Nature. 407 (6805), 770–776.

Jayme, D.W., 2001. Cell Culture Media. Life Technologies Inc., New York.

Karaoz, E., Aksoy, A., Ayhan, S., Sariboyaci, A.E., Kaymaz, F., Kasap, M., 2009. Characterization of mesenchymal stem cells from rat bone marrow: ultrastructural properties, differentiation potential and immunophenotypic markers. Histochem. Cell. Biol. 132 (5), 533–546. Available from: https://doi.org/10.1007/s00418-009-0629-6.

Karaoz, E., Dogan, B.N., Aksoy, A., Gacar, G., Akyuz, S., Ayhan, S., et al., 2010. Isolation and in vitro characterisation of dental pulp stem cells from natal teeth. Histochem. Cell. Biol. 133 (1), 95–112. Available from: https://doi.org/10.1007/s00418-009-0646-5.

Laboratory biosafety level criteria. In: Chosewood, L.C., Wilson, D.E. (Eds.), Biosafety in Microbiological and Biomedical Laboratories. 5th Ed. US Department of Health and Human Services Public Health Service Centers for Disease Control and Prevention National Institutes of Health HHS Publication No. (CDC) 21-1112, USA.

Langdon, S.P., 2004. Cell culture contamination. In: Langdon, S.P. (Ed.), Cancer Cell Culture: Methods and Protocols. Springer, New York, pp. 309–317.

Lanza, R., Gearhart, J., Hogan, B., Melton, D., Pedersen, R., Thomas, E.D., et al., 2009. Essentials of Stem Cell Biology. Elsevier Science, London.

Li, F., Vijayasankaran, N., Shen, A., Kiss, R., Amanullah, A., 2010. Cell culture processes for monoclonal antibody production. MAbs 2 (5), 466–477. Taylor & Francis.

Marks, D.M., 2003. Equipment design considerations for large scale cell culture. Cytotechnology. 42 (1), 21–33.

Mather, J.P., 2013. Introduction to Cell and Tissue Culture: Theory And Technique: Theory and Culture. Springer, New York.

Nelson-Rees, W., Daniels, D., Flandermeyer, R., 1981. Cross-contamination of cells in culture. Science 212 (4493), 446–452.

Ng, K.W., Schantz, J.T., 2010. A Manual for Primary Human Cell Culture. World Scientific, Singapore.

Orlando, G., Wood, K.J., Stratta, R.J., Yoo, J.J., Atala, A., Soker, S., 2011. Regenerative medicine and organ transplantation: past, present, and future. Transplantation. 91 (12), 1310–1317.

Ozturk, S., Hu, W.-S., 2005. Cell Culture Technology for Pharmaceutical and Cell-Based Therapies. CRC Press, Florida.

Phelan, K., May, K.M., 2007. Basic techniques in mammalian cell tissue culture. Curr. Protoc. Cell Biol. 1.1.1–1.1.22.

Pollard, J.W., Walker, J.M., 1997. Basic Cell Culture Protocols. Humana Press, New York.

Przyborski, S., 2016. Technology Platforms for 3D Cell Culture: A Users Guide. Wiley-Blackwell, New York.

Ross, M.H., Pawlina, W., 2014. Histology: A Text and Atlas, with Correlated Cell and Molecular Biology, sixth ed. Lippincott, New York.

Ryan, J.A., 1994. Understanding and managing cell culture contamination. Corning Incorporated.

Saitama, H., 2007. New Cell Differentiation Research Topics. Nova Biomedical Books, New York.

Salzman, N.P., 1961. Animal cell cultures: tissue culture is a powerful tool in the study of nutrition, physiology, virology, and genetics. Science 133 (3464), 1559–1565.

Sewell, D.L., 1995. Laboratory-associated infections and biosafety. Clin. Microbiol. Rev. 8 (3), 389–405.

Shay, J.W., Wright, W.E., 2000. Hayflick, his limit, and cellular ageing. Nat. Rev. Mol. Cell. Biol. 1 (1), 72–76.

Sinha, B.K., 2008. Principles of Animal Cell Culture: Student Compendium. Textbook Library Edition. International Book Distributing Company, Lucknow, India.

Tuschl, G., Mueller, S.O., 2006. Effects of cell culture conditions on primary rat hepatocytes—cell morphology and differential gene expression. Toxicology 218 (2), 205–215.

Turksen, K., 2013. Stem Cells: Current Challenges and New Directions. Springer, New York.

Ulrich, H., Negraes, P.D., 2016. Working with Stem Cells. Springer Verlag, New York.

Wilson, L., Matsudaira, P.T., Mather, J.P., Barnes, D., 1998. Animal Cell Culture Methods. Elsevier Science, London.

World Health Organization, 2004. General principles, Laboratory Biosafety Manual, 3rd ed. WHO, Geneva, Switzerland.

Yadav, P.R., 2008. Cell Culture. Discovery Publishing House, New Delhi, India.

Yilmaz, I., Eker Sariboyaci, A., Subasi, C., Karaoz, E., 2016. Differentiation potential of mouse embryonic stem cells into insulin producing cells in pancreatic islet microenvironment. Exp. Clin. Endocrinol. Diabetes 124 (2), 120–129. Available from: https://doi.org/10.1055/s-0035-1554720.

Zhou, W., Kantardjieff, A., 2014. Mammalian Cell Cultures for Biologics Manufacturing. Springer Verlag, New York.

In Vitro and In Vivo Animal Models: The Engineering Towards Understanding Human Diseases and Therapeutic Interventions

Azka Khan, Kinza Waqar, Adeena Shafique, Rija Irfan and Alvina Gul

National University of Sciences and Technology, Islamabad, Pakistan

18.1 INTRODUCTION

In the history of biomedical sciences, almost all of the medical breakthroughs occurred in past hundred years that have remarkably revolutionized this field. The ultimate goals of biomedical research is to discover advanced approaches that help in making human health better, lessening the existing ailments and increasing life spans. Such goals can only be achieved by carrying out research on experimental basis. Usually scientific research entails systemic investigation of a hypothesis, and in case of biomedical research these systemic investigations involve observational or experimental studies. Research conducted solely on observational basis without involvement of living systems provides significant information but it makes very less contribution towards science. On the contrary, experimental studies are of more importance in biomedical sciences because they provide robust and authentic results. Use of animal models has enabled biologists to access the knowledge that was earlier impossible to perceive. All of these inventions in biomedical field became possible only when biologists introduced animals models for experimentation of living functions.

Experimental studies are important for preclinical studies, and they are either carried out on in vitro or in vivo biological systems. In *in vitro* system, experiments are carried out in microorganisms, isolated cells, biological molecules, cell culture systems, tissue slice preparations, or isolated organs in optimum conditions outside their normal biological context. *In vitro* systems are quite advantageous for early phase research studies such as identification of potential therapeutic candidates. While *in vivo* systems, whole living animal is used to investigate the effect of any therapeutic candidate or to study any biological process. Virtually *in vivo* biological systems provide significant indications about how a certain phenomenon will behave under pathophysiological conditions. *In vivo* studies are extremely important as they ensure proof of principle research because absolute inference cannot be predicted only from *in vitro* data.

Nowadays, various animal models have been extensively used for development of novel and effective ways for diagnostic purposes and therapeutic interventions for both humans as well as animals. These models are a major source for learning about disease pathologies and are suitable means to ensure the safety of new pharmaceutical drugs for treatment of various diseases. Animal models have been delivering novel insights and in-depth understanding of the initiation and prognosis of disease pathologies in humans. They have been used as test subjects for the development and testing of new drugs, vaccines, and other biological materials (such as hormones, antibodies, and components of vaccines) to improve human health.

Animal models of human diseases are deemed relevant only if they are useful in recapitulating disease pathogenesis and assisting in developing approaches to intervention or therapy. Thus, to ensure full utilization, a model needs to reliably mimic the normal anatomy and physiology of human organs and tissues of interest, as well as accurately reflect the morphological and biochemical aspects of the pathogenesis. Models in biomedical research can be analogous or homologous. Analogous model organisms share very little resemblance to the organism under study and not used much in research studies. Homologous animal models are more important in biomedical research because they share more gene sequence homology with the human genome.

As models, scientists aim to produce artificially a condition in an animal in a laboratory, which may resemble the human equivalent of a medical disease or injury. A variety of animals provide very useful models for the study of diseases afflicting both humans and animals. Both vertebrates and invertebrates animals have been used as models. Scientists have been using invertebrate animal models for research in the fields of neuroscience, genetics, and cancer studies such as zebrafish, *Caenorhabditis elegans*, and *Drosophila melanogaster*. A range of vertebrate animal models are important for translation research in biomedical sciences. These include animals such as rats, mice, rabbits, guinea pigs, dogs, cats, and primates.

Development of the methods in molecular and cellular biology now allows specific genetic manipulations of laboratory animals leading to the introduction of an exogenous gene to their genome or elimination of a particular endogenous gene. In all fields of biomedical research, transgenic and knockout (KO) animals have contributed greatly in understanding the molecular cause of several human diseases and allowed production of their animal models that serve as very useful tools for development of new medical drugs and therapeutic procedures for the treatment of human diseases. More recent models signal the incorporation of newer technologies such as genomics and stem cell biology that may prove even more powerful in developing effective vaccines, drugs, and medical devices. In this chapter, in vitro and in vivo animal models being used in cancer, genetics, cardiovascular, metabolic, and neurodegenerative diseases will be discussed in detail.

18.2 MOUSE MODEL

Advances in genetic engineering have revolutionized the biomedical sciences. Genetic manipulation of mouse genome opened a new window for molecular biologists. Over the past few decades, mouse model has been an invaluable tool, extensively used for modeling a wide range of human diseases. Transgenic mouse models with knock-in and KO technologies have been helpful in finding the answer to fundamental questions in basic and applied biomedical sciences (Table 18.1). Moreover, more sophisticated mouse models are required to carry out cutting edge research.

18.2.1 LDLR$^{-/-}$ Mice

The LDLR$^{-/-}$ mice (low-density lipoprotein, LDL, receptor) are the models for studying familial hypocholestrolemia. These mice have a mutation that affects the LDLR; therefore, they resemble humans in the plasma lipoprotein profile. The genetic defect causes a delayed clearance of very low-density lipoprotein (VLDL) and LDL from the plasma and therefore results in an increased plasma level of cholesterol on the normal chow diet (Bentzon and Falk, 2010). High-fat and high-cholesterol diet increases the severity of atherosclerotic lesions and hypercholesterolemia in LDLR$^{-/-}$ mice (Knowles and Maeda 2000).

TABLE 18.1 List of Some Mouse Models Being Used in Research of Different Diseases

Mouse Models	Diseases	References
LDLR$^{-/-}$ mice	Familial hypocholestrolemia	Jawien et al. (2004)
ApoE$^{-/-}$ mice	Atherogenesis	Plump et al. (1992)
Mutant E3L, ApoE mice	Hyperlipidemia and atherosclerosis	Leppanen et al. (1998)
the LDLR$^{-/-}$ and ApoE$^{-/-}$ mouse	Diabetes-associated cardiomyopathy and atherosclerosis	Hayek et al. (2005)
Calcium Chloride–Induced AAA	Hypercholesterolemia	Freestone et al. (1997)
Spontaneous mouse mutants	Aneurysm	Brophy et al. (1998)
Human CRC cell lines in mice	Colon cancer	Rashidi and Gamagami (2000)
C57Bl/6 mice using MCA cells	Colon cancer	de Jong and Aarts (2009)
Mouse model of fatty liver disease	Liver diseases	Chung et al. (2010)

The LDLR$^{-/-}$ and ApoE double-deficient mice (LDLR$^{-/-}$ApoE$^{-/-}$) have the capacity to develop severe atherosclerosis and hyperlipidemia on normal chow diet. These models therefore provide the ease of studying the diseases without putting the burden to feed the mice with atherogenic diet (Jawien et al., 2004).

18.2.2 ApoE$^{-/-}$ Mice

Two different research groups working on embryonic stem cells for the development of ApoE mice came up with the models simultaneously in 1992 by using the method of homogeneous recombination (Zhang et al., 1992; Plump et al., 1992). The plasma levels of VLDL and LDL increase as a result of homogenous deficiency of the ApoE gene, which causes a failure in the function of LDL receptor and related proteins. The model was the first mouse model to mimic lesion similar to humans as it depicts the entire range of lesions observed in the atherogenesis (Plump et al., 1992).

18.2.3 Transgenic Mice of Cardiovascular Diseases

Transgenic mice are widely used for the study of hyperlipidemia and atherosclerosis. The mice with mutant forms of ApoE3 Leiden (E3L) and ApoE (Arg 112→Cys→142) are more commonly used in the studies. The lipoprotein profile of these mice is similar to the patients with dysbetalipoproteinemia (Hofker et al., 1998). The characteristics of human vasculopathy in atherosclerotic lesions of mild, moderate, and severe plaques are presented by the E3L mice (Leppanen et al., 1998).

18.2.4 Diabetes-Accelerated Atherosclerosis Mouse Model

Diabetes is one of the leading causes of cardiovascular diseases. To study diabetes-associated cardiomyopathy and atherosclerosis, the LDLR$^{-/-}$ and ApoE$^{-/-}$ mouse models are frequently used. Type 1 diabetes is introduced in the models by injecting them with viral injections or streptozotocin (Shen and Bornfeldt, 2007). The mice injected with streptozotocin exhibit calcification in proximal aorta and atherosclerosis in the aortic sinus, abdominal aorta, and carotid artery (Hayek et al., 2005).

18.2.5 Calcium Chloride−Induced AAA

The model was first developed in rabbits and then mice were also used. The development of the model involves the peritoneal administration of calcium chloride in the segment between the iliac bifurcation and the renal artery. After 14 days, a significant dilation of the aorta occurs, which results in the development of aneurysm; the severity can be increased by adding calcium chloride with thioglycolate. Similar results can be obtained by feeding the animals on high-cholesterol diet (Freestone et al., 1997).

18.2.6 Spontaneous Mouse Mutants

A spontaneous mutation in the X chromosome causes the abnormal change in the absorption rate of intestinal copper in the blotchy mouse model strain. These models develop aneurysm in the thoracic aorta, abdominal aorta, aortic arch, and abdominal aorta due to failed cross linking between the elastin and collagen fibers, which makes the elastic tissue weak and aneurysm occurs. The interpretations drawn from the mouse model are difficult because mutation produces many other effects in addition to the aneurysm (Brophy et al., 1998).

18.2.7 Mouse Model for Liver Metastases

Liver metastasis that occurs in the colorectal area happens in about 50%–60% of the patients with liver metastasis. There is an immediate need for better treatment strategies to improve the life span of the patients affected with this disease.

For this purpose a strategy was developed for animal trials on rodents (de Jong and Aarts, 2009). A five-point criteria was established, which is as follows:

1. The tumor cells needed to have the potential for metastasis.
2. The cells of the tumor had to have reached and should have carried out an invasion of the liver parenchyma and as a result should have settled and grown out into tumor nodules.
3. The model needs to be efficient (the majority of the animals that are being used for the experimentation need to be affected with the above stated condition).
4. The model under study needs to be reproducible.
5. The model under study should be able to be practically applied for testing further if positive results are observed.

For this study immunocompetent rodents were used because of the advantage they provide to the study as they are similar to the normal immune system in the patients affected with colorectal carcinomas that develop metastases. These animals also show a stable health condition when their immune system is being examined for the effects of immunotherapy, which is being investigated as an option for treatment of the disease. The rodents will then undergo the transplantation of human colorectal cancer (CRC) cells by inoculation to create liver metastases in five areas of the rodent, which include the subcutaneous layer, the intrasplenic, the intraportal, the intrahepatic, and the colonic wall (Kobaek-Larsen and Thorup, 2000).

The benefit of using the human CRC cell lines in the rodents is the belief that they show pathological behavior in rodents, which is quite similar to human pathological behavior. These CRC cell lines are obtained from primary colon tumor or rectum tumor and metastasis (Rashidi and Gamagami, 2000).

18.2.8 Mouse Model of Colon Cancer

One of the most important criteria for the creation of a model with resectable hepatic metastases is the ability to reproduce solitary tumors. Experimentation has established that subcapsular injections in the liver or the intraparenchymal implantation of a small piece of tumor show 100% establishments of hepatic tumors (Tong and Russell, 1983). This method although being quite efficient has a drawback because it does not mimic the spread of tumor in human cells. Therefore, to establish a clinical setting that is the closest to human setting, scientists have used the orthotropic injection of tumors in the cecal walls. Injection of the tumor cells in the spleen or portal system is similar to the hematogenous spread of the cells of tumor in the liver. Using these models it is plausible to obtain macroscopic metastases in all the cells of the animal body. The most successful models have been C57Bl/6 mice using MCA cells and Wistar, WAG/Rij, or BDIX rats with N-methyl-N-nitrosoguanidine-induced adenocarcinoma cells, CC531, or DHDK12/TR colon carcinoma cells. The only disadvantage of intrasplenic injection is that it has shown a mortality rate of 10% because of operational complications and development of splenic, pancreatic, and lung metastases (Wood and Korkola, 2002). Therefore, it can be concluded that the injection of heterotopic syngeneic tumor cells in immunocompetent rodents covers majority of the desired characteristics (de Jong and Aarts, 2009).

18.2.9 Mouse Model of Fatty Liver Disease

Nonalcoholic steatohepatitis is now recognized as a complication of a metabolic syndrome (Zivkovic et al., 2007). The mouse models of nonalcoholic steatohepatitis are fed on diet deficient in choline and methionine. The special diet produces increased hepatic triglycerides, increased serum concentrations of total bilirubin and hepatic steatosis, and fibrosis. These mice show significant reduction in liver weight, body weight, and total protein concentration. These mice develop nonalcoholic steatohepatitis but do not show other signs of metabolic syndrome (Chung et al., 2010).

18.2.10 Mouse Models for Neurodegenerative Diseases

Primary neuronal cultures and neuronal cell lines derived from rodents are widely used to study basic physiological properties of neurons and represent a useful tool to study the potential neurotoxicity of chemicals. While short-term culturing of neurons can be a very straightforward process, long-term cultures of relatively pure neuronal populations require more effort (Giordano and Costa, 2011).

18.2.11 Primary Neuronal Cultures and Neuronal Cell Lines

The effect of environmental toxins on oxidative posttranslational modification of parkin protein has been studied using *in vitro* and *in vivo* mouse model. It has been demonstrated that S-nitrosylation of parkin decreased its activity as a repressor of p53 gene expression, leading to upregulation of p53. Primary mesencephalic cultures isolated from ventral mesencephalon of the mouse have been used to study the posttranslational modification of parkin protein. In primary mesencephalic cultures and in a mouse model of pesticide-induced Parkinson's disease (PD), S-nitrosylation of parkin resulted in p53-mediated

TABLE 18.2 Murine Cell Lines and Their Application in Neurodegenerative Diseases

Cell Lines	Origin	Applications	References
Primary mesencephalic cultures	Ventral mesencephalon of mouse	Posttranslational modification of parkin protein in PD	Sunico et al. (2013)
Primary microglia cultures	Cortex of rat or mouse	Neuroinflammation, oxidative stress–related neurodegenerative disorders	Stansley et al. (2012)

neuronal cell death and contributed to the pathophysiology of sporadic PD (Sunico et al., 2013) (Table 18.2).

18.2.12 Primary Microglia Cultures

For the research of neuroinflammation, primary microglia cultures have been extensively used because they possess same phenotype. These cultures contain neuronal cells obtained from cortex of a rat or mouse before or early after birth. These cells have been shown to produce nitric oxide by upregulation of inducible nitric oxide synthase and superoxide anions via activation of NADPH oxidase complexes when they are stimulated by different cytokines and lipopolysaccharides. Such cell lines are useful to study oxidative stress–related neurodegenerative disorders (Stansley et al., 2012) (Table 18.2).

18.2.13 Transgenic Mouse of PD

Mouse models have been a vital tool for research in neurodegenerative diseases. They have been proved as an effective model organism for PD. Both *in vitro* and *in vivo* mouse models have been extensively used. Many transgenic mouse models have been generated to study PD; α-synuclein protein has very important role in the pathology of this disease. KO mice and some transgenic mice with the ability to overexpress α-synuclein possess familial A53T or A30P mutations. α-Synuclein KO mice are viable and fertile, and they support a significant role of α-synuclein in regulation of dopaminergic neurotransmission, synaptic plasticity, and presynaptic vesicular release and recycling (Janus and Welzl, 2010).

18.2.14 Transgenic Mouse Model for Alzheimer's Disease

Another novel transgenic mouse model has been developed with a C57BL/6 J genetic background. This transgenic mouse harbors KM670/671NL-mutated amyloid precursor protein (APP) and L166P-mutated presenilin 1 and co-expresses them under the influence of neuron-specific Thy1 promoter element. APPPS1 mouse models are well suited for using as research tool for Alzheimer's disease (AD) because of early onset of amyloid lesions, their facile breeding, and defined genetic background (Francis et al., 2009). Transgenic mouse models of AD and PD are summarized in Table 18.3.

TABLE 18.3   Transgenic Mouse Models With Gene Manipulation for PD and AD

Disease	Model Names	Target Genes	References
Transgenic mouse model of PD	KO mice	Overexpression of α-synuclein harboring familial A53T or A30P mutations	Janus and Welzl (2010)
Transgenic mouse model for AD	APPPS1	KM670/671NL-mutated amyloid precursor protein and L166P-mutated presenilin 1	Francis et al. (2009)

18.2.15 Animal Models of Heart Failure

One of the common methods of inducing cardiac damage in mouse models is the ligation of left coronary artery. This method is adapted from the procedure used for inducing myocardial damage in rat heart. Mostly the protocols use permanently occluded arteries, but recent studies have demonstrated that partially occluded arteries can provide with the same results (Michael et al., 1995). Recently, cryoinjuries are used as the means of inducing myocardial damage in rat and mouse models; this method provides promising results (Ryu et al., 2010).

18.2.15.1 Localized Aortic Perfusion With Elastase

This model is developed exposing a segment of abdominal aorta to elastase. An inflammatory response is triggered by the degradation of the elastic fibers and then it develops into an aneurysm. The severity of the induced aortic aneurysms can be enhanced by administration of plasmin into the infusion. This model has been adapted for several other species such as mouse, rabbits, and large animals (Anidjar et al., 1992).

18.2.15.2 Decellularized Xenografts

This model is based on the observation from rejection of arterial allografts and xenografts that eventually results in the rupturing of the arterial wall. The animal model is developed by decellularizing a section from the abdominal aorta of one species, for example, pig and mice, and then grafting the extracellular matrix into another species, i.e., rat (Allaire et al., 1994).

18.3 RAT MODELS

Rat models have dominated the research in cardiovascular diseases. In addition to their low cost and easy handling, they greatly enhance the research on surgical procedures because of their large size. Three processes, electrical, pharmacological, and surgical, are commonly used to induce myocardial damage in rat heart. The first procedure was developed by Pfeffer and colleagues, they ligated the left coronary artery of rat heart. Anesthetized rats were subjected to left thoracotomy and gentle pressure was applied on the right side of the thorax to exteriorize the heart; the left pulmonary artery is heat cauterized or ligated. Heart is finally returned to the normal position and thorax is closed immediately (Pfeffer et al., 1979).

To induce pharmacological damage in myocardial tissue, β-1 adrenergic receptor agonist isoproterenol was first utilized in 1963. When used prior to ischemia, administration of isoproterenol provides a cardioprotective action but the administration of the right dose causes necrosis of myocytes and extensive hypertrophy and left ventricular dilation. This process has been utilized to study underlying mechanism of heart attack (Zbinden and Bagdon, 1963).

For electrically induced myocardial damage in rat heart, 2 mm tipped soldering is applied to exposed rat heart. Electric shock is applied to the left ventricle of the heart. Although this method has high degree of validity to produce heart damage, the same method when used in different laboratories does not provide reproducible results (Adler et al., 1976).

18.3.1 Rat Models for Celiac Disease

Celiac disease is described to be an "immune-mediated small intestinal enteropathy," which occurs because of the consumption of gluten in the diet. This reaction to gluten occurs only in individuals genetically predisposed to this disease. Although statistically the occurrence of this disease is approximated to only 0.5%−2% in the Caucasian people, it is very alarming to note that the disease has increased by two- to fourfold in the industrialized countries in the last 40 years (Di Sabatino and Corazza, 2009). The diagnosis of this disease is dependent on detecting serum antibodies produced as a result of the body's action to the enzyme tissue transglutaminase 2 (TG2) and histological features present in the biopsies of the duodenum, combining villus atrophy, crypt elongation, and accumulation of intraepithelial lymphocytes.

For testing purposes, rat models with gluten sensitization are used to study gluten-dependent enteropathies. The rat models are divided into two categories, HLA-independent and HLA-dependent models. One of the HLA-independent models was created on the basis of the T-cell transfer colitis model used to study the chronic inflammatory bowel disease in the colon (Freitag and Rietdijk, 2009). The development of crypt hyperplasia and villus atrophy was induced in RAG1 mice by the adoptive transfer of the in vitro gliadin primer and orally giving gluten. Moreover, Wistar rats showed smaller villus height and increased TNF-α levels and cellular infiltrates in the small intestinal lamina propria when they were given oral gluten and intraperitoneal injection of INF-γ (Laparra and Olivares, 2012). Therefore, the development of the disease features in the variety of rat models has given us an array of new therapeutic targets and the various pathways of testing that would eventually lead to counter the onset of the celiac disease and help in finding out the missing links that lead to the series of events responsible for the disease in humans (Costes and Meresse, 2015).

18.3.2 Nile Grass Rats

Contrary to normal practice of using laboratory animals for research, two types of wild rodents, sand rat (*Psammomys obesus*) and the Nile rat (African grass rat; *Arvicanthis niloticus*), have been used as animal models for the study of obesity and diabetes. These rats do not develop diabetes in the wild condition but develop the metabolic disorder when fed on high-fat diet in the laboratory (Noda et al., 2010). At the age of 1 year, these rats develop dyslipidemia and hyperglycemia. They also develop other conditions such as abdominal fat deposition, hyperinsulinemia, hypertension, and liver steatosis. They show promising results in metabolic disorders research when they are fed normal diet, contrary to high-fat and carbohydrate diet in humans (Chaabo et al., 2010).

18.4 PORCINE MODELS

Pigs are very important animal models being used in biomedical research such as translational research as surgical model for brain disease, and they are more favorable alternatives to other nonrodent animal models. Pigs represent suitable animal models for neurobiology research because they share more homology with humans in physiological and anatomical features. Transgenic pigs are providing biologists very promising tools to model human genetic disorders in the form of large animal model. Various genetically engineered pigs have been developed through different advanced techniques such as microinjection of DNA into pronuclei of zygotes collected from superovulated female, sperm-mediated gene transfer, and lentivirus and retrovirus-mediated gene transfer into the porcine oocytes and nuclear transfer and cloning (Dolezalova et al., 2014).

Pigs provide researchers with a unique and best recreation of plaques for the study of cardiovascular diseases. They mimic human plaque instability and are an excellent model for the study of accelerated atherosclerosis in a combination of hypercholesterolemia and diabetes (Gerrity et al., 2001). The porcine models of coronary atherosclerosis facilitate the research on vascular remodeling, adventitial neovascularization, and composition of atherosclerotic plaque (Alviar et al., 2010).

Porcine models are frequently used for the study of stent deployment responses and the physiological changes that occur after induction of AAAs. A study reports the development of porcine models by the combination of balloon angioplasty mechanical dilation and enzymatic degradation by the administration of elastase/collagen solution. There is a gradual expansion of abdominal aortic aneurysm as a result of aortic wall degradation and persistent loss of smooth muscles (Molacek et al., 2008).

Using aforementioned techniques, many transgenic porcine models have been generated and used as neurologic models for different diseases. However, not all of them have shown typical phenotypical and histological characteristics of a true model (Swindle et al., 2012). Transgenic pig models used in biomedical research are summarized in Table 18.4.

18.4.1 Gottingen Miniature Pig Model

Human brain biopsies are most important yet difficult to access prerequisite for neurodegenerative disease research. Porcine models have provided an alternative for this problem. Over the past few years, many pig models containing human genes associated with AD have been developed including amyloid precursor protein gene. In one research study, Gottingen miniature pig models of AD have developed, which have overexpression of a

TABLE 18.4 Transgenic Pig Models for AD, HD, and ALS

Model Name	Mutated Genes	References
Gottingen miniature pig models for AD	A mutated version of the amyloid precursor protein gene	Swindle et al. (2012)
Transgenic HD minipigs	N-terminal mutant HTT (208 amino acids and 105 Q)	Bassols et al. (2014)
Minipig models for PD	Homolog of FBXO7 gene	Swindle et al. (2012)
Transgenic pig model of ALS	Expression of mutant G93A hSOD1 gene	Yang et al. (2014)

mutated version of the amyloid precursor protein gene. Another pig model of PD, which has utilized the dopaminergic neurotoxin 1-methyl-4-phenyl-1,2,3,6-tetrahydropyridine in a Gottingen minipig has been developed. Research on this model has showed that neurotoxin has significantly reduced striatal dopamine levels. Recently, porcine models containing homolog of FBXO7 gene have been developed to investigate the Parkinsonian pyramidal syndrome (Swindle et al., 2012).

18.4.2 Transgenic Huntington Disease Minipigs

Cloned transgenic Huntington disease minipigs containing N-terminal mutant HTT (208 amino acids and 105 Q) have been generated through somatic cell nuclear transfer technology. But these models are found to be unstable, and some of them did not last due to extremely high expression of transgene. These cloned transgenic Huntington disease minipigs exhibited classic apoptosis of neurons characterized by activated Caspase-3 and DNA fragmentation in brain region. It has been suggested that mutant HTT is more toxic to large animal models as compared to the mostly used murine HD model. Another transgenic HD porcine model has been successfully developed by lentiviral vector system. For this purpose, a lentiviral vector bearing N-terminal-truncated (548 aa) HTT gene with mixed 145 CAG/CAA repeats expressing under control of human huntingtin promoter was used and injected into one-cell stage embryo. First set of HIV1-HD-548aaHTT-145Q genetically manipulated piglets were produced after standard period of gestation. Stable transgenic mutant HTT protein levels (548aa-124Q) analogous to wild-type pig HTT levels were identified in all other organs specifically in all region of brain, and they are shown to be germline transmissible (Bassols et al., 2014).

18.4.3 Transgenic Pig Model of Amyotrophic Lateral Sclerosis

Transgenic pigs have also been used in research studies of amyotrophic lateral sclerosis (ALS). In one study, transgenic pigs have been generated, which successfully expressed mutant G93A hSOD1 gene and results reflected motor defects in hind limbs. This mutation is shown to have germline transmission in consecutive generation and degeneration occurs in motor neuron of such pigs in dose- and age-dependent manner. Investigation in this model showed that SOD1 binds to PCBP1 (nuclear poly(rC) binding protein 1) in pig brain but not in brain of murine model. This suggested that SOD1—PCBP1 interaction may be responsible for SOD1 accumulation in nucleus and can be further studied to identify the species-specific targets for ALS pathology in primate models (Yang et al., 2014).

18.4.4 Pig Models for Ataxia Telangiectasia

Porcine model have been used to study ataxia telangiectasia (AT), which is characterized by progressive multisystem disorder with motor impairment due to mutations in the AT-mutated gene. A novel pig AT model has been engineered, which depicted better phenotype of disease. Initially this model showed quick cerebellar lesions including loss of Purkinje cells and altered cytoarchitecture. Pig AT model with early developmental etiology can be potentially useful for therapies in early stages of AT patients (Beraldi et al., 2015).

18.4.5 Pig Models for Myocardial Infarction

Pigs have predominated as animal models for study of myocardial infarction. They provide similar arterial anatomy and collateral coronary circulation to humans, and therefore they better present the degree of infarct size and pathophysiology (Kim and Kaelin, 2003). One of the common methods of development of myocardial infarction model of pig is the use of angioplasty balloon inflation at the femoral artery. The surgical equipment and researcher skills are the major limitations of the procedure (Suzuki et al., 2008).

18.5 ZEBRAFISH MODEL

Zebrafish model for neurobiological research has become widely popular in the past decade. To increase the understanding of complicated mechanisms occurring in brain, both adult and larval zebrafish have been used for this purpose. Zebrafish model has provided knowledge about brain both in normal and pathological conditions. Table 18.5 shows some of the most common zebrafish models used in biomedical research.

TABLE 18.5 Zebrafish Models for Neurodegenerative Diseases

Model Name	Genes Involved	References
Zebrafish models of epilepsy	Expression of early proto-oncogenes, e.g., asc-fos *Mind-bomb* mutant zebrafish Zebrafish Nav1.1 mutants	Kalueff et al. (2013); Hortopan and Baraban (2011)
Zebrafish model of AD	Two homologs of APP	Newman et al. (2007)
Zebrafish model of PD	*DJ-1* gene expression (DJ-1 knockdown zebrafish)	Best and Alderton (2008)

18.5.1 Zebrafish Models of Epilepsy

Zebrafish has been used as model organism for epilepsy studies. Through a variety of experimentations, physiological and seizure-like behavior responses have been observed in zebrafish model. This model has shown significant results upon electroencephalographic stimulation recorded in both larval and adult stage. Change in swimming pattern, e.g., circular swimming and other epilepsy-like responses such as seizure-like behavior and hyperactivity have been observed during experimentation. Zebrafish has been used for studying the upregulated expression of early proto-oncogenes, e.g., *asc-fos*, which is an important marker for neuronal activation. The expression of *asc-fos* proto-oncogenes is usually high during seizures in mouse and zebrafish model (Kalueff et al., 2013).

Mind-bomb mutant zebrafish, with dysregulated *Notch* signaling and disturbed E3 ubiquitin ligase activity have been modeled, which depicts defects during development of brain and spontaneous seizures. Research on this model using agar-immobilized mutant zebrafish larvae revealed abnormal motor responses and EEG activity showing seizures along with modified expression of some CNS genes. Zebrafish is also sensitive to antiepileptic pharmacological agent used for therapeutic purpose. Zebrafish Nav1.1 mutants with genetically engineered *scn1Lab* gene have been developed. *scn1Lab* gene encodes a voltage-gated sodium channel, and mutations in this gene cause Dravet syndrome with severe intellectual disability, impaired social development, and drug-resistant seizures in human. Clinical findings have shown that zebrafish exhibits same similar neurological phenotype as observed in epileptic patient (Hortopan and Baraban, 2011).

Zebrafish has performed its role in studying pediatric epilepsy; its genetically engineered models have found to be sensitive to antiepileptic drugs for its treatments. Supporting the use of zebrafish in antiepileptic treatment, this model represents a promising high-throughput screening (HTS) agent for many drugs such as anticonvulsant drugs. During this study, groups of experimental and control models are exposed to a standard proepileptic drug, and groups that are more resistant to evoked seizure-related physiological and behavioral symptoms are used for further research. Recently devices designed for automated drug delivery and medium change in multiwell screening panels can be used as a potential epilepsy-related HTS using larval zebrafish (Stewart et al., 2014).

18.5.2 Zebrafish Model of AD

Zebrafish has become a powerful model organism for AD research as it possesses two homologs of APP with 70% homology to human APP. Research has shown that zebrafish harbors functional γ-secretase machinery that produces Aβ. Presence of tau protein has also been reported. In one study, microinjection of four repeat human tau GFP constructs into 1−2 cell stage embryos exhibited disruption of cytoskeleton and trafficking of tau protein after 48 hours of injection. Ultimately this resulted in hyperphosphorylated fibrillar tau showing same staining properties with neurofibrillary tangles as seen in AD pathology. Zebrafish model for AD offers promising experimental tool to identify novel therapeutics that will be able to reduce Aβ load and hyperphosphorylation in tauopathies (Newman et al., 2007).

18.5.3 Zebrafish Model of PD

Patients with autosomal recessively inherited DJ-1 mutations typically develop early onset PD. Zebrafish has been used to investigate the role of DJ-1 mutations. *DJ-1* gene is expressed in excessively in both embryonic and adult zebrafish. DJ-1 protein of zebrafish has 83% homology with than mouse DJ-1 protein which is 80%. It has been observed that knocking down of DJ-1 in zebrafish has not affected the number of dopaminergic neurons in a same way to mouse DJ-1 null mutant. Keeping in mind the proposed function of DJ-1, knockdown zebrafish larvae are seemed to be more susceptible to oxidative stress and show remarkably high levels of SOD1 (Best and Alderton, 2008).

In most animal models of PD,1-methythl-4-phenyl-1, 2,3,6-tetrahydropyridine (MPTP) is used to develop some of the effects of idiopathic PD. In one of the recent researches, it has shown that Zebrafish embryos treated with MPTP led to loss of tyrosine hydrolase and dopamine transporter expressing neurons. Moreover, monoamine oxidase-B inhibitor deprenyl could be used to reverse loss of these neurons. Using this paradigm of zebrafish models will help neurobiologists in isolating novel compounds for both form of hereditary and the idiopathic forms of PD (Best and Alderton, 2008).

18.6 RABBIT MODELS

Initially diet high in cholesterol was used to mimic atherosclerosis in rabbit models. Studies conducted back in 1913 found that the arterial intima of rabbit presented with similar disease condition as in humans caused by cholesterol. In cardiovascular disease research, rabbit models have primarily been used to establish the effects of statin or diet in lowering lipid levels and the effect on the formation of plaques. These studies contributed towards the understanding of mechanisms involved in atheroma inflammation such as macrophage accumulation and lipid lowering (Aikawa et al., 2002)

18.6.1 Rabbit Model of Inflammation-Associated Atherosclerosis

An experimental study reported the development of rabbit model of inflammation-associated atherosclerosis in which the hyperlipidemic animal had severe vascular lesions with a combination of induced knee arthritis and femoral injury (Largo et al., 2008).

The aortic arteries in rabbit are smaller in the diameter as compared to carotid artery in humans yet they are often used for the study of endovascular therapeutic devices. In addition to that, the MRI quantification studies use rabbit as animal models for the determination and imaging of atherosclerotic aortic component (Helft et al., 2001). Rabbit model for the study of plaque rupture has also been developed. The histological findings of the plaque reveal that it consists of a thin fibrous cap, lipid-rich core, and macrophage accumulation (Shimizu 2009).

Several similar interventions that are used in the development of mouse model are also used in the development of rabbit models. One of the commonly used methods in the infusion of elastase and calcium chloride is, in the abdominal aorta of the rabbit. Rabbits provide several advantages as models for the study of cardiovascular disease, the prominent reasons being the high similarity in the presentation of the aneurysm by rabbits with the occurrence of aneurysm in humans. Rabbit aneurysm can be easily monitored in the femoral artery therefore are excellent models for the study of endovascular therapies (Dai et al., 2008).

18.6.2 Rabbit Model for Myocardial Damage

Rabbits are used as models for the study of myocardial damage for several reasons. They have similar sarcomere protein composition as in humans. The WHHLMI strain of rabbits is the spontaneous myocardial infarction model that does not need any surgical procedure. The strain was developed by a selective breeding of coronary atherosclerosis-prone WHHL rabbits. The limitation of this model is the lack of plaque formation which is contrary to human myocardial infraction and is associated with intravascular thrombosis and coronary plaque rupture (Kuge et al., 2010).

18.7 CONCLUSION

Animal models are of utmost importance in the field of biomedical sciences and have been performing fundamental role in investigating the crucial mechanisms occurring in many fatal human diseases. These models have significantly helped the biologists to study the molecular and genetic causes of many diseases. Transgenic models have been used to develop novel drugs and serve as an important medium for their testing. Animal models with more homology to human genome have proved to be very promising in the research of genetic disorders. The ability to manipulate the genetic makeup of animal models has made it facile to study molecular mechanisms of human genome in animal models, providing potential insights to these sophisticated mechanisms. Unquestionably they are an essential tool in biomedical research; however, there is need of improved animal models that mimic the pathological outlook as much as human diseases to overcome the present shortcomings faced during biomedical research.

References

Adler, N., Camin, L.L., Shulkin, P., 1976. Rat model for acute myocardial infarction: application to technetium-labeled glucoheptonate, tetracycline, and polyphosphate. J. Nucl. Med. 17 (3), 203−207.

Aikawa, M., Sugiyama, S., Hill, C.C., et al., 2002. Lipid lowering reduces oxidative stress and endothelial cell activation in rabbit atheroma. Circulation 106 (11), 1390−1396.

Allaire, E., Guettier, C., Bruneval, P., Plissonnier, D., Michel, J.B., 1994. Cell-free arterial grafts:morphologic characteristics of aortic isografts, allografts, and xenografts in rats. J. Vasc. Surg. 19 (3), 446−456.

Alviar, C.L., Tellez, A., Wallace-Bradley, D., et al., 2010. Impact of adventitial neovascularisation on atherosclerotic plaque composition and vascular remodelling in a porcine model of coronary atherosclerosis. EuroIntervention 5 (8), 981−988.

Anidjar, S., Dobrin, P.B., Eichorst, M., Graham, G.P., Chejfec, G., 1992. Correlation of inflammatory infiltrate with the enlargement of experimental aortic aneurysms. J. Vasc. Surg. 16 (2), 139−147.

Bassols, A., Costa, C., Eckersall, P.D., Osada, J., Sabrià, J., Tibau, J., 2014. The pig as an animal model for human pathologies: a proteomics perspective. Proteomics Clin. Appl. 8 (9−10), 715−731.

Bentzon, J.F., Falk, E., 2010. Atherosclerotic lesions in mouse and man: is it the same disease? Curr. Opin. Lipid. 21 (5), 434–440.

Beraldi, R., Chan, C.-H., Rogers, C.S., Kovács, A.D., Meyerholz, D.K., Trantzas, C., et al., 2015. A novel porcine model of ataxia telangiectasia reproduces neurological features and motor deficits of human disease. Hum. Mol. Genet. 24 (22), 6473–6484.

Best, J., Alderton, W.K., 2008. Zebrafish: an in vivo model for the study of neurological diseases. Neuropsychiatr. Dis Treat. 4 (3), 567.

Brophy, C.M., Netzer, D., Forster, D., 1998. Detonation studies of JP-10 with oxygen and air for pulse detonation engine development. AIAA Paper 98, 4003.

Chaabo, F., Pronczuk, A., Maslova, E., Hayes, K., 2010. Nutritional correlates and dynamics of diabetes in the Nile rat (Arvicanthis niloticus): a novel model for diet-induced type 2 diabetes and the metabolic syndrome. Nutr. Metab. 7, article 29.

Chung, S., Yao, H., Caito, S., Hwang, J.W., Arunachalam, G., Rahman, I., 2010. Regulation of SIRT1 in cellular functions: role of polyphenols. Arch. Biochem. Biophys. 501 (1), 79–90.

Costes, L.M., Meresse, B., 2015. The role of animal models in unravelling therapeutic targets in coeliac disease. Best Pract. Res. Clin. Gastroenterol. 29 (3), 437–450.

Dai, D, Ding, Y.H., Danielson, M.A., et al., 2008. Endovascular treatment of experimental aneurysms with use of fibroblast transfected with replication-deficient adenovirus containing bone morphogenetic protein-13 gene. Am. J. Neuroradiol. 29 (4), 739–744.

de Jong, G.M., Aarts, F., 2009. Animal models for liver metastases of colorectal cancer: research review of preclinical studies in rodents. J. Surg. Res. 29, 167–176.

Di Sabatino, A, Corazza, GR., 2009. Coeliac disease. Lancet 373 (9673), 1480–1493.

Dolezalova, D., Hruska-Plochan, M., Bjarkam, C.R., Sørensen, J.C.H., Cunningham, M., Weingarten, D., et al., 2014. Pig models of neurodegenerative disorders: utilization in cell replacement-based preclinical safety and efficacy studies. J. Comp. Neurol. 522 (12), 2784–2801.

Francis, Y.I., Fà, M., Ashraf, H., Zhang, H., Staniszewski, A., Latchman, D.S., et al., 2009. Dysregulation of histone acetylation in the APP/PS1 mouse model of Alzheimer's disease. J. Alzheimer Dis. 18 (1), 131–139.

Freestone, T., Turner, R.J., Higman, D.J., Lever, M.J., Powell, J.T., 1997. Influence of hypercholesterolemia and adventitial inflammation on the development of aortic aneurysm in rabbits. Arterioscler. Thromb. Vasc. Biol. 17 (1), 10–17.

Freitag, TL, Rietdijk, S., 2009. Gliadin-primed CD4þCD45RBlowCD25-T cells drive gluten-dependent small intestinal damage after adoptive transfer into lymphopenic mice. Gut 58 (12), 1597–1605.

Gerrity, R.G., Natarajan, R., Nadler, J.L., Kimsey, T., 2001. Diabetes-induced accelerated atherosclerosis in swine. Diabetes 50 (7), 1654–1665.

Giordano, G., Costa, L.G., 2011. Primary neurons in culture and neuronal cell lines for in vitro neurotoxicological studies. In Vitro Neurotoxicology: Methods and Protocols 758, 13–27.

Hayek, T., Hussein, K., Aviram, M., et al., 2005. Macrophagefoam cell formation in streptozotocin-induced diabetic mice: stimulatory effect of glucose. Atherosclerosis 183 (1), 25–33.

Helft, G., Worthley, S.G., Fuster, V., et al., 2001. Atherosclerotic aortic component quantification by noninvasive magnetic resonance imaging: an in vivo study in rabbits. J. Am. Coll. Cardiol. 37 (4), 1149–1154.

Hofker, M.H., Van Vlijmen, B.J.M., Havekes, L.M., 1998. Transgenic mouse models to study the role of APOE in hyperlipidemia and atherosclerosis. Atherosclerosis 137 (1), 1–11.

Hortopan, G.A., Baraban, S.C., 2011. Aberrant expression of genes necessary for neuronal development and notch signaling in an epileptic mind bomb zebrafish. Dev. Dynam. 240 (8), 1964–1976.

Janus, C., Welzl, H., 2010. Mouse models of neurodegenerative diseases: criteria and general methodology. Methods Mol. Biol. 602, 323–345.

Jawien, J., Nastałek, P., Korbut, R., 2004. Mouse models of experimental atherosclerosis. J. Physiol. Pharmacol. 55 (3), 503–517.

Kalueff, A.V., Gebhardt, M., Stewart, A.M., Cachat, J.M., Brimmer, M., Chawla, J.S., et al., 2013. Towards a comprehensive catalog of zebrafish behavior 1.0 and beyond. Zebrafish 10 (1), 70–86.

Kim, W., Kaelin, W.G., 2003. The von Hippel–Lindau tumor suppressor protein: new insights into oxygen sensing and cancer. Curr. Opin. Genet. Dev. 13 (1), 55–60.

Knowles, J.W., Maeda, N., 2000. Genetic modifiers of atherosclerosis in mice. Arterioscler. Thromb. Vasc. Biol. 20 (11), 2336–2345.

Kobaek-Larsen, M, Thorup, I., 2000. Review of colorectal cancer and its metastases in rodent models: comparative aspects with those in humans. Comp. Med. 50 (1), 16.

Kuge, Y., Takai, N., Ogawa, Y., et al., 2010. Imaging with radiolabelled anti-membrane type 1 matrix metalloproteinase (MT1-MMP) antibody: potentials for characterizingg atherosclerotic plaques. Eur. J. Nucl. Med. Mol. Imaging 37, 2093–2104.

Laparra, JM, Olivares, M., 2012. Bifidobacterium longum CECT 7347 modulates immune responses in a gliadinin-duced enteropathy animal model. PLoS One 7 (2), e30744.

Largo, R., Sanchez-Pernaute, O., Marcos, M.E., et al., 2008. Chronic arthritis aggravates vascular lesions in rabbits with atherosclerosis: a novel model of atherosclerosis associated with chronic inflammation. Arthritis Rheum. 58 (9), 2723–2734.

Leppanen, P., Luoma, J.S., Hofker, M.H, Havekes, L.M, Yla-Herttuala, S., 1998. Characterization of atherosclerotic lesions in apo E3-leiden transgenic mice. Atherosclerosis 136 (1), 147–152.

Michael, L.H, Entman, M.L., Hartley, C.J., et al., 1995. Myocardial ischemia and reperfusion: a murine model. Am. J. Physiol. 269 (6), H2147–H2154.

Molacek, J., Treska, V., Kobr, J., et al., 2008. Optimization of the model of abdominal aortic aneurysm—experiment in an animal model. J. Vasc. Res. 46 (1), 1–5.

Newman, M., Musgrave, F., Lardelli, M., 2007. Alzheimer disease: amyloidogenesis, the presenilins and animal models. Biochim. Biophys. Acta. 1772 (3), 285–297.

Noda, K., Melhorn, M.I., Zandi, S., et al., 2010. An animal model of spontaneous metabolic syndrome: nile grass rat. FASEB J. 24 (7), 2443–2453.

Pfeffer, M.A., Pfeffer, J.M., Fishbein, M.C., 1979. Myocardial infarct size and ventricular function in rats. Circ. Res. 44 (4), 503–512.

Plump, A.S., Smith, J.D., Hayek, T., et al., 1992. Severe hypercholesterolemia and atherosclerosis in apolipoprotein E- deficient mice created by homologous recombination in ES cells. Cell 71 (2), 343–353.

Rashidi, B, Gamagami, R., 2000. An orthotopic mouse. Clin. Cancer Res. 6 (6), 2556–2561.

Ryu, J.U.H., Kim, I.L.K., Cho, S.W., et al., 2010. Implantation of bone marrow mononuclear cells using injectable fibrin matrix enhances neovascularization in infarcted myocardium. Biomaterials 26 (3), 319–326.

Shen, X., Bornfeldt, K.E., 2007. Mouse models for studies of cardiovascular complications of type 1 diabetes. Ann. N. Y. Acad. Sci. 1103, 202–217.

Shimizu, T., 2009. Lipid mediators in health and disease: enzymes and receptors as therapeutic targets for the regulation of immunity and inflammation. Ann. Rev. Pharmacol. Toxicol. 49, 123–150.

Stansley, B., Post, J., Hensley, K., 2012. A comparative review of cell culture systems for the study of microglial biology in Alzheimer's. J. Neuroinflamm. 9, 115.

Stewart, A.M., Braubach, O., Spitsbergen, J., Gerlai, R., Kalueff, A.V., 2014. Zebrafish models for translational neuroscience research: from tank to bedside. Trends Neurosci. 37 (5), 264–278.

Sunico, C.R., Nakamura, T., Rockenstein, E., Mante, M., Adame, A., Chan, S.F., et al., 2013. S-Nitrosylation of parkin as a novel regulator of p53-mediated neuronal cell death in sporadic Parkinson's disease. Mol. Neurodegener. 8 (29), 105–116.

Suzuki, Y., Matsunami, H., Sakaguchi, Y., Ikeda, M., Nakayama, T., Kamata, K., et al. (2008). *U.S. Patent Application No. 12/745,458.*

Swindle, M., Makin, A., Herron, A., Clubb, F., Frazier, K., 2012. Swine as models in biomedical research and toxicology testing. Vet. Pathol. 49 (2), 344–356.

Tong, D, Russell, A.H., 1983. Second laparotomy for proximal colon cancer. Sites of recurrence and implications for adjuvant therapy. Am. J. Surg. 145, 382–386.

Wood, GA, Korkola, J.E., 2002. Tissue-specific resistance to cancer development in the rat: phenotypes of tumor-modifier genes. Carcinogenesis 23 (1), 1–9.

Yang, H., Wang, G., Sun, H., Shu, R., Liu, T., Wang, C.-E., et al., 2014. Species-dependent neuropathology in transgenic SOD1 pigs. Cell Res. 24 (4), 464–481.

Zbinden, G., Bagdon, R.E., 1963. Isoproterenol-induced heart necrosis, an experimental model for the study of angina pectoris and myocardial infarct. Rev. Can. Biol. 22, 257–263.

Zhang, S.H., Reddick, R.L., Piedrahita, J.A., Maeda, N., 1992. Spontaneous hypercholesterolemia and arterial lesions in mice lacking apolipoprotein E. Science 258 (5081), 468–471.

Zivkovic, M., German, J.B., Sanyal, A.J., 2007. Comparative review of diets for the metabolic syndrome: implications for nonalcoholic fatty liver disease. Am. J. Clin. Nutr. 86 (2), 285–300.

Further Reading

Annambhotla, S., Bourgeois, S., Wang, X., Lin, P.H., Yao, Q., Chen, C., 2008. Recent advances in molecular mechanisms of abdominal aortic aneurysm formation. World J. Surg. 32 (6), 976–986.

Zadelaar, S., Kleemann, R., Verschuren, L., et al., 2007. Mouse models for atherosclerosis and pharmaceutical modifiers. Arterioscler. Thromb. Vasc. Biol. 27 (8), 1706–1721.

Medical Biotechnology: Techniques and Applications

Phuc V. Pham

University of Science, VNUHCM, Ho Chi Minh city, Vietnam

19.1 INTRODUCTION

Medical biotechnology is defined as the application of biotechnology tools for producing medical products that can be used for the diagnosis, prevention, and treatment of diseases. The best-known products of medical biotechnology are antibiotics that are used to treat bacterial infections. Similar products that are applied to crops are called biopesticides.

The most remarkable product of medical biotechnology is human insulin that can now be produced outside of the human body. This product highlights the growth of biomedical technology in recent years. Similar to growth hormones, insulin is also considered a gift of modern biotechnology. These medical products were generated using recombinant DNA technology, allowing for production in high quantities.

With breakthroughs in gene manipulation and cell culture, several tools in medical biotechnology were invented since 2000, including stem cell therapy and tissue engineering (Table 19.1). Stem cell therapy is rapidly being translated to the clinic, with nearly 1000 clinical trials (clinicaltrials.gov) and several approved stem cell–based products being sold in some countries. Other applications of stem cells, including stem cell–based gene therapy, and tissue engineering have also been developed in recent years with dramatic results.

Omics Technologies and Bio-engineering: Towards Improving Quality of Life
DOI: https://doi.org/10.1016/B978-0-12-804659-3.00019-1

TABLE 19.1   Timeline of Medical Biotechnology

Year	Milestone
1882	Chromosomes discovered in salamander larvae
1944	DNA is a hereditary material
1963	Genetic materials decoded
1971	The world's first biotech company is founded in California, United States
1979	First biotech product human growth hormone
1980	Genetech becomes the first biotech company to go public, generating 35 million USD
1983	Stanford Scholl of medicine becomes the first to screen blood to prevent AIDS transmission
1984	World's first DNA fingerprinting techniques is developed
1984	Chiron Corporation announced the first cloning and sequencing of the entire HIV virus genome
1984	Genetech obtains US FDA approval to market human growth hormone, the first recombinant product to be sold by a biotechnology company
1986	The US FDA awards Chiron Corporation, a license for the production of first recombinant vaccine to battle the hepatitis B virus
1988	The "Harvard Mouse" becomes the first mammal patented in the United States
1991	Cancer patients are treated with a gene therapy that produces the tumor necrosis factor, a natural tumor fighting protein
1995	The first full gene sequence of a living organism other than a virus is completed for the bacterium hemophilic influenza
1997	A sheep "Dolly" becomes the first mammal cloned
1998	Thomson's Lab was the first to report the successful isolation of human ESCs
2002	First vaccine against cervical cancer development
2007	Yamanaka and his colleagues found IPSCs with germ line transmission (via selecting for Oct4 or Nanog gene). Also in 2007, they were the first to produce human IPSCs
2007	The 2007 Nobel Prize in physiology or medicine is awarded to Drs Mario R. Capecchi, Martin J. Evans, and Oliver Smithies for their discoveries of principles for introducing specific gene modifications in mice by the use of ESCs
2011	Nobel Prize in Physiology or Medicine to Bruce A. Beutler, Jules A. Hoffmann, and Ralph M. Steinman for discoveries regarding activation of the immune system and the development of new vaccines and the treatment of infectious diseases and cancer
2012	The Nobel Prize in Physiology or Medicine was awarded jointly to Sir John B. Gurdon and Shinya Yamanaka "for the discovery that mature cells can be reprogrammed to become pluripotent"
2015	The Nobel Prize in Physiology or Medicine to William C. Campbell and Satoshi Ōmura "for their discoveries concerning a novel therapy against infections caused by roundworm parasites," and Youyou Tu "for her discoveries concerning a novel therapy against Malaria"

19.2 BACKGROUND OF MEDICAL BIOTECHNOLOGY

19.2.1 Techniques

19.2.1.1 Polymerase Chain Reaction

Polymerase chain reaction (PCR) is a revolutionary method that was developed by Kary Mullis in the 1980s. PCR is based on the in vitro activity of DNA polymerase to synthesize new strands of DNA. This reaction is performed in a tube containing a mixture of the DNA template, nucleotides, primers (forward and reverse primers), and DNA polymerase. Different from an in vivo reaction, the high temperature during an in vitro reaction causes the denaturation of DNA double strands into two single strands. Then, primers bind to specific sequences of target DNA that are homologous to the primers. DNA polymerase then elongates the primer that is homologous with the target DNA. At the end of the PCR reaction, a specific sequence will accumulate numbering in billions of copies (amplicons).

The main components of PCR include:

- DNA template: The sample DNA that contains the target sequence. At the beginning of the reaction, a high temperature is used to separate the double-stranded DNA (dsRNA) molecule to a single strand.
- DNA polymerase: An enzyme used to synthesize new strands of DNA complementary to the target sequence. The first and most commonly used enzyme is *Taq* DNA polymerase (from *Thermophilus aquaticus*). *Pfu* DNA polymerase (from *Pyrococcus furiosus*) is also used widely because of its higher fidelity when copying DNA. These enzymes are resistant to high temperatures such that they stably function during the PCR cycles that range from 50°C to 95°C.
- Primers: Single-stranded DNA complementary to the target sequence. The polymerase begins synthesizing new DNA from one end of the primer.
- Nucleotides (dNTPs or deoxynucleotide triphosphates): Single units of nucleic acid bases, including Adenine (A), Thymine (T), Guanine (G), and Cytosine (C), which are essentially the "building blocks" for new DNA strands.

The basic PCR technique has been modified for amplifying both DNA and RNA templates and is now used for quantitative measurements. Popular types of PCR-based methods are listed in Table 19.2. PCR has become an indispensable tool for both clinical and diagnostic medicine and for biomedical research (Fig. 19.1).

TABLE 19.2 Some Kinds of PCR and Their Properties

Kinds of PCR	Information and properties
AFLP PCR	Involves the digestion of DNA into fragments using restriction enzymes, amplification of the fragments using PCR, and then analysis of the fragments using gel electrophoresis
Allele-specific PCR	This uses allele-specific primers that are designed to match a mutation
Alu PCR	Uses primers that select the Alu elements (commonly repeated sections of DNA) and amplify the sections between these in DNA fingerprinting

(Continued)

TABLE 19.2   (Continued)

Kinds of PCR	Information and properties
Assembly PCR	Also known as polymerase cycling assembly or PCA—Uses PCR to build longer pieces of DNA from fragments using overlapping primers
Asymmetric PCR	Amplifies just one strand of the target DNA
Colony PCR	Used to screen colonies of bacteria after transformation with a plasmid
Conventional PCR	The basic PCR process that produces up to a billion copies of a DNA or RNA strand; the results are only seen at the end of the process
Hot start PCR	Inactivates the Taq polymerase until the reaction starts, using antibodies that are denatured by heat
In situ PCR	PCR that takes place in cells or in fixed tissue on a slide
Inverse PCR	Amplifies DNA next to a known sequence, using primers placed in the reverse direction to normal
ISSR PCR (intersequence-specific PCR)	Amplifies DNA between sequences that are repeated through the genome
LATE PCR (linear after the exponential PCR)	Creates single strands of DNA using PCR techniques
Long-range PCR	Uses a mixture of polymerases to amplify longer stretches of DNA
Methylation-specific PCR	Uses two primer pairs that bind to methylated and unmethylated DNA
Multiplex PCR	Uses a number of pairs of primers to allow analysis of a number of fragments in a single sample
Nested PCR	After an initial 30–35 cycles of PCR, an additional round of PCR uses new primers "nested" within the original primers, making the process more sensitive because it reduces the risk of unwanted products from primers binding to incorrect regions
qPCR	In quantitative PCR (also known as real-time PCR, RT PCR, or qRT-PCR), the DNA or RNA molecules are tagged using fluorescent probes, so that the concentration of amplified products can be monitored and quantified in real time by tracking the level of fluorescence
Repetitive sequence-based PCR	PCR using primers that are complementary to repeated noncoding sequences in the genome
Reverse transcriptase PCR	Confusingly, also known as RT PCR, creates cDNA (complementary DNA) by reverse transcribing RNA to DNA using reverse transcriptase
Single-cell PCR	Exactly as it sounds—PCR on a single, isolated, lysed cell
TD PCR (touchdown PCR)	This begins with an annealing temperature higher than optimum and reduces it each cycle to the optimum or "touchdown" temperature, and this can improve the outcomes
Immuno-PCR	
Digital PCR	

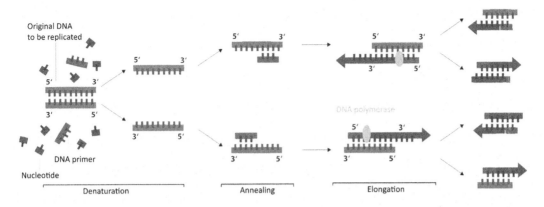

Denaturation   Annealing   Elongation

FIGURE 19.1 PCR principles. Each PCR reaction has three steps: denaturation, annealing, and elongation.

19.2.1.2 *Fluorescence In Situ Hybridization*

Fluorescence in situ hybridization (FISH) is a common cytogenetic technique. This technique, which was first demonstrated by Gall and Pardue (1969), allows the user to localize the position of a specific DNA sequence on the chromosome. To date, several FISH-based procedures have been developed to increase sensitivity and accessibility. However, most procedures use single-stranded DNA conjugated with a fluorescent probe to detect a specific DNA sequence on chromosome(s).

Initially, Joseph Gall and Mary Lou Pardue used radioactively labeled DNA sequences, called a probe, to identify and quantify the target DNA sequence. Soon after, fluorescent labels quickly replaced radioactive labels owing to their greater safety, stability, and ease of detection (Rudkin and Stollar, 1977).

The first step in FISH is the production of a fluorescent copy of the probe sequences. Then, both probes and targets are denatured using heat or chemicals. This step aims to expose hydrogen bonds from targets and probes that can then bind together during the hybridization step. Then, both targets and probes are mixed together. In this mixture, the probe binds to the complementary sequence(s) on the chromosome (target). Finally, the samples are washed to remove excess probe. If the probe is conjugated to a fluorophore, it is visualized on a fluorescent microscope.

Presently, FISH uses different types of probes, including locus specific probes, alphoid or centromeric repeat probes, and whole chromosome probes. Locus-specific probes bind to a particular region on a chromosome. This type of probe is useful for determining a specific locus on a chromosome and for identifying the number of copies of a gene on the chromosomes. Alphoid or centromeric repeat probes are used to determine whether an individual has the correct number of chromosomes. This type of probe can also be used in combination with locus-specific probes to determine whether an individual is missing genetic material.

Whole chromosome probes are collections of smaller probes, each of which can bind to unique sequences on every chromosome. With this technique, each chromosome can be labeled with a color, and the karyotype can be exhibited as a full-color map of all the chromosomes, and is called a spectral karyotype. This technique can be used to determine chromosomal abnormalities.

19.2.1.3 Sequencing

DNA sequencing is the process of determining the order of nucleotides within a DNA molecule. Because DNA is the basis of genetic information, the knowledge of DNA sequences has become an important tool in basic biological research and has applications in medical diagnosis, biotechnology, forensic biology, virology, and biological systematics. DNA sequencing was first developed by Fred Sanger in 1955, wherein he determined the sequence of amino acids in an insulin molecule. Sanger's success greatly influenced X-ray crystallographers, including Watson and Crick (https://en.wikipedia.org/wiki/DNA_sequencing).

The earliest form of nucleotide sequencing is RNA sequencing. In 1972 and 1976, Fiers et al. successfully identified the genome sequence of the bacteriophage, MS2 (Min Jou et al., 1972; Fiers et al., 1976). DNA sequencing methodology was first suggested by Ray Wu at Cornell University in 1970. This method is related to location-specific, primer extension strategy. From 1970 to 1973, Wu and Padmanabhan employed this method to determine DNA sequences, using synthetic location-specific primers (Padmanabhan and Wu, 1972; Wu et al., 1973; Jay et al., 1974). Frederick Sanger applied this strategy to develop a more rapid DNA sequencing method in 1977 (Sanger et al., 1977). Walter Gilbert and Allan Maxam also developed novel sequencing methods. In 1973, they identified a sequence of 24 bp by using a method known as wandering spot analysis.

Current next-generation sequencing (NGS) methods were invented in the 1990s. In October 1990, Roger Tsien, Pepi Ross, Margaret Fahnestock, and Allan J. Johnston patented a novel NGS technology that involved removable 3' blockers on DNA arrays. In 1996, Pal Nyren and Mostafa Ronaghi published another NGS modification called pyrosequencing. In 1997, Pascal Mayer and Laurent Farinelli submitted a method described as DNA colony sequencing. Illumina's Hi-seq genome sequencers implement a mix of these techniques. In 2000, Lynx Therapeutics published and marketed a new NGS system, named as "massively parallel signature sequencing". In 2004, 454 Life Sciences marketed a paralleled version of pyrosequencing (Table 19.3).

DNA sequencing has now become a powerful tool for basic research and translational research. In particular, DNA sequencing has been tested as a tool for diagnosis or prognosis in medicine. Major applications of DNA sequencing include (1) understanding and comprehending the internal structure of genes, (2) understanding which sequence codes for what kind of proteins, (3) predicting gene mutations related to diseases, (4) preparing proteins based on the knowledge of the sequence, and (5) curing diseases based on the knowledge of the sequence.

19.2.1.4 Microarrays

A microarray is a multiplexed lab-on-a-chip. In principle, a microarray is a 2D array on a solid substrate that assays large amounts of biological material using high-throughput screening, multiplexed, and parallel processing and detection methods.

TABLE 19.3 Comparison of NGS Methods (https://en.wikipedia.org/wiki/DNA_sequencing)

Method	Read length	Accuracy (single read not consensus)	Reads per run	Advantages	Disadvantages
Single-molecule real-time sequencing (Pacific Biosciences)	10,000–15,000 bp average (14,000 bp N50); maximum read length >40,000 bases	87% single-read accuracy	50,000 per SMRT cell or 500–1000 megabases	Longest read length. Fast. Detects 4 mC, 5 mC, 6 mA	Moderate throughput. Equipment can be very expensive
Ion semiconductor (Ion Torrent sequencing)	Up to 400 bp	98%	Up to 80 million	Less expensive equipment. Fast	Homopolymer errors
Pyrosequencing (454)	700 bp	99.9%	1 million	Long read size. Fast	Runs are expensive. Homopolymer errors
Sequencing by synthesis (Illumina)	50–300 bp	99.9% (Phred 30)	Up to 6 billion (TruSeq paired-end)	Potential for high-sequence yield, depending upon sequencer model and desired application	Equipment can be very expensive. Requires high concentrations of DNA
Sequencing by ligation (SOLiD sequencing)	50 + 35 or 50 + 50 bp	99.9%	1.2–1.4 billion	Low cost per base	Slower than other methods. Has issues sequencing palindromic sequences
Chain termination (Sanger sequencing)	400–900 bp	99.9%	N/A	Long individual reads. Useful for many applications	More expensive and impractical for larger sequencing projects. This method also requires the time-consuming step of plasmid cloning or PCR

This method was first introduced by Tse Wen Chang in 1983 who used antibody microarrays. This method was then significantly developed when companies such as Affymetrix, Agilent, Applied Microarrays, Arrayit, and Illumina were established.

Although microarrays initially started with DNA, they now include the following: DNA microarrays (cDNA microarrays, oligonucleotide microarrays, BAC microarrays, and SNP microarrays), MMChips (for surveillance of microRNA populations), protein microarrays, peptide microarrays (for analysis of protein–protein interactions), tissue microarrays, cellular microarrays, chemical compound microarrays, antibody microarrays, carbohydrate arrays (glycoarrays), phenotype microarrays, and reverse-phase protein microarray (https:/en.wikipedia.org/wiki/Microarray).

Among these, DNA microarray (also called DNA chip or biochip) is most commonly used. This technology has a long history, from invention to application. DNA chip is a collection of microscopic DNA spots attached to a solid surface. A DNA spot is a specific DNA sequence (called probe). Similar to FISH, the DNA or RNA samples (target sequences) hybridize with a probe that is located in an array on the DNA chip. The probe-target hybridization can be detected or quantified by detection of the fluorophore-, silver-, or chemiluminescence-labeled targets.

Therefore, the core principle of DNA chip is similar to FISH, that is, hybridization occurs between two DNA strands, based on complementary nucleic acid sequences, via hydrogen bonds formed between targets and probes. There are two differences between FISH and DNA chips, including (1) probes are located in an array on a DNA chip, whereas targets are located on the slide in FISH, (2) more probes are used in DNA chip as compared to fewer probes in FISH. Based on the intensity of the signal in the array, the amount of target can be quantified. A DNA chip is commonly used to detect gene mutations and differences in gene expression to diagnose diseases.

19.2.1.5 Cell Culture

Cell culture is an enormous achievement of cell biology. Cell culture is the process wherein cells in vivo are grown outside the body in controlled conditions. The term, "cell culture," is applied to all types of cultures including plant cells, animal cells, microorganisms, and fungi. However, the advent of animal and human cell culture methods leads to more breakthroughs in medical biotechnology, including the production of vaccine, recombinant proteins, monoclonal antibodies (mAbs), and cells for transplantation. The in vitro cell culture method has become more robust in the mid-20th century.

The idea that tissues and cells can be maintained outside the body was formed when the 19th-century English physiologist, Sydney Ringer, could maintain isolated animal hearts in salt solutions that contained sodium, potassium, calcium, and magnesium salts. Then, in 1885, Wilhelm Roux removed and maintained a portion of the medullary plate of an embryonic chicken in warm saline solution. With this experiment, Roux first established the cell culture principle for growing animal cells, namely that animal cells could be maintained with saline in warm conditions. Later, Ross Granville Harrison published some of his experiments from 1907 to 1910, further establishing the methodology for tissue culture.

Cell culture techniques significantly developed in the 1940s and 1950s. With these breakthroughs in animal cell culture, viral culture and production was also developed. The first vaccine based on a virus was successfully generated by Jonas Salk (injectable polio vaccine). This vaccine was developed using results of a cell culture technique pioneered by John Franklin Enders, Thomas Huckle Weller, and Frederick Chapman Robbins, who were awarded a Nobel Prize for culturing virus in monkey kidney cells.

There are two platforms for culturing cells, namely, adherent cell cultures and cell suspension cultures. Efficiency of cell culture increases with become supportive equipment and preprepared media. Using currently available CO_2 incubators, animal cells can be stably maintained at required temperatures (such as 37°C) with a bicarbonate pH buffer that maintains a stable pH value. The culture medium varies and premixed media are available for the culture of general or specific cell types. The use of fetal bovine serum as a supplement for media has now been replaced with chemically defined components, resulting in serum-free media. Defined medium or non-animal component medium is useful in several applications, especially cell therapy. In the 21st century, cell culture has become an essential technique for proliferating stem cells for cell therapy. Since 2010, 3D cell culture has been attempted to generate tissue for research or transplantation. In 3D cell culture, cells are grown in a 3D scaffold in which cells differentiate into specific cell types that are essential for forming a tissue. 3D cell culture has brought about a revolution in tissue engineering.

As a core technique, cell culture is commonly used in the production of biological products, including enzymes, synthetic hormones, mAbs, interleukins, lymphokines, anticancer agents, and vaccines.

19.2.1.6 Interference RNA

RNA interference (RNAi) is a biological process in which mRNA molecules are degraded. This technology was developed by Andrew Fire and Craig C. Mello, who won the Nobel Prize in Physiology or Medicine in 2006. RNAi is found in many eukaryotes, including animals. This process is started by an enzyme, Dicer, which cuts long, dsRNA into short double-stranded fragments of approximately 20 nucleotides, called siRNA. RNA is an essential tool for studying gene functions. Currently, RNAi is used as a tool for disrupting genes involved in cancer. Morpholino is an example of this clinical application (Dirin and Winkler, 2013; Heald et al., 2014).

19.2.1.7 Genome Editing

Recently, genome editing has become a major technique used in medical biotechnology. Genome editing is a technique that is used to change the structure of a gene or genome using engineered nucleases. With this technique, a gene or multiple genes can be inserted, replaced, or removed from the genome. In principle, this technique is initiated with a specific nuclease that causes a double-strand break at desired locations in the genome. Then, the endogenous repair mechanism in cells repairs the breaks by one of two ways: homologous recombination or nonhomologous end joining. To date, there are four families of engineered nucleases that are used: zinc finger nucleases, transcription activator-like effector-based nucleases, the CRISPR/Cas system, and engineered meganuclease reengineered homing endonucleases.

19.2.2 Emerging Trends

19.2.2.1 Stem Cells

Stem cells are nonspecific cells that can differentiate into specific cell types. A cell is considered as a stem cell when it exhibits two important properties, including self-renewal and differentiation. Self-renewal is the capacity of stem cells to produce copies via symmetric or asymmetric division. Differentiation is the capacity of stem cells to achieve function after changes in gene expression. To date, stem cells are classified based on their origins as well as their differentiation potentials. In the most popular classification, stem cells are divided into embryonic stem cells (ESCs) that are isolated from inner cell mass (ICM) of blastocyst embryos, and adult stem cells that are isolated from somatic tissues of fetus or adult.

Recently, other types of stem cells have been demonstrated, including induced pluripotent stem cells (IPSCs), and cancer stem cells (CSCs). IPSCs are stem cells that are artificially produced by inducing pluripotent properties in somatic adult cells. Initially, IPSCs were produced by transfection of four genes, including *Oct-3/4*, *Sox-2*, *Klf4*, and *c-Myc* into fibroblast cells (Takahashi and Yamanaka, 2006). However, other methods have been suggested for inducing pluripotency, including chemicals (Feng et al., 2009). CSCs are defined as cancer cells that exhibit stem cell properties. Therefore, CSCs are also called as tumor-initiating cells.

Mouse ESCs were first isolated from the ICM of murine blastocysts by Evans and Kaufman (1981). Human ESCs were first isolated from the ICM of human blastocysts by Thomson et al. (1998). Although ESCs are better for clinical applications, they can differentiate into multiple cell types in the human body, including cell types from all the three germ layers. However, ESCs are isolated from blastocysts using invasive techniques that can destroy the blastocysts; therefore, this is an ethically controversial issue. To overcome this issue, improvements in isolation techniques of ESCs were suggested, such as using dead embryos (Landry and Zucker, 2004; Gavrilov et al., 2011) or discarded human embryos (Sun et al., 2014; Wang et al., 2012). A better solution is to replace ESCs with IPSCs. Although ESCs and IPSCs have minor differences in properties (Narsinh et al., 2011), IPSCs are pluripotent stem cells that can be used for replacing ESCs in most applications.

Adult stem cells are groups of multiple types of stem cells isolated from a fetus or from adults. To date, adult stem cells can be isolated from almost all somatic adult tissues and extraembryonic tissues, including umbilical cord blood or placenta. However, the most commonly used adult stem cells are hematopoietic stem cells (HSCs) and mesenchymal stem cells (MSCs). HSCs have a long history from research to application. The first reports identifying HSCs were by Ford et al. (1956) and McCulloch and Till (1960). However, HSCs were used in transplantation before the 1950s. At that time, HSCs were regarded as bone marrow cells. The first bone marrow transfusion was carried out in 1939 to treat aplastic anemia (Osgood et al., 1939). The first allogeneic HSC transplantation was pioneered by E. Donnall Thomas and reported in the *New England Journal of Medicine* on September 12, 1957 (Thomas et al., 1957). With this study, in 1990, E. Donnall Thomas won a Nobel Prize for his discoveries aiding in the treatment of human disease. Currently, HSCs can be isolated from various sources, such as the bone marrow, umbilical cord blood, and peripheral blood.

MSCs are multipotent stem cells that can differentiate into a variety of cell types, e.g., osteoblasts (bone cells), chondrocytes (cartilage cells), and adipocytes (fat cells). They were first discovered by Alexander Maximow. McCulloch and James later revealed the clonal nature of marrow cells in 1963 (Becker et al., 1963; Siminovitch et al., 1963). An ex vivo assay for examining the potential of multipotent marrow clonogenic cells was reported in the 1970s by Friedenstein and colleagues (Friedenstein et al., 1974; Friedenstein et al., 1976). The first identified source of MSCs was bone marrow. MSCs are currently isolated from many different tissues in the body, such as adipose tissue, peripheral blood, umbilical cord blood, banked umbilical cord blood, umbilical cord, umbilical cord membrane,

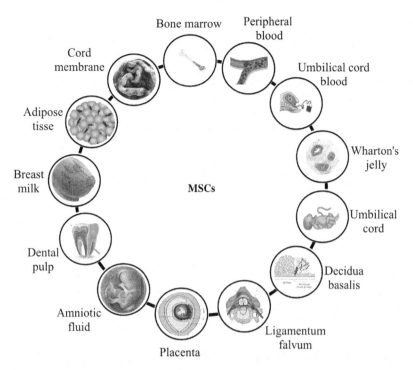

FIGURE 19.2 Sources of MSCs. MSCs can be derived from several tissues in the adult or infant human body.

umbilical cord vein, Wharton's jelly of the umbilical cord, placenta, decidua basalis, ligamentum flavum, amniotic fluid, amniotic membrane, dental pulp, chorionic villi of human placenta, fetal membranes, menstrual blood, breast milk, and urine (Fig. 19.2).

CSCs are a specific type of stem cells that can cause cancer. The idea regarding the existence of CSCs was suggested in 1875 by Julius Cohnheim. The link between cancer and stem cells was first observed by Leroy Stevens and Barry Pierce in 1964. This finding was confirmed by Kleinsmith and Pierce (1964). In recent years, the definition of CSCs has been widely accepted by the science community. CSCs have been discovered in breast cancer, colon cancer, liver cancer, and gliomas. The origins of CSCs in the human body have also been hypothesized and confirmed by experiments. CSCs were validated by the results of genetic and epigenetic modifications of mature cells, progenitor cells, or stem cells. CSCs can be formed by different cells but possess two properties of stem cells, including self-renewal and differentiation potential.

Stem cells bring to medicine a new age of regenerative medicine. Moreover, with the breakthroughs in stem cell research, stem cell-based therapies will truly become a backbone of healthcare in the near future.

19.2.2.2 *The Human Genome Project*

The Human Genome Project (HGP) was an international, collaborative research program that aimed to understand all genes in a human being. All genes are cumulatively known as the "genome." The first full DNA genome to be sequenced was of the bacteriophage φX174 in 1977. Then, in 1984, Medical Research Council scientists completely sequenced the Epstein–Barr virus with 172,282 nucleotides. The first genome of the free-living bacterium, *Haemophilus influenzae*, was published by Venter, Hamilton Smith, and colleagues at The Institute for Genomic Research in 1995. The technology used to analyze the whole genome of this bacterium was called shotgun sequencing that was used to produce a draft sequence of the human genome in 2001.

The HGP was proposed and funded by US government. Planning started in 1984 and got underway in 1990, with the first draft published in 2001, and a completed draft in 2003. This project was performed by 20 universities and research centers in the United States, United Kingdom, Japan, France, Germany, and China.

The International Human Genome Sequencing Consortium published the first draft of the human genome in *Nature* in February 2001. This draft showed that the entire human genome contained approximately three billion base pairs with about 50,000 genes. The complete sequence of the human genome was published in April 2003.

Some findings from the HGP are as noted. Briefly (1) there are approximately 20,500 genes in human beings, (2) the human genome has significantly more segmental duplications than had been previously suspected, and (3) at the time when the draft sequence was published, fewer than 7% of protein families appeared to be vertebrate-specific. With these findings, the human genome holds several benefits for all fields ranging from molecular medicine to human evolution. In fact, the sequencing of human DNA has helped us understand and treat diseases, such as genotyping for specific viruses to direct suitable treatment, identifying of mutations relating to cancer, and other applications in forensic sciences. At present, the human gene sequences are stored at GenBank (US National Center for Biotechnology Information, NCBI) and are available to everyone with access to the Internet.

19.2.2.3 Recombinant DNA Technology

Recombinant DNA technology is a major DNA-based tool that opens a new age for modern biotechnology. With this technology, a gene or multiple genes can be identified, cut, and inserted into the genome of another organism. Using this technology, the first drugs of medical biotechnology were produced, namely human insulin.

The first concept for recombinant DNA technology came from Werner Arber's discovery of restriction enzymes in bacteria that degrade foreign viral DNA molecules. From this discovery, geneticists learned to "cut" and "paste" DNA molecules, and novel restriction enzymes for cutting, and pasting were discovered or invented. The development of recombinant DNA technology was advanced by the collaboration of Stanley Cohen and Herbert Boyer in 1972. They also established the first company that focused on recombinant DNA technology (Genentech) in 1976.

Recombinant DNA technology relates to the usage of three main tools: (1) enzymes (restriction enzymes, polymerases, and ligases); (2) vectors; and (3) host organism. The enzymes will help cut (restriction enzymes), synthesize (polymerases), and bind (ligases) DNA. The restriction enzymes play an important role in this technology. The restriction enzyme will cut at a specific site within the DNA molecule called a restriction site. Usually, the restriction enzyme will produce sticky ends in the DNA sequence that will help it bind specifically to the desired gene. The vector will carry the desired gene. Vectors are important parts of the recombinant DNA technology. They are considered as the final vehicles that carry genes of interest into the host organism. Several types of vectors have been developed to date; however, the most commonly used vectors are plasmids and bacteriophages. A vector must contain the same restriction sites that are found within the desired gene to facilitate gene integration. Besides, vectors should also carry sequences for selection and identification, such as a gene for ampicillin resistance and cloning sites. The host organism is the cell in which the recombinant DNA is introduced. To date, host organisms include bacteria, fungi, and animal cells. To introduce vectors into hosts, techniques involving microinjection, biolistics, gene gun, alternate cooling and heating, and calcium phosphate ions have been used.

In generally, a recombinant DNA technology has five steps: (1) cutting the desired DNA by restriction sites, (2) amplifying the gene copies by PCR, (3) inserting the genes into the vectors, (4) transferring the vectors into host organism, and (5) obtaining the products of recombinant genes (Fig. 19.3).

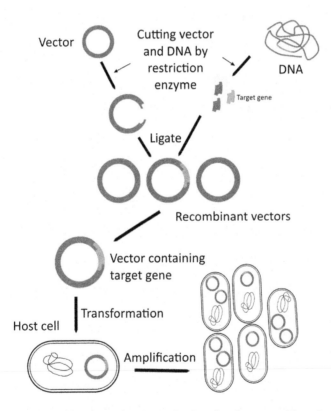

FIGURE 19.3 Recombinant DNA technology. This technology has five steps: (1) cutting the desired DNA by restriction sites, (2) amplifying the gene copies by PCR, (3) inserting the genes into the vectors, (4) transferring the vectors into host organism, and (5) obtaining the products of recombinant genes.

Currently, recombinant DNA technology is widely applied not only in recombinant protein production but also in gene therapy, clinical diagnosis, and in animal and plant transgenesis.

19.3 PRODUCTS OF MEDICAL BIOTECHNOLOGY

19.3.1 Antibiotics

Antibiotics are types of antimicrobial products used for the treatment and prevention of bacterial infections. Antibiotics can kill or inhibit bacterial growth. Although the target of an antibiotic is bacteria, some antibiotics also attack fungi and protozoans, as well as animals and humans. However, antibiotics rarely have an effect on viruses. The major mechanism underlying antibiotics is either interference with the structure of the cell wall of the bacteria or parasites or prevention of them from multiplying.

The first antibiotic, penicillin, was discovered by Alexander Fleming in 1928. To date, several types of antibiotics have been discovered and widely applied in medicine. In addition to naturally occurring compounds, antibiotics have also been synthesized and commercialized. The first synthesized antibiotic was arsphenamine by Alfred Bertheim and Paul Ehrlich in 1907. The first systemically active antibacterial drug was Prontosil that was discovered by Gerhard Domagk in 1933, who was awarded the Nobel Prize in Medicine in 1939.

Antibiotics are mainly used for the treatment and prevention of infection. Antibiotics can be used to treat bacterial infection, protozoan infection, and immunomodulation. They can also be used for preventing infections after surgical wounds and dental work. Antibiotics are categorized based on their mechanism of function. Antibiotics function either against specific bacteria or against parasites. Major types of antibiotics include:

- Penicillins—for example, phenoxymethylpenicillin, flucloxacillin, and amoxicillin
- Cephalosporins—for example, cefaclor, cefadroxil, and cefalexin
- Tetracyclines—for example, tetracycline, doxycycline, and lymecycline
- Aminoglycosides—for example, gentamicin and tobramycin
- Macrolides—for example, erythromycin, azithromycin, and clarithromycin
- Clindamycin
- Sulfonamides and trimethoprim—for example, co-trimoxazole
- Metronidazole and tinidazole
- Quinolones—for example, ciprofloxacin, levofloxacin, and norfloxacin.

19.3.2 Recombinant Proteins

Recombinant proteins provided important breakthroughs in biomedical biotechnology. They are not only used in biomedical research but also in treatment, as drugs. The first recombinant protein used in treatment was recombinant human insulin in 1982. The recombinant protein industry has rapidly grown. To date, more than 130 recombinant proteins are approved by the US FDA for clinical use. However, more than 170 recombinant proteins are produced and used in medicine worldwide.

Recombinant proteins used in the clinic include recombinant hormones, interferons, interleukins, growth factors, tumor necrosis factors, blood clotting factors, thrombolytic drugs, and enzymes for treating major diseases such as diabetes, dwarfism, myocardial infarction, congestive heart failure, cerebral apoplexy, multiple sclerosis, neutropenia, thrombocytopenia, anemia, hepatitis, rheumatoid arthritis, asthma, Crohn's disease, and cancers therapies.

As yet, therapeutic recombinant proteins have undergone three generations, with noticeable improvement in the third generation. The first generation of recombinant proteins included proteins with native structures, while the second generation involved proteins with improved properties, especially PK, biodistribution, specificity, efficacy, and minimal side effects. The third generation will include recombinant proteins that have been improvised for novel routes of administration, include new formulations, exhibit higher efficiency and increased safety (Table 19.4).

TABLE 19.4 Some Recombinant Proteins Were Produced and Commercialized in Medicine

Brand	Generic	Company	Therapeutic category	Indications
THE FIRST GENERATION OF THERAPEUTIC PROTEINS				
Humulin	Insulin	Eli Lilly	Diabetes	Diabetes
Hematrope	Recombinant somatropin	Eli Lilly	Hormones	Growth failure
Genotropin	Somatropin	Pfizer	Hormones	Growth failure
Saizen	Somatropin	Serono	Hormones	Growth failure
Nutropin/ Protropin	Somatropin/ Somatrem	Genetech	Hormones	Growth failure
Intron A	Interferon alpha 2b	Schering-Plough	Anti-infective	Viral infections
Avonex	Interferon beta-1a	Biogen Idec	Multiple sclerosis	Chronic inflammatory demyelinating polyneurophathy
Betaseron/ Betaferon	Interferon beta-1b	Schering AG	Multiple sclerosis	Multiple sclerosis
Procrit/Eprex	Epoetin alpha	J&J	Blood modifier	Anemia
Epogen	Epoetin alpha	Amgen	Blood modifier	Anemia
NeoRecormon	Epoetin beta	Roche	Blood modifier	Anemia
Kogenate	Factor VIII	Bayer	Blood modifier	Hemophilia
NovoSeven	Factor VIIa	Novo Nordisk	Blood modifier	Hemophilia
Benefix	Factor IX	Wyeth	Blood modifier	Hemophilia
Fabrazyme	Agalsidase beta	Genzyme	Enzymes	Fabry disease
Replagal	Agalsidase alfa	TKT Europe	Enzymes	Fabry disease

(Continued)

TABLE 19.4   (Continued)

Brand	Generic	Company	Therapeutic category	Indications
Pulmozyme	Domase alpha	Genetech	Enzymes	Cystic fibrosis
Activase/ Acitlyse	Alteplase	Genetech	Blood factor	Myocardial infarction

THE SECOND GENERATION OF THERAPEUTIC PROTEINS

Humalog/ Liprolog	Insulin Lispro	Eli Lilly	Diabetes	Diabetes
Lantus	Glargine insulin	Sanofi-Aventis	Diabetes	Diabetes
Levemir	Detemir insulin	Novo Nordisk	Diabetes	Diabetes
Pegasys	Pegylated interferon alpha -2a	Roche	Interferon	Hepatitis C
Peg-Intron	Pegylated interferon alpha -2a	Schering Plough	Interferon	Hepatitis C
Aranesp	Darbepoetin alpha	Amgen	Blood modifier	Anemia
Neulasta	PEG-Filgrastim	Amgen	Blood modifier	Neutropenia
Refacto	Factor VIII	Wyeth	Blood modifier	Hemophilia
Amevive	Alefacept	Biogen Idec	Inflammation/ Bone	Plaque psoriasis
Enbrel	Etanercept	Amgen	Anti-arthritic	Arthritis
Ontak	rIL-2-diptheria toxin	Ligand Pharmaceuticals	Cancer	Cancer

19.3.3 Hybridoma and MAb

Hybridoma is a culture of hybrid cells that results from the fusion of B cells and myeloma cells. Hybridoma technology produces hybridomas. This technology was developed to produce mAbs. Hybridomas possess two important properties of B cells, production of antibodies, and immortalization of myeloma cells. MAbs are monospecific antibodies that are produced by a clone of B cells. Therefore, an mAb is understood to be an antibody for a unique epitope.

This technology was invented by Cesar Milstein and Georges J. F. Kohler in 1975. For this invention, they received the Nobel Prize for Medicine and Physiology with Niels Kaj Jerne in 1984. However, the term "hybridoma" was suggested by Leonard Herzenberg in 1976/1977 (Milstein, 1999). Two important inventions include immortalization of the cells and the development of selection techniques for target cells.

Naturally, when an antigen attacks the body (both animal and human), the antigen is phagocytosed by antigen presenting cells. Next, this antigen is presented to B cells for antibody production. An antigen with multiple epitopes can activate several B cells to produce different antibodies. A mixture of these antibodies is referred to as polyclonal antibodies. Antibodies isolated from a B cell line are referred to as mAbs. However, life of B cells is short and the quality of the mAb produced is low and hence not suitable for multiple applications.

Using hybridoma technology, B cells of interest can be immortalized by fusing them with cancer cells (myeloma) to produce hybridomas that are immortal. Methods such electrofusion and chemical fusion can be used to fuse B cells with myeloma cells. To separate B cells (immortal cells) and hybridoma cells (also immortal cells), specific types of myeloma cells are used. Myeloma cells do not secrete any antibody themselves and lack the hypoxanthine-guanine phosphoribosyltransferase (*HGPRT*) gene, making them sensitive to a medium containing hypoxanthine−aminopterin−thymidine (HAT). Fused cells are incubated in HAT medium for roughly 10−14 days. Aminopterin blocks the signaling pathway that is required for nucleotide synthesis. Hence, unfused myeloma cells that lack *HGPRT* die, as they cannot produce nucleotides by the de novo or salvage pathways. Removal of the unfused myeloma cells is necessary, because they are potentially capable of outgrowing other cells, especially weakly established hybridomas. Unfused B cells also die as they have a short life span. Thus, only the B cell-myeloma hybrids survive because the *HGPRT* gene in B cells is functional.

MAbs are widely used for the prevention, diagnosis, and treatment of diseases. In diagnosis, these antibodies are essential components of enzyme-linked immunosorbent assay−based techniques, immunohistochemistry, immunocytochemistry, flow cytometry, and western blots. For example, antibodies to prostate-specific antigen, placental alkaline phosphatase, human chorionic gonadotropin, α-fetoprotein, and other organ-associated antigens are used for diagnosis of primary tumors (Nelson et al., 2000). MAbs for cytokeratin 18 and prostate-specific antigen have also been developed for the diagnosis of prostate cancer (Riesenberg et al., 1993).

MAbs have also been used for the treatment of cancer and autoimmune diseases. The major mechanisms underlying these treatments include blocking of targeted molecule functions, inducing apoptosis and modulating signaling pathways (Breedveld, 2000). During cancer treatment, mAbs can bind to the cancer cell−specific antigen, activate apoptosis and induce an immunological response. They can also be used to deliver a toxin, radioisotope, cytokine, or other conjugate. In autoimmune diseases, they are used to neutralize interleukins such as TNF-alpha, IL-2, or antibodies such as immunoglobulin E.

19.3.4 Vaccines

A vaccine is a biological product that aims to provide acquired immunity for a particular disease. Generally, a vaccine contains the entire disease-causing pathogen as a weakened or killed pathogen or only their toxins or their surface proteins. These agents will actively stimulate the host immune system that then produces a protective response to these agents. Types of vaccines developed thus far include the following:

- Inactivated vaccine: This type of vaccine contains microorganisms that were inactivated or destroyed by chemicals, heat, radiation, or antibiotics. Examples include influenza, cholera, bubonic plague, polio, hepatitis A, and rabies vaccines.

- Attenuated vaccines: This type of vaccine contains live microorganisms that were attenuated. Most vaccines in this type are attenuated viruses such as yellow fever, measles, rubella, and mumps. These vaccines may not be safe for use in immunocompromised individuals.
- Toxoid vaccines: These are made from inactivated toxic compounds. For example, tetanus and diphtheria vaccines. Not all toxoid vaccines are against microorganisms; some are generated for animals, for example, *Crotalus atrox* toxoid is used to vaccinate dogs against rattlesnake bites.
- Subunit vaccines: These vaccines are derived from protein subunits and not the whole microorganism. Subunit vaccines can include surface proteins (vaccine against Hepatitis B virus), virus-like particles (vaccine against human papillomavirus) that contains viral major capsid protein, or the hemagglutinin and neuraminidase subunits (vaccine against the influenza virus).
- Conjugated vaccine: Some bacteria contain outer coats made of poorly immunogenic polysaccharides. By linking these outer coats to proteins (e.g., toxins), the immunogenicity of these bacteria can be enhanced. This approach is used in the *Haemophilus influenzae* type B vaccine.
- DNA vaccine: These vaccines contain DNA sequences derived from viral or bacterial DNA. DNA vaccines cannot directly induce an immune response. When the DNA sequences are translated to a functional protein in animal or human cells, they can evoke an immune response. The immune system recognizes these proteins and attacks the virus or bacteria containing these proteins. However, this type of vaccine is currently undergoing experimental research and is not available for clinical applications.
- Dendritic cell vaccine: This type of vaccine contains dendritic cells primed with particular antigens. Dendritic cells will directly present these antigens to T, B, and NK cells that stimulate the host immune system. Dendritic cell vaccines have been approved for diseases such as prostate cancer and melanoma.
- Heterotypic vaccines: These are specific vaccines that are pathogens of other animals and either do not cause disease or cause mild disease in the organism being treated. A classic example is Jenner's use of cowpox to protect against smallpox. A current example is the use of BCG vaccine made from *Mycobacterium bovis* to protect against human tuberculosis.

19.3.5 Stem Cell Therapy

As one of the four bases of healthcare science, stem cell therapy offers advanced treatment for degenerative diseases as well as for some genetic disorders. Initially, bone marrow was used as source of HSCs. To date, stem cells used include HSCs, MSCs, neural stem cells, epidermal stem cells, endothelial progenitor cells, limbal stem cells, ESCs, and IPSCs. Their use in clinical trials strongly increased approximately 10 years ago. As per clinicaltrials.gov, more than 5000 clinical trials use stem cells for treatment of more than 50 different diseases. More importantly, from 2010 onwards, approximately 12 stem cell–based products have been approved for treatments, some of them as stem cell drugs (Table 19.5).

TABLE 19.5   Some Stem Cell–Based Products Approved for Treatment

Names of products	Kinds of stem cells	Indication	Type of transplantation	Company of production	Country of approval
Cartistem	HSCs from UCB	Osteoarthritis	Allo	Medipost	Korea
MPC	MSCs	N/A	Allo	Mesoblast	Australia
Cupistem	MSCs from AT	Crohn's disease	Auto	Anterogen	Korea
Prochymal	MSCs from BM	GVHD	Allo	Osiris Therapeutics	Canada
AlloStem	MSCs from BM	Bone and cartilage degeneration	Allo	AlloSource	United States
HeartiCellGram-AMI	MSCs from BM	Post heart infarction	Auto	FCB PharmiCell	Korea
Osteocel Plus	MSCs from BM	Bone and cartilage degeneration	Allo	NuVasive	United States
Trinity Evolution	MSCs from BM	Bone and cartilage degeneration	Allo	Orthofix	United States
CardioRel	MSCs from BM	Post heart infarction	Auto	Reliance Life Science	India
HoloClar	Limbal stem cells	Injured cornea	Auto	HoloStem	EU
HemaCord	HSCs	N/A	Allo	New York Blood Center	United States
DuCord	HSCs	N/A	Allo	Duke University and Carolinas Blood Bank	United States

N/A: non indication, GVHD, graft versus host disease; AT, adipose tissue; BM, bone marrow; UCB, umbilical cord blood.

HSC transplantation is now considered to be standard treatment for some types of abnormal hematological conditions, including leukemia, multiple myeloma, non-Hodgkin's lymphoma, Hodgkin's lymphoma, β-thalassemia, and sickle cell anemia (Giralt et al., 2014). HSCs from bone marrow, peripheral blood, and cord blood are used to clinically treat these hematological diseases (Cheuk, 2013). For the past five years, MSC-based therapies have been widely used in clinical applications as one of two main approaches: (1) approved MSC-based products and (2) clinical trials. Some MSC-based products have been approved for clinical applications in several countries for treating diseases involving both autologous and allogeneic transplantation. Besides HSCs and MSCs, other stem cells such as ESCs, IPSCs, limbal stem cells, neural stem cells, and endothelial progenitor cells have also been used to treat specific diseases.

19.3.6 Tissue Engineering

Tissue engineering is the study of growth of new tissues or organs in vitro. Tissue engineering is the product of a combination of stem cells, biomaterials, and signals. Due to strong growth in stem cell–based technologies as well as in biomaterials and recombinant proteins that provide growth factors and cytokines for cell differentiation, tissue engineering is focused to become the future of medicine. At present, multiple ongoing studies are attempting to generate tissues such as bone, cartilage, blood vessels, bladders, skin, and muscle.

The first part of tissue engineering involves stem cells. Stem cells are living parts of tissues. Stem cells can be used to produce tissues in vitro, as observed with adult stem cells (adipose-derived stem cells, bone marrow–derived stem cells, umbilical cord blood–derived stem cells), progenitor cells (as endothelial progenitor cells), and IPSCs. Biomaterials are used to generate the scaffold or 3D structure of the tissue. With 3D printing technology, it has become easier to generate the 3D structure of tissues. Finally, growth factors and cytokines are signals that induce biological effects in stem cells, including stem cell differentiation and stem cell self-renewal. By controlling the effects of these signaling molecules, stem cells can be induced to attach and differentiate within the scaffold to form functional tissue.

19.4 CONCLUSION

Medical biotechnology has provided multiple technologies and products for health care. With discoveries of genes, proteins, and stem cells, medical biotechnology has become an important bridge between biotechnology, medicine, and pharmacology. Molecular medicine, personalized medicine, and regenerative medicine have branched out of medical biotechnology to generate a new era of healthcare science. With these breakthroughs, medical biotechnology is positioned to become the most significant field that improves human health and the quality of life.

References

Becker, A.J., Mc, C.E., Till, J.E., 1963. Cytological demonstration of the clonal nature of spleen colonies derived from transplanted mouse marrow cells. Nature 197, 452–454.
Breedveld, F.C., 2000. Therapeutic monoclonal antibodies. Lancet 355 (9205), 735–740.
Cheuk, D.K., 2013. Optimal stem cell source for allogeneic stem cell transplantation for hematological malignancies. World. J. Transplant. 3 (4), 99–112.
Dirin, M., Winkler, J., 2013. Influence of diverse chemical modifications on the ADME characteristics and toxicology of antisense oligonucleotides. Expert. Opin. Biol. Ther. 13 (6), 875–888.
Evans, M.J., Kaufman, M.H., 1981. Establishment in culture of pluripotential cells from mouse embryos. Nature 292 (5819), 154–156.
Feng, B., et al., 2009. Molecules that promote or enhance reprogramming of somatic cells to induced pluripotent stem cells. Cell Stem Cell 4 (4), 301–312.
Fiers, W., et al., 1976. Complete nucleotide sequence of bacteriophage MS2 RNA: primary and secondary structure of the replicase gene. Nature 260 (5551), 500–507.
Ford, C.E., et al., 1956. Cytological identification of radiation-chimaeras. Nature 177 (4506), 452–454.

Friedenstein, A.J., et al., 1974. Precursors for fibroblasts in different populations of hematopoietic cells as detected by the in vitro colony assay method. Exp. Hematol. 2 (2), 83–92.

Friedenstein, A.J., Gorskaja, J.F., Kulagina, N.N., 1976. Fibroblast precursors in normal and irradiated mouse hematopoietic organs. Exp. Hematol. 4 (5), 267–274.

Gall, J.G., Pardue, M.L., 1969. Formation and detection of RNA–DNA hybrid molecules in cytological preparations. Proc. Natl Acad. Sci. USA 63, 378–383.

Gavrilov, S., et al., 2011. Derivation of two new human embryonic stem cell lines from nonviable human embryos. Stem Cells Int. 2011, 765378.

Giralt, S., et al., 2014. Optimizing autologous stem cell mobilization strategies to improve patient outcomes: consensus guidelines and recommendations. Biol. Blood. Marrow. Transplant. 20 (3), 295–308.

Heald, A.E., et al., 2014. Safety and pharmacokinetic profiles of phosphorodiamidate morpholino oligomers with activity against ebola virus and marburg virus: results of two single-ascending-dose studies. Antimicrob. Agents Chemother. 58 (11), 6639–6647.

Jay, E., et al., 1974. DNA sequence analysis: a general, simple and rapid method for sequencing large oligodeoxyribonucleotide fragments by mapping. Nucleic Acids Res. 1 (3), 331–353.

Kleinsmith, L.J., Pierce, G.B., 1964. Multipotentiality of single embryonal carcinoma cells. Cancer research 24.9, 1544–1551.

Landry, D.W., Zucker, H.A., 2004. Embryonic death and the creation of human embryonic stem cells. J. Clin. Invest. 114 (9), 1184–1186.

McCulloch, E.A., Till, J.E., 1960. The radiation sensitivity of normal mouse bone marrow cells, determined by quantitative marrow transplantation into irradiated mice. Radiat. Res. 13 (1), 115–125.

Milstein, C., 1999. The hybridoma revolution: an offshoot of basic research. Bioessays 21 (11), 966–973.

Min Jou, W., et al., 1972. Nucleotide sequence of the gene coding for the bacteriophage MS2 coat protein. Nature 237 (5350), 82–88.

Narsinh, K.H., Plews, J., Wu, J.C., 2011. Comparison of human induced pluripotent and embryonic stem cells: fraternal or identical twins? Mol. Ther. 19 (4), 635–638.

Nelson, P.N., et al., 2000. Monoclonal antibodies. Mol. Pathol. 53 (3), 111–117.

Osgood, E.E., Riddle, M.C., Mathews, T.J., 1939. Aplastic anemia treated with daily transfusions and intravenous marrow: case report. Ann. Intern. Med. 13 (2), 357–367.

Padmanabhan, R., Wu, R., 1972. Nucleotide sequence analysis of DNA. IX. Use of oligonucleotides of defined sequence as primers in DNA sequence analysis. Biochem. Biophys. Res. Commun. 48 (5), 1295–1302.

Riesenberg, R., et al., 1993. Immunocytochemical double staining of cytokeratin and prostate specific antigen in individual prostatic tumour cells. Histochemistry 99 (1), 61–66.

Rudkin, G.T., Stollar, B.D., 1977. High resolution detection of DNA-RNA hybrids in situ by indirect immunofluorescence. Nature 265, 472–473.

Sanger, F., Nicklen, S., Coulson, A.R., 1977. DNA sequencing with chain-terminating inhibitors. Proc. Natl. Acad. Sci. USA 74 (12), 5463–5467.

Siminovitch, L., McCulloch, E.A., Till, J.E., 1963. The distribution of colony-forming cells among spleen colonies. J. Cell. Physiol. 62, 327–336.

Sun, B., et al., 2014. Effects of group culture on the development of discarded human embryos and the construction of human embryonic stem cell lines. J. Assist. Reprod. Genet. 31 (10), 1369–1376.

Takahashi, K., Yamanaka, S., 2006. Induction of pluripotent stem cells from mouse embryonic and adult fibroblast cultures by defined factors. Cell 126 (4), 663–676.

Thomas, E.D., et al., 1957. Intravenous infusion of bone marrow in patients receiving radiation and chemotherapy. N. Engl. J. Med. 257 (11), 491–496.

Thomson, J.A., et al., 1998. Embryonic stem cell lines derived from human blastocysts. Science 282 (5391), 1145–1147.

Wang, F., et al., 2012. Analysis of blastocyst culture of discarded embryos and its significance for establishing human embryonic stem cell lines. J. Cell. Biochem. 113 (12), 3835–3842.

Wu, R., Tu, C.D., Padmanabhan, R., 1973. Nucleotide sequence analysis of DNA. XII. The chemical synthesis and sequence analysis of a dodecadeoxynucleotide which binds to the endolysin gene of bacteriophage lambda. Biochem. Biophys. Res. Commun. 55 (4), 1092–1099.

Tissue Engineering: Towards Development of Regenerative and Transplant Medicine

Mustafa S. Elitok[1,2], *Esra Gunduz*[1,2], *Hacer E. Gurses*[1,2] *and Mehmet Gunduz*[1,2]

[1]Turgut Ozal University, Ankara, Turkey [2]Yunus Emre Mahallesi, Ankara, Turkey

20.1 INTRODUCTION

The definition of tissue engineering, according to International Union of Pure and Applied Chemistry (IUPAC), is "to use of a combination of cells, engineering and materials, and suitable biochemical and physiochemical factors to improve or replace biological functions" (Griffin et al., 2009). Tissue engineering was initially defined by attendees of the first Natural Science Foundation, United States, sponsored meeting in 1988 as the "application of the principles and methods of engineering and life sciences toward fundamental understanding of structure-function relationship in normal and pathological mammalian tissues and the development of biological substitutes for the repair or regeneration of tissue or organ function" (Chapekar, 2000). With these statements, tissue engineering is considered as one of the major approaches of regenerative medicine, which may offer new treatment alternatives for organ replacement or repair damaged organs. By the understanding of tissue structure, biology, physiology, and cell culture techniques, it is believed that conditions such as end organ failure or the problems after organ transplantation, e.g., life-long immunosuppression, chronic organ rejection, can be overcome.

Today, tissue engineering offers several opportunities including the creation of functional grafts suitable for implantation and the repair of failing tissues, studying stem

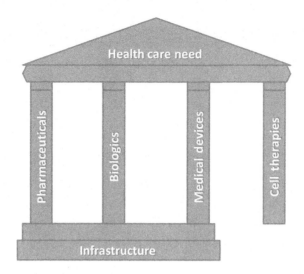

FIGURE 20.1   The platform of healthcare sector. The columns stand for individual sector technologies; the steps represent the essential infrastructure, i.e., scalable manufacturing, appropriate regulation, clinical adoption, and the pediment (roof) symbolize total medical need. In spite of the impressive success of the first three technologies in is the medical sector so far, there are still untreatable and severe diseases and conditions waiting for extensive health care and cure. This unmet medical need can only be satisfied by the contribution of CTI and the synergistic hard work of all four pillars for the benefits of the patients. The lack of steps under the cell therapies tells us the necessary infrastructure for the last pillar is not yet in place and there are still works to do.

cell behavior and developmental processes in the context of controllable three-dimensional (3D) models of engineered tissues, and lastly, the utilization of engineered tissues as models for studies of physiology, diseases, and the medical treatment (Grayson et al., 2010; Godier et al., 2008; Grayson et al., 2009).

Over the past few decades, the creation of implantable tissues, such as skin, cartilage, bone, blood vessels, skeletal muscle, bladder, trachea, and cardiac tissue, have been conducted and several of them have entered into clinical trials (Khademhosseini et al., 2009).

Tissue engineering is also regarded as a part of cell therapy industry (CTI), which is accepted as the fourth and final pillar of global healthcare, beside the pharmaceuticals, biologics, and medical devices (Fig. 20.1). CTI is a distinct healthcare sector being rapidly developed the capability and capacity to be a highly competitive, sustainable, multibillion dollar 21st century industry (Mason et al., 2011).

The main approaches of tissue engineering can be juxtaposed as:

I. Use of an instructive environment (e.g., bioactive material) to recruit and guide host cells to regenerate a tissue;

II. Delivery of repair cells and/or bioactive factors into the damaged area; and

III. Cultivation of cells on a biomaterial scaffold in a culture system (bioreactor), under conditions designed to engineer a functional tissue for implantation (Discher et al., 2009).

In this context, the features of tissue engineering will be summarized and the basic approaches to design and implant an artificial tissue/organ will be explained briefly.

20.2 TYPES OF CELLS (PROLIFERATION AND DIFFERENTIATION)

Three main approaches of tissue engineering have been mentioned above briefly as the need of attraction and guidance of autologous cells into the tissue, providing the conveyance of biological compounds and cells necessary for the proliferation and differentiation mechanisms and, lastly, the incubation of cells and scaffold environment within the needed physical, mechanical, and biological conditions being prepared by a culture system called bioreactors.

As it is seen, the use of appropriate cell type is a critical part of tissue regeneration. Strategies for cellularization of tissue matrix need to account for the unique characteristics of the parenchymal and nonparenchymal cell types of each organ. Depending on the need of cell type, the procedure may alter. Whether it is expected to start tissue cellularization with fully differentiated cells, it is better to obtain a cell biopsy from the patient. Assuming that the cells are unable to divide effectively, diseased or expanding ex vivo insufficiently, endogenous organ-specific progenitor cells might be suitable. As another option, multipotent stem cells (e.g., mesenchymal stromal cells and endothelial progenitor cells; EPCs) can be directed into desired differentiation. Lastly, autologous induced pluripotent stem (iPS) cells are a good alternative choice for the source of required cell types. In general, adult human mesenchymal stem cells have documented capacity to form a number of tissues (including cartilage, bone, fat, and blood vessels) but not necessarily cardiac muscle, nerves, hepatocytes, or Langerhans islets. Although the presence of some evidence that EPCs (or precursor cells) are the most promising among those contributing to the vessel structure because of their success on surviving and proliferation better than terminally differentiated endothelial cells (ECs) (Xu et al., 2012) embryonic-like stem cells, like the newly derived iPS, have essentially unlimited potential for expansion in vitro which might open the possibility of creating autologous embryonic-like cells that could solve the matter of assembling more complex tissues including cardiac regeneration (Jakab et al., 2010).

One obvious requirement is that of immune tolerance of the repaired cells. It is essential to use patient-specific autologous cells, i.e., cells obtained from the patient, in the transplantation to avoid disease transmission, life-long immunosuppression, or the possibility of rejection of transplanted organ. Cell source options mentioned above eliminate the need of immunosuppression as long as the acquired cells are sourced by the patient.

Despite all the information and scientific background we have, there are still unclear aspects about the use of appropriate cell type when constructing a tissue. As an example, it is crucial to improve the cell therapy approaches in the skin regeneration therapy. Keratinocytes alone do not result in fully functional skin, and in vitro cell culture or graft engineering can take weeks. There are also some problems about the functionality and survival of cells after the in situ implantation (Guenou et al., 2009). It is best to say that a mix of different types of cells will be needed, and these cell types will vary by the organs and tissues required, due to the advantages and challenges of various potential cell sources (Badylak et al., 2012).

20.3 SCAFFOLDS

Throughout the experimental approaches to tissue engineering, typically, three main features have been investigated that can be summarized as cell-based therapies, tissue-inducing factors, and biocompatible scaffolds. "Scaffolding" term is first introduced by Barth in 1893 (Barth, 1893) as to use this notion like a porous matrix or an implant allowing cells to infiltrate and regenerate the local tissue. The term, eventually, became possessing alternative concepts (Jakab et al., 2010) such as using natural and synthetic substrates, nanocomposite materials, or decellularized extracellular matrix (ECM), and additionally, maturing the concept towards cell—material interactions, release of biological factors, and design of shape and functionality for specific purposes (Gloria et al., 2010). Early artificial scaffolds were designed to provide cell structural integrity on a macroscopic level, but only achieved moderate success. It is now widely accepted that to facilitate proper tissue functionality, scaffolds should also establish a tissue-specific microenvironment to maintain and regulate cell behavior and function (Khademhosseini et al., 2009). The scaffold, onto which cells are seeded, enabling them to attach and colonize, is therefore a key element for tissue engineering.

As a result of the increasing speed of researching the dynamics and features of scaffolds, it is becoming more complex and sophisticated when differentiated cells and stem cells are included to test their vitality and functionality in the scaffold. Moreover, scaffolds can cause problems owing to their degradation, evoking immunogenic reactions and other unforeseen complications, which arouse the importance of further development and research about scaffolds.

The properties of scaffolds consist of several parameters, including biological substances used, porosity, elasticity, stiffness, and specific anatomical shapes. It is also needed to reinstate the tissue-specific structure, activity, and physical behavior. Ultimately, scaffold should also provide repopulated cell-specific topological features (nano- or micro- and macroscale), mechanical environment, surface ligands, and facility to release chemical compounds (angiogenic factors or cytokines) (Freytes et al., 2009). That reveals the need of dynamic reciprocity between scaffolds and cells that will shape the structure of scaffold and affect the functions of it.

Here, the basic designing approaches of scaffolds will be pointed out and the novel methods will be cited, briefly. It can be separated into four parts as follows: ECM scaffolds, natural substances, synthetic biomaterials, and scaffold-free strategies.

20.3.1 ECM Scaffolds

It is good to start with using naturally occurring ECM as a biological scaffold. All tissues and organs are architected in the design of mutual interaction of cells and ECM. ECM is actually a secreted product of the resident cells, composed of tissue-specific 3D environment of structural and functional molecules. Furthermore, ECM is considered as a dynamic interchangeable media with the resident cell population which turns out that ECM can affect genetic profile, proteome, and protein functionality of cells depending on the parameters of pH, oxygen concentration, mechanical forces, and its biochemical milieu.

All of these properties influence the mitogenic and chemotactic mechanisms of endogenous stem and progenitor cells, regulate the immune response, and impact cell phenotype and function (Tottey et al., 2011). Thus, native ECM is a logical and ideal scaffold for organ and tissue reconstruction.

To establish intact 3D ECM scaffolds, varying allogeneic, xenogeneic tissues and organs have been manufactured. They are prepared by the process of decellularization. The strategy contains the principles of decellularizing a site-specific ECM and repopulating it with autologous fully differentiated, progenitor or stem cells of the current patient to prevent adverse effects of implantation or immune responses. It is demonstrated that these scaffolds can maintain or promote site-appropriate cell phenotypes during the repopulation, through presentation of the ligands and bioactive molecules that are necessary to self-assemble into functional groupings of cells (Ott et al., 2008). To give organ-specific examples, it is shown that ECM successfully directed cells to their target region and supported growth and differentiation of local stem and progenitor cells in skin graft that are potential cure for scar-free healing (Guenou et al., 2009). Akin to skin, there are some evidence that ECM scaffolds induce strong in vivo angiogenic response and could be used for partial or total implantation of larynx and trachea (Baiguera et al., 2011). For the liver, in spite of the potentiality to transplant a recellularized whole-liver scaffold, there are long-term restrictions consisting of hemorrhage and thrombotic complication risks. Including liver, for all complex organs, long-term in vivo perfusion and functional endothelialized vasculature must be formed to allow whole-organ engineering, a main challenge at the moment. Finally, the use of acellular ECM scaffolds have been widely reported in successful clinical studies containing the functional reconstruction of musculotendinous tissues (Borschel et al., 2004), lower urinary tract structures (Chen et al., 1999), esophagus (Nieponice et al., 2006), cardiovascular structures (Bader et al., 1998), and skin (Hodde et al., 2005). Especially for skeletal muscle, scaffold does not need an ex vivo recellularization step, but rather relies on recruitment of endogenous stem or progenitor cells through naturally occurring cryptic peptides derived from the native ECM. These cells proliferate and differentiate in situ consequently (Agrawal et al., 2011).

One additional advantage of using decellularized ECM scaffolds is its ability to serve 3D architecture of the capillary bed, besides its biomechanical properties of the constituent fibers. To empower the repopulation and differentiation processes of implanted cells, and, to maintain life-long restoration of tissue, enough oxygen and nutrient support must be conducted, as well as the transfer of bioactive molecules throughout the tissue. In a recent study, it is stated that after the decellularization process, the first three—four branches of capillary vasculature are preserved, which allows enough perfusion and nutrition conveyance to the tissue (Sarig et al., 2012). This peculiarity of ECM scaffold makes it promising to tackle one of the main difficulties confronted in tissue engineering, capillary network of tissues, and organs.

In addition to ECM scaffolds, amniotic membrane (AM) is an alternative source of construction dock for the repopulating cells. AM is located in the inner membrane of fetal membranes, including the monolayer epithelium, thick basement membrane, and avascular stroma. In literature, acellular human AM is tested as a substrate for culturing corneal epithelial cells (ECs) and encouraging results are obtained (Koizumi et al., 2000).

20.3.2 Natural Biomaterials

Using ECM scaffolds relies on the decellularization technique, which takes quite long time and projects some unwilling consequences afterwards. Additionally, conducting a proper and fully desired tissue construct is not possible yet with this method, regarding the difficulty to accomplish physically and biologically sufficient tissue or organ. With the additional ambition to modify tissues along with specific purposes such as regulated drug or chemical release, employing natural and synthetic biomaterials became inevitable.

So far, natural, synthetic, semisynthetic, and hybrid materials have been proposed and tested as scaffolds for tissue regeneration. Natural polymers used in tissue engineering consist of collagen, alginate, agarose, chitosan, chitin, fibrin, silk fibroin, and hyaluronic acid (or hyaluronan). Collagen is a natural protein material that is commonly used in tissue engineering because environment of the cell metabolism by collagen scaffold is close to physiological conditions. More interestingly, it is stated that collagen, as natural polymers, can promote cell adhesion and proliferation better than synthetic polymers (Wang et al., 2013). Fibrin, alternatively, can be degraded completely after the implantation and can be gathered from autologous plasma of the body. It has the advantages of low price, availability, and good tolerance to cells.

To date, natural biopolymers have been used successfully in various applications of tissue engineering such as skin, cartilage, bone, meniscus, cornea, and nerve tissue engineering. However, the growth and differentiation of progenitor—stem cells have not been conducted as expected by using only natural biopolymer-based scaffolds. Therefore, this result paved the way that synthetic products can be constructed or even natural and synthetic materials can be combined to obtain maximum efficiency.

20.3.3 Synthetic Biomaterials

Unlike natural polymers, synthetic polymers are man-made polymers that may present several advantages as well as more flexibility and manipulability into different size and shapes (Cheung et al., 2007). By incorporating functional groups and side chains, synthetic polymers can also be activated with specific molecules and they are biodegradable.

In tissue engineering, commonly used synthetic polymers are polyglycolic acid (PGA), polylactic acid (PLA) and their copolymers such as poly (lactic-co-glycolic) acid, and polycaprolactone (PCL). When these polymers are degraded, the products of that reaction can be removed by the body cells. On contrary to brittleness of natural biopolymers, synthetic material—based scaffolds are more flexible, which improves the robustness of the tissue.

Consequently, new materials that are the combination of natural substances and synthetic polymers have been established to take the advantages of two groups and construct even more flexible and robust tissues and organs. These chemical constituents, called composite materials, allowed researches to control the degradation rates of scaffolds and bioactivity by fusing natural and synthetic monomers (Mathieu et al., 2006). The term "composite material" generally refers to the combination, on a macroscopic scale, of two or more materials that differ in composition or morphology to obtain particular chemical, physical, and mechanical properties. For example, collagen deposition and alignment during construct maturation is crucial for achieving adequate mechanical strength needed for implantation. Nevertheless, using biodegradable scaffolds can lead to the residual presence of polymer fragments disrupting the normal organization of the vascular wall. For such a mechanically demanding application, balancing the degradation rate of the scaffold material must be calculated and new strategies should be developed.

Up to the present, with attaining specific advantages, various fields of study have been developed new applications by using polymer-based composite biomaterials that are also called "biocomposites" (Ramakrishna et al., 2001). Lately, the interest in applications for biocomposites has extended to nanocomposites, which will be mentioned in Section 20.10.

20.3.4 Scaffold-Free Strategies

Although there have been very exciting developments in tissue engineering based on using natural or synthetic scaffolds, the main problems of constructing a fully functional tissue or organ still remains challenging such as a proper vasculature. It is also being realized that ultimately the best approach is to rely on the self-assembly and self-organizing properties of cells and tissues and the innate regenerative capability of the organism itself, not just simply preparing tissue and organ structures in vitro followed by their implantation.

In parallel to all these strategies, scaffold-free (or self-assembly) approaches demonstrate that fully biological tissues can be engineered with specific compositions and shapes, by exploiting cell–cell adhesion and the ability of cultured cells to grow their own ECM, and thereby help to reduce and mediate inflammatory responses. As an example, human smooth muscle cells are grown on culture plates with fibroblasts to form a "cell sheet" in a porous, tubular mandrel. Consequently, these cells construct the equivalents of media and adventitia of blood vessels. After several weeks of maturation, ECs are seeded into the lumen of sheets (L'Heureux et al., 1998). Eventual engineered blood vessel successfully shows the similar organization and mechanical properties of real blood vessel and its ECM. Now, autologous small-diameter vascular grafts engineered by using this method are currently in clinical trials for hemodialysis access (McAllister et al., 2009). There are alternative methods that can project versatility to use various cell types, including epidermal keratinocytes, kidney ECs, and periodontal ligaments, to establish fully functional tissues. There are clinical trials using the latter method involving corneal transplantation, which promoted the recovery of weakened vision (Nishida et al., 2004).

One of the main advantages to use cell sheets with respect to scaffold-supported cells is that the former can more easily communicate electrically via gap junction formation. It is also believed that self-assembly approach may bring some useful solution to the problem of constructing tissue with "properly working capillary bed or intermediate vasculature," which is the "missing link" between two perfused vascular trees (arteries and veins). This seems possible with the help of manufacturing endothelialized microtissues to form macrotissues (McGuigan and Sefton, 2006). In this approach, modular constructs are assembled from sub-millimeter-sized cylindrical modules of collagen or gelatin seeded with cells and endothelialized. Afterwards, these modules randomly self-assembled into tissue constructs with interstitial spaces enable perfusion of medium or blood. However, the stability and the results of direct anastomosis to the host vasculature still remain unclear and new researches are required (Fig. 20.2).

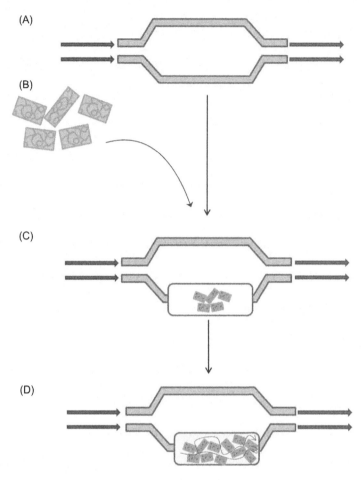

FIGURE 20.2   The self-assembly approach depends on the natural ability of cells to unite and generate new natural environment by secreting extracellular matrix components. (A) The basic construct of vascular graft to provide nutrient support to engineered tissues. (B) Prepared multicellular blocks are seeded into the desired position in the tissue. (C) The blood flow is conducted through cellular blocks, while the conditions of incubation for the cells have been prepared. (D) Self-assembling cells proliferate and generate new ways to provide maximum perfusion among them. These unnatural spaces between the cells become the new coronary bed for the engineered tissue.

Cellular self-assembly approaches represent an alternative and offer a complement to scaffold-based tissue engineering. They allow establishing high cell density, controlled deposition of ECM, and positional specificity of cell patterning (Jakab et al., 2010).

20.4 BIOMOLECULES IMPORTANCE IN TISSUE ENGINEERING

The main aim of tissue engineering, of course, is to build a tissue or organ that is desired to show almost same properties of normal tissue in the aspects of mechanical, functional, physical, and biological conditions. To conduct this, many fields of study involve in and produce new ideas on how to make it possible to engineer a natural construct in different stages. Biochemistry and biology is now two very important fields to study the cell behavior and its relationship with compounds currently in the environment.

To enhance the healthy growth of a novel engineered tissue, physiological mechanisms of cell proliferation and differentiation must be clearly manifested. Today, it is known that enabling a convenient biological, physical, and chemical environment that achieves almost the same condition of body is the key process. Such "biomimetic" environment, as result of biology and engineering interacting at multiple levels, should be suitable to direct the cells to differentiate at the right time, in the right place, and into the right phenotype and eventually to assemble functional tissues by using biologically derived design requirements (Godier et al., 2008; Cimetta et al., 2009). The controlled delivery of biological factors in 3D scaffolds represents another key for tissue regeneration and growth such that vascular endothelial growth factors and basic fibroblast growth factors can specifically enhance vascularization essential for maintaining continuous blood supply to developing tissues (Langer, 2009).

It must also be mentioned that ECM and its chemical properties are very important to direct cells to their target region and to support their development and differentiation. It has been stated before that degradation products of ECM filaments such as collagen lure the progenitor or stem cells of the host body to empower the healing or development process of engineered tissue (Tottey et al., 2011).

20.5 ASSEMBLY METHODS OF A TISSUE CULTURE AND ITS MAINTENANCE

To structure a legitimate engineered tissue, correct cell type must be selected, an appropriate scaffold should be carried out (or a self-assembly approach may be addressed), and by using a conventional assembly technique, cells and scaffolds must be joined together. Consequently, to reinforce the healthy development of a tissue, a bioreactor environment that can support the cells from sides of mechanical, physical, and chemical properties of the natural milieu must be provided. Here, the engineering and biomechanical aspects of assembly methods will be explained briefly.

20.5.1 Engineering Design Aspects

The choice of repair cells is central to any of the many different modalities of tissue engineering—injection of repair cells (with or without biomaterial), implantation of a fully formed cell-based graft or mobilization of the host cells into the site of injury. Completely matured cells, progenitor, or stem cells may alter the required scaffold or environment and their chemical components because of the differing need of growth, differentiation, and attracting factors to the tissue site (Table 20.1). Eventually, after the novel developments happened in molecular biology and materials science in recent years, cell guidance concept has been adapted in tissue engineering (Nair and Laurencin, 2006). According to

TABLE 20.1   Some Cell Sources and the Needed Types of Scaffolds, Bioreactors or Other Conditions.

Cell source	Scaffold/ cell sheets	Bioreactors/other devices	Biological effects
HUVEC	Collagen	Intact microvascular segments gelled in polyethylene tubes	Patent vascularization
HUVEC	Poly(ethylene glycol)-dicrylate hydrogel scaffold	Device micropatterned with cell adhesive ligands	Spatial regulation of the angiogenic response
HUVEC	Alginate	Multishear perfusion bioreactor	Expression of the intercellular adhesion molecule (ICAM-1) and endothelial nitric oxide synthase (eNOS)
MSC	Hyaluronan	Unidirectional cyclic stretch bioreactor	Cell multilayer organization and invasion of the 3D mesh of the scaffold/muscle protein expression
Endothelial cell line	Polycaprolactone porous scaffold	Shear perfusion bioreactor	Rapid endothelialization method to create preformed vascular networks
hESC/hiPSC/ HUVEC/MSC	Collagen	Uniaxial stress bioreactor	Cardiomyocyte and matrix fiber alignment/increased vessel-like structures
HUVEC/human skeletal myoblast	Cell sheets	HUVECs sandwiched between sheets of myoblasts	Cell sheets sprouted a capillary-like network and connected to the host vessels
Human microvascular endothelial cell/human dermal fibroblast	Collagen	System for guided tubulogenesis coupled with 3D organotypic culture	Endothelial tube formation and endothelial vessels surrounded by collagen type IV

HUVEC, human umbilical vein embryonic cell; MSC, mesenchymal stem cell; hESC, human embryonic stem cell; hiPSC, human induced pluripotent stem cell.
Adapted from Muscari, C., Giordano, E., Bonafè, F., Govoni, M., Guarnieri, C., 2014. Strategies affording prevascularized cell-based constructs for myocardial tissue engineering. Stem Cells Int. 2014, 434169. As It Is Seen, Depending on the Type of Cell Conditions, the Type of Scaffold and Its Constructing Method Varies Strikingly. It is the Best to Say That the Procedure and Its Components Must Be Compatible with the Organ or Tissue We Want to Design and Engineer.

this concept, natural, synthetic, or composite biomaterials can be used to form a functional scaffold that can also affect progenitor or stem cells being seeded into it and attract them to their appropriate sites in the tissue. The more we know about the complex features of cell—material interaction, the more we are getting close to a successful tissue graft by using this method.

Based on this concept, several scaffold construction techniques have been published by using biopolymers. Bioresorbable scaffolds would slowly resorb, leaving no foreign substances in the body, making them a good candidate as a template for tissue regeneration. To enhance better quality, as previously defined, composite material consisting of two or more materials has been produced. As an example of bone tissue, hydroxyapatite particles can be considered the inorganic reinforcing phase of the composite scaffold, whereas polymers like PGA or PLA function as the organic matrix (Guarino et al., 2009). In some methods, addition of more supporting compounds may shape the scaffolds into porous and may intensify the interaction between composite materials. Such an approach have been conducted in bone tissue engineering by reinforcing the scaffold with PCL and alpha—tricalcium phosphate that synergistically boost the mechanical response in compression (Guarino and Ambrosio, 2008). Although it is stated in the course of cornea tissue engineering, generally, methods of forming composite scaffolds are all critical to achieve an artificial tissue or replacements for different portions of organs for transplantation (Wang et al., 2013). Aside from the current methods, nanotechnology is also expected to play an important role in creating novel tissue regeneration strategies (e.g., cell sheet engineering) (Elloumi-Hannachi et al., 2010), thus, it should also be taken into consideration when tissue is designed. Benefits of nanotechnology will take place in Section 20.10.

The techniques that produce such composite materials include phase inversion and salt leaching for bone tissue engineering (Guarino et al., 2008); custom braiding techniques, which allows the fabrication of substrates with specific geometry, pore diameter, porosity, and mechanical aspects for ligaments (Cooper et al., 2005); air—liquid technique for cornea (Alaminos et al., 2006); and computed tomography, computer numerical control machining methods for mold preparation, solvent casting, freeze—drying, and lamination for meniscus (Kon et al., 2008). Additionally, well-defined magnetic PCL/iron oxide scaffolds have been fabricated by 3D fiber deposition technique with the help of nanotechnology. With this method, through magnetic driving to attract and take up in vivo growth factors, stem cells or other bioagents bound to magnetic particles, which can result as magnetic nanocomposite scaffold (Bock et al., 2010).

As it has been mentioned earlier, using decellularized ECM is another option to construct a scaffold. Although significant scientific and ethical challenges remain as this approach advances to clinical use, successful proof of principle for organs such as liver, heart, lung, trachea, esophagus, and skeletal muscle has been shown (Badylak et al., 2012). Principle of decellularization depends on harvesting of ECM from an organ or tissue by removing the cell population while restricting changes in structure, composition, and ligand background of the native matrix, including those components that provide the vascular and lymphatic networks. Especially, protection of 3D architecture of the capillary bed is an excellent advantage of ECM scaffold, beside the benefit of proposing biomechanical properties of the constituent fibers it contains. However, removal of cells from their integrin-bound anchors and intercellular adhesion complexes while protecting ECM surface topography and resident ligands is challenging. It is ideal to remove the cell population without disturbing the structure, composition, or ligand background of the native matrix. Failure to effectively and thoroughly remove cellular remnants can cause a proinflammatory response in the recipient, which is a result of interference with structure of recellularized organ and recipient immune system. Although complete removal of all cell remnants is not possible, a combination of qualitative and quantitative strategies avoids such adverse responses. Techniques vary strikingly and total procedure times for the decellularization processes range from 5 hours to 7 weeks (Badylak et al., 2012).

After the decellularization process, autologous cells are seeded into the ECM scaffold. In this procedure, natural characteristics of the parenchymal and nonparenchymal cell types of each organ must be taken into account. Depending on the health level of cells; dividing, maturing, and differentiation capacities and amount of them after obtaining from patients; fully differentiated, progenitor, natural stem, or iPS cells can be used to repopulate the ECM scaffold. The vasculature and airway structure of scaffold can also help to repopulate the whole construct although the complete recellularization of 3D scaffold is not the objective of the initial cell delivery effort. Instead, the main goal for now is to distribute enough number of cells spatially and allow them to contact with other cells, which consequently end up with self-assembly, proliferation, and differentiation within the scaffold.

Lastly, sheet-based self-assembly techniques and bioprinting approaches widen the applications and production procedures of tissue engineering. Cell sheets that have their own specific compositions and shapes can build up their own ECM and differentiate into desired cell population in the tissue without disturbing host immunity and inflammatory environment. Generally, sheet-based tissue engineering uses the mechanisms of cell sorting to the repair site, fusion between distinct tissues, and conducting the appropriate tissue liquidity (Jakab et al., 2010).

Bioprinting, instead of that, can engineer current 3D living structures. Two different approaches have been described in bioprinting technique. Inkjet printing can design individual cells or small clusters rapidly, versatile, and cheaply (Campbell and Weiss, 2007; Nakamura et al., 2005; Nishiyama et al., 2009). Disadvantage of this method is the difficulty of assuring high cell density for the fabrication of solid organ structures. Moreover, because of the high speed of cell deposition, considerable damage is caused to cells. The other method, called extruder bioprinting, uses "bioink" particles to assemble "biopapers," which are the multicellular aggregates of definite composition, to form an organoid by the postprinting fusion. Cells sort within the bioink particles afterwards (Norotte et al., 2009; Jakab et al., 2004). The method consists of three main steps: fabrication of multicellular building blocks, bioink deposition, and the self-assembly of blocks. Cellular blocks are prepared from cell suspensions, either spherical or cylindrical in shape. The deposition is done by delivering spherical or cylindrical multicellular blocks, bioink, together with hydrogel, biopaper, according to a computer-generated template (Fig. 20.3). With the help of computer sciences, a biological construct, which is close to native tissue with aspects of structure and function, is almost possible. In the last step, fusion of bioink—biopaper is incubated to achieve its final 3D structure and maturation (Norotte et al., 2009; Jakab et al., 2008).

The advantage of this method is that the bioink particles represent small 3D tissue fragments, which arrange adhesive contacts between cells allowing the transmission of vital molecular signals. It is also shown that building straight sheets, complex topologies like branched tubes, or vessel formation is possible with this method. This also means small diameter vessels can be produced fully biological (scaffold-free). However, the method is relatively expensive due to the high cost of the printers.

20.5.2 Biomechanical Aspects of Design (Bioreactors)

Bioreactors are closed mechanical chambers providing suitable environmental conditions for cellular activity (Fig. 20.4). To assemble a functional tissue compatible with

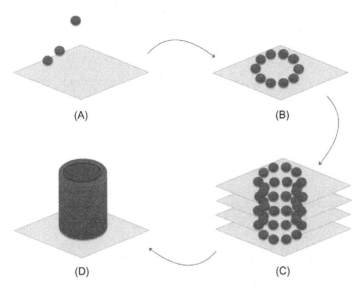

(A)

(B)

(D)

(C)

FIGURE 20.3 Bioprinting processes while using bioink method. (A) Biocompatible hydrogel, also known as biopaper, is prepared, and the building blocks consisting of cells, bioink, is embedded into it. (B) Bioink–biopaper complex is assembled with the shape of circle to design a tubular construct. (C) The deposition of circled layers of bioink and biopaper according to the predefined 3D structure, here a tubular construct. (D) Fusion of the blocks and layers. After the fusion, removal of the hydrogel results in a hollow tube a few days later.

its native form, it is critical to design a surrounding that has the same properties as its original anatomy. As an example, cardiac regeneration and vessel formation need to expose a pulsatile blood flow to maximize its maturation and differentiation. It is also the same with the interaction between lung tissue and airflow (Govoni et al., 2014; Shimizu et al., 2009).

Especially for ECM-based scaffolds, bioreactors are necessary for optimum physiological repopulation of cells by providing a flow of nutrient medium via the vasculature, physical, and environmental stimuli such as electrical activity to induce cardiac maturation or breathing movements for lung cells. The maturation process can take weeks or, even, months inside a bioreactor (Asnaghi et al., 2009).

Bioreactors that can induce shear stress, which is the frictional force of blood flow onto the surface of the cells while being perfused, are used to influence cell behavior according to the phenotype. Shear stress is accepted as an important regulator of the fate of ECs and EPCs, especially when designing arteries and veins. When engineering a vessel, inner surface must be antithrombogenic. Bioreactor that can generate shear stress inducing epithelialization of the designed vessel structure, by this, makes the inner surface antithrombogenic (Kang et al., 2013). Additionally, there are bioreactors to generate pulsatile blood flow and mechanical stress to enhance the growth of cardiomyocytes for heart tissue engineering.

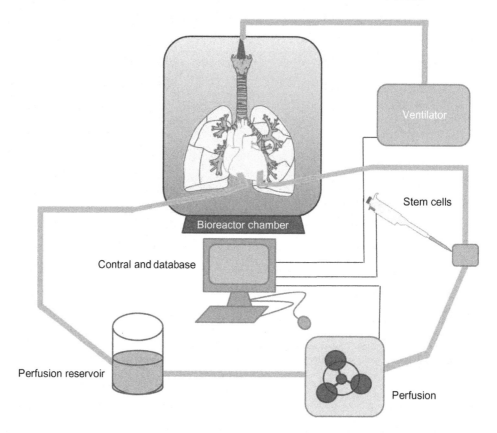

FIGURE 20.4   The basic components of a bioreactor. The bioreactor chamber allows engineered tissues or organs maximum incubation condition by providing similar biological, mechanical, and physical properties of natural milieu. Stem cells can be conveyed to the tissue through natural pathways such as vasculature or airway. The basic needs of cell like nutrients and oxygen support must be provided with the help of enough perfusion and ventilator reinforcement.

20.6 REGENERATION OF A DAMAGED TISSUE USING TISSUE ENGINEERING

So far, several tissues have been successfully regenerated by using the design aspects and methods explained above. By means of regenerative medicine principles, these tissues have been prepared to replace or repair the damaged structures on the body. Here are some major examples about the reflections of tissue engineering on real life.

20.6.1 Blood Vessels

Constructing blood vessels has a key importance in tissue engineering because of the difficulties being confronted during the implantation processes of other engineered tissues such as poor perfusion. To overcome this matter, prevascularization procedure may conduct enough blood flow into the constructs. That also means generating additional blood vessels into the desired tissues. Today, conducting a capillary bed inside the tissue is one of the main researching areas in tissue engineering. Sheet-based self-assembly techniques can generate a nonnatural blood flow path between multicellular blocks after the assembly phase. However, during the implementation, the anastomosis of host vasculature and constructed vessels still stand as a major problem.

Apart from this, repairing a damaged vessel still remains as a problem. At the moment, the best method to design a blood vessel is scaffold-free technique that includes sheet-based tissue engineering and bioprinting. Vessel sheets consisting of epithelial and EPCs are assembled randomly and automatically or printed by the plan of texture entered into the computer. Bioreactors providing pulsatile flow and shear stress intensify the differentiation and maturation of ECs. This mechanical support also empowers the biological conditions of vessel such as being antithrombogenic.

20.6.2 Skin

A bioengineered skin graft must enable the regrowth of a functional and physiological dermal layer of skin, including adnexal characteristics, such as hairs or pigment, to avoid adverse effects happening consequently to implantation like severe scarring and contractures. Skin restoration is accomplished with artificial skin substitutes, cell-based therapies, and their combination.

So far, it is cleared that using ECM-based scaffolds is ideal for skin grafting due to its success on supporting cell engraftment, proliferation, and differentiation in vitro and in vivo. ECM also provides collagen and other substances, which are beneficial to attract autologous progenitor and stem cells into the injury site, too. The popular approach is to implant ECM scaffold into the damaged site and lure host stem cells by the substances being provided from ECM to finish the regeneration (Priya et al., 2008). Cell therapy approaches for skin regeneration and grafting need to be improved in terms of the type of cells used and their survival after in situ delivery. The use of stem cells and progenitor cells with ECM scaffolds would be the best choice until now, to prevent development of scar while repairing the damage.

20.6.3 Bone and Cartilage

Bone can be considered as a natural nanostructured composite consisting of intertwined inorganic and organic compounds. The ECM of bone includes hydroxyapatite crystals (which can be considered as the inorganic part like synthetic polymers in scaffolds), Type I collagen (as the organic part of scaffolds), nanocrystals (as nanocomposites used in scaffolding), and water (Yamashita et al., 2001). With these features, scaffold techniques are based on matching the similarity of bone ECM by performing polymer-based and nanotechnology-supported methods. Besides, due to the hierarchical construct of the bone, with the lowest level belonging to the nanoscale range, materials with nanometer structure appear to be natural choices for creating better scaffolds for bone tissue engineering

(Tran et al., 2009). Comparing to conventional composite materials, nanocomposites induce cell response better due to its similarity to natural structure of bone and presenting better osteoconductivity. Additionally, it is reported that polymer/ceramic composites with nanometer particle size showed better mechanical function in contrast to micro-sized ones (McManus et al., 2005). Recently, carbon nanofibers (CNFs) and carbon nanotubes (CNTs) are used to enhance the properties of bone scaffolds such as weight, strength, and mechanical function. CNFs can be used to empower the functions of selected cells in the scaffold—for example, only the osteoblasts—and to decrease the others that is shown to increase osteoblast adhesion in literature (Price et al., 2003). Lastly, magnetic PCL/iron oxide scaffolds fabricated to lead magnetic nanocomposite scaffolds that can attract and take up in vivo growth factors, stem cells, and other biological agents. This method also conducted better mechanical and biological properties of bones, intensified the adhered number and a more evident spreading of human mesenchymal stem cells comparing to the methods using only PCL fibers (De Santis et al., 2010).

20.6.4 Cardiovascular Diseases

One of the main goals of tissue engineers is to make it possible to repair the damaged heart tissue, especially after the acute myocardial infarction or, in other word, heart attack. Up to now, stem cells transplanted alone or carried onboard of polymeric devices have shown a short survival in the heart because of poor perfusion through myocardium by coronary vessels. To provide efficient oxygen and nutrients, metabolically active grafts should be supplied by a vascular network. Thus, prevascularization process has been described, which represents the maturation of vasculature of construction before the implantation procedure by using specific techniques, generally based on self-assembly (Karam et al., 2012).

To be successfully grafted, a cardiac construct must be sufficiently thick to contract with enough strength and compliance, beat synchronously with neighboring cells, not generate inflammatory reactions, produce a suitable ECM and, ultimately, improve cardiac function in a relevant manner. Unfortunately, the perfusion of a thick construct still remains a challenge in tissue engineering because of the difficulty of architecting capillary network throughout the tissue.

Strategies of prevascularization can be summarized as the transplantation of microvascularized gelled material into the infect region/border zone of the heart, suture/application of the patch on the infarcted tissue, and insertion of the construct into a "pouch" made in the thickness of the cardiac wall and anastomosis of the inlet/outlet perfusion conduits of the construct, often described as arteriovenous loops, with branches of coronary vessels (Muscari et al., 2014).

20.7 THE FORMATION OF BIOARTIFICIAL ORGANS USING CELL-BASED TISSUE ENGINEERING

Besides repairing damaged tissues, tissue engineering also makes it possible to redesign and reassemble a fully functional organ. Until now, musculotendinous tissues, skin, trachea, esophagus, small intestine, bladder, and larynx have been entered into clinical trials, and advancements in liver, heart, and lung have been continuing.

20.7.1 Liver

To construct a fully functional liver, ECM scaffolding methods have been used. In vivo transplantation of a recellularized whole-liver scaffold is possible; but hemorrhage and thrombotic complications have restricted long-term assessment (Uygun et al., 2010). The ability to sustain long-term in vivo perfusion until a functional endothelialized vasculature can be formed is the limiting step at present in the whole-organ engineering approach, not just for liver but for all complex organs. However, there are complex problems about the blood load of the native liver. The healthy liver has two blood supplies, including a low pressure, nutrient-rich, nonpulsatile portal circulation and a high pressure, pulsatile hepatic artery source. It is still unknown how to encourage liver regeneration of implanted engineered liver construct by providing healthy circulatory environment.

20.7.2 Bladder

Akin to liver, decellularized ECM scaffolds are ideal to construct a lower urinary tract structures and urinary bladder. Differently, repopulating cells are not seeded into the scaffolds to generate the novel structure of bladder. Instead of this, scaffold materials depend on endogenous recruitment of cells that have the capacity to form site-specific functional tissue like in skeletal muscle engineering. Degradation of ECM with the release of bioactive cryptic factors is essential for this effect on stem cells, and modulation of the innate immune system has a crucial role in constructive remodeling (Valentin et al., 2009; Brown et al., 2009). Clinical applications for both of tissues (bladder and lower urinary tract) have been performed (Chen et al., 1999). Nevertheless, using endogenous stem cells method still needs to be improved considering the regulatory, financial, and practical cost of such an approach that is probably prohibitive for the incremental advantages it provides.

20.7.3 Kidney

Kidney is one of the necessary organs that lots of patients with end-stage renal disease are waiting for donor. So far, cell therapy with progenitor or multipotent stem cells is the only option for the regeneration of damaged renal tissue. This approach may be effective to acute renal failure but not for the chronic diseases of kidney. To tackle this problem, whole-organ engineering, such as reseeding of ECM has been investigated (Ross et al., 2009). Using decellularized ECM scaffolds may represent promising results; nonetheless, at present, much work needs to be done for ex vivo kidney regeneration with ECM scaffolds and cell therapy approaches to become a viable clinical option.

20.8 CELL THERAPIES IN TISSUE ENGINEERING

Tissue engineering is one of the application areas of CTI such as cell cancer vaccines, cord blood banking, xenografts, and advanced reproductive techniques (Mason et al., 2011). CTI became its international base, with a billion dollar per year turnover and broad spectrum of proven therapies. To enhance better strategies for the novel remedies of diseases, CTI structured treatment options must be reinforced with pharmaceuticals, biological background, and medical devices (Mason and Manzotti, 2009).

One of the early cell therapy applications of tissue engineering is injectable cells onto the damaged organ. Cardiac muscle cells are injected into the injury area to repair and restore the infarction site. The conveyance of the repair cells can be performed via various techniques such as direct injection, systemic intravenous infusion, intracoronary cell infusion, and NOGA mapping system, which is enabled during the open heart surgery by a catheter with a 3D electromechanical mapping system.

20.9 USE OF STEM CELLS AND THERAPEUTIC CLONING IN TISSUE ENGINEERING

The type of cell source is critical for successful treatment in tissue engineering as mentioned before. Native stem cells and, recently, iPS cells propose a promising role in the central of several methods. These methods include the injection of repair cells (with or without biomaterial), implantation of a fully formed cell-based graft, or mobilization of the host cells into the site of injury.

The first method is widely used in bladder and skin graft tissue engineering. Decellularized ECM scaffolds are implanted onto the site regarding the attraction of progenitor or stem cells of the host by ECM degradation products and bioactive factors. It is stated that the use of stem or progenitor cells and engineered scaffolds from preserved ECM is the best solution for such organs (Badylak et al., 2012). However, recellularization procedure of ECM scaffolds is usually performed by using autologous multipotent stem cells with subsequent directed differentiation along organ-specific or tissue-specific lineages. IPS cells are another option to use in such approach. Other methods are cell sheet—based tissue engineering and cell injection technique such as injectable cardiac muscle cells.

The extensive growth and differentiation of EPCs also need shear stress producing bioreactors, which addresses the reality that the importance of providing biomimetic environment including attracting biofactors, mechanical properties of tissue, and, even, nanoscale and microscale shape/adhesion features or neighboring cells. Chitosan-collagen composite-based scaffolds suitable for the proliferation of adipose tissue—derived stem cells, which also support the fact that stem cells are very sensitive to the environmental conditions (Zhu et al., 2009). Hypoxic preconditioning has also declared as a natural process that allows stem cells to survive in microenvironments with a low-oxygen tension that shows the necessity to understand the biological background of stem cells better to have better quality engineered tissues (Muscari et al., 2013).

However, stem cells are not yet proven to differentiate into all of desired matured cells presenting fully accomplishment of its native form such as cardiomyocytes. Furthermore, most allogeneic cells derived from pluripotent cells could be tumorigenic (Sng and Lufkin, 2012).

20.10 AID OF NANOTECHNOLOGY IN TISSUE ENGINEERING

Nanotechnology, a state-of-the-art field of study, proposes unique solutions and novel applications to various fields including drug delivery, in vitro diagnostics, in vivo imaging, therapy techniques, biomaterials, and, eventually, tissue engineering. Especially in tissue engineering, nanotechnology enables the design and fabrication of biocompatible scaffolds at the nanoscale and control the spatiotemporal release of biological factors to direct cell behaviors (Shi et al., 2010).

Recently, nanotechnology makes it possible to construct biomimetic microenvironment at the nanoscale, providing an analog to native ECM (Goldberg et al., 2007). Notably, these technologies have been applied to design nanotopographic surfaces, to encapsulate and control the release of drugs and biofactors. By doing this, nanodevices can direct cellular interactions such as adhesion or, even, gene expression. Living cells are sensitive to nanoscale topography that encouraged scientists to control cell function via this method (Stevens and George, 2005). Today, we can design gratings, pillars and pits onto the cells or scaffold, as well as, nanospheres, vertical nanotubes, and nanofibers. These advances provided opportunity to test cell—nanotopography interaction and the manipulation of cell morphology, signaling, orientation, adhesion, migration, proliferation, and differentiation.

Furthermore, some unique scaffold types have been described by using nanofibers. Nanofibrous scaffolds are now under wide investigation as they exhibit a very similar physical structure to protein nanofibers in ECM (Goldberg et al., 2007). Electrospinning and phase separation techniques are popular to generate continuous fiber network fabrication with tunable pore structure, which results as sponge-like scaffolding. Additionally, nanocomposites-based scaffolds have been researched especially for bone tissue engineering. It is shown that nanocomposites can mimic the constituents of natural bone better than the individual components and induce better cell response because of their similarity with natural structure comparing to conventional composites (Tran et al., 2009). This similarity phenomenon is also shown for tracheal and bronchial scaffolds constructed with nanocomposites (Baiguera et al., 2010). In addition, generating CNTs (single-walled carbon nanotubes) allows the fabrication of lighter scaffolds with mechanical strength and electrical conductivity (Tran et al., 2009).

The integration of fine nanofeatures into microfabricated 3D scaffolds can enhance control of cell function via cell—nanotopography interactions. Microscale scaffolds can be further decorated with these features and patterns, such as grooves ad ridges, to better match the nano-architecture of ECM (Borenstein et al., 2007). The more we approach the native structure of ECM, the more we are getting close successful generation of tissues and organs.

Lastly, nanotechnology allows controlling the release of chemical compounds within the scaffold. Such mechanism is used for angiogenesis by monitoring the release of vascular endothelial growth factor and basic fibroblast growth factor to enhance vascularization for maintaining continuous perfusion. This can be done by affecting the particle formulation, entrapping the compound within nanoparticles, or adjusting the thickness and porosity of the nanopolymers (Ma, 2008). Some of these nanocarriers even have the unique ability to trigger drug release by responding to environmental changes such as pH, temperature, light, and mechanical stress. However, these efforts are still in its infancy compared to the other advancements of nanotechnology in other fields (Zhang and Uludağ, 2009). Aside from these benefits, nanotechnology is also expected to play an important role in establishing novel tissue regeneration strategies like cell-sheet engineering and handle the obstacles presenting today. One exciting opportunity lying ahead is the idea of "organ-on-a-chip" which, in the foreseeable future, will replace the expensive and life-costing animal testing being used for drug development. This feature will be mentioned in Section 20.12 in more detail.

20.11 THE CHALLENGES OF TISSUE ENGINEERING

Throughout the whole advancements, there are some key barriers remaining. The challenges remained depends on the technique related with. The obstacles of EMC scaffolding are the need of the identification of the optimal cell source for different organs, the loss of an effective method for recellularization of denuded vascular structure in whole-organ scaffolds, and problems on the identification of the appropriate population of patients (Badylak et al., 2012). In addition, scaffold-based tissue engineering still faces the problems of immunogenicity, acute and chronic inflammatory response resulting from the host response to the scaffolds, and its biodegradation products, mechanical mismatch with the surrounding tissue, difficulties in incorporating high numbers of cells uniformly distributed within the scaffold, and the limitation in introducing multiple cell types with positional specificity (Langer, 2007). Another critical problem present in the field is to provide sufficient vascular supply to thick constructs, as molecular diffusion can assure the exchange of nutrients and oxygen within limited ranges (Ko et al., 2008). This could only be done by conducting proper capillary bed within the scaffold in nanoscale.

Another problem waiting for being addressed is the transition and the integration of a tissue from an in vitro to an in vivo setting. According to the surgical point of view, macro- and microvascular tree must be connected successfully to provide enough perfusion among the tissue. Three main components of vasculature, capillary, intermediate microvessels, and microvasculature, must work together. However, constructing intermediate microvessels is still a challenge in the field, and it is unclear how the microvasculature can be connected to the capillary tree.

Bioprinting also faces some difficulties. Aside from the expense of the printers, it is hard to assure high cell density to build up solid organs. Sometimes, high-speed deposition of cells may damage the construction. Additionally, the success of printing depends on the control of the gelation state of the collagen layers. To integrate the final structure of tissue, collagen must be removed, which is very difficult. Also, printing larger and more complex patterns like branching tubes limits the use of printing, and it is excessively time-consuming. Establishing compatible bioreactors is also difficult. The maturation of the tissue in bioreactor takes really long time and new designs of bioreactors are certainly needed.

Economical and financial aspects of the field are in their infancy. CTI, in which tissue engineering is one of the fields, has some problems to market its products and to make them approved by FDA. It is also stated that the necessary infrastructure, including appropriate regulation, reimbursement regimes, scalable manufacturing, robust business models, and clinical outlets are not yet in place (Mason et al., 2011).

Last, but not the least, it is still possible to construct very complex tissues such as pancreas, brain, and eyes. Hopefully, if progress continues, this might also be possible in the future.

20.12 CONCLUSION AND FUTURE PROSPECTS

It is considered that the development of treatment modalities for reestablishing tissue structure and function is getting close instead of replacing parts and tissues. Consequently, the major focus will be on the cells in next years. Scientists describe their real goal to make the cells ultimate "tissue engineer" that completes the effort by providing the conditions to the cells, so that the cells do the engineering. And with the help of novel technological developments, especially on stem cell biology, "instructing" stem cells to regenerate defective tissues by providing highly engineered environmental milieu would change the scheme of the future (Jakab et al., 2010).

The spectrum of cell therapy—based industries, including tissue engineering, is highly diverse from permanent cell therapies such as the replacement of limbal cells for damaged corneas to transient cell therapies such as the immunomodulation provided by adult stem cells for the treatment of graft-versus-host disease. In the future, the number of different cell types used for therapies will expand rapidly like bespoke designer cells, synthetically engineered cells, cell fragments, cell hybrids, substitute tissues, and the gradual emergence of enhancement therapies rather than mere health restoration. As tissue engineering is described as the fourth and the final pillar of healthcare sector, to achieve providing the best possible treatments for all patients, all of the pillars must work together. In the example of cardiovascular diseases, small molecule drugs (aspirin or digitalis), biologics (glycoprotein IIB/IIA inhibitors), devices (synthetic bypass grafts or stents), and cell therapies (cardiomyocyte tissue engineering and blood vessel formation) must work synergistically (Mason et al., 2011).

Organ-on-a-chip phenomenon contains designing an organ concept of natural system on a microenvironment called "chip" with its substituted vasculature and other biological substances. This method uses the benefits of nanotechnology to be assembled. It allows testing any chemical effects on that system or organ and the optimization of nanoparticulate systems for drug delivery, for example. Similarly, a lung-on-a-chip device was used to assess how nanoparticles and bacteria enter the lungs (Huh et al., 2010). Considering the relevance of effects of drugs on patients, testing the drugs on chips before clinical trials may enhance choosing the right drug form at the right time to give the right patient to achieve maximum success and relief of disease, which is called "personalized medicine." This could be possible by means of tissue engineering.

In conclusion, it is clearly stated that establishment of "biomimetic" approach in the field and unlocking the full biological potential of stem cells (both in vitro and in vivo) could be the best solution. This necessitates an environment mimicking the native development. For this to happen, it is needed to continue to cross the boundaries between several disciplines, to take advantage of the synergism of developmental, stem cell and adult biology, biomaterials, biomedical engineering, and medicine.

20.12.1 Conclusion

Regenerative medicine offers a better life quality and proposes properties to solve major obstacles of nowadays medicine such as organ transplantation, scar-free healing, and repairing damaged internal tissues, and, even, new researching approaches for drug development and pharmaceutical delivery. Tissue engineering is the pioneer of this field with devastating developmental speed and the importance that may resolve the great surgical difficulties of organ transplantation such as life-long immunosuppression and adverse inflammatory reactions like graft-versus-host disease. Tissue engineering also helps the new study areas by providing useful information about promising strategies such as using stem cells or cell-sheet technology by conducting application fields for novel scientific knowledge. That means, it is an interdisciplinary science field, which serves others reciprocally and mutually while it is gaining some benefits from them. However, further developmental approaches depending on subfields of sciences such as biomaterials, stem cell biology, and nanotechnology and better understanding of constructing tissue structure and environment might make it possible to mimic the native design of the body and pave the way to serve better and qualified treatment options in the future.

References

Agrawal, V., Tottey, S., Johnson, S.A., Freund, J.M., Siu, B.F., Badylak, S.F., 2011. Recruitment of progenitor cells by an ECM cryptic peptide in a mouse model of digit amputation. Tissue Eng. Part A 17, 2335–2343.

Alaminos, M., Del Carmen Sánchez-Quevedo, M., Muñoz-Avila, J.I., Serrano, D., Medialdea, S., Carreras, I., et al., 2006. Construction of a complete rabbit cornea substitute using a fibrin-agarose scaffold. Invest. Ophthalmol. Vis. Sci. 47 (8), 3311–3317.

Asnaghi, M.A., Jungebluth, P., Raimondi, M.T., Dickinson, S.C., Rees, L.E., Go, T., et al., 2009. A double-chamber rotating bioreactor for the development of tissue-engineered hollow organs: from concept to clinical trial. Biomaterials 30, 5260–5269.

Bader, A., Schilling, T., Teebken, O.E., Brandes, G., Herden, T., Steinhoff, G., et al., 1998. Tissue engineering of heart valves—human endothelial cell seeding of detergent acellularized porcine valves. Eur. J. Cardiothorac. Surg. 14, 279–284.

Badylak, S.F., Weiss, D.J., Caplan, A., Macchiarini, P., 2012. Engineered whole organs and complex tissues. Lancet. 379, 943–952.

Baiguera, S., Jungebluth, P., Burns, A., Mavilia, C., Haag, J., De Coppi, P., et al., 2010. Tissue engineered human tracheas for in vivo implantation. Biomaterials 31, 8931–8938.

Baiguera, S., Gonfiotti, A., Jaus, M., Comin, C.E., Paglierani, M., Del Gaudio, C., et al., 2011. Development of bioengineered human larynx. Biomaterials 32, 4433–4442.

Barth, A., 1893. Ueber histologische befunde nach knochen-implantationen. Ach. Klein. Chir. 46, 409–417.

Bock, N., Riminucci, A., Dionigi, C., Russo, A., Tampieri, A., Landi, E., et al., 2010. A novel route in bone tissue engineering: magnetic biomimetic scaffolds. Acta. Biomater. 6, 786–796.

Borenstein, J.T., Weinberg, E.J., Orrick, B.K., Sundback, C., Kaazempur-Mofrad, M.R., Vacanti, J.P., 2007. Tissue Eng. 13, 1837–1844.

Borschel, G.H., Dennis, R.G., Kuzon Jr., W.M., 2004. Contractile skeletal muscle tissue-engineered on an acellular scaffold. Plast. Reconstr. Surg. 113, 595–602.

Brown, B.N., Valentin, J.E., Stewart-Akers, A.M., McCabe, G.P., Badylak, S.F., 2009. Macrophage phenotype and remodeling outcomes in response to biologic scaffolds with and without a cellular component. Biomaterials 30, 1482–1491.

Campbell, P.G., Weiss, L.E., 2007. Tissue engineering with the aid of inkjet printers. Expert Opin. Biol. Ther. 7, 1123–1127.

Chapekar, M.S., 2000. Tissue engineering: challenges and opportunities. J. Biomed. Mater. Res. B Appl. Biomater. 53, 617–620.

Chen, F., Yoo, J.J., Atala, A., 1999. Acellular collagen matrix as a possible "off the shell" biomaterial for urethral repair. Urology 54, 407–410.

Cheung, H.-Y., Lau, K.-T., Lu, T.-P., Hui, D., 2007. A critical review on polymer-based bio-engineered materials for scaffold development. Compos. Part B Eng. 38, 291–300.

Cimetta, E., Figallo, E., Cannizzaro, C., Elvassore, N., Vunjak-Novakovic, G., 2009. Micro-bioreactor arrays for controlling cellular environments: design principles for human embryonic stem cell applications. Methods 47, 81–89.

Cooper, J.A., Lu, H.H., Ko, F.K., Freeman, J.W., Laurencin, C.T., 2005. Fiber-based tissue engineered scaffold for ligament replacement: design considerations and in vitro evaluation. Biomaterials 26, 1523–1532.

De Santis, et al., 2010. An approach in developing 3D fiber-deposited magnetic scaffolds for tissue engineering. In: D'Amore, et al., (Eds.), Vth International Conference on Times of Polymers (TOP) and Composites. American Institute of Physics, Melville, New York, pp. 420–422.

Discher, D.E., Mooney, D.J., Zandstra, P.W., 2009. Growth factors, matrices, and forces combine and control stem cells. Science 324, 1673–1677.

Elloumi-Hannachi, I., Yamato, M., Okano, T., 2010. Cell sheet engineering: a unique nanotechnology for scaffold-free tissue reconstruction with clinical applications in regenerative medicine. J. Intern. Med. 267, 54–70.

Freytes, D.O., Wan, L.Q., Vunjak-Novakovic, G., 2009. Geometry and force control of cell function. J. Cell. Biochem. 108, 1047–1058.

Gloria, A., De Santis, R., Ambrosio, L., 2010. Polymer-based composite scaffolds for tissue engineering. J. Appl. Biomat. Biomech. 8, 57–67.

Godier, A.F., Marolt, D., Gerecht, S., Tajnsek, U., Martens, T.P., Vunjak-Novakovic, G., 2008. Engineered microenvironments for human stem cells. Birth Defect. Res. C Embryo Today 84, 335–347.

Goldberg, M., Langer, R., Jia, X., 2007. Nanostructured materials for applications in drug delivery and tissue engineering. Biomater. Sci. 18, 241–268.

Govoni, M., Lotti, F., Biagiotti, L., Lannocca, M., Pasquinelli, G., Valente, S., et al., 2014. An innovative stand-alone bioreactor for the highly reproducible transfer of cyclic mechanical stretch to stem cells cultured in a 3D scaffold. J. Tissue Eng. Reg. Med. 8 (10), 787–793.

Grayson, W.L., Martens, T.P., Eng, G.M., Radisic, M., Vunjak-Novakovic, G., 2009. Biomimetic approach to tissue engineering. Semin. Cell. Dev. Biol. 20, 665–673.

Grayson, W.L., Fröhlich, M., Yeager, K., Bhumiratana, S., Chan, M.E., Cannizzaro, C., et al., 2010. Regenerative medicine special feature: engineering anatomically shaped human bone grafts. Proc. Natl. Acad. Sci. USA 107, 3299–3304.

Griffin, J.P., O'Grady, J., Baber, N., Hutt, P.B., Bendall, C., 2009. The Textbook of Pharmaceutical Medicine. Wiley-Blackwell.

Guarino, V., Ambrosio, L., 2008. The synergic effect of polylactide fiber and calcium phosphate particle reinforcement in poly-ε-caprolactone composite saffolds. Acta. Biomater. 4, 1778–1787.

Guarino, V., Causa, F., Netti, P.A., Ciapetti, G., Pagani, S., Martini, D., et al., 2008. The role of hydroxyapatite as solid signal on performance of PCL porous scaffolds for bone tissue regeneration. J. Biomed. Mater. Res. B Appl. Biomater. 86, 548–557.

Guarino, V., Taddei, P., Di Foggia, M., Fagnano, C., Ciapetti, G., Ambrosio, L., 2009. The influence of hydroxyapatite particles on in vitro degradation behavior of poly-ε-caprolactone based composite scaffolds. Tissue Eng. Part A 15, 3655–3668.

Guenou, H., Nissan, X., Larcher, F., Feteira, J., Lemaitre, G., Saidani, M., et al., 2009. Human embryonic stem cell derivatives for full reconstruction of the pluristratified epidermis: a preclinical study. Lancet 374, 1745–1753.

Hodde, J.P., Ernst, D.M., Hiles, M.C., 2005. An investigation of the long-term bioactivity of endogenous growth factor in OASIS wound matrix. J. Wound Care 14, 23–25.

Huh, D., Matthews, B.D., Mammoto, A., Montoya-Zavala, M., Hsin, H.Y., Ingber, D.E., 2010. Reconstructing organ-level lung functions on a chip. Science 328, 1662–1668.

Jakab, K., Neagu, A., Mironov, V., Markwald, R.R., Forgacs, G., 2004. Engineering biological structure of prescribed shaped using self-assembling multicellular systems. Proc. Natl. Acad. Sci. USA 101, 2864–2869.

Jakab, K., Norotte, C., Damon, B., Marga, F., Neagu, A., Besch-Williford, C.L., et al., 2008. Tissue engineering by self-assembly of cells printed into topologically defined structures. Tissue Eng. Part A 14, 413–421.

Jakab, K., Norotte, C., Marga, F., Murphy, K., Vunjak-Novakovic, G., Forgacs, G., 2010. Tissue engineering by self-assembly and bio-printing of living cells. Biofabrication 2 (2), 022001.

Kang, T.Y., Hong, J.M., Kim, B.J., Cha, H.J., Cho, D.W., 2013. Enhanced endothelialization for developing artificial vascular networks with a natural vessel mimicking the luminal surface in scaffolds. Acta Biomater. 9 (1), 4716–4725.

Karam, J.P., Muscari, C., Montero-Menei, C.N., 2012. Combining adult stem cells and polymeric devices for tissue engineering in infarcted myocardium. Biomaterials 33, 5683–5695.

Khademhosseini, A., Vacanti, J.P., Langer, R., 2009. Progress in tissue engineering. Sci. Am. 300, 64–71.

Ko H.C. et al., Engineering thick tissues—the vascularization problem, Eur Cell Mater. 2007 Jul 25;14:1-18; discussion 18-9.

Koizumi, N., Inatomi, T., Quantock, A.J., Fullwood, N.J., Dota, A., Kinoshita, S., 2000. Amniotic membrane as a substrate for cultivating limbal corneal epithelial cells for autologous transplantation in rabbits. Cornea 19 (1), 65–71.

Kon, E., Chiari, C., Marcacci, M., Delcogliano, M., Salter, D.M., Martin, I., et al., 2008. Tissue engineering approach to meniscus regeneration in a sheep model. Tissue Eng. Part A 14, 1067–1080.

Langer, R., 2007. Tissue engineering: perspectives, challenges and future directions. Tissue Eng. 13, 1–2.

Langer, R., 2009. A conversation with Robert Langer: pioneering biomedical scientist and engineer. Interview by Paul S. Weiss. ACS Nano 3, 756–761.

L'Heureux, N., Pâquet, S., Labbé, R., Germain, L., Auger, F.A., 1998. A completely biological tissue-engineered human blood vessel. FASEB J 12, 47–56.

Ma, P.X., 2008. Biomimetic materials for tissue engineering. Drug Deliv. Rev. 60, 184–198.

Mason, C., Manzotti, E., 2009. Regen: the industry responsible for cell-based therapies. Regen. Med. 4 (6), 783–785.

Mason, C., Brindley, D.A., Culme-Seymour, E.J., Davie, N.L., 2011. Cell therapy industry: billion dollar global business with unlimited potential. Regen. Med. 6, 265–272.

Mathieu, L.M., Mueller, T.L., Bourban, P.E., Pioletti, D.P., Müller, R., Månson, J.A., 2006. Architecture and properties of anisotropic polymer composite scaffolds for bone tissue engineering. Biomaterials 27, 905–916.

McAllister, T.N., Maruszewski, M., Garrido, S.A., Wystrychowski, W., Dusserre, N., Marini, A., et al., 2009. Effectiveness of hemodialysis access with an autologous tissue-engineered vascular graft: a muticentre cohort study. Lancet 373, 1440–1446.

McGuigan, A.P., Sefton, M.V., 2006. Vascularized organoid engineered by modular assembly enables blood perfusion. Proc. Natl. Acad. Sci. USA 103, 11461–11466.

McManus, A.J., Doremus, R.H., Siegel, R.W., Bizios, R., 2005. Evaluation of cytocompatibility and bending modulus of nanoceramic/polymer composites. J. Biomed. Mater. Res. A 72, 98–106.

Muscari, C., Giordano, E., Bonafè, F., Govoni, M., Pasini, A., Guarnieri, C., 2013. Molecular mechanisms of ischemic pre-conditioning and post-conditioning as putative therapeutic targets to reduce tumor survival and malignancy. Med. Hypotheses 81, 1141–1145.

Muscari, C., Giordano, E., Bonafè, F., Govoni, M., Guarnieri, C., 2014. Strategies affording prevascularized cell-based constructs for myocardial tissue engineering. Stem Cells Int. 2014, 434169.

Nair, L.S., Laurencin, C.T., 2006. Polymers as biomaterials for tissue engineering and controlled drug delivery. Adv. Biochem. Eng. Biotechnol. 102, 47–90.

Nakamura, M., Kobayashi, A., Takagi, F., Watanabe, A., Hiruma, Y., Ohuchi, K., et al., 2005. Biocompatible inkjet printing technique for designed seeding of individual living cells. Tissue Eng. 11, 1658–1666.

Nieponice, A., Gilbert, T.W., Badylak, S.F., 2006. Reinforcement of esophageal anastomoses with an extracellular matrix scaffold in a canine model. Ann. Thorac. Surg. 82, 2050–2058.

Nishida, K., Yamato, M., Hayashida, Y., Watanabe, K., Yamamoto, K., Adachi, E., et al., 2004. Corneal reconstruction with tissue-engineered cell sheets composed of autologous oral mucosal epithelium. N. Engl. J. Med. 351, 1187–1196.

Nishiyama, Y., Nakamura, M., Henmi, C., Yamaguchi, K., Mochizuki, S., Nakagawa, H., et al., 2009. Development of a three-dimensional bioprinter: construction of cell supporting structures using hydrogel and state-of-the-art inkjet technology. J. Biomech. Eng. 131, 035001.

Norotte, C., Marga, F.S., Niklason, L.E., Forgacs, G., 2009. Scaffold-free vascular tissue engineering using bioprinting. Biomaterials 30, 5910–5917.

Ott, H.C., Matthiesen, T.S., Goh, S.K., Black, L.D., Kren, S.M., Netoff, T.I., et al., 2008. Perfusion-decellularized matrix: using nature's platform to engineer a bioartifical heart. Nat. Med. 14, 213–221.

Price, R.L., Waid, M.C., Haberstroh, K.M., Webster, T.J., 2003. Selective bone cell adhesion on formulations containing carbon nanofibers. Biomaterials 24, 1877–1887.

Priya, S.G., Jungvid, H., Kumar, A., 2008. Skin tissue engineering for tissue repair and regeneration. Tissue Eng. Part B Rev. 14, 105–118.

Ramakrishna, S., Mayer, J., Wintermantel, E., Leong, K.W., 2001. Biomedical applications of polymer-composite materials: a review. Comp. Sci. Tech. 61, 1189–1224.

Ross, E.A., Williams, M.J., Hamazaki, T., Terada, N., Clapp, W.L., Adin, C., et al., 2009. Embryonic stem cells proliferate and differentiate when seeded into kidney scaffolds. J. Am. Soc. Nephrol. 20, 2338–2347.

Sarig, U., Au-Yeung, G.C., Wang, Y., Bronshtein, T., Dahan, N., Boey, F.Y., et al., 2012. Thick acellular heart extracellular matrix with inherent vasculature: a potential platform for myocardial tissue regeneration. Tissue Eng. A 18, 2125–2137.

Shi, J., Votruba, A.R., Farokhzad, O.C., Langer, R., 2010. Nanotechnology in drug delivery and tissue engineering: from discovery to applications. Nano Lett. 10 (9), 3223–3230.

Shimizu, T., Sekine, H., Yamato, M., Okano, T., 2009. Cell sheet-based myocardial tissue engineering: new hope for damaged heart rescue. Current Pharma. Design 15, 2807–2814.

Sng, J., Lufkin, T., 2012. Emerging stem cell therapies: treatment, safety and biology. Stem Cells Int. 2012, 521343.

Stevens, M.M., George, J.H., 2005. Exploring and engineering the cell surface interface. Science 310, 1135–1138.

Tottey, S., Corselli, M., Jeffries, E.M., Londono, R., Peault, B., Badylak, S.F., 2011. Extracellular matrix degradation products and low-oxygen conditions enhance the regenerative potential of perivascular stem cells. Tissue Eng. Part A 17, 661–680.

Tran, N., et al., 2009. Nanotechnology for bone materials. WIRES Nanomed. Nanobiotech. 24, 2133–2151.

Tran, P.A., Zhang, L., Webster, T.J., 2009. Carbon nanofibers and carbon nanotubes in regenerative medicine. Adv. Drug. Deliv. Rev. 61, 1097–1114.

Uygun, B.E., Soto-Gutierrez, A., Yagi, H., Izamis, M.L., Guzzardi, M.A., Shulman, C., et al., 2010. Organ reengineering through development of a transplantable recellularized liver graft using decellularized liver matrix. Nat. Med. 16, 814–820.

Valentin, J.E., Stewart-Akers, A.M., Gilbert, T.W., Badylak, S.F., 2009. Macrophage participation in the degradation and remodeling of extracellular matrix scaffolds. Tissue Eng. Part A 15, 1687–1694.

Wang, H.Y., Wei, R.H., Zhao, S.Z., 2013. Evaluation of corneal cell growth on tissue engineering materials as artificial cornea scaffolds. Int. J. Ophthalmol. 6, 873–878.

Xu, S., Zhu, J., Yu, L., Fu, G., 2012. Endothelial progenitor cells: current development of their paracrine factors in cardiovascular therapy. J. Cardiovasc. Pharmacol. 59, 387–396.

Yamashita, J., Furman, B.R., Rawls, H.R., Wang, X., Agrawal, C.M., 2001. The use of dynamic mechanical analysis to assess the viscoelastic properties of human cortical bone. J. Biomed. Mater. Res. B Apple Biomater. 59, 47–53.

Zhang, S., Uludağ, H., 2009. Nanoparticulate systems for growth factor delivery. Pharm. Res. 26, 1561–1580.

Zhu, Y., Liu, T., Song, K., Jiang, B., Ma, X., Cui, Z., 2009. Collagen-chitosan polymer as a scaffold for the proliferation of human adipose tissue-derived stem cells. J. Mater. Sci. Mater. Med. 20 (3), 799–808.

21

Therapeutic Aspects of Stem Cells in Regenerative Medicine

Alok Mishra[1] and Mukesh Verma[2]

[1]National Institutes of Health (NIH), Bethesda, MD, United States
[2]National Institutes of Health (NIH), Rockville, MD, United States

21.1 INTRODUCTION

Stem cell—based regenerative medicine is one of the most modern areas of life sciences. According to NIH Fact sheet definition, the discipline of regenerative medicine is the process of creating living, functional tissues to repair or replace tissue or organ function lost due to age, disease, damage, or congenital defects (https://report.nih.gov/nihfactsheets/viewfactsheet.aspx?csid=62). Either stimulation of self-cells in humans, such as the healing of a wound, or transplanting appropriately conditioned/engineered cells from outside can reach regeneration. Potential of differentiation in any lineage and their proliferation are the two fundamental hallmarks of stem cells that offer the possibility to renew, replenish, and replace any types of cells as a source towards therapeutics.

Increasing demand of donated organs and tissues is more than their supply to fulfill the malfunctioned and damaged tissue arises due to several pathological conditions prevailing in human population such as diseases and man-made injuries. Therefore, stem cells offer an attractive repertoire towards lifesaving clinical interventions.

21.2 CHARACTERISTICS OF STEM CELLS SUITABLE FOR REGENERATIVE MEDICINE

An ideal stem cell for potential use in regenerative medicine must proliferate and mitotically divide efficiently for ample supply, can be directed to differentiate into the desired lineage, must survive, integrate in a transplanted niche without any immunological

Omics Technologies and Bio-engineering: Towards Improving Quality of Life
DOI: https://doi.org/10.1016/B978-0-12-804659-3.00021-X

rejection, and eventually exhibit the requisite contextual functionality. At the same time, component of regenerative medicine focuses on choice and source of cells, biocompatibility of scaffold material, the decision to add a supplement for patient, and transplant safety compliances. The role of enhancer of zeste homolog 2, a histone methyl transferase, is needed for the maintenance and lineage specification of stem cells (Lin, 2015). Those factors that are required for the proper growth of stem cells include leukemia inhibitory factor and fibroblast growth factor along with activation of the AKT pathway (Su et al., 2015).

21.3 POLICIES OF UNITED STATES AND OTHER NATIONS IN THE FIELD OF STEM CELL RESEARCH

Many clinical trials are underway in testing and approving stem cell–based therapies for different diseases (Dulamea, 2015, Savitz et al., 2014, Song et al., 2013). In the United States, Food and Drug Administration's (FDA's) Center for Biologics Evaluation and Research oversees the stem cell–based product under the guidance "Tissue rules" OCTGT-CFR-127. NIH guidelines and policies for stem cell research and therapy can be found at http://stemcells.nih.gov/research/pages/newcell_qa.aspx. The existing US government policy states that "The governing Federal statute on human fetal tissue research and transplantation is found in sections 498A and 498B of the PHS Act, 42 U.S.C. 298g-1 and 298g-2. The statute specifically prohibits any person from knowingly acquiring, receiving, or transferring any human fetal tissue for valuable consideration, and requires certain safeguards in transplantation research." In the European Union (EU), the Europeans Medicine Agency follows the Advance Therapy Medicine product policies. Japanese organization PMDA (http://www.pmda.go.jp/english/) plays a significant role in approval of policies. ISSR and CRIM are the two non-federal key organizations, which also actively engage in policies and awareness of stem cell therapies. The US FDA cautions strictly the general public, consumer, and practitioners about the use of stem cell–based therapies (http://www.fda.gov/ForConsumers/ConsumerUpdates/ucm286155.htm). In general, FDA states that no such therapy except one, discussed below, is approved for clinic. So far, the US FDA has approved only one stem cell product, Hemacord, a cord blood–derived product manufactured by the New York Blood Center. This product is used for patients with specific disorders related to blood-forming system.

21.4 STEM CELLS IN REGENERATIVE MEDICINE OF DIFFERENT DISEASES

21.4.1 Spinal Cord Injuries

Spinal cord injuries due to mechanical/traumatic stressors result in severe dysfunction of neuromotor system. The limitations associated with the intrinsic mechanism of stem cell activation and proliferation in spinal cord need transplantation. Sanchez-Ramos and colleagues have shown that bone marrow–derived stromal cells (BMSCs) supply a stem cell population that can differentiate in neuronal lineages (Sanchez-Ramos et al., 2000). In rats, the advantages of human umbilical cord blood—derived stem cell are reported by Saporta et al. (2003). Later on in clinics, Sykova et al. (2006) have shown autologous bone marrow transplant (BMT) in patients with subacute and chronic spinal cord injury. Later on Yoon et al. (2007) have also demonstrated complete spinal cord injury treatment using autologous bone marrow cell transplantation and bone marrow stimulation with Phase I/II clinical trial.

21.4.2 Heart and Vascular System

Makino et al. (1999) had reported that BMSCs can differentiate into cardiomyocytes in vitro condition. The source of cardiomyocytes can be different kinds of stem cells derived from embryonic or bone marrow (Hassink et al., 2003, Povsic, 2016). Orlic and colleagues seminal works reported that bone marrow cells could regenerate infarcted heart (Orlic, 2003, 2004, 2005, Orlic et al., 2002, 2003). In 2010, a randomized, placebo-treatment study of adult mesenchymal stem cells' intravenous injection post-myocardial infarction (MI) was reported by Hare et al. (2009). Cardiovascular Cell Therapy Research Network (CCTRN) of the Center for Cell and Gene Therapy, Baylor College of Medicine, Texas, United States, conducted multicenter cellular therapy clinical trials with 60 patients (Gee et al., 2010). A clinical trial showing the implication of transendocardial injection of bone marrow and mesenchymal stem cell in the patients of Chronic Ischemic Left Ventricular Dysfunction and Heart Failure Secondary to Myocardial Infarction (TAC-HFT) was coordinated and published by the Interdisciplinary Stem Cell Institute, University of Miami Leonard Miller School of Medicine, Miami, FL, United States (Trachtenberg et al., 2011). A US-based company Osiris therapeutics successfully ran a Phase II clinical trial evaluating Prochymal (MSC) drip injection following MI in 220 patients in the United States and Canada. For the peripheral heart diseases, most notably the critical limb ischemia, two different companies, Harvest Tech Co (United States) and Pluristem Therapeutics (Israel), had conducted clinical trials with significant success without any side effects utilizing stem cells from the placenta.

21.4.3 Periodontal Diseases

The complex periodontal structure is made of different kinds of tissues bone, cementum, and ligament, and so it poses challenges for regeneration. One of the recent approaches to recreate periodontal apparatus is to use stem cells. Bone marrow skeletal stem cell, adipose tissue stem cell, periodontal ligament stem cell (PDSLC), dental pulp stem cell, exfoliated teeth—derived stem cell, dental follicle stem cell, dental apical papilla stem cell, dental socket stem cell, and pluripotent stem cells are some of the studied sources for regenerative medicine of periodontal diseases. The PDSLC and spermatogonial stem cells are most promising candidates among these. Interestingly pigs, sheep, and dogs are models together with mice and rats for these studies. Yamada and colleagues in 2006 and later Feng and colleagues in 2010 have conducted important clinical studies successfully (Irokawa et al., 2010, Kitamura et al., 2011, Feng et al., 2010). Seo et al. have also reported multipotent stem cell from human periodontal ligament (Seo et al., 2004).

21.4.4 Ocular Diseases

Transplantation of human bone marrow−derived mesenchymal stem cell could ameliorate retinal degeneration in rats (Tzameret et al., 2014). The same cells could also protect glaucoma in aging rats (Hu et al., 2013). The most studied human pathologies related to the eyes are macular degeneration, hereditary dystrophy, and glaucoma.

21.4.5 Diabetes

Stem cell regeneration has been studied in diabetes by several groups (Mansouri, 2012, Masjkur et al., 2016, Wolfe-Coote et al., 1998). Human embryonic stem cells (ESCs) have been also used in vitro to study differentiation in pancreatic lineages or insulin secretive cells (Kroon et al., 2008). Stem cells of beta cells were used successfully for Type 1 diabetes by one group of investigators (Goland and Egli, 2014, Li and Ikehara, 2014).

21.4.6 Skin Wounds and Burns

The basement membrane of the epidermis has a repertoire of stem cells. Skin-derived Schwann cells are also used in regeneration therapies of spinal cord and sciatic nerve. Serial cultivation of skin cells is used to generate a sheet to use as an autograft to cover burns. Induction of vasculogenesis in non-self-dermal layers and alloderms is still a challenge for skin grafting (Pellegrini et al., 1998). MSCs have been utilized for tissue regeneration in thermal burns as stem cell therapy (Leclerc et al., 2011). A human skin equivalent, Apligraf, is an example of tissue engineering and regeneration (Eaglstein and Falanga, 1998).

In 2013, the US FDA cleared fish-skin technology of Kerecis Limited to heal human wounds, and released the statement "When this product is inserted into or onto damaged human tissue, protease activity is modulated, the fish skin is vascularized and populated by the patient's own cells, and ultimately converted into living tissue" (http://www.kerecis.com/technology/29-fda-approval).

21.4.7 Ischemic Limb Disease

Cell populations of $CD34^+/CD31^-$ residing in stromal−vascular fraction of human fats have been tested in hindlimb ischemia model for neovascularization (Miranville et al., 2004). Human cord blood with $CD34^+$ cells could also be differentiated into endothelial and skeletal muscle cells in ischemic limb tissue regeneration model in mice (Pasche et al., 2002).

21.4.8 Bone, Cartilage, and Muscle-Related Abnormalities

Bone marrow stromal cells with or without biomaterials are established for treating bone-related abnormalities (viz. size defects, arthritis, and cleft palate) in mouse and larger animal models too (Kon et al., 2000, Ohgushi et al., 1989). Examples of human patients treated with such materials are also reported by many investigators for tibia, ulna, humorous, mandible, femur, and distal phalanx. Biomaterials—hydroxpartite, CaCl, and titanium—were used with bone marrow stromal cells (Jingchao et al., 2011, Plander et al., 2011, Shinagawa et al., 2013, and Tsai et al., 2011). Wakitani and colleagues have reported mesenchymal cell-based repair of articular cartilage (Wakitani et al., 1994). The same lead author later reported bone marrow stromal cell—based therapy on 12 patients suffering from osteoarthritis.

21.4.9 Neurological Disorder

Clinical trials based on the cells derived from the bone marrow and umbilical cord for Parkinson's disease, Alzheimer's disease, and amyotrophic lateral sclerosis are undergoing (Venkataramana et al., 2010, Shin et al., 2014). Geron Corp, a California-based company, got an FDA approval for using ESC-derived oligodendrocyte. It has been demonstrated that ESC can integrate and differentiate when delivered at injured regions.

21.4.10 Cancers

Injection/grafting/transplantation of hematopoietic stem cell (HSC)/progenitor cells is a therapeutic hallmark for hematological malignancies. Historically, first human intravenous infusion was reported by Ferrebee and colleagues in 1957 (Thomas et al., 1957). Cord blood transplantation offered a good alternative for BMT (Rocha et al., 1998). Mesenchymal stem cells are the main source, which supply cells required for proper functioning of the hematopoietic system. CD34$^+$ HSC are implicated in many such therapies.

Human bone marrow—derived mesenchymal stem cells are reported for the treatment of gliomas (Nakamizo et al., 2005). Nakamura and colleagues have published a report on the antitumor effect of genetically engineered mesenchymal stem cells in a rat glioma model (Nakamura et al., 2004). Mesenchymal stem cells and neuronal stem cells derived from human/mouse/rat are used in experimental models of glioma.

21.4.11 Liver Injury and Cirrhosis

Liver transplantation is the standard care for many end-stage liver diseases. For regeneration of damaged liver, iPSC (Rashid et al., 2010, Si-Tayeb et al., 2010), umbilical cord mesenchymal stromal cells (Zhao et al., 2009), ESCs (Ramasamy et al., 2013), mesenchymal stem cell (Mohamadnejad et al., 2013), and bone marrow—derived stem cells (Amer et al., 2011) were employed. All these cells either alone or in combination with biomaterials have shown promising results (Tsolaki and Yannaki, 2015). The common challenges in these studies include complicated surgeries and donor-tissue rejection. Chen et al. demonstrated treatment of radiation-induced liver injury by mesenchymal stem cells (Chen et al., 2015).

21.4.12 Muscular Dystrophy

Emerging studies have suggested that significant heterogeneity exists within the satellite cell niche, and satellite cells operate like muscle-specific adult stem cells

(Schultz, 1996). These cells are the basis for transplantation studies for muscular dystrophy. Rudnicki and colleagues reviewed the implications of muscle satellite cells in treating muscular diseases (Kuang et al., 2008).

21.5 CHALLENGES IN STEM CELL RESEARCH

Despite the progress in theoretical understanding of stem cell and medicine, still we have many challenges before applying those in clinical practice. In a laboratory, extremely low abundance, purification and characterization, apoptosis, inefficient expansion, and enormous cost of maintenance are the important limitations of generation of stem cell repertoire for therapy (Vacanti et al., 2001). The development of teratoma and limited achievement of vascular niche are other major problems at transplantation stage. Eventually, scientists and clinician might face socio-ethical issues too. Well-controlled large animal models with follow-up and randomized controlled clinical trials are needed before extensive use of stem cells in regenerative medicine.

21.6 CONCLUDING REMARKS

Among the different types of stem cells, bone marrow—derived mesenchymal stem cells offer the widest avenues for therapeutics in human regenerative medicine. ESCs pose risk for teratoma as well as socio-religious issues. Future of stem cell—based therapies will be using genetically modified multipotent stem cells with nanotechnology-based delivery systems for the contextual, spatial, and temporal release of biomolecules from stem cells with appropriate balance in proliferation—differentiation, targeted homing in niche (Yin et al., 2016). Increase yield and biocompatible grafting materials are some of the issues that can be resolved by biomedical engineering. Fundamental mechanism of differentiation—proliferation balance in the context of stemness and niche is an important area to explore. More randomized and control clinical trials as well as clinical and regulatory guidelines are needed to establish the discipline of stem cells in regenerative medicine.

References

Amer, M.E., El-Sayed, S.Z., El-Kheir, W.A., Gabr, H., Gomaa, A.A., El-Noomani, N., et al., 2011. Clinical and laboratory evaluation of patients with end-stage liver cell failure injected with bone marrow-derived hepatocyte-like cells. Eur. J. Gastroenterol. Hepatol. 23 (10), 936—941. Available from: https://doi.org/10.1097/MEG.0b013e3283488b00.
Chen, Y.X., Zeng, Z.C., Sun, J., Zeng, H.Y., Huang, Y., Zhang, Z.Y., 2015. Mesenchymal stem cell-conditioned medium prevents radiation-induced liver injury by inhibiting inflammation and protecting sinusoidal endothelial cells. J. Radiat. Res. 56 (4), 700—708. Available from: https://doi.org/10.1093/jrr/rrv026.
Dulamea, A., 2015. Mesenchymal stem cells in multiple sclerosis—translation to clinical trials. J. Med. Life. 8 (1), 24—27.
Eaglstein, W.H., Falanga, V., 1998. Tissue engineering and the development of Apligraf a human skin equivalent. Adv. Wound Care 11 (4 Suppl), 1—8.
Feng, F., Akiyama, K., Liu, Y., Yamaza, T., Wang, T.M., Chen, J.H., et al., 2010. Utility of PDL progenitors for in vivo tissue regeneration: a report of 3 cases. Oral Dis. 16 (1), 20—28.

Gee, A.P., Richman, S., Durett, A., McKenna, D., Traverse, J., Henry, T., et al., 2010. Multicenter cell processing for cardiovascular regenerative medicine applications: the Cardiovascular Cell Therapy Research Network (CCTRN) experience. Cytotherapy 12 (5), 684−691. Available from: https://doi.org/10.3109/14653249.2010.487900.

Goland, R., Egli, D., 2014. Stem cell-derived beta cells for treatment of type 1 diabetes?. EBioMedicine 1 (2-3), 93−94. Available from: https://doi.org/10.1016/j.ebiom.2014.10.018.

Hare, J.M., Traverse, J.H., Henry, T.D., Dib, N., Strumpf, R.K., Schulman, S.P., et al., 2009. A randomized, double-blind, placebo-controlled, dose-escalation study of intravenous adult human mesenchymal stem cells (prochymal) after acute myocardial infarction. J. Am. Coll. Cardiol. 54 (24), 2277−2286. Available from: https://doi.org/10.1016/j.jacc.2009.06.055.

Hassink, R.J., Dowell, J.D., Brutel de la Riviere, A., Doevendans, P.A., Field, L.J., 2003. Stem cell therapy for ischemic heart disease. Trends Mol. Med. 9 (10), 436−441.

Hu, Y., Tan, H.B., Wang, X.M., Rong, H., Cui, H.P., Cui, H., 2013. Bone marrow mesenchymal stem cells protect against retinal ganglion cell loss in aged rats with glaucoma. Clin. Interv. Aging 8, 1467−1470. Available from: https://doi.org/10.2147/CIA.S47350.

Irokawa, D., Ota, M., Yamamoto, S., Shibukawa, Y., Yamada, S., 2010. Effect of beta tricalcium phosphate particle size on recombinant human platelet-derived growth factor-BB-induced regeneration of periodontal tissue in dog. Dent. Mater. J. 29 (6), 721−730.

Jingchao, H., Rong, S., Zhongchen, S., Lan, C., 2011. Human amelogenin up-regulates osteogenic gene expression in human bone marrow stroma cells. Biochem. Biophys. Res. Commun. 408 (3), 437−441. Available from: https://doi.org/10.1016/j.bbrc.2011.04.042.

Kitamura, M., Akamatsu, M., Machigashira, M., Hara, Y., Sakagami, R., Hirofuji, T., et al., 2011. FGF-2 stimulates periodontal regeneration: results of a multi-center randomized clinical trial. J. Dent. Res. 90 (1), 35−40. Available from: https://doi.org/10.1177/0022034510384616.

Kon, E., Muraglia, A., Corsi, A., Bianco, P., Marcacci, M., Martin, I., et al., 2000. Autologous bone marrow stromal cells loaded onto porous hydroxyapatite ceramic accelerate bone repair in critical-size defects of sheep long bones. J. Biomed. Mater. Res. 49 (3), 328−337.

Kroon, E., Martinson, L.A., Kadoya, K., Bang, A.G., Kelly, O.G., Eliazer, S., et al., 2008. Pancreatic endoderm derived from human embryonic stem cells generates glucose-responsive insulin-secreting cells in vivo. Nat. Biotechnol. 26 (4), 443−452. Available from: https://doi.org/10.1038/nbt1393.

Kuang, S., Gillespie, M.A., Rudnicki, M.A., 2008. Niche regulation of muscle satellite cell self-renewal and differentiation. Cell Stem Cell 2 (1), 22−31. Available from: https://doi.org/10.1016/j.stem.2007.12.012.

Leclerc, T., Thepenier, C., Jault, P., Bey, E., Peltzer, J., Trouillas, M., et al., 2011. Cell therapy of burns. Cell Prolif. 44 (Suppl 1), 48−54. Available from: https://doi.org/10.1111/j.1365-2184.2010.00727.x.

Li, M., Ikehara, S., 2014. Stem cell treatment for type 1 diabetes. Front. Cell Dev. Biol. 2, 9. Available from: https://doi.org/10.3389/fcell.2014.00009.

Lin, S.Z., 2015. Era of stem cell therapy for regenerative medicine and cancers: an introduction for the special issue of Pan Pacific Symposium on Stem Cells and Cancer Research. Cell Transplant. 24 (3), 311−312. Available from: https://doi.org/10.3727/096368915X686814.

Makino, S., Fukuda, K., Miyoshi, S., Konishi, F., Kodama, H., Pan, J., et al., 1999. Cardiomyocytes can be generated from marrow stromal cells in vitro. J. Clin. Invest. 103 (5), 697−705. Available from: https://doi.org/10.1172/JCI5298.

Mansouri, A., 2012. Development and regeneration in the endocrine pancreas. ISRN Endocrinol. 2012, 640956. Available from: https://doi.org/10.5402/2012/640956.

Masjkur, J., Poser, S.W., Nikolakopoulou, P., Chrousos, G., McKay, R.D., Bornstein, S.R., et al., 2016. Endocrine pancreas development and regeneration: noncanonical ideas from neural stem cell biology. Diabetes 65 (2), 314−330. Available from: https://doi.org/10.2337/db15-1099.

Miranville, A., Heeschen, C., Sengenes, C., Curat, C.A., Busse, R., Bouloumie, A., 2004. Improvement of postnatal neovascularization by human adipose tissue-derived stem cells. Circulation 110 (3), 349−355. Available from: https://doi.org/10.1161/01.CIR.0000135466.16823.D0.

Mohamadnejad, M., Alimoghaddam, K., Bagheri, M., Ashrafi, M., Abdollahzadeh, L., Akhlaghpoor, S., et al., 2013. Randomized placebo-controlled trial of mesenchymal stem cell transplantation in decompensated cirrhosis. Liver Int. 33 (10), 1490−1496. Available from: https://doi.org/10.1111/liv.12228.

Nakamizo, A., Marini, F., Amano, T., Khan, A., Studeny, M., Gumin, J., et al., 2005. Human bone marrow-derived mesenchymal stem cells in the treatment of gliomas. Cancer Res. 65 (8), 3307–3318. Available from: https://doi.org/10.1158/0008-5472.CAN-04-1874.

Nakamura, K., Ito, Y., Kawano, Y., Kurozumi, K., Kobune, M., Tsuda, H., et al., 2004. Antitumor effect of genetically engineered mesenchymal stem cells in a rat glioma model. Gene Ther. 11 (14), 1155–1164. Available from: https://doi.org/10.1038/sj.gt.3302276.

Ohgushi, H., Goldberg, V.M., Caplan, A.I., 1989. Repair of bone defects with marrow cells and porous ceramic. Experiments in rats. Acta. Orthop. Scand. 60 (3), 334–339.

Orlic, D., 2003. Adult bone marrow stem cells regenerate myocardium in ischemic heart disease. Ann. N. Y. Acad. Sci. 996, 152–157.

Orlic, D., 2004. The strength of plasticity: stem cells for cardiac repair. Int. J. Cardiol. 95 (Suppl 1), S16–S19.

Orlic, D., 2005. BM stem cells and cardiac repair: where do we stand in 2004? Cytotherapy 7 (1), 3–15. Available from: https://doi.org/10.1080/14653240510018028.

Orlic, D., Hill, J.M., Arai, A.E., 2002. Stem cells for myocardial regeneration. Circ. Res. 91 (12), 1092–1102.

Orlic, D., Kajstura, J., Chimenti, S., Bodine, D.M., Leri, A., Anversa, P., 2003. Bone marrow stem cells regenerate infarcted myocardium. Pediatr. Transplant. 7 (Suppl 3), 86–88.

Pasche, B., Mulcahy, M., Benson III, A.B., 2002. Molecular markers in prognosis of colorectal cancer and prediction of response to treatment. Best Pract. Res. Clin. Gastroenterol. 16 (2), 331–345. Available from: https://doi.org/10.1053/bega.2002.0289.

Pellegrini, G., Bondanza, S., Guerra, L., De Luca, M., 1998. Cultivation of human keratinocyte stem cells: current and future clinical applications. Med. Biol. Eng. Comput. 36 (6), 778–790.

Plander, M., Ugocsai, P., Seegers, S., Orso, E., Reichle, A., Schmitz, G., et al., 2011. Chronic lymphocytic leukemia cells induce anti-apoptotic effects of bone marrow stroma. Ann. Hematol. 90 (12), 1381–1390. Available from: https://doi.org/10.1007/s00277-011-1218-z.

Povsic, T.J., 2016. Current state of stem cell therapy for ischemic heart disease. Curr. Cardiol. Rep. 18 (2), 17. Available from: https://doi.org/10.1007/s11886-015-0693-6.

Ramasamy, T.S., Yu, J.S., Selden, C., Hodgson, H., Cui, W., 2013. Application of three-dimensional culture conditions to human embryonic stem cell-derived definitive endoderm cells enhances hepatocyte differentiation and functionality. Tissue Eng. Part A 19 (3-4), 360–367. Available from: https://doi.org/10.1089/ten.tea.2012.0190.

Rashid, S.T., Corbineau, S., Hannan, N., Marciniak, S.J., Miranda, E., Alexander, G., et al., 2010. Modeling inherited metabolic disorders of the liver using human induced pluripotent stem cells. J. Clin. Invest. 120 (9), 3127–3136. Available from: https://doi.org/10.1172/JCI43122.

Rocha, V., Chastang, C., Souillet, G., Pasquini, R., Plouvier, E., Nagler, A., et al., 1998. Related cord blood transplants: the Eurocord experience from 78 transplants. Eurocord Transplant group. Bone Marrow Transplant. 21 (Suppl 3), S59–S62.

Sanchez-Ramos, J., Song, S., Cardozo-Pelaez, F., Hazzi, C., Stedeford, T., Willing, A., et al., 2000. Adult bone marrow stromal cells differentiate into neural cells in vitro. Exp. Neurol. 164 (2), 247–256. Available from: https://doi.org/10.1006/exnr.2000.7389.

Saporta, S., Kim, J.J., Willing, A.E., Fu, E.S., Davis, C.D., Sanberg, P.R., 2003. Human umbilical cord blood stem cells infusion in spinal cord injury: engraftment and beneficial influence on behavior. J. Hematother. Stem Cell Res. 12 (3), 271–278. Available from: https://doi.org/10.1089/152581603322023007.

Savitz, S.I., Cramer, S.C., Wechsler, L., Steps Consortium, 2014. Stem cells as an emerging paradigm in stroke 3: enhancing the development of clinical trials. Stroke 45 (2), 634–639. Available from: https://doi.org/10.1161/STROKEAHA.113.003379.

Schultz, E., 1996. Satellite cell proliferative compartments in growing skeletal muscles. Dev. Biol. 175 (1), 84–94. Available from: https://doi.org/10.1006/dbio.1996.0097.

Seo, B.M., Miura, M., Gronthos, S., Bartold, P.M., Batouli, S., Brahim, J., et al., 2004. Investigation of multipotent postnatal stem cells from human periodontal ligament. Lancet 364 (9429), 149–155. Available from: https://doi.org/10.1016/S0140-6736(04)16627-0.

Shin, J.Y., Park, H.J., Kim, H.N., Oh, S.H., Bae, J.S., Ha, H.J., et al., 2014. Mesenchymal stem cells enhance autophagy and increase beta-amyloid clearance in Alzheimer disease models. Autophagy 10 (1), 32–44. Available from: https://doi.org/10.4161/auto.26508.

Shinagawa, K., Kitadai, Y., Tanaka, M., Sumida, T., Onoyama, M., Ohnishi, M., et al., 2013. Stroma-directed imatinib therapy impairs the tumor-promoting effect of bone marrow-derived mesenchymal stem cells in an orthotopic transplantation model of colon cancer. Int. J. Cancer 132 (4), 813–823. Available from: https://doi.org/10.1002/ijc.27735.

Si-Tayeb, K., Noto, F.K., Nagaoka, M., Li, J., Battle, M.A., Duris, C., et al., 2010. Highly efficient generation of human hepatocyte-like cells from induced pluripotent stem cells. Hepatology 51 (1), 297–305. Available from: https://doi.org/10.1002/hep.23354.

Song, P., Inagaki, Y., Sugawara, Y., Kokudo, N., 2013. Perspectives on human clinical trials of therapies using iPS cells in Japan: reaching the forefront of stem-cell therapies. Biosci. Trends 7 (3), 157–158.

Su, Y.J., Lin, W.H., Chang, Y.W., Wei, K.C., Liang, C.L., Chen, S.C., et al., 2015. Polarized cell migration induces cancer type-specific CD133/integrin/Src/Akt/GSK3beta/beta-catenin signaling required for maintenance of cancer stem cell properties. Oncotarget 6 (35), 38029–38045. Available from: https://doi.org/10.18632/oncotarget.5703.

Sykova, E., Homola, A., Mazanec, R., Lachmann, H., Konradova, S.L., Kobylka, P., et al., 2006. Autologous bone marrow transplantation in patients with subacute and chronic spinal cord injury. Cell Transplant. 15 (8-9), 675–687.

Thomas, E.D., Lochte Jr., H.L., Lu, W.C., Ferrebee, J.W., 1957. Intravenous infusion of bone marrow in patients receiving radiation and chemotherapy. N. Engl. J. Med. 257 (11), 491–496. Available from: https://doi.org/10.1056/NEJM195709122571102.

Trachtenberg, B., Velazquez, D.L., Williams, A.R., McNiece, I., Fishman, J., Nguyen, K., et al., 2011. Rationale and design of the transendocardial injection of autologous human cells (bone marrow or mesenchymal) in chronic ischemic left ventricular dysfunction and heart failure secondary to myocardial infarction (TAC-HFT) trial: a randomized, double-blind, placebo-controlled study of safety and efficacy. Am. Heart J. 161 (3), 487–493. Available from: https://doi.org/10.1016/j.ahj.2010.11.024.

Tsai, M.Y., Shyr, C.R., Kang, H.Y., Chang, Y.C., Weng, P.L., Wang, S.Y., et al., 2011. The reduced trabecular bone mass of adult ARKO male mice results from the decreased osteogenic differentiation of bone marrow stroma cells. Biochem. Biophys. Res. Commun. 411 (3), 477–482. Available from: https://doi.org/10.1016/j.bbrc.2011.06.113.

Tsolaki, E., Yannaki, E., 2015. Stem cell-based regenerative opportunities for the liver: state of the art and beyond. World J. Gastroenterol. 21 (43), 12334–12350. Available from: https://doi.org/10.3748/wjg.v21.i43.12334.

Tzameret, A., Sher, I., Belkin, M., Treves, A.J., Meir, A., Nagler, A., et al., 2014. Transplantation of human bone marrow mesenchymal stem cells as a thin subretinal layer ameliorates retinal degeneration in a rat model of retinal dystrophy. Exp. Eye Res. 118, 135–144. Available from: https://doi.org/10.1016/j.exer.2013.10.023.

Vacanti, C.A., Bonassar, L.J., Vacanti, M.P., Shufflebarger, J., 2001. Replacement of an avulsed phalanx with tissue-engineered bone. N. Engl. J. Med. 344 (20), 1511–1514. Available from: https://doi.org/10.1056/NEJM200105173442004.

Venkataramana, N.K., Kumar, S.K., Balaraju, S., Radhakrishnan, R.C., Bansal, A., Dixit, A., et al., 2010. Open-labeled study of unilateral autologous bone-marrow-derived mesenchymal stem cell transplantation in Parkinson's disease. Transl. Res. 155 (2), 62–70. Available from: https://doi.org/10.1016/j.trsl.2009.07.006.

Wakitani, S., Goto, T., Pineda, S.J., Young, R.G., Mansour, J.M., Caplan, A.I., et al., 1994. Mesenchymal cell-based repair of large, full-thickness defects of articular cartilage. J. Bone Joint Surg. Am. 76 (4), 579–592.

Wolfe-Coote, S., Louw, J., Woodroof, C., du Toit, D.F., 1998. Development, differentiation, and regeneration potential of the Vervet monkey endocrine pancreas. Microsc. Res. Tech. 43 (4), 322–331. Available from: https://doi.org/10.1002/(SICI)1097-0029(19981115)43:4%3C322::AID-JEMT6%3E3.0.CO;2-7.

Yin, P.T., Shah, S., Pasquale, N.J., Garbuzenko, O.B., Minko, T., Lee, K.B., 2016. Stem cell-based gene therapy activated using magnetic hyperthermia to enhance the treatment of cancer. Biomaterials 81, 46–57. Available from: https://doi.org/10.1016/j.biomaterials.2015.11.023.

Yoon, S.H., Shim, Y.S., Park, Y.H., Chung, J.K., Nam, J.H., Kim, M.O., et al., 2007. Complete spinal cord injury treatment using autologous bone marrow cell transplantation and bone marrow stimulation with granulocyte macrophage-colony stimulating factor: phase I/II clinical trial. Stem Cells 25 (8), 2066–2073. Available from: https://doi.org/10.1634/stemcells.2006-0807.

Zhao, Q., Ren, H., Li, X., Chen, Z., Zhang, X., Gong, W., et al., 2009. Differentiation of human umbilical cord mesenchymal stromal cells into low immunogenic hepatocyte-like cells. Cytotherapy 11 (4), 414–426. Available from: https://doi.org/10.1080/14653240902849754.

Genetic Engineering: Towards Gene Therapy and Molecular Medicine

Shailendra Dwivedi[1,2], Purvi Purohit[1], Yogesh Mittal[1], Garima Gupta[1], Apul Goel[2], Rakesh C. Verma[3], Sanjay Khattri[1], Praveen Sharma[1], Sanjeev Misra[1,2] and Kamlesh K. Pant[1]

[1]All India Institute of Medical Sciences (AIIMS), Jodhpur, India [2]King George Medical University (KGMU), Lucknow, India [3]UP Rural Institute of Medical Sciences & Research Saifai, Etawah, India

22.1 INTRODUCTION: GENE THERAPY AND MOLECULAR MEDICINE

Research advances in the fields of genetics, molecular biology, clinical medicine, and human genomics has led to the development of the new stream of gene therapy. It involves the transfer of healthy genes into cells with faulty genes, of an individual to treat or prevent a disease. In times to come, this technique would allow clinicians to treat a disorder by inserting a gene into a patient's cells instead of using drugs or surgery. Gene therapy offers the potential of a one-time cure for devastating inherited disorders, for which current therapeutic approaches are either ineffective or have deficient prospects for effective treatment. It can be accomplished by changing a mutated, disease-causing gene with a healthy copy of the gene or "knocking out," a mutated gene or introducing a new gene into the body to help fight a disease. Molecular medicine incorporates the revelation of the genetic basis of disease, diagnosis of the disease, the design of an appropriate approach to disease management or therapy, the application of approved therapeutic protocols, and monitoring of clinical outcomes. Experimental gene therapy research breakthroughs observed in animal model systems are modified for clinical or bedside use, forming the emerging practice of molecular medicine (Kay et al., 1997). There are currently two types of gene therapy proposed for intervention in human beings.

1. Germ line gene therapy
2. Somatic gene therapy

22.1.1 Germinal Gene Therapy

Germinal gene therapy involves the introduction of corrective genes into germ cells (sperm and eggs) or zygotes, with the effect being not just limited to the individual but also genetic change being transmitted to the progeny. This therapy has been fruitful in mice, but the protocols developed for its use in mice have not been effective in humans till date. One protocol that might be possible in human germinal gene therapy includes the elimination of an embryo during the blastocyst stage and injection with transgenic cells. The injected cells then become integrated into tissues of the developing embryo, including the germ line and eventually the gametes of that individual (Griffiths et al., 1996).

22.1.2 Somatic Gene Therapy

Somatic gene therapy pinpoints on the improvement of genetic disease by handling of nonreproductive or somatic tissues. This method of gene therapy includes the exclusion of some of the dysfunctional cells and introducing them with a cloned wild-type gene. The transgenic cells are then implanted into the patient's body where they offer the rectified gene function. Unlike germinal gene therapy, somatic gene therapy technology has progressed swiftly and is presently being assessed in frequent clinical trials. This therapy currently focuses on recovering diseases that are restricted to the one tissue, including CF (Cystic fibrosis) and adenosine deaminase (ADA). In near future, somatic therapy may be tried on multigenic diseases (Griffiths et al., 1996). However, the limitations of somatic gene therapy have been depicted in the delivery of the wild-type gene and nourishing expression (Verma and Somia, 1997).

22.2 HISTORICAL SIGNIFICANCE

The emergence of idea of gene therapy is not new, but it was dictated well by the molecular biologist Joshua Lederberg. In 1963, Joshua Lederberg wrote, "We might anticipate the in vitro culture of germ cells and such manipulations as interchange of chromosomes and segments. The ultimate application of molecular biology would be the direct control of nucleotide sequences in human chromosomes, coupled with recognition, selection and integration of the desired genes."

In 1985, Drs W. French Anderson and Michael Blaese in the National Heart, Lung, and Blood Institute and the National Cancer Institute (NCI) worked in collaboration to explore the possibilities of ADA deficiency corrections in cultured cells of patients. They utilized retroviral vectors to carry the correct human ADA gene to the cells. Furthermore, in 1986, this team fruitfully transformed the modified gene into bone marrow cells of the animals.

In 1988, the researchers successfully transformed the white blood cells (T cells), substituting the bone marrow cells. This transformation ultimately increased the output of corrected cell numbers than bone barrow. Thus this experiment actually paves the way to test the delivery system in humans (Anderson, 1990).

Furthermore, in 1989 Dr Steven Rosenberg tested safety and effectiveness of this gene therapy in a cancer patient by isolating malignant melanoma cells (tumor infiltrating lymphocyte or TIL cells) and then transformed these cells by adding DNA markers with the help of viral vectors. Thus they concluded that TIL cells have an effective role in treatment and engineered viral vectors can be safely utilized in cancer treatment (Rosenberg et al., 1990).

Ashanthi DeSilva, an 8-year-old girl, made a medical history in September 1990 at the NIH Clinical Center when she received her first authorized human gene therapy. This girl by birth had a defective gene that codes for an essential enzyme ADA, which is known to cause malfunctioning of the immune system. White blood cells were extracted from her blood and normal genes for making ADA were inserted into them. The transformed cells were reintroduced into her blood. After taking the doses, she is doing well and no adverse effect and abnormalities were recorded. Thus in 1993 several researchers utilized gene therapy to correct ADA deficiency in newborn babies by delivering ADA gene in isolated immature blood cells from their umbilical cords.

22.3 GENE TRANSFER STRATEGY: DELIVERY VEHICLE

An obvious therapeutic option for treating hereditary disorders would be gene transfer, by replacing a mutated gene by a corrected gene copy. The execution of successful gene therapy in humans has been a challenge despite a number of successful animal models (like mice and other higher animals).

There are two basic strategies of gene transfer for a hereditary disorder, ex vivo and in vivo gene transfer. The various types vectors currently utilized in gene transfer are shown in Table 22.1.

The ex vivo approach is restricted to disorders in which the significant cell population can be removed from the affected individual, modified genetically, and then replaced. Ex vivo strategies are best suited to applications in which the corrected cells can be easily obtained (for example, bone marrow), or when the corrected cells have a discerning benefit when they are back to the patient (for example, the rectification of hematopoietic stem cells (HSCs) for severe combined immunodeficiency, SCID). Other ex vivo applications for hereditary disorders include the use of corrected cells as sources of a secreted protein (for example, factor VIII, F8, gene transfer to autologous fibroblasts in hemophilia A) or to treat a complication from other therapies (for example, transfer of a suicide gene to T lymphocytes to control graft rejection in nonautologous bone marrow transplants).

The in vivo method is the most straight tactic for gene transfer, and human in vivo studies for monogenic disorders have directed the vector containing the therapeutic DNA either directly to the organ of interest or into blood vessels that nourish the organ. Table 22.2 represents some of the common method by which we can insert gene directly into host.

Feasibility of in vivo gene therapy has been demonstrated in animal models at least partially, with target viral gene transfer vectors to different organs. However, targeting

TABLE 22.1   Common Types of Vectors Used in Gene Therapy

Nonviral			Viral
Modes	In vitro	In vitro and In vivo	In vitro and In vivo
Chemical	Calcium phosphate transfection	Liposomes naked plasmid DNA insertion	
Physical	Electroporation Particle bombardment		
Biological			Retrovirus Lentivirus Adenovirus Herpes virus Vaccinia virus Measles virus Epstein—Barr virus

TABLE 22.2   Most Common Direct Methods of Integrating Gene Directly Into Host Cells

Method	Detail	Advantage/disadvantage
Jet gun	Solution of DNA powered by gas such as CO_2 fired into host cell	Good for local treatment of skin/breast cancers but causes bleeding and bruising
Hydrodynamic gene transfer	High-pressure solution of DNA pumped into veins or through a catheter	offer quicker and more convenient, but volumes are too large for human use at present
Electroporation	Electricity used to alter permeability of cell walls and cause gene transfer	Versatility (works with any cell type), efficiency, very low DNA requirements, and the ability to operate in living organisms. Disadvantages include potential cell damage and the nonspecific transport of molecules into and out of the cell
Gene gun	Inert gas used to fire heavy metal particles (gold) coated with DNA at the site and transfer DNA on impact	Good for muscle and tumor treatment gets higher expression rates than needles, especially useful in neurological diseases but expensive and sometimes inserted without genetic materials and hence it damages cells
Ultrasound	Combined ultrasound energy and micro bubbles used to increase permeability of cell membrane to plasmid DNA	Safe and efficient, good gene expression in vascular cells and muscles

strategies have not been used clinically for hereditary disorders. Effective gene transfer strategies must involve:

- The insertion of the gene to an in vivo cell population, with a sufficient number of target cells affected for a corrected clinical phenotype.

TABLE 22.3   General Characteristics of Gene Therapy Vectors

Gene Transfer	Advantage	Disadvantage	Clinical application
Viruses Retrovirus Adenovirus Adeno-associated herpes poxvirus HIV-1	Efficient entry into cells stable integration Biology known	Need higher titer Limited payload Immunogenetic Difficult to control and stabilize Expression can induce adverse events Random insertion	Suitable for permanent correction Extensive use in marking studied specific virus for specific disease, e.g. herpes neurology
Liposomes	Commercially availableeasy to use targetable large payload	Entry into cells Integration rate	
Naked DNA	Ease in preparation Safe No size limitation No moderate application Extraneous genes	Inefficient entry into cells not stable	Topic application
Complexed DNA	More efficient uptake than naked DNA: protected from degradation targetable unlimited construct size	Not stable Inefficient cell entry Limited targetability	Limited clinical use Vaccination
Artificial chromosomes	Autonomous vectors No insertion required Regulatable tissue and temporal	Unpredictable chromosome formation Centromere formation	Experimental: only in human transformed cells
Artificial cells	Designer potential	Complexity	Conceptual

- By contrast, if the phenotype results from a secreted protein, well-regulated sufficient levels of post-transnationally modified protein needs to be produced to correct the phenotype. If these circumstances come across, it does not mean to which cells or organ the gene is transferred. However, if the phenotype to be corrected is in organs that have barriers to the route of proteins, for example, eye and brain, the gene coding for the secreted protein expression needs to be directly delivered to the suffered organ.

Vectors are vehicles assisting the transfer of genetic information into a cell, and most common characteristics of the vectors have been shown in Table 22.3. They can be divided into viral and nonviral delivery systems (Kresina and Branch, 2000).

22.3.1 Viral Vectors: Gene Therapy

Infectious virions are very promising in transferring genetic information. Most gene therapy experiments have utilized viral vectors comprising elements of a virus that result in a replication-incompetent virus. These viral vectors replaced one or more viral genes with a promoter and coding sequence of interest and could potentially undergo recombination to produce a wild-type virus capable of multiple rounds of replication. Several viruses have been used to generate viral vectors for its application in gene therapy as shown in Table 22.4. Important

TABLE 22.4 Types of Viral Vector Used in Gene Therapy

Viral vector	Types	Advantage	Disadvantage
Retrovirus	Integrates with host chromatin	Effective over long periods Efficient transfection ex vivo Low immune response in host	Small, max 8 kb insert size inefficient transfection in vivo Relies of target cell mitosis safety concerns
Lentivirus	Integrates with host chromatin	Transfects proliferating and nonproliferating hosts and hemo-stem cells New generations are self-inactivating for safety	Need active transport into cell small, max 8 kb insert size Technologically challenging safety concern simmunod eficiency origins
Adeno-associated virus	Either	Very good length of expression especially in vivo Low immune response in host	Safety problems owing to potential insertional mutagenesis Small, max 4.5 kb insert size High immune response Technologically challenging
Adenovirus	Extra chromosomal DNA	High efficient transfection in vivo and ex vivo Transfects proliferating and nonproliferating hosts	Repeat treatments ineffective due to strong immune response Small, max 7.5 kb insert size Technologically challenging short expression duration
Herpes simplex virus	Extra chromosomal DNA	Very good length of expression especially in vivo Safe for use in immune-compromised patients Large insert size to 30 kbEffective on many cell types	Difficult to produce in large quantities

features that distinguish the different viral vectors include the size of the gene length accepted, the duration of expression, target cell infectivity, and integration of the vector into the genome.

22.3.1.1 Retroviral Vectors

Retroviruses include two copies of a positive single-stranded RNA genome of 7−10 kb. Their RNA genome is converted into double-stranded DNA, which incorporates stably into the host cell chromosome. They were originally isolated since they induced tumors in susceptible animals by the transfer of cellular oncogenes into cells.

Retroviruses membrane envelope protein is a virus-encoded glycoprotein precise to host range or types of cells that can be infected by binding to a cellular receptor. The envelope protein endorses fusion with a cellular membrane on one side of the cell surface or in an endosomal compartment. To date, retroviruses based on the Mouse Moloney Leukemia virus have been used most commonly in clinical trials. These vectors are enveloped into viral particles, have all viral genes deleted except some of the viral controlling sequences, and will only transduce dividing cells, because movement of the DNA to the nucleus only occurs during the breakdown of nuclear membrane during mitosis (Rosenberg et al., 1990).

22.3.1.2 *Adenovirus-Based Vectors*

The human adenoviruses are known for mild sicknesses such as upper respiratory infections. The more known serotypes (2 and 5) have been assessed for its usefulness in clinical trials for cystic fibrosis and cancer. Human adenoviruses are viruses having 36-kb double-stranded DNA that contain genes that code for more than 50 gene products during its life cycle. In the removal of the E1 region of the vector, space is created for placing therapeutic expression sequences and, in the absence of the trans-activating E1a protein, the virus failed to replicate. Thus, after gene transfer, no viral spread will occur making first-generation adenoviral vectors to be very efficient at transferring genes into most tissues after in vivo administration. These vectors were utilized to partially cure a number of animal models with typical genetic diseases like hemophilia and hypercholesterolemia. Passion for these vectors was moderated by the finding that low-level production of viral antigens from the vector caused a vigorous immunologic response that excluded transduced cells and transgene product and or inhibited repeat administration.

22.3.2 Nonviral Vectors: Liposome

Liposomes or DNA/lipid complexes are very easy in preparation, and there is no restriction in the size of genes that can be transfected. As they are protein-deficient, they may induce negligible immunogenic responses. Furthermore, they also have a lesser risk of generating the infectious form or inducing tumorigenic mutations as genes delivered have low integration frequency and cannot replicate or recombine. During the last decades, two classes of cationic lipids have been manufactured and shown decent transfection activity in vitro with established cell lines: One with two alkyl chains in each cationic lipid molecule and the other type uses cholesterol as the backbone, which interact with DNA via electrostatic interactions (Kresina and Branch, 2000).

22.4 CLINICAL TRIALS (IN VIVO AND EX VIVO): AN UPDATE

The first therapeutic human gene therapy clinical trial was approved in 1990 and involved two children suffering from SCID. Since then, several gene therapy trials have been performed and currently about 90–100 new trials are registered a year all over the world, involving around 31 countries from all continents.

22.5 GENE THERAPY AND DISEASE/DISORDERS

Gene therapy was primarily regarded as a tool for treating inherited diseases, but now it has open new vistas for noninherited ones also. With all the paradoxical dilemmas, gene therapy has made significant medical advances in less than two decades. It has changed from theoretical to technological progression and laboratory research to clinical translational trials for many diseases. The major diseases and disorders for which gene therapy or gene transfer permitted are shown in Table 22.5. Some of the most notable advancements are discussed below.

TABLE 22.5 Common Diseases and Disorders in Which Gene Therapy—Based Trials Have Been Recommended

Monogenetic disorders	Cancer
Adrenoleukodystrophy	Gynecological: breast, ovary, cervix, vulva
α-1-antitrypsin deficiency	Nervous system: glioblastoma, astrocytoma, neuroblastoma, retinoblastoma
Becker muscular dystrophy	Gastrointestinal: colon, colorectal, liver metastases, gall bladder
β-Thalassemia	Genitourinary: prostate, renal, bladder
Cystic fibrosis	Skin: melanoma
Duchene muscular dystrophy	Head and neck: nasopharyngeal carcinoma, squamous cell carcinoma,
Fabry disease	esophageal carcinoma
Familial adenomatous polyposis	Lung: adenocarcinoma, small cell/non-small cell
Familial hypercholesterolemia	Hematological: Leukemia, lymphoma
Fanconi anemia	Sarcoma
Gaucher's disease	Li—Fraumeni syndrome
Hemophilia A and B	
Hurler syndrome	
Hunter syndrome	
Huntington's chorea	
Pompe disease	
Sickle cell disease	
Tay—Sachs disease	

Cardiovascular disease	Neurological disease
Anemia of end-stage renal disease	Alzheimer's disease
Angina pectoris	Amylotrophic lateral sclerosis
Coronary artery stenosis	Carpel tunnel syndrome
Critical limb ischemia	Diabetic nephropathy
Heart failure	Epilepsy
Myocardial ischemia	Multiple sclerosis
Pulmonary hypertension	Myasthenia gravis
Venous ulcers	Parkinson's disease

Ocular diseases	Inflammatory disease
Age-related macular degeneration	Arthritis
Diabetic macular edema	Degenerative joint disease
Glaucoma	Ulcerative colitis
Retinitis pigmentosa	

Infective disease	Other disease
Adenovirus infection	Chronic renal disease
Cytomegalovirus infection	Erectile dysfunction
Epstein—Barr virus	Oral mucositis
Hepatitis B and C	Fractures
HIV/AIDS	Type 1 diabetes
Influenza	Diabetic ulcers
Japanese encephalitis	Graft-versus-host disease
Malaria	
Tetanus	
Tuberculosis	

22.5.1 Gene Therapy and Hemophilia

Hemophilia were supposed to be perfect applicants for gene therapy due to lack of a single protein that circulates in little quantity in the plasma, and an increase of 1%−2% in circulating levels of the deficient clotting factor can significantly alter the bleeding diathesis. Several preclinical studies have shown continued expression of factor VIII (FVIII) and factor IX (FIX), using a wide range of gene transfer technologies targeted at diverse tissues. This led the foundation to six different Phase I/II clinical trials that resulted in limited efficacy and, importantly, minimal toxicity. Currently, recombinant adeno-associated virus (rAAV) vectors have shown emerging potential to be utilized in gene therapy of hemophilia B. These vectors have a good safety profile among gene transfer vectors of viral class, as wild-type AAV has never been linked with human disease. Long-lasting expression of human FIX in several animal models has been demonstrated using rAAV, after delivery of the vector to the liver or muscle. A dose escalation study of rAAV with intramuscular injection failed to show any severe effects. Human FIX was identified immunohistochemically at the site of injection in all patients although systemic levels were below the therapeutic range. Study with higher doses of rAAV-2 vector shall not be practiced as it may possibly aggravate a neutralizing antibody response to human FIX, as assessed in animal models (Mingozzi et al., 2003). Another clinical trial with AAV vectors pinpointed on liver-directed delivery of rAAV-2 at a dose ranging from 2 to 50 … 1011 2 to 50 • 1011 vector genomes (vg)/kg. All patients had shown vector genomes momentarily in the semen, although there was no proof of germ line transmission. Therapeutic levels of human FIX has thus far only been noticed only transiently (up to 6 weeks after gene transfer) in two patients who were under the highest dose (50 … 1011 vg/kg). One of these patients had an episode of reversible subclinical elevation (fivefold) of liver transaminase at 4 weeks after gene transfer (High et al., 2003) possibly because of a 'recall' T-cell-mediated immune response to the viral capsid proteins, leading to the termination of this trial (High et al., 2004). The new generation of rAAV vectors based on alternative serotypes will avoid these immunologic difficulties.

Gene therapy (based on vector innovation) for hematologic disease and disorders have been cutting-edge extraordinarily in the past 10 years, with 2014 perceiving constant improvement in certain inherited diseases. One notable example of technologically claimed and fruitful gene therapy is that of X-linked SCID (Xl-SCID). A recent revolutionary study by Dr Salima Hacein-Bey-Abina research group in Paris and in the United States applied a novel safety-modified self-inactivating γ-retrovirus encoding the interleukin-2 receptor γ-chain to transfect the normal gene into pediatric patients. They removed the Moloney murine leukemia virus LTR U3 enhancer from this altered vector, thereby dropping the mutagenic potential. Out of nine children with X1-SCID registered in the study, six exhibited continued lymphocyte reconstitution including T and NK cells, with variable functional B-cell recovery. No leukemia was seen in any patient on the follow-up period of 33 months (median). Another achievement story of 2014 was the success by Dr Amit Nathwani and colleagues on long-term efficacy and safety of gene therapy for severe hemophilia B using a new vector, an adeno-associated virus serotype 8 (AAV8) vector, scAAV2/8-LP1-hFIXco, containing a codon-optimized modified factor IX transgene. In the Annual Meeting of ASH 2014, two abstracts reported the usage of a lentivirus to cure β-thalassemia. Lentiviral vectors are replication-defective and self-inactivating and also comprise an engineered β-globin gene (βA-T87Q) and is called LentiGlobin BB305. Both studies described that some study subjects no longer needed red blood cell transfusions (Nathwani et al., 2005, 2014).

An invention in preclinical studies in the field of gene editing was witnessed in 2014. Gene editing comprises the ability to correct defective mutant genes in situ and also the ability to interrupt gene function. Endonucleases undertake the task of gene editing that can be directed to specific regions of DNA to excise mutated regions (XI-SCID treatment) and exchange them with a normal sequence or to interrupt genes to modify or eliminate unwanted gene function. Dr Fei Xie and colleagues utilizing the endonuclease system CRISPR (clustered, regularly interspaced, palindromic repeats) associated protein 9 (Cas9) and the piggyBac cassette encompassing normal parts of the gene corrected induced pluripotent stem cells from patients with β-thalassemia. The CRISPR/Cas9 endonuclease system was also benefitted in (1) β-2 microglobulin as the auxiliary chain of the major histocompatibility gene class I molecules in order to yield hypoimmunogenic cells for transplantation, and (2) CCR5, the main co-receptor for assured strains of HIV that could potentially interfere with viral entry. These reports signify key milestones in gene editing technology that promise a good future success in the clinic (Xie et al., 2014).

22.5.2 Gene Therapy and Cardiovascular Disorders

The Shh protein is a significant element of the hedgehog signaling. The embryonic hedgehog signaling pathway can be revitalized to combat ischemia in adult mammals as shown in experimental data. Recombinant Shh protein causes aggravation of multiple angiogenic factors, including VEGF, in interstitial mesenchymal cells, thus causing robust angiogenic effect. Also genetic transfer of Shh induces the expression of trophic factors, such as SDF-1, which escalates the recruitment and incorporation of bone marrow–derived progenitor cells into the growing vasculature leading to enhanced regeneration of ischemic myocardium.

Gli3, a transcription factor targeted by Shh during hedgehog signaling, gets strongly up regulated in the ischemic tissue of adult mammals and probably has a favorable effect on my genesis and angiogenesis after an ischemic insult. Shh appears to be particularly effective for angiogenic gene therapy (Roncalli et al., 2010).

22.5.2.1 Angiogenic Gene Therapy

Post-ischemic heart failure is the major contributor to morbidity and mortality in the western world. Thus the main goal for treating heart failure is to get better perfusion in the ischemic region. Therapeutic vascular growth has been accomplished in vivo by the genetic transfer of cytokines by several researchers. The common studied and best developed cytokines used in the clinical setting are VEGF and FGF. The REVASC trial studied 67 patients with coronary artery disease, severe angina, and no conventional options for revascularization. They were randomized to receive direct intra-myocardial gene transfer of adenoviral VEGF-121 (AdVEGF-121) via mini-thoracotomy or to continue receiving maximal medical treatment. The patients showed exercise time, the primary efficacy endpoint, significantly greater when treated with AdVEGF-121 than in the control group

(P = 0.026), and there was no significant difference between the two treatment groups in overall adverse event occurrence.

An enhanced regional wall motion and a favorable anti-ischemic effect was witnessed in the Euroinject One trial, 80 'no-option' patients with severe, stable, ischemic heart disease receiving pVEGF-165 (0.5 mg) or a placebo plasmid; the therapy looked to be safe, but did not advance significantly stress-induced abnormalities in myocardial perfusion. The AGENT trials evaluated the intracoronary injection of AdFGF-4 in patients with stable coronary artery disease and reported positive trends were observed in the two small Phase 1/2 trials (AGENT 1 and 2) (Stewart et al., 2006).

22.5.2.2 Non-Angiogenic Gene Therapy

Gene therapy with rAAV vectors coding for SERCA2a was well accepted in both small and large animal heart failure models. Restoring the levels of SERCA2a to normal led to a substantial enhancement in cardiac function. These findings prompted initiation of the first-in-human Phase 1/2 CUPID trial. In this study, nine patients with advanced heart failure received a single intracoronary infusion of rAAV SERCA2a. A second, randomized, double-blind study (ClinicalTrials.gov identifier: NCT00534703) is also investigating the safety and feasibility of SERCA2a gene therapy when delivered with the AAV6 vector and driven by the cytomegalovirus promoter (AAV6-CMV-SERCA2a). Sixteen patients with advanced heart failure who have received a left ventricular assist device will be randomized to receive AAV6-CMV-SERCA2a or placebo infusion into the coronary arteries, followed by the evaluation of recovery of contractile function during attempts to wean patients from the left ventricular assist device. A clinical study (ClinicalTrials.gov identifier: NCT00787059) is presently underway to determine whether a Type 5 adenovirus encoding this gene can be administered safely and is potentially beneficial in patients with congestive heart failure (Kastrup et al., 2005).

22.5.2.3 Combination Therapy

With the characterization of individual gene therapies becoming almost complete, preclinical investigations are being designed to recognize the potential complementary or synergistic effects achieved with combinations of therapies. The results of these studies would depend upon the same variables that prompt the effectiveness of single-gene therapy, including the model species, the delivery vector, the organ and disease treated, and the genes delivered. At times, two (or more) co-injected genes may not be expressed in the intended ratio, due to preferential silencing or removal of one of the vectors. Thus, a therapeutic approach needs to be believed such that it ensures the stable co-expression of both genes.

One system for inducing stable gene expression involves the use of IRESs. IRESs are structural elements located in the 5'-untranslated region of several mRNAs. These elements can assist to construct expression cassettes that code for combinations of genes within a single mRNA sequence (Jaski et al., 2009).

22.5.3 Gene Therapy and Diabetes

The 1970s observed a revolution in the medicine sector and gene therapy, when several experiments led to cloning and expression of insulin in the cultures cells and was proposed as a possible cure for diabetes. The somatic ex vivo therapy poised at the generation of cells which retain the properties of β cells—the insulin-generating cells. It has also been utilized to generate β cells for transplantation. However, it is of apprehension that surgically removed tissue from the patient and its reimplantation post-genetic modification, back into the body of the patients, might produce immune response. Type 1 diabetes that happens due to autoimmune destruction of insulin synthesizing pancreatic β cells is being evaluated for islet transplantation as a possible treatment strategy.

In vivo gene therapy is nontoxic and more beneficial. Existing tactics for in vivo therapy are:

- genetic transfer of noninsulin glucose dropping genes,
- an enhancer of glucose utilization by liver or skeletal muscles, and
- an inhibitor of glucose production by the liver.

For example, glucokinase as a transgene is found to have glucose dropping effect in the liver, probably by enhancing glucose utilization by the body.

The genetic transfer of glucokinase had been practiced as an adjuvant therapy in the treatment of diabetes. In another method, it was to regulate glucose production in the liver via a gene known as protein targeting to glycogen (PTG) that changes glucose to glycogen. The PTG protein belongs to the family of glycogen targeting subunits of protein phosphatase-1. Adenoviral-mediated PTG transfer in rats has demonstrated to stimulate glycogen synthesis in the liver and drops blood glucose levels. This has been deliberated as a therapeutic approach for diabetes. Some other areas of genetic engineering include transfer of genes that show response to glucose and to induce β cell production in the hepatocytes using gene therapy.

In addition, the insulin gene can be altered to encode single-chain insulin with 20%—40% activity of normal mature insulin. Induction of hepatocytes for synthesis of β cells has been described by Kojima et al., by inserting islets-specific transcription factors to hepatocytes. The regulation of insulin production and its control is still unclear. Thus induced β cell neogenesis appears to be an encouraging tactic as a therapeutic for diabetes, as it provides a solution for the autoimmunity in Type 1 diabetes (Tiwari, 2015).

22.5.4 Gene Therapy and Neurological Disorders

Human neurological disorders are complex, regardless of intracerebral grafts of fetal and/or adult-derived cells, and are beneficial in somatic gene therapy. Cells of varied origin survive transplantation into the brain and can replace or supplement deficient molecules. Many studies have reconnoitered somatic gene therapy focusing on the ex vivo approach (Blömer et al., 1996).

22.5.4.1 Parkinson's Disease

Parkinson's disease (PD) is one of the most common neurodegenerative disorders with prevalence of 0.1%—1%. Bulk amount of cases are noticed every year, and its biological cause is generally unidentified but may be linked to oxidative stress, lack of neurotropic support, or exposure to toxins. The disease is characterized by tremor, rigidity, and movement disorder due to the loss of inhibitory input on the extra-pyramidal system and loss of dopaminergic neurons of the substantia nigra that project into the striatum. The effect of oral L-Dopa shows that improvement does not necessitate the restoration of the neuronal circuitry. L-Dopa is synthesized from tyrosine by enzyme tyrosine hydroxylase (TH). Local L-Dopa supply may thus be enhanced by the introduction of a single-gene TH into the cells. A rodent model permits testing of the proficiency of gene therapy in PD. 6-Hydroxydopamine, a neurotoxin, destroys nigro-striatal dopaminergic neurons and eliminate nigral dopaminergic input and upregulation of dopamine receptors in the lesioned striatum, still maintaining striatal dopamine receptor density in the unlesioned side. Application of apomorphine (dopaminergic agent) results in rotational behavior, i.e., asymmetry due to differential postsynaptic receptor sensitivities between denervated and intact striatum.

Several viral vectors have been used for the direct TH gene transfer into denervated striatum and in vivo expression using AAV vectors showed long-term expression in vivo in lesioned animals. Earlier reports mostly using adenoviral vectors were not able to retain long-term transgene expression (Kaplitt et al., 1994).

22.5.4.2 *Alzheimer's disease*

Alzheimer's disease (AD) belongs to a large group of degenerative brain disorders which are a common dementia (0.02%−5% prevalence) of older patients. AD has a large number of acquired cases as compared to inherited ones. It is characterized by progressive dementia due to cortical atrophy, neuronal loss, neurofibrillary tangles, senile plaques, and vascular deposits of β-amyloid in various regions of the cerebral cortex and the hippocampus, with β-amyloid and its precursor playing a critical role in the pathogenesis of AD. The cause of degeneration of forebrain cholinergic neurons responsible for memory acquisition and retention is not known. A fimbria-fornix lesion that disconnects the cholinergic neurons of medial septum to their nerve growth factor (NGF) supply is a well-established rodent model for degeneration of cholinergic neurons. There are reports that exogenous substitution of NGF in the rodent model circumvents cholinergic neurons degeneration and ameliorates some forms of memory deficit. Measurable generation of neurons in AD patients has been shown in response to experimental gene therapy which involved NGF injection into their brains. This report is claimed by the researchers at University of California, San Diego School of Medicine, in the current issue of *JAMA Neurology*.

Prof. Mark H. Tuszynski (MD, PhD), Professor and Director, UC San Diego Translational Neuroscience Institute, and a neurologist at VA Medical Center, San Diego, reported in 10 AD patients in a Phase I trial (2001) whether injected NGF (a protein essential to cellular growth, maintenance, and survival) safely slows or prevents neuronal degeneration that affected neurons displaying heightened growth, axonal sprouting, and activation of functional markers. The reports are derived from post mortem analyses. This was the first study of direct transfer of NGF into the brain, since NGF is too large to cross the blood−brain barrier. Besides, circulating NGF causes adverse effects such as pain and weight loss. In the direct injection of NGF into targeted regions of the brain, the protein was actually introduced to surrounding degenerating neurons.

The participants of the Phase I trials (March 2001 to October 2012 at UC San Diego Medical Center) survived 1–10 years (Scott, 2015; La Spada and Ranum, 2010) and since then the studies have moved onto Phase II trials. However, the results have not been published as yet.

22.5.5 Gene Therapy and HIV Infection

The principle goal of gene therapy in HIV is to limit the replication of the HIV-1.

22.5.5.1 Genetic Approaches to Inhibit HIV Replication

There are three major gene therapy categories against HIV replication: (1) transdominant negative proteins (TNPs) and single-chain antibodies as the protein approaches; (2) antisense DNA/RNA, RNA decoys, and catalytic RNA moieties (ribozymes) that are the nucleic acid–based gene therapies; and (3) genetic vaccines or pathogen-specific lymphocytes included as immunotherapeutic approaches. These techniques may be used concurrently impeding multiple stages of the viral life cycle, along with other approaches, such as hematopoietic stem cell transplantation or vaccination. The effectiveness of the gene therapy against HIV-1 depends upon (1) appropriate target cell selection; (2) gene delivery efficiency; (3) expression, regulation, and stability of the gene product(s); and (iv) potency of gene therapy against viral replication.

22.5.5.2 Transdominant Negative Proteins

TNPs inhibit replication as they are mutant versions of regulatory or structural proteins with a dominant negative phenotype. The major hindrance with the use of viral TNPs is the possible immunogenicity when expressed by the transduced. This might result in their own destruction reducing the efficacy of antiviral gene therapy. The most thoroughly studied HIV-1 TNP is Mutant Rev Protein denoted RevM10, although other potential targets include HIV-1 regulatory (Tat and Rev) and structural proteins (Env and Gag).

22.5.5.3 Single-Chain Antibodies (Intrabodies)

Intrabodies are single-chain antibodies expressed intracellularly, which are used in antimicrobial gene therapies. The variable fragment of the single-chain antibody holds antigen specificity and binding capabilities of the parental antibody. Intracellular expression of the intrabody is guided by these genes, and an immunoglobulin heavy chain leader sequence targets the intrabody to the endoplasmic reticulum (ER). An interchain linker connects the heavy and light chain variable regions.

In HIV-1 gene therapy, the use of an intrabody specific for the CD4 binding region of the gp120 (Env) markedly reduced the HIV-1 replication. The intrabody inhibited replication by entrapping gp160 in the ER, thus preventing its maturation into the gp120/gp41 proteins.

22.5.5.4 Endogenous Cellular Proteins as Anti-HIV Agents

Certain endogenous cellular protein genes have been identified to have specific gene inhibitory activity. These proteins act by checking attachment of HIV to cells, by direct binding to the regulatory/structural proteins, or indirect induction or repression of cellular factors that in turn affect viral gene expression. Soluble version of the HIV receptor CD4 (sCD4) is the most successful in vitro endogenous cellular proteins inhibiting an infectious agent.

22.5.5.5 Nucleic Acid–Based Gene Therapy Approaches: RNA Decoys

This approach employs the use of decoy short RNA molecules which disrupt normal interaction of the HIV regulatory proteins with their cis-acting regulatory elements by competing with viral RNA elements for binding of proteins that are required for virus replication. Two such viral regulatory elements found in HIV are TAR (transactivation response) and RRE (Rev response element). They are the binding sites for the transactivating proteins Tat and Rev, respectively. A retroviral-mediated gene transfer in T-cell lines demonstrated the antiviral activity of the TAR element decoys in vitro.

22.5.5.6 Antisense DNA and RNA

Antisense nucleic acid technology involves the introduction of an antisense RNA or single-stranded DNA moiety (oligodeoxynucleotide) into the cell or tissue, which is complementary to a target mRNA. Chemically modified DNA oligonucleotides are used as one of the major approaches for application of antisense RNA. These chemically modified DNA oligonucleotides have higher stability in target cells. A number of artificial antisense oligonucleotides have been designed for inhibiting replication of HIV-1.

22.5.5.7 Ribozymes (Catalytic Antisense RNA)

Catalytic RNA molecules with antisense oligonucleotide nature are ribozymes. They bind to the target RNA moiety by antisense sequence-specific hybridization. They act by cleavage of the phosphodiester backbone and inactivating the target HIV-1. The hammerhead and hairpin ribozymes are two most thoroughly studied ribozyme types. Ribozymes are advantageous over traditional antisense technology since they are not consumed during target cleavage reaction and several target molecules are inactivated by single ribozyme. These ribozymes are highly efficient catalytic molecules even at low concentrations. Multiple ribozymes can be packed into single vector owing to the small transcription unit, facilitating transfer of ribozymes targeted to several HIV-1 regions to be delivered into the same cell.

22.5.5.8 DNA Vaccines

Two groups using plasmid DNA for the first time demonstrated generation of an immune response to marker proteins encoded by plasmids by introduction into the skin of mice using biolistic gene delivery approach. The advantages of the DNA-based vaccination technology as compared to traditional vaccine strategies include (1) convenient production and preparation of plasmid DNA, (2) efficient generation of cytotoxic and helper T cells owing to the expression of antigens in their native form, (3) provides long-term immunity so less number of doses of vaccine required, and (4) the cells need not be the target cells that are normally infected by the infectious agent.

22.5.5.9 HIV-Specific Cytotoxic T Lymphocytes

Passive restoration of the immune system function by using an infected individual's own cells is another technique of gene therapies for containing HIV infection. This involves an ex vivo expansion of either CD4 or CD8 lymphocytes, which are then reinfused into the HIV-1-infected individual. In this context, CD8 cells have been mainly used as adoptive cell therapy for HIV-1 infection. Since early in the infection there is a rise in HIV-specific CD8 cells, this correlates well with the improvement of viremia. Thus MHC class 1—restricted CD8 cells play a role in containing infection during the acute phase of infection (Dwivedi et al., 2013a,b).

22.5.6 Gene Therapy and Various Cancers

Abnormal gene expression, mutations, nonfunctional or missing tumor suppressor genes, and activated proto-oncogenes are associated with cancers of veritable forms. Gene therapy raises hope by inserting a correct gene into the tumor or cancer cells.

22.5.6.1 Gene Therapy and Hematological Malignancy

Graft-versus-leukemia (GvL) effect, i.e., alloreactivity of the donor lymphocytes against cancer cells following allogeneic bone marrow transplantation is the major form of immunotherapy utilized. But the host tissue alloreactivity evades this effect manifested by donor lymphocytes and causes severe complication of graft-versus-host disease (Ennis and Dingli, 2010). However, in patients with hemtological malignancies, gene therapy helps enhance the GvL effect. Bonini et al. (1997) reported the use of onco-retroviral vector transduced T cells in eight patients with diverse post-transplant malignancies. The onco-retrovirus sensitive to the prodrug ganciclovir (GCV) encoded the herpes simplex thymidine kinase gene that caused inactivation of these cells. Five out of the eight patients were reported to have a clinical antitumor response. Tiberghien et al. (2001) showed alloreactive capacity of gene-modified T cells in Epstein—Barr virus—induced lymphoproliferative disease. Oncolytic viruses are the new focus of gene cancer gene therapy today. Viruses usually express variable forms of proteins, e.g., adenoviral E1A and E1B capable of altering normal cellular defenses, including pRB or p53 to enhance their likelihoods of replication. Phase I/II studies show promise and are underway with these vectors in patients with solid tumors (Mullen and Tanabe, 2002). Currently more lymphoproliferative disorder treatments using vectors based on viruses that have a natural tropism for lymphoid tissues are being explored.

One of the major neoplastic disorders of hematopoietic origin is acute myeloid leukemia (AML). Sindbis virus (SIN) replicons were tested for therapy of AML (Tan et al., 2010) SIN, a prototypic alpha virus that is transmitted to humans by arthropods, is a member of the Togavirus family. SIN has a ~12 kb positive strand RNA genome. SIN is an attractive vector for gene therapy owing to its fast viral life cycle, high viral genome amplification, and the lack of integration, especially when -level transgene expression is the motive. The self-limiting nature of the vector makes it an attractive mode of gene therapy in humans. SIN vector is produced by transfection of cells with a genomic RNA that codes for the nonstructural and structural proteins and a therapeutic gene of interest. The structural portion is eliminated. Meruleo's group was the first to report the oncolytic activity of SIN. A luciferase reporter gene inserted with SIN replicon when the injected intraperitoneally was able to infect subcutaneous BHK tumors. Other cancer models showed successful infection on intravenous injection of the vector. The extra-therapeutic benefit of GALV.fus expression was determined by using it as a reporter gene in one of the replicons and comparing it with another replicon carrying DsRed, another reporter gene. Successful infection of the AML cell line HL-60 with either replicon was achieved, and GALV.fus caused formation of syncytia and a higher cell death rate. There was reduced colony formation from CD34 + AML progenitor cells, however, not from normal progenitors, thus supporting the relative tumor specificity of these replicons. Reduced tumor burden was also observed in sublethally irradiated NOD/SCID mice harboring primary human AML cells. Interestingly, probably expression of GALV.fus inhibits NF-κB signaling, a master regulator of many genes, so an important therapeutic target in AML with other hematologic and nonhematologic malignancies. Consequently, SIN/GALV.fus can kill AML cells because of (1) elevated level expression of the laminin receptor favoring entry, (2) strong effect is possibly due to formation of syncytia and these are nonviable, (3) due to blocking of NF-κB signaling, and (4) apoptosis stimulation due to expression of nsP1-4 (Tan et al., 2010).

22.5.6.2 *Gene Therapy and Oral Cancer*

The availability of primary and recurrent lesions of oral squamous cell carcinoma (OSCC) makes it an interesting candidate for gene therapy. The OSCC lesions are swiftly available for injection or application of the therapeutic agent. *Addiction gene therapy* is the present-day gene therapy advancement. It involves insertion of tumor suppressor genes that inactivate carcinogenic cells and inhibit the growth of tumors. There are reports of p53 alteration as the primary event in oral cavity carcinogenesis and mutation of p53 expression as reported in noncancerous epithelium adjacent to OCSS (Ishiguro et al., 1997).

Presently, the Phase III study on adenovirus vector Ad5CMV-p53 suggests that application as intramucosal injection followed 2 hours later by a mouthwash BD for 2−5 days and repeated its cycle after 28 days has a reduction in progression of precancerous lesions with no adverse effects. Similarly pro-apoptotic and antitumor effect was seen on insertion of Rb (retinoblastoma gene) and mda-7 (melanoma differentiation-associated gene-7) tumor suppressor gene in to cancerous cells. Furthermore, p27 gene showed similar antitumor and pro-apoptotic effect when transfected into tongue cancer.

Allovectin-7 (Vical Inc., San Diego, CA, USA) suppressed the growth of head and neck tumors on intratumoral injection in animal models. It is produced due to the co-expression of the human gene of the HLA-B7 leukocyte antigen with beta2-microglobulin gene. This seems to be relatively well tolerated, and only mild or moderate adverse reactions around the insertion site are observed (Yoo et al., 2009).

Another viral vector in OSCC is adenovirus ONYX-015 with deleted region of the E1B. It is a unique viral vector that shows negligible viral replication in cells with normal p53 function but showed high levels in cells with p53 alterations or mutations. There is significant tumor regression following intravenous injection of this vector, improving survival in the presence of metastasis. Oncolytic viruses used along with chemotherapy have shown enhanced activity. Thus the efficacy of ONYX-015 was enhanced by 6% on usage with

5-fluorouracil (5-FU). Suicide genes cause expression of enzymes transforming nontoxic drugs into cytotoxic substances such as thymidine kinase gene of Herpes Simplex Virus (HSV) transforms GCV into GCV phosphate.

A vaccine against FLK-1 that stimulates T lymphocytes, represses this receptor, and vascularization is said to be effective. It is also useful in the tongue metastasis of OSCC, with enhanced immune response at 10 months of the inoculation. NF-kB inhibition is a possibly useful co-adjuvant treatment in oral cancer as NF-kB is an important contributor to the progression and metastasis of various cancers, including OSCC. Besides, the insertion of IL-2 gene showed a high antitumor effect and major necrotic changes due to intratumoral injection of an adenovirus and an RGD peptide (Adv-F/RGD) in mutated fibroblast.

22.5.6.3 *Gene Therapy and Breast Cancer*

There are reports of clinical studies using a combination systemic chemotherapy and of local injection of p53-adenoviral vector into skin metastatic lesions or locally advanced breast cancer. The p53 gene transfer has been reported to have no effect on normal PBSC (Baynes et al., 2007). ErbB2/HER2 gene is amplified in 20%−30% of breast cancer patients and causative of poor prognosis, and hormone therapy resistance may be suppressed on introduction of monoclonal humanized murine antibody of ErbB2/HER2 protein (trastuzumab/Herceptin) in ErbB2/HER2-overexpressing breast cancer patients. This leads to tumor reduction and enhanced sensitivity to in vivo and in vitro chemotherapy. Furthermore, anti-sense mRNA with mammary tumor virus promoters in retroviral vectors suppresses c-myc and c-fos gene in animal models (Arteaga and Holt, 1996). Growth suppression and apoptosis is caused by insertion of melanoma differentiation associated protein 7 (MDA-7). But normal cell lines on inoculation with MDA-7 gene do not respond, triggering a clinical trial that inserts MDA-7−adenoviral vector (Ad-mda7, ISGN 241) into tumor cells. Phase I trials showed no adverse effects and so a combination Phase I/II study is underway. Breast cancer patients have shown considerable reduction in tumor size and enhanced immune response on injection of IL-2 and IL-12 gene expression vectors. Furthermore, activation of systemic immune reaction has been reported on retroviral transfer of GMCSF gene into tumor cells. Similarly, T-cell growth and immune response has been shown in tumor cells on transfection of T-cell co-stimulatory molecule CD80 (B7.1) gene by lipofection and injection of those tumor cells into subcutaneous tissue (Urba) or direct CD80-adenoviral vector injection of into cancerous tissue.

Suicide gene therapy, a mode of drug-activating enzyme gene transfer into tumor cells creates a high concentration of the activated drug from a prodrug in the tumor tissue activating apoptosis of tumor cells. Currently, the use of GCV in combination with retroviral herpes simplex virus thymidine kinase (HSV-TK) gene transfected into breast cancer tumor tissues is under trial. A Phase I trial for the metastatic skin lesions of breast cancer treatment with a combination of prodrug (fluorocytosine) and of injection of HER2 promoter−driven cytosine deaminase (CD) gene plasmid has been successfully tested. The CD gene transform fluorocytosine transforms into 5-FU. There was 81.8% expression at Day 2 and 30% expression at Day 7 in the transfected tumor cells. There were 33.33% cases that showed tumor reduction. The researchers at MD Anderson Cancer Centre primarily conducted retroviral gene transfer without using cytokines, in suspension or with autologous stromal cells. Using the in situ PCR, they reported an in vitro transduction efficiency of 2.8% with the solution method and 5.6% with the stromal method. This study reported insufficient transduction efficiency in the absence of cytokines as peripheral blood leukocytes had nil positive results with solution method and 62.5% results with stromal method after 3 weeks of therapy. Furthermore, retroviral MDR1 gene therapy has been reported by the NCI and Columbia University researchers. By transfecting MDR1 genes into bone marrow mononuclear cells or peripheral blood stem cells, they stimulated by IL-3, IL-6, and SCF (Stem Cell Factor). There was a difference in the transfection efficiency of NCI experiments (0.2%–0.5%) and of Columbia University (20%–70%) for BFUE or CFU-GM (Colony forming unit-granulocyte, macrophage) colonies from transferred CD34 + cells that were positive for MDR1 by PCR (Takahashi et al., 2006; Dwivedi et al., 2014a,b).

22.5.6.4 Gene Therapy and Ovarian Cancer

Ovarian cancer is one of the common reasons of mortality from gynecologic malignancies, even though conventional chemotherapy offers some positive assistance, but drug resistance and adverse drug response still pose as obstacles in this cancer. So extra competent and sensitive tactics are in need of time, and few researchers have utilized virus gene therapy as one probable path to add-on conventional chemotherapy. Adenovirus vectors have shown promises for gene therapy because of their high infection efficiency in dividing and nondividing cells and the wide prevalence of the coxsackie adenovirus receptors (CARs) in a variety of cells and tissues. Though, one of the reasons accountable for this disappointment appears to be the lack of CARs in primary tumor tissues, leading to poor virus uptake and gene expression, but this problem have been resolved to some extent by inserting another ligand (RGD) into the virus adherent protein. Thus the oncolytic virus can target the more ubiquitous integrins as a portal of entry.

Hemminki and Alvarez (2002) presented the success of this modification in targeting ovarian cancer in vivo of animal models. Furthermore, after introducing an extra somatostatin receptor into the virus, they succeeded to carry out noninvasive imaging using a 99mTc-somatostatin analogue to trace the progress of the virus after infection. Furthermore, this working group confirmed that the vector used in their experiments will moderately ablate neutralization of the virus. This has made a noteworthy effect exclusively in ovarian cancers, in which malignant ascites, often a characteristic clinical feature of the disease, can have substantial levels of adenovirus-neutralizing antibodies (Hemminki and Alvarez, 2002).

Adenoviral vector therapy also has revealed interference with the PI-3/Akt pathway, which arises to be overactive in ovarian cancer cells. This pathway is generally controlled by the tumor suppressor PTEN, which encodes a phosphatidylinositol phosphatase, which is often defective in ovarian cancer (Russell, 2002).

22.5.6.5 Gene Therapy and Lung Cancer

Like other cancers in lung cancer, cellular and animal studies have informed that replacement of the normal p53 tumor suppressor gene in tumor cells induced anticancerous effect and apoptosis. In the first Phase I study, a retrovirus vector carrying wild-type p53 was inserted into seven patients with lung cancer by direct intratumoral injection. In six cases, there was confirmation of augmented apoptosis, and tumor regression in three of the patients. This was the primary study that showed the possibility of tumor suppressor gene mediating tumor regression. Numerous Phase I studies on Ad.p53 gene transfer, when pooled with chemotherapy, confirmed safety and evidence of increased apoptosis (histologically examination) in transduced tumors. Though, in a Phase II study, no alteration in response rates for Ad.p53/chemotherapy-treated lesions were conveyed for primary tumor lesions versus comparison lesions treated with chemotherapy alone, suggesting that Ad.p53 provided slight local benefit over chemotherapy. Preclinical studies displayed that DC transformed with CCL21 on insertion into tumors has shown potent anticancerous activity against lung cancers.

Antisense therapy also offers the opportunity of downregulating a wide variety of molecules like the first utilized aprinocarsen, a 20-mer oligonucleotide that binds to the mRNA for protein kinase C-a and inhibits its expression. Though, a Phase III trial of chemotherapy with or without aprinocarsen as first-line therapy failed to enhanced survival and did show some toxicity. A third set of studies targeted Bcl-2, an apoptotic inhibitor which is over expressed by several tumors (around 80%−90% of small-cell lung cancer) and is linked with increased resistance to chemotherapy. Though two Phase I trials informed promising results, subsequently a Phase II of standard chemotherapy with or without a bcl-2 antisense oligonucleotide (oblimersen) demonstrated worst survival in the experimental arm and greater hematologic toxicity.

Preclinical studies disclosed that transfection of tumor cells with the GMCSF gene amazingly increased the ability of these cells to induce antitumor immune responses. In the first trial of metastatic non-small-cell lung cancer (NSCLC), GMCSF was introduced into autologous tumor cells with the utilization of adenoviral vector prior to irradiation and patient vaccination. Some clinical responses were perceived and numerous reports recommended a strong immune response. B7.1/HLA vaccination has been tested as B7.1 (CD80) is accountable for co-stimulation of T cells during priming by an antigen-presenting cell. Tumor cells transfected with B7.1 and foreign HLA molecules have been proven to excite an immune response leading to T-cell activation. The mode of treatment with an allogeneic lung cancer cell line vaccine transfected with B7.1 and HLA A1 or A2 was examined in a Phase I trial of 19 patients with advanced NSCLC. On the whole, one patient had a partial response, and five had stable disease; though, the median overall survival in this group with a very poor prognosis was an impressive 18 months. In the six responders, the CD8 T-cell titers to tumor cell stimulations remained elevated up to 150 weeks after cessation of therapy. Based on the promising results of these trials in heavily pretreated patients, a trial is presently ongoing in stage IIIB/intravenous patients who fail first-line chemotherapy (ClinicalTrials.gov Identifier: NCT00534209).

Furthermore, vaccination of MUC-1 is also in test, and it is a tumor-associated mucin-type surface antigen usually found on epithelial cells in several tissues. Attempting MUC-1 in lung cancer has been implemented in different ways, including both gene and non-gene therapy tactics. A vaccinia virus has been designed that comprises the coding sequences for MUC-1 and IL-2 (TG4010), and it was examined in a two-arm Phase II trial of 65 patients with IIIB/intravenous NSCLC. Arm 1 composed of up-front combination therapy of TG4010 with cisplatin/vinorelbine, whereas arm 2 used TG4010 monotherapy followed by combination therapy at progression. In Arm 1 (44 patients), partial response (29.5%) has been documented in within 1 year with a survival rate of 53%. While Arm 2 had shown two patients (of 21 total) with stable disease for more than 6 months who were under TG4010 monotherapy, but efficacy was in question and so it was terminated.

L523S vaccination also attempted in lung cancer, as L523S is an immunogenic lung cancer antigen expressed in approximately 80% of lung cancer cells. In a Phase I study, 13 patients with early-stage NSCLC (stage 1B, IIA, and IIB) were provided with two doses of intramuscular recombinant DNA (pVAX/L523S) followed by two doses of Ad.L523S given for 4 weeks apart. This vaccination was in fact an effort to enhance immune response against the recombinant protein and thereby achieve a more substantial immune response to L523S. However, the regimen was tolerated and only one patient showed an L523S-specific antibody response (Vachani et al., 2010; Nasu et al., 1999a,b).

22.5.6.6 Gene Therapy and Prostate Cancer

Prostate cancer is also one of the prevalent cancers of the male; several mode of treatment have been verified but still management is not up to the mark, emerging gene therapy also has shown some potential to treat this cancer. The most extensively studied gene therapy tactic for managing prostate cancer is of the cytoreductive approach. As discussed earlier, cytoreductive therapy by utilizing herpes simplex virus thymidine kinase (HSV-tk) gene and its role in conversion of GCV into a phosphorylated compound, GCV triphosphate is also used to treat cancer cells. This ultimately causes DNA chain termination and its fragmentation or necrosis death of cancerous cells. This method also shows a beneficial secondary effect, the "bystander effect," by which nontransduced adjacent cancer cells are also killed.

Researchers from Baylor College of Medicine have constructed adenovirus-mediated HSV-tk + GCV suicide gene therapy for the treatment of prostate cancer patient. After explorative preclinical study of both in vivo and in vitro prostate cancer model, this group now demonstrated the results of the first gene therapy trial for human prostate cancer, in which they reported the feasibility and safety of their strategy (Herman et al., 1999). Furthermore, the gene therapy with standard hormonal therapy (Hall et al., 1998) has shown promise in modulation of the antitumor cell immune response (Nasu et al., 1999a, b). In fact, remarkable response rates have not yet been described in any single gene therapy protocol for cancer, signifying that upcoming hard work will focus on combining different types of gene-based strategies with more predictable therapies.

Hall et al. (1999) further reported that with the combination of androgen ablation and HSV-tk + GCV instead of using a replication-defective recombinant adenovirus (ADV) under the transcriptional control of the Rous sarcoma virus, they utilized an androgen-sensitive mouse prostate cancer cell line (RM-2) from a mouse prostate reconstitution model system to establish subcutaneous and orthotopic prostate tumors. The subcutaneous tumors were introduced on Day 2 and orthotopic tumors on Day 6 were treated with HSV-tk, followed by intraperitoneal administration of GCV (10 mg/kg twice daily for 6 days). The combination of castration and HSV-tk + GCV treatment showed significantly higher growth suppression than castration alone or HSV-tk + GCV alone. After 14 days of treatment, tumors were significantly reduced in size and volume (Hall et al., 1999).

Several study demonstrated the cytokine especially interleukins (IL-12, IL-18, and IL-2) have the ability to enhance the immune response, and this may be true as many reports have claimed their higher levels during carcinoma (Dwivedi et al., 2011, 2012, 2013a, b, 2014c, 2015a, b, c). Therefore, to enhance this antitumor immune response, the researchers have utilized an ADV expressing human cytomegalovirus promoter−driven recombinant murine IL-12 (AdmIL-12). These studies have shown significant growth suppression (more than 50% reduction of tumor weight) on Day 14 after vector insertions as well as increased mean survival time (28.9 ± 1.2 days vs 23.4 ± 0.8 days, $P < 0.0001$) were observed in the AdmIL-12 group as compared with controls. Cytolytic natural killer cell activity was also aggravated within splenocytes shortly after virus injection. Moreover, the intratumoral infiltration of CD4 + and CD8 + T cells, as well as nitric oxide synthase—positive macrophages, was increased in AdmIL-12—treated animals (Shariat and Slawin, 2000).

22.6 OBSTACLES AND BARRIERS

Gene therapy is not only a novel field, but still it is fronting several problems over the last few decades. One of the main concerns being the lack of knowledge about the long-term effects of the therapy and ethical issues associated with trials. The brief explanation of shortcomings as in activation and delivery, controlled gene expression, etc., is discussed below.

22.6.1 Activation and Delivery of Gene

Gene therapy works only if we are able to deliver a normal gene to a large number of cells in a tissue. Once the gene reaches its destination, it must be activated, to make the protein it encodes, and has to remain "on" as cells have a habit of shutting down genes that are too active or exhibiting other unusual behaviors.

For the accomplishment of any gene therapy treatment, "Targeting" a gene to the correct cells is crucial. If a gene is delivered to a wrong cell, it could cause health problems to the individual or could be inefficient. For example, if a therapeutic gene is mistakenly transferred to germ cells instead of somatic cells, it might be expressed in the progeny of the individual.

22.6.2 Controlled Gene Expression

One of the most essential issues is that of controlled gene expression. For those strategies requiring long-term expression and those inducing inflammation or utilizing growth factors, the ability to turn "on" and "off" the expression of a therapeutic gene will be essential. For treating certain disease like cancer, induction of inflammation may be useful, but once the cancer is cured, the inflammation will continue if the cells continue to express the inciting transgene, which is undesirable.

22.6.3 Activation of Immune Response

An excessive immune response can cause serious illness, even pertaining to death. There is also a risk of inducing tumor growth, a concept referred to as insertional mutagenesis; if the inserted DNA is incorrectly placed, such as in a tumor suppressor gene, then a tumor may originate. These factors have led to three reported deaths during gene therapy trials. One of the most notable was the death of 18 year old Jesse Gelsinger. He died of complications resulting from an inflammatory response against the experimental adenovirus vector.

22.6.4 Commercially Unviable

Many of the genetic disorders, which have possibility to be treated with gene therapy, are extremely rare. Gene therapy, although, can be life-saving for these patients, but the high cost of developing a treatment makes it unappealing for pharmaceutical companies, as there are limited number of patients to recover from those expenses.

22.6.5 Safety Issues

Over 3000 patients have been cured with gene therapy since the authorization of the first clinical gene therapy trial in 1988 and its commencement in 1989. Still, many of the initial safety considerations elevated with early trials remain today, which can be generally classified as the ones pertaining to the delivery vector or the ones related to expression of the transferred gene.

A vast majority of clinical trials use viruses to transfer genetic material to cells. Administration of a virus can result in inflammation or active infection. Active uncontrolled infection also can occur either through multiple recombination events or through the contamination of replication incompetent viral stocks with a helper virus. Insertional mutagenesis and malignant transformation are also relevant issues. The expression of various types of therapeutic genes predisposes patients to various adverse effects. For example, tumor growth can be endorsed by the growth factors used for neurodegenerative disease or the proangiogenic molecules used for the treatment of CAD.

22.7 CONCLUSIONS AND FUTURE PROSPECTIVE

As we know that our fortune is indeed in our genes and gene therapy, it deals a great promise in treating disease for which the existing modalities are either ineffective or no options are available. Novel and perceptive techniques are emerging continuously in this field day by day. These trials ultimately will successfully treat or even cure patients. As this occurs, it will be essential for all physicians to understand the basic concepts encompassing this complex field. In the future, this technique will eventually be an everyday word used in our households. It has tremendous potential to change the field of medicine from what it is today. With the discovery of more genes and their functions, the potential of this treatment is unlimited. The key to our future is locked in our genome, which is the blueprint of our body. As the researchers start to understand this blueprint, our lives will be forever changed.

Gene therapy has the potential to become the strategy of choice in a wide variety of clinical settings in future prospects. It aims to direct the biological processes towards the goal of correction of disease, tissue repair, and regeneration. However, certain major challenges are still required to be addressed like improvements in the efficiency of gene transfer into target cells and in the maintenance of expression from the relevant transferred gene. Efficient gene transfer requires further researches to improve the delivery systems and vector constructions as well as to better understand the target cell biology (Kerr and Mulé, 1994).

Acknowledgments

The authors are thankful to Mrs Shashi Dwivedi, Virendra Kumar Saini, and Jatin Joshi for their support in the manuscript preparation and editing.

References

Anderson, W.F., 1990. Genetics and human malleability. Hastings Center Report 20, 21–24. Available from: http://dx.doi.org/10.2307/3562969.

Arteaga, C.L., Holt, J., 1996. Tissue-targeted antisense c-fos retroviral vector inhibits established breast cancer xenografts in nude mice. Cancer Res. 56, 1098–1103.

Baynes, C., Healey, C.S., Pooley, K.A., Scollen, S., Luben, R.N., Thompson, D.J., et al., 2007. Common variants in the ATM, BRCA1, BRCA2, CHEK2 and TP53 cancer susceptibility genes are unlikely to increase breast cancer risk. Breast Cancer Res. 9 (2), R27.

Blömer, U., Naldini, L., Verma, I.M., Trono, D., Gage, F.H., 1996. Applications of gene therapy to the CNS. Hum. Mol. Genet. 5, 1397–1404.

Bonini, C., Ferrari, G., Verzeletti, S., Servida, P., Zappone, E., Ruggieri, L., et al., 1997. HSV-TK gene transfer into donor lymphocytes for control of allogeneic graft-versus-leukemia. Science 276, 1719–1724.

Dwivedi, S., Goel, A., Natu, S.M., Mandhani, A., Khattri, S., Pant, K.K., 2011. Diagnostic and prognostic significance of prostate specific antigen and serum interleukin 18 and 10 in patients with locally advanced prostate cancer: a prospective study. Asian Pac. J. Cancer Prev. 12, 1843–1848.

Dwivedi, S., Khattri, S., Pant, K.K., 2012. In: Tiwari, S.P., Sharma, R., Singh, R.K. (Eds.), Recent Advances in Molecular Diagnostic Approaches for Microbial Technology, 133–154. Nova Science Publishers, Inc., USA, Hauppauge, NY.

Dwivedi, S., Shukla, K.K., Gupta, G., Sharma, P., 2013a. Non-invasive biomarker in prostate cancer: a novel approach. Indian J. Clin. Biochem. 28 (2), 107–109.

Dwivedi, S., Yadav, S.S., Singh, M.K., Shukla, S., Khattri, S., Pant, K.K., 2013b. Pharmacogenomics of viral diseases. In: Barh, D. (Ed.), Omics for Personalized Medicine. Springer India, New Delhi, pp. 637–676. Available from: http://dx.doi.org/10.1007/978-81-322-1184-6_28.

Dwivedi, S., Goel, A., Sadashiv, Verma, A., Shukla, S., Sharma, P., et al., 2014a. Molecular diagnosis of metastasizing breast cancer based upon liquid biopsy. In: Barh, D. (Ed.), Omics Approaches in Breast Cancer. Springer India, New Delhi, pp. 425–460. Available from: http://dx.doi.org/10.1007/978-81-322-0843-3_22.

Dwivedi, S., Shukla, S., Goel, A., Sharma, P., Khattri, S., Pant, K.K., 2014b. Nutrigenomics in cancer. In: Barh, D. (Ed.), Omics Approaches in Breast Cancer. Springer India, New Delhi, pp. 105–126. Available from: http://dx.doi.org/10.1007/978-81-322-0843-3_6.

Dwivedi, S., Goel, A., Khattri, S., Mandhani, A., Sharma, P., Pant, K.K., 2014c. Tobacco exposure by various modes may alter pro-inflammatory (IL-12) and anti-inflammatory (IL-10) levels and affects the survival of prostate carcinoma patients: an explorative study in North India. Biomed. Res. Int. 2014, 2014; 158530. 11 https://doi.org/10.1155/2014/158530.

Dwivedi, S., Goel, A., Khattri, S., Mandhani, A., Sharma, P., Misra, S., et al., 2015a. Genetic variability at promoters of IL-18 (pro-) and IL-10 (anti-) inflammatory gene affects susceptibility and their circulating serum levels: An explorative study of prostate cancer patients in North Indian populations. Cytokine 74, 117–122.

Dwivedi, S., Goel, A., Mandhani, A., Khattri, S., Sharma, P., Misra, S., et al., 2015b. Functional genetic variability at promoters of pro-(IL-18) and anti-(IL-10) inflammatory affects their mRNA expression and survival in prostate carcinoma patients: five year follow-up study. Prostate 75, 1737–1746.

Dwivedi, S., Singh, S., Goel, A., Khattri, S., Mandhani, A., Sharma, P., et al., 2015c. Pro-(IL-18) and Anti-(IL-10) inflammatory promoter genetic variants (intrinsic factors) with tobacco exposure (extrinsic factors) may influence susceptibility and severity of prostate carcinoma: a prospective study. Asian Pac. J. Cancer Prev. 16 (8), 3173–3181. 2015.

Ennis, M.K., Dingli, D., 2010. Death by fusion for acute leukemia cells. Cancer Biol. Ther. 9 (5), 358–361. Available from: http://dx.doi.org/10.4161/cbt.9.5.11140.

Griffiths, A.J.E., Miller, J., Suzuki, D.T., Lewontin, R.C., Gelbart, W.M., 1996. An Introduction to Genetic Analysis, sixth ed. W. H. Freeman Company, New York.

Hall, S.J., Sanford, M.A., Atkinson, G., Chen, S.H., 1998. Induction of potent antitumor natural killer cell activity by herpes simplex virus-thymidine kinase and ganciclovir therapy in an orthotopic mouse model of prostate cancer. Cancer Res. 58, 3221–3225.

Hall, S.J., Mutchnik, S.E., Yang, G., Timme, T.L., Nasu, Y., Bangma, C.H., et al., 1999. Cooperative therapeutic effects of androgen ablation and adenovirus-mediated herpes simplex virus thymidine kinase gene and ganciclovir therapy in experimental prostate cancer. Cancer Gene Ther. 6, 54–63.

Hemminki, A., Alvarez, R.D., 2002. Adenoviruses in oncology: a viable option? BioDrugs. 16 (2), 77–87.

Herman, J.R., Adler, H.L., Aguilar-Cordova, E., Rojas-Martinez, A., Woo, S., Timme, T.L., et al., 1999. In situ gene therapy for adenocarcinoma of the prostate: a phase I clinical trial. Hum. Gene Ther. 10, 1239–1249.

High, K.A., Manno, C.S., Sabatino, D.E., Hutchison, S., Dake, M., Razavi, M., et al., 2003. Immune responses to AAV and to factor IX in a Phase I study of AAV-mediated liver-directed gene transfer for hemophilia B. Blood 102, 154a–155a (abstract).

High, K.A., Manno, C.S., Sabatino, D.E., Hutchison, S., Dake, M., Razavi, M., et al., 2004. Immune response to AAV and to factor IX in a Phase I study of AAV-mediated liver directed gene transfer for hemophilia B. Mol. Ther. 9, S383–S384 (abstract).

Ishiguro, N., Brown, G.D., Meruelo, D., 1997. Activation transcription factor 1 involvement in the regulation of murine H-2Dd expression. J. Biol. Chem. 272 (25), 15993–16001.

Jaski, B.E., Jessup, M.L., Mancini, D.M., Cappola, T.P., Pauly, D.F., Greenberg, B., et al., 2009. Calcium upregulation by percutaneous administration of gene therapy in cardiac disease (CUPID Trial), a first-in-human phase 1/2 clinical trial. J. Card. Fail. 15, 171–181.

Kaplitt, M.G., Leone, P., Samulski, R.J., Xiao, X., Pfaff, D.W., O'Malley, K.L., et al., 1994. Long-term gene expression and phenotypic correction using adeno-associated virus vectors in the mammalian brain. Nat. Genet. 8 (2), 148–154.

Kastrup, J., Jørgensen, E., Rück, A., Tägil, K., Glogar, D., Ruzyllo, W., et al., 2005. Direct intramyocardial plasmid vascular endothelial growth factor-A165 gene therapy in patients with stable severe angina pectoris. A randomized double-blind placebo-controlled study: the Euroinject One trial. J. Am. Coll. Cardiol. 45, 982–988.

Kay, M.A., Liu, D., Hoogerbrugge, P.M., 1997. Gene therapy. PNAS 94, 12744–12746.

Kerr, W.G., Mulé, J.J., 1994. Gene therapy: current status and future prospects. J. Leukoc. Biol. 56 (2), 210–214.

Kresina, T.F., Branch, A.D., 2000. Molecular medicine and gene therapy: an introduction 1. In: Kresina, T.F. (Ed.), An Introduction to Molecular Medicine and Gene Therapy. John Wiley & Sons, New York, USA.

La Spada, A., Ranum, L.P., 2010. Molecular genetic advances in neurological disease: special review issue. Hum. Mol. Genet. 19, R1–R3. Available from: http://dx.doi.org/10.1093/hmg/ddq193.

Mingozzi, F., Liu, Y.L., Dobrzynski, E., Kaufhold, A., Liu, J.H., Wang, Y., et al., 2003. Induction of immune tolerance to coagulation factor IX antigen by in vivo hepatic gene transfer. J. Clin. Invest. 111, 1347–1356.

Mullen, J.T., Tanabe, K.K., 2002. Viral oncolysis. Oncologist 7 (2), 106–119.

Nasu, Y., Djavan, B., Marberger, M., Kumon, H., 1999a. Prostate cancer gene therapy: outcome of basic and clinical trials. Tech. Urol. 5, 185–190.

Nasu, Y., Bangma, C.H., Hull, G.W., Lee, H.M., Hu, J., Wang, J., et al., 1999b. Adenovirus-mediated interleukin-12 gene therapy for prostate cancer: suppression of orthotopic tumor growth and pre-established lung metastases in an orthotopic model. Gene Ther. 6, 338–349.

Nathwani, A.C., Davidoff, A.M., Linch, D.C., 2005. A review of gene therapy for haematological disorders. Br. J. Haematol. 128 (1), 3–17.

Nathwani, A.C., Reiss, U.M., Tuddenham, E.G., Rosales, C., Chowdary, P., McIntosh, J., et al., 2014. Long-term safety and efficacy of factor IX gene therapy in hemophilia B. N. Engl. J. Med. 371 (21), 1994–2004.

Roncalli, J., Tongers, J., Losordo, D.W., 2010. Update on gene therapy for myocardial ischaemia and left ventricular systolic dysfunction or heart failure. Arch. Cardiovasc. Dis. 103, 469–476.

Rosenberg, S.A., Aebersold, P., Cornetta, K., Kasid, A., Morgan, R.A., Moen, R., et al., 1990. Gene transfer into humans—immunotherapy of patients with advanced melanoma, using tumor-infiltrating lymphocytes modified by retroviral gene transduction. N. Engl. J. Med. 323 (9), 570–578.

Russell, W., 2002. Adenovirus gene therapy for ovarian cancer. J. Natl. Cancer Inst. 94 (10), 706–707.

Scott L., 2015. Degenerating neurons respond to gene therapy treatment for Alzheimer's disease. Postmortem brain studies suggest nerve growth factor safely triggered functional cell growth. ENews.

Shariat, S.F., Slawin, K.M., 2000. Gene therapy for prostate cancer. Rev. Urol. 2 (2), 81–87.

Stewart, D.J., Hilton, J.D., Arnold, J.M., Gregoire, J., Rivard, A., Archer, S.L., et al., 2006. Angiogenic gene therapy in patients with nonrevascularizable ischemic heart disease: a phase 2 randomized, controlled trial of AdVEGF(121) (AdVEGF-121) versus maximum medical treatment. Gene Ther. 13, 1503–1511.

Takahashi, S., Ito, Y., Hatake, K., Sugimoto, Y., 2006. Gene therapy for breast cancer—review of clinical gene therapy trials for breast cancer and MDR1 gene therapy trial in Cancer Institute Hospital. Breast Cancer 13 (1), 8–15.

Tan, Li, Xu, Bing, Liu, Ranyi, Liu, Haibo, Tan, Huo, Huang, Wenlin, 2010. Gene therapy for acute myeloid leukemia using Sindbis vectors expressing a fusogenic membrane glycoprotein. Cancer Biol. Ther. 9 (5), 350–357.

Tiberghien, P., Ferrand, C., Lioure, B., Milpied, N., Angonin, R., Deconinck, E., et al., 2001. Administration of herpes simplex-thymidine kinase-expressing donor T cells with a T-cell-depleted allogeneic marrow graft. Blood 97 (1), 63–72.

Tiwari, P., 2015. Recent trends in therapeutic approaches for diabetes management: a comprehensive update. J. Diabetes Res. 340838. Available from: https://doi.org/10.1155/2015/340838.

Vachani, A., Moon, E., Wakeam, E., Albelda, S.M., 2010. Gene therapy for mesothelioma and lung cancer. Am. J. Respir. Cell. Mol. Biol. 42 (4), 385–393.

Verma, I., Somia, N., 1997. Gene therapy—promises, problems and prospects. Nature 389, 239–242.

Xie, F., Ye, L., Chang, J.C., Beyer, A.I., Wang, J., Muench, M.O., et al., 2014. Seamless gene correction of ß-thalassemia mutations in patient-specific iPSCs using CRISPR/Cas9 and piggyBac. Genome Res. 24 (9), 1526–1533.

Yoo, G.H., Moon, J., Leblanc, M., Lonardo, F., Urba, S., Kim, H., et al., 2009. A phase 2 trial of surgery with perioperative INGN 201 (Ad5CMV-p53) gene therapy followed by chemoradiotherapy for advanced, resectable squamous cell carcinoma of the oral cavity, oropharynx, hypopharynx, and larynx: report of the Southwest Oncology Group. Arch. Otolaryngol. Head Neck Surg. 135 (9), 869–874.

23

Biotechnology for Biomarkers: Towards Prediction, Screening, Diagnosis, Prognosis, and Therapy

Dipali Dhawan

PanGenomics International Pvt. Ltd., Ahmedabad, Gujarat, India

23.1 INTRODUCTION

Biomarkers are already benefiting many therapeutic areas; however, even greater value may be realized as the industry begins to combine multiple biomarker technologies. Biomarker is a term often used to refer to a protein measured in blood whose concentration reflects the severity or presence of some disease state. More generally a biomarker can be used as an indicator of a particular disease state or some other biological state of an organism. Biomarkers can be specific cells, molecules, or genes, gene products, enzymes, or hormones (Biomarkers Consortium, http://www.biomarkersconsortium.org). Biomarkers of all types have been used by generations of epidemiologists, physicians, and scientists to study human disease (Houlka, 1990).

Application of biomarkers in the field of human health is improving our understanding of disease and will provide new knowledge of disease mechanisms and processes providing a means for improved health management through the earlier diagnosis of disease and the delivery of more efficacious and safer therapies. Novel molecular biomarkers have the potential to transform much of the current health care model, shifting the focus from a reactive "one-size-fits-all" system to one that is more proactive and precise. In this new, proactive approach, disease or disease susceptibility may be diagnosed earlier, and disease may be controlled or possibly prevented before it starts; and when disease is detected, new biomarker-based diagnostics may be used to develop treatment strategies that are tailored to the characteristics of individual patients. Over the long term, the use of biomarkers may improve patient welfare by delivering better health outcomes.

Omics Technologies and Bio-engineering: Towards Improving Quality of Life
DOI: https://doi.org/10.1016/B978-0-12-804659-3.00023-3

23.2 EVOLUTION OF BIOMARKERS

The idea of using biomarkers to detect disease and improve treatment goes back to the very beginnings of medical treatment. Biomarkers—a measure of a normal biological process in the body, a pathological process, or the response of the body to a therapy—may offer information about the mechanism of action of the drug, its efficacy, its safety, and its metabolic profile.

23.3 CLASSIFICATION OF BIOMARKERS

Biomarkers can be classified into susceptibility biomarkers, diagnostic and prognostic biomarkers, and therapeutic biomarkers. Disease-related biomarkers give an indication of whether there is a risk of a particular disease (predictive or susceptibility biomarkers), if a disease already exists (diagnostic biomarkers), or how such a disease may develop in an individual scenario (prognostic biomarker). Although, therapeutic biomarkers indicate whether a drug will be effective in a specific patient and how the patient's body will process it.

23.3.1 Susceptibility Biomarkers

Susceptibility to infection and many other human diseases arises from the complex interaction of environmental and host genetic factors. In general, many genetic loci make modest contributions to human disease susceptibility, and most of the focus in the field has been on identifying these loci and their effects in infection and in other conditions.

23.3.1.1 Infectious Diseases

There is a strong evidence that high frequencies of genetic disorders, predominately those involving hemoglobin and red blood cell metabolic pathways and membranes, reflect relative resistance to malaria. The strongest signal of association was located close to the hemoglobin beta (*HBB*) gene, which contains the classic sickle cell hemoglobin variant hemoglobin S (HbS) polymorphism. The causal SNP for HbS (rs334) results in the nonsynonymous replacement of glutamic acid with valine at amino acid residue 6 of the ß-globin chain. Despite rs334 homozygotes experiencing life-threatening sickle cell disease, rs334 heterozygotes have a 10-fold reduced risk of severe malaria (Hill, 2006; Allison, 1954). It has been seen that in case of HIV-1, individuals who are heterozygous for a 32 bp deletion in the cytoplasmic tail of *CCR5* (known as CCR5Δ32) progress more slowly to AIDS (Dean et al., 1996; de Roda Husman et al., 1997). Furthermore, individuals homozygous for CCR5Δ32 are highly resistant to acquiring HIV-1 infection, even after repeated exposure (Dean et al., 1996; Liu et al., 1996; Samson et al., 1996). Conversely, a promoter variant in *CCR5* has been associated with increased CCR5 expression, a higher viral load set point and faster disease progression (Martin et al., 1998; Fellay et al., 2009; Salkowitz et al., 2003).

23.3.1.2 Cardiovascular Diseases

C-reactive protein (CRP) is a protein found in serum or plasma at elevated levels during an inflammatory process. It is a sensitive marker of acute and chronic inflammation and infection and, in such cases, is increased several hundredfold. More than 15 studies have shown CRP to be associated with short-term and long-term mortality risk not only for patients with acute and chronic ischemic heart disease but also for those at risk for atherosclerosis (Scirica et al., 2007). Myeloperoxidase, released from activated neutrophils, is a leukocyte enzyme possessing powerful prooxidative and proinflammatory properties that play important roles in the pathogenesis of destabilization of coronary artery disease. A number of studies have shown that extracellular matrix degradation by matrix metalloproteinases (MMPs), specifically MMP-9, are involved in the pathogenesis of a wide spectrum of cardiovascular disorders, including atherosclerosis, restenosis, cardiomyopathy, congestive heart failure, myocardial infarction (MI), and aortic aneurysm (Dollery et al., 1995; Lindsay et al., 2002). Troponin is a complex of three regulatory proteins, namely TnC (18 kD), TnI (26 kD), and TnT (39 kD), which form the thin filaments of muscle fibers and regulate the movement of contractile proteins in muscle tissue (Frey et al., 1998). cTnT and cTnI assays can detect heart muscle injury with great sensitivity and specificity. They appear in serum within 4−8 hours after symptom onset, but unlike CK-MB, it remains elevated for as long as 7−10 days post-MI but a much lower increase and elevation duration in cases of "microinfarction" (Antman, 2002). Assays for cTnT have a sensitivity of up to 100% for myocardial damage within 4−6 hours after an acute myocardial infarction (AMI), and assays for cTnI have a sensitivity of up to 100% by 6 hours after an AMI (Wong, 1996). The cTn together with clinical information from the patient history and the electrocardiogram are the gold standard for diagnosing AMI (Thygesen et al., 2007). Troponin is the preferred biomarker but others, such as the MB fraction of creatinine kinase, suffice when troponin testing is unavailable. Both cTnT and cTnI provide independent prognostic information with regard to cardiac death and AMI (Luscher et al., 1997).

23.3.1.3 Rheumatoid Arthritis

There are several risk genes for rheumatoid arthritis (RA); however, given their function may also play important roles in response to therapy. CD226 is a membrane protein expressed on the surface of NK, T-cell, and B-cell subsets and plays a role in their activation and inhibition. A Gly307Ser substitution in the CD226 gene caused by SNP (rs763361) has been associated with autoimmune susceptibility, including RA (Hafler et al., 2009; Tan et al., 2010). Protein tyrosine phosphatase receptor-C is an RA susceptibility gene (Raychaudhuri et al., 2009) and also plays a dual role in both risk and response to anti-tumor necrosis factor (anti-TNF) therapy in RA.

23.3.2 Diagnostic and Prognostic Biomarkers

Tumor classification, stage, and sometimes grade are used to assess prognosis. Oncotype DX assay is a molecular diagnostic test, which uses multiple biomarkers to provide information-rich diagnostics for breast cancer. By analyzing a panel of 21 genes, Oncotype DX provides information on the type of breast cancer, information regarding likelihood of recurrence after 10 years, and information on the utility of chemotherapy as a treatment option.

23.3.2.1 Cancer

Antigen CA 125 has been used as a serum marker for ovarian cancer diagnosis and monitoring responses to chemotherapy; its application is severely limited due to poor sensitivity in early detection (van Haaften-Day et al., 2001; Bast et al., 2005). Several other tumor biomarkers, such as CEA and CA19-9, are currently utilized for the detection of ovarian cancer in clinical practice. However, these detection methods also lack the sensitivity and specificity required for early detection of potentially curable lesions and are not suitable for screening of population (Akdogan et al., 2001; Yurkovetsky et al., 2006, 2007, 2010).

23.3.2.2 Rheumatoid Arthritis

The association of RA with the major histocompatibility complex (MHC) has been well documented, in particular the HLA-DRB1 locus is believed to account for around 30% of RA genetic risk (Morel and Combe, 2005; Gregersen et al., 1987). RA patients with the HLA-DRB1*0401 and DRB1*0404 alleles have been shown to have increased radiological erosions and joint replacement compared to individuals without these alleles (Combe et al., 2001). Multiple studies of genes within and outside the MHC region have been carried out to identify and validate RA risk markers. These concerted efforts in more than 40,000 samples have led to confirmation of 31 loci associated with RA risk (Stahl et al., 2010).

23.3.3 Therapeutic Biomarkers

Biomarker expression often supplants or complements tumor classification, stage, and grade when biologically targeted therapeutics are under consideration. Some important examples include CD20 positivity for treatment of lymphomas with rituximab, HER2/ NEU positivity for treatment of breast cancer with trastuzumab, BCR-ABL translocation for treatment of chronic myelogenous leukemia with imatinib, and KIT or platelet-derived growth factor receptor-α (PDGFRA) positivity for treatment of gastrointestinal stromal tumors with imatinib (Ludwig and Weinstein, 2005).

Prediction of response is necessary for the selection of neoadjuvant or adjuvant chemotherapy. Biomarkers can also be used to avoid idiosyncratic drug toxicity such as the sustained, life-threatening leukocyte suppression seen when mercaptopurine is given to leukemia patients with homozygous mutations of the thiopurine methyltransferase (*TPMT*) gene (Relling and Dervieux, 2001).

In addition to biomarker expression in resected specimens or biopsy samples, further emphasis should be placed on the role of circulating serum biomarkers. Assessment of molecular biomarkers in serum and other body fluids including urine may allow formulation of preoperative prognostic criteria to identify patients most likely to benefit from particular therapies, such as hepatic resection and transplantation, as well as predict those most likely to respond to different chemotherapeutic agents. Serum and urinary biomarkers may also have a potential role in screening for recurrent disease after treatment.

CRP is released by inflamed atherosclerotic plaques in the arteries of individuals with coronary heart disease, and increased levels of CRP are associated with a greater risk of the plaque rupturing, forming a clot, and causing a heart attack. CRP is currently being used as a biomarker to measure drug efficacy, in particular whether rosuvastatin reduces the risk of cardiovascular morbidity and mortality in apparently healthy individuals with low-density lipoprotein cholesterol levels but elevated CRP.

23.3.3.1 Infectious Diseases

The major toxicity to abacavir, which is a nucleoside analog reverse transcriptase inhibitor, manifests as a hypersensitivity reaction that occurs in approximately 5%−8% of recipients within 6 weeks of commencing therapy (Hetherington et al. 2001). Mallal et al. (2008) showed that patients carrying the HLA-B5701 allele are more likely to experience this hypersensitivity reaction to abacavir. This marker had a negative predictive value of 100%, indicating that genetic tests can provide unequivocal information that can help to predict and prevent otherwise unpredictable drug reactions. Another example of the use of genomics to guide therapy is in the case of the antiretroviral drug, maraviroc, which is a CCR5-specific HIV entry inhibitor. This drug is only effective in patients who are infected with the CCR4-tropic viruses. Therefore, by screening for the specific tropism of the strain of HIV with which the patient is infected, clinicians can more specifically target drug therapy; for example, by using maraviroc only in patients who are infected with the susceptible strain of HIV (Genebat et al., 2009).

23.3.3.2 Cardiology

Warfarin is used to reduce the risk of death, heart attack, or stroke after a patient has a heart attack. It is also used to treat and prevent venous thrombosis (blood clots) and pulmonary embolism associated with atrial fibrillation or heart valve replacement surgery. Dosing of warfarin is difficult as the effective dose may vary depending on the patient. It has now been shown that the doses on which people are eventually stabilized are related to genetic profiles. Variation in two genes (*VKORC1* and *CYP2C9*) account for 30%−50% of individual variation in response to warfarin among patients. As the relationship between these biomarkers and clinical outcomes of warfarin become known, these biomarkers may be used to guide warfarin treatment significantly reducing costs associated with adverse drug reactions (Epstein et al., 2010).

Most of the biomarker-based diagnostics have been developed to guide drug delivery. These companion diagnostics may be used to identify patients likely to respond well to certain drugs or treatment options. These diagnostics are often used in conjunction with a specific drug. For example, the Trofile assay is used to determine the trophism of the HIV and the likelihood a patient will respond well to prescription of Selzentry.

23.3.3.3 Rheumatoid Arthritis

There are lot of data regarding single nucleotide polymorphisms (SNPs), which effect the *TNF* promoter such as 308G > A, and 857C > T. Reports on *TNF*-308G/A have indicated that a genetic predisposition toward increased TNF expression (carrying the A allele) can be associated with nonresponse to anti-TNF therapy (Mugnier et al., 2003; Balog et al., 2004; Criswell et al., 2004; Martinez et al., 2004). In addition to studies of TNF, SNPs within other cytokine genes such as *IL-10*, *TGF-B1*, *IL-1B*, and IL-1 receptor antagonist *IL-RA* have also been examined where opposing associations as biomarkers for response and nonresponse are reported (Marotte et al., 2006; Padyukov et al., 2003). SNPs located within the TLR and NF-κB pathways such as rTLR signaling protein MyD88 (rs7744) and kinase CHUK (rs11591741) have been demonstrated to be associated with good response to anti-TNF therapy (Potter et al., 2010). Interestingly, these two SNPs (rs7744 and rs11591741) remained significant against DAS28 or European League Against Rheumatism response criteria, where others lost significance. Response to anti-TNF therapy has been associated with SNPs within the genes coding for *MAP3K1* (rs96844), *MAP3K14* (rs4792847) (Bowes et al., 2009), and *MAP2K6* (rs11656130) (Coulthard et al., 2011) proteins.

23.4 FOOD AND DRUG ADMINISTRATION–APPROVED BIOMARKERS

The chemotherapy drug irinotecan is one example of personalized medicine, using a biomarker to guide both clinical practice and subsequent clinical trials. Irinotecan is used to treat advanced colorectal cancer. Once administered, it is activated to the metabolite SN-38 and then eventually inactivated in the body by the UGT1A1 enzyme. In 2005, the US Food and Drug Administration (FDA) added a warning to the label of the drug, stating that patients homozygous for a particular version of the *UGT1A1* gene—the UGT1A1*28 allele, associated with decreased UGT1A1 enzyme activity—should be given a reduced dose. Because patients with this allele clear the drug less quickly from their body than the rest of the population, they effectively receive a greater exposure to the drug from the same dose. As a consequence, they are at higher risk of potentially life-threatening side effects such as neutropenia (a decrease in white blood cells) and diarrhea (O'Dwyer and Catalano, 2006). The toxicity of irinotecan has long been a concern, and this biomarker now allows clinicians to better identify those patients who are at high risk of serious side effects (about 10% of the population are homozygous for UGT1A1*28).

Although this pharmacogenomics information has helped improve the clinical use and efficacy of irinotecan, it has also fed back into the development of other drugs; almost immediately, this new understanding prompted the use of the *UGT1A1* biomarker to guide other studies ongoing at the time, including several new irinotecan and oxaliplatin-based chemotherapy regimens (Pfizer, 2004).

23.5 BIOMARKERS FOR DRUG DISCOVERY AND DEVELOPMENT

Biomarkers are useful throughout the drug discovery and development process. Biomarkers appear in drug development programs as opportunists—taking advantage of spare samples and leftover money in the budget—often resulting in incomplete or inadequate data. Nowadays biomarkers have become an integral part of all the stages of drug development (Fig. 23.1) including target discovery, evaluation of drug activity,

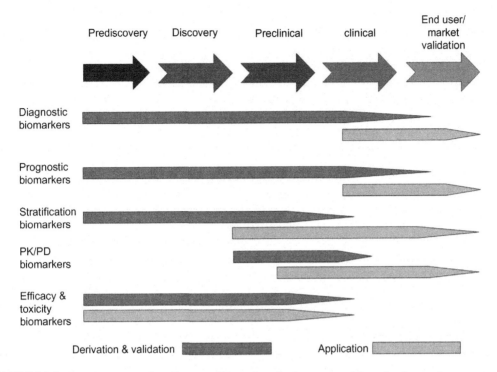

FIGURE 23.1 Development and application of biomarkers in the process of drug development.

understanding mechanisms of action, toxicity and safety evaluation, internal decision-making, clinical study design, diagnostic tools, and understanding disease processes. The ultimate aim is the development of more effective drugs at a lower cost.

In recent times the conventional approach to drug discovery and development has been courting failure more than it has success. The attrition rate for drugs in the clinical development is high: the percentage of tested products entering Phase I trials that eventually gain regulatory approval has been estimated at a paltry 8% (O'Connell and Roblin, 2006). Many of these failures happen late in clinical trials, with the consequence that expenditure in clinical drug development—already a mammoth effort requiring a huge amount of money, time, and patients—is increasing. Another consequence is that very few drugs are making it out of the clinical research pipeline. Because biomarkers can predict drug efficacy more quickly than conventional clinical endpoints, they hold the potential to substantially accelerate product development in certain disease areas (FDA Critical Path Opportunities Report, 2006). And because they help identify earlier those candidates that are likely to fail, they reduce drug development costs, giving life to the concept of "fail early, fail cheap."

The field of oncology is leading the way in the use of biomarkers in drug development. An excellent example of a biomarker in use in oncology is circulating tumor cells (CTCs), a biomarker present in the blood of cancer patients. At the moment, CTCs are used in the development of anticancer drugs as an objective and direct measurement of the response of the cancer to a novel agent. Although still in development, this biomarker holds further promise: the number of CTCs in the blood of patients with breast cancer, for example, is potentially a prognostic biomarker, and there is currently research ongoing to find out whether a decrease in CTCs associated with treatment can be used as a predictive biomarker of long-term benefit. This use of biomarkers as alternatives to clinical endpoints in drug development, alongside an increased understanding of disease and the availability of funds for research into cancer, has meant that oncology has not experienced the downturn in drug development experiences by many other therapeutic areas.

23.6 REGULATORY ISSUES

According to the FDA's draft guidance for pharmacogenomic data submissions, a validated biomarker is one that is "measured in an analytical test system with well-established performance characteristics with an established scientific framework or body of evidence that elucidates the physiologic, pharmacologic, toxicologic, or clinical significance of the test results" (FDA, 2005a,b).

Herceptin is indicated for treating only certain breast cancer tumors—those overexpressing the HER-2 protein. Although Herceptin's clinical trials utilized a laboratory designed test to measure expression of the HER-2 protein, FDA required that a test kit be approved with Herceptin. The Dako HercepTest was approved on the same day as the drug in 1998 (Press Release, HHS, 2001). Additional diagnostics were approved subsequently for use with Herceptin: the current labeling also references the Pathway HER-2/neu, PathVysion, and HER2 FISH pharmDx assays (Herceptin Prescribing Information, 2009). In 2008, FDA approved another test, the SPOT-Light HER2 CISH Kit, for use with Herceptin, but Herceptin's labeling does not mention this assay by name (Approval Letter for P050040, 2008). Tykerb (lapatinib), which also treats breast cancer tumors overexpressing HER-2, was approved in 2007 for use after failure of Herceptin (Approval Letter from Richard Padzur, 2007). Presumably patients already treated with Herceptin will have had the HER-2 test prior to treatment with Tykerb, but the clinical studies of Tykerb did require the use of a HER-2 diagnostic prior to entrance (Tykerb Prescribing Information, 2007).

Erbitux is indicated for treatment of colorectal tumors expressing the epidermal growth factor receptor (EGFR) protein. Its companion diagnostic, the DakoCytomation EGFR pharmDx, determines whether a given patient's tumors express this protein (Erbitux Prescribing Information, 2009a). Pretreatment screening for EGFR expression is not required for use of Erbitux in treating head and neck cancer. Because expression of EGFR has been detected in nearly all squamous cell head and neck cancers, clinical studies for the head/neck cancer indication did not have entry criteria based on EGFR tumor expression (Erbitux Prescribing Information, 2009b).

FDA has approved drug labeling that "recommends" a genetic test where the diagnostic provides information regarding adverse events associated with the therapy (Table 23.1). Testing also is recommended for certain mutations that can cause nonresponse to Vectibix and Erbitux.

The labeling of Camptosar (irinotecan) deems testing "recommended" for detecting whether the individual has a *UGT1A1*28* polymorphism (which can cause the adverse

TABLE 23.1 Biomarkers for Therapeutics and Their Implications in Drug Labels

Drug	Therapeutic Area	HUGO Symbol	Referenced Subgroup	Labeling Sections
Abacavir	Infectious diseases	HLA-B	HLA-B*5701 allele carriers	Boxed Warning, Contraindications, Warnings and Precautions, Patient Counseling Information
Ando-Trastuzumab Emtansine	Oncology	ERBB2	HER2 protein overexpression or gene amplification positive	Indications and Usage, Warnings and Precautions, Adverse Reactions, Clinical Pharmacology, Clinical Studies
Afatinib	Oncology	EGFR	EGFR exon 19 deletion or exon 21 substitution (L858R) mutation positive	Indications and Usage, Dosage and Administration, Adverse Reactions, Clinical Pharmacology, Clinical Studies, Patient Counseling Information
Amitriptyline	Psychiatry	CYP2D6	CYP2D6 poor metabolizers	Precautions
Anastrozole	Oncology	ESR1, PGR	Hormone receptor positive	Indications and Usage, Clinical Pharmacology, Clinical Studies
Aripiprazole	Psychiatry	CYP2D6	CYP2D6 poor metabolizers	Clinical Pharmacology, Dosage and Administration
Arsenic Trioxide	Oncology	PML/RARA	PML/RARα (t(15;17)) gene expression positive	Boxed Warning, Clinical Pharmacology, Indications and Usage, Warnings
Atomoxetine	Psychiatry	CYP2D6	CYP2D6 poor metabolizers	Dosage and Administration, Warnings and Precautions, Drug Interactions, Clinical Pharmacology
Atorvastatin	Endocrinology	LDLR	Homozygous familial hypercholesterolemia	Indications and Usage, Dosage and Administration, Warnings and Precautions, Clinical Pharmacology, Clinical Studies
Azathioprine	Rheumatology	TPMT	TPMT intermediate or poor metabolizers	Dosage and Administration, Warnings and Precautions, Drug Interactions, Adverse Reactions, Clinical Pharmacology
Boceprevir	Infectious diseases	IFNL3	IL28B rs12979860 T allele carriers	Clinical Pharmacology

(Continued)

II. ANIMAL AND MEDICAL BT

TABLE 23.1   (Continued)

Drug	Therapeutic Area	HUGO Symbol	Referenced Subgroup	Labeling Sections
Bosutinib	Oncology	BCR/ABL1	Philadelphia chromosome (t(9;22)) positive	Indications and Usage, Adverse Reactions, Clinical Studies
Brentuximab Vedotin	Oncology	TNFRSF8	CD30 Positive	Indications and Usage, Description, Clinical Pharmacology
Busulfan	Oncology	Ph Chromosome	Ph Chromosome negative	Clinical Studies
Capecitabine	Oncology	DPYD	DPD deficient	Contraindications, Warnings and Precautions, Patient Information
Carbamazepine (1)	Neurology	HLA-B	HLA-B*1502 allele carriers	Boxed Warning, Warnings and Precautions
Carbamazepine (2)	Neurology	HLA-A	HLA-A*3101 allele carriers	Boxed Warning, Warnings and Precautions
Carglumic Acid	Metabolic disorders	NAGS	N-acetylglutamate synthase deficiency	Indications and Usage, Warnings and Precautions, Special Populations, Clinical Pharmacology, Clinical Studies
Carisoprodol	Rheumatology	CYP2C19	CYP2C19 poor metabolizers	Clinical Pharmacology, Special Populations
Carvedilol	Cardiology	CYP2D6	CYP2D6 poor metabolizers	Drug Interactions, Clinical Pharmacology
Celecoxib	Rheumatology	CYP2C9	CYP2C9 poor metabolizers	Dosage and Administration, Drug Interactions, Use in Specific Populations, Clinical Pharmacology
Cetuximab (1)	Oncology	EGFR	EGFR protein expression positive	Indications and Usage, Warnings and Precautions, Description, Clinical Pharmacology, Clinical Studies
Cetuximab (2)	Oncology	KRAS	KRAS codon 12 and 13 mutation negative	Indications and Usage, Dosage and Administration, Warnings and Precautions, Adverse Reactions, Clinical Pharmacology, Clinical Studies
Cevimeline	Dermatology	CYP2D6	CYP2D6 poor metabolizers	Drug Interactions

(Continued)

TABLE 23.1   (Continued)

Drug	Therapeutic Area	HUGO Symbol	Referenced Subgroup	Labeling Sections
Chloroquine	Infectious diseases	G6PD	G6PD deficient	Precautions
Chlorpropamide	Endocrinology	G6PD	G6PD deficient	Precautions
Cisplatin	Oncology	TPMT	TPMT intermediate or poor metabolizers	Clinical Pharmacology, Warnings, Precautions
Citalopram (1)	Psychiatry	CYP2C19	CYP2C19 poor metabolizers	Drug Interactions, Warnings
Citalopram (2)	Psychiatry	CYP2D6	CYP2D6 poor metabolizers	Drug Interactions
Clobazam	Neurology	CYP2C19	CYP2C19 poor metabolizers	Clinical Pharmacology, Dosage and Administration, Use in Specific Populations
Clomipramine	Psychiatry	CYP2D6	CYP2D6 poor metabolizers	Drug interactions
Clopidogrel	Cardiology	CYP2C19	CYP2C19 intermediate or poor metabolizers	Boxed Warning, Dosage and Administration, Warnings and Precautions, Drug Interactions, Clinical Pharmacology
Clozapine	Psychiatry	CYP2D6	CYP2D6 poor metabolizers	Drug Interactions, Clinical Pharmacology
Codeine	Anesthesiology	CYP2D6	CYP2D6 Ultra-rapid metabolizers	Boxed Warnings, Warnings and Precautions, Use in Specific Populations, Clinical Pharmacology, Patient Counseling Information
Crizotinib	Oncology	ALK	ALK gene rearrangement positive	Indications and Usage, Dosage and Administration, Drug Interactions, Warnings and Precautions, Adverse Reactions, Clinical Pharmacology, Clinical Studies
Dabrafenib (1)	Oncology	BRAF	BRAF V600E mutation positive	Indications and Usage, Dosage and Administration, Warnings and Precautions, Clinical Pharmacology, Clinical Studies, Patient Counseling Information
Dabrafenib (2)	Oncology	G6PD	G6PD deficient	Warnings and Precautions, Adverse Reactions, Patient Counseling Information

(Continued)

TABLE 23.1 (Continued)

Drug	Therapeutic Area	HUGO Symbol	Referenced Subgroup	Labeling Sections
Dapsone (1)	Dermatology	G6PD	G6PD deficient	Indications and Usage, Precautions, Adverse Reactions, Patient Counseling Information
Dapsone (2)	Infectious diseases	G6PD	G6PD deficient	Precautions, Adverse Reactions, Over Dosage
Dasatinib	Oncology	BCR/ABL1	Philadelphia chromosome (t(9;22)) positive; T315I mutation positive	Indications and Usage, Clinical Studies, Patient Counseling Information
Denileukin diftitox	Oncology	IL2RA	CD25 antigen positive	Indications and Usage, Warnings and Precautions, Clinical Studies
Desipramine	Psychiatry	CYP2D6	CYP2D6 poor metabolizers	Drug interactions
Dexlansoprazole (1)	Gastroenterology	CYP2C19	CYP2C19 poor metabolizers	Clinical Pharmacology, Drug Interactions
Dexlansoprazole (2)	Gastroenterology	CYP1A2	CYP1A2 genotypes	Clinical Pharmacology
Dextromethorphan and Quinidine	Neurology	CYP2D6	CYP2D6 poor metabolizers	Clinical Pharmacology, Warnings and Precautions, Drug Interactions
Diazepam	Psychiatry	CYP2C19	CYP2C19 poor metabolizers	Drug Interactions, Clinical Pharmacology
Doxepin	Psychiatry	CYP2D6	CYP2D6 poor metabolizers	Precautions
Drospirenone and Ethinyl Estradiol	Neurology	CYP2C19	CYP2C19 poor metabolizers	Clinical Pharmacology, Warnings and Precautions, Drug Interactions
Eltromobopag (1)	Hematology	F5	Factor V Leiden carriers	Warnings and Precautions
Eltromobopag (2)	Hematology	SERPINC1	Antithrombin III deficient	Warnings and Precautions
Erlotinib (1)	Oncology	EGFR	EGFR protein expression positive	Clinical Pharmacology
Erlotinib (2)	Oncology	EGFR	EGFR exon 19 deletion or exon 21 substitution (L858R) positive	Indications and Usage, Dosage and Administration, Clinical Pharmacology, Clinical Studies
Esomeprazole	Gastroenterology	CYP2C19	CYP2C19 poor metabolizers	Drug Interactions, Clinical Pharmacology

(Continued)

TABLE 23.1 (Continued)

Drug	Therapeutic Area	HUGO Symbol	Referenced Subgroup	Labeling Sections
Everolimus (1)	Oncology	ERBB2	HER2 protein overexpression negative	Indications and Usage, Boxed Warning, Adverse Reactions, Use in Specific Populations, Clinical Pharmacology, Clinical Studies
Everolimus (2)	Oncology	ESR1	Estrogen receptor positive	Clinical Pharmacology, Clinical Studies
Exemestane	Oncology	ESR1	Estrogen receptor positive	Indications and Usage, Dosage and Administration, Clinical Studies, Clinical Pharmacology
Fluorouracil (1)	Dermatology	DPYD	DPD deficient	Contraindications, Warnings, Patient Information
Fluorouracil (2)	Oncology	DPYD	DPD deficient	Warnings
Fluoxetine	Psychiatry	CYP2D6	CYP2D6 poor metabolizers	Warnings, Precautions, Clinical Pharmacology
Flurbiprofen	Rheumatology	CYP2C9	CYP2C9 poor metabolizers	Clinical Pharmacology, Special Populations
Fluvoxamine	Psychiatry	CYP2D6	CYP2D6 poor metabolizers	Drug Interactions
Fulvestrant	Oncology	ESR1	Estrogen receptor positive	Indications and Usage, Clinical Pharmacology, Clinical Studies, Patient Counseling Information
Galantamine	Neurology	CYP2D6	CYP2D6 poor metabolizers	Special Populations
Glimepiride	Endocrinology	G6PD	G6PD deficient	Warning and Precautions
Glipizide	Endocrinology	G6PD	G6PD deficient	Precautions
Glyburide	Endocrinology	G6PD	G6PD deficient	Precautions
Ibritumomab tiuxetan	Oncology	MS4A1	CD20 positive	Indications and Usage, Clinical Pharmacology, Description
Iloperidone	Psychiatry	CYP2D6	CYP2D6 poor metabolizers	Clinical Pharmacology, Dosage and Administration, Drug Interactions, Specific Populations, Warnings and Precautions
Imatinib (1)	Oncology	KIT	C-KIT D816V mutation negative	Indications and Usage, Dosage and Administration, Clinical Pharmacology, Clinical Studies

(Continued)

II. ANIMAL AND MEDICAL BT

TABLE 23.1 (Continued)

Drug	Therapeutic Area	HUGO Symbol	Referenced Subgroup	Labeling Sections
Imatinib (2)	Oncology	BCR/ABL1	Philadelphia chromosome (t(9;22)) positive	Indications and Usage, Dosage and Administration, Clinical Pharmacology, Clinical Studies
Imatinib (3)	Oncology	PDGFRB	PDGFR gene rearrangement positive	Indications and Usage, Dosage and Administration, Clinical Studies
Imatinib (4)	Oncology	FIP1L1/ PDGFRA	FIP1L1/PDGFRα fusion kinase (or CHIC2 deletion) positive	Indications and Usage, Dosage and Administration, Clinical Studies
Imipramine	Psychiatry	CYP2D6	CYP2D6 poor metabolizers	Drug Interactions
Indacaterol	Pulmonary	UGT1A1	UGT1A1*28 allele homozygotes	Clinical Pharmacology
Irinotecan	Oncology	UGT1A1	UGT1A1*28 allele carriers	Dosage and Administration, Warnings, Clinical Pharmacology
Isosorbide and Hydralazine	Cardiology	NAT1-2	Slow acetylators	Clinical Pharmacology
Ivacaftor	Pulmonary	CFTR	CFTR G551D, G1244E, G1349D, G178R, G551S, S1251N, S1255P, S549N, or S549R carriers	Indications and Usage, Adverse Reactions, Use in Specific Populations, Clinical Pharmacology, Clinical Studies
Lansoprazole	Gastroenterology	CYP2C19	CYP2C19 poor metabolizer	Drug Interactions, Clinical Pharmacology
Lapatinib	Oncology	ERBB2	HER2 protein overexpression positive	Indications and Usage, Clinical Pharmacology, Patient Counseling Information
Lenalidomide	Hematology	del (5q)	Chromosome 5q deletion	Boxed Warning, Indications and Usage, Clinical Studies, Patient Counseling
Letrozole	Oncology	Esr1, PGR	Hormone receptor positive	Indications and Usage, Adverse Reactions, Clinical Studies, Clinical Pharmacology

(Continued)

TABLE 23.1 (Continued)

Drug	Therapeutic Area	HUGO Symbol	Referenced Subgroup	Labeling Sections
Lomitapide	Endocrinology	LDLR	Homozygous familial hypercholesterolemia and LDL receptor mutation deficient	Indication and Usage, Adverse Reactions, Clinical Studies
Mafenide	Infectious diseases	G6PD	G6PD deficient	Warnings, Adverse Reactions
Maraviroc	Infectious diseases	CCR5	CCR5 positive	Indications and Usage, Warnings and Precautions, Clinical Pharmacology, Clinical Studies, Patient Counseling Information
Mercaptopurine	Oncology	TPMT	TPMT intermediate or poor metabolizers	Dosage and Administration, Precautions, Adverse Reactions, Clinical Pharmacology
Methylene blue	Hematology	G6PD	G6PD deficient	Precautions
Metoclopramide	Gastroenterology	CYB5R1-4	NADH cytochrome b5 reductase deficient	Precautions
Metoprolol	Cardiology	CYP2D6	CYP2D6 poor metabolizers	Precautions, Clinical Pharmacology
Mipomersen	Endocrinology	LDLR	Homozygous familial hypercholesterolemia and LDL receptor mutation deficient	Indication and Usage, Clinical Studies, Use in Specific Populations
Modafinil	Psychiatry	CYP2D6	CYP2D6 poor metabolizers	Drug Interactions
Mycophenolic acid	Transplantation	HPRT1	HGPRT deficient	Precautions
Nalidixic acid	Infectious diseases	G6PD	G6PD deficient	Precautions, Adverse Reactions
Nefazodone	Psychiatry	CYP2D6	CYP2D6 poor metabolizers	Drug Interactions
Nilotinib (1)	Oncology	BCR/ABL1	Philadelphia chromosome (t(9;22)) positive	Indications and Usage, Patient Counseling Information
Nilotinib (2)	Oncology	UGT1A1	UGT1A1*28 allele homozygotes	Warnings and Precautions, Clinical Pharmacology
Nitrofurantoin	Infectious diseases	G6PD	G6PD deficient	Warnings, Adverse Reactions
Nortriptyline	Psychiatry	CYP2D6	CYP2D6 poor metabolizers	Drug Interactions

(Continued)

II. ANIMAL AND MEDICAL BT

TABLE 23.1 (Continued)

Drug	Therapeutic Area	HUGO Symbol	Referenced Subgroup	Labeling Sections
Obinutuzumab	Oncology	Ms4A1	CD20 positive	Indication and Usage, Warnings and Precautions, Description, Clinical Pharmacology, Clinical Studies
Ofatumumab	Oncology	Ms4A1	CD20 positive	Indications and Usage, Clinical Pharmacology
Omacetaxine	Oncology	BCR/ABL1	BCR-ABL T315I	Clinical Pharmacology
Omeprazole	Gastroenterology	CYP2C19	CYP2C19 poor metabolizers	Dosage and Administration, Warnings and Precautions, Drug Interactions
Panitumumab (1)	Oncology	EGFR	EGFR protein expression	Indications and Usage, Warnings and Precautions, Clinical Pharmacology, Clinical Studies
Panitumumab (2)	Oncology	KRAS	KRAS codon 12 and 13 mutant negative	Indications and Usage, Clinical Pharmacology, Clinical Studies
Pantoprazole	Gastroenterology	CYP2C19	CYP2C19 poor metabolizers	Clinical Pharmacology, Drug Interactions, Special Populations
Paroxetine	Psychiatry	CYP2D6	CYP2D6 poor metabolizers	Clinical Pharmacology, Drug Interactions
Pazopanib	Oncology	UGT1A1	(TA)7/(TA)7 genotype (UGT1A1*28/*28)	Clinical Pharmacology, Warnings and Precautions
PEG-3350. Sodium Chloride, Potassium Chloride, Sodium Ascorbate, and Ascorbic Acid	Gastroenterology	G6PD	G6PD deficient	Warnings and Precautions
Peginterferon alfa-2b	Infectious diseases	IFNL3	IL28B rs12979860 T allele carriers	Clinical Pharmacology
Pegloticase (1)	Rheumatology	G6PD	G6PD deficient	Contraindications, Patient Counseling Information
Pegloticase (2)	Rheumatology	G6PD	G6PD deficient	Contraindications, Patient Information counseling
Perphenazine	Psychiatry	CYP2D6	CYP2D6 poor metabolizers	Clinical Pharmacology, Drug Interactions

(Continued)

TABLE 23.1 (Continued)

Drug	Therapeutic Area	HUGO Symbol	Referenced Subgroup	Labeling Sections
Pertuzumab	Oncology	ERBB2	HER2 protein overexpression positive	Indications and Usage, Warnings and Precautions, Adverse Reactions, Clinical Studies, Clinical Pharmacology
Phenytoin	Neurology	HLA-B	HLA-B*1502 allele carriers	Warnings
Pimozide	Psychiatry	CYP2D6	CYP2D6 poor metabolizers	Warnings, Precautions, Contraindications, Dosage and Administration
Ponatinib	Oncology	BCR-ABL T315I	BCR-ABL T315I mutation	Indications and Usage, Adverse Reactions, Clinical Pharmacology, Clinical Studies
Prasugrel	Cardiology	CYP2C19	CYP2C19 poor metabolizers	Use in Specific Populations, Clinical Pharmacology, Clinical Studies
Pravastatin	Endocrinology	LDLR	Homozygous familial hypercholesterolemia and LAL receptor deficient	Clinical Studies, Use in Specific Populations
Primaquine	Infectious diseases	G6PD	G6PD deficient	Warnings and Precautions, Adverse Reactions
Propafenone	Cardiology	CYP2D6	CYP2D6 poor metabolizers	Clinical Pharmacology
Propranolol	Cardiology	CYP2D6	CYP2D6 poor metabolizers	Precautions, Drug Interactions, Clinical Pharmacology
Protriptyline	Psychiatry	CYP2D6	CYP2D6 poor metabolizers	Precautions
Quinidine	Cardiology	CYP2D6	CYP2D6 poor metabolizers	Precautions
Quinine Sulfate	Infectious diseases	G6PD	G6PD deficient	Contraindications, Patient Counseling Information
Rabeprazole	Gastroenterology	CYP2C19	CYP2C19 poor metabolizers	Drug Interactions, Clinical Pharmacology
Rasburicase	Oncology	G6PD	G6PD deficient	Boxed Warning: Contraindications
Rifampin, Isoniazid, and Pyrazinamide	Infectious diseases	NAT1-2	Slow inactivators	Adverse Reactions, Clinical Pharmacology

(Continued)

TABLE 23.1 (Continued)

Drug	Therapeutic Area	HUGO Symbol	Referenced Subgroup	Labeling Sections
Risperidone	Psychiatry	CYP2D6	CYP2D6 poor metabolizers	Drug Interactions, Clinical Pharmacology
Rituximab	Oncology	Ms4A1	CD20 positive	Indication and Usage, Clinical Pharmacology, Description, Precautions
Rosuvastatin	Endocrinology	LDLR	Homozygous and Heterozygous familial hypercholesterolemia	Indications and Usage, Dosage and Administration, Clinical Pharmacology, Clinical Studies
Simeprevir	Infectious diseases	IFNL3	IL28B allele carriers	Clinical studies, Clinical Pharmacology
Sodium Nitrite	Antidotal therapy	G6PD	G6PD deficient	Warnings and Precautions
Sofosbuvir	Infectious diseases	IFNL3	IL28B allele carriers	Clinical Pharmacology, Clinical Studies
Succimer	Hematology	G6PD	G6PD deficient	Clinical Pharmacology
Sulfamethoxazole and Trimethoprim	Infectious diseases	G6PD	G6PD deficient	Precautions
Tamoxifen (1)	Oncology	ESR1, PGR	Hormone receptor positive	Indications and Usage, Precautions, Medication Guide
Tamoxifen (2)	Oncology	F5	Factor V Leiden carriers	Warnings
Tamoxifen (3)	Oncology	F2	Prothrombin mutation G20210A	Warnings
Telaprevir	Infectious diseases	IFNL3	IL28B rs12979860 T allele carriers	Clinical Pharmacology
Terbinafine	Infectious diseases	CYP2D6	CYP2D6 poor metabolizers	Drug Interactions
Tetrabenazine	Neurology	CYP2D6	CYP2D6 poor metabolizers	Dosage and Administration, Warnings, Clinical Pharmacology
Thioguanine	Oncology	TPMT	TPMT poor metabolizer	Dosage and Administration, Precautions, Warnings
Thioridazine	Psychiatry	CYP2D6	CYP2D6 poor metabolizers	Precautions, Warnings, Contraindications
Ticagrelor	Cardiology	CYP2C19	CYP2C19 poor metabolizers	Clinical Studies

(Continued)

TABLE 23.1   (Continued)

Drug	Therapeutic Area	HUGO Symbol	Referenced Subgroup	Labeling Sections
Tolterodine	Genitourinary	CYP2D6	CYP2D6 poor metabolizers	Warnings and Precautions, Drug Interactions, Use in Specific Populations, Clinical Pharmacology
Tositumomab	Oncology	Ms4A1	CD20 antigen positive	Indications and Usage, Clinical Pharmacology
Tramadol	Analgesic	CYP2D6	CYP2D6 poor metabolizers	Clinical Pharmacology
Trametinib	Oncology	BRAF	BRAF V600E/K mutation positive	Indications and Usage, Dosage and Administration, Adverse Reactions, Clinical Pharmacology, Clinical Studies, Patient Counseling Information
Trastuzumab	Oncology	ERBB2	HER2 protein overexpression positive	Indications and Usage, Warnings and Precautions, Clinical Pharmacology, Clinical Studies
Tretinoin	Oncology	PML/RARA	PML/RARα (t(15;17)) gene expression positive	Clinical Studies, Indications and Usage, Warnings
Trimipramine	Psychiatry	CYP2D6	CYP2D6 poor metabolizers	Drug Interactions
Valproic Acid (1)	Neurology	POLG	POLG mutation positive	Boxed Warning, Contraindications, Warnings and Precautions
Valproic Acid (2)	Neurology	NAGS, CPS1, ASS1, OTC, ASL, ABL2	Urea cycle enzyme deficient	Contraindications, Warnings, and Precautions, Adverse Reactions, Medication Guide
Velaglucerase Alfa	Metabolic disorders	GBA	Lysosomal glucocerebrosidase enzyme	Indication and Usage, Description, Clinical Pharmacology, Clinical Studies
Vemurafenib	Oncology	BRAF	BRAF V600E mutation positive	Indications and Usage, Warning and Precautions, Clinical Pharmacology, Clinical Studies, Patient Counseling Information
Venlafaxine	Psychiatry	CYP2D6	CYP2D6 poor metabolizers	Drug Interactions
Voriconazole	Infectious diseases	CYP2C19	CYP2C19 intermediate or poor metabolizers	Clinical Pharmacology, Drug Interactions

(Continued)

II. ANIMAL AND MEDICAL BT

TABLE 23.1   (Continued)

Drug	Therapeutic Area	HUGO Symbol	Referenced Subgroup	Labeling Sections
Vortioxetine	Neurology	CYP2D6	CYP2D6 poor metabolizers	Dosage and Administration, Drug Interactions, Clinical Pharmacology
Warfarin (1)	Cardiology or hematology	CYP2C9	CYP2C9 intermediate or poor metabolizers	Dosage and Administration, Drug Interactions, Clinical Pharmacology
Warfarin (2)	Cardiology or hematology	VKORC1	VKORC1 rs9923231 A allele carriers	Dosage and Administration, Clinical Pharmacology
Warfarin (3)	Cardiology or hematology	PROC	Protein C deficient	Warning and precautions

event of neutropenia upon treatment with the drug) advising that "a reduction in the starting dose by at least one level…should be considered for patients known to be homozygous for the *UGT1A1*28* allele" (Frueh et al., 2008). FDA approved Camptosar in 1996 and later cleared the Invader *UGT1A1* Molecular Assay for performing the relevant genetic testing (Talk Paper, FDA, 1996; Summary for Invader *UGT1A1* Molecular Assay, 2005).

The Imuran (azathioprine) labeling recommends testing to determine whether patients have low or absent thiopurine 5-methyltransferase (TPMT) activity, which poses an enhanced risk for development of severe myelotoxicity with standard Imuran doses (Imuran Prescribing Information, 2008).

Prior to administration of valproic acid, testing for urea cycle enzyme disorder (UCD) is recommended. The labeling recommends that patients with a family history of UCD or signs or symptoms of UCD be evaluated for UCD prior to initiation of valproic acid therapy (Nat'l Urea Cycle Disorders Foundation, 2005).

Several "information-only" labels provide background regarding rare side effects. As one example, Xeloda (capecitabine) is contraindicated for patients who are known to lack the enzyme dihydropyrimidine dehydrogenase, which converts the 5-FU metabolite of capecitabine to a much less toxic substance (Xeloda Prescribing Information, 2009). The labeling for Rifater (rifampin, isoniazid, and pyrazinamide) indicates that the rate of acetylation of the drug's isoniazid component is "genetically determined" (Rifater Prescribing Information, 2008).

Imaging and diagnostic measures will need to be validated against certain standards and metrics for use in internal BioPharma decision-making, clinical practice, clinical diagnostic claims, as well as for regulatory purposes.

23.7 FUTURE PROSPECTS

While there is a widespread recognition of biomarker value, scientific progress continues to outpace acceptance. Over the years, biomarkers have sometimes been the center of excessive "hype," promoting excessive expectations. Also, biomarkers as surrogate end-points have had some public failures when they were believed to be falsely reassuring or too alarming, creating general skepticism among some scientists. Resistance still hindering biomarker acceptance includes:

- Resistance to sharing data across independent efforts: Organizations may work on similar research or discover keystone advances yet resist sharing knowledge because they believe that doing so will jeopardize their competitive advantage. However, sharing information could help companies achieve greater overall progress or reduce costs by not having to work on efforts independently.
- Limited biomarker validation: Validation is critical for establishing biomarkers as reliable tools to support development, medical care, health policy such as the FDA's critical path, and BioPharma investment decisions (FDA, 2004). The biomarker development and validation process is necessary but costly for one company to do in isolation. Innovation takes place in many organizations, and so stakeholders work redundantly on the same effort. Many collaborative forums exist, but these usually involve sharing "safe" information that really does not hasten overall progress.
- Need for new R&D models with greater precision and flexibility: The industry needs an R&D model with greater precision to improve pipelines, leveraging active clinical knowledge to offset the declining success in new drug development (Martin et al., 2006). Some R&D leaders are concerned that using an approach that targets treatment for limited patient groups decreases profits and increases research costs. Others recognize that this direction has already created value beyond costs and are building these capabilities into their new R&D models and tactics. For example, Herceptin is considered an effective targeted treatment for breast cancer (Burstein, 2005). Targeted treatments could actually increase both medical and economic success of a therapeutic (Trusheim et al., 2007).

One of the major challenges to innovation is our ability to discover new biomarkers. Part of the answer lies in gaining a better understanding of the pathophysiology of disease, thereby uncovering potential drug targets and biomarkers in the disease pathway.

23.8 CONCLUSION

Biomarkers are becoming a foundation for information-based medicine in determining who should be treated, how, and with what. For biomarkers to reach their full potential, alone or in combination, they must undergo their own development process. Biomarker development has lagged significantly behind therapeutic development because of scientific, economic, and regulatory factors (Woodcock, 2007). It is important to accelerate biomarker development and to help biomarkers advance in parallel with therapeutics (Mills, 2007).

Applying biomarkers to the traditional blockbuster model may have its benefits, but biomarkers offer a new level of precision and value in a transformed R&D model. They can make trials more economically feasible by improving patient selection and not including patients that have more known risk of toxicity or lack of efficacy (Greener, 2006).

Biobanks and associated epidemiology studies are playing an important role in the discovery of new biomarkers and the understanding of disease of new biomarkers and the understanding of disease mechanisms. In turn, biomarkers are being applied in an expanding range of clinical and therapeutic areas. Combining information from diverse biomarker sources is increasing its usefulness and importance in making critical patient and investment decisions. However, the complete utilization of biomarkers is still lagging in drug development.

Many pharmaceutical companies have begun to invest in "omics"—genomics, proteomics, and metabonomics—to begin to sort through this huge number of biomarkers and characterize them on the basis of a molecular understanding of disease. The "omics" approach enables the detection of small changes in tissue composition through protein profiling technologies such as mass spectrometry and gel electrophoresis. Essentially, it is about capturing a molecular profile from a clinical sample and converting into information about a clinical condition, for example, the stage of disease or what players are involved in the disease pathways.

By finding molecular biomarkers of the disease, diagnosis could be improved and could reveal new information about the disease.

References

Akdogan, M., Sasmaz, N., Kayhan, B., Biyikoglu, I., Disibeyaz, S., Sahin, B., 2001. Extraordinarily elevated CA19-9 in benign conditions: a case report and review of literature. Tumori 87, 337–339.

Allison, A.C., 1954. Protection afforded by sickle-cell trait against subtertian + malarial infection. BMJ 1, 290–294.

Antman, E., 2002. Decision making with cardiac troponin tests. N. Engl. J. Med. 346, 2079–2082.

Approval Letter from Richard Padzur, M.D., CDER, FDA, to SmithKline Beecham Corp. (March 13, 2007).

Balog, A., Klausz, G., Gál, J., Molnár, T., Nagy, F., Ocsovszky, I., et al., 2004. Investigation of the prognostic value of TNF-alpha gene polymorphism among patients treated with infliximab, and the effects of infliximab therapy on TNF-alpha production and apoptosis. Pathobiology 71, 274–280.

Bast Jr., R.C., Badgwell, D., Lu, Z., Marquez, R., Rosen, D., Liu, J., et al., 2005. New tumor markers: CA 125 and beyond. Int. J. Gynecol. Cancer 15 (Suppl. 3), 274–281.

Bowes, J.D., Potter, C., Gibbons, L.J., Hyrich, K., Plant, D., Morgan, A.W., et al., 2009. Investigation of genetic variants within candidate genes of the TNFRSF1B signalling pathway on the response to anti-TNF agents in a UK cohort of rheumatoid arthritis patients. Pharmacogenet. Genomics 19, 319–323.

Burstein, H.J., 2005. The distinctive nature of HER2-positive breast cancer. N. Engl. J. Med. 353 (16), 1652–1654.

Combe, B., Dougados, M., Gouille, P., Cantagrel, A., Eliaou, J.F., Sibilia, J., et al., 2001. Prognostic factors for radiographic damage in early rheumatoid arthritis: a multiparameter prospective study. Arthritis Rheum. 44, 1736–1743.

Coulthard, L.R., Taylor, J.C., Eyre, S., Biologics in rheumatoid arthritis genetics and genomics, Robinson, J.I., Wilson, A.G., et al., 2011. Genetic variants within the MAP kinase signalling network and antiTNF treatment response in rheumatoid arthritis patients. Ann. Rheum. Dis. 70, 98–103.

Criswell, L.A., Lum, R.F., Turner, K.N., Woehl, B., Zhu, Y., Wang, J., et al., 2004. The influence of genetic variation in the HLA-DRB1 and LTA-TNF regions on the response to treatment of early rheumatoid arthritis with methotrexate or etanercept. Arthritis Rheum. 50, 2750–2756.

Dean, M., Carrington, M., Winkler, C., Huttley, G.A., Smith, M.W., Allikmets, R., et al., 1996. Genetic restriction of HIV-1 infection and progression to AIDS by a deletion allele of the CKR5 structural gene. Science 273, 1856–1862.

de Roda Husman, A.M., Koot, M., Cornelissen, M., Keet, I.P., Brouwer, M., Broersen, S.M., et al., 1997. Association between CCR5 genotype and the clinical course of HIV-1 infection. Ann. Intern. Med. 127, 882–890.

Dollery, C.M., McEwan, J.R., Henney, A.M., 1995. Matrix metalloproteinases and cardiovascular disease. Circ. Res. 77, 863–868.

Epstein, R.S., Toyer, T.P., Aubert, R.E., O'Kane, D.J., Xia, F., Verbrugge, R.R., et al., 2010. Warfarin genotyping reduces hospitalisation rates: Results from the MM-WES (Medco-Mayo Effectiveness Study). J. Am. Coll. Cardiol. 55 (25), 2804–2812.

Erbitux Prescribing Information, 1.2, 2009a. Approval letter for P030044 from Steven I. Gutman, M.D., M.B.A., CDRH, FDA, to Ronald F. Lagerquist, DakoCytomation California, Inc. (Feb. 12, 2004).

Erbitux Prescribing Information, 5.7, 2009b.

FDA, 2004. Innovation or Stagnation: Challenge and Opportunity on the Critical Path to New Medical Products. US Department of Health and Human Services. http://www.fda.gov/oc/initiatives/criticalpath.

FDA, 2005a. Guidance for Industry: Pharmacogenomic Data Submissions. US Department of Health and Human Services. http://www.fda.gov/cber/gdlns/pharmdtasub.pdf.

FDA, 510(k) Substantial Equivalence Determination Decision Summary for K042884, DakoCytomation ER/PR pharmDx™ Kit, section H (2005b).

FDA Critical Path Opportunities Report 2006, 18.

F.W. Frueh et al., Pharmacogenomic Biomarker Information in Drug Labels Approved by the United States Food and Drug Administration: Prevalence of Related Drug Use, 28 Pharmacotherapy 992, 997 (2008); Camptosar Prescribing Information, Dosage and Administration & Warnings sections (July 2008). Lower Drug Dose Spells Trouble for Pharmacogenetic Test, The Gray Sheet (June 25, 2007), at 10.

Frey, N., Muller-Bardorff, M., Katus, H.A., 1998. Myocardial damage: the role of troponin T. In: Kaski, JC, Holt, DW (Eds.), Myocardial Damage. Early Detection by Novel Biochemical Markers. Kluwer Academic Publishers, Boston, MA, pp. 27–40.

Genebat, M., Ruiz-Mateos, E., Leon, J.A., Gonzalez-Serna, A., Pulido, I., Rivas, I., et al., 2009. Correlation between the Trofile test and virological response to a short-term maraviroc exposure in HIV-infected patients. J. Antimicrob. Chemother. 64, 845–849.

Greener, M., 2006. SNPs: Driving Variability and Tailoring Treatments. Drug Discovery and Development. http://www.dddmag.com/ShowPR_Print.aspx?PUBCODE=016&ACCT=1600000100&ISSUE=0407&RELTYPE=PR&ORIGRELTYPE=GPF&PRODCODE=00000000&PRODLETT=Y&CALLFROM=RELPGM.

Gregersen, P.K., Silver, J., Winchester, R.J., 1987. The shared epitope hypothesis: an approach to understanding the molecular genetics of susceptibility to rheumatoid arthritis. Arthritis Rheum. 30, 1205–1213.

Hafler, J.P., Maier, L.M., Cooper, J.D., Plagnol, V., Hinks, A., Simmonds, M.J., et al., 2009. CD226, Gly307Ser association with multiple autoimmune diseases. Genes Immun. 10, 5–10.

Herceptin Prescribing Information, 5.5 2009.

Hetherington, S., McGuirk, S., Powell, G., Cutrell, A., Naderer, O., Spreen, B., et al., 2001. Hypersensitivity reactions during therapy with the nucleoside reverse transcriptase inhibitor abacavir. Clin. Ther. 23, 1603–1614.

Hill, A.V., 2006. Aspects of genetic susceptibility to human infectious diseases. Annu. Rev. Genet. 40, 469–486.

Houlka, B.S., 1990. Overview of biological markers. Biological Markers in Epidemiology. Oxford University Press, New York, pp. 3–15. Available from: http://www.accessdata.fda.gov/cdrh_docs/reviews/K042884.pdf.

Id: Approval Letter for P050040 from Maria M. Chau, Ph.D., Center for Devices and Radiological Health (CDRH), FDA, to Kelli L. Tanzella, Ph.D., Invitrogen Corp. (July 1, 2008).

Imuran Prescribing Information, Clinical Pharmacology Section (2008).

Lindsay, M.M., Maxwell, P., Dunn, F.G., 2002. TIMP-1: a marker of left ventricular diastolic dysfunction and fibrosis in hypertension. Hypertension 40, 136–141.

Liu, R., Paxton, W.A., Choe, S., Ceradini, D., Martin, S.R., Horuk, R., et al., 1996. Homozygous defect in HIV-1 coreceptor accounts for resistance of some multiply-exposed individuals to HIV-1 infection. Cell 86, 367–377.

Ludwig, J.A., Weinstein, J.N., 2005. Biomarkers in cancer staging, prognosis and treatment selection. Nat. Rev. Cancer. 5 (11), 845–856.

Luscher, M.S., Thygesen, K., Ravkilde, J., Heickendorff, L., 1997. Applicability of cardiac troponin T and I for early risk stratification in unstable coronary artery disease. Circulation 96, 2578–2585.

Mallal, S., Phillips, E., Carosi, G., Molina, J.M., Workman, C., Tomazic, J., et al., 2008. HLA-B*5701 screening for hypersensitivity to abacavir. N. Engl. J. Med. 358, 568–579.

Marotte, H., Pallot-Prades, B., Grange, L., Tebib, J., Gaudin, P., Alexandre, C., et al., 2006. The shared epitope is a marker of severity associated with selection for, but not with response to, infliximab in a large rheumatoid arthritis population. Ann. Rheum. Dis. 65, 342–347.

Martin, K., M. Hammond and S. Henderson., 2006. The eClinical equation: Part 2—bridging connections for innovation. IBM Institute for Business Value. http://www-935.ibm.com/services/us/index.wss/ibvstudy/gbs/a1025940?cntxt = a1000060.

Martin, M.P., Dean, M., Smith, M.W., Winkler, C., Gerrard, B., Michael, N.L., et al., 1998. Genetic acceleration of AIDS progression by a promoter variant of CCR5. Science 282, 1907–1911.

Martinez, A., Salido, M., Bonilla, G., Pascual-Salcedo, D., Femandez-Arquero, M., de Miguel, S., et al., 2004. Association of the major histocompatibility complex with response to infliximab therapy in rheumatoid arthritis patients. Arthritis Rheum. 50, 1077–1082.

Mills, G., January 2007. FDA Update. IBM (Imaging) Biomarker Summit III.

Morel, J., Combe, B., 2005. How to predict prognosis in early rheumatoid arthritis. Best. Prac. Res. Clin. Rheumatol. 19, 137–146.

Mugnier, B., Balandraud, N., Darque, A., Roudier, C., Roudier, J., Reviron, D., 2003. Polymorphism at position-308 of the tumour necrosis factor alpha gene influences outcome of infliximab therapy in rheumatoid arthritis. Arthritis Rheum. 48, 1849–1852.

Nat'l Urea Cycle Disorders Foundation, What is a Urea Cycle Disorder? (2005), http://www.nucdf.org/ucd.htm.

O'Connell, D., Roblin, D., 2006. Translational research in the pharmaceutical industry: from bench to bedside. Drug Discov. Today 11 (17/18), 833–838.

O'Dwyer, P.J., Catalano, R.B., 2006. UGT1A1 and irinotecan: practical pharmacogenomics arrives in cancer therapy. J. Clin. Oncol. 24 (28), 4534–4538.

Padyukov, L., Lampa, J., Heimbürger, M., Ernestam, S., Cederholm, T., Lundkvist, I., et al., 2003. Genetic markers for the efficacy of tumour necrosis factor blocking therapy in rheumatoid arthritis. Ann. Rheum. Dis. 62, 526–529.

Pfizer, 2004. Background document on the UGT1A1 polymorphisms and irinotecan toxicity. ACPS November 2004 Advisory Committee Meeting, published on the FDA website at http://www.fda.gov/ohrms/dockets/ac/04/briefing/2004-4079B1_07_Pfizer-UGT1A1.pdf.

Potter, C., Cordell, H.J., Barton, A., Daly, A.K., Hyrich, K.L., Mann, D.A., et al., 2010. Association between anti-tumor necrosis factor treatment response and genetic variants within the TLR and NF[kappa]B signalling pathways. Ann. Rheum. Dis. 69, 1315–1320.

Press Release, HHS, New Monoclonal Antibody Approved for Advanced Breast Cancer, supra note 100; Center for Biologics Evaluation and Research, FDA, Clinical Review Briefing Document: sBLA STN: 103792/5008/0, Trastuzumab, at 4 (November 5, 2001); see also Herceptin Prescribing Information, Clinical Studies section (Immunohistochemical Detection)1 (September 1998).

Raychaudhuri, S., Thomson, B.P., Remmers, E.F., Eyre, S., Hinks, A., Guiducci, C., et al., 2009. Genetic variants at CD28, PRDM1 and CD2/CD58 are associated with rheumatoid arthritis risk. Nat. Genet. 41, 1313–1318.

Relling, M., Dervieux, T., 2001. Pharmacogenetics and cancer therapy. Nat. Rev. Cancer 1, 99–108.

Rifater Prescribing Information, Clinical Pharmacology Section (April 2008).

Salkowitz, J.R., Bruse, S.E., Meyerson, H., Valdez, H., Mosier, D.E., Harding, C.V., et al., 2003. CCR5 promoter polymorphisms determines macrophage CCR5 density and magnitude of HIV-1 propagation in vitro. Clin. Immunol. 108, 234–240.

Samson, M., Libert, F., Doranz, B.J., Rucker, J., Liesnard, C., Farber, C.M., et al., 1996. Resistance to HIV-1 infection in Caucasian individuals bearing mutant alleles of the CCR-5 chemokine receptor gene. Nature 382, 722–725.

Scirica, B.M., Morrow, D.A., Cannon, C.P., de Lemos, J.A., Murphy, S., Sabatine, M.S., et al., 2007. Clinical application of C-reactive protein across the spectrum of acute coronary syndromes. Clin. Chem. 53, 1800–1807.

Stahl, E.A., Raychaudhuri, S., Remmers, E.F., Xie, G., Eyre, S., Thomson, B.P., et al., 2010. Genome-wide association study meta-analysis identifies seven new rheumatoid arthritis risk loci. Nat. Genet. 42, 508–514.

Talk Paper, FDA, FDA Approves Drug for Advanced Colorectal Cancer, T96-42 (June 17, 1996); 510(k) Summary for Invader UGT1A1 Molecular Assay 11 (Aug. 18, 2005).

Tan, R.J., Gibbons, L.J., Potter, C., Hyrich, K.L., Morgan, A.W., Wilson, A.G., et al., 2010. Investigation of rheumatoid arthritis susceptibility gene identifies association of AFF3 and CD226 variants with response to anti-tumor necrosis factor alpha therapy. Ann. Rheum. Dis. 69, 1029–1035.

Thygesen, K., Alpert, J.S., White, H.D., Jaffe, A.S., Apple, F.S., Galvani, M., et al., 2007. Joint ESC/ACCf/AHA/ WHF task force for the redefinition of myocardial infarction, universal definition of myocardial infarction. Circulation 116, 2634−2653.

Trusheim, M.R., Berndt, E.R., Douglas, F.L., 2007. Stratified medicine: strategic and economic implications of combining drugs and clinical biomarkers. Nat. Rev. Drug Discov. 6, 287−293.

Tykerb Prescribing Information, March 14, 2007.

van Haaften-Day, C., Shen, Y., Xu, F., Yu, Y., Berchuck, A., Havrilesky, L.J., et al., 2001. OVX1, macrophage-colony stimulating factor, and CA-125-II as tumor markers for epithelial ovarian carcinoma: a critical appraisal. Cancer 92, 2837−2844.

Wong, S.S., 1996. Strategic utilization of cardiac markers for the diagnosis of acute myocardial infarction. Ann. Clin. Lab. Sci. 26, 301−312.

Woodcock, J., 2007. Biomarkers: Physiological and Laboratory Markers of Drug Effect. National Institutes of Health Clinical Center. http://clinicalcenter.nih.gov/researchers/training/principles/ppt/woodcock_2006-2007.ppt.

Xeloda Prescribing Information, Contraindications, Precautions, & Clinical Pharmacology Sections, November 2009; see also Carac (fluorouracil cream) Prescribing Information, Contraindications and Warnings sections (November 2006) noting this drug also should not be given to patients with DPD deficiency due to risk of serious toxicity.

Yurkovetsky, Z., Ta'asan, S., Skates, S., Rand, A., Lomakin, A., Linkov, F., et al., 2007. Development of multimarker panel for early detection of endometrial cancer. High diagnostic power of prolactin. Gynecol. Oncol. 107, 58−65.

Yurkovetsky, Z., Skates, S., Lomakin, A., Nolen, B., Pulsipher, T., Modugno, F., et al., 2010. Development of a multimarker assay for early detection of ovarian cancer. J. Clin. Oncol. 28, 2159−2166.

Yurkovetsky, Z.R., Linkov, F.Y., E Malehorn, D., Lokshin, A.E., 2006. Multiple biomarker panels for early detection of ovarian cancer. Future Oncol. 2, 733−741.

Further Reading

Fellay, J., Ge, D., Shianna, K.V., Colombo, S., Ledergerber, B., Cirulli, E.T., et al., 2009. Common genetic variation and the control of HIV-1 in humans. PLoS Genet. 5, e1000791.

Tykerb Prescribing Information, January 1, 2010.

Omics Approaches in In Vitro Fertilization

Aiman Tanveer, Neha Malviya and Dinesh Yadav

D.D.U. Gorakhpur University, Gorakhpur, India

24.1 INTRODUCTION

Infertility is a prevalent condition that is suffered by more than 10% of all couples worldwide. It can lead to depression, social isolation, and psychological stress to most of the patients. Before the introduction of assisted reproductive technologies, very little medical help was available to infertile individuals (Elder and Dale, 2000). In vitro fertilization (IVF) is a type of assisted reproductive technology in which a man's sperm and a woman's eggs are combined outside the body under laboratory conditions. One or more fertilized eggs (embryos) may be transferred into the woman's uterus, where they may be implanted in the uterine lining and finally develop. Excess embryos may be cryopreserved (frozen) for future use. IVF is an important technique used to cure both male and female infertility. The basic steps associated with an IVF treatment cycle are ovarian stimulation, egg retrieval, fertilization, embryo culture, and embryo transfer. These are discussed in the following sections.

24.2 HISTORICAL ASPECTS

In the present scenario, IVF has emerged as a well-established technology finding its application in both basic and applied sciences. Apart from the treatment of human infertility, the technology is also currently being applied routinely in a wide variety of species. A large number of experiments in reproductive biology since the last few decades have immensely contributed to the success of IVF. Artificial insemination was first reported in 1884. For the first time in 1906 Surgeon Robert Tuttle Morris reported successful partial "ovarian transplantations" from healthy women to those unable to have children. In 1934

Harvard scientist Gregory Pincus performed IVF experiments on rabbits and indicated that similar fertilization procedure can also be applied for humans (Bavister, 2002).

In 1938 John Rock attempted similar fertilization in the humans assisted by former technician of Gregory Pincus, Miriam Menkin. After several years of hard work, in 1944 John Rock and Miriam Menkin were successful in fertilizing four ova after long-term incubation of egg and sperm resulting in first successful IVF of human eggs though they did not attempt its implantation. Their experiments were duplicated by the Physician Landrum Shettles in 1951. Later in 1960 Shettles published a book entitled "Ovum Humanum" containing more than 1000 photographs showing the development of human eggs. British scientist Robert Edwards who later moved to America had several unsuccessful attempts with IVF of human eggs, but finally succeeded in 1965 with the help of Howard and Georgeanna Jones. In an attempt to implant fertilized eggs back into previously infertile women, Edwards collaborated with Patrick Steptoe, a gynecologist in Oldham, England, who has developed a new technique of abdominal surgery called laparoscopy that may allow the retrieval of a mature human egg. They together succeeded in fertilizing human eggs in vitro and their successful IVF experiment was published in Journal *Nature* in 1969. Although Robert Edwards and Patrick Steptoe have the first successful IVF pregnancy among their patients, it was an ectopic pregnancy, whereby the implantation was done in the fallopian tubes instead of the uterus, and the baby is lost.

In 1975 focus was laid down on the surgical retrieval of a single egg that would be then fertilized. In November 1977 Steptoe surgically removes an egg from an infertile patient, Lesley Brown's ovaries. Two days later the fertilized egg has developed into an eight-cell embryo, which was then implanted back into Lesley's uterus by him. By December in the same year, Edwards and Steptoe discovered that Lesley is pregnant and the egg that was fertilized in vitro has become the very first one to grow in utero. In 1978 they announced birth of a healthy, fit, and normal baby by successful IVF. This led to the emergence of a technique with immense potential for treating infertility patients.

24.3 HUMAN IVF

IVF has finally emerged as a promising technique to cure infertility. It comes to the rescue of infertile patients when other methods of assisted reproductive technology are unsuccessful. It involves fertilization of an egg by a sperm outside the body in in vitro conditions. The procedure involves monitoring and stimulating a woman's ovulatory process followed by removal of ovum or ova (egg or eggs) from the woman's ovaries. The eggs are allowed to fertilize with the sperm in a fluid medium under laboratory conditions. After the eggs get fertilized, the zygotes are cultured for 2–6 days in a growth medium and then transferred to the patient's uterus for successful pregnancy.

IVF is used to deal with both male and female infertility problems. It helps to overcome male infertility manifested by the sperm quality of the patient and female infertility resulting due to malfunctioning of the fallopian tubes. IVF has also paved way for the development of preimplantation genetic diagnostics whereby the in vitro fertilized embryo cells are tested for genetic disorders and chromosomal abnormality so that these anomalies are not carried over to nest generation. The detailed protocol for carrying out the IVF procedure is discussed in the later section of the chapter.

24.4 BENEFITS OF IVF

The technique of IVF offers the advantage of increasing the chances of becoming pregnant. It provides opportunity of getting pregnant in women having blocked fallopian tubes. It is quite helpful in conditions where the male partner has a reduced sperm count or other various issues that influence his capability of fertilizing the egg. IVF and the implantation of already created embryos eliminate the need for sperms to navigate the reproductive tract of the woman. IVF is also useful in case of patients having issues with the egg supply. In such cases donor eggs can be used and embryos are created using the sperms of the male partner. The embryos are later implanted in the uterus of the woman at the correct time in her menstrual cycle. It increases the opportunity of having healthier children as genetic anomaly can be monitored at the embryo stage (Hart and Norman, 2013). This is very beneficial since some women have a higher risk for having a child affected by various chromosomal complications. Embryos are checked for their viability before implantation.

24.5 OMICS APPROACHES IN IVF

The prime issue faced in the treatment of infertility is the lack of understanding of reproductive components at the molecular level. Advancement in the omics technology provides better overview of the biological system. Omics technologies can be used in screening of best spermatozoa and oocyte for improved success in assisted reproduction.

24.5.1 Genomics in IVF

Aneuploidy is the prime phenomenon limiting the success rate of IVF. Fluorescence in situ hybridization has been used in the screening of aneuploid embryos but all the chromosomes cannot be screened at the same time. However, through comparative genome hybridization (CGH), all the 24 chromosomes can be screened along with the typing of copy number variations (Wilton, 2005). CGH integrated with the microarray technology provides better understanding of the ploidy status of zygotes. It is also applied to determine the copy number of both first and second polar bodies. Single nucleotide polymorphism (SNP) microarray has also been used for the assisted reproductive techniques whereby each parent and zygote DNA are screened in parallel. It also provides DNA fingerprint for the identification of the parental origin. Various SNPs have been identified having direct associations with the pregnancy and IVF outcomes. Table 24.1 gives a list of SNPs associated with various IVF and fertility-related traits, and screening of those SNPs could lead to better IVF success.

24.5.2 Transcriptomics in IVF

Referring to the transcriptome, mRNA microarray is an invasive technique requiring disruption of the oocyte which may have an adverse effect on the resulting embryo. Researchers have come up with a noninvasive microarray in which granulosa cells and/or

TABLE 24.1 SNPs Associated Fertility and IVF Associated Traits (Lledo et al., 2014)

Fertility and IVF Associated Traits	Gene Name	SNP
Deep vein thrombosis	SERPINC1/ATIII	rs2227589
IVF outcome	GDF9	rs10491279
IVF outcome	SOD2	rs4880
Spontaneous abortion	F2	rs1799963
Spontaneous abortion	MTHFR	rs1801131
Spontaneous abortion	MTHFR	rs1801133
Spontaneous abortion	F13	rs5985
Spontaneous abortion	F5	rs6025
Spontaneous abortion	EPCR	rs867186
Spontaneous abortion	EPCR	rs9574
Ovarian hyperstimulation syndrome	LHCGR	rs4073366
Ovarian hyperstimulation syndrome	BMP15	rs58995369
Ovarian hyperstimulation syndrome	BMP15	rs3810682
Ovarian hyperstimulation syndrome	BMP15	rs3897937
Ovarian hyperstimulation syndrome	BMP15	rs41308602
Ovarian response during IVF	FSHR	rs6166
Ovarian response during IVF	FSHR	rs1394205
Ovarian response during IVF	ESR1	rs9340799
Ovarian response during IVF	ESR1	rs2234693
Ovarian response during IVF	ESR2	rs4986938
Ovarian response during IVF	FSHR	rs6165
Ovarian response during IVF	LHB	rs1800447
Ovarian response during IVF	LHB	rs34349826
Protein C deficiency	PROC	rs121918143
Protein C deficiency	PROC	rs146922325
Protein C deficiency	PROC	rs121918143
Recurrent pregnancy loss	TP53	rs1042522
Recurrent pregnancy loss	VEGF	rs1570360
Recurrent pregnancy loss	VEGF	rs3025039
Recurrent pregnancy loss	TNF-alpha	rs1800629
Recurrent pregnancy loss	IL10	rs1800871

(Continued)

TABLE 24.1   (Continued)

Fertility and IVF Associated Traits	Gene Name	SNP
Recurrent pregnancy loss	IL10	rs1800872
Recurrent pregnancy loss	IL10	rs1800896
Recurrent pregnancy loss	APOE	rs429358
Recurrent pregnancy loss	APOE	rs7412
Male infertility	MTHFR	rs1801133
Male infertility	FSHR	rs1394205
Male infertility	FSHR	rs6166
Male infertility	FSHB	rs10835638
Male infertility	CFTR	rs113993960
Male infertility	CFTR	rs78655421

cumulus cells (CCs) are used as marker in evaluation of embryo potential. Given that the expression pattern of the CCs reflects the quality of the oocyte. Reports have suggested that abundance of CC cathepsin mRNA is indicative of embryo quality (Bettegowda et al., 2008). In addition to the oocytes, comparative mRNA profiling can be performed for fertile and infertile males to understand molecular basis of male infertility.

Transcriptome profiling could be helpful in the determination of the implantation time in the endometrium. Expression profile of the biomarkers in the endometrium can be monitored to increase the success rate of IVF.

24.5.3 Proteomics in IVF

Proteomics provides an overview regarding all the protein expressed from a single genome (Venter et al., 2001). After the hyperstimulation protocol, the protein profiling can be performed by high resolution two-dimensional protein electrophoresis after metabolic labeling with [^{35}S]-methionine. Proteomics study of an embryo is a challenging task because of large number of dynamic molecules but of smaller quantity. Currently, the proteomics is performed by two technologies, first separation of the proteins and then their identification. Most commonly used separation techniques are surface-enhanced laser desorption ionizing time of flight, two-dimensional gel electrophoresis, and liquid chromatography. The identification is performed with the mass spectrophotometer. The mass spectrophotometry technique can even analyze the secretome of single oocye and embryos. Protein microarray has also been a useful technique for proteomics study. Some endometrial proteins have also been used as marker for the study of embryo implantation.

24.5.4 Metabolomics in IVF

Metabolomics refers to the study of all the metabolites present in a cell, tissue, and organism. Metabolites are the end products of many biological processes and using tools of metabolomics, an insight into metabolic status and biochemical processes in any living entity could be deciphered. It refers to the comprehensive profiling of all the metabolites present for a particular system. In recent years, this study has been applied to the assisted reproductive technology to analyze human embryo and the oocytes. Study of metabolites is a challenging task owing to the wide spectrum of molecules present in a sample. The small quantities of all these molecules make the study comparatively challenging as they cannot be amplified at par with the DNA or RNA molecules. Recent technological advancements in the nuclear magnetic resonance and mass spectroscopy have facilitated the study of metabolites. Studies on the metabolic profiling of embryo culture media have been performed (Seli et al., 2008) but extensive studies need to be done to get the real picture of the relatedness between the embryo culture media profile and success rate of IVF.

24.5.5 Pharmacogenomics of IVF

Pharmacogenomics refers to the establishment of relationship between genetic variability in organism and related medical response which is needed to study the action of the drug and define therapy accordingly. The pharmacogenomics aspect has also been investigated in IVF. Follicle-stimulating hormone (FSH) and its receptor (FSHR) play a major role in follicular development and regulation of steroidogenesis in the ovary (Lledo et al., 2014). Almost 1000 SNPs have been located in the FSHR gene out of which only two SNPs (rs6165) and 680 (rs6166) have been found to be related to ovarian response. FSHR genotype also plays an important factor for determining the prognosis of controlled ovarian stimulation cycles. Polymorphism has also been identified in luteinizing hormone (LH) β subunit gene. Female patients carrying the variants W8R (rs1800447) and I15T (rs3439826) require higher FSH doses while fewer oocytes are retrieved. Further polymorphism in gene encoding LH receptor, estrogen receptors, 5-methyltetrahydrofolate receptor, p53, and some other genes have been investigated. Despite the current information further studies need to be done to analyze the difference between different polymorphisms and their clinical significance. Genome wide analysis would give better picture of the polymorphism associated with these genes for their application in pharmacogenomics.

24.6 TECHNIQUES AND PROTOCOLS INVOLVED IN DIFFERENT STEPS OF IVF AND EMBRYO TRANSFER

The following steps are associated with IVF:

1. Ovarian stimulation
2. Oocyte collection
3. Sperm processing and intracytoplasmic sperm injection (ICSI)
4. In vitro interaction of sperm and oocyte

5. Embryo culture
6. Embryo quality selection
7. Preimplantation genetic diagnosis
8. Cryopreservation of extra embryos for future transfer
9. Embryo implantation

The following steps are shown in Fig. 24.1.

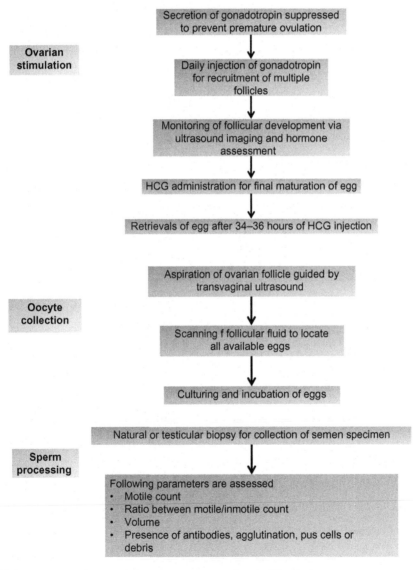

FIGURE 24.1 Flowchart showing different steps associated with IVF technique.

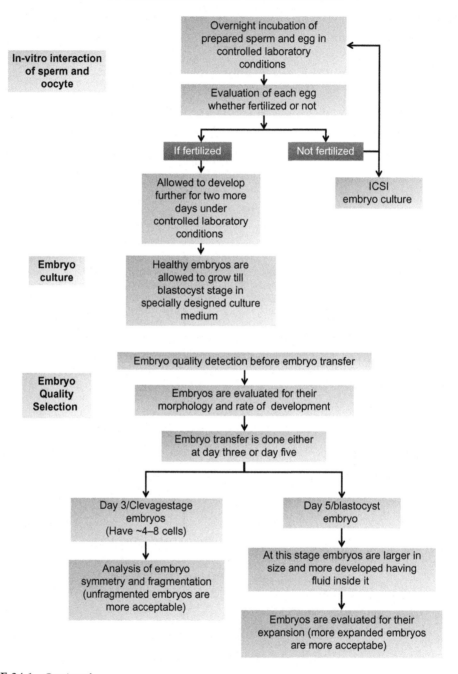

FIGURE 24.1   Continued

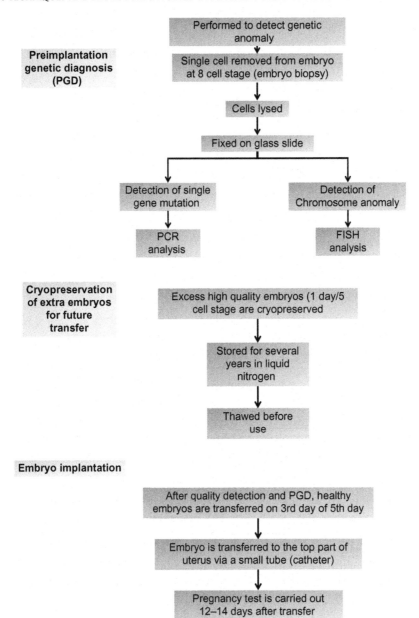

FIGURE 24.1 Continued

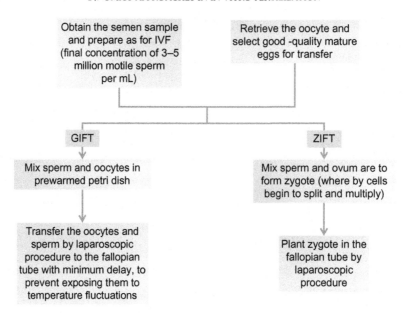

FIGURE 24.2 Flowchart showing differences in the two variants of IVF procedure namely GIFT and ZIFT.

24.7 VARIATIONS OF IVF

In general, there exits two variants of IVF referred as gamete intrafallopian transfer (GIFT) and zygote intrafallopian transfer (ZIFT) (Meldrum, 1992; Pelinck et al., 2002; Seibel and Weissman, 2008). Both these methods are the modified variations of the traditional IVF process. The methodologies used in both these techniques are summarized in Fig. 24.2. The difference is manifested in terms of degree of manipulation and time elapsed between retrieval of eggs from the woman's body and its reimplantation. The main disadvantage with GIFT is that it involves the use of laparoscopy which increases the risk and expenses as compared to the normal IVF procedure (Meldrum, 1992). Also on GIFT one cannot make out whether the pregnancy has occurred or not. Typically, more eggs are used for better chances which ultimately results in a higher risk of multiples (Dickens and Cook, 2008). The disadvantage of the ZIFT procedure is that it includes the need for two separate operations within a short time. Firstly for the aspiration of oocytes and another for the plantation of zygote back into the patient. Like GIFT, it also suffers with an increased risk of multiple and ectopic pregnancy.

24.8 SUCCESS RATES OF IVF

The technique of IVF serves as an important procedure to cure infertility when all other assisted reproductive technology cannot provide fruitful results. Success rate of IVF depends on a host of factors which ascertains further success and pregnancy rates (Pelinck et al., 2002). Patient's age is an important factor determining the success rate of IVF. Success rate decreases with the increasing age of woman. A data collected from year 2010 is a clear representation of decrease in fertility rate with increase in woman's age:

- 32.2% for women aged under 35
- 27.7% for women aged between 35 and 37
- 20.8% for women aged between 38 and 39
- 13.6% for women aged between 40 and 42
- 5.0% for women aged between 43 and 44
- 1.9% for women aged 45 and over.

To further enhance the success rate of IVF, pregnancy rate can be enhanced by freezing extra preembryos or attempting the process several times. The success rate of IVF is presented either as pregnancy rate or live birth rate. The pregnancy rate is determined by the ultrasound tests on the patients while the percentage of all the IVF cycle that leads to live birth is termed as live birth rate except miscarriage or stillbirth and multiple-order births. Birth rate decreases with the increase in patient's age. Necessarily, pregnancy rates are not the same as live birth rates.

The status of woman's health is an important parameter determining the success rate of IVF including the patient's age, lifestyle, and the status of woman's uterus. Tobacco smoking reduces the chances of IVF producing a live birth by 34% and increases the risk of an IVF pregnancy miscarrying by 30%. Patients having body mass index over 27 have reduced likelihood of successful IVF procedure. Low alcohol/caffeine intake also increases success rate. Obese patients have higher rates of miscarriage, gestational diabetes, hypertension, thromboembolism, and problems during delivery, as well as leading to an increased risk of fetal congenital abnormality.

Success rate is also determined by quality, number, and maturity levels of eggs and number of embryos transferred and their quality. IVF is also affected by a number of biomarkers such as antral follicle count, anti-Müllerian hormone levels, semen quality, progesterone elevation on the day of induction of final maturation and characteristics of cells from the cumulus oophorus and the membrana granulosa, which were aspirated during oocyte retrieval (La Marca and Sunkara, 2013).

24.9 COMPLICATIONS OF IVF

Although IVF has proved to be boon for numerous infertility patients, there are many complications associated with it. IVF may lead to multiple births as multiple embryos may be transferred at the time of embryo transfer (Dickens and Cook, 2008; Fauser et al., 2005). Multiple births may lead to pregnancy loss, obstetrical complications, prematurity, and neonatal morbidity with the potential for long-term damage (Fauser et al., 2005). The egg donor is also prone to several risks. During egg retrieval by laproscopy, there are chances of difficulty in breathing, chest infection, allergic reactions, or nerve damage. Ovarian stimulation for inducing final oocyte maturation may cause ovarian hyperstimulation syndrome (La Marca and Sunkara, 2013; Fauser et al., 2010). Around 30% of the patients develop this syndrome resulting in swollen and painful ovaries. Under severe conditions, it may cause sudden excess abdominal pain, nausea, vomiting, and patient has to be hospitalized. There is also possibility of ectopic pregnancy where fetus has to be immediately destroyed. A negative pregnancy test after IVF is also associated with an increased risk for depression in women. A recent review has reported that that infants resulting from IVF (with or without ICSI) have a relative risk of birth defects of 1.32 (95% confidence interval 1.24–1.42) compared to naturally conceived infants.

24.10 COST AND CONVENIENCE

The cost associated with the technique of IVF is an important consideration for its feasibility and popularity, and number of factors influencing the cost. The important factors are discussed in this section.

24.10.1 Clinical Factors

Success of live birth rate specific to a clinic is an important factor determining the cost of IVF. If the success rate is low, the procedure has to be repeated resulting in the increase of overall cost.

24.10.2 Patient Factors

Patient's age is an important factor that affects the cost of IVF as in the case of older patients higher dose of medication is needed for longer durations (Armstrong and Akande, 2013). Physical condition of woman's uterus and endometrium (uterine lining) also influences the treatment cost. The type of IVF cycle greatly affects the cost of treatment. Using donor eggs typically requires agency fees, donor fees, and other laboratory and screening fees which ultimately add significant cost to the IVF cycle. Cycles utilizing a surrogate for carrying the baby also include additional charges.

24.10.3 Medications

Medications for IVF treatment are used for suppression, stimulation, and supplementation. In general, dose and duration of medication increases with the increase in patient's age.

24.10.4 Precycle Costs

Following procedures are carried out before the IVF cycle which ultimately adds to the cost of the procedure. These includes

1. *Sonohysterogram*—ultrasound procedure to detect possible abnormalities inside the uterus lining which may adversely affect pregnancy.

2. *Trial Transfer*—This procedure typically consists placement of a tiny catheter inside the uterus to determine the direction and length of the uterine cavity prior to the IVF cycle.
3. *Semen Analysis*—Determination of sperm quality in terms of volume, sperm concentration, sperm motility, and morphology to determine if more advanced techniques such as ICSI should be used for successful fertilization.
4. *Semen Culture*—To identify any bacterial contamination in semen sample.
5. *FSH Levels*—This test is used to measure the baseline FSH present in a woman's body which is indicative of egg quality.
6. *Initial Visit*—The initial visit or consultation at an IVF clinic to review patient's medical history and to perform a physical examination.
7. *E2 Level*—Determination of patient's estradiol level to validate the accuracy of the timing of the FSH.
8. *TSH and Free T4*—To check proper functioning of thyroid hormone.
9. *Prolactin*—The level of prolactin hormone is determined as high levels of prolactin can interfere with conceiving.
10. *Prepregnancy Screening*—Infectious disease screening.

24.10.5 Cycle Costs

This includes

1. *Base Cycle Cost*—The base fee paid to the medical center performing the IVF treatment.
2. *Anesthesia*—The fee paid to the surgery center or anesthetist for anesthesia during egg retrieval.

24.10.5.1 *Embryo Freezing Costs*

Cost covered in freezing of additional viable embryos that can be implanted in future frozen embryo cycles. The total IVF cost can be summarized Table 24.2.

24.11 CHALLENGES AND ISSUES

The technique of IVF has a great potential for alleviating the problem of infertility but suffers from several moral, ethical, and religious controversies since its development (Brezina and Zhao, 2012).

24.11.1 Ethical

IVF had faced ethical issues from the members of almost all the religions but the major opposition arose from the Roman Catholic Church, which in 1987, issued a doctrinal statement opposing IVF. It expressed concerns over three major issues: namely, the destruction of human embryos not used for implantation, IVF by a donor other than the husband, and severing of an essential connection between the conjugal act and procreation. The unusually high rate of multiple births associated with IVF is also a major concern as it increases

TABLE 24.2 Details of the Percent IVF Cost Coverage

Procedure	Detail	Percentage of the Basic IVF Treatment Cost
Initial consultation	Consultation and initial check-up by the doctor and discussion with the financial consultant	6%–10%
Basic IVF treatment	Includes doctor's fees for egg collection and embryo transfer, embryologist's fees for IVF, anesthetics, ultrasound scans, theater charges	Varies from country to country
ICSI	Creation of embryos by direct injection of sperm into egg, procedure used for male infertility but increasingly common in standard IVF	28%–40%
Hormonal drugs	Precise cost will depend on which drugs are prescribed for stimulating the ovaries	20%–80%
Embryo freezing	Freezing of additional viable embryos which can be implanted in future frozen embryo cycles	Up to 20% plus around 10% of the basic IVF cost per year for storage between 20% and 100% of the basic IVF treatment cost for later thawing and transfer
Other investigations • Immunological therapies • Preimplantation genetic screening	To pick the best embryos	Up to 50% 55%–120%
Regulatory fee		Be 4%–5%

the risk of spontaneous miscarriages. This issue is being addressed through the development of better techniques which use fewer fertilized embryos to achieve pregnancy. Although technique for the SET is available, very few patients opt for SET owing to its low success rate over multiple embryo transfer. Moreover, all patients are unable to afford additional cycles of IVF using SET, hence they prefer multiple embryo transfer. IVF has also raised a number of unresolved moral issues concerning the freezing (cryopreservation) of ovaries, eggs, sperm, or embryos for future pregnancies.

24.11.2 Social

Major social concern regarding IVF is its high costs of treatment and research which is not affordable by many needy patients (Fasouliotis and Schenker, 1999). Moreover, having multiple IVF births from the one pregnancy has a sociological impact on the couple. The destruction of embryos can be arguably not good for society when many couples are childless and desiring children. Screening and destruction of embryos found to have a disability or disorder of some kind is not good on social ground as it sets an example of refining human race in terms of physical characteristics or intellectual capacity. It is a debatable issue that the lives of embryonic humans be sacrificed for the good of other human beings. Meanwhile, sufficient funding should be provided so that adequate research can be conducted upon adult stem cells.

24.11.3 Religious

The technique of IVF has faced several objections from all religious bodies. The Catholic Church is arguably the most outspoken religious authority opposing IVF. The United Methodist Church, a mainstream Protestant denomination, passed a resolution on the subject of IVF in 2004. According to the resolution, IVF is an acceptable option as long as care is taken by couples and clinicians to prevent the overproduction of embryos. Sunni Islam also permits IVF, with some limitations. Caution must be taken that the sperm and eggs used must be from the parents wishing to conceive, and a physician must carry out the procedure. Use of donor eggs and sperms is not permitted. Hinduism permits donor sperm on the grounds that it comes from one of the husband's close relatives.

24.11.4 Psychological and Emotional

IVF is generally considered as a final option when other methods of assisted reproduction have failed. Hence, IVF procedure may cause anxiety to the couple seeking treatment. Since the patient has no control over the progression and outcome of the procedure, during the course of treatment, the patient may feel depressed. If the treatment fails, the couple grief over the child that was never born. This eventually leads to the loss of self-esteem and self-confidence. On the high end, it may lead to the ending of relationship and loss of social status. Multiple failures related to the IVF treatment may result in depressive symptoms and grief. It is recommended that the patient's undergo infertility counseling to explore, understand, and cope with the issues related to infertility and treatment.

24.12 LEGAL ISSUES

As with any medical procedure, patients must provide informed consent to fertility treatments in IVF. The informed consent paperwork contains the details of procedures, benefits and risks, and issues such as the establishment and relinquishment of parental rights (Singer and Wells, 1983). If the consent forms are not properly filled, it may lead to devastating consequences in terms of many important legal issues. Other legal concerns that may arise are the custody issues and inheritance rights over the ovum or egg. Parentage issue is also a major concern when the fertilized egg is replanted in the donor or when the donor sperm or egg is used for fertilization. Legal issue regarding the parentage may arise also in the case of surrogate mother.

24.13 CONCLUSION AND FUTURE PROSPECTIVE

In the present time, IVF has emerged as one of the most widely adopted technology for infertility treatment. Although it gives hope to a number of childless couples, it has come up with new ethical, legal, and social questions that society must address. Society and the funding agencies must come together to provide better treatment options and generate better facilities to increase access to a larger lot of people. Additionally, the numerous issues regarding gamete and embryo donation must be addressed. IVF is a dynamic technique with possibilities for new advancements regularly. Due to the rapidly evolving nature of the technique, legislation is often unable to keep pace and address relevant ethical and legal issues that are constantly emerging in the field. Therefore the physicians must take responsibility to continually monitor these issues and ensure that the technology is offered and delivered in a manner that balances patient care with social and moral responsibility.

Today, IVF is practically a well-established technology for curing infertility with certain inherent social, moral, and legal complications which needs to be addressed. The various limitations of the technique mentioned earlier needs to be mitigated by undergoing research to refine the technology as per the needs of the patients. The advent of SET has substantially reduced the chances of multiple birth to a large extent and further focus should be on minimally stimulated and unstimulated IVF. Advances in genetics and genome sequencing can be used to find out genes that correlate with implantation and normal development. Likewise genes that correlate with early or late implantation failure can also be detected to improve outcome of procedure. Many scientific groups are working in this direction in hope of finding more impressive outcome. With the assistance of genetic screening, genetic diseases can be defined more easily and screening can be done against a greater number of serious conditions. In order to increase the popularity of IVF, there is a need for curtailing the inherent cost of the technology so that it is accessible to common people. One way of bringing down the cost of treatment is automating the IVF laboratory. The current modern technologies such as robotics, microfluidics, and incubators with cameras can assist in making the IVF technology a lot cheaper.

References

Armstrong, S., Akande, V., 2013. What is the best treatment option for infertile women aged 40 and over? J. Assist. Reprod. Genet. 30 (5), 667–671.

Bavister, B.D., 2002. Early history of in vitro fertilization. Reproduction 124, 181–196.

Bettegowda, A., Patel, O.V., Lee, K.B., Park, K.E., Salem, M., Yao, J., et al., 2008. Identification of novel bovine cumulus cell molecular markers predictive of oocyte competence: functional and diagnostic implications. Biol. Reprod. 79, 301–309.

Brezina, P.R., Zhao, Y., 2012. The ethical, legal, and social issues impacted by modern assisted reproductive technologies. Obstet. Gynecol. Int. 2012.

Dickens, B.M., Cook, R.J., 2008. Multiple pregnancy: legal and ethical issues. Int. J. Gynecol. Obstet. 103, 270–274.

Elder, K., Dale, B., 2000. In vitro fertilization, Second edition Cambridge University Press, Cambridge.

Fasouliotis, S.J., Schenker, J.G., 1999. Social aspects in assisted reproduction. Hum. Reprod. Update 5 (1), 26–39.

Fauser, B.C., Devroey, P., Macklon, N.S., 2005. Multiple births resulting from ovarian stimulation for subfertility treatment. Lancet 365 (943), 1807–1816.

Fauser, B.C., Nargund, G., Andersen, A.N., Norman, R., Tarlatzis, B., Boivin, J., et al., 2010. Mild ovarian stimulation for IVF: 10 years later. Hum. Reprod. 25 (11), 2678−2684.

Hart, R., Norman, R.J., 2013. The longer-term health outcomes for children born as a result of IVF treatment. Part II-Mental health and development outcomes. Hum. Reprod. Update 19 (3), 244−250.

La Marca, A., Sunkara, S.K., 2013. Individualization of controlled ovarian stimulation in IVF using ovarian reserve markers: from theory to practice. Hum. Reprod. Update 20, 124.

Lledo, B., Ortiz, J.A., Llacer, J., Bernabeu, R., 2014. Pharmacogenetics of ovarian response. Pharmacogenomics 15 (6), 885−893.

Meldrum, D., 1992. IVF, GIFT, or ZIFT. J. Assist. Reprod. Genet 9 (5), 427−428.

Pelinck, M.J., Hoek, A., Simons, A.H.M., Heineman, MJ., 2002. Efficacy of natural cycle IVF: a review of the literature. Hum. Reprod. Update 8 (2), 129−139.

Seibel, M.M., Weissman, A., 2008. Gamete intrafallopian transfer (GIFT) and zygote intrafallopian transfer (ZIFT). Textbook of Assisted Reproductive Technologies. Informa Healthcare, London.

Seli, E., Botros, L., Sakkas, D., Burns, DH., 2008. Noninvasive metabolomic profiling of embryo culture media using proton nuclear magnetic resonance correlates with reproductive potential of embryos in women undergoing in vitro fertilization. Fertil. Steril. 90, 2183−2189.

Singer, P., Wells, D., 1983. In vitro fertilization: the major issues. J. Med. Ethics 9, 192−195.

Venter, J.C., Adams, M.D., Myers, E.W., Li, P.W., Mural, R.J., Sutton, G.G., et al., 2001. The sequence of the human genome. Science 291, 1304−1351.

Wilton, L., 2005. Preimplantation genetic diagnosis and chromosome analysis of blastomeres using comparative genomic hybridization. Hum. Reprod. Update 11, 33−41.

25

Safety and Ethics in Biotechnology and Bioengineering: What to Follow and What Not to

Anjana Munshi[1] and Vandana Sharma[2]

[1]Central University of Punjab, Bathinda, India [2]Indraprastha Apollo Hospitals, New Delhi, India

25.1 INTRODUCTION

Biotechnology is a scientific discipline, which consists of development of living organisms to make new products/services or to modify products or processes for specific use and benefit of mankind. Genetic engineering differs from biotechnology because it uses isolation and insertion of genetic material or DNA, thereby it reprograms the life of an organism or better known as genetically modified organisms (GMO). Both these fields often overlap with each other depending upon the use of techniques and their applications. Omics-based approaches such as proteomics, metabolomics, and pharmacogenomics have added more weight to advances in these fields (Jeanette and Emon, 2016). A wide range of applications of biotechnology and bioengineering involve the areas of health care and medicine, agriculture, transgenic plants (edible vaccines), industrial, and environmental uses. These developments have significantly reduced the global constraints and have benefitted the society by providing novel resources. Many lifesaving drugs and vaccines have been developed and are under development with the use of recombinant technology, monoclonal antibodies, and protein engineering techniques.

At the same time, the developments of biotechnology and bioengineering have raised eyes of researchers, policy makers, economic representatives, law, politics, and even common man. Some issues of ethical and social concerns have been triggered from different corners of the society including public, researchers, government, and nongovernment organizations. Especially, genetic engineering of living beings, plants, animals, and human has generated issues and safety concerns. Diverging views and many ethical discussions have been expressed with mediatic announcement of creation of genetically engineered (GE) crops (soya and tomato), cloning of sheep and other animals, or research on human embryos. The guidelines pertaining to animal and human research have been established and documented for proper utilization of benefits of these two fields (Shankar and Simmons 2009).

25.1.1 Addressing Ethical Issues

The word "Ethics" has been derived from "Ethikos" in Greek language, which means "arising from a Habit" to analyze good or right in human conduct. Ethics is an act of defining right or wrong within moral limits. Ethical aspects cover crops animals, medicine, stem cells, and medicines. The use of biotechnology and bioengineering has raised a number of ethical and social issues such as:

- Who owns GMO? Can such organisms be patented like inventions?
- Are genetically modified (GM) crops and foods safe to eat and are without any harmful effect?
- Are GE crops safe for the environment and ecosystem?
- Who controls an individual's genetic information, its privacy, and safeguarding issues? Whether this will categorize them on the basis of genetic mutations.
- How far should we go to ensure that we will be free of mutations? What will be the influence on future generations?
- Should a pregnancy be ended if the fetus has a mutation for a serious genetic disorder?
- Storing, Collection, and Safe processing of samples, all these issues need to be addressed.
- What about the risks obtained by an individual/animal during a research study?

Till date the technically sound answer to these concerns is not available and depends on moral values, dignity, justice, safety, welfare, humanity, biodiversity, health, and security (www.oecd.org/futures/long-termtechnologicalsocietalchallenges/40926844.pdf). Usually the ethical values or code of conduct conflict with each other and produce dilemmas for public, policy makers, and researchers. Awash in predictions and uncertainties, we need to devise and follow appropriate controls and should frame precise laws for using research and development in the favor of society.

25.1.2 Organizations Framing Research Ethics in Biotechnology and Bioengineering

Several organizations all over the world like US Food and Drug Regulatory agency that regulates the FDA regulates the safety of food and drugs for humans and animals. A number of organizations, e.g., World Medical Association (Helsinki Declaration), Nuffield Council of Bioethics (2002), Council for International Organizations of Medical Sciences (CIOMS, 2002), Indian Council for Medical Research (in India) and The United Nations

Educational, Scientific and Cultural Organization's declaration on bioethics and human rights (2005) have developed framework of regulatory guidelines to help and guide researchers in the controversial arena of human using advanced biotechnology (Ethical Guidelines for Biomedical Research on Human, 2006).

25.1.3 Ethical Concern in Agriculture

Biotechnology has contributed to sustainable agriculture in many ways, such as increased resistance against stresses including pests and diseases, against abiotic stresses, bioremediation of polluted soils, improvement in quality and productivity, enhanced nitrogen fixation, improvement in fermentation technology, and improved levels of nutrients in crops. It has increased sustainability in agriculture by reducing the dependence on agro chemicals. Pyramiding of genes in a systematic manner helps to improve the quality and quantity of crops such as tolerance to stresses, productivity, and nutritional quality. GE crops or GM crops differ from conventional crops because of genetic modification due to genetic engineering involving one or a few well-characterized genes into a plant species or it can introduce genes from any species into a plant. Whereas most of the conventional methods of genetic modification used to create new varieties (e.g., artificial selection, forced interspecific transfer, random mutagenesis, marker-assisted selection, and grafting of two species) to introduce many uncharacterized genes into the same species. It can transfer genes between species, such as wheat and rye or barley and rye.

Statistical results in year 2008 showed that approximately 30 GE crops were grown on almost 300 million acres in 25 countries (including 15 developing countries) (James, 2009; Ronald, 2011). According to an estimate in year 2015, >120 GE crops (including potato and rice) have been grown worldwide (Stein and Rodriguez-Cerezo, 2009).

There is no doubt that GM crops have created greater possibilities to ensure future food security; especially for small-scale agriculture, associated concerns towards negative and ethical concerns have been increasing (Azadi and Ho, 2010). A diversity of views and opinions related to risks and benefits is available on whether GM crops should be used. There are other questions: Will there be adverse effects of GM crops to obtain medicinal product will endanger their original species on Earth? What will be the effects on environment or ecological balance? What will be the effect on economic condition of farmers? It has been observed that technological advances often bring disproportionate disadvantages to small-scale farmers (Belcher et al., 2005). There will be financial growth of the firms, such as Dupoint and Monsanto, which invest in genetically contrived herbicide-resistant crops (Comstock, 1989; http://www.biofortified.org/2011/02/organic-infighting-over-ge-alfalfa/; https://www.ethics.iit.edu/publication/ethics&biotechnology.pdf). The most debated ethical issue is of market monopoly by the big companies and threatening of small farms (Amin, 2009). Major impact will be on rural economies of the developing countries with the redistribution of benefits in a pattern from small to large and better-off farmers. All these will result in increasing commercialization of science only for profit and will reduce the social benefit of mankind. This will also put a question on scientific purity and public trust on scientists (Thompson, 1997). Moreover, the burden of justification falls on scientists who introduce innovations bringing foreseeable adverse effects. These situations are needed to be addressed by governing bodies at the international and national level to distribute equal benefit of modern biotechnology and bioengineering, so that fruits of innovations and products in these fields should be accessible to all regardless of economic status and scientific purity should be maintained (Amin, 2009).

25.1.3.1 Sustainable Agriculture

Bt (*Bacillus thuringiensis*) is a soil bacterium, which is used to produce an insecticidal toxin within plants and microbial pesticide. Bt (*Bacillus thuringiensis*), a type of cotton crop, was developed and used after conducting several toxicity and allergenicity tests (Thompson, 1997; European Food Safety Authority, 2004). The genes encoding insecticidal toxins were transferred to many other crops such as cotton, soybean, and rape, to breed pest-resistant crops. As a result of this, fewer chemical pesticides were used and this provided environmental and economic benefits leading to sustainable agriculture production and increased biological diversity. Marvier et al. (2007) found that along with Bt crops nontargeted invertebrates (such as insects, spiders, mites, and related species that are not pests targeted by Bt crops) were more abundant in Bt cotton and Bt corn fields when compared to conventional fields managed with insecticides (Marvier et al., 2007). This crop when commercially introduced in 2002 as a trial in India (especially in Maharashtra and Andhra Pradesh) failed to achieve the target. The environmentalists and the Green Peace Corp observed that as these crops produced the insecticidal toxins later in their life cycle, the Bt transgenic crops planted on a large-scale might increase the resistance of pests to the Bt toxins (Uzogara, 2000). Ultimately, Bt pesticides may lose their efficacy and can result into heavy economic losses. The US Environmental Protection Agency (EPA) was sued in year 1997 by a coalition of consumers, environmentalists, and farmers using organic farming methods for allowing the genetically modified planting of crops (corn, cotton, and tomato) to produce an insecticidal Bt toxin. They demanded that EPA revoke 11 registrations issued to five companies to cease granting such permissions and complete a statement to find out the environmental impact of Bt crops so far (Ronald, 2011; Wadman, 1997).

25.1.3.2 Transgenic Plants

Transgenic plants are able to produce medicine of desired therapeutic value. They consist of elements of two different species. Transgenics helps scientists to develop organisms that express a novel trait not found in a species normally; for example, potatoes (protein rich) or rice that has elevated levels of vitamin A (known as "golden rice") (James, 2009; http://www.goldenrice.org). These may be used to save endangered species, e.g., American Chestnut tree, currently being repopulated by Chinese–American chestnut hybrids specifically engineered with a genetic resistance to the chestnut blight (the deadly fungus that nearly decimated native populations in the early 1900s) (http://www.nytimes.com). Transgenic have also been used to develop novel vaccines such as edible vaccines. A transgenic plant project, known as the "glowing plant project," incorporated a gene from a firefly into a houseplant, resulting in plants displaying a soft illumination in the darkness. This was to create trees that could illuminate streets and pathways, thereby saving energy and reducing our dependence upon limited energy resources. However, this has sparked a heated debate related to potential environmental consequences via using highly GE plants into natural ecosystems (The Daily Beast, 2013).

25.1.4 Impact of Biotechnology and Bioengineering on Animals

The use of cutting edge technology in these fields has created revolution and has touched the field of veterinary science. These are evidenced from the use of research methods for enhancing livestock production, to develop disease models, to develop diagnostic procedures, and for preclinical evaluation of adverse effects, vaccines, growth hormones via manipulating or modifying the animal genome to bring out desired traits (Kinter and Valentin, 2002; Berger, 2005). Selective breeding has produced physiologically and genetically stable strains to be used as models for certain human disorders such as cancer (Berger, 2005). Some disease are created in animals by special surgical procedures, via administration of toxins, by molecular biology techniques, e.g., knock in and generalized knock out (Levine et al., 2004; Rees and Alcolado, 2005). Many countries have imposed legal restrictions on animal experimentation (Kromka, 2003). Several times unfavorable impact of these advances of biotechnology have been observed on socioeconomic status, e.g., bovine growth hormone (BGH), a product of genetic engineering used in dairy, increases the milk yield of a cow by 30%. Therefore, this undoubtedly has major influence on the dairy industry (Feenstra, 1993; Liu and Xue, 1998). Then the question is regarding quality and safety of milk produced using this method and "inhumane" treatment of cows. The use of BGH threatened to displace a disproportionate number of disadvantaged farmers. The most affected were farmers of small- and medium-sized with small herbs and high debts loads, without highly mechanized and intensively managed operations (Comstock, 1989).

Scientists have claimed that consumption of GE food is safe, based on the view that this food gets destroyed in stomach by acid and enzymes. However, studies have shown that GE food is not completely destroyed or acted upon by stomach acids and enters into systemic circulation after absorption and digestion. Moreover, our defense system is not capable of removing these GE food from the cells.

25.1.4.1 Transgenic Animals

The major ethical issues in transgenic animals are whether it is ethical or unethical alteration of the natural order of the universe. The genetic make-up of an individual is modified for a specific purpose, without predicting in advance whether there are any side-effects that may cause disease to the animal (http://www.bbc.co.uk/ethics/animals). Can we use animals or creatures as a commodity? Disease model or suffering may last for a long time in these animals as researchers want to conduct long-term investigations.

Another transgenic product, BioSteel, is a silent silk product created by inserting the genes from a silk-spinning spider into the genome of a goat's egg prior to fertilization (http://www.bbc.co.uk/ethics/animals). Transgenic goats are obtained by pronuclear microinjection and somatic cell nuclear transfer. After the maturation of the transgenic female goats, they produce milk containing the protein from which spider silk is made. The fiber artificially created from this silk protein has several potentially valuable uses, e.g., bulletproof vests. Industrial and medical applications include stronger automotive and aerospace components, stronger and more biodegradable sutures, and bioshields, which can protect army personnel and first responders from chemical threats such as Sarin gas (Satya, 2007).

The recombinant protein in the milk of transgenic goats is alpha-fetoprotein used in autoimmune diseases, malaria vaccine antigen, antithrombin III with antiinflammatory and anticoagulant properties, tissue plasminogen activator used in stroke, butyrylcholinesterase in treating neurodegenerative diseases, human growth hormone, human granulocyte colony-stimulating factor, and many others have been already developed and used and some are under development stages in clinical trials (Moura et al., 2011).

Transgenic combinations may include plant−animal−human transgenes. The DNA of human tumor fragments is inserted into tobacco plants to develop a vaccine against non-Hodgkin's lymphoma (Dador, 2013). Researchers have developed a flu vaccine using human DNA and tobacco plants (Swaminathan, 2008). Incorporating a human protein into bananas, potatoes, and tomatoes, researchers have successfully developed edible vaccines for hepatitis B, cholera, and rotavirus, the latter of which can cause fatal bouts of diarrhea (Thomas et al., 2002; http://www.actionbioscience.org).

25.1.5 Ethical Concerns of Xenotransplantation

Xenotransplantation is a procedure, which is used to create model of a disease and sometimes might offer a potential solution to organ/tissue shortages for human recipients. Xenotransplantation, or the transplantation of living tissues or organs from one species to another, alleviates the shortage of human organs such as heart and kidney. Pigs have a similar physiology and organ size, making porcine (pig) organs ideal candidates for transplantation into human recipients. Research studies are going on to explore the use of cell transplantation therapy for patients with spinal cord injury or Parkinson's disease. Due to rapid scientific progress in the field of xenotransplantation, it has overcome the major problem of hyperacute rejection, i.e., massive destruction of transplanted organ within 24 hours (Ekser and Cooper, 2010). The pigs have been introduced with the deletion of gene alpha 1, 3, galactosyltransferase, so that the endothelium of blood vessels of pigs no longer expresses Galα1,3Gal antigen. Humans have naturally formed antibodies against this antigen (Kolber-Simonds et al., 2004; Smetanka and Cooper, 2005; Zeyland et al., 2013).

A number of ethical concerns have been raised regarding xenotransplantation (Smetanka and Cooper, 2005; Melo et al., 2001). These involve risk of transfer of infection, such as cytomegalovirus, Epstein−Barr virus, and hepatitis B or C, and sometimes even HIV illegal trade in sale and purchase of organs, in case of allotransplantation where the organ is transplanted from an individual to other (same species). In case of xenotransplantation, informed consent is difficult to take, because the person (in case vegetarian) with end organ failure, after knowing that organ is from pig, may deny for the procedure. Second is the risk of transfer of infection from porcine to human and the possibility of the transmission of a porcine endogenous retrovirus, cytomegalovirus may persist, and the person in clinical trials is required to undergo lifetime surveillance that may be extended to the family members. This will invade the privacy of an individual along with health risk. A high level of care has to be observed in xenotransplantation; first the animal used should be genetically engineered to avoid the rejection of immune response and should be kept in a healthy and safe environment to avoid risk of transfer of infection. Concerns have been expressed from different communities indicating that animals will be placed at an "additional risk" in case of the use of this technology.

The UK Advisory Group on the Ethics of Xenotransplantation stated that "some degree of genetic modification is ethically acceptable" but that "there are limits to the extent to which an animal should be genetically modified" (Smetanka and Cooper, 2005; Nuffield Council on Bioethics, 1996).

25.1.6 Ethical Issues in Stem Cell Research

Stem cell research has promised a new life to medical science and treatment strategies. The pluripotent stem cell lines, which are derived from oocytes and embryos, are fraught with disputes about the onset of human personhood (Lo and Parham, 2009). Genetic manipulation of stem cells now includes the growth of tissues on scaffolding, or a 3-D printer, which then can be used as a temporary skin substitute for healing wounds or burns. This has become a viable alternative in procedures that involve replacement of cartilage, heart valves, cerebrospinal shunts, burns, grafting of skin, and other organs. However, the reprogramming of somatic cells to produce induced pluripotent stem cells avoids the ethical problems related to embryonic stem cell research (Lo and Parham, 2009). Stem cells from adult and umbilical cord blood cells do not create any ethical concerns and are widely used in research. The use of embryonic stem cells is ethically and politically challenged.

Several other ethical concerns include counseling, collection, banking, informed consent, confidentiality, and privacy. Also, the issue related to patent of stem cells derived from human embryos has been under debate in scientific and legal communities since years. These have been discussed in detail and reviewed by Lo and Parham (2009). Many countries have established guidelines to address these ethical issues and have made rules for working on embryo transplantation, embryo research, surrogate motherhood, and other issues (Saxena et al., 2012). Cloning of human embryos only to be used for therapeutic purpose was made legal in year 2001, after making an amendment in "Human Embryology Act." However, cloning humans for reproductive purposes is illegal and punishable under the act (Asmatulu et al., 2010).

25.1.7 Ethical Issues in Aquaculture Industry

In aquaculture industry, the transgenic fish is a perfect example to be mentioned here. GM food in aquaculture industry has made it a significant contributor of food production in many countries. Almost 35 species of fish have been genetically modified, e.g., trout, catfish, tilapia, and salmon. These are genetically engineered to grow fast, develop large muscles, and tolerate temperature. Increased efficiency and high production levels are offsetting conventional practices in this field particularly in developing countries. The transgenic fish is able to breed with existing species in an uncontained environment. Some are already marketed in China (Carp), the United States (Salmon), and Cuba (Tilapia). However, Costco, the second largest retailer of GM Salmon fish has stopped selling Salmon.

Genetic modification represents certain unanswerable and sometime uncontrollable uncertainties that represent risks for human health, animal welfare, and environment (Weaver and Morris, 2005). Approximately, 2 million people, involving scientists, fisherman, businessman, and consumers, have opposed the US FDA's approval for GM Salmon due to the risks related to health (as it continuously releases growth hormone), environment, and wild salmon species. Center for Food Safety in November 2015 announced to sue US FDA for this approval (http://www.centerforfoodsafety.org).

25.1.8 Ethics in Biobanking

Human tissues have been derived and stored, distributed, and used for forensic, educational, therapeutic, and research purposes. These have been known under various names such as biobanks, biolibraries, tissue repositories, genetic databases, or DNA banks (Hoeyer, 2008). These biobanks and other sources contain important samples that contain personal health information of an individual may be owned by a private owner or a profit or nonprofit organization. This type of diversity has raised many ethical and legal questions in biobanking. Many ethical dilemmas exist such as ownership, use of donation, processing of sample, storage, and results of research. Commercialization of biobanks will create many issues such as prevention of data exploitation, ensuring justice to participants, balancing costs and benefits, and ensuring and sustaining public trust (Rothstein, 2005; Budimir et al., 2011). The role of ethical review boards working under ethical and legal framework and national legislations is very important, which provide protection to participants and ensure proper use of their sample. Proper coding and storage of biobank samples, with restricted access to personal information, must be ensured to promote the safety and personal integrity of sample donors. Optimal storage conditions should be followed. Specific consideration applies in case its donor is not alive. These samples should not be used in case if the deceased participants did not wish to carry on research on his/her sample after death, respect should be maintained, and sample should be destructed and disposed off. The review discussing ethical concerns in biobanking has been provided by Budimir et al (2011).

25.1.8.1 Biosafety Issues of Modern Biotechnology and Bioengineering

Biosafety can be defined as an asset of actions or concerns to minimize or erase the potential risks to environment or living beings derived from development, implementation, or commercialization of biotechnology or bioengineering or to counteract the negative impact of these two fields. The valuable knowledge of these two fields if used inappropriately can cause serious disaster to mankind and living beings on our planet. For example, GM crops and food are improved in quality than naturally occurring but are associated with some allergic reactions, and consumption of these for long time may lead to development of some fatal diseases such as cancers. The treatments of weeds such as herbicide, stress resistance and pest resistance may escape plants from cultivation system. These modified crops may create a great threat to normal food chain, which may influence beneficial insects, birds, mammals, and microbes (Buiatti and Christou, 2013). If used for a long time and in large scale, these transgenic plants or crops may harm natural ecological balance more than a nuclear reaction can cause. These biosafety concerns have hindered the growth of biotechnology and bioengineering. Many countries have formulated and documented the laws regarding biosafety issues, e.g., chapter 16 of Agenda 21 adopted at United Nations Conference in 1992 and "International Biosafety protocol" to ensure the safety of development, application, exchange, and transfer of biotechnology (Buiatti and Christou, 2013). International biosafety project report has focused on the benefit and prioritization of biosafety in the developing world for the benefit of biodiversity and to enable consumers to use their right to choose a healthy sustainable environment and to be informed. The precautionary measure requires an integrated system of risk management in biosafety that includes active participation of representatives of public, government, scientific researchers, policy makers, stakeholders, and manufacturers from biotech companies. The other ingredients of this system are education, information, proper waste management of toxic and contaminating products, and adaptive management (http://www.consumersinternational.org). Above all there should be proper analysis of risk and benefit ratio before directly implementing/using any benefit for society. A number of organizations have come into existence across world such as UN Framework Convention on Biological Diversity controls biodiversity loss and Biological Weapons Convention regulates the use of biological weapons in a proper way.

We can use biotechnology and bioengineering to significantly increase the quality of life and can affect the world around us—for better or for worse. If used negatively, it could cause unprecedented pandemic from lapse of biosafety or bioterrorism (Baum and Wilson, 2013).

25.1.8.2 Adverse Effect on Health of People/Environment

These include increased disease burden, increased rate of pathogenicity, emergence of a new disease or allergic reaction, pest or weed, and adverse effect on species and ecosystem.

25.1.8.3 Unpredictable and Unintended effects

Horizontal gene transfer may transfer the desired gene from a genetically modified organism to potential pests or pathogens and many yet to be identified organisms (Praksh et al., 2011). This may lead to alteration of ecological niche or ecological potential of the recipient organisms (Praksh et al., 2011). The other possibility of gene transfer is that it may insert at variable sites of the recipient gene, resulting into a novel gene that may cause unintended effects.

25.1.8.4 Impacts on Socioeconomic Welfare of Countries and of Communities

Application of genetic engineering has enabled to diagnose and cure some genetically inherited diseases (e.g., some cancers and hemophilia). Molecular diagnosis also has impacts on the employment and marriage prospects of human beings. For example, in a research carried out by geneticists at Johns Hopkins University, it was found that Ashkenazi Jews have double the risk of colon cancer. This work led the Jewish groups being targeted as a potential market for commercial genetic tests because of their susceptibility to disease.

25.1.8.5 *Impact of Traditional Values and Culture*

Because the genetic modification interferes with nature's creation, the concerns have been raised that identification of certain people susceptible to a certain disease or condition might make descendent selection and will disturb the evolution in a normal way. Because it interferes with fundamental laws of nature.

25.1.8.6 *Ethical Issues in Medical Biotechnology*

The advances in biotechnology and bioengineering have created a revolution by manipulating genetic heritage in the field of manufacturing medicines and in medical sciences such as personalized medicine. The ability to modify and fix mutant genes has provided unbeatable treatment of disease. Also the diagnosis has become precise after identifying the related genetic variants causing the disease. Animal models of human diseases have been developed to understand the molecular basis of disease. Clinical trials have shown the successful use of gene therapy to attack tumor cells of patients with recurrent B-cell acute lymphoblastic leukemia (Baum and Wilson, 2013; Brenner et al., 2013). However, it has raised many ethical issues that need close attention (Kuszler, 2006). Because we can modify living beings for required benefit, there is great potential of ethical regulation (Persson, 2006). Research costs, storage of DNA samples, risk/benefit to the study subject insurance, privacy, employers, providers, and third-party payers are the other issues bubbled to the surface. To address all the concerned issues, scientific communities, medical professional people, related government officers, policy makers, stakeholders along with public should develop an infrastructure so that biotechnogical development should be used only for the benefit of mankind.

25.1.8.7 *Protecting Human Beings in Clinical Trials*

This major issue has already created debate in year 1999. With the death of Jesse Gelsinger (18 years of age) after participating in a gene therapy trial in Pennsylvania, the institution faced great criticism for failing to disclose crucial information on informed consent documents, relaxing criteria for accepting volunteers, and eligibility criteria (Silverman, 2004). This issue prompted a great deal of curiosity among researchers and regulators, and many institutions came forward to implement new standards.

Each trial/study should be carefully planned necessary for the success of the scientific research, and the study participants should be aware of the risks and benefits of the research study to which they are participating. Patient's privacy and confidentiality is a must during and after the study. After decoding the human genome, the scientists are adept at deciphering genetic composition that has increasingly alarmed the privacy of an individual's information including prospective employer and insurers or others who can mislead this information.

25.1.8.8 *Accountability*

The accountability to frame and regulate the "public policy and regulatory ethics" in biotechnology is of government and other research communities. Ethical committees should be framed and maintained at institutional and central level. They should report time to time to the central body.

25.1.8.9 *Affordability*

The sky-rising costs of health care and cost of medicines are a political hot potato and will remain so (Silverman, 2004). As compared to pills and tablets, the cost of a biologic will be 10 times more or even more costlier. Sometimes a patient cannot afford the out-of-pocket treatment, and situation becomes worse if insurers decline to add a biologic to its formulary because of cost. The high pricing of biological/medicines, which are the result of biotechnology or bioengineering provides additional burden, e.g., in case of a chronic disease or the case where an insurer does not cover a particular treatment.

25.1.8.10 *Privacy*

Protecting privacy is another concerned issue. After decoding human genome, the whole personal information is available just like a Kundli (A document prepared at birth in Indian Hindu religion to forecast the future based on birth), the single nucleotide polymorphisms, susceptibility to disease, the adverse drug reactions that one can experience if undergoes treatment with particular drugs, and many others predictabilities. This creates enormous problems in case an employer comes or an insurer comes to know the details. This should be kept highly confidential to avoid any misuse, or in other words identifiable patient information should not be shared without the patient's consent.

25.1.8.11 *Intellectual Property Rights*

The inventions made in the fields of biotechnology and bioengineering are patentable. Patents are used to provide exclusivity for the inventor in commercializing his invention. There is an agreement under World Trade Organization known as The Agreement on Trade-Related Aspects of Intellectual Property Rights (IPR) that requires countries to grant patents for inventions, in all fields of technology provided that the invention is new and industrially applicable (Singh, 2000). IPR encourages inventions and their disclosers. The global market of biotechnology agricultural products has increased approximately to 20 from 0.5 billion dollars in 1996. The Organization of Economic Cooperation and Development (OECD) has expressed 7 billion dollars and accounts for about half the world's entire agricultural research investment (Singh, 2000). Issues in patenting GMO allow industries with big setup to have monopoly of genetically modified plants and animals and violate the sanctity of life (Uzogara, 2000). The patent system can also be used to usurp intellectual ownership of "native technologies." For example, Neem tree, known as blessed tree, used from ancient times in India for medicinal values. A US company did some research on Neem tree and got the patent for extracting "Azadirachtin," a potential pesticide. There are more than 40 patents on Neem only in the United States. A number of Patent controversies and court cases have been going on globally on different aspects. These have been discussed by Fialho and Chakrabarty (2012).

25.2 CONCLUSION

The use of emerging techniques in the field of biotechnology and bioengineering is very important to meet the challenging demands of world around us and to improve existing conditions prevalent. We are at the juncture where on one side we have capability to change the scenario by meeting all the demands of food and medicine, whereas on the other side, there is unprecedented threat to originality of species, their existence, human health, and environment. Inventions are made every day but the critical approval requires proper risk assessment and management with appropriate monitoring methods before it is applicable to society/general public. A broader base is required for discussion on GMO and their impact on ecosystem. The international biosafety regulatory frameworks are sufficient to be stringent to protect against genuine ascertainable risks. Interdisciplinary training and education should be provided to the students, scientists and professors, doctors, engineers, social scientists, workers, lawyers who are working in the research-related areas with main focus on biosafety and bioethics. The best way is consideration of ethical social and economic issues related to research and its implementation in the fields of biotechnology and bioengineering. The precautionary approach will provide new avenues for future research and development in these areas.

References

Amin, L., 2009. Modern Biotechnology: ethical issues, ethical principles, and guidelines. http://www.ukm.my/jmalim/images/vol_10_2009/a1%20latifah%20amin.pdf.

Asmatulu R., Khan W.S., Asmatulu E., Ceylan M., 2010. Biotechnology and bioethics in engineering education. In: Proceedings of the 2010 Midwest Section Conference of the American Society for Engineering Education. September 22–24, 2010.

Azadi, H., Ho, P., 2010. Genetically modified and organic crops in developing countries: a review of options for food security. Biotechnology Adv. 28, 160–168.

Baum, S.D., Wilson, G.S., 2013. The ethics of global catastrophic risk from dual use bioengineering. Ethics Biol. Eng. Med. 4, 59–72.

Belcher, K., Nolana, J., Phillips, P.W.B., 2005. Genetically modified crops and agricultural landscapes: spatial patterns of contamination. Ecol. Econ. 53, 387–401.

Berger, J., 2005. Current ethical problems in cell biology. J. Appl. Biomed. 3, 109–113.

Brenner, M.K., Gottschalk, S., Leen, A.M., Vera, J.F., 2013. Is cancer gene therapy an empty suit? Lancet Oncol. 14 (11), e447–e456.

Budimir, D., Polasek, O., Marusić, A., Kolcić, I., Zemunik, T., Boraska, V., Jeroncić, A., Boban, M., Campbell, H., Rudan, I., 2011. Ethical aspects of human biobanks: a systematic review. Croat. Med. J. 52 (3), 262–279.

Buiatti, M., Christou, P., 2013. The application of GMOs in agriculture and in food production for a better nutrition: two different scientific points of view. Genes Nutr. 8 (3), 255–270.

Comstock, G., 1989. Genetically engineered herbicide resistance part one. J. Agriculture Ethics. 2, 263–306.

Dador, D., 2013. New flu vaccine made from tobacco plant in the works. http://abclocal.go.com/kabc/story?section=news/health/your_health&id=9115757.

Ekser, B., Cooper, D.K.C., 2010. Overcoming the barriers to xenotransplantation: prospects for the future. Expert Rev. Clin. Immunol. 6 (2), 219–230.

Indian Council of Medical Research, 2006. Ethical Guidelines for Biomedical Research on Human. icmr.nic.in/ethical_guidelines.pdf.

European Food Safety Authority, 2004. Opinion of the scientific panel on genetically modified organisms. EFSA J. 1–25.

Feenstra, G., 1993. Is BGH sustainable? The consumer perspective. In: Liebhardt, W.C. (Ed.), The Dairy Debate: Consequences of Bovine Growth Hormone and Rotational Grazing Technologies. University of California, Davis, CA, 1, 20–27.

Fialho, AM, Chakrabarty, AM., 2012. Patent controversies and court cases. Cancer diagnosis, therapy and prevention. Cancer Biol. Ther. 13 (13), 1229–1234.

Hoeyer, K., 2008. The ethics of research biobanking: a critical review of the literature. Biotechnol. Genet. Eng. Rev. 25, 429–452.

James, C., 2009. Global Status of Commercialized Biotech/GM Crops. International Service for the Acquisition of Agri-biotech Applications (ISAAA). Southeast Asia Center, Manila, The Philippines; ISAAA Africa Center, Nairobi, Kenya; ISAAA American Center, New York.

Jeanette, M., Emon, V., 2016. The omics revolution in agricultural research. J. Agricul Food Chem. 64 (1), 36−44. Available from: http://dx.doi.org/10.1021/acs.jafc5b04515.

Kinter, L.B., Valentin, J.P., 2002. Safety pharmacology and risk assessment. Fund. Clin. Pharmacol. 16, 175−182.

Kolber-Simonds, D., Lai, L., Watt, S.R., et al., 2004. α1,3-galactosyltransferase null pigs via nuclear transfer with fibroblasts bearing loss of heterozygosity mutations. Proc. Natl. Acad. Sci. USA 19, 7335−7340.

Kromka, F., 2003. Equality for man and animal—a common sense criticism of this animal rights concept. Ber. Landwirt. 81, 150−158.

Kuszler, P.C., 2006. Biotechnology entrepreneurship and ethics: principles, paradigms, and products. Med. Law. 25 (3), 491−502.

Levine, M.S., Cepeda, C., Hickey, M.A., Flemming, S.M., Chesselet, M.F., 2004. Genetic mouse models of Huntington's and Parkinson's diseases: illuminating but imperfect. Trends Neurosci. 27, 691−697.

Liu, B., Xue, D., 1998. Progress and the focal points of the international legislation on biosafety. Rural Eco-Environment 14 (2), 45−48.

Lo, B., Parham, L., 2009. Ethical issues in stem cell research. Endocr. Rev. 30 (3), 204−213.

Marvier, M., McCreedy, C., Regetz, J., Kareiva, P., 2007. A meta-analysis of effects of Bt cotton and maize on non-target invertebrates. Science 316, 1475−1477.

Melo, H., Brandao, C., Rego, G., Nunes, R., 2001. Ethical and legal issues in xenotransplantation. Bioethics. 15 (5-6), 427−442.

Moura, R.R., Melo, L.M., de, V.J., Freitas, F., 2011. Production of recombinant proteins in milk of transgenic and non-transgenic goats. Human and Animal Health. Braz. Arch. Biol. Technol. 54, 5.

Nuffield Council on Bioethics, 1996. Animal-to-Human Transplants: The Ethics of Xenotransplantation. Nuffield Council on Bioethics, London, p. 147.

Persson, A., 2006. Research ethics and the development of medical biotechnology. Xenotransplantation 13 (6), 511−513.

Praksh, D., Verma, S., Bhatia, R., Tiwary, B.N., 2011. Review article: Risks and precautions of genetically modified organisms. Internat. Sch. Res. Network 2011. Article ID 369573, 13 pp. doi:10.5402/2011/369573.

Rees, D.A., Alcolado, J.C., 2005. Animal models of diabetes mellitus. Diab. Med. 22, 359−370.

Ronald, P., 2011. Plant genetics, sustainable agriculture and global food security. Genetics 188 (1), 11−20.

Rothstein, M.A., 2005. Expanding the ethical analysis of biobanks. J. Law Med. Ethics 33, 89−101.

Satya, P., 2007. Genomics and Genetic Engineering. New India Publishing Agency, New Delhi, India, p. 208.

Saxena, P., Mishra, A., Malik, S., 2012. Surrogacy: ethical and legal issues. Indian J. Community Med. 37 (4), 211−213.

Shankar, G., Simmons, A., 2009. Understanding ethics guidelines using an internet-based expert system. J. Med. Ethics 35 (1), 65−68.

Silverman, E.D., 2004. The 5 most pressing ethical issues in biotech medicine. Biotechnol. Health 1 (6), 41−45. 46.

Singh R.B., 2000. Biotechnology, Biodiversity, and Sustainable Agriculture: A Contradiction. http://www.bic.searca.org/seminar_proceedings/bangkok-2000/H-plenary_papers/singh.pdf.

Smetanka, C., Cooper, D.K.C., 2005. The ethics debate in relation to xenotransplantation. Rev. Sci. Tech. Off. Int. Epiz. 24 (1), 335−342.

Stein, A.J., Rodriguez-Cerezo, E., 2009. In: Joint Research Centre European Commission (Ed.), The Global Pipeline of New GM Crops: Implications of Asynchronous Approval for International Trade, JRC Scientific and Technical Reports. Institute for Prospective Technological Studies Joint Research Centre, Institute for Prospective Technological Studies, Seville, Spain, pp. 1−114.

Swaminathan, N., 2008. Good and Evil: A Cancer Vaccine From Tobacco Plants. http://www.scientificamerican.com/article.cfm?id = cancer-vaccine-tobacco-plants.

The Daily Beast, 2013. Plants that glow in the dark spark heated debate. http://www.thedailybeast.com/articles/2013/08/18/plants-that-glow-in-the-dark-spark-heated-debate.html.

Thomas, B., Van Deynze, A., & Bradford, K., 2002. Production of Therapeutic Proteins in Plants. University of California Division of Agriculture and Natural Resources: Agricultural Biotechnology in California Series, Publication 8078. http://anrcatalog.ucdavis.edu/pdf/8078.pdf.

Thompson, P.B., 1997. Food biotechnology's challenge to cultural integrity and individual consent. Hastings Cent. Rep. 27 (4), 34–38.

Uzogara, S.G., 2000. The impact of genetic modification of human foods in the 21st century: a review. Biotechnol. Adv. 18, 179–206.

Wadman, M., 1997. EPA to be sued over gene-modified crops. Nature 389, 317.

Weaver, S.A., Morris, M.C., 2005. Risks associated with genetic modification: an annotated bibliography of peer reviewed natural science publications. J. Agricul. Environ. Eth. 18 (2), 157–189.

Zeyland, J., Gawrońska, B., Juzwa, W., Jura, J., Nowak, A., Słomski, R., et al., 2013. Transgenic pigs designed to express human α-galactosidase to avoid humoral xenograft rejection. J. Appl. Genet. 54 (3), 293–303.

Further Reading

Van der Geer, J., Hanraads, J.A.J., Lupton, R.A., 2010. The art of writing a scientific article. J. Sci. Commun. 163, 51–59.

Index

Note: Page numbers followed by "f" and "t" refer to figures and tables, respectively.

Printed in the United States
By Bookmasters